W9-AXL-460

KEY EXPERIMENTS	MOLECULAR MEDICINE FEATURES

STUDENT CD-ROM

Adopters have the option of bundling *The Cell* with a special edition of the CD-ROM *Molecules, Cells, and Genes* (ISBN 0-87893-074-4).* This invaluable resource is keyed to the textbook and combines the essential features of a Study Guide and a Problems book. The CD-ROM includes:

- multiple-choice quiz modules for each chapter, with feedback provided for every possible response
- several problems with answers for each textbook chapter
- over 30 minicourses, keyed to text chapters and covering over 1,000 topics in basic cell biology, molecular biology, genetics, immunology, terminology, cancer, and more
- more than 100 animated sequences, with narration
- 800 original illustrations
- 17 video microscopy sequences
- over 125 micrographs
- an extensive library of spoken pronunciations
- an Index that allows the student to instantly locate every Key Term
- endless hypertext links that provide easy navigation throughout the program
- web links that guide the student to useful and effective web sites to further augment the learning process

*Students who wish to order the CD-ROM directly can do so at the following web site: www.sinauer.com

THE CELL

A MOLECULAR APPROACH

SECOND EDITION

THE CELL

A MOLECULAR APPROACH

SECOND EDITION

Geoffrey M. Cooper

Boston University

ASM Press
Washington, D.C.

ASM PRESS

Sinauer Associates, Inc.
Sunderland, Massachusetts

The Cover
A nuclear pore complex (center) is shown with the cytoplasm above and the nucleus below. The illustration includes all macromolecules, including ribosomes (purple) and actin filaments (blue) in the cytoplasm, and DNA double helices (red) being transcribed into messenger RNA (white) in the nucleus. Illustration by David S. Goodsell, The Scripps Research Institute.

Part One opener image
Confocal micrograph of mouse embryo cells.
(Courtesy of David Albertini, Tufts University School of Medicine)

Part Two opener image
Light micrograph of stained salivary gland polytene chromosomes in *Drosophila*.
(Peter J. Bryant/Biological Photo Service)

Part Three opener image
Micrograph of macrophages ingesting red blood cells.
(Courtesy of Joel Swanson)

Part Four opener image
Fluorescence micrograph of the telophase stage of mitosis in newt lung cells.
(Conly L. Rieder/Biological Photo Service)

The Cell: A Molecular Approach, Second Edition

Copyright © 2000 by Geoffrey M. Cooper. All rights reserved.
This book may not be reproduced in whole or in part without permission.

Address editorial correspondence to ASM Press, c/o The American Society for Microbiology, 1752 N Street NW, Washington, DC 20036 U.S.A.

Address orders and requests for examination copies to Sinauer Associates, Inc., P.O. Box 407, 23 Plumtree Road, Sunderland, MA 01375 U.S.A.
Phone: 413-549-4300
FAX: 413-549-1118
email: orders@sinauer.com
www.sinauer.com

Library of Congress Cataloging-in-Publication Data

Cooper, Geoffrey M.
    The cell : a molecular approach / Geoffrey M. Cooper.--2nd ed.
        p. cm.
    Includes bibliographical references and index.
    ISBN 0-87893-106-6 (cloth : alk. paper)
1. Cytology. 2. Molecular biology. I. Title.

QH581.2 .C66 2000
571.6--dc21   00–036528

Printed in U.S.A.

5 4 3 2 1

This book is dedicated to
Howard M. Temin (1934–1994),
whom I was privileged to know as
a teacher, mentor, and friend.

Contents in Brief

Contents

PART I Introduction

1

An Overview of Cells and Cell Research 3

2

The Chemistry of Cells 41

3

Fundamentals of Molecular Biology 89

PART II *The Flow of Genetic Information*

4

The Organization of Cellular Genomes 137

5

Replication, Maintenance, and Rearrangements of Genomic DNA 177

6

RNA Synthesis and Processing 227

7

Protein Synthesis, Processing, and Regulation 273

PART III Cell Structure and Function

8

The Nucleus 315

9

Protein Sorting and Transport: The Endoplasmic Reticulum, Golgi Apparatus, and Lysosomes 347

10

Bioenergetics and Metabolism: Mitochondria, Chloroplasts, and Peroxisomes 387

11

The Cytoskeleton and Cell Movement 421

15

Cancer 609

Preface

Cell and molecular biology is one of the most rapidly moving areas of the life sciences, and a number of substantial advances have been made since the first edition of *The Cell* was published in 1997. Particularly noteworthy progress has been made in the area of genome sequencing: We now know the complete sequences not only of yeast and bacterial genomes but also of *C. elegans* and *Drosophila melanogaster*, as well as much of the genomic sequence of the model plant *Arabidopsis thaliana*. Projects directed toward sequencing the human genome have also moved forward rapidly, and completion of the human genome sequence can be confidently anticipated in the near future. These advances in genome sequencing are opening whole new approaches to understanding the regulation and function of cells, and they will undoubtedly provide a new database upon which much of the future research in molecular and cell biology will be built.

There have been significant advances in the understanding of several aspects of the molecular biology of eukaryotic cells—the mechanisms that control the initiation of DNA replication have been elucidated, studies of histone acetylation have provided a clear link between chromatin structure and transcriptional regulation, and new mechanisms of translational regulation have been described. Substantial progress has similarly been made in understanding a variety of aspects of protein trafficking, including nuclear import and export, the pathways of protein import to the endoplasmic reticulum and transport to the Golgi apparatus, the molecular mechanisms of vesicular transport, and the mechanisms of protein import into mitochondria and chloroplasts. The exciting areas of cell signaling and cell cycle regulation have also continued to rapidly move forward, with particularly noteworthy progress being made in understanding the regulation of programmed cell death and the signaling pathways that control cell fate during embryonic development. These advances in basic science have been coupled with progress in medically-related areas, including the potential use of stem cells to replace damaged human tissue and the development of new anticancer drugs targeted against specific oncogene proteins.

The second edition of *The Cell* has been completely updated to include these exciting advances and to present the most current information available. The goals and distinguishing features of *The Cell*, however, remain unchanged from the first edition. *The Cell* continues to be a basic text that provides an accessible introduction for undergraduate or medical students who are taking a first course on cell and molecular biology. My principal goal has been to provide an accessible and readable book that can be approached and mastered by undergraduate students, while still being intellectually gratifying and conveying the excitement and challenges of research in contemporary molecular and cellular biology.

As in the first edition, *The Cell* is focused on the molecular biology of cells as a unifying theme, with specialized topics discussed throughout the book as examples of more general principles. Aspects of developmental biology, the immune system, the nervous system, and plant biology are thus discussed in their broader biological context in chapters covering areas such as genome structure, gene expression, DNA rearrangements, the plasma membrane, cell signaling, and the cell cycle. Relationships between cell biology and medicine are similarly discussed throughout the text, as well as being highlighted in the Molecular Medicine essays that are included as a special feature in each chapter. These discussions illustrate the striking impact of molecular and cellular biology on human health, and are intended to stimulate as well as inform those students interested in medicine.

Since *The Cell* deals with one of the most rapidly progressing areas of biology, it remains critical for students not only to have the most current information available, but also to understand the experimental nature of contemporary research in cell and molecular biology. With this goal in mind, representative experiments are discussed throughout the text to illustrate the kinds of approaches upon which our current understanding has been built. In addition, each chapter contains a Key Experiment essay that describes a seminal paper and its background in detail, with the intent of giving the reader a sense of "doing science." Finally, I have tried to point out the limits of our knowledge and to identify outstanding unanswered questions throughout the text. I hope that, together, these approaches give the student an appreciation of the ways in which progress in understanding the cell has been made, and convey the excitement and challenges of participating in this frontier of scientific exploration.

Geoffrey M. Cooper
May 2000

ORGANIZATION AND FEATURES OF
The Cell

The Cell has been designed to be an approachable and teachable text that can be covered in a single semester while allowing students to master the material in the entire book. It is assumed that most students will have had introductory biology and general chemistry courses, but will not have had previous courses in organic chemistry, biochemistry, or molecular biology. Several aspects of the organization and features of the book will help students to approach and understand its subject matter.

ORGANIZATION

The Cell is divided into four parts, each of which is self-contained, so that the order and emphasis of topics can be easily varied according to the needs of individual courses. In covering this vast subject matter, however, I developed an organizational overview of the book, as described below.

Part I of the book provides background chapters on the evolution of cells, methods for studying cells, the chemistry of cells, and the fundamentals of modern molecular biology. For those students who have a strong background from either a comprehensive introductory biology course or a previous course in molecular biology, various parts of these chapters can be skipped or used for review.

Part II focuses on the molecular biology of cells and contains chapters dealing with genome organization; DNA replication, repair, and recombination; transcription and RNA processing; and the synthesis, processing, and regulation of proteins. The order of chapters follows the flow of genetic information (DNA → RNA → protein) and provides a concise but up-to-date overview of these topics.

Part III contains the core block of chapters on cell structure and function, including chapters on the nucleus, cytoplasmic organelles, the cytoskeleton, and the cell surface. This part of the book starts with coverage of the nucleus, which puts the molecular biology of Part II within the context of the eukaryotic cell, and then works outward through cytoplasmic organelles and the cytoskeleton to the plasma membrane. These chapters are relatively self-contained, however, and could be used in a different order should that be more appropriate for a particular course.

Finally, Part Four focuses on the exciting and fast-moving area of cell regulation, including coverage of topics such as cell signaling, the cell cycle, and programmed cell death. This part of the book concludes with a chapter on cancer, my own field of research, which synthesizes the consequences of defects in basic cell regulatory mechanisms.

FEATURES

Several pedagogical features have been incorporated into *The Cell* in order to help students master and integrate its contents. These features are reviewed below as a guide to students studying from this book.

Chapter organization. Each chapter is divided into four or five major sections, which are further divided into a similar number of subsections. An outline listing the major sections at the beginning of each chapter provides a brief overview of its contents.

Key Terms and Glossary. Key terms are identified as boldfaced words when they are introduced in each chapter. These key terms are reiterated in the chapter summary and defined in the glossary at the end of the book.

Illustrations and micrographs. An illustration program of full-color art and micrographs has been carefully developed to complement and visually reinforce the text.

Key Experiment and Molecular Medicine essays. Each chapter contains one Key Experiment and one Molecular Medicine feature. These essays are designed to provide the student with a sense of both the experimental basis of cell and molecular biology and its applications to modern medicine.

Chapter Summaries. Chapter summaries are organized in outline form corresponding to the major sections and subsections of each chapter. This section-by-section format is coupled with a list of the key terms introduced in each section, providing a succinct but comprehensive review of the material.

Questions and Answers. Questions at the end of each chapter (with answers in the back of the book) are designed to further facilitate review by calling for students to understand and integrate the material presented in the chapter and to use this material to predict or interpret experimental results.

References. Comprehensive lists of references at the end of each chapter provide access to both reviews and selected papers from the primary literature. In order to help the student identify articles of interest, the references are organized according to chapter sections. Review articles and primary papers are distinguished by [R] and [P] designations, respectively.

Acknowledgments

Many people made critical contributions to both the first and second editions of *The Cell*, and it is a pleasure to thank them for their efforts.

The project was initiated by Patrick Fitzgerald at ASM Press. His efforts were carried on for the second edition by Jeff Holtmeier, who provided a continued source of encouragement and support. The help of Karen Jones and Norma Davis in obtaining reviews of the second edition is also greatly appreciated.

Both the first and second editions of *The Cell* benefited greatly from the many scientists who gave generously of their time to read and critique the text. The reviewers of both editions are listed separately on the following pages, and I am deeply grateful for their efforts and advice. It is a special pleasure to acknowledge the help of Elaine Lai (Brandeis University), who read and critiqued the entire text of the second edition, providing extremely valuable advice from her perspective as both a scientist and a teacher of cell biology.

The production of the second edition of *The Cell* was again undertaken by Andy Sinauer and his colleagues at Sinauer Associates. I am delighted to thank Andy and Dean Scudder for their continued help and encouragement throughout all stages of the project. It is also a pleasure to acknowledge the superb efforts of Joyce Zymeck, who did a wonderful job editing the text, David McIntyre, who obtained the many new micrographs included in the book, and Christopher Small, who was responsible for production of the completed volume. The artwork is the product of Patrick Lane of the J/B Woolsey studio, who once again succeeded in turning my rough sketches into an outstanding art program. The cover illustration was created especially for the second edition of *The Cell* by David S. Goodsell of the Scripps Research Institute, and I am most grateful for his contribution.

Finally, it is a pleasure to acknowledge the careful and cheerful work of Sean Valles, who compiled and copied the many research papers I used as references to bring the second edition of *The Cell* up to date.

Reviewers of the Second Edition

Dr. Randy Bennett
Brigham Young University

Dr. James T. Bradley
Auburn University

Dr. Ulla Hansen
Boston University

Dr. Richard Holdeman
Indiana University, Bloomington

Dr. Elaine Lai
Brandeis University

Dr. Umadas Maitra
Albert Einstein College of Medicine

Dr. Harry Nickla
Creighton University

Dr. Richard Nuccitelli
University Of California, Davis

Dr. Clare M. O'Connor
Boston College

Dr. Alan Wildeman
University of Guelph
Ontario, Canada

Reviewers of the First Edition

Gerald M. Adair
M. D. Anderson Cancer Center
University of Texas

Ibrahim Ades
University of Maryland, College Park

Ronald Berezney
State University of New York at Buffalo

Niels Bols
University of Waterloo
Ontario, Canada

Joan S. Brugge
ARIAD Pharmaceuticals, Inc.
Cambridge, MA

Brian Burke
The University of Calgary
Alberta, Canada

David A. Clayton
Stanford University School of Medicine

Roger I. Davis
Howard Hughes Medical Institute
University of Massachusetts
Medical Center

Michael Edidin
Johns Hopkins University

Gary Felsenfeld
National Institutes of Health
Bethesda, MD

Michael Forman
Purdue University

Errol C. Friedberg
University of Texas Southwestern
Medical Center

David Fromson
California State University, Fullerton

Larry Gerace
Scripps Research Institute
La Jolla, CA

Reid Gilmore
University of Massachusetts
Medical Center

Dennis Goode
University of Maryland

Michael M. Gottesman
National Cancer Institute
National Institutes of Health

Michael E. Greenberg
Children's Hospital
Harvard Medical School

Barry Gumbiner
Memorial Sloan-Kettering Cancer Center

Alan Hall
MRC Laboratory for Molecular Cell Biology
University College London

Joyce L. Hamlin
University of Virginia, Charlottesville

Ulla Hansen
Dana-Farber Cancer Institute
Harvard Medical School

F. Ulrich Hartl
Howard Hughes Medical Institute
Memorial Sloan-Kettering Cancer Center

Lawrence Hightower
University of Connecticut, Storrs

Tony Hunter
Salk Institute
La Jolla, CA

James. N. Ihle
St. Jude's Children's Research Hospital
Memphis, TN

Judith A. Jaehning
Indiana University, Bloomington

Haig H. Kazazian, Jr.
University of Pennsylvania

Kenneth Keegstra
MSU-DOE Plant Research Laboratory
Michigan State University

Robert A. Koch
California State University, Fullerton

Elaine Lai
Brandeis University

Jerry B. Lingrel
University of Cincinnati Medical Center

Elizabeth J. Luna
Worcester Foundation for Experimental
Biology
Shrewsbury, MA

Carolyn Machamer
Johns Hopkins University School of
Medicine

Umadas Maitra
Albert Einstein College of Medicine

Russell Malmberg
University of Georgia, Athens

James Manley
Columbia University

J. Richard McIntosh
University of Colorado, Boulder

Donald Miles
University of Missouri, Columbia

David A. Mullin
Tulane University

Eva Neer
Harvard Medical School

Carol S. Newlon
UMDNJ/New Jersey Medical School

John Newport
University of California, San Diego

Danton O'Day
University of Toronto, Erindale Campus

Donata Oertel
University of Wisconsin, Madison

Barbara Pearse
MRC Laboratory of Molecular Biology
Cambridge, UK

Hugh Pelham
Medical Research Council
Cambridge, UK

Jonathon Pines
The Wellcome Cancer Research Centre
Cambridge, UK

T. A. Rapoport
Max-Delbrück-Centrum
Berlin, Germany

Marilyn D. Resh
Memorial Sloan-Kettering Cancer Center

Joel D. Richter
Worcester Foundation for Biomedical
Research
Shrewsbury, MA

David A. Rintoul
Kansas State University, Manhattan

Thomas M. Roberts
Florida State University, Tallahassee

Alan Sachs
University of California, Berkeley

Shelley Sazer
Baylor College of Medicine

Gottfried Schatz
University of Basel
Switzerland

Randy Schekman
University of California, Berkeley

Tom Shrader
Albert Einstein College of Medicine

David Schlessinger
Washington University, St. Louis

Kai Simons
European Molecular Biology Laboratory
Heidelberg, Germany

Roger Sloboda
Dartmouth College

Timothy A. Springer
Harvard Medical School

Joseph M. Steffen
University of Louisville

Thomas P. Stossel
Brigham and Women's Hospital
Harvard Medical School

Suresh Subramani
University of California, San Diego

John Taylor
Wayne State University

Patricia Wadsworth
University of Massachusetts, Amherst

Jonathan R. Warner
Albert Einstein College of Medicine

Larry G. Williams
Kansas State University, Manhattan

Fred Wilt
University of California, Berkeley

Debra J. Wolgemuth
Columbia University

SUPPLEMENTS TO ACCOMPANY
The Cell

FOR THE STUDENT

Student CD-ROM (ISBN 0-87893-074-4)

Adopters have the option of bundling *The Cell* with a special edition of the CD-ROM *Molecules, Cells, and Genes*. This invaluable resource is keyed to the textbook and combines the essential features of a Study Guide and a Problems book. The CD-ROM includes:

- multiple-choice quiz modules for each chapter, with feedback provided for every possible response
- several problems with answers for each textbook chapter
- over 30 minicourses, keyed to text chapters and covering over 1,000 topics in basic cell biology, molecular biology, genetics, immunology, terminology, cancer, and more
- more than 100 animated sequences, with narration
- 800 original illustrations
- 17 video microscopy sequences
- over 125 micrographs
- an extensive library of spoken pronunciations
- an Index that allows the student to instantly locate every Key Term
- endless hypertext links that provide easy navigation throughout the program
- web links that guide the student to useful and effective web sites to further augment the learning process

World Wide Web Site [www.sinauer.com/cooper/]

This site provides links, for each chapter, to resources on the World Wide Web.

FOR THE INSTRUCTOR

Instructor's Resource CD-ROM (ISBN 0-87893-072-8)

This CD-ROM includes:

- all of the full-color figures and tables from the text
- video clips that include: cell signaling proteins localized in relation to cell behavior; interflagellar transport (GFP-labeled); Ca^{++} wave during frog egg fertilization; 4-cell-stage embryo; lysosomes and mitochondrial transport; lamellipodia and wound closure; actin nucleation and retrograde flow; epidermal migration; endoplasmic reticulum extension; stationary actin; and nucleation
- animations from the Student CD-ROM
- over 200 electron micrographs
- all problems and solutions from the Student CD-ROM
- the computerized test file
- PowerPoint® presentations
- PowerPoint® tutorial for new users

Overhead Transparencies (ISBN 0-87893-104-X)

A set of 150 full-color figures from the textbook is available as overhead transparencies.

Test File (ISBN 0-87893-103-1)

A collection of over 500 multiple-choice and true/false questions is available in printed form, or in Windows or Macintosh computer formats (on the Instructor's Resource CD-ROM).

Other Media

The following videos and CD-ROMs are available to qualified adopters of the text:

- Sardet, Larsonneur, and Koch: *Voyage Inside the Cell* (ISBN 0-87893-763-3, NTSC Video)
- Pickett-Heaps and Pickett-Heaps: *Living Cells: Structure and Diversity* (ISBN 0-646-29291-9, NTSC Video)
- Pickett-Heaps and Pickett-Heaps: *From Egg to Tadpole: Early Morphogenesis in* Xenopus (ISBN 09586081-1-3, NTSC Video)
- Pickett-Heaps and Pickett-Heaps: *Predatory Tactics: Survival in the Microcosmos* (ISBN 0-646-29290-0, NTSC Video)
- Fink: *CELLebration* (ISBN 0-87893-166-X, NTSC Video)
- Tyler, Schnetzer, and Tartaglia: *Fly Cycle: The Lives of a Fly*, Drosophila melanogaster (ISBN 0-87893-837-0, NTSC Video)
- Tyler, Kozlowski, and Iten: *FlyCycle* CD (ISBN 0-87893-839-7, CD-ROM)
- Tyler and Kozlowski: *Vade Mecum: An Interactive Guide to Developmental Biology* (ISBN 0-87893-842-7, CD-ROM)

PART I

Introduction

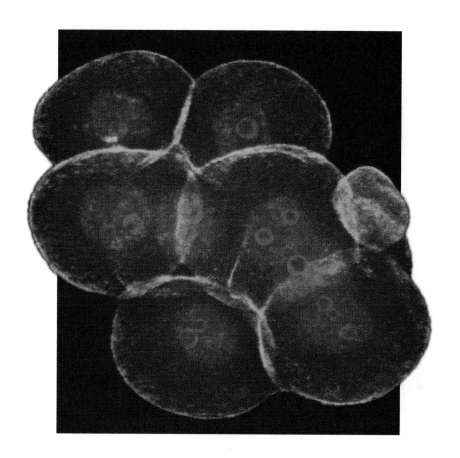

1

An Overview of Cells and Cell Research

UNDERSTANDING THE MOLECULAR BIOLOGY OF CELLS is an active area of research that is fundamental to all of the biological sciences. This is true not only from the standpoint of basic science, but also with respect to a growing number of practical applications in agriculture, biotechnology, and medicine. Medical applications provide some particularly exciting examples, with new approaches to prevention and treatment being made possible by an increased understanding of the cellular and molecular basis of many human diseases.

Because cell and molecular biology is a rapidly growing field of research, this chapter will focus on how cells are studied, as well as reviewing some of their basic properties. Appreciating the similarities and differences between cells is particularly important to understanding cell biology. The first section of this chapter therefore discusses both the unity and the diversity of present-day cells in terms of their evolution from a common ancestor. On the one hand, all cells share common fundamental properties that have been conserved throughout evolution. For example, all cells employ DNA as their genetic material, are surrounded by plasma membranes, and use the same basic mechanisms for energy metabolism. On the other hand, present-day cells have evolved a variety of different lifestyles. Many organisms, such as bacteria, amoebas, and yeasts, consist of single cells that are capable of independent self-replication. More complex organisms are composed of collections of cells that function in a coordinated manner, with different cells specialized to perform particular tasks. The human body, for example, is composed of more than 200 different kinds of cells, each specialized for such distinctive functions as memory, sight, movement, and digestion. The diversity exhibited by the many different kinds of cells is striking; for example, consider the differences between bacteria and the cells of the human brain.

The fundamental similarities between different types of cells provide a unifying theme to cell biology, allowing the basic principles learned from experiments with one kind of cell to be extrapolated and generalized to other cell types. Several kinds of cells and organisms are widely used to study different aspects of cell and molecular biology; the second section of

3

this chapter discusses some of the properties of these cells that make them particularly valuable as experimental models. Finally, it is important to recognize that progress in cell biology depends heavily on the availability of experimental tools that allow scientists to make new observations or conduct novel kinds of experiments. This introductory chapter therefore concludes with a discussion of some of the experimental approaches used to study cells, as well as a review of some of the major historical developments that have led to our current understanding of cell structure and function.

The Origin and Evolution of Cells

Cells are divided into two main classes, initially defined by whether they contain a nucleus. **Prokaryotic cells** (bacteria) lack a nuclear envelope; **eukaryotic cells** have a nucleus in which the genetic material is separated from the cytoplasm. Prokaryotic cells are generally smaller and simpler than eukaryotic cells; in addition to the absence of a nucleus, their genomes are less complex and they do not contain cytoplasmic organelles or a cytoskeleton (Table 1.1). In spite of these differences, the same basic molecular mechanisms govern the lives of both prokaryotes and eukaryotes, indicating that all present-day cells are descended from a single primordial ancestor. How did this first cell develop? And how did the complexity and diversity exhibited by present-day cells evolve?

The First Cell

It appears that life first emerged at least 3.8 billion years ago, approximately 750 million years after Earth was formed (Figure 1.1). How life originated and how the first cell came into being are matters of speculation, since these events cannot be reproduced in the laboratory. Nonetheless, several types of experiments provide important evidence bearing on some steps of the process.

It was first suggested in the 1920s that simple organic molecules could form and spontaneously polymerize into macromolecules under the conditions thought to exist in primitive Earth's atmosphere. At the time life arose, the atmosphere of Earth is thought to have contained little or no free oxygen, instead consisting principally of CO_2 and N_2 in addition to smaller amounts of gases such as H_2, H_2S, and CO. Such an atmosphere provides reducing conditions in which organic molecules, given a source of energy such as sunlight or electrical discharge, can form spontaneously. The spontaneous formation of organic molecules was first demonstrated experimentally in the 1950s, when Stanley Miller (then a graduate stu-

TABLE 1.1 Prokaryotic and Eukaryotic Cells

Characteristic	Prokaryote	Eukaryote
Nucleus	Absent	Present
Diameter of a typical cell	$\approx 1\ \mu m$	$10–100\ \mu m$
Cytoskeleton	Absent	Present
Cytoplasmic organelles	Absent	Present
DNA content (base pairs)	1×10^6 to 5×10^6	1.5×10^7 to 5×10^9
Chromosomes	Single circular DNA molecule	Multiple linear DNA molecules

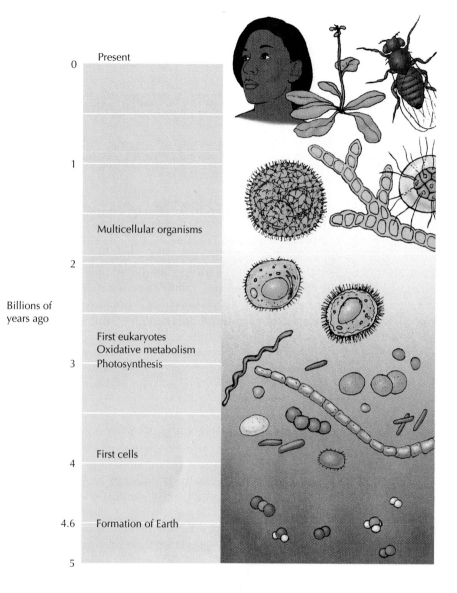

Present

0

1

Multicellular organisms

2

Billions of years ago

First eukaryotes
Oxidative metabolism

3 Photosynthesis

First cells

4

4.6 Formation of Earth

5

dent) showed that the discharge of electric sparks into a mixture of H_2, CH_4, and NH_3, in the presence of water, led to the formation of a variety of organic molecules, including several amino acids (Figure 1.2). Although Miller's experiments did not precisely reproduce the conditions of primitive Earth, they clearly demonstrated the plausibility of the spontaneous synthesis of organic molecules, providing the basic materials from which the first living organisms arose.

The next step in evolution was the formation of macromolecules. The monomeric building blocks of macromolecules have been demonstrated to polymerize spontaneously under plausible prebiotic conditions. Heating dry mixtures of amino acids, for example, results in their polymerization to form polypeptides. But the critical characteristic of the macromolecule from

Organic molecules
- Alanine
- Aspartic acid
- Glutamic acid
- Glycine
- Urea
- Lactic acid
- Acetic acid
- Formic acid

Figure 1.2
Spontaneous formation of organic molecules Water vapor was refluxed through an atmosphere consisting of CH_4, NH_3, and H_2, into which electric sparks were discharged. Analysis of the reaction products revealed the formation of a variety of organic molecules, including the amino acids alanine, aspartic acid, glutamic acid, and glycine.

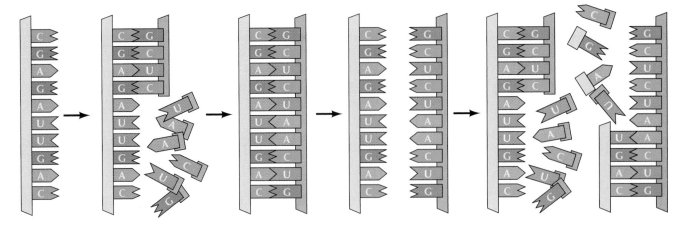

Figure 1.3
Self-replication of RNA
Complementary pairing between nucleotides (adenine [A] with uracil [U] and guanine [G] with cytosine [C]) allows one strand of RNA to serve as a template for the synthesis of a new strand with the complementary sequence.

which life evolved must have been the ability to replicate itself. Only a macromolecule capable of directing the synthesis of new copies of itself would have been capable of reproduction and further evolution.

Of the two major classes of informational macromolecules in present-day cells (nucleic acids and proteins), only the nucleic acids are capable of directing their own self-replication. Nucleic acids can serve as templates for their own synthesis as a result of specific base pairing between complementary nucleotides (Figure 1.3). A critical step in understanding molecular evolution was thus reached in the early 1980s, when it was discovered in the laboratories of Sid Altman and Tom Cech that RNA is capable of catalyzing a number of chemical reactions, including the polymerization of nucleotides. RNA is thus uniquely able both to serve as a template for and to catalyze its own replication. Consequently, RNA is generally believed to have been the initial genetic system, and an early stage of chemical evolution is thought to have been based on self-replicating RNA molecules—a period of evolution known as the **RNA world**. Ordered interactions between RNA and amino acids then evolved into the present-day genetic code, and DNA eventually replaced RNA as the genetic material.

The first cell is presumed to have arisen by the enclosure of self-replicating RNA in a membrane composed of **phospholipids** (Figure 1.4). As discussed in detail in the next chapter, phospholipids are the basic components of all present-day biological membranes, including the plasma

Figure 1.4
Enclosure of self-replicating RNA in a phospholipid membrane   The first cell is thought to have arisen by the enclosure of self-replicating RNA and associated molecules in a membrane composed of phospholipids. Each phospholipid molecule has two long hydrophobic tails attached to a hydrophilic head group. The hydrophobic tails are buried in the lipid bilayer; the hydrophilic heads are exposed to water on both sides of the membrane.

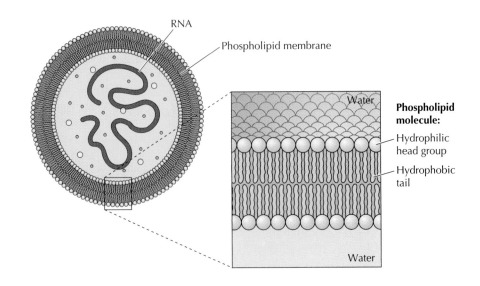

membranes of both prokaryotic and eukaryotic cells. The key characteristic of the phospholipids that form membranes is that they are **amphipathic** molecules, meaning that one portion of the molecule is soluble in water and another portion is not. Phospholipids have long, water-insoluble (**hydrophobic**) hydrocarbon chains joined to water-soluble (**hydrophilic**) head groups that contain phosphate. When placed in water, phospholipids spontaneously aggregate into a bilayer with their phosphate-containing head groups on the outside in contact with water and their hydrocarbon tails in the interior in contact with each other. Such a phospholipid bilayer forms a stable barrier between two aqueous compartments—for example, separating the interior of the cell from its external environment.

The enclosure of self-replicating RNA and associated molecules in a phospholipid membrane would thus have maintained them as a unit, capable of self-reproduction and further evolution. RNA-directed protein synthesis may already have evolved by this time, in which case the first cell would have consisted of self-replicating RNA and its encoded proteins.

The Evolution of Metabolism

Because cells originated in a sea of organic molecules, they were able to obtain food and energy directly from their environment. But such a situation is self-limiting, so cells needed to evolve their own mechanisms for generating energy and synthesizing the molecules necessary for their replication. The generation and controlled utilization of metabolic energy is central to all cell activities, and the principal pathways of energy metabolism (discussed in detail in Chapter 2) are highly conserved in present-day cells. All cells use **adenosine 5′-triphosphate** (**ATP**) as their source of metabolic energy to drive the synthesis of cell constituents and carry out other energy-requiring activities, such as movement (e.g., muscle contraction). The mechanisms used by cells for the generation of ATP are thought to have evolved in three stages, corresponding to the evolution of glycolysis, photosynthesis, and oxidative metabolism (Figure 1.5). The development of these metabolic pathways changed Earth's atmosphere, thereby altering the course of further evolution.

In the initially anaerobic atmosphere of Earth, the first energy-generating reactions presumably involved the breakdown of organic molecules in the absence of oxygen. These reactions are likely to have been a form of present-day **glycolysis**—the anaerobic breakdown of glucose to lactic acid, with the net energy gain of two molecules of ATP. In addition to using ATP

Glycolysis

$C_6H_{12}O_6$ $\longrightarrow$ $2\ C_3H_6O_3$ Generates 2 ATP

Glucose   Lactic acid

Photosynthesis

$6\ CO_2 + 6\ H_2O \longrightarrow C_6H_{12}O_6 + 6\ O_2$

Glucose

Oxidative metabolism

$C_6H_{12}O_6 + 6\ O_2 \longrightarrow 6\ CO_2 + 6\ H_2O$ Generates 36–38 ATP

Glucose

Figure 1.5
Generation of metabolic energy Glycolysis is the anaerobic breakdown of glucose to lactic acid. Photosynthesis utilizes energy from sunlight to drive the synthesis of glucose from CO_2 and H_2O, with the release of O_2 as a by-product. The O_2 released by photosynthesis is used in oxidative metabolism, in which glucose is broken down to CO_2 and H_2O, releasing much more energy than is obtained from glycolysis.

as their source of intracellular chemical energy, all present-day cells carry out glycolysis, consistent with the notion that these reactions arose very early in evolution.

Glycolysis provided a mechanism by which the energy in preformed organic molecules (e.g., glucose) could be converted to ATP, which could then be used as a source of energy to drive other metabolic reactions. The development of **photosynthesis** is generally thought to have been the next major evolutionary step, which allowed the cell to harness energy from sunlight and provided independence from the utilization of preformed organic molecules. The first photosynthetic bacteria, which evolved more than 3 billion years ago, probably utilized H_2S to convert CO_2 to organic molecules—a pathway of photosynthesis still used by some bacteria. The use of H_2O as a donor of electrons and hydrogen for the conversion of CO_2 to organic compounds evolved later and had the important consequence of changing Earth's atmosphere. The use of H_2O in photosynthetic reactions produces the by-product free O_2; this mechanism is thought to have been responsible for making O_2 abundant in Earth's atmosphere.

The release of O_2 as a consequence of photosynthesis changed the environment in which cells evolved and is commonly thought to have led to the development of **oxidative metabolism**. Alternatively, oxidative metabolism may have evolved before photosynthesis, with the increase in atmospheric O_2 then providing a strong selective advantage for organisms capable of using O_2 in energy-producing reactions. In either case, O_2 is a highly reactive molecule, and oxidative metabolism, utilizing this reactivity, has provided a mechanism for generating energy from organic molecules that is much more efficient than anaerobic glycolysis. For example, the complete oxidative breakdown of glucose to CO_2 and H_2O yields energy equivalent to that of 36 to 38 molecules of ATP, in contrast to the 2 ATP molecules formed by anaerobic glycolysis. With few exceptions, present-day cells use oxidative reactions as their principal source of energy.

Present-Day Prokaryotes

Present-day prokaryotes, which include all the various types of bacteria, are divided into two groups—the **archaebacteria** and the **eubacteria**—which diverged early in evolution. Some archaebacteria live in extreme environments, which are unusual today but may have been prevalent in primitive Earth. For example, thermoacidophiles live in hot sulfur springs with temperatures as high as 80°C and pH values as low as 2. The eubacteria include the common forms of present-day bacteria—a large group of organisms that live in a wide range of environments, including soil, water, and other organisms (e.g., human pathogens).

Most bacterial cells are spherical, rod-shaped, or spiral, with diameters of 1 to 10 μm. Their DNA contents range from about 0.6 million to 5 million base pairs, an amount sufficient to encode about 5000 different proteins. The largest and most complex prokaryotes are the **cyanobacteria**, bacteria in which photosynthesis evolved.

The structure of a typical prokaryotic cell is illustrated by *Escherichia coli* (*E. coli*), a common inhabitant of the human intestinal tract (Figure 1.6). The cell is rod-shaped, about 1 μm in diameter and about 2 μm long. Like most other prokaryotes, *E. coli* is surrounded by a rigid **cell wall** composed of polysaccharides and peptides. Within the cell wall is the **plasma membrane**, which is a bilayer of phospholipids and associated proteins. Whereas the cell wall is porous and readily penetrated by a variety of molecules, the plasma membrane provides the functional separation between the inside of the cell and its external environment. The DNA of *E. coli* is a

Figure 1.6
Electron micrograph of *E. coli*
The cell is surrounded by a cell wall, within which is the plasma membrane. DNA is located in the nucleoid. (Menge and Wurtz/Biozentrum, University of Basel/Science Photo Library/Photo Researchers, Inc.)

Plasma membrane

Cell wall

Nucleoid

0.5 μm

single circular molecule in the nucleoid, which, in contrast to the nucleus of eukaryotes, is not surrounded by a membrane separating it from the cytoplasm. The cytoplasm contains approximately 30,000 **ribosomes** (the sites of protein synthesis), which account for its granular appearance.

Eukaryotic Cells

Like prokaryotic cells, all eukaryotic cells are surrounded by plasma membranes and contain ribosomes. However, eukaryotic cells are much more complex and contain a nucleus, a variety of cytoplasmic organelles, and a cytoskeleton (Figure 1.7). The largest and most prominent organelle of eukaryotic cells is the **nucleus**, with a diameter of approximately 5 μm. The nucleus contains the genetic information of the cell, which in eukaryotes is organized as linear rather than circular DNA molecules. The nucleus is the site of DNA replication and of RNA synthesis; the translation of RNA into proteins takes place on ribosomes in the cytoplasm.

In addition to a nucleus, eukaryotic cells contain a variety of membrane-enclosed organelles within their cytoplasm. These organelles provide compartments in which different metabolic activities are localized. Eukaryotic cells are generally much larger than prokaryotic cells, frequently having a cell volume at least a thousandfold greater. The compartmentalization provided by cytoplasmic organelles is what allows eukaryotic cells to function efficiently. Two of these organelles, **mitochondria** and **chloroplasts**, play critical roles in energy metabolism. Mitochondria, which are found in almost all eukaryotic cells, are the sites of oxidative metabolism and are thus responsible for generating most of the ATP derived from the breakdown of organic molecules. Chloroplasts are the sites of photosynthesis and are found only in the cells of plants and green algae. **Lysosomes** and **peroxisomes** also provide specialized metabolic compartments for the digestion of macromolecules and for various oxidative reactions, respectively. In addition, most plant cells contain large **vacuoles** that perform a variety of functions, including the digestion of macromolecules and the storage of both waste products and nutrients.

Because of the size and complexity of eukaryotic cells, the transport of proteins to their correct destinations within the cell is a formidable task. Two cytoplasmic organelles, the **endoplasmic reticulum** and the **Golgi apparatus**, are specifically devoted to the sorting and transport of proteins destined for secretion, incorporation into the plasma membrane, and incorporation into lysosomes. The endoplasmic reticulum is an extensive network of intracellular membranes, extending from the nuclear membrane throughout the cytoplasm. It functions not only in the processing and transport of proteins, but also in the synthesis of lipids. From the endoplasmic reticulum, proteins are transported within small membrane vesicles to the Golgi apparatus, where they are further processed and sorted for transport to their final destinations. In addition to this role in protein transport, the Golgi apparatus serves as a site of lipid synthesis and (in plant cells) as the site of synthesis of some of the polysaccharides that compose the cell wall.

Eukaryotic cells have another level of internal organization: the **cytoskeleton**, a network of protein filaments extending throughout the cytoplasm. The cytoskeleton provides the structural framework of the cell, determining cell shape and the general organization of the cytoplasm. In addition, the cytoskeleton is responsible for the movements of entire cells (e.g., the contraction of muscle cells) and for the intracellular transport and positioning of organelles and other structures, including the movements of chromosomes during cell division.

Animal cell

Figure 1.7
Structures of animal and plant cells
Both animal and plant cells are surrounded by a plasma membrane and contain a nucleus, a cytoskeleton, and many cytoplasmic organelles in common. Plant cells are also surrounded by a cell wall and contain chloroplasts and large vacuoles.

The eukaryotes developed at least 2.7 billion years ago, following some 1 to 1.5 billion years of prokaryotic evolution. Studies of their DNA sequences indicate that the archaebacteria and eubacteria are as different from each other as either is from present-day eukaryotes. Therefore, a very early event in evolution appears to have been the divergence of three lines of descent from a common ancestor, giving rise to present-day archaebacteria, eubacteria, and eukaryotes. Interestingly, many archaebacterial genes are more similar to those of eukaryotes than to those of eubacteria, indicating that the archaebacteria and eukaryotes share a common line of evolutionary descent and are more closely related to each other than either is to the eubacteria (Figure 1.8).

A critical step in the evolution of eukaryotic cells was the acquisition of membrane-enclosed subcellular organelles, allowing the development of the complexity characteristic of these cells. The organelles are thought to have been acquired as a result of the association of prokaryotic cells with the ancestor of eukaryotes.

The hypothesis that eukaryotic cells evolved from a symbiotic association of prokaryotes—**endosymbiosis**—is particularly well supported by studies of mitochondria and chloroplasts, which are thought to have evolved from bacteria living in large cells. Both mitochondria and chloroplasts are similar to bacteria in size, and like bacteria, they reproduce by dividing in two. Most important, both mitochondria and chloroplasts contain their own DNA, which encodes some of their components. The mitochondrial and chloroplast DNAs are replicated each time the organelle

Plant cell

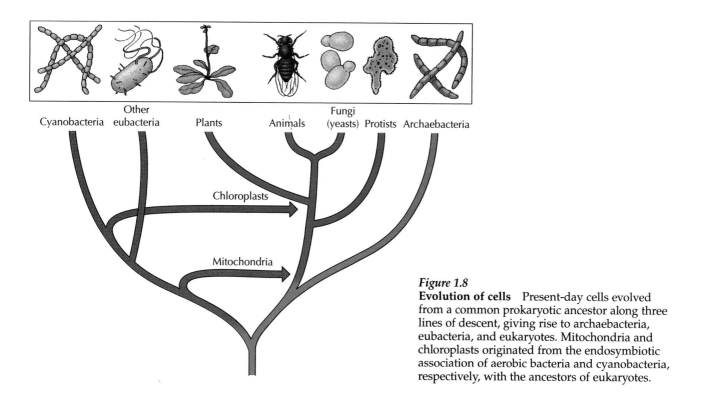

Figure 1.8
Evolution of cells   Present-day cells evolved from a common prokaryotic ancestor along three lines of descent, giving rise to archaebacteria, eubacteria, and eukaryotes. Mitochondria and chloroplasts originated from the endosymbiotic association of aerobic bacteria and cyanobacteria, respectively, with the ancestors of eukaryotes.

TABLE 1.2 DNA Content of Cells

Organism	Haploid DNA content (millions of base pairs)
Bacteria	
Mycoplasma	0.6
E. coli	4.6
Unicellular eukaryotes	
Saccharomyces cerevisiae (yeast)	12
Dictyostelium discoideum	70
Euglena	3000
Plants	
Arabidopsis thaliana	130
Zea mays (corn)	5000
Animals	
Caenorhabditis elegans (nematode)	97
Drosophila melanogaster (fruit fly)	180
Chicken	1200
Zebrafish	1700
Mouse	3000
Human	3000

Figure 1.9
Scanning electron micrograph of *Saccharomyces cerevisiae* Artificial color has been added to the micrograph. (Andrew Syed/Science Photo Library/Photo Researchers, Inc.)

divides, and the genes they encode are transcribed within the organelle and translated on organelle ribosomes. Mitochondria and chloroplasts thus contain their own genetic systems, which are distinct from the nuclear genome of the cell. Furthermore, the ribosomes and ribosomal RNAs of these organelles are more closely related to those of bacteria than to those encoded by the nuclear genomes of eukaryotes.

An endosymbiotic origin for these organelles is now generally accepted, with mitochondria thought to have evolved from aerobic bacteria and chloroplasts from photosynthetic bacteria, such as the cyanobacteria. The acquisition of aerobic bacteria would have provided an anaerobic cell with the ability to carry out oxidative metabolism. The acquisition of photosynthetic bacteria would have provided the nutritional independence afforded by the ability to perform photosynthesis. Thus, these endosymbiotic associations were highly advantageous to their partners and were selected for in the course of evolution. Through time, most of the genes originally present in these bacteria apparently became incorporated into the nuclear genome of the cell, so only a few components of mitochondria and chloroplasts are still encoded by the organelle genomes.

The Development of Multicellular Organisms

Many eukaryotes are unicellular organisms that, like bacteria, consist of only single cells capable of self-replication. The simplest eukaryotes are the **yeasts**. Yeasts are more complex than bacteria, but much smaller and simpler than the cells of animals or plants. For example, the commonly studied yeast *Saccharomyces cerevisiae* is about 6 μm in diameter and contains 12 million base pairs of DNA (Figure 1.9). Other unicellular eukaryotes, however, are far more complex cells, some containing as much DNA as human cells have (Table 1.2). They include organisms specialized to perform a variety of tasks, including photosynthesis, movement, and the capture and ingestion of other organisms as food. *Amoeba proteus*, for example, is a large, complex cell. Its volume is more than 100,000 times that of *E. coli*, and its length can exceed 1 mm when the cell is fully extended (Figure 1.10). Amoebas are highly mobile organisms that use cytoplasmic extensions,

Figure 1.10
Light micrograph of *Amoeba proteus* (M. I. Walker/Photo Researchers, Inc.)

0.2 mm

called **pseudopodia**, to move and to engulf other organisms, including bacteria and yeasts, as food. Other unicellular eukaryotes (the green algae) contain chloroplasts and are able to carry out photosynthesis.

Multicellular organisms evolved from unicellular eukaryotes at least 1.7 billion years ago. Some unicellular eukaryotes form multicellular aggregates that appear to represent an evolutionary transition from single cells to multicellular organisms. For instance, the cells of many algae (e.g., the green alga *Volvox*) associate with each other to form multicellular colonies (Figure 1.11), which are thought to have been the evolutionary precursors of present-day plants. Increasing cell specialization then led to the transition from colonial aggregates to truly multicellular organisms. Continuing cell specialization and division of labor among the cells of an organism have led to the complexity and diversity observed in the many types of cells that make up present-day plants and animals, including human beings.

Plants are composed of fewer cell types than are animals, but each different kind of plant cell is specialized to perform specific tasks required by the organism as a whole (Figure 1.12). The cells of plants are organized into

Figure 1.11
Colonial green algae   Individual cells of *Volvox* form colonies consisting of hollow balls in which hundreds or thousands of cells are embedded in a gelatinous matrix. (Cabisco/Visuals Unlimited.)

Figure 1.12
Light micrographs of representative plant cells   (A) Parenchyma cells, which are responsible for photosynthesis and other metabolic reactions. (B) Collenchyma cells, which are specialized for support and have thickened cell walls. (C) Epidermal cells on the surface of a leaf. Tiny pores (stomata) are flanked by specialized cells called guard cells. (D) Vessel elements and tracheids are elongated cells that are arranged end to end to form vessels of the xylem. (A, Jack M. Bastsack/Visuals Unlimited; B, A. J. Karpoff/Visuals Unlimited; C, Alfred Owczarzak/Biological Photo Service; D, Biophoto Associates/Science Source/Photo Researchers Inc.)

(A) (B)

(C) (D)

50 μm

three main tissue systems: ground tissue, dermal tissue, and vascular tissue. The ground tissue contains **parenchyma cells**, which carry out most of the metabolic reactions of the plant, including photosynthesis. Ground tissue also contains two specialized cell types (**collenchyma cells** and **sclerenchyma cells**) that are characterized by thick cell walls and provide structural support to the plant. Dermal tissue covers the surface of the plant and is composed of **epidermal cells**, which form a protective coat and allow the absorption of nutrients. Finally, several types of elongated cells form the vascular system (the xylem and phloem), which is responsible for the transport of water and nutrients throughout the plant.

The cells found in animals are considerably more diverse than those of plants. The human body, for example, is composed of more than 200 dif-

(A)i Mouth

(A)ii Bile duct

(A)iii Intestine

(B)

(C)

Erythrocyte

Granulocyte

Lymphocyte

Monocyte

Figure 1.13
Light micrographs of representative animal cells   (A) Epithelial cells of the mouth (a thick, multilayered sheet), bile duct, and intestine. (B) Fibroblasts are connective tissue cells characterized by their elongated spindle shape. (C) Erythrocytes, granulocytes, lymphocytes, and monocytes in human blood. [(A)i and (A)ii, G. W. Willis/Biological Photo Service; (A)iii, Biophoto Associates/Photo Researchers, Inc.; B, Don W. Fawcett/Visuals Unlimited; C, G. W. Willis/Biological Photo Service.]

ferent kinds of cells, which are generally considered to be components of five main types of tissues: epithelial tissue, connective tissue, blood, nervous tissue, and muscle (Figure 1.13). **Epithelial cells** form sheets that cover the surface of the body and line the internal organs. There are many different types of epithelial cells, each specialized for a specific function, including protection (the skin), absorption (e.g., the cells lining the small intestine), and secretion (e.g., cells of the salivary gland). Connective tissues include bone, cartilage, and adipose tissue, each of which is formed by different types of cells (osteoblasts, chondrocytes, and adipocytes, respectively). The loose connective tissue that underlies epithelial layers and fills the spaces between organs and tissues in the body is formed by another cell type, the **fibroblast**. Blood contains several different types of cells, which function in oxygen transport (red blood cells, or **erythrocytes**), inflammatory reactions (**granulocytes**, **monocytes**, and **macrophages**), and the immune response (**lymphocytes**). Nervous tissue is composed of nerve cells, or **neurons**, which are highly specialized to transmit signals throughout the body. Various types of sensory cells, such as cells of the eye and ear, are further specialized to receive external signals from the environment. Finally, several different types of muscle cells are responsible for the production of force and movement.

The evolution of animals clearly involved the development of considerable diversity and specialization at the cellular level. Understanding the mechanisms that control the growth and differentiation of such a complex array of specialized cells, starting from a single fertilized egg, is one of the major challenges facing contemporary cell and molecular biology.

Cells As Experimental Models

The evolution of present-day cells from a common ancestor has important implications for cell and molecular biology as an experimental science. Because the fundamental properties of all cells have been conserved during evolution, the basic principles learned from experiments performed with one type of cell are generally applicable to other cells. On the other hand, because of the diversity of present-day cells, many kinds of experiments can be more readily undertaken with one type of cell than with another. Several different kinds of cells and organisms are commonly used as experimental models to study various aspects of cell and molecular biology. The features of some of these cells that make them particularly advantageous as experimental models are discussed in the sections that follow.

E. coli

Because of their comparative simplicity, prokaryotic cells (bacteria) are ideal models for studying many fundamental aspects of biochemistry and molecular biology. The most thoroughly studied species of bacteria is *E. coli*, which has long been the favored organism for investigation of the basic mechanisms of molecular genetics. Most of our present concepts of molecular biology—including our understanding of DNA replication, the genetic code, gene expression, and protein synthesis—derive from studies of this humble bacterium.

E. coli has been especially useful to molecular biologists because of both its relative simplicity and the ease with which it can be propagated and studied in the laboratory. The genome of *E. coli*, for example, consists of approximately 4.6 million base pairs and encodes about 4000 different proteins. The human genome is nearly a thousand times more complex (approximately 3 billion base pairs) and encodes about 100,000 different proteins (see Table 1.2). The small size of the *E. coli* genome provides obvi-

Figure 1.14
Bacterial colonies Photograph of colonies of *E. coli* growing on the surface of agar-containing medium. (A. M. Siegelman/Visuals Unlimited.)

ous advantages for genetic analysis, and the sequence of the entire *E. coli* genome has been determined.

Molecular genetic experiments are further facilitated by the rapid growth of *E. coli* under well-defined laboratory conditions. Depending on the culture conditions, *E. coli* divide every 20 to 60 minutes. Moreover, a clonal population of *E. coli*, in which all cells are derived by division of a single cell of origin, can be readily isolated as a colony grown on semisolid agar-containing medium (Figure 1.14). Because bacterial colonies containing as many as 10^8 cells can develop overnight, selecting genetic variants of an *E. coli* strain—for example, mutants that are resistant to an antibiotic, such as penicillin—is easy and rapid. The ease with which such mutants can be selected and analyzed was critical to the success of experiments that defined the basic principles of molecular genetics, discussed in Chapter 3.

The nutrient mixtures in which *E. coli* divide most rapidly include glucose, salts, and various organic compounds, such as amino acids, vitamins, and nucleic acid precursors. However, *E. coli* can also grow in much simpler media consisting only of salts, a source of nitrogen (such as ammonia), and a source of carbon and energy (such as glucose). In such a medium, the bacteria grow a little more slowly (with a division time of about 40 minutes) because they must synthesize all their own amino acids, nucleotides, and other organic compounds. The ability of *E. coli* to carry out these biosynthetic reactions in simple defined media has made them extremely useful in elucidating the biochemical pathways involved. Thus, the rapid growth and simple nutritional requirements of *E. coli* have greatly facilitated fundamental experiments in both molecular biology and biochemistry.

Yeasts

Although bacteria have been an invaluable model for studies of many conserved properties of cells, they obviously cannot be used to study aspects of cell structure and function that are unique to eukaryotes. Yeasts, the simplest eukaryotes, have a number of experimental advantages similar to those of *E. coli*. Consequently, yeasts have provided a crucial model for studies of many fundamental aspects of eukaryotic cell biology.

The genome of the most frequently studied yeast, *Saccharomyces cerevisiae*, consists of 12 million base pairs of DNA and contains about 6000 genes. Although the yeast genome is approximately three times larger than that of *E. coli*, it is far more manageable than the genomes of more complex eukaryotes, such as humans. Yet even in its simplicity, the yeast cell exhibits the typical features of eukaryotic cells (Figure 1.15): It contains a distinct nucleus surrounded by a nuclear membrane, its genomic DNA is organized as 16 linear chromosomes, and its cytoplasm contains a cytoskeleton and subcellular organelles.

Yeasts can be readily grown in the laboratory and can be studied by many of the same molecular genetic approaches that have proved so successful with *E. coli*. Although yeasts do not replicate as rapidly as bacteria, they still divide as frequently as every 2 hours and can easily be grown as colonies from a single cell. Consequently, yeasts can be used for a variety of genetic manipulations similar to those that can be performed using bacteria.

These features have made yeast cells the most approachable eukaryotic cells from the standpoint of molecular biology. Yeast mutants have been important in understanding many fundamental processes in eukaryotes, including DNA replication, transcription, RNA processing, protein sorting, and the regulation of cell division, as will be discussed in subsequent chapters. The unity of molecular cell biology is made abundantly clear by the fact that the general principles of cell structure and function revealed by studies of yeasts apply to all eukaryotic cells.

Figure 1.15
Electron micrograph of *Saccharomyces cerevisiae* (David Scharf/Peter Arnold, Inc.)

2 μm

Dictyostelium discoideum

Dictyostelium discoideum is a cellular slime mold, which, like yeast, is a comparatively simple unicellular eukaryote. The genome of *Dictyostelium* is approximately ten times larger than that of *E. coli*—more complex than the yeast genome but considerably simpler than the genomes of higher eukaryotes. Moreover, *Dictyostelium* can be readily grown in the laboratory and is amenable to a variety of genetic manipulations.

Under conditions of plentiful food, *Dictyostelium* lives as a single-celled amoeba, feeding on bacteria and yeasts. It is a highly mobile cell, and this property has made *Dictyostelium* an important model for studying the molecular mechanisms responsible for animal cell movements (Figure 1.16). For example, introducing the appropriate mutations into *Dictyostelium* has revealed the roles of several genes in cell motility.

An additional interesting feature of *Dictyostelium* is the ability of single cells to aggregate into multicellular structures. If an adequate supply of food is not available, the cells associate to form wormlike structures called slugs, each consisting of up to 100,000 cells that function as a unit. *Dictyostelium* thus appears to straddle the border between unicellular and multicellular organisms, providing an important model for studies of cell signaling and cell–cell interactions.

Caenorhabditis elegans

The unicellular eukaryotes *Saccharomyces* and *Dictyostelium* are important models for studies of eukaryotic cells, but understanding the development of multicellular organisms requires the experimental analysis of plants and animals, organisms that are more complex. The nematode *Caenorhabditis elegans* (Figure 1.17) possesses several notable features that make it one of the most widely used models for studies of animal development and cell differentiation.

Although the genome of *C. elegans* (approximately 100 million base pairs) is larger than those of unicellular eukaryotes, it is simpler and more manageable than the genomes of most animals. Its complete sequence has been determined, revealing that the genome of *C. elegans* contains approximately 19,000 genes—about three times the number of genes in yeast, and one-fifth the number of genes predicted in humans. Biologically, *C. elegans* is also a relatively simple multicellular organism: Adult worms consist of only 959 **somatic cells**, plus 1000 to 2000 **germ cells**. In addition, *C. elegans* can be easily grown and subjected to genetic manipulations in the laboratory.

The simplicity of *C. elegans* has enabled the course of its development to be studied in detail by microscopic observation. Such analyses have successfully traced the embryonic origin and lineage of all the cells in the adult

10 µm

Figure 1.16
Dictyostelium discoideum These photographs show the movement of two amoebas over the course of about 40 seconds. (Courtesy of David Knecht, University of Connecticut.)

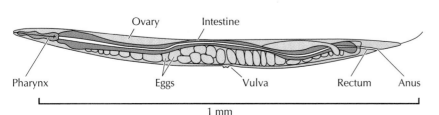

Ovary   Intestine

Pharynx   Eggs   Vulva   Rectum   Anus

1 mm

Figure 1.17
Caenorhabditis elegans   (From J. E. Sulston and H. R. Horvitz, 1977. *Dev. Biol.* 56: 110.)

Figure 1.18
Drosophila melanogaster   (Darwin Dale/Photo Researchers, Inc.)

worm. Genetic studies have also identified some of the mutations responsible for developmental abnormalities, leading to the isolation and characterization of critical genes that control nematode development and differentiation. Importantly, similar genes have also been found to function in complex animals (including humans), making *C. elegans* an important model for studies of animal development.

Drosophila melanogaster

Like *C. elegans*, the fruit fly **Drosophila melanogaster** (Figure 1.18) has been a crucial model organism in developmental biology. The genome of *Drosophila* is similar in size to that of *C. elegans*, and *Drosophila* can be easily maintained and bred in the laboratory. Furthermore, the short reproductive cycle of *Drosophila* (about 2 weeks) makes it a very useful organism for genetic experiments. Many fundamental concepts of genetics—such as the relationship between genes and chromosomes—were derived from studies of *Drosophila* early in the twentieth century (see Chapter 3).

Extensive genetic analysis of *Drosophila* has uncovered many genes that control development and differentiation, and current methods of molecular biology have allowed the functions of these genes to be analyzed in detail. Consequently, studies of *Drosophila* have led to striking advances in understanding the molecular mechanisms that govern animal development, particularly with respect to formation of the body plan of complex multicellular organisms. As with *C. elegans*, similar genes and mechanisms exist in vertebrates, validating the use of *Drosophila* as a major experimental model in contemporary developmental biology.

Arabidopsis thaliana

The study of plant molecular biology and development is an active and expanding field of considerable economic importance as well as intellectual interest. Since the genomes of plants cover a range of complexity comparable to that of animal genomes (see Table 1.2), an optimal model for studies of plant development would be a relatively simple organism with some of the advantageous properties of *C. elegans* and *Drosophila*. The small flowering plant **Arabidopsis thaliana** (Figure 1.19) meets these criteria and is therefore widely used as a model to study the molecular biology of plants.

Arabidopsis is notable for its genome of only about 130 million base pairs—a complexity similar to that of *C. elegans* and *Drosophila*. In addition, *Arabidopsis* is relatively easy to grow in the laboratory, and methods for molecular genetic manipulations of this plant have been developed. These studies have led to the identification of genes involved in various aspects of plant development, such as the development of flowers. Analysis of these genes points to clear similarities between the mechanisms that control the development of plants and animals, further emphasizing the fundamental unity of cell and molecular biology.

Figure 1.19
Arabidopsis thaliana   (Jeremy Burgess/Photo Researchers, Inc.)

Vertebrates

The most complex animals are the vertebrates, including humans and other mammals. The human genome is approximately 3 billion base pairs—about 30 times larger than the genomes of *C. elegans*, *Drosophila*, or *Arabidopsis*. Moreover, the human body is composed of more than 200 different kinds of specialized cell types. This complexity makes the vertebrates difficult to study from the standpoint of cell and molecular biology, but much of the interest in biological sciences nonetheless stems from the desire to understand the human organism. Moreover, an understanding of many questions of immediate practical importance (e.g., in medicine) must be based directly on studies of human (or closely related) cell types.

One important approach to studying human and other mammalian cells is to grow isolated cells in culture, where they can be manipulated under controlled laboratory conditions. The use of cultured cells has allowed studies of many aspects of mammalian cell biology, including experiments that have elucidated the mechanisms of DNA replication, gene expression, protein synthesis and processing, and cell division. Moreover, the ability to culture cells in chemically defined media has allowed studies of the signaling mechanisms that normally control cell growth and differentiation within the intact organism.

The specialized properties of some highly differentiated cell types have made them important models for studies of particular aspects of cell biology. Muscle cells, for example, are highly specialized to undergo contraction, producing force and movement. Because of this specialization, muscle cells are a crucial model for studying cell movement at the molecular level. Another example is provided by nerve cells (neurons), which are specialized to conduct electrochemical signals over long distances. In humans, nerve cell axons may be more than a meter long, and some invertebrates, such as the squid, have giant neurons with axons as large as 1 mm in diameter. Because of their highly specialized structure and function, these giant neurons have provided important models for studies of ion transport across the plasma membrane, and of the role of the cytoskeleton in the transport of cytoplasmic organelles.

The frog *Xenopus laevis* is an important model for studies of early vertebrate development. *Xenopus* eggs are unusually large cells, with a diameter of approximately 1 mm (Figure 1.20). Because those eggs develop outside of the mother, all stages of development from egg to tadpole can be readily studied in the laboratory. In addition, *Xenopus* eggs can be obtained in large numbers, facilitating biochemical analysis. Because of these technical advantages, *Xenopus* has been widely used in studies of developmental biology and has provided important insights into the molecular mechanisms that control development, differentiation, and embryonic cell division.

The **zebrafish** (Figure 1.21) possesses a number of advantages for genetic studies of vertebrate development. These small fish are easy to maintain in the laboratory and they reproduce rapidly. In addition, the embryos develop outside of the mother and are transparent, so that early stages of development can be easily observed. Powerful methods have been developed to facilitate the isolation of mutations affecting zebrafish development, and

1 mm

Figure 1.20
Eggs of the frog *Xenopus laevis*
(Courtesy of Michael Danilchik and Kimberly Ray.)

(A)

Figure 1.21
Zebrafish (A) A 24-hour-old embryo. (B) An adult fish. (A, courtesy of Charles Kimmel, University of Oregon; B, courtesy of S. Kondo)

(B)

Figure 1.22
The mouse as a model for human development A child and a mouse show similar defects in pigmentation (piebaldism) as a result of mutations in a gene required for normal migration of melanocytes (the cells responsible for skin pigmentation) during embryonic development. (Courtesy of R. A. Fleischman, Markey Cancer Center, University of Kentucky.)

several thousand such mutations have now been identified. Because the zebrafish is an easily studied vertebrate, it promises to bridge the gap between humans and the simpler invertebrate systems, such as *C. elegans* and *Drosophila*.

Among mammals, the mouse is the most suitable for genetic analysis. Although the technical difficulties in studying mouse genetics (compared, for example, to the genetics of yeasts or *Drosophila*) are formidable, several mutations affecting mouse development have been identified. Most important, recent advances in molecular biology have enabled the production of **transgenic mice**, in which specific mutant genes have been introduced into the mouse germ line, so that their effects on development or other aspects of cell function can be studied in the context of the whole animal. The suitability of the mouse as a model for human development is illustrated by the fact that mutations in homologous genes result in similar developmental defects in both species; piebaldism is a striking example (Figure 1.22).

Tools of Cell Biology

As in all experimental sciences, research in cell biology depends on the laboratory methods that can be used to study cell structure and function. Many important advances in understanding cells have directly followed the development of new methods that have opened novel avenues of investigation. An appreciation of the experimental tools available to the cell biologist is thus critical to understanding both the current status and future directions of this rapidly moving area of science. Some of the important general methods of cell biology are described in the sections that follow. Other experimental approaches, including the methods of biochemistry and molecular biology, will be discussed in later chapters.

Light Microscopy
Because most cells are too small to be seen by the naked eye, the study of cells has depended heavily on the use of microscopes. Indeed, the very discovery of cells arose from the development of the microscope: Robert Hooke first coined the term "cell" following his observations of a piece of cork with

a simple light microscope in 1665 (Figure 1.23). Using a microscope that magnified objects up to about 300 times their actual size, Antony van Leeuwenhoek, in the 1670s, was able to observe a variety of different types of cells, including sperm, red blood cells, and bacteria. The proposal of the cell theory by Matthias Schleiden and Theodor Schwann in 1838 may be seen as the birth of contemporary cell biology. Microscopic studies of plant tissues by Schleiden and of animal tissues by Schwann led to the same conclusion: All organisms are composed of cells. Shortly thereafter, it was recognized that cells are not formed *de novo* but arise only from division of preexisting cells. Thus, the cell achieved its current recognition as the fundamental unit of all living organisms because of observations made with the light microscope.

The light microscope remains a basic tool of cell biologists, with technical improvements allowing the visualization of ever-increasing details of cell structure. Contemporary light microscopes are able to magnify objects up to about a thousand times. Since most cells are between 1 and 100 μm in diameter, they can be observed by light microscopy, as can some of the larger subcellular organelles, such as nuclei, chloroplasts, and mitochondria. However, the light microscope is not sufficiently powerful to reveal fine details of cell structure, for which **resolution**—the ability of a microscope to distinguish objects separated by small distances—is even more important than magnification. Images can be magnified as much as desired (for example, by projection onto a large screen), but such magnification does not increase the level of detail that can be observed.

The limit of resolution of the light microscope is approximately 0.2 μm; two objects separated by less than this distance appear as a single image, rather than being distinguished from one another. This theoretical limitation of light microscopy is determined by two factors—the wavelength (λ) of visible light and the light-gathering power of the microscope lens (numerical aperture, *NA*)—according to the following equation:

$$\text{Resolution} = \frac{0.61\lambda}{NA}$$

The wavelength of visible light is 0.4 to 0.7 μm, so the value of λ is fixed at approximately 0.5 μm for the light microscope. The numerical aperture can be envisioned as the size of the cone of light that enters the microscope lens after passing through the specimen (Figure 1.24). It is given by the equation

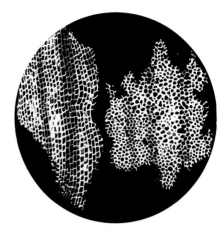

Figure 1.23
The cellular structure of cork
A reproduction of Robert Hooke's drawing of a thin slice of cork examined with a light microscope. The "cells" that Hooke observed were actually only the cell walls remaining from cells that had long since died.

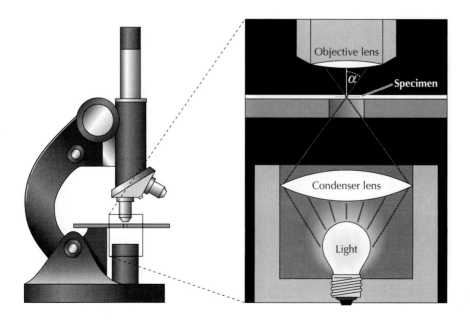

Figure 1.24
Numerical aperture Light is focused on the specimen by the condenser lens and then collected by the objective lens of the microscope. The numerical aperture is determined by the angle of the cone of light entering the objective lens (α) and by the refractive index of the medium (usually air or oil) between the lens and the specimen.

$$NA = \eta \sin \alpha$$

where η is the refractive index of the medium through which light travels between the specimen and the lens. The value of η for air is 1.0, but it can be increased to a maximum of approximately 1.4 by using an oil-immersion lens to view the specimen through a drop of oil. The angle α corresponds to half the width of the cone of light collected by the lens. The maximum value of α is 90°, at which $\sin \alpha = 1$, so the highest possible value for the numerical aperture is 1.4.

The theoretical limit of resolution of the light microscope can therefore be calculated as follows:

$$\text{Resolution} = \frac{0.61 \times 0.5}{1.4} = 0.22 \ \mu\text{m}$$

Microscopes capable of achieving this level of resolution had been made already by the end of the nineteenth century; further improvements in this aspect of light microscopy cannot be expected.

Several different types of light microscopy are routinely used to study various aspects of cell structure. The simplest is **bright-field microscopy**, in which light passes directly through the cell and the ability to distinguish different parts of the cell depends on contrast resulting from the absorption of visible light by cell components. In many cases, cells are stained with dyes that react with proteins or nucleic acids in order to enhance the contrast between different parts of the cell. Prior to staining, specimens are usually treated with fixatives (such as alcohol, acetic acid, or formaldehyde) to stabilize and preserve their structures. The examination of fixed and stained tissues by bright-field microscopy is the standard approach for the analysis of tissue specimens in histology laboratories (Figure 1.25). Such staining procedures kill the cells, however, and therefore are not suitable for many experiments in which the observation of living cells is desired.

Without staining, the direct passage of light does not provide sufficient contrast to distinguish many parts of the cell, limiting the usefulness of bright-field microscopy. However, optical variations of the light microscope

Figure 1.25
Bright-field micrograph of stained tissue Cross section of a hair follicle in human skin, stained with hematoxylin and eosin. (G. W. Willis/ Biological Photo Service.)

Figure 1.26
Microscopic observation of living cells Photomicrographs of human cheek cells obtained with (A) bright-field, (B) phase-contrast, and (C) differential interference-contrast microscopy. (Courtesy of Mort Abramowitz, Olympus America, Inc.)

50 μm

can be used to enhance the contrast between light waves passing through regions of the cell with different densities. The two most common methods for visualizing living cells are **phase-contrast microscopy** and **differential interference-contrast microscopy** (Figure 1.26). Both kinds of microscopy use optical systems that convert variations in density or thickness between different parts of the cell to differences in contrast that can be seen in the final image. In bright-field microscopy, transparent structures (such as the nucleus) have little contrast because they absorb light poorly. However, light is slowed down as it passes through these structures so that its phase is altered compared to light that has passed through the surrounding cytoplasm. Phase-contrast and differential interference-contrast microscopy convert these differences in phase to differences in contrast, thereby yielding improved images of live, unstained cells.

The power of the light microscope has been considerably expanded by the use of video cameras and computers for image analysis and processing. Such electronic image-processing systems can substantially enhance the contrast of images obtained with the light microscope, allowing the visualization of small objects that otherwise could not be detected. For example, **video-enhanced differential interference-contrast microscopy** has allowed visualization of the movement of organelles along microtubules, which are cytoskeletal protein filaments with a diameter of only 0.025 μm (Figure 1.27). However, this enhancement does not overcome the theoretical limit of resolution of the light microscope, approximately 0.2 μm. Thus, although video enhancement allows the visualization of microtubules, the microtubules appear as blurred images at least 0.2 μm in diameter and an individual microtubule cannot be distinguished from a bundle of adjacent structures.

Light microscopy has been brought to the level of molecular analysis by methods for labeling specific molecules so that they can be visualized within cells. Specific genes or RNA transcripts can be detected by hybridization with nucleic acid probes of complementary sequence, and proteins can be detected using appropriate antibodies (see Chapter 3). Both nucleic acid probes and antibodies can be labeled with a variety of tags that allow their visualization in the light microscope, making it possible to determine the location of specific molecules within individual cells.

Fluorescence microscopy is a widely used and very sensitive method for studying the intracellular distribution of molecules (Figure 1.28). A fluorescent dye is used to label the molecule of interest within either fixed or living cells. The fluorescent dye is a molecule that absorbs light at one wavelength and emits light at a second wavelength. This fluorescence is detected by illuminating the specimen with a wavelength of light that excites the fluorescent dye and then using appropriate filters to detect the specific wavelength of light that the dye emits. Fluorescence microscopy

Figure 1.27
Video-enhanced differential interference-contrast microscopy Electronic image processing allows the visualization of single microtubules. (Courtesy of E. D. Salmon, University of North Carolina, Chapel Hill.)

2.5 μm

(A)

(B)

⊢——————⊣
10 μm

Figure 1.28
Fluorescence microscopy (A) Light passes through an excitation filter to select light of the wavelength (e.g., blue) that excites the fluorescent dye. A dichroic mirror then deflects the excitation light down to the specimen. The fluorescent light emitted by the specimen (e.g., green) then passes through the dichroic mirror and a second filter (the barrier filter) to select light of the wavelength emitted by the dye. (B) Fluorescence micrograph of a newt lung cell in which the DNA is stained blue and microtubules in the cytoplasm are stained green. (Conly S. Rieder/Biological Photo Service.)

can be used to study a variety of molecules within cells. One frequent application is to label antibodies directed against a specific protein with fluorescent dyes, so that the intracellular distribution of the protein can be determined. Proteins in living cells can be visualized by using the **green fluorescent protein (GFP)** of jellyfish as a fluorescent label. GFP can be fused to a wide range of proteins using standard methods of recombinant DNA, and the GFP-tagged protein can then be introduced into cells and detected by fluorescence microscopy.

Confocal microscopy combines flourescence microscopy with electronic image analysis to obtain three-dimensional images. A small point of light, usually supplied by a laser, is focused on the specimen at a particular depth. The emitted flourescent light is then collected using a detector, such as a video camera. Before the emitted light reaches the detector, however, it must pass through a pinhole aperture (called a confocal aperture) placed at precisely the point where light emitted from the chosen depth of the specimen comes to a focus (Figure 1.29). Consequently, only light emitted from the plane of focus is able to reach the detector. Scanning across the specimen generates a two-dimensional image of the plane of focus, a much sharper image than that obtained with standard flourescence microscopy (Figure 1.30). Moreover, a series of images obtained at different depths can be used to reconstruct a three-dimensional image of the sample.

Two-photon excitation microscopy is an alternative to confocal microscopy that can be applied to living cells. The specimen is illuminated with a wavelength of light such that excitation of the fluorescent dye requires the simultaneous absorption of two photons (Figure 1.31). The probability of two photons simultaneously exciting the fluorescent dye is only significant at the point in the specimen upon which the input laser beam is focused, so

Figure 1.29
Confocal microscopy A pinpoint of light is focused on the specimen at a particular depth, and emitted fluorescent light is collected by a detector. Before reaching the detector, the fluorescent light emitted by the specimen must pass through a confocal aperture placed at the point where light emitted from the chosen depth of the specimen comes into focus. As a result, only in-focus light emitted from the chosen depth of the specimen is detected.

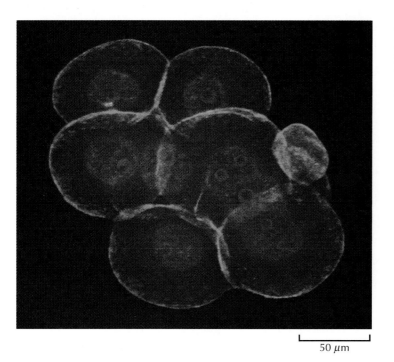

Figure 1.30
Confocal micrograph of mouse embryo cells
Nuclei are stained red and actin filaments under-lying the plasma membrane are stained green. (Courtesy of David Albertini, Tufts University School of Medicine.)

50 μm

fluorescence is only emitted from the plane of focus of the input light. This highly localized excitation automatically provides three-dimensional reso-lution, without the need for passing the emitted light through a pinhole aperture, as in confocal microscopy. Moreover, the localization of excitation minimizes damage to the specimen, allowing three-dimensional imaging of living cells.

Electron Microscopy

Because of the limited resolution of the light microscope, analysis of the details of cell structure has required the use of more powerful microscopic techniques—namely electron microscopy, which was developed in the 1930s and first applied to biological specimens by Albert Claude, Keith Porter, and George Palade in the 1940s and 1950s. The electron microscope can achieve a much greater resolution than that obtained with the light microscope because the wavelength of electrons is shorter than that of light. The wavelength of electrons in an electron microscope can be as short as 0.004 nm—about 100,000 times shorter than the wavelength of visible

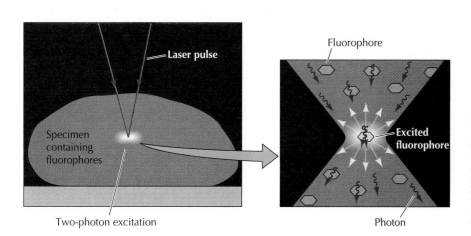

Figure 1.31
Two-photon excitation microscopy
Simultaneous absorption of two pho-tons is required to excite the fluores-cent dye. This only occurs at the point in the specimen upon which the input light is focused, so fluorescent light is only emitted from the chosen depth of the specimen.

5 µm

Figure 1.32
Positive staining   Transmission electron micrograph of a positively stained white blood cell. (Don W. Fawcett/ Visuals Unlimited.)

light. Theoretically, this wavelength could yield a resolution of 0.002 nm, but such a resolution cannot be obtained in practice, because resolution is determined not only by wavelength, but also by the numerical aperture of the microscope lens. Numerical aperture is a limiting factor for electron microscopy because inherent properties of electromagnetic lenses limit their aperture angles to about 0.5 degrees, corresponding to numerical apertures of only about 0.01. Thus, under optimal conditions, the resolving power of the electron microscope is approximately 0.2 nm. Moreover, the resolution that can be obtained with biological specimens is further limited by their lack of inherent contrast. Consequently, for biological samples the practical limit of resolution of the electron microscope is 1 to 2 nm. Although this resolution is much less than that predicted simply from the wavelength of electrons, it represents more than a hundredfold improvement over the resolving power of the light microscope.

Two types of electron microscopy—transmission and scanning—are widely used to study cells. In principle, **transmission electron microscopy** is similar to the observation of stained cells with the bright-field light microscope. Specimens are fixed and stained with salts of heavy metals, which provide contrast by scattering electrons. A beam of electrons is then passed through the specimen and focused to form an image on a fluorescent screen. Electrons that encounter a heavy metal ion as they pass through the sample are deflected and do not contribute to the final image, so stained areas of the specimen appear dark.

Specimens to be examined by transmission electron microscopy can be prepared by either positive or negative staining. In positive staining, tissue specimens are cut into thin sections and stained with heavy metal salts (such as osmium tetroxide, uranyl acetate, and lead citrate) that react with lipids, proteins, and nucleic acids. These heavy metal ions bind to a variety of cell structures, which consequently appear dark in the final image (Figure 1.32). Alternative positive-staining procedures can also be used to identify specific macromolecules within cells. For example, antibodies labeled with electron-dense heavy metals (such as gold particles) are frequently used to determine the subcellular location of specific proteins in the electron microscope. This method is similar to the use of antibodies labeled with fluorescent dyes in fluorescence microscopy.

Negative staining is useful for the visualization of intact biological structures, such as bacteria, isolated subcellular organelles, and macromolecules (Figure 1.33). In this method, the biological specimen is deposited on a supporting film, and a heavy metal stain is allowed to dry around its surface. The unstained specimen is then surrounded by a film of electron-dense

Figure 1.33
Negative staining   Transmission electron micrograph of negatively stained actin filaments. (Courtesy of Roger Craig, University of Massachusetts Medical Center.)

Figure 1.34
Metal shadowing Electron micrograph of actin/myosin filaments of the cytoskeleton prepared by metal shadowing. (Don W. Fawcett, J. Heuser/ Photo Researchers, Inc.)

stain, producing an image in which the specimen appears light against a stained dark background.

Metal shadowing is another technique used to visualize the surface of isolated subcellular structures or macromolecules in the transmission electron microscope (Figure 1.34). The specimen is coated with a thin layer of evaporated metal, such as platinum. The metal is sprayed onto the specimen from an angle so that surfaces of the specimen that face the source of evaporated metal molecules are coated more heavily than others. This differential coating creates a shadow effect, giving the specimen a three-dimensional appearance in electron micrographs.

The preparation of samples by **freeze fracture**, in combination with metal shadowing, has been particularly important in studies of membrane structure. Specimens are frozen in liquid nitrogen (at –196°C) and then fractured with a knife blade. This process frequently splits the lipid bilayer, revealing the interior faces of a cell membrane (Figure 1.35). The specimen

(A)

Proteins

Phospholipids

(B)

Figure 1.35
Freeze fracture (A) Freeze fracture splits the lipid bilayer, leaving proteins embedded in the membrane associated with one of the two membrane halves. (B) Micrograph of freeze-fractured plasma membranes of two adjacent cells. Proteins that span the bilayer appear as intramembranous particles (arrow). (Don W. Fawcett/Photo Researchers, Inc.)

is then shadowed with platinum, and the biological material is dissolved with acid, producing a metal replica of the surface of the sample. Examination of such replicas in the electron microscope reveals many surface bumps, corresponding to proteins that span the lipid bilayer. A variation of freeze fracture called **freeze etching** allows visualization of the external surfaces of cell membranes in addition to their interior faces.

The second type of electron microscopy, **scanning electron microscopy**, is used to provide a three-dimensional image of cells (Figure 1.36). In scanning electron microscopy the electron beam does not pass through the specimen. Instead, the surface of the cell is coated with a heavy metal, and a beam of electrons is used to scan across the specimen. Electrons that are scattered or emitted from the sample surface are collected to generate a three-dimensional image as the electron beam moves across the cell. Because the resolution of scanning electron microscopy is only about 10 nm, its use is generally restricted to studying whole cells rather than subcellular organelles or macromolecules.

Subcellular Fractionation

Although the electron microscope has allowed detailed visualization of cell structure, microscopy alone is not sufficient to define the functions of the various components of eukaryotic cells. To address many of the questions concerning the function of subcellular organelles, it has proven necessary to isolate the organelles of eukaryotic cells in a form that can be used for biochemical studies. This is usually accomplished by **differential centrifugation**—a method developed largely by Albert Claude, Christian de Duve, and their colleagues in the 1940s and 1950s to separate the components of cells on the basis of their size and density.

The first step in subcellular fractionation is the disruption of the plasma membrane under conditions that do not destroy the internal components of the cell. Several methods are used, including sonication (exposure to high-frequency sound), grinding in a mechanical homogenizer, or treatment with a high-speed blender. All these procedures break the plasma membrane and the endoplasmic reticulum into small fragments while leaving other components of the cell (such as nuclei, lysosomes, peroxisomes, mitochondria, and chloroplasts) intact.

The suspension of broken cells (called a lysate or homogenate) is then fractionated into its components by a series of centrifugations in an **ultracentrifuge**, which rotates samples at very high speeds (up to 100,000 rpm) to produce forces up to 500,000 times greater than gravity. This force causes cell components to move toward the bottom of the centrifuge tube and form a pellet (a process called sedimentation) at a rate that depends on their size and density, with the largest and heaviest structures sedimenting most rapidly (Figure 1.37). Usually the cell homogenate is first centrifuged at a low speed, which sediments only unbroken cells and the largest subcellular structures—the nuclei. Thus, an enriched fraction of nuclei can be recovered from the pellet of such a low-speed centrifugation while the other cell components remain suspended in the supernatant (the remaining solution). The supernatant is then centrifuged at higher speed to sediment mitochondria, chloroplasts, lysosomes, and peroxisomes. Recentrifugation of the supernatant at still higher speed sediments fragments of the plasma membrane and the endoplasmic reticulum. A fourth centrifugation at still higher speed sediments ribosomes, leaving only the soluble portion of the cytoplasm (the cytosol) in the supernatant.

The fractions obtained from differential centrifugation correspond to enriched, but still not pure, organelle preparations. A greater degree of

Figure 1.36
Scanning electron microscopy
Scanning electron micrograph of a macrophage. (David Phillips/Visuals Unlimited.)

5 μm

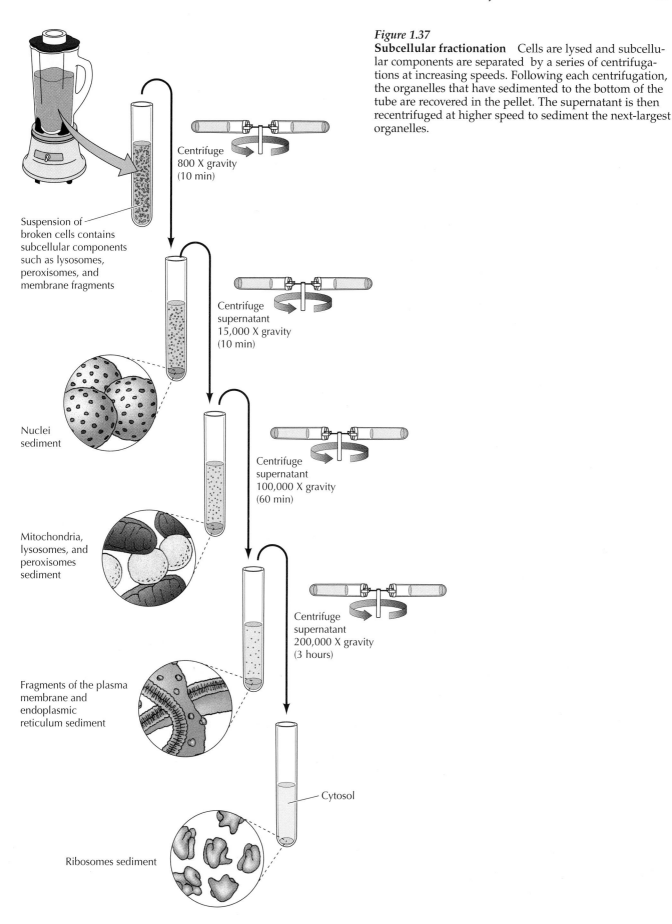

Figure 1.37
Subcellular fractionation Cells are lysed and subcellular components are separated by a series of centrifugations at increasing speeds. Following each centrifugation, the organelles that have sedimented to the bottom of the tube are recovered in the pellet. The supernatant is then recentrifuged at higher speed to sediment the next-largest organelles.

Centrifuge
800 X gravity
(10 min)

Suspension of broken cells contains subcellular components such as lysosomes, peroxisomes, and membrane fragments

Centrifuge
supernatant
15,000 X gravity
(10 min)

Nuclei sediment

Centrifuge
supernatant
100,000 X gravity
(60 min)

Mitochondria, lysosomes, and peroxisomes sediment

Centrifuge
supernatant
200,000 X gravity
(3 hours)

Fragments of the plasma membrane and endoplasmic reticulum sediment

Cytosol

Ribosomes sediment

purification can be achieved by **density-gradient centrifugation**, in which organelles are separated by sedimentation through a gradient of a dense substance, such as sucrose. In **velocity centrifugation**, the starting material is layered on top of the sucrose gradient (Figure 1.38). Particles of different sizes sediment through the gradient at different rates, moving as discrete bands. Following centrifugation, the collection of individual fractions of the gradient provides sufficient resolution to separate organelles of similar size, such as mitochondria, lysosomes, and peroxisomes.

Equilibrium centrifugation in density gradients can be used to separate subcellular components on the basis of their buoyant density, independent of their size and shape. In this procedure, the sample is centrifuged in a gradient containing a high concentration of sucrose or cesium chloride. Rather than being separated on the basis of their sedimentation velocity, the sample particles are centrifuged until they reach an equilibrium position at which their buoyant density is equal to that of the surrounding

Figure 1.38
Velocity centrifugation in a density gradient The sample is layered on top of a gradient of sucrose, and particles of different sizes sediment through the gradient as discrete bands. The separated particles can then be collected in individual fractions of the gradient, which can be obtained simply by puncturing the bottom of the centrifuge tube and collecting drops.

sucrose or cesium chloride solution. Such equilibrium centrifugations are useful in separating different types of membranes from one another and are sufficiently sensitive to separate macromolecules that are labeled with different isotopes. A classic example, discussed in Chapter 3, is the analysis of DNA replication by separating DNA molecules containing heavy and light isotopes of nitrogen (^{15}N and ^{14}N) by equilibrium centrifugation in cesium chloride gradients.

Growth of Animal Cells in Culture

The ability to study cells depends largely on how readily they can be grown and manipulated in the laboratory. Although the process is technically far more difficult than the culture of bacteria or yeasts, a wide variety of animal and plant cells can be grown and manipulated in culture. Such *in vitro* cell culture systems have enabled scientists to study cell growth and differentiation, as well as to perform genetic manipulations required to understand gene structure and function.

Animal cell cultures are initiated by the dispersion of a piece of tissue into a suspension of its component cells, which is then added to a culture dish containing nutrient media. Most animal cell types, such as fibroblasts and epithelial cells, attach and grow on the plastic surface of dishes used for cell culture (Figure 1.39). Because they contain rapidly growing cells, embryos or tumors are frequently used as starting material. Embryo fibroblasts grow particularly well in culture and consequently are one of the most widely studied types of animal cells. Under appropriate conditions, however, some specialized cell types can also be grown in culture, allowing their differentiated properties to be studied in a controlled experimental environment.

The culture media required for the propagation of animal cells are much more complex than the minimal media sufficient to support the growth of bacteria and yeasts. Early studies of cell culture utilized media consisting of undefined components, such as plasma, serum, and embryo extracts. A major advance was thus made in 1955, when Harry Eagle described the first defined media that supported the growth of animal cells. In addition to salts and glucose, the media used for animal cell cultures contain various amino acids and vitamins, which the cells cannot make for themselves. The growth media for most animal cells in culture also include serum, which serves as a source of polypeptide growth factors that are required to stimulate cell division. Several such growth factors have been identified. They serve as critical regulators of cell growth and differentiation in multicellular organisms, providing signals by which different cells communicate with each other. For example, an important function of skin fibroblasts in the intact animal is to proliferate when needed to repair damage resulting from a cut or wound. Their division is triggered by a growth factor released from platelets during blood clotting, thereby stimulating proliferation of fibroblasts in the neighborhood of the damaged tissue. The identification of individual growth factors has made possible the culture of a variety of cells in serum-free media (media in which serum has been replaced by the specific growth factors required for proliferation of the cells in question).

Figure 1.39
Animal cells in culture Scanning electron micrograph of human fibroblasts attached to the surface of a culture dish. (David M. Phillips/Visuals Unlimited.)

10 μm

Key Experiment

Animal Cell Culture

Nutrition Needs of Mammalian Cells in Tissue Culture

Harry Eagle
National Institutes of Health, Bethesda, MD
Science, Volume 122, 1955, pages 501–504

Harry Eagle

The Context

The earliest cell cultures involved the growth of cells from fragments of tissue that were embedded in clots of plasma—a culture system that was far from suitable for experimental analysis. In the late 1940s, a major advance was the establishment of cell lines that grew from isolated cells attached to the surface of culture dishes. But these cells were still grown in undefined media consisting of varying combinations of serum and embryo extracts. For example, a widely used human cancer cell line (called HeLa cells) was initially established in 1952 by growth in a medium consisting of chicken plasma, bovine embryo extract, and human placental cord serum. The use of such complex and undefined culture media made analysis of the specific growth requirements of animal cells impossible. Harry Eagle was the first to solve this problem, by carrying out a systematic analysis of the nutrients needed to support the growth of animal cells in culture.

The Experiments

Eagle studied the growth of two established cell lines: HeLa cells and a mouse fibroblast line called L cells. He was able to grow these cells in a medium consisting of a mixture of salts, carbohydrates, amino acids, and vitamins, supplemented with serum protein. By systematically varying the components of this medium, Eagle was able to determine the specific nutrients required for cell growth. In addition to salts and glu-

cose, these nutrients included 13 amino acids and several vitamins. A small amount of serum protein was also required. The basal medium developed by Eagle is described in the accompanying table, reprinted from his 1955 paper.

The Impact

The medium developed by Eagle is still the basic medium used for animal cell culture. Its use has enabled scientists to grow a wide variety of cells under defined experimental conditions, which has been critical to studies of animal cell growth and differentiation, including identification of the growth factors present in serum—now known to include polypeptides that control the behavior of individual cells within intact animals.

Table 4. Basal media for cultivation of the HeLa cell and mouse fibroblast (*10*)

L-Amino acids* (mM)		Vitamins‡ (mM)		Miscellaneous	
Arginine	0.1	Biotin	10^{-3}	Glucose	5mM§
Cystine	0.05 (0.02)†	Choline	10^{-3}	Penicillin	0.005%#
Glutamine	2.0 (1.0)‖	Folic acid	10^{-3}	Streptomycin	0.005%#
Histidine	0.05 (0.02)†	Nicotinamide	10^{-3}	Phenol red	0.0005%#
Isoleucine	0.2	Pantothenic acid	10^{-3}		
Leucine	0.2 (0.1)†	Pyridoxal	10^{-3}		
Lysine	0.2 (0.1)†	Thiamine	10^{-3}	For studies of cell nutrition	
Methionine	0.05	Riboflavin	10^{-4}	Dialyzed horse serum, 1%†	
Phenylalanine	0.1 (0.05)†			Dialyzed human serum, 5%	
Threonine	0.2 (0.1)†	**Salts§** (mM)			
Tryptophan	0.02 (0.01)†			For stock cultures	
Tyrosine	0.1	NaCl	100	Whole horse serum, 5%†	
Valine	0.2 (0.1)†	KCl	5	Whole human serum, 10%	
		NaH$_2$PO$_4$ · H$_2$O	1		
		NaHCO$_3$	20		
		CaCl$_2$	1		
		MgCl$_2$	0.5		

* Conveniently stored in the refrigerator as a single stock solution containing 20 times the indicated concentration of each amino acid.
† For mouse fibroblast.
‡ Conveniently stored as a single stock solution containing 100 or 1000 times the indicated concentration of each vitamin; kept frozen.
§ Conveniently stored in the refrigerator in two stock solutions, one containing NaCl, KCl, NaH$_2$PO$_4$, NaHCO$_3$, and glucose at 10 times the indicated concentration of each, and the second containing CaCl$_2$ and MgCl$_2$ at 20 times the indicated concentration.
‖ Conveniently stored as a 100mM stock solution; frozen when not in use.
Conveniently stored as a single stock solution containing 100 times the indicated concentrations of penicillin, streptomycin, and phenol red.

The initial cell cultures established from a tissue are called **primary cultures** (Figure 1.40). The cells in a primary culture usually grow until they cover the culture dish surface. They can then be removed from the dish and replated at a lower density to form secondary cultures. This process can be repeated many times, but most normal cells cannot be grown in culture indefinitely. For example, normal human fibroblasts can usually be cultured for 50 to 100 population doublings, after which they stop growing and die. In contrast, cells derived from tumors frequently proliferate indefinitely in culture and are referred to as **immortal cell lines**. In addition, a number of immortalized rodent cell lines have been isolated from cultures of normal fibroblasts. Instead of dying as most of their counterparts do, a few cells in these cultures continue proliferating indefinitely, forming cell lines like those derived from tumors. Such permanent cell lines have been particularly useful for many types of experiments because they provide a continuous and uniform source of cells that can be manipulated, cloned, and indefinitely propagated in the laboratory.

Even under optimal conditions, the division time of most actively growing animal cells is on the order of 20 hours—ten times longer than the division time of yeasts. Consequently, experiments with cultured animal cells are more difficult and take much longer than those with bacteria or yeasts. For example, the growth of a visible colony of animal cells from a single cell takes a week or more, whereas colonies of *E. coli* or yeast develop from single cells overnight. Nonetheless, genetic manipulations of animal cells in culture have been indispensable to our understanding of cell structure and function.

Culture of Plant Cells

Plant cells can also be cultured in nutrient media containing appropriate growth regulatory molecules. In contrast to the polypeptide growth factors that regulate the proliferation of most animal cells, the growth regulators of plant cells are small molecules that can pass through the plant cell wall. When provided with appropriate mixtures of these growth regulatory molecules, many types of plant cells proliferate in culture, producing a mass of undifferentiated cells called a callus (Figure 1.41).

A striking feature of plant cells that contrasts sharply to the behavior of animal cells is the phenomenon called **totipotency**. Differentiated animal cells, such as fibroblasts, cannot develop into other cell types, such as nerve cells. Many plant cells, however, are capable of forming any of the different cell types and tissues ultimately needed to regenerate an entire plant. Consequently, by appropriate manipulation of nutrients and growth regulatory molecules, undifferentiated plant cells in culture can be induced to form a variety of plant tissues, including roots, stems, and leaves. In many cases, even an entire plant can be regenerated from a single cultured cell. In addition to its theoretical interest, the ability to produce a new plant from a single cell that has been

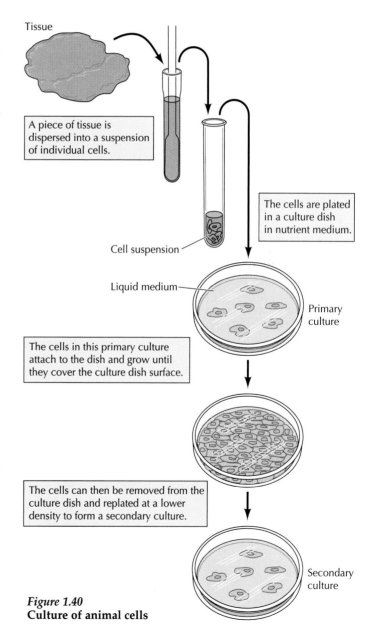

Tissue

A piece of tissue is dispersed into a suspension of individual cells.

Cell suspension

The cells are plated in a culture dish in nutrient medium.

Liquid medium

Primary culture

The cells in this primary culture attach to the dish and grow until they cover the culture dish surface.

The cells can then be removed from the culture dish and replated at a lower density to form a secondary culture.

Secondary culture

Figure 1.40
Culture of animal cells

Figure 1.41
Plant cells in culture An undifferentiated mass of plant cells (a callus) growing on a solid medium. (John N. A. Lott/Biological Photo Service.)

manipulated in culture makes it easy to introduce genetic alterations into plants, opening important possibilities for agricultural genetic engineering.

Viruses

Viruses are intracellular parasites that cannot replicate on their own. They reproduce by infecting host cells and usurping the cellular machinery to produce more virus particles. In their simplest forms, viruses consist only of genomic nucleic acid (either DNA or RNA) surrounded by a protein coat (Figure 1.42). Viruses are important in molecular and cellular biology because they provide simple systems that can be used to investigate the functions of cells. Because virus replication depends on the metabolism of the infected cells, studies of viruses have revealed many fundamental aspects of cell biology. Studies of bacterial viruses contributed substantially to our understanding of the basic mechanisms of molecular genetics, and experiments with a plant virus (tobacco mosaic virus) first demonstrated the genetic potential of RNA. Animal viruses have provided particularly sensitive probes for investigations of various activities of eukaryotic cells.

The rapid growth and small genome size of bacteria make them excellent subjects for experiments in molecular biology, and bacterial viruses (**bacteriophages**) have simplified the study of bacterial genetics even further. One of the most important bacteriophages is T4, which infects and replicates in

(A)

DNA   Capsid proteins

(B)

50 nm

Figure 1.42
Structure of an animal virus
(A) Papillomavirus particles contain a small circular DNA molecule enclosed in a protein coat (the capsid). (B) Electron micrograph of human papillomavirus particles. Artificial color has been added. (B, Alfred Pasieka/Science Photo Library/Photo Researchers, Inc.)

Molecular Medicine

Viruses and Cancer

The Disease

Cancer is a family of diseases characterized by uncontrolled cell proliferation. The growth of normal animal cells is carefully regulated to meet the needs of the complete organism. In contrast, cancer cells grow in an unregulated manner, ultimately invading and interfering with the function of normal tissues and organs. Cancer is the second most common cause of death (next to heart disease) in the United States. Approximately one out of every three Americans will develop cancer at some point in life and, in spite of major advances in treatment, nearly one out of every four Americans ultimately die of this disease. Understanding the causes of cancer and developing more effective methods of cancer treatment therefore represent major goals of medical research.

Molecular and Cellular Basis

Cancer is now known to result from mutations in the genes that normally control cell proliferation. The major insights leading to identification of these genes came from studies of viruses that cause cancer in animals, the prototype of which was isolated by Peyton Rous in 1911. Rous found that sarcomas (a cancer of connective tissues) in chickens could be transmitted by a virus, now known as Rous sarcoma virus, or RSV. Because RSV is a retrovirus with a genome of only 10,000 base pairs, it can be subjected to molecular analysis much more readily than the complex genomes of chickens or other animal cells can. Such studies eventually led

to identification of a specific cancer-causing gene (oncogene) carried by the virus, and to the discovery of related genes in normal cells of all vertebrate species, including humans. Some cancers in humans are now known to be caused by viruses; others result from mutations in normal cell genes similar to the oncogene first identified in RSV.

Prevention and Treatment

The human cancers that are caused by viruses include cervical and other anogenital cancers (papilloma viruses), liver cancer (hepatitis B and C viruses), and some types of lymphomas (Epstein-Barr virus and human T-cell lymphotropic virus). Together, these virus-induced cancers account for about 20% of worldwide cancer incidence. In principle, these cancers could be prevented by vaccination against the responsible viruses, and considerable progress in this area has been made by the development of an effective vaccine against hepatitis B virus.

Other human cancers are caused by mutations in normal cell genes, most of which occur during the lifetime of the individual rather than being inherited. Studies of cancer-causing viruses have led to the identification of many of the genes responsible for non-virus-induced cancers, and to an understanding of the molecular mechanisms responsible for cancer development. Major efforts are now under way to use these insights into the molecular and cellular biology of cancer to develop new approaches to cancer treatment.

Reference

Rous, P. 1911. A sarcoma of the fowl transmissible by an agent separable from the tumor cells. *J. Exp. Med.* 13: 397–411.

The transplantable tumor from which Rous sarcoma virus was isolated.

E. coli. Infection with a single particle of T4 leads to the formation of approximately 200 progeny virus particles in 20 to 30 minutes. The initially infected cell then bursts (lyses), releasing progeny virus particles into the medium, where they can infect new cells. In a culture of bacteria growing on agar medium, the replication of T4 leads to the formation of a clear area

Figure 1.43
Bacteriophage plaques   T4 plaques are visible on a lawn of *E. coli*. Each plaque arises by the replication of a single virus particle. (E. C. S. Chen/ Visuals Unlimited.)

TABLE 1.3 Examples of Animal Viruses

Virus family	Representative member	Genome size (thousands of base pairs)
RNA genomes		
Picornaviruses	Poliovirus	7–8
Togaviruses	Rubella virus	12
Flaviviruses	Yellow fever virus	10
Paramyxoviruses	Measles virus	16–20
Orthomyxoviruses	Influenza virus	14
Retroviruses	Human immunodeficiency virus	9
DNA genomes		
Hepadnaviruses	Hepatitis B virus	3.2
Papovaviruses	Human papillomavirus	5–8
Adenoviruses	Adenovirus	36
Herpesviruses	Herpes simplex virus	120–200
Poxviruses	Vaccinia virus	130–280

of lysed cells (a plaque) in the lawn of bacteria (Figure 1.43). Just as infectious virus particles are easy to grow and assay, viral mutants—for example, viruses that will grow in one strain of *E. coli* but not another—are easy to isolate. Thus, T4 is manipulated even more readily than *E. coli* for studies of molecular genetics. Moreover, the genome of T4 is 20 times smaller than that of *E. coli*—approximately 0.2 million base pairs—further facilitating genetic analysis. Some other bacteriophages have even smaller genomes—the simplest consisting of RNA molecules of only about 3600 nucleotides. Bacterial viruses have thus provided extremely facile experimental systems for molecular genetics. Studies of these viruses are largely what have led to the elucidation of many fundamental principles of molecular biology.

Because of the increased complexity of the animal cell genome, viruses have been even more important in studies of animal cells than in studies of bacteria. Many animal viruses replicate and can be assayed by plaque formation in cell cultures, much as bacteriophages can. Moreover, the genomes of animal viruses are similar in complexity to those of bacterial viruses (ranging from approximately 3000 to 300,000 base pairs), so animal viruses are far more manageable than are their host cells.

There are many diverse animal viruses, each containing either DNA or RNA as their genetic material (Table 1.3). One family of animal viruses—the **retroviruses**—contain RNA genomes in their virus particles but synthesize a DNA copy of their genome in infected cells. These viruses provide a good example of the importance of viruses as models, because studies of the retroviruses are what first demonstrated the synthesis of DNA from RNA templates—a fundamental mode of genetic information transfer now known to occur in both prokaryotic and eukaryotic cells. Other examples in which animal viruses have provided important models for investigations of their host cells include studies of DNA replication, transcription, RNA processing, and protein transport and secretion.

It is particularly noteworthy that infection by some animal viruses, rather than killing the host cell, converts a normal cell into a cancer cell. Studies of such cancer-causing viruses, first described by Peyton Rous in 1911, not only have provided the basis for our current understanding of cancer at the level of cell and molecular biology, but also have led to the elucidation of many of the molecular mechanisms that control animal cell growth and differentiation.

SUMMARY	KEY TERMS

THE ORIGIN AND EVOLUTION OF CELLS

The First Cell: All present-day cells, both prokaryotes and eukaryotes, are descended from a single ancestor. The first cell is thought to have arisen at least 3.8 billion years ago as a result of enclosure of self-replicating RNA in a phospholipid membrane.

prokaryotic cell, eukaryotic cell, RNA world, phospholipid, amphipathic, hydrophobic, hydrophilic

The Evolution of Metabolism: The earliest reactions for the generation of metabolic energy were a form of anaerobic glycolysis. Photosynthesis then evolved, followed by oxidative metabolism.

adenosine 5'-triphosphate (ATP), glycolysis, photosynthesis, oxidative metabolism

Present-Day Prokaryotes: Present-day prokaryotes are divided into two groups, the archaebacteria and the eubacteria, which diverged early in evolution.

archaebacteria, eubacteria, cyanobacteria, *Escherichia coli* (*E. coli*), cell wall, plasma membrane, ribosome

Eukaryotic Cells: Eukaryotic cells, which are larger and more complex than prokaryotic cells, contain a nucleus, cytoplasmic organelles, and a cytoskeleton. They are thought to have evolved from symbiotic associations of prokaryotes.

nucleus, mitochondrion, chloroplast, lysosome, peroxisome, vacuole, endoplasmic reticulum, Golgi apparatus, cytoskeleton, endosymbiosis

The Development of Multicellular Organisms: The simplest eukaryotes are unicellular organisms, such as yeasts and amoebas. Multicellular organisms evolved from associations between such unicellular eukaryotes, and division of labor led to the development of the many kinds of specialized cells that make up present-day plants and animals.

yeast, *Saccharomyces cerevisiae*, pseudopodium, parenchyma cell, collenchyma cell, schlerenchyma cell, epidermal cell, epithelial cell, fibroblast, erythrocyte, granulocyte, monocyte, macrophage, lymphocyte, neuron

CELLS AS EXPERIMENTAL MODELS

E. coli: Because of their genetic simplicity and ease of study, bacteria such as *E. coli* are particularly useful for investigation of fundamental aspects of biochemistry and molecular biology.

Yeasts: As the simplest eukaryotic cells, yeasts are an important model for studying various aspects of eukaryotic cell biology.

Dictyostelium discoideum: The unicellular eukaryote *Dictyostelium* is widely used for experimental analysis of cell movement.

Dictyostelium discoideum

Caenorhabditis elegans: The nematode *C. elegans* is a simple multicellular organism that serves as an important model in developmental biology.

Caenorhabditis elegans, somatic cell, germ cell

Drosophila melanogaster: Because of extensive genetic analysis, studies of the fruit fly *Drosophila* have led to major advances in understanding animal development.

Drosophila melanogaster

Arabidopsis thaliana: The small flowering plant *Arabidopsis* is widely used as a model for studies of plant molecular biology and development.

Arabidopsis thaliana

Vertebrates: Many kinds of vertebrate cells can be grown in culture, where they can be studied under controlled laboratory conditions. Specialized cell types, such as neurons and muscle cells, provide useful models for investigating particular aspects of cell biology. The frog *Xenopus laevis* and zebrafish are important models for studies of early vertebrate development, and the mouse is a mammalian species suitable for genetic analysis.

Xenopus laevis, zebrafish, transgenic mouse

resolution, bright-field microscopy, phase-contrast microscopy, differential interference-contrast microscopy, video-enhanced differential interference-contrast microscopy, fluorescence microscopy, green flourescent protein (GFP), confocal microscopy, two-photon excitation microscopy

transmission electron microscopy, metal shadowing, freeze fracture, freeze etching, scanning electron microscopy

differential centrifugation, ultra-centrifuge, density-gradient centrifugation, velocity centrifugation, equilibrium centrifugation

primary culture, immortal cell line

totipotency

bacteriophage, retrovirus

TOOLS OF CELL BIOLOGY

Light Microscopy: A variety of methods are used to visualize cells and subcellular structures and to determine the intracellular localization of specific molecules using the light microscope.

Electron Microscopy: Electron microscopy, with a resolution that is approximately a hundredfold greater than that of light microscopy, is used to analyze details of cell structure.

Subcellular Fractionation: The organelles of eukaryotic cells can be isolated for biochemical analysis by differential centrifugation.

Growth of Animal Cells in Culture: The propagation of animal cells in culture has allowed studies of the mechanisms that control cell growth and differentiation.

Culture of Plant Cells: Cultured plant cells can differentiate to form specialized cell types and, in some cases, can regenerate entire plants.

Viruses: Viruses provide simple models for studies of cell function.

Questions

1. Why is the evolution of photosynthesis thought to have favored the subsequent evolution of oxidative metabolism?

2. Yeasts have been used as a model for the study of many aspects of the biology of eukaryotic cells. Why are they not a suitable model for analysis of animal cell movements?

3. What resolution can be obtained with a light microscope if the specimen is viewed through air rather than through oil? Assume that the wavelength of visible light is 0.5 μm.

4. You need to separate two organelles that have the same buoyant density but differ in size. Would you use velocity centrifugation or equilibrium centrifugation in density gradients?

5. One advantage of the T4 bacteriophage in the study of molecular genetics is the fact that its genome is 20 times smaller than that of *E. coli*. How much smaller is the genome of Rous sarcoma virus than that of its host?

References and Further Reading

The Origin and Evolution of Cells

Andersson, S. G. E., A. Zomorodipour, J. O. Andersson, T. Sicheritz-Ponten, U. C. M. Alsmark, R. M. Podowski, A. K. Naslund, A.-S. Eriksson, H. H. Winkler and C. G. Kurland. 1998. The genome sequence of *Rickettsia prowazekii* and the origin of mitochondria. *Nature* 396: 133–140. [P]

Brocks, J. J., G. A. Logan, R. Buick and R. E. Summons. 1999. Archean molecular fossils and the early rise of eukaryotes. *Science* 285: 1033-1036. [P]

Bult, C. J. and 39 others. 1996. Complete genome sequence of the methanogenic Archaeon, *Methanococcus jannaschii*. *Science* 273: 1058–1073. [P]

Cech, T. R. 1986. A model for the RNA-catalyzed replication of RNA. *Proc. Natl. Acad. Sci. USA* 83: 4360–4363. [P]

Crick, F. H. C. 1968. The origin of the genetic code. *J. Mol. Biol.* 38: 367–379. [P]

Darnell, J. E. and W. F. Doolittle. 1986. Speculations on the early course of evolution. *Proc. Natl. Acad. Sci. USA* 83: 1271–1275. [P]

Doolittle, W. F. 1999. Phylogenetic classification and the universal tree. *Science* 284: 2124–2128. [R]

Ekland, E. H. and D. P. Bartel. 1996. RNA-catalysed RNA polymerization using nucleoside triphosphates. *Nature* 382: 373–376. [P]

Fawcett, D. W. 1994. *Bloom and Fawcett: A Textbook of Histology.* 12th ed. New York: Chapman and Hall.

Gesteland, R. F., T. R. Cech and J. F., Atkins (eds.). 1999. *The RNA World.* 2nd ed.

Plainview, NY: Cold Spring Laboratory Press.

Gilbert, W. 1986. The RNA world. *Nature* 319: 618. [R]

Joyce, G. F. 1989. RNA evolution and the origins of life. *Nature* 338: 217–224. [R]

Kasting, J. F. 1993. Earth's early atmosphere. *Science* 259: 920–926. [R]

Knoll, A. H. 1992. The early evolution of eukaryotes: A geological perspective. *Science* 256: 622–627. [R]

Margulis, L. 1992. *Symbiosis in Cell Evolution.* 2nd ed. New York: W.H. Freeman.

Miller, S. L. 1953. A production of amino acids under possible primitive earth conditions. *Science* 117: 528–529. [P]

Mojzsis, S. J., G. Arrhenius, K. D. McKeegan, T. M. Harrison, A. P. Nutman and C. R. L. Friend. 1996. Evidence for life on Earth before 3,800 million years ago. *Nature* 384: 55–59. [P]

Orgel, L. E. 1998. The origin of life—a review of facts and speculations. *Trends Biochem. Sci.* 23: 491–495. [R]

Pace, N. R. 1997. A molecular view of microbial diversity and the biosphere. *Science* 276: 734–740. [R]

Shixing, Z. and C. Huineng. 1995. Megascopic multicellular organisms from the 1700-million-year-old Tuanshanzi formation in the Jixian area, north China. *Science* 270: 620–622. [P]

Strickberger, M. W. 1996. *Evolution.* 2nd ed. Boston: Jones and Bartlett.

Woese, C. R., O. Kandler and M. L. Wheelis. 1990. Towards a natural system of organisms: Proposal for the domains Archae, Bacteria, and Eucarya. *Proc. Natl. Acad. Sci. USA* 87: 4576–4579. [P]

Cells as Experimental Models

Blattner, F.R., G. Plunkett III., C. A. Bloch, N. T. Perna, V. Burland, M. Riley, J. Collado-Vides, J. D. Glasner, C. K. Rode, G. F. Mayhew, J. Gregor, N. W. Davis, H. A. Kirkpatrick, M. A. Goeden, D. J. Rose, B. Mau and Y. Shao. 1997. The complete genome sequence of *Escherichia coli* K-12. *Science* 277: 1453–1462. [P]

Botstein, D., S. A. Chervitz and J. M. Cherry. 1997. Yeast as a model organism. *Science* 277: 1259–1260. [R]

Broach, J. R., J. R. Pringle and E. W. Jones (eds.). 1991. *The Molecular and Cellular Biology of the Yeast* Saccharomyces. Plainview, NY: Cold Spring Harbor Laboratory Press.

Condeelis, J. 1993. Understanding the cortex of crawling cells: Insights from *Dictyostelium*. *Trends Cell Biol.* 3: 371–376. [R]

Demerec, M. (ed.). 1994. *Biology of* Drosophila. Plainview, NY: Cold Spring Harbor Laboratory Press.

Eisen, J. S. 1996. Zebrafish make a big splash. *Cell* 87: 969–977. [R]

Goffeau, A. and 15 others. 1996. Life with 6000 genes. *Science* 274: 546–567. [P]

Hogan, B., R. Beddington, F. Costantini and E. Lacey. 1994. *Manipulating the Mouse Embryo.* 2nd ed. Plainview, NY: Cold Spring Harbor Laboratory Press.

Maliga, P., D. F. Klessig, A. R. Cashmore, W. Gruissem and J. E. Varner (eds.). 1995. *Methods in Plant Molecular Biology.* Plainview, NY: Cold Spring Harbor Laboratory Press.

Meinke, D. W., J. M. Cherry, C. Dean, S. D. Rounsley and M. Koornneef. 1998. *Arabidopsis thaliana*: A model plant for genome analysis. *Science* 282: 662–682. [R]

Meyerowitz, E. M. and C. R. Somerville (eds.). 1995. *Arabidopsis.* Plainview, NY: Cold Spring Harbor Laboratory Press.

Miklos, G. L. G., and G. M. Rubin. 1996. The role of the genome project in determining gene function: insights from model organisms. *Cell* 86: 521–529. [R]

Neidhardt, F. C., R. Curtiss III., J. L. Ingraham, E. C. C. Lin, K. B. Low Jr., B. Magasanik, W. Reznikoff, M. Riley, M. Schaechter and H. E. Umbarger (eds.). 1996. Escherichia coli *and* Salmonella: *Cellular and Molecular Biology.* 2nd ed. Washington, DC: ASM Press.

Sive, H. L., R. M. Grainger and R. M. Harland (eds.). 1999. *Early Development of* Xenopus laevis: *A Course Manual.* Plainview, NY: Cold Spring Harbor Laboratory Press.

The *C. elegans* Sequencing Consortium. 1998. Genome sequence of the nematode *C. elegans*: A platform for investigating biology. *Science* 282: 2012–2018. [P]

Tools of Cell Biology

Bowers, W. E. 1998. Christian de Duve and the discovery of lysosomes and peroxisomes. *Trends Cell Biol.* 8: 330–333 [R]

Cairns, J., G. S. Stent and J. D. Watson (eds.). 1992. *Phage and the Origins of Molecular Biology.* Plainview, NY: Cold Spring Harbor Laboratory Press.

Claude, A. 1975. The coming of age of the cell. *Science* 189: 433–435. [R]

De Duve, C. 1975. Exploring cells with a centrifuge. *Science* 189: 186–194. [R]

Eagle, H. 1955. Nutrition needs of mammalian cells in tissue culture. *Science* 235: 442–447. [P]

Flint, S. J., L. W. Enquist, R. M. Krug, V. R. Racaniello and A. M. Skalka. 1999. *Principles of Virology: Molecular Biology, Pathogenesis, and Control.* Washington, DC: ASM Press.

Graham, J. and D. Rickwood (ed.). 1997. *Subcellular Fractionation: A Practical Approach.* New York: Oxford University Press.

Lacey, A. J. (ed.) 1999. *Light Microscopy in Biology: A Practical Approach.* New York: Oxford University Press.

Palade, G. 1975. Intracellular aspects of the process of protein synthesis. *Science* 189: 347–358. [R]

Piston, D. W. 1999. Imaging living cells and tissues by two-photon excitation microscopy. *Trends Cell Biol.* 9: 66–69 [R]

Porter, K. R., A. Claude and E. F. Fullam. 1945. A study of tissue culture cells by electron microscopy. *J. Exp. Med.* 81: 233–246. [P]

Rous, P. 1911. A sarcoma of the fowl transmissible by an agent separable from the tumor cells. *J. Exp. Med.* 13: 397–411. [P]

Salmon, E. D. 1995. VE-DIC light microscopy and the discovery of kinesin. *Trends Cell Biol.* 5: 154–158. [R]

Slayter, E. M. and H. S. Slayter. 1992. *Light and Electron Microscopy.* Cambridge, England: Cambridge University Press.

Spector, D. L., R. Goldman and L. Leinwand. 1998. *Cells: A Laboratory Manual.* Plainview, NY: Cold Spring Harbor Laboratory Press.

Tsien, R. Y. and A Miyawaki. 1998. Seeing the machinery of live cells. *Science* 280: 1954–1955. [R]

2

The Chemistry of Cells

C ELLS ARE INCREDIBLY COMPLEX AND DIVERSE STRUCTURES, capable not only of self-replication—the very essence of life—but also of performing a wide range of specialized tasks in multicellular organisms. Yet cells obey the same laws of chemistry and physics that determine the behavior of nonliving systems. Consequently, modern cell biology seeks to understand cellular processes in terms of chemical and physical reactions.

This chapter considers the fundamental principles of biological chemistry that govern the lives of cells. It is intended neither to be a comprehensive discussion of biochemistry nor to chart all the metabolic reactions within cells. Rather, the chapter will focus on five major topics: the types of molecules within cells, the central role of proteins as biological catalysts, the generation and utilization of metabolic energy, the biosynthesis of major cell constituents, and the structure of biological membranes. An appreciation of these chemical foundations forms the basis for understanding the diverse aspects of cell structure and function that will be discussed throughout the rest of this text.

The Molecular Composition of Cells

Cells are composed of water, inorganic ions, and carbon-containing (organic) molecules. Water is the most abundant molecule in cells, accounting for 70% or more of total cell mass. Consequently, the interactions between water and the other constituents of cells are of central importance in biological chemistry. The critical property of water in this respect is that it is a polar molecule, in which the hydrogen atoms have a slight positive charge and the oxygen has a slight negative charge (Figure 2.1). Because of their polar nature, water molecules can form hydrogen bonds with each other or with other polar molecules, as well as interacting with positively or negatively charged ions. As a result of these interactions, ions and polar molecules are readily soluble in water (hydrophilic). In contrast, nonpolar molecules, which cannot interact with water, are poorly soluble in an aqueous environment (hydrophobic). Consequently, nonpolar molecules tend to minimize their contact with water by associating closely with each other

(A)

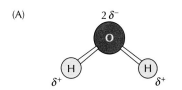

Figure 2.1
Characteristics of water (A) Water is a polar molecule, with a slight negative charge (δ^-) on the oxygen atom and a slight positive charge (δ^+) on the hydrogen atoms. Because of this polarity, water molecules can form hydrogen bonds (dashed lines) either with each other or with other polar molecules (B), in addition to interacting with charged ions (C).

(B)

Hydrogen bond

(C)

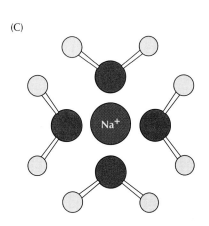

instead. As discussed later in this chapter, such interactions of polar and nonpolar molecules with water and with each other play crucial roles in the formation of biological structures, such as cell membranes.

The inorganic ions of the cell, including sodium (Na^+), potassium (K^+), magnesium (Mg^{2+}), calcium (Ca^{2+}), phosphate (HPO_4^{2-}), chloride (Cl^-), and bicarbonate (HCO_3^-), constitute 1% or less of the cell mass. These ions are involved in a number of aspects of cell metabolism, and thus play critical roles in cell function.

It is, however, the organic molecules that are the unique constituents of cells. Most of these organic compounds belong to one of four classes of molecules: carbohydrates, lipids, proteins, and nucleic acids. Proteins, nucleic acids, and most carbohydrates (the polysaccharides) are macromolecules formed by the joining (polymerization) of hundreds or thousands of low-molecular-weight precursors: amino acids, nucleotides, and simple sugars, respectively. Such macromolecules constitute 80 to 90% of the dry weight of most cells. Lipids are the other major constituent of cells. The remainder of the cell mass is composed of a variety of small organic molecules, including macromolecular precursors. The basic chemistry of cells can thus be understood in terms of the structures and functions of four major classes of organic molecules.

Carbohydrates

The **carbohydrates** include simple sugars as well as polysaccharides. These simple sugars, such as glucose, are the major nutrients of cells. As discussed later in this chapter, their breakdown provides both a source of cellular energy and the starting material for the synthesis of other cell constituents. Polysaccharides are storage forms of sugars and form structural components of the cell. In addition, polysaccharides and shorter polymers of sugars act as markers for a variety of cell recognition processes, including the adhesion of cells to their neighbors and the transport of proteins to appropriate intracellular destinations.

The structures of representative simple sugars (**monosaccharides**) are illustrated in Figure 2.2. The basic formula for these molecules is $(CH_2O)_n$, from which the name carbohydrate is derived (C = "carbo" and H_2O = "hydrate"). The six-carbon ($n = 6$) sugar glucose ($C_6H_{12}O_6$) is especially important in cells, since it provides the principal source of cellular energy. Other simple sugars have between three and seven carbons, with three- and five-carbon sugars being the most common. Sugars containing five or more carbons can cyclize to form ring structures, which are the predominant forms of these molecules within cells. As illustrated in Figure 2.2, the cyclized sugars exist in two alternative forms (called α or β), depending on the configuration of carbon 1.

Monosaccharides can be joined together by dehydration reactions, in which H_2O is removed and the sugars are linked by a **glycosidic bond** between two of their carbons (Figure 2.3). If only a few sugars are joined together, the resulting polymer is called an **oligosaccharide**. If a large number (hundreds or thousands) of sugars are involved, the resulting polymers are macromolecules called **polysaccharides**.

Triose sugars (C₃H₆O₃)

Glyceraldehyde

Dihydroxyacetone

Figure 2.2
Structure of simple sugars Representative sugars containing three, five, and six carbons (triose, pentose, and hexose sugars, respectively) are illustrated. Sugars with five or more carbons can cyclize to form rings, which exist in two alternative forms (α and β) depending on the configuration of carbon 1.

Pentose sugars (C₅H₁₀O₅)

Ribose   Linear form

Ring forms

α β

Hexose sugars (C₆H₁₂O₆)

Glucose   Linear form

Ring forms

α β

Two common polysaccharides—**glycogen** and **starch**—are the storage forms of carbohydrates in animal and plant cells, respectively. Both glycogen and starch are composed entirely of glucose molecules in the α configuration (Figure 2.4). The principal linkage is between carbon 1 of one glucose and carbon 4 of a second. In addition, both glycogen and one form of starch (amylopectin) contain occasional α (1→6) linkages, in which carbon 1 of one glucose is joined to carbon 6 of a second. As illustrated in Figure 2.4, these linkages lead to the formation of branches resulting from the joining of two separate α (1→4) linked chains. Such branches are present in glycogen and amylopectin, although another form of starch (amylose) is an unbranched molecule.

The structures of glycogen and starch are thus basically similar, as is their function: to store glucose. **Cellulose**, in contrast, has a quite distinct function as the principal structural component of the plant cell wall. Perhaps surprisingly, then, cellulose is also composed entirely of glucose molecules. The glucose residues in cellulose, however, are in the β rather than the α configuration, and cellulose is an unbranched polysac-

α (1→ 4) glycosidic bond

Figure 2.3
Formation of a glycosidic bond Two simple sugars are joined by a dehydration reaction (a reaction in which water is removed). In the example shown, two glucose molecules in the α configuration are joined by a bond between carbons 1 and 4, which is therefore called an α (1→4) glycosidic bond.

Amylopectin (starch)

Glycogen

Cellulose

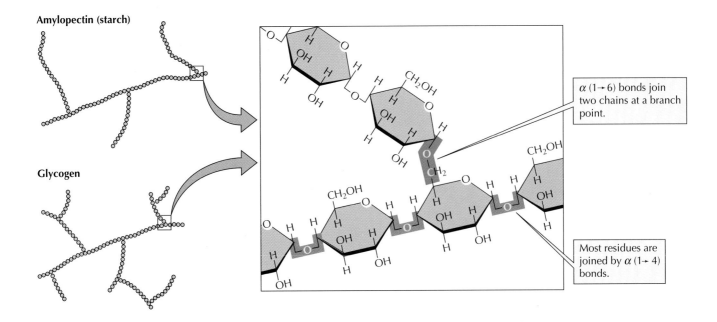

Figure 2.4
Structure of polysaccharides
Polysaccharides are macromolecules consisting of hundreds or thousands of simple sugars. Glycogen, starch, and cellulose are all composed entirely of glucose residues, which are joined by α (1→4) glycosidic bonds in glycogen and starch but by β (1→4) bonds in cellulose. Glycogen and one form of starch (amylopectin) also contain occasional α (1→6) bonds, which serve as branch points by joining two separate α (1→4) chains.

charide (see Figure 2.4). The linkage of glucose residues by β (1→4) rather than α (1→4) bonds causes cellulose to form long extended chains that pack side by side to form fibers of great mechanical strength.

In addition to their roles in energy storage and cell structure, oligosaccharides and polysaccharides are important in a variety of cell signaling processes. For example, oligosaccharides are frequently linked to proteins, where they serve as markers to target proteins for transport to the cell surface or incorporation into different subcellular organelles. Oligosaccharides and polysaccharides also serve as markers on the surface of cells, playing important roles in cell recognition and the interactions between cells in tissues of multicellular organisms.

Lipids

Lipids have three major roles in cells. First, they provide an important form of energy storage. Second, and of great importance in cell biology, lipids are the major components of cell membranes. Third, lipids play important roles in cell signaling, both as steroid hormones (e.g., estrogen and testosterone) and as messenger molecules that convey signals from cell surface receptors to targets within the cell.

The simplest lipids are **fatty acids**, which consist of long hydrocarbon chains, most frequently containing 16 or 18 carbon atoms, with a carboxyl group (COO⁻) at one end (Figure 2.5). Unsaturated fatty acids contain one

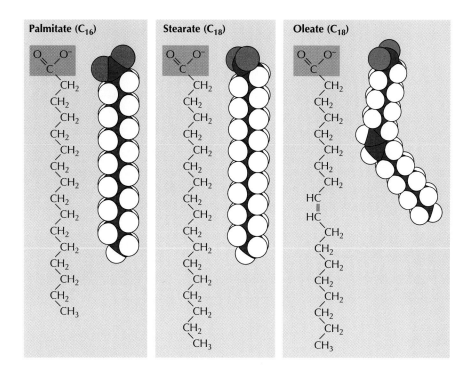

Palmitate (C$_{16}$)   Stearate (C$_{18}$)   Oleate (C$_{18}$)

Figure 2.5
Structure of fatty acids Fatty acids consist of long hydrocarbon chains terminating in a carboxyl group (COO$^-$). Palmitate and stearate are saturated fatty acids consisting of 16 and 18 carbons, respectively. Oleate is an unsaturated 18-carbon fatty acid containing a double bond between carbons 9 and 10. Note that the double bond introduces a kink in the hydrocarbon chain.

or more double bonds between carbon atoms; in saturated fatty acids all of the carbon atoms are bonded to the maximum number of hydrogen atoms. The long hydrocarbon chains of fatty acids contain only nonpolar C—H bonds, which are unable to interact with water. The hydrophobic nature of these fatty acid chains is responsible for much of the behavior of complex lipids, particularly in the formation of biological membranes.

Fatty acids are stored in the form of **triacylglycerols,** or **fats**, which consist of three fatty acids linked to a glycerol molecule (Figure 2.6). Triacylglycerols are insoluble in water and therefore accumulate as fat droplets in the cytoplasm. When required, they can be broken down for use in energy-yielding reactions discussed later in this chapter. It is noteworthy that fats are a more efficient form of energy storage than carbohydrates, yielding more than twice as much energy per weight of material broken down. Fats therefore allow energy to be stored in less than half the body weight that would be required to store the same amount of energy in carbohydrates—a particularly important consideration for animals because of their mobility.

Phospholipids, the principal components of cell membranes, consist of two fatty acids joined to a polar head group (Figure 2.7). In the **glycerol phospholipids**, the two fatty acids are bound to carbon atoms in glycerol, as in triacylglycerols. The third carbon of glycerol, however, is bound to a phosphate group, which is in turn frequently attached to another small polar molecule, such as choline, serine, inositol, or ethanolamine. **Sphingomyelin**, the only nonglycerol phospholipid in cell membranes, contains two hydrocarbon chains linked to a polar head group formed from serine rather than from glycerol. All phospholipids have hydrophobic tails, con-

Glycerol

Fatty acids

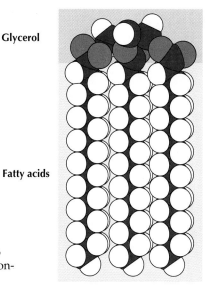

Glycerol

Fatty acids

Figure 2.6
Structure of triacylglycerols Triacylglycerols (fats) contain three fatty acids joined to glycerol. In this example, all three fatty acids are palmitate, but triacylglycerols often contain a mixture of different fatty acids.

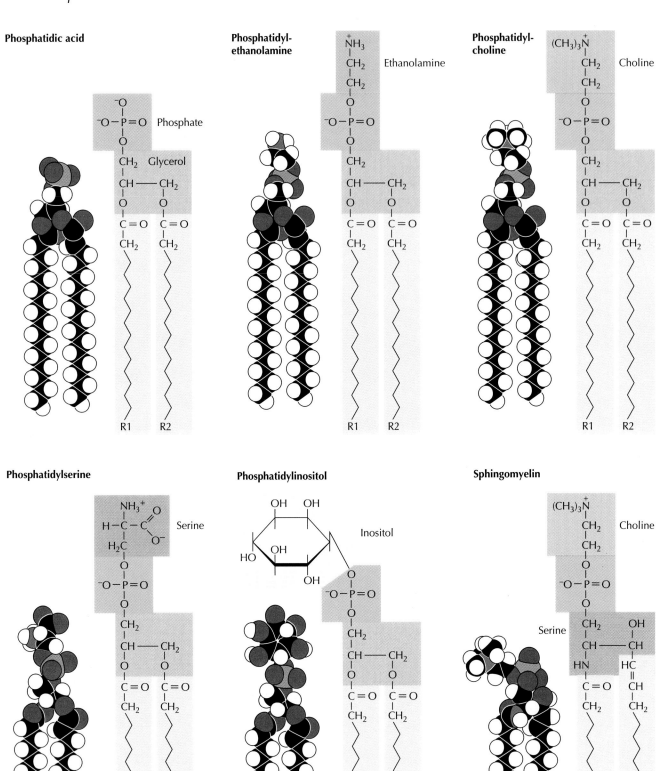

Phosphatidic acid

⁻O
|
⁻O—P=O   Phosphate
|
O
|
CH₂   Glycerol
|
CH——CH₂
| |
O   O
| |
C=O   C=O
| |
CH₂   CH₂

R1   R2

Phosphatidyl-ethanolamine

⁺NH₃
|
CH₂
|
CH₂   Ethanolamine
|
O
|
⁻O—P=O
|
O
|
CH₂
|
CH——CH₂
| |
O   O
| |
C=O   C=O
| |
CH₂   CH₂

R1   R2

Phosphatidyl-choline

(CH₃)₃N⁺
|
CH₂
|
CH₂   Choline
|
O
|
⁻O—P=O
|
O
|
CH₂
|
CH——CH₂
| |
O   O
| |
C=O   C=O
| |
CH₂   CH₂

R1   R2

Phosphatidylserine

NH₃⁺   O
| ‖
H—C—C
| \
H₂C   O⁻   Serine
|
O
|
⁻O—P=O
|
O
|
CH₂
|
CH——CH₂
| |
O   O
| |
C=O   C=O
| |
CH₂   CH₂

R1   R2

Phosphatidylinositol

OH   OH

HO

OH   Inositol

OH
|
O
|
⁻O—P=O
|
O
|
CH₂
|
CH——CH₂
| |
O   O
| |
C=O   C=O
| |
CH₂   CH₂

R1   R2

Sphingomyelin

(CH₃)₃N⁺
|
CH₂
|
CH₂   Choline
|
O
|
⁻O—P=O
|
O
|
CH₂   OH
| |
Serine   CH——CH
| |
HN   HC
| ‖
C=O   CH
| |
CH₂   CH₂

R1   R2

◀ *Figure 2.7*
Structure of phospholipids Glycerol phospholipids contain two fatty acids joined to glycerol. The fatty acids may be different from each other and are designated R1 and R2. The third carbon of glycerol is joined to a phosphate group (forming phosphatidic acid), which in turn is frequently joined to another small polar molecule (forming phosphatidylethanolamine, phosphatidylcholine, phosphatidylserine, or phosphatidylinositol). In sphingomyelin, two hydrocarbon chains are bound to a polar head group formed from serine instead of glycerol.

Figure 2.8
Structure of glycolipids Two hydrocarbon chains are joined to a polar head group formed from serine and containing carbohydrates (e.g., glucose).

sisting of the two hydrocarbon chains, and hydrophilic head groups, consisting of the phosphate group and its polar attachments. Consequently, phospholipids are **amphipathic** molecules, part water-soluble and part water-insoluble. This property of phospholipids is the basis for the formation of biological membranes, as discussed later in this chapter.

In addition to phospholipids, many cell membranes contain **glycolipids** and **cholesterol**. Glycolipids consist of two hydrocarbon chains linked to polar head groups that contain carbohydrates (Figure 2.8). They are thus similar to the phospholipids in their general organization as amphipathic molecules. Cholesterol, in contrast, consists of four hydrocarbon rings rather than linear hydrocarbon chains (Figure 2.9). The hydrocarbon rings are strongly hydrophobic, but the hydroxyl (OH) group attached to one end of cholesterol is weakly hydrophilic, so cholesterol is also amphipathic.

In addition to their roles as components of cell membranes, lipids function as signaling molecules, both within and between cells. The **steroid hormones** (such as estrogens and testosterone) are derivatives of cholesterol (see Figure 2.9). These hormones are a diverse group of chemical messengers, all of which contain four hydrocarbon rings to which distinct functional groups are attached. Derivatives of phospholipids also serve as messenger molecules within cells, acting to convey signals from cell surface receptors to intracellular targets (see Chapter 13).

Figure 2.9
Cholesterol and steroid hormones Cholesterol, an important component of cell membranes, is an amphipathic molecule because of its polar hydroxyl group. Cholesterol is also a precursor to the steroid hormones, such as testosterone and estradiol (a form of estrogen). The hydrogen atoms bonded to the ring carbons are not shown in this figure.

Nucleic Acids

The nucleic acids—DNA and RNA—are the principal informational molecules of the cell. **Deoxyribonucleic acid (DNA)** has a unique role as the genetic material, which in eukaryotic cells is located in the nucleus. Different types of **ribonucleic acid (RNA)** participate in a number of cellular activities. **Messenger RNA (mRNA)** carries information from DNA to the ribosomes, where it serves as a template for protein synthesis. Two other types of RNA (**ribosomal RNA** and **transfer RNA**) are involved in protein synthesis. Still other kinds of RNAs are involved in the processing and transport of both RNAs and proteins. In addition to acting as an informational molecule, RNA is also capable of catalyzing a number of chemical reactions. In present-day cells, these include reactions involved in both protein synthesis and RNA processing.

Figure 2.10
Components of nucleic acids
Nucleic acids contain purine and pyrimidine bases linked to phosphorylated sugars. A nucleic acid base linked to a sugar alone is a nucleoside. Nucleotides additionally contain one or more phosphate groups.

Figure 2.11
Polymerization of nucleotides A phosphodiester bond is formed between the 3′ hydroxyl group of one nucleotide and the 5′ phosphate group of another. A polynucleotide chain has a sense of direction, one end terminating in a 5′ phosphate group (the 5′ end) and the other in a 3′ hydroxyl group (the 3′ end).

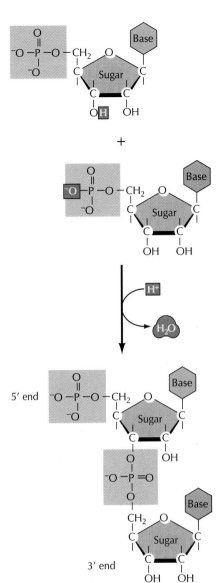

DNA and RNA are polymers of **nucleotides**, which consist of **purine** and **pyrimidine** bases linked to phosphorylated sugars (Figure 2.10). DNA contains two purines (**adenine** and **guanine**) and two pyrimidines (**cytosine** and **thymine**). Adenine, guanine, and cytosine are also present in RNA, but RNA contains **uracil** in place of thymine. The bases are linked to sugars (**2′-deoxyribose** in DNA, or **ribose** in RNA) to form **nucleosides**. Nucleotides additionally contain one or more phosphate groups linked to the 5′ carbon of nucleoside sugars.

The polymerization of nucleotides to form nucleic acids involves the formation of **phosphodiester bonds** between the 5′ phosphate of one nucleotide and the 3′ hydroxyl of another (Figure 2.11). **Oligonucleotides** are small polymers containing only a few nucleotides; the large **polynucleotides** that make up cellular RNA and DNA may contain thousands or millions of nucleotides, respectively. It is important to note that a polynucleotide chain has a sense of direction, with one end of the chain terminating in a 5′ phosphate group and the other in a 3′ hydroxyl group. Polynucleotides are always synthesized in the 5′ to 3′ direction, with a free nucleotide being added to the 3′ OH group of a growing chain. By convention, the sequence of bases in DNA or RNA is also written in the 5′ to 3′ direction.

The information in DNA and RNA is conveyed by the order of the bases in the polynucleotide chain. DNA is a double-stranded molecule consisting of two polynucleotide chains running in opposite directions (see Chapter 3). The bases are on the inside of the molecule, and the two chains are joined by hydrogen bonds between complementary base pairs—adenine pairing with thymine and guanine with cytosine (Figure 2.12). The important consequence of such complementary base pairing is that one strand of DNA (or RNA) can act as a template to direct the synthesis of a complementary strand. Nucleic acids are thus uniquely capable of directing their own self-replication, allowing them to function as the fundamental informational molecules of the cell. The information carried by DNA and RNA directs the synthesis of specific proteins, which control most cellular activities.

Nucleotides are not only important as the building blocks of nucleic acids; they also play critical roles in other cell processes. Perhaps the most prominent example is adenosine 5′-triphosphate (ATP), which is the prin-

Figure 2.12
Complementary pairing between nucleic acid bases The formation of hydrogen bonds between bases on opposite strands of DNA leads to the specific pairing of guanine (G) with cytosine (C) and adenine (A) with thymine (T).

cipal form of chemical energy within cells. Other nucleotides similarly function as carriers of either energy or reactive chemical groups in a wide variety of metabolic reactions. In addition, some nucleotides (e.g., cyclic AMP) are important signaling molecules within cells (see Chapter 13).

Proteins

While nucleic acids carry the genetic information of the cell, the primary responsibility of **proteins** is to execute the tasks directed by that information. Proteins are the most diverse of all macromolecules, and each cell contains several thousand different proteins, which perform a wide variety of functions. The roles of proteins include serving as structural components of cells and tissues, acting in the transport and storage of small molecules (e.g., the transport of oxygen by hemoglobin), transmitting information between cells (e.g., protein hormones), and providing a defense against infection (e.g., antibodies). The most fundamental property of proteins, however, is their ability to act as enzymes, which, as discussed in the following section, catalyze nearly all the chemical reactions in biological systems. Thus, proteins direct virtually all activities of the cell. The central importance of proteins in biological chemistry is indicated by their name, which is derived from the Greek word *proteios*, meaning "of the first rank."

Proteins are polymers of 20 different **amino acids**. Each amino acid consists of a carbon atom (called the α carbon) bonded to a carboxyl group (COO^-), an amino group (NH_3^+), a hydrogen atom, and a distinctive side chain (Figure 2.13). The specific chemical properties of the different amino acid side chains determine the roles of each amino acid in protein structure and function.

The amino acids can be grouped into four broad categories according to the properties of their side chains (Figure 2.14). Ten amino acids have nonpolar side chains that do not interact with water. Glycine is the simplest amino acid, with a side chain consisting of only a hydrogen atom. Alanine, valine, leucine, and isoleucine have hydrocarbon side chains consisting of up to four carbon atoms. The side chains of these amino acids are hydrophobic and therefore tend to be located in the interior of proteins, where they are not in contact with water. Proline similarly has a hydrocarbon side chain, but it is unique in that its side chain is bonded to the nitrogen of the amino group as well as to the α carbon, forming a cyclic structure. The side chains of two amino acids, cysteine and methionine, contain sulfur atoms. Methionine is quite hydrophobic, but cysteine is less so because of its sulfhydryl (SH) group. As discussed later, the sulfhydryl group of cysteine plays an important role in protein structure because disulfide bonds can form between the side chains of different cysteine residues. Finally, two nonpolar amino acids, phenylalanine and tryptophan, have side chains containing very hydrophobic aromatic rings.

Five amino acids have uncharged but polar side chains. These include serine, threonine, and tyrosine, which have hydroxyl groups on their side chains, as well as asparagine and glutamine, which have polar amide ($O=C-NH_2$) groups. Because the polar side chains of these amino acids can form hydrogen bonds with water, these amino acids are hydrophilic and tend to be located on the outside of proteins.

The amino acids lysine, arginine, and histidine have side chains with charged basic groups. Lysine and arginine are very basic amino acids, and their side chains are positively charged in the cell. Consequently, they are very hydrophilic and are found in contact with water on the surface of proteins. Histidine can be either uncharged or positively charged at physiological pH, so it frequently plays an active role in enzymatic reactions involving

Figure 2.13
Structure of amino acids   Each amino acid consists of a central carbon atom (the α carbon) bonded to a hydrogen atom, a carboxyl group, an amino group, and a specific side chain (designated R). At physiological pH, both the carboxyl and amino groups are ionized, as shown.

Nonpolar amino acids

Glycine (Gly) G   Alanine (Ala) A   Valine (Val) V   Leucine (Leu) L   Isoleucine (Ile) I   Proline (Pro) P

Cysteine (Cys) C   Methionine (Met) M   Phenylalanine (Phe) F   Tryptophan (Trp) W

Polar amino acids

Serine (Ser) S   Threonine (Thr) T   Tyrosine (Tyr) Y   Asparagine (Asn) N   Glutamine (Gln) Q

Basic amino acids

Lysine (Lys) K   Arginine (Arg) R   Histidine (His) H

Acidic amino acids

Aspartic acid (Asp) D   Glutamic acid (Glu) E

Figure 2.14
The amino acids The three-letter and one-letter abbreviations for each amino acid are indicated. The amino acids are grouped into four categories according to the properties of their side chains: nonpolar, polar, basic, and acidic.

the exchange of hydrogen ions, as illustrated in the example of enzymatic catalysis discussed in the following section.

Finally, two amino acids, aspartic acid and glutamic acid, have acidic side chains terminating in carboxyl groups. These amino acids are negatively charged within the cell and are therefore frequently referred to as aspartate and glutamate. Like the basic amino acids, these acidic amino acids are very hydrophilic and are usually located on the surface of proteins.

Amino acids are joined together by **peptide bonds** between the α amino group of one amino acid and the α carboxyl group of a second (Figure 2.15). **Polypeptides** are linear chains of amino acids, usually hundreds or thousands of amino acids in length. Each polypeptide chain has two distinct ends, one terminating in an α amino group (the amino, or N, terminus) and the other in an α carboxyl group (the carboxy, or C, terminus). Polypeptides

Key Experiment

The Folding of Polypeptide Chains

Reductive Cleavage of Disulfide Bridges in Ribonuclease

Michael Sela, Frederick H. White, Jr., and Christian B. Anfinsen
National Institutes of Health, Bethesda, MD
Science, Volume 125, 1957, pages 691–692

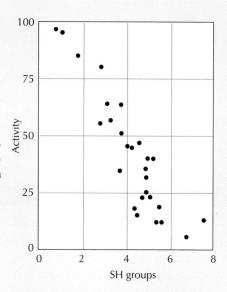

A summary of results of the renaturation experiments. The enzymatic activity of ribonuclease is plotted as a function of the number of sulfhydryl groups present after various treatments. Activity is expressed as percent activity of the native enzyme.

The Context

Functional proteins are structurally far more complex than linear chains of amino acids. The formation of active enzymes or other proteins requires the folding of polypeptide chains into precise three-dimensional conformations. This difference between proteins and polypeptide chains raises critical questions with respect to understanding protein structure and function. How is the correct protein conformation chosen from the many possible conformations that could be adopted by a polypeptide, and what is the nature of the information that directs protein folding?

Classic experiments done by Christian Anfinsen and his colleagues provided the answer to these questions. Studying the enzyme ribonuclease, Anfinsen and his collaborators were able to demonstrate that denatured proteins can spontaneously refold into an active conformation. Therefore, all the information required to specify the correct three-

dimensional conformation of a protein is contained in its primary amino acid sequence. A series of such experiments led Anfinsen to the conclusion that the native three-dimensional structure of a protein corresponds to its thermodynamically most stable conformation, as determined by interactions between its constituent amino acids. The original observations leading to the formulation of this critical principle were reported in this 1957 paper by Anfinsen, Michael Sela, and Fred White.

The Experiments

The protein studied by Sela, White, and Anfinsen was bovine ribonuclease, a small protein of 124 amino acids that contains four disulfide (S—S) bonds between side chains of cysteine residues. The enzymatic activity of ribonuclease could be determined by its ability to degrade RNA to nucleotides, providing a ready assay for function of the native protein. This enzymatic activity was completely lost when ribonuclease

was subjected to treatments that both disrupted noncovalent bonds (e.g., hydrogen bonds) and cleaved the disulfide bonds by reducing them to sulfhydryl (SH) groups. The denatured protein thus appeared to be in a random inactive conformation.

Sela, White, and Anfinsen made the crucial observation that enzymatic activity reappears if the dena-

---

Peptide bond

are synthesized from the amino to the carboxy terminus, and the sequence of amino acids in a polypeptide is written (by convention) in the same order.

The defining characteristic of proteins is that they are polypeptides with specific amino acid sequences. In 1953 Frederick Sanger was the first to determine the complete amino acid sequence of a protein, the hormone insulin. Insulin was found to consist of two polypeptide chains, joined by disulfide bonds between cysteine residues (Figure 2.16). Most important, Sanger's experiment revealed that each protein consists of a specific amino

Figure 2.15
Formation of a peptide bond The carboxyl group of one amino acid is linked to the amino group of a second.

The Folding of Polypeptide Chains (continued)

tured protein is incubated under conditions that allowed the polypeptide chain to refold and disulfide bonds to re-form. In these experiments the denaturing agents were removed, and the inactive enzyme was incubated at room temperature in a physiological buffer in the presence of O_2. This procedure led to oxidation of the sulfhydryl groups and the re-formation of disulfide bonds. During this process, the enzyme regained its catalytic activity, indicating that it had refolded to its native conformation. Because no other cellular components were present, all the information required for proper protein folding appeared to be contained in the primary amino acid sequence of the polypeptide chain.

The Impact

Further experiments defined conditions under which denatured ribonuclease completely regains its native structure and enzymatic activity, establishing the "thermodynamic hypothesis" of protein folding—that is, the native three-dimensional structure of a protein corresponds to the thermodynamically most stable state under physiological conditions. Thermodynamic stability is governed by interactions of the constituent amino acids, so the three-dimensional conformations of proteins are determined directly by their primary amino acid sequences. Since the order of nucleotides in DNA specifies the amino acid sequence of a polypeptide, it follows that the nucleotide sequence of a gene contains all the information needed to determine the three-dimensional structure of its protein product.

Although Anfinsen's work established the thermodynamic basis of protein folding, understanding the mechanism of this process remains an active area of research. Protein

Christian Anfinsen

folding is extremely complex, and we are still unable to deduce the three-dimensional structure of a protein directly from its amino acid sequence. It is also important to note that the spontaneous folding of proteins *in vitro* is much slower than protein folding within the cell, which is assisted by enzymes (see Chapter 7). The protein folding problem thus remains a central challenge in biological chemistry.

acid sequence. Proteins are currently sequenced using automated methods, and the complete amino acid sequences of over 100,000 proteins are now known. Each consists of a unique sequence of amino acids, determined by the order of nucleotides in a gene (see Chapter 3).

The amino acid sequence of a protein is only the first element of its structure. Rather than being extended chains of amino acids, proteins adopt distinct three-dimensional conformations that are critical to their function. These three-dimensional conformations of proteins are the result of interactions between their constituent amino acids, so the shapes of proteins are determined by their amino acid sequences. This was first demonstrated by experiments of Christian Anfinsen in which he disrupted the three-dimensional structures of proteins by treatments, such as heating,

Figure 2.16
Amino acid sequence of insulin   Insulin consists of two polypeptide chains, one of 21 and the other of 30 amino acids (indicated here by their one-letter codes). The side chains of three pairs of cysteine residues are joined by disulfide bonds, two of which connect the polypeptide chains.

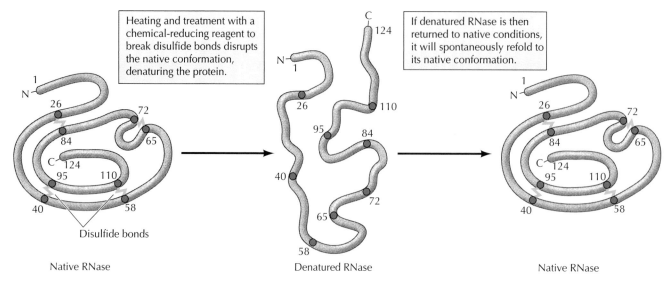

Heating and treatment with a chemical-reducing reagent to break disulfide bonds disrupts the native conformation, denaturing the protein.

If denatured RNase is then returned to native conditions, it will spontaneously refold to its native conformation.

Disulfide bonds

Native RNase   Denatured RNase   Native RNase

Figure 2.17
Protein denaturation and refolding
Ribonuclease (RNase) is a protein of 124 amino acids (indicated by numbers). The protein is normally folded into its native conformation, which contains four disulfide bonds (indicated as paired circles representing the cysteine residues).

that break noncovalent bonds—a process called denaturation (Figure 2.17). Following incubation under milder conditions, such denatured proteins often spontaneously returned to their native conformations, indicating that these conformations were directly determined by the amino acid sequence.

The three-dimensional structure of proteins is most frequently analyzed by **X-ray crystallography,** a high-resolution technique that can determine the arrangement of individual atoms within a molecule. A beam of X rays is directed at crystals of the protein to be analyzed, and the pattern of X rays that pass through the protein crystal is detected on X-ray film. As the X rays strike the crystal, they are scattered in characteristic patterns determined by the arrangement of atoms in the molecule. The structure of the molecule can therefore be deduced from the pattern of scattered X rays (the diffraction pattern).

In 1958 John Kendrew was the first to determine the three-dimensional structure of a protein, myoglobin—a relatively simple protein of 153 amino acids (Figure 2.18). Since then, the three-dimensional structures of several thousand proteins have been analyzed. Most, like myoglobin, are globular proteins with polypeptide chains folded into compact structures, although some (such as the structural proteins of connective tissues) are long fibrous molecules. Analysis of the three-dimensional structures of these proteins has revealed several basic principles that govern protein folding, although protein structure is so complex that predicting the three-dimensional structure of a protein directly from its amino acid sequence is impossible.

Protein structure is generally described as having four levels. The **primary structure** of a protein is the sequence of amino acids in its polypeptide chain. The **secondary structure** is the regular arrangement of amino acids within localized regions of the polypeptide. Two types of secondary structure, which were first proposed by Linus Pauling and Robert Corey in 1951, are particularly common: the α **helix** and the β **sheet**. Both of these secondary structures are held together by hydrogen bonds between the CO and NH groups of peptide bonds. An α helix is formed when a region of a

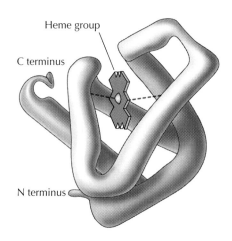

Heme group

C terminus

N terminus

Figure 2.18
Three-dimensional structure of myoglobin   Myoglobin is a protein of 153 amino acids that is involved in oxygen transport. The polypeptide chain is folded around a heme group that serves as the oxygen-binding site.

α helix

β sheet

Figure 2.19
Secondary structure of proteins
The most common types of secondary structure are the α helix and the β sheet. In an α helix, hydrogen bonds form between CO and NH groups of peptide bonds separated by four amino acid residues. In a β sheet, hydrogen bonds connect two parts of a polypeptide chain lying side by side. The amino acid side chains are not shown.

polypeptide chain coils around itself, with the CO group of one peptide bond forming a hydrogen bond with the NH group of a peptide bond located four residues downstream in the linear polypeptide chain (Figure 2.19). In contrast, a β sheet is formed when two parts of a polypeptide chain lie side by side with hydrogen bonds between them. Such β sheets can be formed between several polypeptide strands, which can be oriented either parallel or antiparallel to each other.

Tertiary structure is the folding of the polypeptide chain as a result of interactions between the side chains of amino acids that lie in different regions of the primary sequence (Figure 2.20). In most proteins, combinations of α helices and β sheets, connected by loop regions of the polypeptide chain, fold into compact globular structures called **domains**, which are the basic units of tertiary structure. Small proteins, such as ribonuclease or myoglobin, contain only a single domain; larger proteins may contain a number of different domains, which are frequently associated with distinct functions.

A critical determinant of tertiary structure is the localization of hydrophobic amino acids in the interior of the protein and of hydrophilic amino acids on the surface, where they interact with water. The interiors of folded proteins thus consist mainly of hydrophobic amino acids arranged in α helices and β sheets; these secondary structures are found in the hydrophobic cores of proteins because hydrogen bonding neutralizes the polar character of the CO and NH groups of the polypeptide backbone. The loop regions connecting the elements of secondary structure are found on the surface of folded proteins, where the polar components of the peptide bonds form hydrogen bonds with water or with the polar side chains of hydrophilic amino acids. Interactions between polar amino

Figure 2.20
Tertiary structure of ribonuclease
Regions of α-helix and β-sheet secondary structures, connected by loop regions, are folded into the native conformation of the protein. In this schematic representation of the polypeptide chain as a ribbon model, α helices are represented as spirals and β sheets as wide arrows.

Heme group   β chains

α chains

Figure 2.21
Quaternary structure of hemoglobin
Hemoglobin is composed of four poly-peptide chains, each of which is bound to a heme group. The two α chains and the two β chains are identical.

acid side chains (hydrogen bonds and ionic bonds) on the protein surface are also important determinants of tertiary structure. In addition, the covalent disulfide bonds between the sulfhydryl groups of cysteine residues stabilize the folded structures of many cell-surface or secreted proteins.

The fourth level of protein structure, **quaternary structure**, consists of the interactions between different polypeptide chains in proteins composed of more than one polypeptide. Hemoglobin, for example, is composed of four polypeptide chains held together by the same types of interactions that maintain tertiary structure (Figure 2.21).

The distinct chemical characteristics of the 20 different amino acids thus lead to considerable variation in the three-dimensional conformations of folded proteins. Consequently, proteins constitute an extremely complex and diverse group of macromolecules, suited to the wide variety of tasks they perform in cell biology.

The Central Role of Enzymes as Biological Catalysts

A fundamental task of proteins is to act as enzymes—catalysts that increase the rate of virtually all the chemical reactions within cells. Although RNAs are capable of catalyzing some reactions, most biological reactions are catalyzed by proteins. In the absence of enzymatic catalysis, most biochemical reactions are so slow that they would not occur under the mild conditions of temperature and pressure that are compatible with life. Enzymes accelerate the rates of such reactions by well over a million-fold, so reactions that would take years in the absence of catalysis can occur in fractions of seconds if catalyzed by the appropriate enzyme. Cells contain thousands of different enzymes, and their activities determine which of the many possible chemical reactions actually take place within the cell.

The Catalytic Activity of Enzymes

Like all other catalysts, **enzymes** are characterized by two fundamental properties. First, they increase the rate of chemical reactions without themselves being consumed or permanently altered by the reaction. Second, they increase reaction rates without altering the chemical equilibrium between reactants and products.

These principles of enzymatic catalysis are illustrated in the following example, in which a molecule acted upon by an enzyme (referred to as a **substrate** [S]) is converted to a **product** (P) as the result of the reaction. In the absence of the enzyme, the reaction can be written as follows:

$$S \rightleftharpoons P$$

The chemical equilibrium between S and P is determined by the laws of thermodynamics (as discussed further in the next section of this chapter) and is represented by the ratio of the forward and reverse reaction rates ($S{\rightarrow}P$ and $P{\rightarrow}S$, respectively). In the presence of the appropriate enzyme, the conversion of S to P is accelerated, but the equilibrium between S and P is unaltered. Therefore, the enzyme must accelerate both the forward and reverse reactions equally. The reaction can be written as follows:

$$S \overset{E}{\rightleftharpoons} P$$

Note that the enzyme (E) is not altered by the reaction, so the chemical equilibrium remains unchanged, determined solely by the thermodynamic properties of S and P.

The effect of the enzyme on such a reaction is best illustrated by the energy changes that must occur during the conversion of S to P (Figure

2.22). The equilibrium of the reaction is determined by the final energy states of *S* and *P*, which are unaffected by enzymatic catalysis. In order for the reaction to proceed, however, the substrate must first be converted to a higher energy state, called the **transition state**. The energy required to reach the transition state (the **activation energy**) constitutes a barrier to the progress of the reaction, limiting the rate of the reaction. Enzymes (and other catalysts) act by reducing the activation energy, thereby increasing the rate of reaction. The increased rate is the same in both the forward and reverse directions, since both must pass through the same transition state.

The catalytic activity of enzymes involves the binding of their substrates to form an enzyme-substrate complex (*ES*). The substrate binds to a specific region of the enzyme, called the **active site**. While bound to the active site, the substrate is converted into the product of the reaction, which is then released from the enzyme. The enzyme-catalyzed reaction can thus be written as follows:

$$S + E \rightleftharpoons ES \rightleftharpoons E + P$$

Note that *E* appears unaltered on both sides of the equation, so the equilibrium is unaffected. However, the enzyme provides a surface upon which the reactions converting *S* to *P* can occur more readily. This is a result of interactions between the enzyme and substrate that lower the energy of activation and favor formation of the transition state.

Mechanisms of Enzymatic Catalysis

The binding of a substrate to the active site of an enzyme is a very specific interaction. Active sites are clefts or grooves on the surface of an enzyme, usually composed of amino acids from different parts of the polypeptide chain that are brought together in the tertiary structure of the folded protein. Substrates initially bind to the active site by noncovalent interactions, including hydrogen bonds, ionic bonds, and hydrophobic interactions. Once a substrate is bound to the active site of an enzyme, multiple mechanisms can accelerate its conversion to the product of the reaction.

Although the simple example discussed in the previous section involved only a single substrate molecule, most biochemical reactions involve interactions between two or more different substrates. For example, the formation of a peptide bond involves the joining of two amino acids. For such reactions, the binding of two or more substrates to the active site in the proper position and orientation accelerates the reaction (Figure 2.23). The enzyme provides a template upon which the reactants are brought together and properly oriented to favor the formation of the transition state in which they interact.

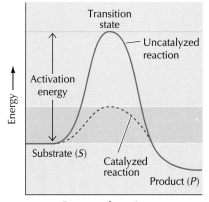

Figure 2.22
Energy diagrams for catalyzed and uncatalyzed reactions   The reaction illustrated is the simple conversion of a substrate *S* to a product *P*. Because the final energy state of *P* is lower than that of *S*, the reaction proceeds from left to right. For the reaction to occur, however, *S* must first pass through a higher energy transition state. The energy required to reach this transition state (the activation energy) represents a barrier to the progress of the reaction and thereby determines the rate at which the reaction proceeds. In the presence of a catalyst (e.g., an enzyme), the activation energy is lowered and the reaction proceeds at an accelerated rate.

Figure 2.23
Enzymatic catalysis of a reaction between two substrates   The enzyme provides a template upon which the two substrates are brought together in the proper position and orientation to react with each other.

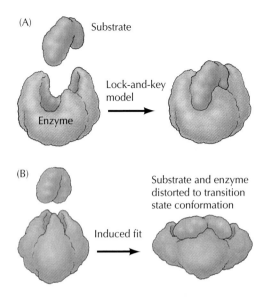

Enzymes accelerate reactions also by altering the conformation of their substrates to approach that of the transition state. The simplest model of enzyme-substrate interaction is the **lock-and-key model**, in which the substrate fits precisely into the active site (Figure 2.24). In many cases, however, the configurations of both the enzyme and substrate are modified by substrate binding—a process called **induced fit**. In such cases the conformation of the substrate is altered so that it more closely resembles that of the transition state. The stress produced by such distortion of the substrate can further facilitate its conversion to the transition state by weakening critical bonds. Moreover, the transition state is stabilized by its tight binding to the enzyme, thereby lowering the required energy of activation.

In addition to bringing multiple substrates together and distorting the conformation of substrates to approach the transition state, many enzymes participate directly in the catalytic process. In such cases, specific amino acid side chains in the active site may react with the substrate and form bonds with reaction intermediates. The acidic and basic amino acids are often involved in these catalytic mechanisms, as illustrated in the following discussion of chymotrypsin as an example of enzymatic catalysis.

Chymotrypsin is a member of a family of enzymes (serine proteases) that digest proteins by catalyzing the hydrolysis of peptide bonds. The reaction can be written as follows:

$$\text{Protein} + \text{H}_2\text{O} \rightarrow \text{Peptide}_1 + \text{Peptide}_2$$

The different members of the serine protease family (including chymotrypsin, trypsin, elastase, and thrombin) have distinct substrate specificities; they preferentially cleave peptide bonds adjacent to different amino acids. For example, whereas chymotrypsin digests bonds adjacent to hydrophobic amino acids, such as tryptophan and phenylalanine, trypsin digests bonds next to basic amino acids, such as lysine and arginine. All the serine proteases, however, are similar in structure and use the same mechanism of catalysis. The active sites of these enzymes contain three critical amino acids—serine, histidine, and aspartate—that drive hydrolysis of the peptide bond. Indeed, these enzymes are called serine proteases because of the central role of the serine residue.

Substrates bind to the serine proteases by insertion of the amino acid adjacent to the cleavage site into a pocket at the active site of the enzyme (Figure 2.25). The nature of this pocket determines the substrate specificity

Figure 2.24
Models of enzyme-substrate interaction   (A) In the lock-and-key model, the substrate fits precisely into the active site of the enzyme. (B) In the induced-fit model, substrate binding distorts the conformations of both substrate and enzyme. This distortion brings the substrate closer to the conformation of the transition state, thereby accelerating the reaction.

Figure 2.25
Substrate binding by serine proteases   The amino acid adjacent to the peptide bond to be cleaved is inserted into a pocket at the active site of the enzyme. In chymotrypsin, the pocket binds hydrophobic amino acids; the binding pocket of trypsin contains a negatively charged aspartate residue that binds basic amino acids via an ionic interaction.

Figure 2.26
Catalytic mechanism of chymotrypsin
Three amino acids at the active site (Ser-195, His-57, and Asp-102) play critical roles in catalysis.

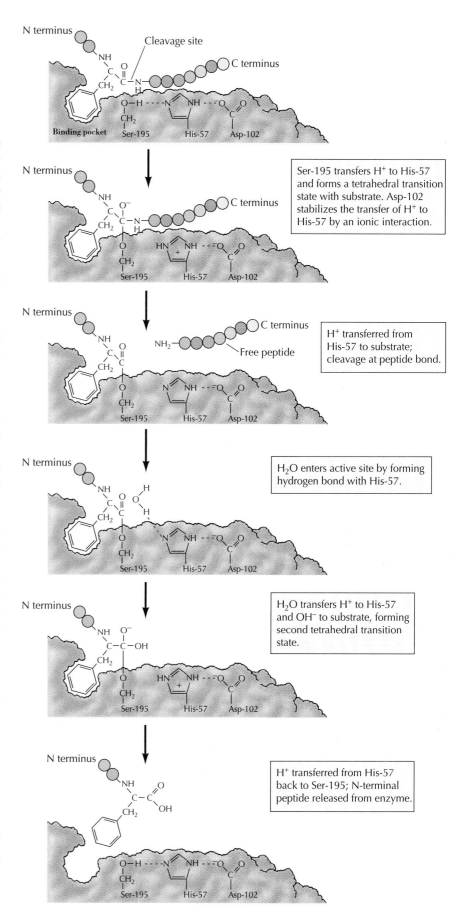

Ser-195 transfers H⁺ to His-57 and forms a tetrahedral transition state with substrate. Asp-102 stabilizes the transfer of H⁺ to His-57 by an ionic interaction.

H⁺ transferred from His-57 to substrate; cleavage at peptide bond.

H_2O enters active site by forming hydrogen bond with His-57.

H_2O transfers H⁺ to His-57 and OH⁻ to substrate, forming second tetrahedral transition state.

H⁺ transferred from His-57 back to Ser-195; N-terminal peptide released from enzyme.

of the different members of the serine protease family. For example, the binding pocket of chymotrypsin contains hydrophobic amino acids that interact with the hydrophobic side chains of its preferred substrates. In contrast, the binding pocket of trypsin contains a negatively charged acidic amino acid (aspartate), which is able to form an ionic bond with the lysine or arginine residues of its substrates.

Substrate binding positions the peptide bond to be cleaved adjacent to the active site serine (Figure 2.26). The proton of this serine is then transferred to the active site histidine. The conformation of the active site favors this proton transfer because the histidine interacts with the negatively charged aspartate residue. The serine reacts with the substrate, forming a tetrahedral transition state. The peptide bond is then cleaved, and the C-terminal portion of the substrate is released from the enzyme. However, the N-terminal peptide remains bound to serine. This situation is resolved when a water molecule (the second substrate) enters the active site and reverses the preceding reactions. The proton of the water molecule is transferred to histidine, and its hydroxyl group is transferred to the peptide, forming a second tetrahedral transition state. The proton is then transferred from histidine back to serine, and the peptide is released from the enzyme, completing the reaction.

This example illustrates several features of enzymatic catalysis; the specificity of enzyme-substrate interactions, the positioning of different substrate molecules in the active site, and the involvement of active-site residues in the formation and stabilization of the transition state. Although the thousands of enzymes in cells catalyze many different types of chemical reactions, the same basic principles apply to their operation.

Coenzymes

In addition to binding their substrates, the active sites of many enzymes bind other small molecules that participate in catalysis. **Prosthetic groups** are small molecules bound to proteins in which they play critical functional roles. For example, the oxygen carried by myoglobin and hemoglobin is bound to heme, a prosthetic group of these proteins. In many cases metal ions (such as zinc or iron) are bound to enzymes and play central roles in the catalytic process. In addition, various low-molecular-weight organic molecules participate in specific types of enzymatic reactions. These molecules are called **coenzymes** because they work together with enzymes to enhance reaction rates. In contrast to substrates, coenzymes are not irreversibly altered by the reactions in which they are involved. Rather, they are recycled and can participate in multiple enzymatic reactions.

Coenzymes serve as carriers of several types of chemical groups. A prominent example of a coenzyme is **nicotinamide adenine dinucleotide (NAD⁺)**, which functions as a carrier of electrons in oxidation–reduction reactions (Figure 2.27). NAD⁺ can accept a hydrogen ion (H⁺) and two elec-

Figure 2.27
Role of NAD⁺ in oxidation–reduction reactions
(A) Nicotinamide adenine dinucleotide (NAD⁺) acts as a carrier of electrons in oxidation–reduction reactions by accepting electrons (e⁻) to form NADH. (B) For example, NAD⁺ can accept electrons from one substrate (S1), yielding oxidized S1 plus NADH. The NADH formed in this reaction can then transfer its electrons to a second substrate (S2), yielding reduced S2 and regenerating NAD⁺. The net effect is the transfer of electrons (carried by NADH) from S1 (which becomes oxidized) to S2 (which becomes reduced).

TABLE 2.1 Examples of Coenzymes and Vitamins

Coenzyme	Related vitamin	Chemical reaction
NAD$^+$, NADP$^+$	Niacin	Oxidation–reduction
FAD	Riboflavin (B$_2$)	Oxidation–reduction
Thiamine pyrophosphate	Thiamine (B$_1$)	Aldehyde group transfer
Coenzyme A	Pantothenate	Acyl group transfer
Tetrahydrofolate	Folate	Transfer of one-carbon groups
Biotin	Biotin	Carboxylation
Pyridoxal phosphate	Pyridoxal (B$_6$)	Transamination

trons (e$^-$) from one substrate, forming NADH. NADH can then donate these electrons to a second substrate, re-forming NAD$^+$. Thus, NAD$^+$ transfers electrons from the first substrate (which becomes oxidized) to the second (which becomes reduced).

Several other coenzymes also act as electron carriers, and still others are involved in the transfer of a variety of additional chemical groups (e.g., carboxyl groups and acyl groups; Table 2.1). The same coenzymes function together with a variety of different enzymes to catalyze the transfer of specific chemical groups between a wide range of substrates. Many coenzymes are closely related to vitamins, which contribute part or all of the structure of the coenzyme. Vitamins are not required by bacteria such as *E. coli* but are necessary components of the diets of human and other higher animals, which have lost the ability to synthesize these compounds.

Regulation of Enzyme Activity

An important feature of most enzymes is that their activities are not constant but instead can be modulated. That is, the activities of enzymes can be regulated so that they function appropriately to meet the varied physiological needs that may arise during the life of the cell.

One common type of enzyme regulation is **feedback inhibition**, in which the product of a metabolic pathway inhibits the activity of an enzyme involved in its synthesis. For example, the amino acid isoleucine is synthesized by a series of reactions starting from the amino acid threonine (Figure 2.28). The first step in the pathway is catalyzed by the enzyme threonine deaminase, which is inhibited by isoleucine, the end product of the pathway. Thus, an adequate amount of isoleucine in the cell inhibits threonine deaminase, blocking further synthesis of isoleucine. If the concentration of isoleucine decreases, feedback inhibition is relieved, threonine deaminase is no longer inhibited, and additional isoleucine is synthesized. By so regulating the activity of threonine deaminase, the cell synthesizes the necessary amount of isoleucine but avoids wasting energy on the synthesis of more isoleucine than is needed.

Feedback inhibition is one example of **allosteric regulation**, in which enzyme activity is controlled by the binding of small molecules to regulatory sites on the enzyme (Figure 2.29). The term "allosteric regulation" derives from the fact that the regulatory molecules bind not to the catalytic site, but to a distinct site on the protein (*allo* = "other" and *steric* = "site"). Binding of the regulatory molecule changes the conformation of the protein, which in turn alters the shape of the active site and the catalytic activity of the enzyme. In the case of threonine deaminase, binding of the regulatory molecule (isoleucine) inhibits enzymatic activity. In other cases regulatory molecules serve as activators, stimulating rather than inhibiting their target enzymes.

Figure 2.28
Feedback inhibition The first step in the conversion of threonine to isoleucine is catalyzed by the enzyme threonine deaminase. The activity of this enzyme is inhibited by isoleucine, the end product of the pathway.

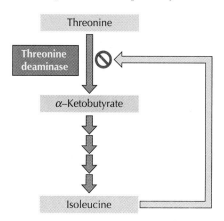

Figure 2.29
Allosteric regulation   In this example, enzyme activity is inhibited by the binding of a regulatory molecule to an allosteric site. In the absence of inhibitor, the substrate binds to the active site of the enzyme and the reaction proceeds. The binding of inhibitor to the allosteric site induces a conformational change in the enzyme and prevents substrate binding. Most allosteric enzymes consist of multiple subunits.

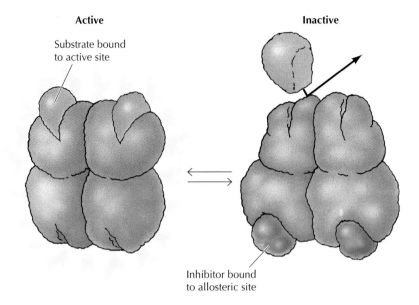

The activities of enzymes can also be regulated by their interactions with other proteins and by covalent modifications, such as the addition of phosphate groups to serine, threonine, or tyrosine residues. **Phosphorylation** is a particularly common mechanism for regulating enzyme activity; the addition of phosphate groups either stimulates or inhibits the activities of many different enzymes (Figure 2.30). For example, muscle cells respond to epinephrine (adrenaline) by breaking down glycogen into glucose, thereby providing a source of energy for increased muscular activity. The breakdown of glycogen is catalyzed by the enzyme glycogen phosphorylase, which is activated by phosphorylation in response to the binding of epinephrine to a receptor on the surface of the muscle cell. Protein phosphorylation plays a central role in controlling not only metabolic reactions but also many other cellular functions, including cell growth and differentiation.

Figure 2.30
Protein phosphorylation   Some enzymes are regulated by the addition of phosphate groups to the side-chain OH groups of serine (as shown here), threonine, or tyrosine residues. For example, the enzyme glycogen phosphorylase, which catalyzes the conversion of glycogen into glucose-1-phosphate, is activated by phosphorylation in response to the binding of epinephrine to muscle cells.

Metabolic Energy

Many tasks that a cell must perform, such as movement and the synthesis of macromolecules, require energy. A large portion of the cell's activities are therefore devoted to obtaining energy from the environment and using that energy to drive energy-requiring reactions. Although enzymes control the rates of virtually all chemical reactions within cells, the equilibrium position of chemical reactions is not affected by enzymatic catalysis. The laws of thermodynamics govern chemical equilibria and determine the energetically favorable direction of all chemical reactions. Many of the reactions that must take place within cells are energetically unfavorable, and are therefore able to proceed only at the cost of additional energy input. Consequently, cells must constantly expend energy derived from the environment. The generation and utilization of metabolic energy is thus fundamental to all of cell biology.

Free Energy and ATP

The energetics of biochemical reactions are best described in terms of the thermodynamic function called **Gibbs free energy (G)**, named for Josiah Willard Gibbs. The change in free energy (ΔG) of a reaction combines the effects of changes in enthalpy (the heat that is released or absorbed during a chemical reaction) and entropy (the degree of disorder resulting from a reaction) to predict whether or not a reaction is energetically favorable. All chemical reactions spontaneously proceed in the energetically favorable direction, accompanied by a decrease in free energy ($\Delta G < 0$). For example, consider a hypothetical reaction in which A is converted to B:

$$A \rightleftharpoons B$$

If $\Delta G < 0$, this reaction will proceed in the forward direction, as written. If $\Delta G > 0$, however, the reaction will proceed in the reverse direction and B will be converted to A.

The ΔG of a reaction is determined not only by the intrinsic properties of reactants and products, but also by their concentrations and other reaction conditions (e.g., temperature). It is thus useful to define the free-energy change of a reaction under standard conditions. (Standard conditions are considered to be a 1-M concentration of all reactants and products, and 1 atm of pressure). The standard free-energy change ($\Delta G°$) of a reaction is directly related to its equilibrium position because the actual ΔG is a function of both $\Delta G°$ and the concentrations of reactants and products. For example, consider the reaction

$$A \rightleftharpoons B$$

The free-energy change can be written as follows:

$$\Delta G = \Delta G° + RT \ln [B]/[A]$$

where R is the gas constant and T is the absolute temperature.

At equilibrium, $\Delta G = 0$ and the reaction does not proceed in either direction. The equilibrium constant for the reaction ($K = [B]/[A]$ at equilibrium) is thus directly related to $\Delta G°$ by the above equation, which can be expressed as follows:

$$0 = \Delta G° + RT \ln K$$

or

$$\Delta G° = - RT \ln K$$

If the actual ratio [B]/[A] is greater than the equilibrium ratio (K), $\Delta G > 0$ and the reaction proceeds in the reverse direction (conversion of B to A).

On the other hand, if the ratio [B]/[A] is less than the equilibrium ratio, ΔG < 0 and A is converted to B.

The standard free-energy change ($\Delta G°$) of a reaction therefore determines its chemical equilibrium and predicts in which direction the reaction will proceed under any given set of conditions. For biochemical reactions, the standard free-energy change is usually expressed as $\Delta G°'$, which is the standard free-energy change of a reaction in aqueous solution at pH = 7, approximately the conditions within a cell.

Many biological reactions (such as the synthesis of macromolecules) are thermodynamically unfavorable ($\Delta G > 0$) under cellular conditions. In order for such reactions to proceed, an additional source of energy is required. For example, consider the reaction

$$A \rightleftharpoons B \qquad \Delta G = +10 \text{ kcal/mol}$$

The conversion of A to B is energetically unfavorable, so the reaction proceeds in the reverse rather than the forward direction. However, the reaction can be driven in the forward direction by coupling the conversion of A to B with an energetically favorable reaction, such as:

$$C \rightleftharpoons D \qquad \Delta G = -20 \text{ kcal/mol}$$

If these two reactions are combined, the coupled reaction can be written as follows:

$$A + C \rightleftharpoons B + D \qquad \Delta G = -10 \text{ kcal/mol}$$

The ΔG of the combined reaction is the sum of the free-energy changes of its individual components, so the coupled reaction is energetically favorable and will proceed as written. Thus, the energetically unfavorable conversion of A to B is driven by coupling it to a second reaction associated with a large decrease in free energy. Enzymes are responsible for carrying out such coupled reactions in a coordinated manner.

The cell uses this basic mechanism to drive the many energetically unfavorable reactions that must take place in biological systems. **Adenosine 5′-triphosphate (ATP)** plays a central role in this process by acting as a store of free energy within the cell (Figure 2.31). The bonds between the phosphates in ATP are known as **high-energy bonds** because their hydrolysis is accompanied by a relatively large decrease in free energy. There is nothing special about the chemical bonds themselves; they are called high-energy bonds only because a large amount of free energy is released when they are hydrolyzed within the cell. In the hydrolysis of ATP to ADP plus phosphate (P_i), $\Delta G°' = -7.3$ kcal/mol. Recall, however, that $\Delta G°'$ refers to "standard conditions," in which the concentrations of all products and reactants are 1 M. Actual intracellular concentrations of P_i are approximately 10^{-2} M, and intracellular concentrations of ATP are higher than those of ADP. These differences between intracellular concentrations and those of the standard state favor ATP hydrolysis, so for ATP hydrolysis within a cell, ΔG is approximately −12 kcal/mol.

Alternatively, ATP can be hydrolyzed to AMP plus pyrophosphate (PP_i). This reaction yields about the same amount of free energy as the hydrolysis of ATP to ADP does. However, the pyrophosphate produced by this reaction is then itself rapidly hydrolyzed, with a ΔG similar to that of ATP hydrolysis. Thus, the total free-energy change resulting from the hydrolysis of ATP to AMP is approximately twice that obtained by the hydrolysis of ATP to ADP. For comparison, the bond between the sugar and phosphate group of AMP, rather than having high energy, is typical of covalent bonds; for the hydrolysis of AMP, $\Delta G°' = -3.3$ kcal/mol.

Figure 2.31
ATP as a store of free energy   The bonds between the phosphate groups of ATP are called high-energy bonds because their hydrolysis results in a large decrease in free energy. ATP can be hydrolyzed either to ADP plus a phosphate group (HPO_4^{2-}) or to AMP plus pyrophosphate. In the latter case, pyrophosphate is itself rapidly hydrolyzed, releasing additional free energy.

Because of the accompanying decrease in free energy, the hydrolysis of ATP can be used to drive other energy-requiring reactions within the cell. For example, the first reaction in glycolysis (discussed in the next section) is the conversion of glucose to glucose-6-phosphate. The reaction can be written as follows:

$$\text{Glucose} + \text{Phosphate } (HPO_4^{2-}) \rightarrow \text{Glucose-6-phosphate} + H_2O$$

Because this reaction is energetically unfavorable as written ($\Delta G^{\circ\prime} = +3.3$ kcal/mol), it must be driven in the forward direction by being coupled to ATP hydrolysis ($\Delta G^{\circ\prime} = -7.3$ kcal/mol):

$$\text{ATP} + H_2O \rightarrow \text{ADP} + HPO_4^{2-}$$

The combined reaction can be written as follows:

$$\text{Glucose} + \text{ATP} \rightarrow \text{Glucose-6-phosphate} + \text{ADP}$$

The free-energy change for this reaction is the sum of the free-energy changes for the individual reactions, so for the coupled reaction $\Delta G^{\circ\prime} = -4.0$ kcal/mol, favoring glucose-6-phosphate formation.

Other molecules, including other nucleoside triphosphates (e.g., GTP), also have high-energy bonds and can be used as ATP is to drive energy-requiring reactions. For most reactions, however, ATP provides the free energy. The energy-yielding reactions within the cell are therefore coupled to ATP synthesis, while the energy-requiring reactions are coupled to ATP hydrolysis. The high-energy bonds of ATP thus play a central role in cell metabolism by serving as a usable storage form of free energy.

The Generation of ATP from Glucose

The breakdown of carbohydrates, particularly glucose, is a major source of cellular energy. The complete oxidative breakdown of glucose to CO_2 and H_2O can be written as follows:

$$C_6H_{12}O_6 + 6\,O_2 \rightarrow 6\,CO_2 + 6\,H_2O$$

The reaction yields a large amount of free energy: $\Delta G°' = -686$ kcal/mol. To harness this free energy in usable form, glucose is oxidized within cells in a series of steps coupled to the synthesis of ATP.

Glycolysis, the initial stage in the breakdown of glucose, is common to virtually all cells. Glycolysis occurs in the absence of oxygen and can provide all the metabolic energy of anaerobic organisms. In aerobic cells, however, glycolysis is only the first stage in glucose degradation.

The reactions of glycolysis result in the breakdown of glucose into pyruvate, with the net gain of two molecules of ATP (Figure 2.32). The initial reactions in the pathway actually consume energy, using ATP to phosphorylate glucose to glucose-6-phosphate and then fructose-6-phosphate to fructose-1,6-bisphosphate. The enzymes that catalyze these two reactions—hexokinase and phosphofructokinase, respectively—are important regulatory points of the glycolytic pathway. The key control element is phosphofructokinase, which is inhibited by high levels of ATP. Inhibition of phosphofructokinase results in an accumulation of glucose-6-phosphate, which in turn inhibits hexokinase. Thus, when the cell has an adequate supply of metabolic energy available in the form of ATP, the breakdown of glucose is inhibited.

The reactions following the formation of fructose-1,6-bisphosphate constitute the energy-producing part of the glycolytic pathway. Cleavage of fructose-1,6-bisphosphate yields two molecules of the three-carbon sugar glyceraldehyde-3-phosphate, which is oxidized to 1,3-bisphosphoglycerate. The phosphate group of this compound has a very high free energy of hydrolysis ($\Delta G°' = -11.5$ kcal/mol), so it is used in the next reaction of glycolysis to drive the synthesis of ATP from ADP. The product of this reaction, 3-phosphoglycerate, is then converted to phosphoenolpyruvate, the second high-energy intermediate in glycolysis. In the hydrolysis of the high-energy phosphate of phosphoenolpyruvate, $\Delta G°' = -14.6$ kcal/mol, its conversion to pyruvate is coupled to the synthesis of ATP. Each molecule of glyceraldehyde-3-phosphate converted to pyruvate is thus coupled to the generation of two molecules of ATP; in total, four ATPs are synthesized from each starting molecule of glucose. Since two ATPs were required to prime the initial reactions, the net gain from glycolysis is two ATP molecules.

In addition to producing ATP, glycolysis converts two molecules of the coenzyme NAD^+ to NADH. In this reaction, NAD^+ acts as an oxidizing agent that accepts electrons from glyceraldehyde-3-phosphate. The NADH formed as a product must be recycled by serving as a donor of electrons for other oxidation–reduction reactions within the cell. In anaerobic conditions, the NADH formed during glycolysis is reoxidized to NAD^+ by the conversion of pyruvate to lactate or ethanol. In aerobic organisms, however, the NADH serves as an additional source of energy by donating its

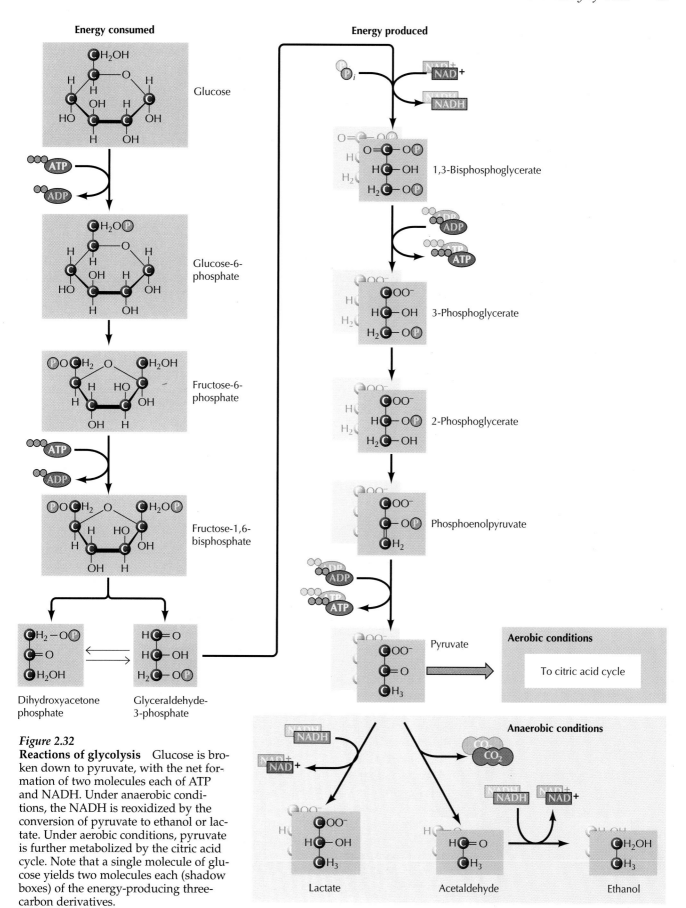

Energy consumed

Glucose

Glucose-6-phosphate

Fructose-6-phosphate

Fructose-1,6-bisphosphate

Dihydroxyacetone phosphate

Glyceraldehyde-3-phosphate

Energy produced

1,3-Bisphosphoglycerate

3-Phosphoglycerate

2-Phosphoglycerate

Phosphoenolpyruvate

Pyruvate

Aerobic conditions

To citric acid cycle

Anaerobic conditions

Lactate

Acetaldehyde

Ethanol

Figure 2.32
Reactions of glycolysis Glucose is broken down to pyruvate, with the net formation of two molecules each of ATP and NADH. Under anaerobic conditions, the NADH is reoxidized by the conversion of pyruvate to ethanol or lactate. Under aerobic conditions, pyruvate is further metabolized by the citric acid cycle. Note that a single molecule of glucose yields two molecules each (shadow boxes) of the energy-producing three-carbon derivatives.

Figure 2.33
Oxidative decarboxylation of pyruvate   Pyruvate is converted to CO_2 and acetyl CoA, and one molecule of NADH is produced in the process. Coenzyme A (CoA-SH) is a general carrier of activated acyl groups in a variety of reactions.

electrons to the electron transport chain, where they are ultimately used to reduce O_2 to H_2O, coupled to the generation of additional ATP.

In eukaryotic cells, glycolysis takes place in the cytosol. Pyruvate is then transported into mitochondria, where its complete oxidation to CO_2 and H_2O yields most of the ATP derived from glucose breakdown. The next step in the metabolism of pyruvate is its oxidative decarboxylation in the presence of **coenzyme A (CoA)**, which serves as a carrier of acyl groups in various metabolic reactions (Figure 2.33). One carbon of pyruvate is released as CO_2, and the remaining two carbons are added to CoA to form acetyl CoA. In the process, one molecule of NAD^+ is reduced to NADH.

The acetyl CoA formed by this reaction enters the **citric acid cycle** or **Krebs cycle** (Figure 2.34), which is the central pathway in oxidative metabolism. The two-carbon acetyl group combines with oxaloacetate (four carbons) to yield citrate (six carbons). Through eight further reactions, two carbons of citrate are completely oxidized to CO_2 and oxaloacetate is regenerated. During the cycle, one high-energy phosphate bond is formed in GTP, which is used directly to drive the synthesis of one ATP molecule. In addition, each turn of the cycle yields three molecules of NADH and one molecule of reduced **flavin adenine dinucleotide (FADH$_2$)**, which is another carrier of electrons in oxidation–reduction reactions.

The citric acid cycle completes the oxidation of glucose to six molecules of CO_2. Four molecules of ATP are obtained directly from each glucose molecule—two from glycolysis and two from the citric acid cycle (one for each molecule of pyruvate). In addition, ten molecules of NADH (two from glycolysis, two from the conversion of pyruvate to acetyl CoA, and six from the citric acid cycle) and two molecules of FADH$_2$ are formed. The remaining energy derived from the breakdown of glucose comes from the reoxidation of NADH and FADH$_2$, with their electrons being transferred through the electron transport chain to (eventually) reduce O_2 to H_2O.

During **oxidative phosphorylation**, the electrons of NADH and FADH$_2$ combine with O_2, and the energy released from the process drives the synthesis of ATP from ADP. The transfer of electrons from NADH to O_2 releases a large amount of free energy: $\Delta G°' = -52.5$ kcal/mol for each pair of electrons transferred. So that this energy can be harvested in usable form, the process takes place gradually by the passage of electrons through a series of carriers, which constitute the **electron transport chain**

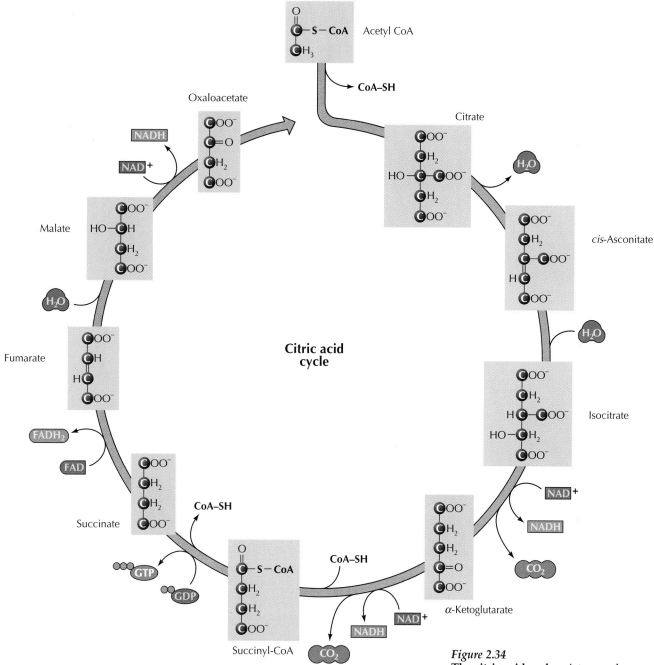

Figure 2.34
The citric acid cycle   A two-carbon acetyl group is transferred from acetyl CoA to oxaloacetate, forming citrate. Two carbons of citrate are then oxidized to CO_2 and oxaloacetate is regenerated. Each turn of the cycle yields one molecule of GTP, three of NADH, and one of $FADH_2$.

(Figure 2.35). The components of the electron transport chain are located in the inner mitochondrial membrane of eukaryotic cells, and oxidative phosphorylation is considered in more detail when mitochondria are discussed in Chapter 10. In aerobic bacteria, which use a comparable system, components of the electron transport chain are located in the plasma membrane. In either case, the transfer of electrons from NADH to O_2 yields sufficient energy to drive the synthesis of approximately three molecules of ATP. Electrons from $FADH_2$ enter the electron transport chain at a lower energy level, so their transfer to O_2 yields less usable free energy, only two ATP molecules.

Figure 2.35
The electron transport chain Electrons from NADH and FADH$_2$ are transferred to O$_2$ through a series of carriers organized into four protein complexes in the mitochondrial membrane. The free energy derived from electron transport reactions at complexes I, III, and IV is used to drive the synthesis of ATP.

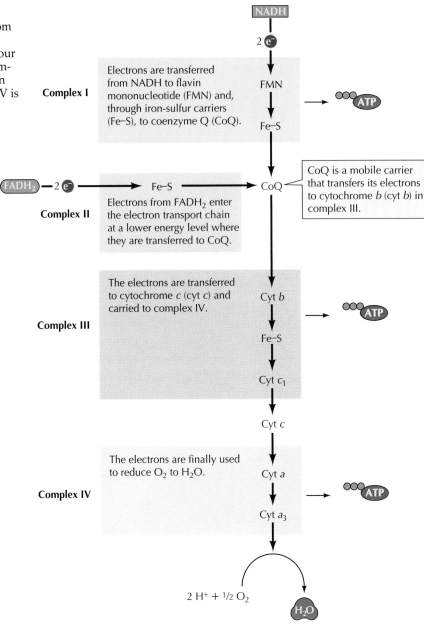

It is now possible to calculate the total yield of ATP from the oxidation of glucose. The net gain from glycolysis is two molecules of ATP and two molecules of NADH. The conversion of pyruvate to acetyl CoA and its metabolism via the citric acid cycle yields two additional molecules of ATP, eight of NADH, and two of FADH$_2$. Assuming that three molecules of ATP are derived from the oxidation of each NADH and two from each FADH$_2$, the total yield is 38 molecules of ATP per molecule of glucose. However, this yield is lower in some cells because the two molecules of NADH generated by glycolysis in the cytosol are unable to enter mitochondria directly. Instead, their electrons must be transferred into the mitochondrion via a shuttle system. Depending on the system used, this transfer may result in these electrons entering the electron transport chain at the level of FADH$_2$. In such cases, the two molecules of NADH derived from glycolysis give rise to two rather than three molecules of ATP, reducing the total yield to 36 rather than 38 ATPs per molecule of glucose.

Figure 2.36
Oxidation of fatty acids The fatty acid (e.g., the 16-carbon saturated fatty acid palmitate) is initially joined to coenzyme A at the cost of one molecule of ATP. Oxidation of the fatty acid then proceeds by stepwise removal of two-carbon units as acetyl CoA, coupled to the formation of one molecule each of NADH and $FADH_2$.

The Derivation of Energy from Other Organic Molecules

Energy in the form of ATP can be derived from the breakdown of other organic molecules, with the pathways involved in glucose degradation again playing a central role. Nucleotides, for example, can be broken down to sugars, which then enter the glycolytic pathway, and amino acids are degraded via the citric acid cycle. The two principal storage forms of energy within cells, polysaccharides and lipids, can also be broken down to produce ATP. Polysaccharides are broken down into free sugars, which are then metabolized as discussed in the previous section. Lipids, however, are an even more efficient energy storage molecule. Because lipids are more reduced than carbohydrates, consisting primarily of hydrocarbon chains, their oxidation yields substantially more energy per weight of starting material.

Fats (triacylglycerols) are the major storage form of lipids. The first step in their utilization is their breakdown to glycerol and free fatty acids. Each fatty acid is joined to coenzyme A, yielding a fatty acyl-CoA at the cost of one molecule of ATP (Figure 2.36). The fatty acids are then degraded in a stepwise oxidative process, two carbons at a time, yielding acetyl CoA plus a fatty acyl-CoA shorter by one two-carbon unit. Each round of oxidation also yields one molecule of NADH and one of $FADH_2$. The acetyl CoA then enters the citric acid cycle, and degradation of the remainder of the fatty acid continues in the same manner.

The breakdown of a 16-carbon fatty acid thus yields seven molecules of NADH, seven of $FADH_2$, and eight of acetyl CoA. In terms of ATP generation, this yield corresponds to 21 molecules of ATP derived from NADH (3×7), 14 ATPs from $FADH_2$ (2×7), and 96 from acetyl CoA (8×12). Since one ATP was used to start the process, the net gain is 130 ATPs per molecule of a 16-carbon fatty acid. Compare this yield with the net gain of 38 ATPs per molecule of glucose. Since the molecular weight of a saturated 16-carbon fatty acid is 256 and that of glucose is 180, the yield of ATP is approximately 2.5 times greater per gram of the fatty acid—hence the advantage of lipids over polysaccharides as energy storage molecules.

Photosynthesis

The generation of energy from oxidation of carbohydrates and lipids relies on the degradation of preformed organic compounds. The energy required for the synthesis of these compounds is ultimately derived from sunlight, which is harvested and used by plants and photosynthetic bacteria to drive the synthesis of carbohydrates. By converting the energy of sunlight to a usable form of chemical energy, photosynthesis is the source of virtually all metabolic energy in biological systems.

The overall equation of photosynthesis can be written as follows:

$$6\ CO_2 + 6\ H_2O \xrightarrow{\text{Light}} C_6H_{12}O_6 + 6\ O_2$$

CHO in chlorophyll *b*

CH₃ in chlorophyll *a*

Porphyrin ring

Hydrocarbon tail

Figure 2.37
The structure of chlorophyll   Chlorophylls consist of porphyrin ring structures linked to hydrocarbon tails. Chlorophylls *a* and *b* differ by a single functional group in the porphyrin ring.

The process is much more complex, however, and takes place in two distinct stages. In the first, called the **light reactions**, energy absorbed from sunlight drives the synthesis of ATP and NADPH (a coenzyme similar to NADH), coupled to the oxidation of H_2O to O_2. The ATP and NADPH generated by the light reactions drive the synthesis of carbohydrates from CO_2 and H_2O in a second set of reactions, called the **dark reactions** because they do not require sunlight. In eukaryotic cells, both the light and dark reactions occur in chloroplasts.

Photosynthetic pigments capture energy from sunlight by absorbing photons. Absorption of light by these pigments causes an electron to move from its normal molecular orbital to one of higher energy, thus converting energy from sunlight into chemical energy. In plants the most abundant photosynthetic pigments are the **chlorophylls** (Figure 2.37), which together absorb visible light of all wavelengths other than green. Additional pigments absorb light of other wavelengths, so essentially the entire spectrum of visible light can be captured and utilized for photosynthesis.

The energy captured by the absorption of light is used to convert H_2O to O_2 (Figure 2.38). The high-energy electrons derived from this process then enter an electron transport chain, in which their transfer through a series of carriers is coupled to the synthesis of ATP. In addition, these high energy electrons reduce $NADP^+$ to NADPH.

In the dark reactions, the ATP and NADPH produced from the light reactions drive the synthesis of carbohydrates from CO_2 and H_2O. One molecule of CO_2 at a time is added to a cycle of reactions—known as the **Calvin cycle** after its discoverer, Melvin Calvin—that leads to the formation of carbohydrates (Figure 2.39). Overall, the Calvin cycle consumes 18 molecules of ATP and 12 of NADPH for each molecule of glucose synthesized. Two electrons are needed to convert each molecule of $NADP^+$ to NADPH, so 24 electrons must pass through the electron transport chain to generate sufficient NADPH to synthesize one molecule of glucose. These electrons are obtained by the conversion of 12 molecules of H_2O to six molecules of O_2, consistent with the formation of six molecules of O_2 for each molecule of glucose. It is not clear, however, whether the passage of the same 24 electrons through the electron transport chain is also sufficient to generate the 18

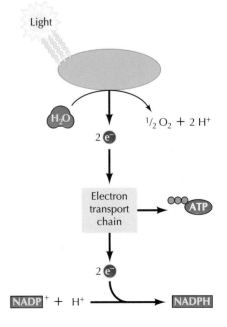

Figure 2.38
The light reactions of photosynthesis   Energy from sunlight is used to split H_2O to O_2. The high-energy electrons derived from this process are then transported through a series of carriers and used to convert $NADP^+$ to NADPH. Energy derived from the electron transport reactions also drives the synthesis of ATP. The details of these reactions are discussed in Chapter 10.

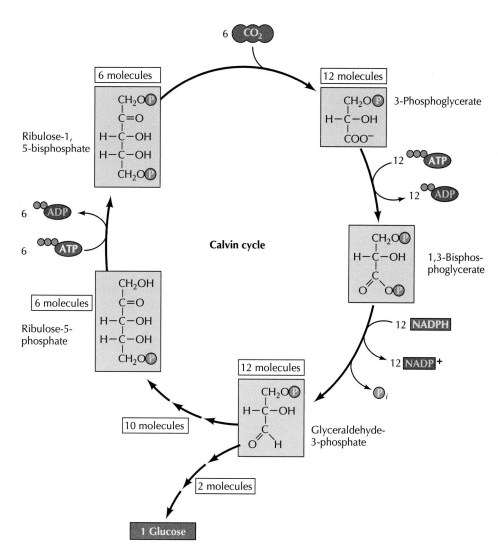

Figure 2.39
The Calvin cycle   Shown here is
the synthesis of one molecule of
glucose from six molecules of CO_2.
Each molecule of CO_2 is added to
ribulose-1,5-bisphosphate to yield
two molecules of 3-phospho-
glycerate. Six molecules of CO_2
thus lead to the formation of 12
molecules of 3-phosphoglycerate,
which are converted to 12 molecules of glyceraldehyde-3-phos-
phate at the cost of 12 molecules
each of ATP and NADPH. Two
molecules of glyceraldehyde-3-
phosphate are then used for synthesis of glucose and ten molecules continue in the Calvin cycle
to form six molecules of ribulose-5-
phosphate. The cycle is then completed by the use of six additional
ATP molecules for the synthesis of
ribulose-1,5-bisphosphate.

ATPs that are required by the Calvin cycle. Some of these ATP molecules
may instead be generated by alternative electron transport chains that use
the energy derived from sunlight to synthesize ATP without the synthesis
of NADPH (see Chapter 10).

The Biosynthesis of Cell Constituents

The preceding section reviewed the major metabolic reactions by which the
cell obtains and stores energy in the form of ATP. This metabolic energy is
then used to accomplish various tasks, including the synthesis of macromolecules and other cell constituents. Thus, energy derived from the breakdown of organic molecules (catabolism) is used to drive the synthesis of
other required components of the cell. Most catabolic pathways involve the
oxidation of organic molecules coupled to the generation of both energy
(ATP) and reducing power (NADH). In contrast, biosynthetic (anabolic)
pathways generally involve the use of both ATP and reducing power (usually in the form of NADPH) for the production of new organic compounds.
One major biosynthetic pathway, the synthesis of carbohydrates from CO_2
and H_2O during the dark reactions of photosynthesis, was discussed in the
preceding section. Additional pathways leading to the biosynthesis of

major cellular constituents (carbohydrates, lipids, proteins, and nucleic acids) are reviewed in the sections that follow.

Carbohydrates

In addition to being obtained directly from food or generated by photosynthesis, glucose can be synthesized from other organic molecules. In animal cells, glucose synthesis (**gluconeogenesis**) usually starts with lactate (produced by anaerobic glycolysis), amino acids (derived from the breakdown of proteins), or glycerol (produced by the breakdown of lipids). Plants (but not animals) are also able to synthesize glucose from fatty acids—a process that is particularly important during the germination of seeds, when energy stored as fats must be converted to carbohydrates to support growth of the plant. In both animal and plant cells, simple sugars are polymerized and stored as polysaccharides.

Gluconeogenesis involves the conversion of pyruvate to glucose—essentially the reverse of glycolysis. However, as discussed earlier, the glycolytic conversion of glucose to pyruvate is an energy-yielding pathway, generating two molecules each of ATP and NADH. Although some reactions of glycolysis are readily reversible, others will proceed only in the direction of glucose breakdown, because they are associated with a large decrease in free energy. These energetically favorable reactions of glycolysis are bypassed during gluconeogenesis by other reactions (catalyzed by different enzymes) that are coupled to the expenditure of ATP and NADH in order to drive them in the direction of glucose synthesis. Overall, the generation of glucose from two molecules of pyruvate requires four molecules of ATP, two of GTP, and two of NADH. This process is considerably more costly than the simple reversal of glycolysis (which would require two molecules of ATP and two of NADH), illustrating the additional energy required to drive the pathway in the direction of biosynthesis.

In both plant and animal cells, glucose is stored in the form of polysaccharides (starch and glycogen, respectively). The synthesis of polysaccharides, like that of all other macromolecules, is an energy-requiring reaction. As noted earlier, the linkage of two sugars by a glycosidic bond can be written as a dehydration reaction, in which H_2O is removed (see Figure 2.3). Such a reaction, however, is energetically unfavorable and therefore unable to proceed in the forward direction. Consequently, the formation of a glycosidic bond must be coupled to an energy-yielding reaction, which is accomplished by the use of nucleotide sugars as intermediates in polysaccharide synthesis (Figure 2.40). Glucose is first phosphorylated in an ATP-driven reaction to glucose-6-phosphate, which is then converted to glucose-1-phosphate. Glucose-1-phosphate reacts with UTP (uridine triphosphate), yielding UDP-glucose plus pyrophosphate, which is hydrolyzed to phosphate with the release of additional free energy. UDP-glucose is an activated intermediate that then donates its glucose residue to a growing polysaccharide chain in an energetically favorable reaction. Thus, chemical energy in the form of ATP and UTP drives the synthesis of polysaccharides from simple sugars.

Lipids

Lipids are important energy storage molecules and the major constituent of cell membranes. They are synthesized from acetyl CoA, which is formed from the breakdown of carbohydrates, in a series of reactions that resemble the reverse of fatty acid oxidation. As with carbohydrate biosynthesis, however, the reactions leading to the synthesis of fatty acids differ from those involved in their degradation and are driven in the biosynthetic direction by being coupled to the expenditure of both energy in the form of ATP and reducing power in the form of NADPH. Fatty acids are synthesized by the

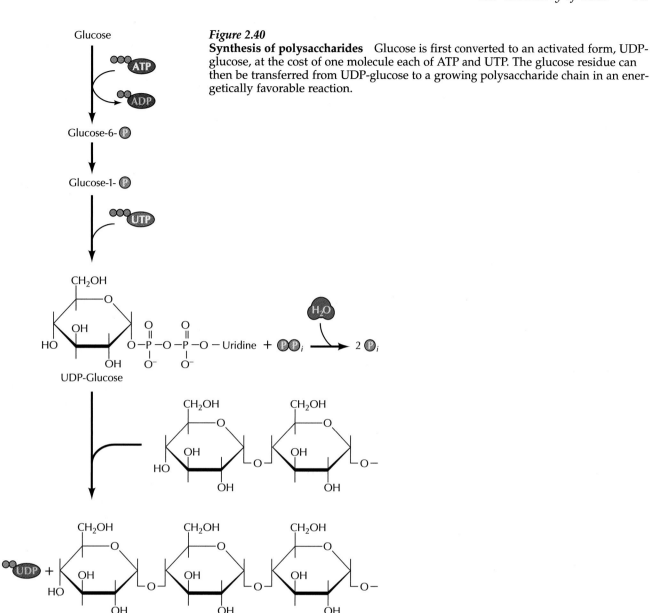

Figure 2.40
Synthesis of polysaccharides   Glucose is first converted to an activated form, UDP-glucose, at the cost of one molecule each of ATP and UTP. The glucose residue can then be transferred from UDP-glucose to a growing polysaccharide chain in an energetically favorable reaction.

stepwise addition of two-carbon units derived from acetyl CoA to a growing chain. The addition of each of these two-carbon units requires the expenditure of one molecule of ATP and two molecules of NADPH.

The major product of fatty acid biosynthesis, which occurs in the cytosol of eukaryotic cells, is the 16-carbon fatty acid palmitate. The principal constituents of cell membranes (phospholipids, sphingomyelin, and glycolipids) are then synthesized from free fatty acids in the endoplasmic reticulum and Golgi apparatus (see Chapter 9).

Proteins

In contrast to carbohydrates and lipids, proteins (as well as nucleic acids) contain nitrogen in addition to carbon, hydrogen, and oxygen. Nitrogen is incorporated into organic compounds from different sources in different organisms (Figure 2.41). Some bacteria can use atmospheric N_2 by a process called **nitrogen fixation**, in which N_2 is reduced to NH_3 at the expense of energy in the form of ATP. Although relatively few species of bacteria are

Figure 2.41
Assimilation of nitrogen into organic compounds   Ammonia is incorporated into organic compounds by all organisms. Some bacteria are capable of converting atmospheric nitrogen to ammonia, and most bacteria, fungi, and plants can utilize nitrate from soil.

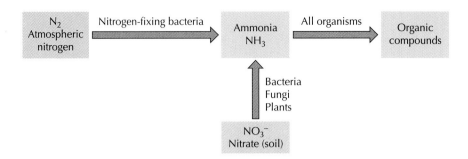

TABLE 2.2   Dietary Requirements for Amino Acids in Humans

Essential	Nonessential
Histidine	Alanine
Isoleucine	Arginine[a]
Leucine	Asparagine
Lysine	Aspartate
Methionine	Cysteine
Phenylalanine	Glutamate
Threonine	Glutamine
Tryptophan	Glycine
Valine	Proline
	Serine
	Tyrosine

The essential amino acids must be obtained from dietary sources; the nonessential amino acids can be synthesized by human cells.

[a] Although arginine is classified as a nonessential amino acid, growing children must obtain additional arginine from their diet.

capable of nitrogen fixation, most bacteria, fungi, and plants can use nitrate (NO_3^-), which is a common constituent of soil, by reducing it to NH_3 via electrons derived from NADH or NADPH. Finally, all organisms are able to incorporate ammonia (NH_3) into organic compounds.

NH_3 is incorporated into organic molecules primarily during the synthesis of the amino acids glutamate and glutamine, which are derived from the citric acid cycle intermediate α-ketoglutarate. These amino acids then serve as donors of amino groups during the synthesis of the other amino acids, which are also derived from central metabolic pathways, such as glycolysis and the citric acid cycle (Figure 2.42). The raw material for amino acid synthesis is thus obtained from glucose, and the amino acids are synthesized at the cost of both energy (ATP) and reducing power (NADPH). Many bacteria and plants can synthesize all 20 amino acids. Humans and other mammals, however, can synthesize only about half of the required amino acids; the remainder must be obtained from dietary sources (Table 2.2).

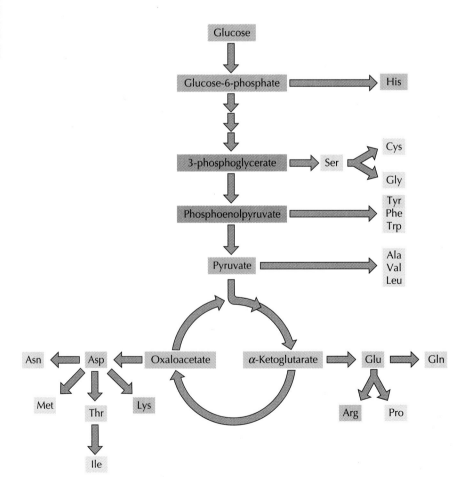

Figure 2.42
Biosynthesis of amino acids   The carbon skeletons of the amino acids are derived from intermediates in glycolysis and in the citric acid cycle.

Molecular Medicine

Phenylketonuria

The Disease

Phenylketonuria, or PKU, is an inborn error of amino acid metabolism with devastating effects. It afflicts approximately one in 10,000 newborn infants and, if untreated, results in severe mental retardation. Fortunately, understanding of the nature of the defect responsible for phenylketonuria has allowed early diagnosis and effective treatment.

Molecular and Cellular Basis

Phenylketonuria is caused by a deficiency of the enzyme phenylalanine hydroxylase, which converts phenylalanine to tyrosine. This deficiency causes phenylalanine to accumulate to very high levels and undergo other reactions, such as conversion to phenylpyruvate. Phenylalanine, phenylpyruvate, and other abnormal metabolites accumulate in the blood and are secreted at high levels in the urine (the name of the disease derives from the high levels of phenylpyruvate, a phenylketone, found in the urine of affected children). Although

the precise biochemical cause is unknown, mental retardation is a critical consequence of the accumulation of these abnormal phenylalanine metabolites.

Prevention and Treatment

The enzyme deficiency causes no difficulties while the fetus is within the uterus, so children with phenylketonuria are normal at birth. If untreated, however, affected children become permanently and severely retarded within the first year of life. Fortunately, newborns with phenyl-

ketonuria can be readily identified by routine screening tests that detect elevated levels of phenylalanine in blood. Mental retardation can be prevented by feeding affected infants a synthetic diet that is low in phenylalanine. This dietary treatment eliminates the accumulation of toxic phenylalanine metabolites and effectively prevents the permanent mental deficiency that would otherwise ensue. Routine screening for phenylketonuria is thus a critical test for all newborn infants.

Abnormal metabolism of phenylalanine in patients with phenylketonuria.

The polymerization of amino acids to form proteins also requires energy. Like the synthesis of polysaccharides, the formation of the peptide bond can be considered a dehydration reaction, which must be driven in the direction of protein synthesis by being coupled to another source of metabolic energy. In the biosynthesis of polysaccharides, this coupling is accomplished through the conversion of sugars to activated intermediates, such as UDP-glucose. Amino acids are similarly activated before being used for protein synthesis.

A critical difference between the synthesis of proteins and that of polysaccharides is that the amino acids are incorporated into proteins in a unique order, specified by a gene. The order of nucleotides in a gene specifies the amino acid sequence of a protein via translation, in which messenger RNA (mRNA) acts as a template for protein synthesis (see Chapter 3). Each amino acid is first attached to a specific transfer RNA (tRNA) molecule in a reaction coupled to ATP hydrolysis (Figure 2.43). The aminoacyl tRNAs then align on the mRNA template bound to ribosomes, and each amino acid is added to the C terminus of a growing peptide chain through a series of reactions that will be discussed in detail in Chapter 7. During the process, two additional molecules of GTP are hydrolyzed, so the incorporation of each amino acid into a protein is coupled to the hydrolysis of one ATP and two GTP molecules.

Figure 2.43
Formation of the peptide bond An amino acid is first activated by attachment to its tRNA in a two-step reaction
involving the hydrolysis of ATP to AMP. The tRNAs serve as
adaptors to align the amino acids on an mRNA template
bound to ribosomes. Each amino acid is then transferred to
the C terminus of the growing peptide chain at the cost of
two additional molecules of GTP.

Nucleic Acids

The precursors of nucleic acids, the nucleotides, are composed of phosphorylated five-carbon sugars joined to nucleic acid bases. Nucleotides can be
synthesized from carbohydrates and amino acids; they can also be obtained from dietary sources or reused following nucleic acid breakdown.
The starting point for nucleotide biosynthesis is the phosphorylated sugar
ribose-5-phosphate, which is derived from glucose-6-phosphate. Divergent
pathways then lead to the synthesis of purine and pyrimidine ribonucleotides, which are the immediate precursors for RNA synthesis. These
ribonucleotides are converted to deoxyribonucleotides, which serve as the
monomeric building blocks of DNA.

RNA and DNA are polymers of nucleoside monophosphates. As for
other macromolecules, however, direct polymerization of nucleoside mono-

Growing polynucleotide chain

Figure 2.44
Synthesis of polynucleotides Nucleoside triphosphates are joined to the 3′ end of a growing polynucleotide chain with the release of pyrophosphate.

phosphates is energetically unfavorable, and the synthesis of polynucleotides instead uses nucleoside triphosphates as activated precursors (Figure 2.44). A nucleoside 5′-triphosphate is added to the 3′ hydroxyl group of a growing polynucleotide chain, with the release and subsequent hydrolysis of pyrophosphate serving to drive the reaction in the direction of polynucleotide synthesis.

Cell Membranes

The structure and function of cells are critically dependent on membranes, which not only separate the interior of the cell from its environment but also define the internal compartments of eukaryotic cells, including the nucleus and cytoplasmic organelles. The formation of biological membranes is based on the properties of lipids, and all cell membranes share a common structural organization: bilayers of phospholipids with associated proteins. These membrane proteins are responsible for many specialized functions; some act as receptors that allow the cell to respond to external signals, some are responsible for the selective transport of molecules across the membrane, and others participate in electron transport and oxidative phosphorylation. In addition, membrane proteins control the interactions between cells of multicellular organisms. The common structural organization of membranes thus underlies a variety of biological processes and specialized membrane functions, which will be discussed in detail in later chapters.

Figure 2.45
A phospholipid bilayer Phospholipids spontaneously form stable bilayers, with their polar head groups exposed to water and their hydrophobic tails buried in the interior of the membrane.

Membrane Lipids

The fundamental building blocks of all cell membranes are phospholipids, which are amphipathic molecules, consisting of two hydrophobic fatty acid chains linked to a phosphate-containing hydrophilic head group (see Figure 2.7). Because their fatty acid tails are poorly soluble in water, phospholipids spontaneously form bilayers in aqueous solutions, with the hydrophobic tails buried in the interior of the membrane and the polar head groups exposed on both sides, in contact with water (Figure 2.45). Such **phospholipid bilayers** form a stable barrier between two aqueous compartments and represent the basic structure of all biological membranes.

Lipids constitute approximately 50% of the mass of most cell membranes, although this proportion varies depending on the type of membrane. Plasma membranes, for example, are approximately 50% lipid and 50% protein. The inner membrane of mitochondria, on the other hand, contains an unusually high fraction (about 75%) of protein, reflecting the abundance of protein complexes involved in electron transport and oxidative phosphorylation. The lipid composition of different cell membranes also varies (Table 2.3). The plasma membrane of *E. coli* consists predominantly of phosphatidylethanolamine, which constitutes 80% of total lipid. Mammalian plasma membranes are more complex, containing four major phospholipids—phosphatidylcholine, phosphatidylserine, phosphatidylethanolamine, and sphingomyelin—which together constitute 50 to 60% of total membrane lipid. In addition to the phospholipids, the plasma membranes of animal cells contain glycolipids and cholesterol, which generally correspond to about 40% of the total lipid molecules.

An important property of lipid bilayers is that they behave as two-dimensional fluids in which individual molecules (both lipids and proteins) are free to rotate and move in lateral directions (Figure 2.46). Such fluidity is a critical property of membranes and is determined by both temperature and lipid composition. For example, the interactions between shorter fatty acid chains are weaker than those between longer chains, so membranes containing shorter fatty acid chains are less rigid and remain fluid at lower temperatures. Lipids containing unsaturated fatty acids similarly increase membrane fluidity because the presence of double bonds introduces kinks in the fatty acid chains, making them more difficult to pack together.

Because of its hydrocarbon ring structure (see Figure 2.9), cholesterol plays a distinct role in determining membrane fluidity. Cholesterol molecules insert into the bilayer with their polar hydroxyl groups close to the hydrophilic head groups of the phospholipids (Figure 2.47). The rigid

Figure 2.46
Mobility of phospholipids in a membrane Individual phospholipids can rotate and move laterally within a bilayer.

TABLE 2.3 Lipid Composition of Cell Membranes[a]

Lipid	Plasma membrane		Rough endoplasmic reticulum	Outer mitochondrial membranes
	E. coli	Erythrocyte		
Phosphatidylcholine	0	17	55	50
Phosphatidylserine	0	6	3	2
Phosphatidyl-ethanolamine	80	16	16	23
Sphingomyelin	0	17	3	5
Glycolipids	0	2	0	0
Cholesterol	0	45	6	<5

Source: Data from P. L. Yeagle, 1993. *The Membranes of Cells*, 2nd ed. San Diego, CA: Academic Press.
[a] Membrane compositions are indicated as the mole percentages of major lipid constituents.

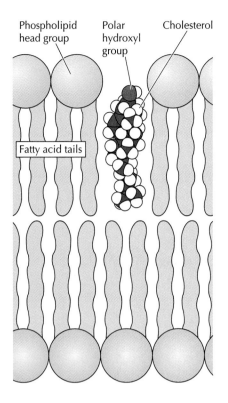

Figure 2.47
Insertion of cholesterol in a membrane Cholesterol inserts into the membrane with its polar hydroxyl group close to the polar head groups of the phospholipids.

hydrocarbon rings of cholesterol therefore interact with the regions of the fatty acid chains that are adjacent to the phospholipid head groups. This interaction decreases the mobility of the outer portions of the fatty acid chains, making this part of the membrane more rigid. On the other hand, insertion of cholesterol interferes with interactions between fatty acid chains, thereby maintaining membrane fluidity at lower temperatures.

Membrane Proteins

Proteins are the other major constituent of cell membranes, constituting 25 to 75% of the mass of the various membranes of the cell. The current model of membrane structure, proposed by Jonathan Singer and Garth Nicolson in 1972, views membranes as a **fluid mosaic** in which proteins are inserted into a lipid bilayer (Figure 2.48). While phospholipids provide the basic structural organization of membranes, membrane proteins carry out the specific functions of the different membranes of the cell. These proteins are divided into two general classes, based on the nature of their association with the membrane. **Integral membrane proteins** are embedded directly within the lipid bilayer. **Peripheral membrane proteins** are not inserted into the lipid bilayer but are associated with the membrane indirectly, generally by interactions with integral membrane proteins.

Many integral membrane proteins (called **transmembrane proteins**) span the lipid bilayer, with portions exposed on both sides of the membrane. The membrane-spanning portions of these proteins are usually α-helical regions of 20 to 25 nonpolar amino acids. The hydrophobic side chains of these amino acids interact with the fatty acid chains of membrane lipids, and the formation of an α helix neutralizes the polar character of the peptide bonds, as discussed earlier in this chapter with respect to protein folding. Like the phospholipids, transmembrane proteins are amphipathic molecules, with their hydrophilic portions exposed to the aqueous environment on both sides of the membrane. Some transmembrane proteins span the membrane only once; others have multiple membrane-spanning regions. Most transmembrane proteins of eukaryotic plasma membranes have been modified by the addition of carbohydrates, which are exposed on the surface of the cell and may participate in cell-cell interactions.

Proteins can also be anchored in membranes by lipids that are covalently attached to the polypeptide chain (see Chapter 7). Distinct lipid modifications anchor proteins to the cytosolic and extracellular faces of the plasma membrane. Proteins can be anchored to the cytosolic face of the membrane either by the addition of a 14-carbon fatty acid (myristic acid) to their amino terminus or by the addition of either a 16-carbon fatty acid (palmitic acid) or 15- or 20-carbon prenyl groups to the side chains of cysteine residues. Alternatively, proteins are anchored to the extracellular face of the plasma membrane by the addition of glycolipids to their carboxy terminus.

Transport across Cell Membranes

The selective permeability of biological membranes to small molecules allows the cell to control and maintain its internal composition. Only small uncharged molecules can diffuse freely through phospholipid bilayers (Figure 2.49). Small nonpolar molecules, such as O_2 and CO_2, are soluble in

Outside   Carbohydrate   Proteins

Inside

Glycolipid

Membrane-
spanning
α helix

Fatty acid or
prenyl group

Lipid
bilayer

Integral
membrane
protein

Peripheral membrane protein

Figure 2.48
Fluid mosaic model of membrane structure   Biological membranes consist of proteins inserted into a lipid bilayer. Integral membrane proteins are embedded in the membrane, usually via α-helical regions of 20 to 25 hydrophobic amino acids. Some transmembrane proteins span the membrane only once; others have multiple membrane-spanning regions. In addition, some proteins are anchored in the membrane by lipids that are covalently attached to the polypeptide chain. These proteins can be anchored to the extracellular face of the plasma membrane by glycolipids and to the cytosolic face by fatty acids or prenyl groups (see Chapter 7 for structures). Peripheral membrane proteins are not inserted in the membrane but are attached via interactions with integral membrane proteins.

Small uncharged molecules

O_2 CO_2 N_2

H_2O Ethanol

Glycerol

Large polar molecules and ions

H^+ Na^+ Ca^{2+}

Glucose Cl^-

Amino acids
(e.g., alanine)

Figure 2.49
Permeability of phospholipid bilayers   Small uncharged molecules can diffuse freely through a phospholipid bilayer. However, the bilayer is impermeable to larger polar molecules (such as glucose and amino acids) and to ions.

the lipid bilayer and therefore can readily cross cell membranes. Small uncharged polar molecules, such as H_2O, also can diffuse through membranes, but larger uncharged polar molecules, such as glucose, cannot. Charged molecules, such as ions, are unable to diffuse through a phospholipid bilayer regardless of size; even H^+ ions cannot cross a lipid bilayer by free diffusion.

Although ions and most polar molecules cannot diffuse across a lipid bilayer, many such molecules (such as glucose) are able to cross cell membranes. These molecules pass across membranes via the action of specific transmembrane proteins, which act as transporters. Such transport proteins determine the selective permeability of cell membranes and thus play a critical role in membrane function. They contain multiple membrane-spanning regions that form a passage through the lipid bilayer, allowing polar or charged molecules to cross the membrane through a protein pore without interacting with the hydrophobic fatty acid chains of the membrane phospholipids.

As discussed in detail in Chapter 12, there are two general classes of membrane transport proteins (Figure 2.50). **Channel proteins** form open pores through the membrane, allowing the free passage of any molecule of the appropriate size. Ion channels, for example, allow the passage of inorganic ions such as Na^+, K^+, Ca^{2+}, and Cl^- across the plasma membrane. Once open, channel proteins form small pores through which ions of the appropriate size and charge can cross the membrane by free diffusion. The pores formed by these channel proteins are not permanently open; rather, they can be selectively opened and closed in response to extracellular signals, allowing the cell to control the movement of ions across the membrane. Such regulated ion channels have been particularly well studied in nerve and muscle cells, where they mediate the transmission of electrochemical signals.

In contrast to channel proteins, **carrier proteins** selectively bind and transport specific small molecules, such as glucose. Rather than forming open channels, carrier proteins act like enzymes to facilitate the passage of specific molecules across membranes. In particular, carrier proteins bind specific molecules and then undergo conformational changes that open channels through which the molecule to be transported can pass across the membrane and be released on the other side.

Figure 2.50
Channel and carrier proteins
(A) Channel proteins form open pores through which molecules of the appropriate size (e.g., ions) can cross the membrane. (B) Carrier proteins selectively bind the small molecule to be transported and then undergo a conformational change to release the molecule on the other side of the membrane.

(A)

(B)

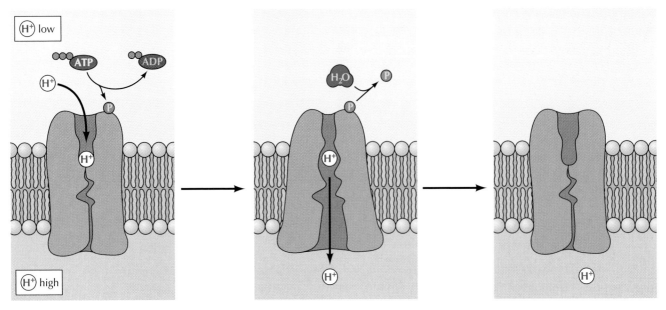

Figure 2.51
Model of active transport   Energy derived from the hydrolysis of ATP is used to transport H^+ against the electrochemical gradient (from low to high H^+ concentration). Binding of H^+ is accompanied by phosphorylation of the carrier protein, which induces a conformational change that drives H^+ transport against the electrochemical gradient. Release of H^+ and hydrolysis of the bound phosphate group then restore the carrier to its original conformation.

As described so far, molecules transported by either channel or carrier proteins cross membranes in the energetically favorable direction, as determined by concentration and electrochemical gradients—a process known as **passive transport**. However, carrier proteins also provide a mechanism through which the energy changes associated with transporting molecules across a membrane can be coupled to the use or production of other forms of metabolic energy, just as enzymatic reactions can be coupled to the hydrolysis or synthesis of ATP. For example, molecules can be transported in an energetically unfavorable direction across a membrane (e.g., against a concentration gradient) if their transport in that direction is coupled to ATP hydrolysis as a source of energy—a process called **active transport** (Figure 2.51). The free energy stored as ATP can thus be used to control the internal composition of the cell, as well as to drive the biosynthesis of cell constituents.

KEY TERMS

carbohydrate, monosaccharide, glycosidic bond, oligosaccharide, polysaccharide, glycogen, starch, cellulose

lipid, fatty acid, triacylglycerol, fat, phospholipid, glycerol phospholipid, sphingomyelin, amphipathic, glycolipid, cholesterol, steroid hormone

SUMMARY

THE MOLECULAR COMPOSITION OF CELLS

Carbohydrates: Carbohydrates include simple sugars and polysaccharides. Polysaccharides serve as storage forms of sugars, structural components of cells, and markers for cell recognition processes.

Lipids: Lipids are the principal components of cell membranes, and they serve as energy storage and signaling molecules. Phospholipids consist of two hydrophobic fatty acid chains linked to a hydrophilic phosphate-containing head group.

Nucleic Acids: Nucleic acids are the principal informational molecules of the cell. Both DNA and RNA are polymers of purine and pyrimidine nucleotides. Hydrogen bonding between complementary base pairs allows nucleic acids to direct their self-replication.

deoxyribonucleic acid (DNA), ribonucleic acid (RNA), messenger RNA (mRNA), ribosomal RNA, transfer RNA, nucleotide, purine, pyrimidine, adenine, guanine, cytosine, thymine, uracil, 2′-deoxyribose, ribose, nucleoside, phosphodiester bond, oligonucleotide, polynucleotide

Proteins: Proteins are polymers of 20 different amino acids, each of which has a distinct side chain with specific chemical properties. Each protein has a unique amino acid sequence, which determines its three-dimensional structure. In most proteins, combinations of α helices and β sheets fold into globular domains with hydrophobic amino acids in the interior and hydrophilic amino acids on the surface.

protein, amino acid, peptide bond, polypeptide, X-ray crystallography, primary structure, secondary structure, α helix, β sheet, tertiary structure, domain, quaternary structure

THE CENTRAL ROLE OF ENZYMES AS BIOLOGICAL CATALYSTS

The Catalytic Activity of Enzymes: Virtually all chemical reactions within cells are catalyzed by enzymes.

enzyme, substrate, product, transition state, activation energy, active site

Mechanisms of Enzymatic Catalysis: Enzymes increase reaction rates by binding substrates in the proper position, by altering the conformation of substrates to approach the transition state, and by participating directly in chemical reactions.

lock-and-key model, induced fit

Coenzymes: Coenzymes function in conjunction with enzymes to carry chemical groups between substrates.

prosthetic group, coenzyme, nicotinamide adenine dinucleotide (NAD$^+$)

Regulation of Enzyme Activity: The activities of enzymes are regulated to meet the physiological needs of the cell. Enzyme activity can be controlled by the binding of small molecules, by interactions with other proteins, and by covalent modifications.

feedback inhibition, allosteric regulation, phosphorylation

METABOLIC ENERGY

Free Energy and ATP: ATP serves as a store of free energy, which is used to drive energy-requiring reactions within cells.

Gibbs free energy (G), adenosine 5′-triphosphate (ATP), high-energy bond

The Generation of ATP from Glucose: The breakdown of glucose provides a major source of cellular energy. In aerobic cells, the complete oxidation of glucose yields 36 to 38 molecules of ATP. Most of this ATP is derived from electron transport reactions in which O_2 is reduced to H_2O.

glycolysis, coenzyme A (CoA), citric acid cycle, Krebs cycle, flavin adenine dinucleotide (FADH$_2$), oxidative phosphorylation, electron transport chain

The Derivation of Energy from Other Organic Molecules: ATP can also be derived from the breakdown of organic molecules other than glucose. Because fats are more reduced than carbohydrates, they provide a more efficient form of energy storage.

Photosynthesis: The energy required for the synthesis of organic molecules is ultimately derived from sunlight, which is harvested by plants and photosynthetic bacteria. In the first stage of photosynthesis, energy from sunlight is used to drive the synthesis of ATP and NADPH, coupled to the oxidation of H_2O to O_2. The ATP and NADPH produced by these reactions are then used to synthesize glucose from CO_2 and H_2O.

light reactions, dark reactions, photosynthetic pigment, chlorophyll, Calvin cycle

THE BIOSYNTHESIS OF CELL CONSTITUENTS

gluconeogenesis

Carbohydrates: Glucose can be synthesized from other organic molecules, using energy and reducing power in the forms of ATP and NADH, respectively. Additional ATP is then needed to drive the synthesis of polysaccharides from simple sugars.

Lipids: Lipids are synthesized from acetyl CoA, which is formed from the breakdown of carbohydrates.

nitrogen fixation

Proteins: The amino acids are synthesized from intermediates in glycolysis and the citric acid cycle. Their polymerization to form proteins requires additional energy in the form of ATP and GTP.

Nucleic Acids: Purine and pyrimidine nucleotides are synthesized from carbohydrates and amino acids. Their polymerization to DNA and RNA is driven by the use of nucleoside triphosphates as activated precursors.

CELL MEMBRANES

phospholipid bilayer

Membrane Lipids: The basic structure of all cell membranes is a phospholipid bilayer. Membranes of animal cells also contain glycolipids and cholesterol.

fluid mosaic, integral membrane protein, peripheral membrane protein, transmembrane protein

Membrane Proteins: Proteins can either be inserted into the lipid bilayer or associated with the membrane indirectly, by protein-protein interactions. Some proteins span the lipid bilayer; others are anchored to one side of the membrane.

channel protein, carrier protein, passive transport, active transport

Transport across Cell Membranes: Lipid bilayers are permeable only to small uncharged molecules. Ions and most polar molecules are transported across cell membranes by specific transport proteins, the action of which can be coupled to the hydrolysis or synthesis of ATP.

Questions

1. What key molecular characteristic of phospholipids makes them the basic components of cell membranes?

2. Where would you expect to find phenylalanine residues in a folded protein? What amino acid residues would you expect to find in the loop regions connecting different α helices?

3. The binding pocket of trypsin contains an aspartate residue. How do you think changing this amino acid to lysine would affect the enzyme's activity?

4. Consider the reaction A $\rightleftharpoons$ B + C, in which $\Delta G^{\circ\prime}$ = +3.5 kcal/mol. Calculate ΔG under intracellular conditions, given that the concentration of A is 10^{-2} M and the concentrations of B and C are each 10^{-3} M. In which direction will the reaction proceed in the cell? For your calculation $R = 1.98 \times 10^{-3}$ kcal/mol/degree and $T = 298$ K (25°C). Note that ln $(x) = 2.3$ log10 (x).

5. Which process releases more free energy: the degradation of saturated fatty acids or the degradation of unsaturated fatty acids?

6. Why are different enzymes needed for gluconeogenesis and glycolysis?

7. Why are the membrane-spanning regions of transmembrane proteins frequently α-helical?

References and Further Reading

General References

Mathews, C. K. and K. E. van Holde. 1996. *Biochemistry*. 2nd ed. Redwood City, CA: Benjamin Cummings.

Stryer, L. 1995. *Biochemistry*. 4th ed. New York: W. H. Freeman.

The Molecular Composition of Cells

Anfinsen, C. B. 1973. Principles that govern the folding of protein chains. *Science* 181: 223–230. [P]

Branden, C. and J. Tooze. 1991. *Introduction to Protein Structure*. New York: Garland.

Chothia, C. and A. V. Finkelstein. 1990. The classification and origins of protein folding patterns. *Ann. Rev. Biochem.* 59: 1007–1039. [R]

Holm, L. and C. Sander. 1996. Mapping the protein universe. *Science* 273: 595–602. [R]

Kendrew, J. C. 1961. The three-dimensional structure of a protein molecule. *Sci. Am.* 205(6): 96–111. [R]

Levitt, M., M. Gerstein, E. Huang, S. Subbiah and J. Tsai. 1997. Protein folding: The endgame. *Ann. Rev. Biochem.* 66: 549–579. [R]

Pauling, L., R. B. Corey, and H. R. Branson. 1951. The structure of proteins: Two hydrogen bonded configurations of the polypeptide chain. *Proc. Natl. Acad. Sci. USA* 37: 205–211. [P]

Richardson, J. S. 1981. The anatomy and taxonomy of protein structure. *Adv. Protein Chem.* 34: 167–339. [R]

Sanger, F. 1988. Sequences, sequences, and sequences. *Ann. Rev. Biochem.* 57: 1–28. [R]

Tanford, C. 1978. The hydrophobic effect and the organization of living matter. *Science* 200: 1012–1018. [R]

The Central Role of Enzymes as Biological Catalysts

Dressler, D. and H. Potter. 1991. *Discovering Enzymes*. New York: Scientific American Library.

Fersht, A. 1985. *Enzyme Structure and Mechanism*. 2nd ed. New York: W. H. Freeman.

Koshland, D. E. 1984. Control of enzyme activity and metabolic pathways. *Trends Biochem. Sci.* 9: 155–159. [R]

Lienhard, G. E. 1973. Enzymatic catalysis and transition-state theory. *Science* 180: 149–154. [R]

Lipscomb, W. N. 1983. Structure and catalysis of enzymes. *Ann. Rev. Biochem.* 52: 17–34. [R]

Monod, J., J.-P. Changeux and F. Jacob. 1963. Allosteric proteins and cellular control systems. *J. Mol. Biol.* 6: 306–329. [P]

Narlikar, G. J. and D. Herschlag. 1997. Mechanistic aspects of enzymatic catalysis: Lessons from comparison of RNA and protein enzymes. *Ann. Rev. Biochem.* 66: 19–59. [R]

Neurath, H. 1984. Evolution of proteolytic enzymes. *Science* 224: 350–357. [R]

Schramm, V. L. 1998. Enzymatic transition states and transition state analog design. *Ann. Rev. Biochem.* 67: 693–720. [R]

Walsh, C. T. 1979. *Enzymatic Reaction Mechanisms*. New York: W. H. Freeman.

Metabolic Energy

Beinert, H., R. H. Holm and E. Munck. 1997. Iron-sulfur clusters: Nature's modular, multipurpose structures. *Science* 277: 653–659. [R]

Bennett, J. 1979. The protein that harvests sunlight. *Trends Biochem. Sci.* 4: 268–271. [R]

Calvin, M. 1962. The path of carbon in photosynthesis. *Science* 135: 879–889. [R]

Deisenhofer, J. and H. Michel. 1991. Structures of bacterial photosynthetic reaction centers. *Ann. Rev. Cell Biol.* 7: 1–23. [R]

Klotz, I. 1967. *Energy Changes in Biochemical Reactions*. New York: Academic Press.

Krebs, H. A. 1970. The history of the tricarboxylic cycle. *Perspect. Biol. Med.* 14: 154–170. [R]

Kuhlbrandt, W., D. N. Wang and Y. Fujiyoshi. 1994. Atomic model of plant light-harvesting complex by electron crystallography. *Nature* 367: 614–621. [P]

Nicholls, D. G. and S. J. Ferguson. 1992. *Bioenergetics 2*. London: Academic Press.

Saraste, M. 1999. Oxidative phosphorylation at the fin de siecle. *Science* 283: 1488–1493. [R]

The Biosynthesis of Cell Constituents

Hers, H. G. and L. Hue. 1983. Gluconeogenesis and related aspects of glycolysis. *Ann. Rev. Biochem.* 52: 617–653. [R]

Jones, M. E. 1980. Pyrimidine nucleotide biosynthesis in animals: Genes, enzymes, and regulation of UMP biosynthesis. *Ann. Rev. Biochem.* 49: 253–279. [R]

Kornberg, A. and T. A. Baker. 1992. *DNA Replication*. 2nd ed. New York: W. H. Freeman.

Tolbert, N. E. 1981. Metabolic pathways in peroxisomes and glyoxysomes. *Ann. Rev. Biochem.* 50: 133–157. [R]

Umbarger, H. E. 1978. Amino acid biosynthesis and its regulation. *Ann. Rev. Biochem.* 47: 533–606. [R]

Van den Bosch, H., R. B. H. Schutgens, R. J. A. Wanders and J. M. Tager. 1992. Biochemistry of peroxisomes. *Ann. Rev. Biochem.* 61: 157–197. [R]

Wakil, S. J., J. K. Stoops and V. C. Joshi. 1983. Fatty acid synthesis and its regulation. *Ann. Rev. Biochem.* 52: 537–579. [R]

Cell Membranes

Bretscher, M. 1985. The molecules of the cell membrane. *Sci. Am.* 253(4): 100–108. [R]

Dowham, W. 1997. Molecular basis for membrane phospholipid diversity: Why are there so many lipids? *Ann. Rev. Biochem.* 66: 199–212. [R]

Englund, P. T. 1993. The structure and biosynthesis of glycosyl phosphatidylinositol protein anchors. *Ann. Rev. Biochem.* 62: 121–138. [R]

Petty, H. R. 1993. *Molecular Biology of Membranes: Structure and Function*. New York: Plenum Press.

Singer, S. J. 1990. The structure and insertion of integral proteins in membranes. *Ann. Rev. Cell Biol.* 6: 247–296. [R]

Singer, S. J. and G. L. Nicolson. 1972. The fluid mosaic model of the structure of cell membranes. *Science* 175: 720–731. [P]

Towler, D. A., J. I. Gordon, S. P. Adams and L. Glaser. 1988. The biology and enzymology of eukaryotic protein acylation. *Ann. Rev. Biochem.* 57: 69–99. [R]

Yeagle, P. L. 1993. *The Membranes of Cells*. 2nd ed. San Diego, CA: Academic Press.

Zhang, F. L. and P. J. Casey. 1996. Protein prenylation: Molecular mechanisms and functional consequences. *Ann. Rev. Biochem.* 65: 241–269. [R]

3

Fundamentals of Molecular Biology

CONTEMPORARY MOLECULAR BIOLOGY is concerned principally with understanding the mechanisms responsible for transmission and expression of the genetic information that ultimately governs cell structure and function. As reviewed in Chapter 1, all cells share a number of basic properties, and this underlying unity of cell biology is particularly apparent at the molecular level. Such unity has allowed scientists to choose simple organisms (such as bacteria) as models for many fundamental experiments, with the expectation that similar molecular mechanisms are operative in organisms as diverse as *E. coli* and humans. Numerous experiments have established the validity of this assumption, and it is now clear that the molecular biology of cells provides a unifying theme to understanding diverse aspects of cell behavior.

Initial advances in molecular biology were made by taking advantage of the rapid growth and readily manipulable genetics of simple bacteria, such as *E. coli*, and their viruses. More recently, not only the fundamental principles but also many of the experimental approaches first developed in prokaryotes have been successfully applied to eukaryotic cells. The development of recombinant DNA has had a tremendous impact, allowing individual eukaryotic genes to be isolated and characterized in detail. Current advances in recombinant DNA technology have made even the determination of the complete sequence of the human genome a feasible project.

Heredity, Genes, and DNA

Perhaps the most fundamental property of all living things is the ability to reproduce. All organisms inherit the genetic information specifying their structure and function from their parents. Likewise, all cells arise from pre-existing cells, so the genetic material must be replicated and passed from parent to progeny cell at each cell division. How genetic information is replicated and transmitted from cell to cell and organism to organism thus represents a question that is central to all of biology. Consequently, elucidation of the mechanisms of genetic transmission and identification of the

89

genetic material as DNA were discoveries that formed the foundation of our current understanding of biology at the molecular level.

Genes and Chromosomes

The classical principles of genetics were deduced by Gregor Mendel in 1865, on the basis of the results of breeding experiments with peas. Mendel studied the inheritance of a number of well-defined traits, such as seed color, and was able to deduce general rules for their transmission. In all cases, he could correctly interpret the observed patterns of inheritance by assuming that each trait is determined by a pair of inherited factors, which are now called **genes**. One gene copy (called an **allele**) specifying each trait is inherited from each parent. For example, breeding two strains of peas—one having yellow seeds, and the other green seeds—yields the following results (Figure 3.1). The parental strains each have two identical copies of the gene specifying yellow (Y) or green (y) seeds, respectively. The progeny plants are therefore hybrids, having inherited one gene for yellow seeds (Y) and one for green seeds (y). All these progeny plants (the first filial, or F$_1$, generation) have yellow seeds, so yellow (Y) is said to be **dominant** and green (y) **recessive**. The **genotype** (genetic composition) of the F$_1$ peas is thus Yy, and their **phenotype** (physical appearance) is yellow. If one F$_1$ offspring is bred with another, giving rise to F$_2$ progeny, the genes for yellow and green seeds segregate in a characteristic manner such that the ratio between F$_2$ plants with yellow seeds and those with green seeds is 3:1.

Mendel's findings, apparently ahead of their time, were largely ignored until 1900, when Mendel's laws were rediscovered and their importance recognized. Shortly thereafter, the role of **chromosomes** as the carriers of genes was proposed. It was realized that most cells of higher plants and animals are **diploid**—containing two copies of each chromosome. Formation of the germ cells (the sperm and egg), however, involves a unique type of cell division (**meiosis**) in which only one member of each chromosome pair is trans-

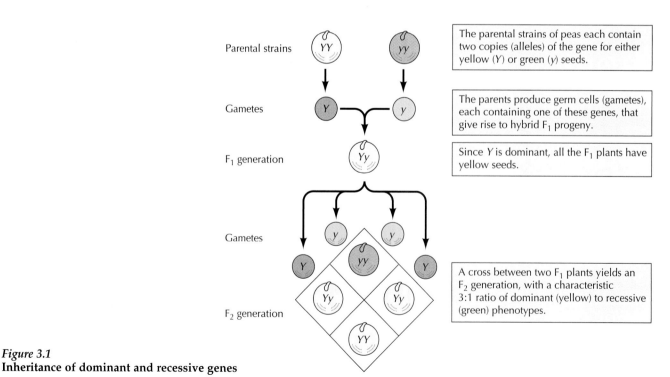

Figure 3.1
Inheritance of dominant and recessive genes

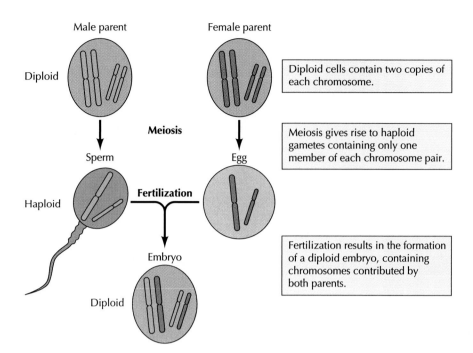

Male parent

Female parent

Diploid

Diploid cells contain two copies of each chromosome.

Meiosis

Sperm

Egg

Meiosis gives rise to haploid gametes containing only one member of each chromosome pair.

Haploid

Fertilization

Embryo

Fertilization results in the formation of a diploid embryo, containing chromosomes contributed by both parents.

Diploid

Figure 3.2
Chromosomes at meiosis and fertilization   Two chromosome pairs of a hypothetical organism are illustrated.

mitted to each progeny cell (Figure 3.2). Consequently, the sperm and egg are **haploid**, containing only one copy of each chromosome. The union of these two haploid cells at fertilization creates a new diploid organism, now containing one member of each chromosome pair derived from the male and one from the female parent. The behavior of chromosome pairs thus parallels that of genes, leading to the conclusion that genes are carried on chromosomes.

The fundamentals of mutation, genetic linkage, and the relationships between genes and chromosomes were largely established by experiments performed with the fruit fly, *Drosophila melanogaster. Drosophila* can be easily maintained in the laboratory, and they reproduce about every two weeks, which is a considerable advantage for genetic experiments. Indeed, these features continue to make *Drosophila* an organism of choice for genetic studies of animals, particularly the genetic analysis of development and differentiation.

In the early 1900s, a number of genetic alterations (**mutations**) were identified in *Drosophila*, usually affecting readily observable characteristics such as eye color or wing shape. Breeding experiments indicated that some of the genes governing these traits are inherited independently of each other, suggesting that these genes are located on different chromosomes that segregate independently during meiosis (Figure 3.3). Other genes, however, are frequently inherited together as paired characteristics. Such genes are said to be linked to each other by virtue of being located on the same chromosome. The number of groups of linked genes is the same as the number of chromosomes (four in *Drosophila*), supporting the idea that chromosomes are carriers of the genes.

Linkage between genes is not complete, however; chromosomes exchange material during meiosis, leading to **recombination** between linked genes (Figure 3.4). The frequency of recombination between two linked genes depends on the distance between them on the chromosome; genes that are close to each other recombine less frequently than do genes farther apart. Thus, the frequencies with which different genes recombine can be used to determine their relative positions on the chromosome, allowing the

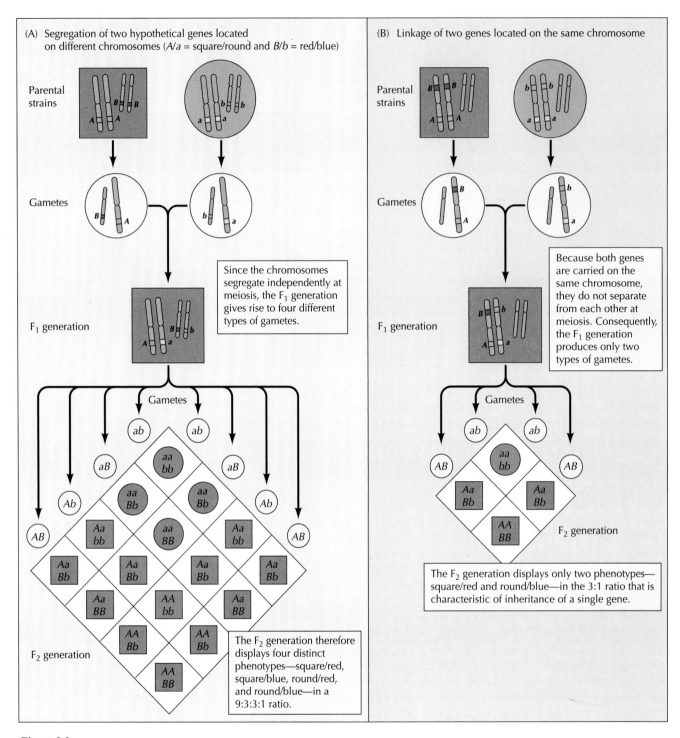

(A) Segregation of two hypothetical genes located on different chromosomes (*A/a* = square/round and *B/b* = red/blue)

Parental strains

Gametes

Since the chromosomes segregate independently at meiosis, the F_1 generation gives rise to four different types of gametes.

F_1 generation

Gametes

The F_2 generation therefore displays four distinct phenotypes—square/red, square/blue, round/red, and round/blue—in a 9:3:3:1 ratio.

F_2 generation

(B) Linkage of two genes located on the same chromosome

Parental strains

Gametes

Because both genes are carried on the same chromosome, they do not separate from each other at meiosis. Consequently, the F_1 generation produces only two types of gametes.

F_1 generation

Gametes

F_2 generation

The F_2 generation displays only two phenotypes—square/red and round/blue—in the 3:1 ratio that is characteristic of inheritance of a single gene.

Figure 3.3
Gene segregation and linkage
(A) Segregation of two hypothetical genes for shape (*A/a* = square/round) and color (*B/b* = red/blue) located on different chromosomes. (B) Linkage of two genes located on the same chromosome.

construction of **genetic maps** (Figure 3.5). By 1915, nearly a hundred genes had been defined and mapped onto the four chromosomes of *Drosophila*, leading to general acceptance of the chromosomal basis of heredity.

Genes and Enzymes

Early genetic studies focused on the identification and chromosomal localization of genes that control readily observable characteristics, such as the eye color of *Drosophila*. How these genes lead to the observed phenotypes,

Figure 3.4
Genetic recombination During meiosis, members of chromosome pairs exchange material. The result is recombination between linked genes.

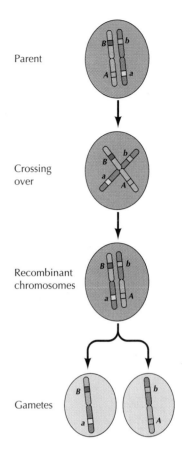

Parent

Crossing over

Recombinant chromosomes

Gametes

however, was unclear. The first insight into the relationship between genes and enzymes came in 1909, when it was realized that the inherited human disease phenylketonuria (see Molecular Medicine in Chapter 2) results from a genetic defect in metabolism of the amino acid phenylalanine. This defect was hypothesized to result from a deficiency in the enzyme needed to catalyze the relevant metabolic reaction, leading to the general suggestion that genes specify the synthesis of enzymes.

Clearer evidence linking genes with the synthesis of enzymes came from experiments of George Beadle and Edward Tatum, performed in 1941 with the fungus *Neurospora crassa*. In the laboratory, *Neurospora* can be grown on minimal or rich media similar to those discussed in Chapter 1 for the growth of *E. coli*. For *Neurospora*, minimal media consist only of salts, glucose, and biotin; rich media are supplemented with amino acids, vitamins, purines, and pyrimidines. Beadle and Tatum isolated mutants of *Neurospora* that grew normally on rich media but could not grow on minimal media. Each mutant was found to require a specific nutritional supplement, such as a particular amino acid, for growth. Furthermore, the requirement for a specific nutritional supplement correlated with the failure of the mutant to synthesize that particular compound. Thus, each mutation resulted in a deficiency in a specific metabolic pathway. Since such metabolic pathways were known to be governed by enzymes, the conclusion from these experiments was that each gene specified the structure of a single enzyme—the **one gene–one enzyme hypothesis**. Many enzymes are now known to consist of multiple polypeptides, so the currently accepted statement of this hypothesis is that each gene specifies the structure of a single polypeptide chain.

Identification of DNA as the Genetic Material

Understanding the chromosomal basis of heredity and the relationship between genes and enzymes did not in itself provide a molecular explanation of the gene. Chromosomes contain proteins as well as DNA, and it was initially thought that genes were proteins. The first evidence leading to the identification of DNA as the genetic material came from studies in bacteria. These experiments represent a prototype for current approaches to defining the function of genes by introducing new DNA sequences into cells, as discussed later in this chapter.

The experiments that defined the role of DNA were derived from studies of the bacterium that causes pneumonia (*Pneumococcus*). Virulent strains of *Pneumococcus* are surrounded by a polysaccharide capsule that protects the bacteria from attack by the immune system of the host. Because the capsule gives bacterial colonies a smooth appearance in culture, encapsulated strains are denoted S. Mutant strains that have lost the ability to make a capsule (denoted R) form rough-edged colonies in culture and are no longer lethal when inoculated into mice. In 1928 it was observed that mice inoculated with nonencapsulated (R) bacteria plus heat-killed encapsulated (S) bacteria developed pneumonia and died. Importantly, the bacteria that were then isolated from these mice were of the S type. Subsequent experiments showed that a cell-free extract of S bacteria was similarly capable of converting (or transforming) R bacteria to the S state. Thus, a substance in the S extract (called the transforming

Figure 3.5
A genetic map Three genes are localized on a hypothetical chromosome based on frequencies of recombination between them (1% recombination between *a* and *b*; 3% between *b* and *c*; 4% between *a* and *c*). The frequencies of recombination are approximately proportional to the distances between genes on the chromosome.

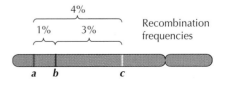

4%

1% 3%

Recombination frequencies

a b   c

principle) was responsible for inducing the genetic **transformation** of R to S bacteria.

In 1944 Oswald Avery, Colin MacLeod, and Maclyn McCarty established that the transforming principle was DNA, both by purifying it from bacterial extracts and by demonstrating that the activity of the transforming principle is abolished by enzymatic digestion of DNA but not by digestion of proteins (Figure 3.6). Although these studies did not immediately lead to the acceptance of DNA as the genetic material, they were extended within a few years by experiments with bacterial viruses. In particular, it was shown that, when a bacterial virus infects a cell, the viral DNA rather than the viral protein must enter the cell in order for the virus to replicate. Moreover, the parental viral DNA (but not the protein) is transmitted to progeny virus particles. The concurrence of these results with continuing studies of the activity of DNA in bacterial transformation led to acceptance of the idea that DNA is the genetic material.

The Structure of DNA

Our understanding of the three-dimensional structure of DNA, deduced in 1953 by James Watson and Francis Crick, has been the basis for present-day molecular biology. At the time of Watson and Crick's work, DNA was known to be a polymer composed of four nucleic acid bases—two purines (adenine [A] and guanine [G]) and two pyrimidines (cytosine [C] and thymine [T])—linked to phosphorylated sugars. Given the central role of DNA as the genetic material, elucidation of its three-dimensional structure appeared critical to understanding its function. Watson and Crick's consideration of the problem was heavily influenced by Linus Pauling's description of hydrogen bonding and the α helix, a common element of the secondary structure of proteins (see Chapter 2). Moreover, experimental data

Figure 3.6
Transfer of genetic information by DNA   DNA is extracted from a pathogenic strain of *Pneumococcus*, which is surrounded by a capsule and forms smooth colonies (S). Addition of the purified S DNA to a culture of nonpathogenic, nonencapsulated bacteria (R for "rough" colonies) results in the formation of S colonies. The purified DNA therefore contains the genetic information responsible for transformation of R to S bacteria.

on the structure of DNA were available from X-ray crystallography studies by Maurice Wilkins and Rosalind Franklin. Analysis of these data revealed that the DNA molecule is a helix that turns every 3.4 nm. In addition, the data showed that the distance between adjacent bases is 0.34 nm, so there are ten bases per turn of the helix. An important finding was that the diameter of the helix is approximately 2 nm, suggesting that it is composed of not one but two DNA chains.

From these data, Watson and Crick built their model of DNA (Figure 3.7). The central features of the model are that DNA is a double helix with the sugar–phosphate backbones on the outside of the molecule. The bases are on the inside, oriented such that hydrogen bonds are formed between purines and pyrimidines on opposite chains. The base pairing is very specific: A always pairs with T and G with C. This specificity accounts for the earlier results of Erwin Chargaff, who had analyzed the base composition of various DNAs and found that the amount of adenine was always equal to that of thymine, and the amount of guanine to that of cytosine. Because

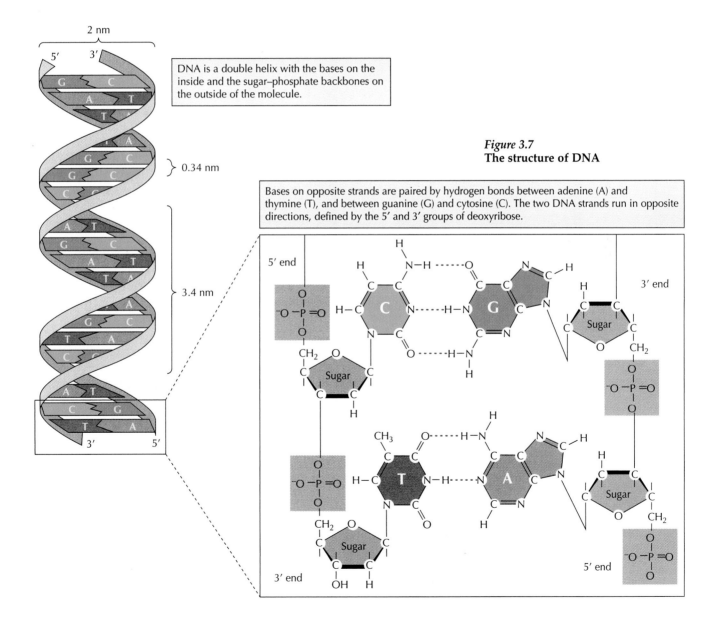

DNA is a double helix with the bases on the inside and the sugar–phosphate backbones on the outside of the molecule.

Figure 3.7
The structure of DNA

Bases on opposite strands are paired by hydrogen bonds between adenine (A) and thymine (T), and between guanine (G) and cytosine (C). The two DNA strands run in opposite directions, defined by the 5′ and 3′ groups of deoxyribose.

of this specific base pairing, the two strands of a DNA molecule are complementary: Each strand contains all the information required to specify the sequences of bases on the other.

Replication of DNA

The discovery of complementary base pairing between DNA strands immediately suggested a molecular solution to the question of how the genetic material could direct its own replication—a process that is required each time a cell divides. It was proposed that the two strands of a DNA molecule could separate and serve as templates for synthesis of new complementary strands, the sequence of which would be dictated by the specificity of base pairing (Figure 3.8). The process is called **semiconservative replication** because one strand of parental DNA is conserved in each progeny DNA molecule.

Direct support for semiconservative DNA replication was obtained in 1958 as a result of elegant experiments, performed by Matthew Meselson and Frank Stahl, in which DNA was labeled with isotopes that altered its density (Figure 3.9). *E. coli* were first grown in media containing the heavy isotope of nitrogen (^{15}N) in place of the normal light isotope (^{14}N). The DNA of these bacteria consequently contained ^{15}N and was heavier than that of bacteria grown in ^{14}N. Such heavy DNA could be separated from DNA containing ^{14}N by equilibrium centrifugation in a density gradient of CsCl. This ability to separate heavy (^{15}N) DNA from light (^{14}N) DNA enabled the study of DNA synthesis. *E. coli* that had been grown in ^{15}N were transferred to media containing ^{14}N and allowed to replicate one more time. Their DNA was then extracted and analyzed by CsCl density gradient centrifugation. The results of this analysis indicated that all of the heavy DNA had been replaced by newly synthesized DNA with a density intermediate between that of heavy (^{15}N) and that of light (^{14}N) DNA molecules. The implication was that during replication, the two parental strands of heavy DNA separated and served as templates for newly synthesized progeny strands of light DNA, yielding double-stranded molecules of intermediate density. This experiment thus provided direct evidence for semiconservative DNA replication, clearly underscoring the importance of complementary base pairing between strands of the double helix.

The ability of DNA to serve as a template for its own replication was further established with the demonstration that an enzyme purified from

Figure 3.8
Semiconservative replication of DNA   The two strands of parental DNA separate, and each serves as a template for synthesis of a new daughter strand by complementary base pairing.

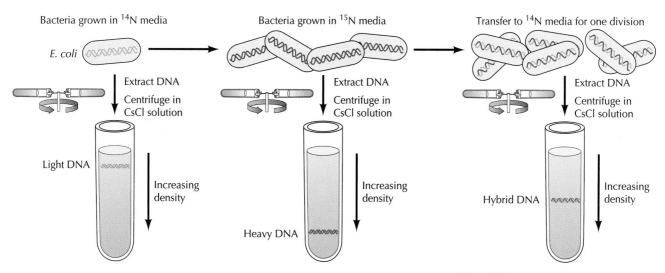

Figure 3.9
Experimental demonstration of semi-conservative replication   Bacteria grown in medium containing the normal isotope of nitrogen (^{14}N) are transferred into medium containing the heavy isotope (^{15}N) and grown in this medium for several generations. They are then transferred back to medium containing ^{14}N and grown for one additional generation. DNA is extracted from these bacteria and analyzed by equilibrium ultracentrifugation in a CsCl solution. The CsCl sediments to form a density gradient, and the DNA molecules band at a position where their density is equal to that of the CsCl solution. DNA of the bacteria transferred from ^{15}N to ^{14}N medium for a single generation bands at a density intermediate between that of ^{15}N DNA and that of ^{14}N DNA, indicating that it represents a hybrid molecule with one heavy and one light strand.

E. coli (**DNA polymerase**) could catalyze DNA replication *in vitro*. In the presence of DNA to act as a template, DNA polymerase was able to direct the incorporation of nucleotides into a complementary DNA molecule.

Expression of Genetic Information

Genes act by determining the structure of proteins, which are responsible for directing cell metabolism through their activity as enzymes. The identification of DNA as the genetic material and the elucidation of its structure revealed that genetic information must be specified by the order of the four bases (A, C, G, and T) that make up the DNA molecule. Proteins, in turn, are polymers of 20 amino acids, the sequence of which determines their structure and function. The first direct link between a genetic mutation and an alteration in the amino acid sequence of a protein was made in 1957, when it was found that patients with the inherited disease sickle-cell anemia had hemoglobin molecules that differed from normal ones by a single amino acid substitution. Deeper understanding of the molecular relationship between DNA and proteins came, however, from a series of experiments that took advantage of *E. coli* and its viruses as genetic models.

Colinearity of Genes and Proteins

The simplest hypothesis to account for the relationship between genes and enzymes was that the order of nucleotides in DNA specified the order of amino acids in a protein. Mutations in a gene would correspond to alterations in the sequence of DNA, which might result from the substitution of one nucleotide for another or from the addition or deletion of nucleotides. These changes in the nucleotide sequence of DNA would then lead to corresponding changes in the amino acid sequence of the protein encoded by the gene in question. This hypothesis predicted that different mutations within a single gene could alter different amino acids in the encoded protein, and that the positions of mutations in a gene should reflect the positions of amino acid alterations in its protein product.

The rapid replication and the simplicity of the genetic system of *E. coli* were of major help in addressing these questions. A variety of mutants of *E. coli* could be isolated, including nutritional mutants that (like the *Neurospora* mutants discussed earlier) require particular amino acids for

Figure 3.10
Colinearity of genes and proteins   A series of mutations (arrowheads) were mapped in the *E. coli* gene encoding tryptophan synthetase (top line). The amino acid substitutions resulting from each of the mutations was then determined by sequence analysis of the proteins of mutant bacteria (bottom line). These studies revealed that the order of mutations in DNA was the same as the order of amino acid substitutions in the encoded protein.

growth. Importantly, the rapid growth of *E. coli* made feasible the isolation and mapping of multiple mutants in a single gene, leading to the first demonstration of the linear relationship between genes and proteins. In these studies, Charles Yanofsky and his colleagues mapped a series of mutations in the gene that encodes an enzyme required for synthesis of the amino acid tryptophan. Analysis of the enzymes encoded by the mutant genes indicated that the relative positions of the amino acid alterations were the same as those of the corresponding mutations (Figure 3.10). Thus, the sequence of amino acids in the protein was colinear with that of mutations in the gene, as expected if the order of nucleotides in DNA specifies the order of amino acids in proteins.

The Role of Messenger RNA

Although the sequence of nucleotides in DNA appeared to specify the order of amino acids in proteins, it did not necessarily follow that DNA itself directs protein synthesis. Indeed, this appeared not to be the case, since DNA is located in the nucleus of eukaryotic cells, whereas protein synthesis takes place in the cytoplasm. Some other molecule was therefore needed to convey genetic information from DNA to the sites of protein synthesis (the ribosomes).

RNA appeared a likely candidate for such an intermediate because the similarity of its structure to that of DNA suggested that RNA could be synthesized from a DNA template (Figure 3.11). RNA differs from DNA in that it is single-stranded rather than double-stranded, its sugar component is ribose instead of deoxyribose, and it contains the pyrimidine base uracil (U) instead of thymine (T) (see Figure 2.10). However, neither the change in sugar nor the substitution of U for T alters base pairing, so the synthesis of RNA can be readily directed by a DNA template. Moreover, since RNA is located primarily in the cytoplasm, it appeared a logical intermediate to convey information from DNA to the ribosomes. These characteristics of RNA suggested a pathway for the flow of genetic information that is known as the **central dogma** of molecular biology:

$$DNA \rightarrow RNA \rightarrow Protein$$

Figure 3.11
Synthesis of RNA from DNA   The two strands of DNA unwind, and one is used as a template for synthesis of a complementary strand of RNA.

According to this concept, RNA molecules are synthesized from DNA templates (a process called **transcription**), and proteins are synthesized from RNA templates (a process called **translation**).

Experimental evidence for the RNA intermediates postulated by the central dogma was obtained by Sidney Brenner, Francois Jacob, and Matthew Meselson in studies of *E. coli* infected with the bacteriophage T4. The synthesis of *E. coli* RNA stops following infection by T4, and the only new RNA synthesized in infected bacteria is transcribed from T4 DNA. This T4 RNA becomes associated with bacterial ribosomes, thus conveying the information from DNA to the site of protein synthesis. Because of their role as intermediates in the flow of genetic information, RNA molecules that serve as templates for protein synthesis are called **messenger RNAs (mRNAs)**. They are transcribed by an enzyme (**RNA polymerase**) that catalyzes the synthesis of RNA from a DNA template.

In addition to mRNA, two other types of RNA molecules are important in protein synthesis. **Ribosomal RNA (rRNA)** is a component of ribosomes, and **transfer RNAs (tRNAs)** serve as adaptor molecules that align amino acids along the mRNA template. The structures and functions of these molecules are discussed in the following section and in more detail in Chapters 6 and 7.

The Genetic Code

How is the nucleotide sequence of mRNA translated into the amino acid sequence of a protein? In this step of gene expression genetic information is transferred between chemically unrelated types of macromolecules—nucleic acids and proteins—raising two new types of problems in understanding the action of genes.

First, since amino acids are structurally unrelated to the nucleic acid bases, direct complementary pairing between mRNA and amino acids during the incorporation of amino acids into proteins seemed impossible. How then could amino acids align on an mRNA template during protein synthesis? This question was solved by the discovery that tRNAs serve as adaptors between amino acids and mRNA during translation (Figure 3.12). Prior to its use in protein synthesis, each amino acid is attached by a specific enzyme to its appropriate tRNA. Base pairing between a recognition sequence on each tRNA and a complementary sequence on the mRNA then directs the attached amino acid to its correct position on the mRNA template.

The second problem in the translation of nucleotide sequence to amino acid sequence was determination of the **genetic code**. How could the information contained in the sequence of four different nucleotides be converted to the sequences of 20 different amino acids in proteins? Because 20 amino acids must be specified by only four nucleotides, at least three nucleotides must be used to encode each amino acid. Used singly, four nucleotides could encode only four amino acids and, used in pairs, four nucleotides could encode only sixteen (4^2) amino acids. Used as triplets, however, four nucleotides could encode 64 (4^3) different amino acids—more than enough to account for the 20 amino acids actually found in proteins.

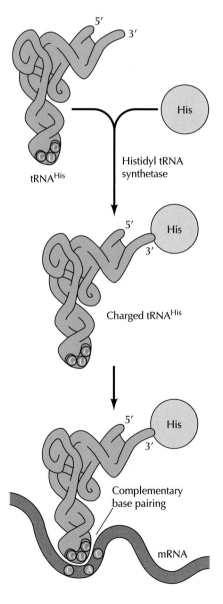

Figure 3.12
Function of transfer RNA   Transfer RNA serves as an adaptor during protein synthesis. Each amino acid (e.g., histidine) is attached to the 3′ end of a specific tRNA by an appropriate enzyme (an aminoacyl tRNA synthetase). The charged tRNAs then align on an mRNA template by complementary base pairing.

Figure 3.13
Genetic evidence for a triplet code
A series of mutations consisting of additions of one, two, or three nucleotides were studied in the *rII* gene of bacteriophage T4. Additions of one or two nucleotides alter the reading frame of the remainder of the gene. Therefore, all the subsequent amino acids are abnormal, and an inactive protein is produced, giving rise to mutant phage. Additions of three nucleotides, however, alter only a single amino acid. The reading frame of the remainder of the gene is normal, and an active protein giving rise to wild-type (WT) phage is produced.

Figure 3.14
The triplet UUU encodes phenylalanine *In vitro* translation of a synthetic RNA consisting of repeated uracils (a poly-U template) results in the synthesis of a polypeptide containing only phenylalanine.

poly-U template

Direct experimental evidence for the triplet code was obtained by studies of bacteriophage T4 bearing mutations in an extensively studied gene called *rII*. Phages with mutations in this gene form abnormally large plaques, which can be clearly distinguished from those formed by wild-type phages. Hence, isolating and mapping a number of *rII* mutants was easy and led to the establishment of a detailed genetic map of this locus. Study of recombinants between *rII* mutants that had arisen by additions or deletions of nucleotides revealed that phages containing additions or deletions of one or two nucleotides always exhibited the mutant phenotype. Phages containing additions or deletions of three nucleotides, however, were frequently wild-type in function (Figure 3.13). These findings suggested that the gene is read in groups of three nucleotides, starting from a fixed point. Additions or deletions of one or two nucleotides would then alter the reading frame of the entire gene, leading to the coding of abnormal amino acids throughout the encoded protein. In contrast, additions or deletions of three nucleotides would lead to the addition or deletion of only a single amino acid; the rest of the amino acid sequence would remain unaltered, frequently yielding an active protein.

Deciphering the genetic code thus became a problem of assigning nucleotide triplets to their corresponding amino acids. This problem was approached using *in vitro* systems that could carry out protein synthesis (***in vitro* translation**). Cell extracts containing ribosomes, amino acids, tRNAs, and the enzymes responsible for attaching amino acids to the appropriate tRNAs (aminoacyl-tRNA synthetases) were known to catalyze the incorporation of amino acids into proteins. However, such protein synthesis depends on the presence of mRNA bound to the ribosomes, and can be greatly enhanced by the addition of purified mRNA. Since added mRNA directs protein synthesis in such systems, the genetic code could be deciphered by study of the translation of synthetic mRNAs of known base sequence.

The first such experiment, performed by Marshall Nirenberg and Heinrich Matthaei, involved the *in vitro* translation of a synthetic RNA polymer containing only uracil (Figure 3.14). This poly-U template was found to direct the incorporation of only a single amino acid—phenylalanine—into a polypeptide consisting of repeated phenylalanine residues. Therefore, the triplet UUU encodes the amino acid phenylalanine. Similar experiments with RNA polymers containing only single nucleotides established that AAA encodes lysine and CCC encodes proline. The remainder of the code was deciphered using RNA polymers containing mixtures of nucleo-

TABLE 3.1 The Genetic Code

First position	Second position				Third position
	U	C	A	G	
U	Phe	Ser	Tyr	Cys	U
	Phe	Ser	Tyr	Cys	C
	Leu	Ser	STOP	STOP	A
	Leu	Ser	STOP	Trp	G
C	Leu	Pro	His	Arg	U
	Leu	Pro	His	Arg	C
	Leu	Pro	Gln	Arg	A
	Leu	Pro	Gln	Arg	G
A	Ile	Thr	Asn	Ser	U
	Ile	Thr	Asn	Ser	C
	Ile	Thr	Lys	Arg	A
	Met	Thr	Lys	Arg	G
G	Val	Ala	Asp	Gly	U
	Val	Ala	Asp	Gly	C
	Val	Ala	Glu	Gly	A
	Val	Ala	Glu	Gly	G

tides, leading to the coding assignment of all 64 possible triplets (called **codons**) (Table 3.1). Of the 64 codons, 61 specify an amino acid; the remaining three (UAA, UAG, and UGA) are stop codons that signal the termination of protein synthesis. The code is degenerate; that is, many amino acids are specified by more than one codon. With few exceptions (discussed in Chapter 10), all organisms utilize the same genetic code, providing strong support for the conclusion that all present-day cells evolved from a common ancestor.

RNA Viruses and Reverse Transcription

With the elucidation of the genetic code, the fundamental principles of the molecular biology of cells appeared to have been established. According to the central dogma, the genetic material consists of DNA, which is capable of self-replication as well as being transcribed into mRNA, which serves in turn as the template for protein synthesis. However, as noted in Chapter 1, many viruses contain RNA rather than DNA as their genetic material, implying the use of other modes of information transfer.

RNA genomes were first discovered in plant viruses, many of which were found to be composed of only RNA and protein. Direct proof that RNA acts as the genetic material of these viruses was obtained in the 1950s by experiments demonstrating that RNA purified from tobacco mosaic virus could infect new host cells, giving rise to infectious progeny virus. The mode of replication of most viral RNA genomes was subsequently determined by studies of the RNA bacteriophages of *E. coli*. These viruses were found to encode a specific enzyme that could catalyze the synthesis of RNA from an RNA template (RNA-directed RNA synthesis), using the same mechanism of base pairing between complementary strands as is employed during DNA replication or transcription of RNA from DNA.

However, RNA-directed RNA synthesis did not appear to account for the replication of certain animal viruses (RNA tumor viruses), which were

Key Experiment

The DNA Provirus Hypothesis

Nature of the Provirus of Rous Sarcoma

Howard M. Temin
McArdle Laboratory, University of Wisconsin, Madison, WI
National Cancer Institute Monographs, Volume 17, 1964, pages 557–570

The Context

Rous sarcoma virus (RSV), the first cancer-causing virus to be described, was of considerable interest as an experimental system for studying the molecular biology of cancer. Howard Temin began his research in this area when, as a graduate student in 1958, he developed the first assay for the transformation of normal cells to cancer cells in culture following infection with RSV. The availability of such a quantitative *in vitro* assay provided the tool needed for further studies of both cell transformation and virus replication. As Temin proceeded with these studies, he made a series of unexpected observations indicating that the replication of RSV was fundamentally different from that of other RNA viruses. These experiments led to Temin's proposal of the DNA provirus hypothesis, which stated that the viral RNA was copied into DNA in infected cells—a proposal that ran directly counter to the universally accepted central dogma of molecular biology.

The Experiments

The DNA provirus hypothesis was based on several different types of experimental evidence. First, studies of cell transformation using mutants of RSV indicated that important characteristics of transformed cells were determined by genetic information of the virus. This information was regularly transmitted to daughter cells following cell division, even in the absence of virus replication. Temin therefore proposed that the viral genome was present in infected cells in a stably inherited form, which he called a provirus.

Evidence that the provirus was DNA was then derived from experiments with metabolic inhibitors. First, actinomycin D, which inhibits the synthesis of RNA from a DNA template, was found to inhibit virus production by RSV-infected cells (see figure). Second, inhibitors of DNA synthesis inhibited early stages of cell infection by RSV. Thus, DNA synthesis appeared to be required early in infection, and DNA-directed RNA synthesis appeared to be needed subsequently for the production of progeny viruses, leading to the proposal that the provirus was a DNA copy of the viral RNA genome. Temin sought further evidence for this proposal by using nucleic acid hybridization to detect viral sequences in infected cell DNA, but the sensitivity of the available techniques was limited and the data were unconvincing.

The Impact

The DNA provirus hypothesis was thus proposed principally on the basis of genetic experiments and the effects of metabolic inhibitors. It was a radical proposal, which contradicted the accepted central dogma of molecular biology. In this setting, Temin's hypothesis that RSV replicated by the transfer of information from RNA to DNA not only failed to win the acceptance of the scientific community, but was met with general derision. Nonetheless, Temin persevered through the 1960s, continuing with experiments to test his

Effect of actinomycin D on RSV replication. RSV-infected cells were cultured with the indicated concentrations of actinomycin D for 8 hours. Actinomycin D was then removed and the amount of virus produced was determined.

of particular interest because of their ability to cause cancer in infected animals. Although these viruses contain genomic RNA in their viral particles, experiments performed by Howard Temin in the early 1960s indicated that their replication requires DNA synthesis in infected cells, leading to the hypothesis that the RNA tumor viruses (now called **retroviruses**) replicate via synthesis of a DNA intermediate, called a DNA provirus (Figure 3.15). This hypothesis was initially met with widespread disbelief because it

The DNA Provirus Hypothesis (continued)

hypothesis and providing increasingly convincing evidence in its support. These efforts culminated in 1970 with the discovery by Temin and Satoshi Mizutani, and at the same time by David Baltimore, of a viral enzyme, now known as reverse transcriptase, that synthesizes DNA from an RNA template—an unambiguous biochemical demonstration that the central dogma could be reversed.

Temin concluded his 1970 paper with the statement that the results "constitute strong evidence that the DNA provirus hypothesis is correct and that RNA tumour viruses have a DNA genome when they are in cells and an RNA genome when they are in virions. This result would have strong implications for theories of viral carcinogenesis and, possibly, for theories of information transfer in other biological systems." As Temin predicted, the discovery of RNA-directed DNA synthesis has led to major advances in our understanding of cancer, human retroviruses, and gene rearrangements. Reverse transcriptase has further provided a critical tool for cDNA cloning, thereby impacting virtually all areas of contemporary cell and molecular biology.

Howard Temin

involves RNA-directed synthesis of DNA—a reversal of the central dogma. In 1970, however, Temin and David Baltimore independently discovered that the RNA tumor viruses contain a novel enzyme that catalyzes the synthesis of DNA from an RNA template. In addition, clear-cut evidence for the existence of viral DNA sequences in infected cells was obtained. The synthesis of DNA from RNA, now called **reverse transcription**, was thus established as a mode of information transfer in biological systems.

Reverse transcription is important not only in the replication of retroviruses, but also in at least two other broad aspects of molecular and cellular biology. First, reverse transcription is not restricted to retroviruses; it also occurs in cells and, as discussed in Chapter 5, is frequently responsible for the transposition of DNA sequences from one chromosomal location to another. Second, enzymes that catalyze RNA-directed DNA synthesis (**reverse transcriptases**) can be used experimentally to generate DNA copies of any RNA molecule. The use of reverse transcriptase has thus allowed mRNAs of eukaryotic cells to be studied using the molecular approaches that are currently applied to the manipulation of DNA, as discussed in the following section.

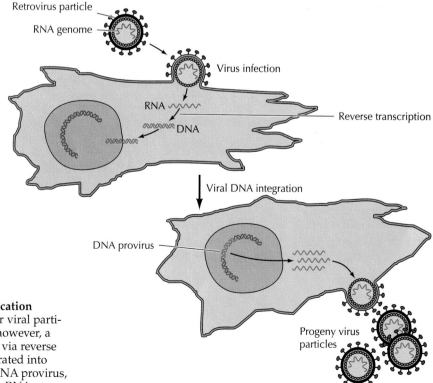

Figure 3.15
Reverse transcription and retrovirus replication
Retroviruses contain RNA genomes in their viral particles. When a retrovirus infects a host cell, however, a DNA copy of the viral RNA is synthesized via reverse transcription. This viral DNA is then integrated into chromosomal DNA of the host to form a DNA provirus, which is transcribed to yield progeny virus RNA.

Recombinant DNA

Classical experiments in molecular biology were strikingly successful in developing our fundamental concepts of the nature and expression of genes. Since these studies were based primarily on genetic analysis, their success depended largely on the choice of simple, rapidly replicating organisms (such as bacteria and viruses) as models. It was not clear, however, how these fundamental principles could be extended to provide a molecular understanding of the complexities of eukaryotic cells, since the genomes of most eukaryotes (e.g., the human genome) are up to a thousand times more complex than that of *E. coli*. In the early 1970s, the possibility of studying such genomes at the molecular level seemed daunting. In particular, there appeared to be no way in which individual genes could be isolated and studied.

This obstacle to the progress of molecular biology was overcome by the development of recombinant DNA technology, which provided scientists with the ability to isolate, sequence, and manipulate individual genes derived from any type of cell. The application of recombinant DNA has thus enabled detailed molecular studies of the structure and function of eukaryotic genes, thereby revolutionizing our understanding of cell biology.

Restriction Endonucleases

The first step in the development of recombinant DNA technology was the characterization of **restriction endonucleases**—enzymes that cleave DNA at specific sequences. These enzymes were identified in bacteria, where they apparently provide a defense against the entry of foreign DNA (e.g., from a virus) into the cell. Bacteria have a variety of restriction endonucleases that cleave DNA at more than a hundred distinct recognition sites, each of which consists of a specific sequence of four to eight base pairs (examples are given in Table 3.2).

Since restriction endonucleases digest DNA at specific sequences, they can be used to cleave a DNA molecule at unique sites. For example, the restriction endonuclease *Eco*RI recognizes the six-base-pair sequence GAATTC. This sequence is present at five sites in DNA of the bacteriophage

TABLE 3.2 Recognition Sites of Representative Restriction Endonucleases

Enzyme[a]	Source	Recognition site[b]
*Bam*HI	*Bacillus amyloliquefaciens* H	GGATCC
*Eco*RI	*Escherichia coli* RY13	GAATTC
*Hae*III	*Haemophilus aegyptius*	GGCC
*Hind*III	*Haemophilus influenzae* Rd	AAGCTT
*Hpa*I	*Haemophilus parainfluenzae*	GTTAAC
*Hpa*II	*Haemophilus parainfluenzae*	CCGG
*Mbo*I	*Moraxella bovis*	GATC
*Not*I	*Nocardia otitidis-caviarum*	GCGGCCGC
*Sfi*I	*Streptomyces fimbriatus*	GGCCNNNNNGGCC
*Taq*I	*Thermus aquaticus*	TCGA

[a] Enzymes are named according to their species of isolation, followed by a number to distinguish different enzymes isolated from the same organism (e.g., *Hpa*I and *Hpa*II).

[b] Recognition sites show the sequence of only one strand of double-stranded DNA. "N" represents any base.

Molecular Medicine

HIV and AIDS

The Disease

Acquired immune deficiency syndrome (AIDS) is a new disease, first described in 1981. It has now become a worldwide pandemic, with more than 47 million people having been infected with HIV and nearly 14 million having died of AIDS. In the United States, AIDS is the leading cause of death among men between the ages of 25 and 44. The clinical manifestations of AIDS result principally from failure of the immune system to function normally. In the absence of normal immunity, AIDS patients are sensitive to opportunistic infections by agents (viruses, bacteria, fungi, and protozoans) against which a healthy individual would be resistant. People with AIDS also suffer a high frequency of some types of cancers, particularly lymphomas and Kaposi's sarcoma, although it is the opportunistic infections that are responsible for most deaths.

Molecular and Cellular Basis

AIDS is caused by a retrovirus (human immunodeficiency virus or HIV) that was discovered by the research groups of Robert Gallo and Luc Montagnier in 1983. HIV infects principally a specific type of lymphocyte (the T4 lymphocyte) that is required for a normal immune response. In contrast to many other retroviruses, such as Rous sarcoma virus, HIV does not cause the cells it infects to become cancerous. Instead, HIV eventually kills the cells in which it replicates, ultimately resulting in the depletion of the population of T4 lymphocytes and the failure of the immune system in infected individuals. This failure of the immune system in turn leads to the opportunistic infections and cancers that represent the clinical manifestations of AIDS.

Prevention and Treatment

At present, the only means of preventing AIDS is to avoid HIV infection. HIV is a fragile virus that quickly loses infectivity outside the body, so it cannot be transmitted by casual contact with an infected person. HIV can be transmitted in three ways: through sexual contact, through contaminated blood products, and from mother to child during pregnancy or breast-feeding. Following the isolation of HIV, screening tests were developed to ensure the safety of clotting factors and blood supplies used for transfusions. Prevention of HIV infection by other routes currently depends on individuals minimizing their personal risk of infection by adhering to safe sexual practices and avoiding sources of contaminated blood, such as shared needles used for intravenous drug injection.

Beyond modifying individual behavior to reduce the risk of infection, the identification of HIV as the cause of AIDS opens possibilities for prevention and treatment. A vaccine to prevent HIV infection is being actively pursued, although several features of the biology of HIV pose difficulties to this approach. Alternatively, drugs that inhibit virus replication are now providing effective therapies for HIV-infected individuals. These drugs either act as inhibitors of the HIV reverse transcriptase or of the HIV protease, which is an enzyme required for processing viral proteins. Combinations of such drugs are now prolonging the lives of AIDS patients, although further work is clearly needed to develop drugs that are not only more effective but also less expensive and more practical for use in developing countries.

Scanning electron micrograph of HIV budding from T lymphocytes. (Cecil Fox/Photo Researchers, Inc.)

0.1 μm

λ, so *Eco*RI digests λ DNA into six fragments ranging from 3.6 to 21.2 kilobases long (1 kilobase, or kb = 1000 base pairs) (Figure 3.16). These fragments can be separated according to size by **gel electrophoresis**—a common method in which molecules are separated based on the rates of their migration in an electric field. A gel, usually formed from agarose or polyacrylamide, is placed between two buffer compartments containing electrodes. The sample (e.g., the mixture of DNA fragments to be analyzed) is then pipetted into preformed slots in the gel, and the electric field is turned on. Nucleic acids are negatively charged (because of their phosphate backbone), so they migrate toward the positive electrode. The gel acts like a sieve, selectively retarding the movement of larger molecules. Smaller molecules therefore move through the gel more rapidly, allowing a mixture of nucleic acids to be separated on the basis of size.

In addition to size, the order of restriction fragments can be determined by a variety of methods, yielding (for example) a map of the *Eco*RI sites in λ DNA. The locations of cleavage sites for multiple different restriction endonucleases can be used to generate detailed **restriction maps** of DNA molecules, such as viral genomes (Figure 3.17). In addition, individual DNA fragments produced by restriction endonuclease digestion can be isolated following electrophoresis for further study—including determination of their DNA sequence. The DNAs of many viruses have been characterized by this approach.

Figure 3.16
***Eco*RI digestion and gel electrophoresis of λ DNA**   *Eco*RI cleaves λ DNA at five sites (arrows), yielding six DNA fragments. These fragments are then separated by electrophoresis in an agarose gel. The DNA fragments migrate toward the positive electrode, with smaller fragments moving more rapidly through the gel. Following electrophoresis, the DNA is stained with a fluorescent dye and photographed. The sizes of DNA fragments are indicated.

Figure 3.17
Restriction maps of λ and adenovirus DNAs
The locations of cleavage sites for *Bam*HI, *Eco*RI, and *Hind*III are shown in the DNAs of *E. coli* bacteriophage λ (48.5 kb) and human adenovirus-2 (35.9 kb).

Restriction endonuclease digestion alone, however, does not provide sufficient resolution for the analysis of larger DNA molecules, such as cellular genomes. A restriction endonuclease with a six-base-pair recognition site (such as *Eco*RI) cleaves DNA with a statistical frequency of once every 4096 base pairs ($1/4^6$). A molecule the size of λ DNA (48.5 kb) would therefore be expected to yield about ten *Eco*RI fragments, consistent with the results illustrated in Figure 3.16. However, restriction endonuclease digestion of larger genomes yields quite different results. For example, the human genome is approximately 3×10^6 kb long and is therefore expected to yield more than 500,000 *Eco*RI fragments. Such a large number of fragments cannot be separated from one another, so agarose gel electrophoresis of *Eco*RI-digested human DNA yields a continuous smear rather than a discrete pattern of DNA fragments. Because it is impossible to isolate single restriction fragments from such digests, restriction endonuclease digestion alone does not yield a source of homogeneous DNA suitable for further analysis. Quantities of such purified DNA fragments, however, can be obtained through molecular cloning.

Generation of Recombinant DNA Molecules

The basic strategy in **molecular cloning** is to insert a DNA fragment of interest (e.g., a segment of human DNA) into a DNA molecule (called a **vector**) that is capable of independent replication in a host cell. The result is a **recombinant molecule** or **molecular clone**, composed of the DNA insert linked to vector DNA sequences. Large quantities of the inserted DNA can be obtained if the recombinant molecule is allowed to replicate in an appropriate host. For example, fragments of human DNA can be cloned in bacteriophage λ vectors (Figure 3.18). These recombinant molecules can then be introduced into *E. coli*, where they replicate efficiently to yield millions of progeny phages containing the human DNA insert. The DNA of these phages can then be isolated, yielding large quantities of recombinant molecules containing a single fragment of human DNA. Whereas this fragment might represent one part in 100,000 of human genomic DNA, it represents approximately one part in 10 after being cloned in the λ vector. Moreover, the fragment can be easily isolated from the rest of the vector DNA by restriction endonuclease digestion and gel electrophoresis, allowing a pure fragment of human DNA to be analyzed and further manipulated.

The DNA fragments used to create recombinant molecules are usually generated by digestion with restriction endonucleases. Many of these enzymes cleave their recognition sequences at staggered sites, leaving

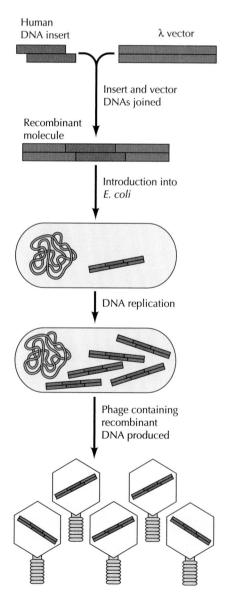

Figure 3.18
Generation of a recombinant DNA molecule   A fragment of human DNA is inserted into a λ DNA vector. The resulting recombinant molecule is then introduced into *E. coli*, where it replicates to yield recombinant progeny phage containing the human DNA insert.

Figure 3.19
Joining of DNA molecules Vector and insert DNAs are digested with a restriction endonuclease (such as *Eco*RI), which cleaves at staggered sites leaving overhanging single-stranded tails. Vector and insert DNAs can then associate by complementary base pairing, and covalent joining of the DNA strands by DNA ligase yields a recombinant molecule.

Insert DNA

*Eco*RI

*Eco*RI

*Eco*RI cleavage

Vector DNA

*Eco*RI

*Eco*RI

*Eco*RI cleavage

*Eco*RI cleavage results in overhanging single-stranded tails.

Complementary base pairing

Joining by DNA ligase

Recombinant molecule

Figure 3.20
cDNA cloning

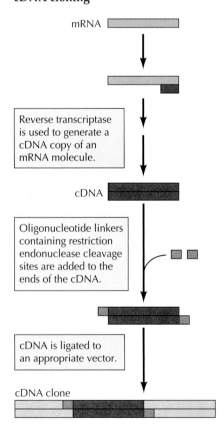

mRNA

Reverse transcriptase is used to generate a cDNA copy of an mRNA molecule.

cDNA

Oligonucleotide linkers containing restriction endonuclease cleavage sites are added to the ends of the cDNA.

cDNA is ligated to an appropriate vector.

cDNA clone

overhanging or cohesive single-stranded tails that can associate with each other by complementary base pairing (Figure 3.19). The association between such paired complementary ends can be established permanently by treatment with **DNA ligase**, an enzyme that seals breaks in DNA strands (see Chapter 5). Thus, two different fragments of DNA (e.g., a human DNA insert and a λ DNA vector) prepared by digestion with the same restriction endonuclease can be readily joined to create a recombinant DNA molecule.

The fragments of DNA that can be cloned are not limited to those that terminate in restriction endonuclease cleavage sites. Synthetic DNA "linkers" containing a variety of restriction endonuclease sites can be added to the blunt ends of any DNA fragment. Linkers are short oligonucleotides that can be readily obtained by chemical synthesis, allowing virtually any fragment of DNA to be prepared for ligation to a vector.

Not only DNA, but also RNA sequences can be cloned (Figure 3.20). The first step is to synthesize a DNA copy of the RNA using the enzyme reverse transcriptase. The DNA product (called a **cDNA** because it is *c*omplementary to the RNA used as a template) can then be ligated to vector DNA as already described. Since eukaryotic genes are usually interrupted by noncoding sequences (introns; see Chapter 4), which are removed from mRNA by splicing, the ability to clone cDNA as well as genomic DNA has been critical for understanding gene structure and function.

Vectors for Recombinant DNA

Depending on the size of the insert DNA and the purpose of the experiment, many different types of cloning vectors can be used for the generation of recombinant molecules. The basic vector systems used for the isolation and propagation of cloned DNAs are reviewed here. Other vectors developed for the expression of cloned DNAs and the introduction of recombinant molecules into eukaryotic cells are discussed in subsequent sections.

Bacteriophage λ vectors are frequently used for the initial isolation of either genomic or cDNA clones from eukaryotic cells (Figure 3.21). In λ cloning vectors, sequences of the bacteriophage genome that are dispensable for virus replication have been removed and replaced with unique restriction sites for insertion of cloned DNA. DNA inserts can be as large as about 15 kb and still yield a recombinant genome that can be packaged into phage particles. To isolate genomic clones of human DNA, for example, random fragments of human DNA with an average size of about 15 kb are ligated to λ vector arms. These recombinant DNA molecules can then be efficiently packaged into phage particles by mixing DNA with λ proteins

Figure 3.21
Cloning in bacteriophage λ vectors   The vector contains a restriction site (e.g., an *Eco*RI site) for insertion of cloned DNA. In addition, cos sites (cohesive ends), which are required for packaging DNA into phage particles, are present on both ends of the vector DNA. Insert DNA (e.g., human DNA) is ligated to the vector, and the recombinant molecules are packaged into phage particles by being mixed with phage proteins. The recombinant phages are then used to infect *E. coli*. Each recombinant phage, which carries a unique insert of cloned DNA, forms a single plaque in the infected bacterial culture. Progeny phage carrying unique DNA inserts can then be isolated from individual plaques and grown in large quantity.

(called packaging extracts) *in vitro*. The phage particles are then used to infect cultures of *E. coli*. Since each recombinant phage forms a single plaque, recombinants carrying unique inserts of human DNA can be isolated. In addition, recombinant phages containing particular genes of interest can be identified by nucleic acid hybridization or other screening methods, as discussed in the next section.

Plasmid vectors (Figure 3.22) allow easier manipulation of cloned DNA sequences than do phage vectors. Plasmids are small circular DNA molecules that can replicate independently—without being associated with chromosomal DNA—in bacteria. All that is required on the plasmid DNA is an **origin of replication**—the DNA sequence that signals the host cell DNA polymerase to

Figure 3.22
Cloning in plasmid vectors The vector is a small circular molecule that contains an origin of replication (*ori*), a gene conferring resistance to ampicillin (*Amp*^r), and a restriction site (e.g., *Eco*RI) which can be used to insert foreign DNA. Insert DNA is ligated to the vector, and the recombinant plasmids are used to transform *E. coli*. The bacteria are plated on medium containing ampicillin, so that only the bacteria that are ampicillin-resistant because they carry plasmid DNAs are able to form colonies. Individual colonies of plasmid-containing bacteria can then be isolated and grown in large quantities for isolation of recombinant plasmids.

replicate the DNA molecule. In addition, plasmid vectors carry genes that confer resistance to antibiotics (e.g., ampicillin resistance), so bacteria carrying the plasmids can be selected. Plasmid vectors usually consist of only 2 to 4 kb of DNA, in contrast to the 30 to 45 kb of phage DNA present in λ vectors, facilitating the analysis of an inserted DNA fragment.

To be cloned into a plasmid vector, a fragment of the insert DNA is ligated to an appropriate restriction site in the vector and the recombinant molecule is used to transform *E. coli*. Antibiotic-resistant colonies, which contain plasmid DNA, are selected. Such plasmid-containing bacteria can then be grown in large quantities and their DNA extracted. The small circular plasmid DNA molecules, of which there are often hundreds of copies per cell, can be separated from the bacterial chromosomal DNA; the result is purified plasmid DNA that is suitable for analysis of the cloned insert.

For some studies involving analysis of genomic DNA, it is desirable to clone even larger fragments of DNA than are accommodated by λ vectors. **Cosmid** and **yeast artificial chromosome (YAC)** vectors can be used for this purpose. Cosmid vectors (Figure 3.23) accommodate inserts of approximately 45 kb. These vectors contain bacteriophage λ sequences that allow efficient packaging of the cloned DNA into phage particles. In addition, cosmids contain origins of replication and the genes for antibiotic resistance that are characteristic of plasmids, so they are able to replicate as plasmids in bacterial cells. Even larger fragments of DNA (hundreds of kilobases) can be cloned in YAC vectors, which replicate as chromosomes in yeast cells. These vectors are particularly useful for chromosome mapping studies, as discussed in Chapter 4.

DNA Sequencing

Molecular cloning allows the isolation of individual fragments of DNA in quantities suitable for detailed characterization, including the determination of nucleotide sequence. Indeed, determination of the nucleotide sequences of many genes has elucidated not only the structure of their protein products, but also the properties of DNA sequences that regulate gene expression. Furthermore, the coding sequences of novel genes are frequently related to those of previously studied genes, and the functions of newly isolated genes can often be correctly deduced on the basis of such sequence similarities.

Current methods of DNA sequencing are both rapid and accurate, and determining the sequence of several kilobases of DNA is a straightforward task for most molecular biology laboratories. Thus, it is now far easier to clone and sequence DNA than it is to determine the amino acid sequence of a protein. Since the nucleotide sequence of a gene can be readily translated into the amino acid sequence of its encoded protein, the easiest way of determining protein sequence is the sequencing of a cloned gene.

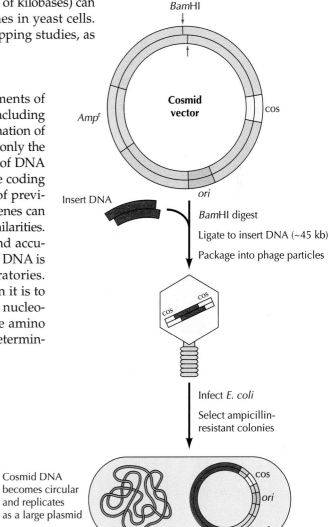

Figure 3.23
Cloning in cosmid vectors　A cosmid is a plasmid containing cos sites, which allow DNA to be packaged into λ phage particles. Large fragments of insert DNA (approximately 45 kb) are ligated to a cloning site (e.g., *Bam*HI) to yield molecules of approximately the size of the λ genome (48.5 kb), which is appropriate for incorporation into phage particles. The recombinant molecules are made linear (so that a cos site is present at each end), packaged into phage, and used to infect *E. coli*. The cosmid DNA becomes circular again within the infected bacteria, yielding molecules that replicate as plasmids and confer ampicillin resistance to infected cells.

Figure 3.24
DNA sequencing by the Sanger procedure Dideoxynucleotides, which lack OH groups at the 3′ as well as the 2′ position of deoxyribose, are used to terminate DNA synthesis at specific bases. These molecules are incorporated normally into growing DNA strands. Because they lack a 3′ OH, however, the next nucleotide cannot be added, so synthesis of that DNA strand terminates. DNA synthesis is initiated with a radioactive primer. Four separate reactions are carried out, each containing one dideoxynucleotide mixed with its normal counterpart as well as the three other normal deoxynucleotides. When the dideoxynucleotide is incorporated, DNA synthesis stops, so each reaction yields a series of products extending from the radioactive primer to the base substituted by a dideoxynucleotide. Products of the four reactions are separated by electrophoresis and analyzed by autoradiography to determine the DNA sequence.

The most common method of DNA sequencing is based on premature termination of DNA synthesis resulting from the inclusion of chain-terminating **dideoxynucleotides** (which do not contain the deoxyribose 3′ hydroxyl group) in DNA polymerase reactions (Figure 3.24). DNA synthesis is initiated from a primer that has been labeled at one end with a radioisotope. Four separate reactions are run, each including one dideoxynucleotide (either A, C, G, or T) in addition to its normal counterpart. Incorporation of a dideoxynucleotide stops further DNA synthesis because no 3′ hydroxyl group is available for addition of the next nucleotide. Thus, a series of labeled DNA molecules is generated, each terminating at the base represented by the dideoxynucleotide in each reaction. These fragments of DNA are then separated according to size by gel electrophoresis and detected by exposure of the gel to X-ray film (**autoradiography**). The size of each fragment is determined by its terminal dideoxynucleotide, so the DNA sequence corresponds to the order of fragments read from the gel.

Large-scale DNA sequencing is frequently performed using automated systems, which use fluorescence-labeled primers in dideoxynucleotide sequencing reactions (Figure 3.25). As the newly syntehsized DNA strands are electrophoresed through a gel, they pass through a laser beam that excites the fluorescent label. The resulting emitted light is then detected by

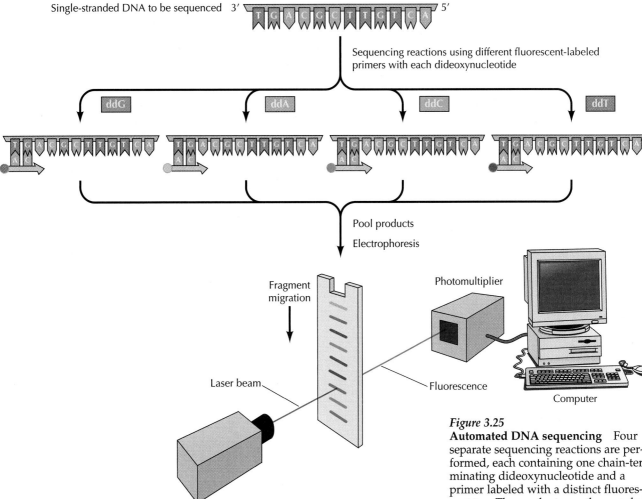

Single-stranded DNA to be sequenced   3′ [T G A C G C T T G T C A] 5′

Sequencing reactions using different fluorescent-labeled primers with each dideoxynucleotide

ddG   ddA   ddC   ddT

Pool products
Electrophoresis

Fragment migration

Laser beam

Photomultiplier

Fluorescence

Computer

Figure 3.25
Automated DNA sequencing Four separate sequencing reactions are performed, each containing one chain-terminating dideoxynucleotide and a primer labeled with a distinct fluorescent tag. The products are then pooled and subjected to gel electrophoresis. As the DNA strands migrate through the gel, they pass through a laser beam that excites the fluorescent label. The emitted light is detected by a photomultiplier, which is connected to a computer that collects and analyzes the data.

a photomultiplier, and a computer collects and analyzes the data. This type of automated DNA sequencing has enabled the large-scale analysis required for determination of the complete genome sequences of bacteria, yeast, *C. elegans*, and *Drosophila*, and is soon expected to yield the complete sequence of the human genome.

Expression of Cloned Genes

In addition to enabling determination of the nucleotide sequences of genes—and hence the amino acid sequences of their protein products—molecular cloning has provided new approaches to obtaining large amounts of proteins for structural and functional characterization. Many proteins of interest are present at only low levels in eukaryotic cells and therefore cannot be purified in significant amounts by conventional biochemical techniques. Given a cloned gene, however, this problem can be solved by the engineering of vectors that lead to high levels of gene expression in either bacteria or eukaryotic cells.

To express a eukaryotic gene in *E. coli*, the cDNA of interest is cloned into a plasmid or phage vector (called an **expression vector**) that contains sequences that drive transcription and translation of the inserted gene in bacterial cells (Figure 3.26). Inserted genes often can be expressed at levels high enough that the protein encoded by the cloned gene corresponds to as much as 10% of the total bacterial protein. Purifying the protein encoded

Figure 3.26
Expression of cloned genes in bacteria   Expression vectors contain promoter sequences (pro) that direct transcription of inserted DNA in bacteria and sequences required for binding of mRNA to bacterial ribosomes (Shine-Delgarno [SD] sequences). A eukaryotic cDNA inserted adjacent to these sequences can be efficiently expressed in *E. coli*, resulting in production of eukaryotic proteins in transformed bacteria.

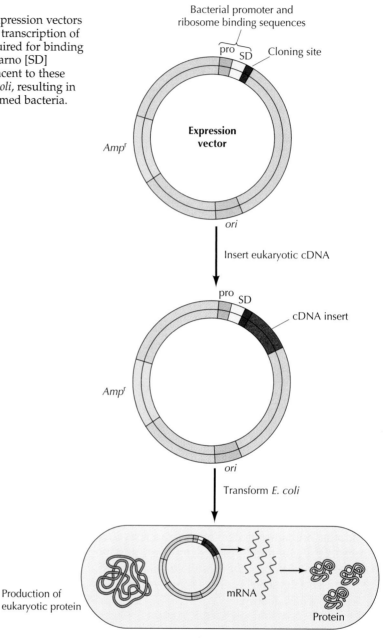

by the cloned gene in quantities suitable for detailed biochemical or structural studies is then a straightforward matter.

It is frequently useful to express high levels of a cloned gene in eukaryotic cells, rather than in bacteria. This mode of expression may be important, for example, to ensure that posttranslational modifications of the protein (such as addition of carbohydrates or lipids) occur normally. Such protein expression in eukaryotic cells can be achieved, as in *E. coli*, by insertion of the cloned gene into a vector (usually derived from a virus) that directs high-level gene expression. One system frequently used for protein expression in eukaryotic cells is infection of insect cells by **baculovirus** vectors, which direct very high levels of expression of genes inserted in place of a viral structural protein. Alternatively, high levels of protein expression can be achieved using appropriate vectors in mammalian cells. Expression of cloned genes in yeast is particularly useful because simple methods of yeast genetics can be employed to identify proteins that interact with other cloned proteins or with specific DNA sequences.

Amplification of DNA by the Polymerase Chain Reaction

Molecular cloning allows individual DNA fragments to be propagated in bacteria and isolated in large amounts. An alternative method to isolating large amounts of a single DNA molecule is the **polymerase chain reaction (PCR)**, which was developed by Kary Mullis in 1988. Provided that some sequence of the DNA molecule is known, PCR can achieve a striking amplification of DNA via reactions carried out entirely *in vitro*. Essentially, DNA polymerase is used for repeated replication of a defined segment of DNA. The number of DNA molecules increases exponentially, doubling with each round of replication, so a substantial quantity of DNA can be obtained from a small number of initial template copies. For example, a single DNA molecule amplified through 30 cycles of replication would theoretically yield 2^{30} (approximately 1 billion) progeny molecules. Single DNA molecules can thus be amplified to yield readily detectable quantities of DNA that can be isolated by molecular cloning or further analyzed directly by restriction endonuclease digestion or nucleotide sequencing.

The general procedure for PCR amplification of DNA is illustrated in Figure 3.27. The starting material can be either a cloned DNA fragment or a mixture of DNA molecules—for example, total DNA from human cells. A specific region of DNA can be amplified from such a mixture, provided that the nucleotide sequence surrounding the region is known so that primers can be designed to initiate DNA synthesis at the desired point. Such primers are usually chemically synthesized oligonucleotides containing 15 to 20 bases of DNA. Two primers are used to initiate DNA synthesis in opposite directions from complementary DNA strands. The reaction is started by heating the template DNA to a high temperature (e.g., 95°C) so that the two strands separate. The temperature is then lowered to allow the primers to pair with their complementary sequences on the template strands. DNA polymerase then uses the primers to synthesize a new strand complementary to each template. Thus in one cycle of amplification, two new DNA molecules are synthesized from one template molecule. The process can be repeated multiple times, with a twofold increase in DNA molecules resulting from each round of replication.

The multiple cycles of heating and cooling involved in PCR are performed by programmable heating blocks called thermocyclers. The DNA polymerases used in these reactions are heat-stable enzymes from bacteria such as *Thermus aquaticus*, which lives in hot springs at temperatures of about 75°C. These polymerases are stable even at the high temperatures used to separate the strands of double-stranded DNA, so PCR amplification can be performed rapidly and automatically. RNA sequences can also be amplified by this method if reverse transcriptase is used to synthesize a cDNA copy prior to PCR amplification.

If enough of the sequence of a gene is known that primers can be specified, PCR amplification provides an extremely powerful method of obtaining readily detectable and manipulable amounts of DNA from starting material that may contain only a few molecules of the desired DNA sequence in a complex mixture of other molecules. For example, defined DNA sequences of up to several kilobases can be readily amplified from total genomic DNA, or a single cDNA can be amplified from total cell RNA. These amplified DNA segments can then be further manipulated or analyzed, for example, to detect mutations within a gene of interest. PCR is thus a powerful addition to the repertoire of recombinant DNA techniques.

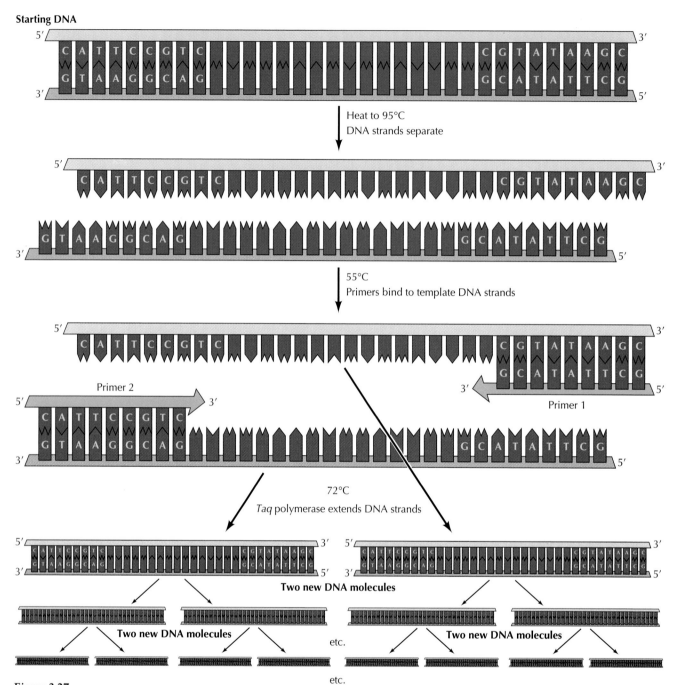

Figure 3.27
Amplification of DNA by PCR
The region of DNA to be amplified is flanked by two sequences used to prime DNA synthesis. The starting double-stranded DNA is heated to separate the strands and then cooled to allow primers (usually oligonucleotides of 15 to 20 bases) to bind to each strand of DNA. DNA polymerase from *Thermus aquaticus* (*Taq* polymerase) is used to synthesize new DNA strands starting from the primers, resulting in the formation of two new DNA molecules. The process can be repeated for multiple cycles, each resulting in a twofold amplification of DNA.

Its power is particularly apparent in applications such as the diagnosis of inherited diseases, studies of gene expression during development, and forensic medicine.

Detection of Nucleic Acids and Proteins

The advent of molecular cloning has enabled the isolation and characterization of individual genes from eukaryotic cells. Understanding the role of genes within cells, however, requires analysis of the intracellular organization and expression of individual genes and their encoded proteins. In this section, the basic procedures currently available for detection of specific nucleic acids and proteins are discussed. These approaches are important for a wide variety of studies, including the mapping of genes to chromosomes, the analysis of gene expression, and the localization of proteins to subcellular organelles. The same general procedures are also used to isolate specific genes as molecular clones.

Nucleic Acid Hybridization

The key to detection of specific nucleic acid sequences is base pairing between complementary strands of RNA or DNA. At high temperatures (e.g., 90 to 100°C), the complementary strands of DNA separate (denature), yielding single-stranded molecules. If such denatured DNA strands are then incubated under appropriate conditions (e.g., 65°), they will renature to form double-stranded molecules as dictated by complementary base pairing—a process called **nucleic acid hybridization**. Nucleic acid hybrids can be formed between two strands of DNA, two strands of RNA, or one strand of DNA and one of RNA.

Nucleic acid hybridization provides a means for detecting DNA or RNA sequences that are complementary to any isolated nucleic acid, such as a viral genome or a cloned DNA sequence (Figure 3.28). The cloned DNA is radiolabeled, usually by being synthesized in the presence of radioactive nucleotides. This radioactive DNA is then used as a **probe** for hybridization to complementary DNA or RNA sequences, which are detected by virtue of the radioactivity of the resulting double-stranded hybrids.

Southern blotting (a technique developed by E. M. Southern) is widely used for detection of specific genes in cellular DNA (Figure 3.29). The DNA to be analyzed is digested with a restriction endonuclease, and the digested DNA fragments are separated by gel electrophoresis. The gel is then overlaid with a nitrocellulose filter or nylon membrane, to which the DNA fragments are transferred (blotted) to yield a replica of the gel. The filter is then incubated with a radiolabeled probe, which hybridizes to the DNA fragments that contain the complementary sequence. These fragments are then visualized by exposure of the filter to X-ray film.

Northern blotting is a variation of the Southern blotting technique (hence its name) that is used for detection of RNA instead of DNA. In this method, total cellular RNAs are extracted and fractionated according to size by gel electrophoresis. As in Southern blotting, the RNAs are transferred to a filter and detected by hybridization with a radioactive probe. Northern blotting is frequently used in studies of gene expression—for example, to determine whether specific mRNAs are present in different types of cells.

Nucleic acid hybridization can be used to detect homologous DNA or RNA sequences not only in cell extracts, but also in chromosomes or intact cells—a procedure called *in situ* hybridization (Figure 3.30). In this case, the hybridization of radioactive or fluorescent probes to specific cells or subcellular structures is analyzed by microscopic examination. For example, labeled probes can be hybridized to intact chromosomes in order to identify the chromosomal regions that contain a gene of interest. *In situ* hybridization can also be used to detect specific mRNAs in different types of cells within a tissue.

Detection of Small Amounts of DNA or RNA by PCR

Amplification of DNA by the polymerase chain reaction is a much more sensitive technique for detecting cellular DNA or RNA sequences than is Southern or Northern blotting. Approximately 100,000 copies of a DNA or RNA sequence are required for detection by blot hybridization. In contrast, PCR can amplify single copies of DNA (or RNA after reverse transcription) to readily detectable levels.

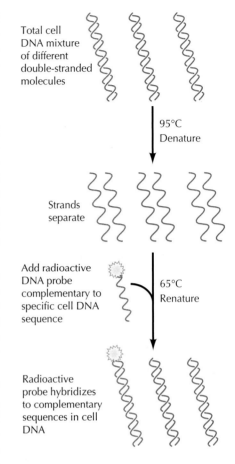

Total cell DNA mixture of different double-stranded molecules

95°C Denature

Strands separate

Add radioactive DNA probe complementary to specific cell DNA sequence

65°C Renature

Radioactive probe hybridizes to complementary sequences in cell DNA

Figure 3.28
Detection of DNA by nucleic acid hybridization   A specific sequence can be detected in total cell DNA by hybridization with a radiolabeled DNA probe. The DNA is denatured by heating to 95°C, yielding single-stranded molecules. The radiolabeled probe is then added and the temperature is lowered to 65°C, allowing complementary DNA strands to renature by pairing with each other. The radioactive probe hybridizes to complementary sequences in cell DNA, which can then be detected as radioactive double-stranded molecules.

DNA

DNA is digested with a restriction endonuclease.

Migration

Restriction fragments of different sizes are separated by gel electrophoresis.

Paper towels
Filter
Gel
Sponge
Salt solution

The DNA is denatured and transferred to a filter by passage of a salt solution through the gel.

DNA fragments

Filter

The filter is hybridized with a radioactive probe, which binds to complementary DNA sequences.

The probe bound to the filter is detected by autoradiography, which reveals the DNA fragment to which the probe hybridized.

X-ray film

Figure 3.29
Southern blotting   Restriction endonuclease fragments of DNA are separated by gel electrophoresis. Specific DNA fragments are then identified by hybridization with an appropriate probe.

Figure 3.30
Fluorescence *in situ* hybridization   Hybridization of human chromosomes with chromosome-specific fluorescent probes that label each of the 24 chromosomes a different color. (Courtesy of Thomas Reid and Hesed Padilla-Nash, National Cancer Institute.)

As discussed earlier, the specificity of amplification in PCR is provided by the use of oligonucleotide primers that hybridize to complementary sequences on the template molecule. Therefore, PCR can selectively amplify a specific DNA molecule from complex mixtures, such as total cell DNA or RNA. Consequently, PCR amplification can be used to detect specific DNA or RNA molecules in very small amounts of starting material, such as extracts of single cells. This extraordinary sensitivity has made PCR an important method for a variety of applications, including the analysis of gene expression in cells available in only limited quantities.

Antibodies as Probes for Proteins

Studies of gene expression and function require the detection not only of DNA and RNA, but also of specific proteins. For these studies, **antibodies** take the place of nucleic acid probes as reagents that can selectively react with unique protein molecules. Antibodies are proteins produced by cells of the immune system (B lymphocytes) that react against molecules (**antigens**) that the host organism recognizes as foreign substances—for example, the protein coat of a virus. The immune systems of vertebrates are capable of producing millions of different antibodies, each of which specifically recognizes a unique antigen, which may be a protein, a carbohydrate, or a nonbiological molecule. An individual lymphocyte produces only a single type of antibody, but the antibody genes of different lymphocytes vary as a result of programmed gene rearrangements during development of the immune system (see Chapter 5). This variation gives rise to an array of lymphocytes with distinct antibody genes, programmed to respond to different antigens.

Antibodies can be generated by inoculation of an animal with any foreign protein. For example, antibodies against human proteins are frequently raised in rabbits. The sera of such immunized animals contain a mixture of antibodies (produced by different lymphocytes) that react against multiple sites on the immunizing antigen. However, single species of antibodies (**monoclonal antibodies**) can also be produced by the culturing of clonal lines of B lymphocytes from immunized animals (usually mice). Because each lymphocyte is programmed to produce only a single antibody, a clonal line of lymphocytes produces a monoclonal antibody that recognizes only a single antigenic determinant, thereby providing a highly specific immunological reagent.

Although antibodies can be raised against proteins purified from cells, other materials may also be used for immunization. For example, animals may be immunized with intact cells to raise antibodies against unknown proteins expressed by a specific cell type (e.g., a cancer cell). Such antibodies may then be used to identify proteins specifically expressed by the cell type used for immunization. In addition, antibodies are frequently raised against proteins expressed in bacteria as recombinant clones. In this way, molecular cloning allows the production of antibodies against proteins that may be difficult to isolate from eukaryotic cells. Moreover, antibodies can be raised against synthetic peptides that consist of only 10 to 15 amino acids, rather than against intact proteins. Therefore, once the sequence of a gene is known, antibodies against peptides synthesized from part of its predicted protein sequence can be produced. Because antibodies against these synthetic peptides frequently react with the intact protein as well, it is possible to produce antibodies against a protein starting with only the sequence of a cloned gene.

Antibodies can be used in a variety of ways to detect proteins in cell extracts. Two common methods are **immunoblotting** (also called **Western blotting**) and **immunoprecipitation**. Western blotting (Figure 3.31) is another variation of Southern blotting. Proteins in cell extracts are first separated according to size by gel electrophoresis. Because proteins have different shapes and charges, however, this process requires a modification of the

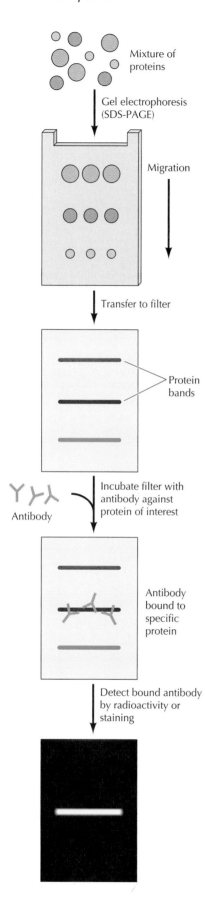

Mixture of proteins

Gel electrophoresis (SDS-PAGE)

Migration

Transfer to filter

Protein bands

Antibody

Incubate filter with antibody against protein of interest

Antibody bound to specific protein

Detect bound antibody by radioactivity or staining

Figure 3.31
Western blotting   Proteins are separated according to size by SDS-polyacrylamide gel electrophoresis and transferred from the gel to a filter. The filter is incubated with an antibody directed against a protein of interest. The antibody bound to the filter can then be detected by reaction with various reagents, such as a radioactive probe that binds to the antibody.

methods used for electrophoresis of nucleic acids. Proteins are separated by a method known as **SDS-polyacrylamide gel electrophoresis (SDS-PAGE)**, in which they are dissolved in a solution containing the negatively charged detergent sodium dodecyl sulfate (SDS). Each protein binds many detergent molecules, which denature the protein and give the protein an overall negative charge. Under these conditions, all proteins migrate toward the positive electrode—their rates of migration determined (like those of nucleic acids) only by size. Following electrophoresis, the proteins are transferred to a filter, which is then allowed to react with antibodies against the protein of interest. The antibody bound to the filter can be detected by various methods, thereby identifying the protein against which the antibody is targeted.

In immunoprecipitation, antibodies are used to isolate the proteins against which they are directed (Figure 3.32). Typically, cells are incubated with radioactive amino acids to label their proteins. Such a radiolabeled cell extract is then incubated with an antibody, which binds to its antigenic target protein. The resulting antigen-antibody complexes are isolated and subjected to electrophoresis, allowing detection of the radioactive antigen by autoradiography.

As discussed in Chapter 1, antibodies can also be used to visualize proteins within cells, as well as in cell lysates. For example, cells can be stained with antibodies labeled with fluorescent dyes, and the subcellular localization of the antigenic proteins can be visualized by fluorescence microscopy (see Figure 1.28). Antibodies can also be labeled with tags that are visible in the electron microscope, such as heavy metals, allowing visualization of antigenic proteins at the ultrastructural level.

Probes for Screening Recombinant DNA Libraries

The same basic methods for detecting nucleic acids and proteins in cell extracts are used for identifying molecular clones that contain specific cellular DNA inserts. For example, nucleic acid hybridization can be used to identify genomic or cDNA clones that contain DNA sequences for which a probe is available. Cloned cDNAs in expression vectors can also be identified by the use of antibodies against their encoded proteins.

The first step in isolation of either genomic or cDNA clones is frequently the preparation of **recombinant DNA libraries**—collections of clones that contain all the genomic or mRNA sequences of a particular cell type (Figure 3.33). A **genomic library** of human DNA, for example, might be prepared by the cloning of random DNA fragments with average sizes of about 15 kb in a λ vector. Since the size of the human genome is about 3×10^6 kb, the DNA equivalent of one genome would be represented by 200,000 such λ clones. Because of statistical fluctuations in sampling, however, many genes will not be represented in a library of 200,000 recombinants, while other genes will be represented by multiple clones. Therefore, larger libraries, consisting of approximately 1 million recombinant phages are usually prepared to ensure a high likelihood that any gene of interest is represented in the collection.

Any gene for which a probe is available can readily be isolated from such a recombinant library. The recombinant phages are plated on *E. coli*, and each phage replicates to produce a plaque on the lawn of bacteria. The plaques are then blotted onto filters in a process similar to the transfer of DNA from a gel to a filter during Southern blotting, and the filters are

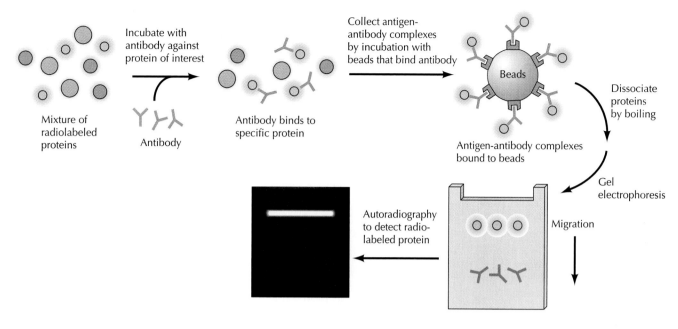

Figure 3.32
Immunoprecipitation Radiolabeled proteins are incubated with an antibody, which forms complexes with the protein against which it is directed (the antigen). These antigen-antibody complexes are collected on beads that bind the antibody. The beads are then boiled to dissociate the antigen-antibody complexes, and the recovered proteins are analyzed by SDS-polyacrylamide gel electrophoresis. The radioactive protein that was immunoprecipitated is detected by autoradiography.

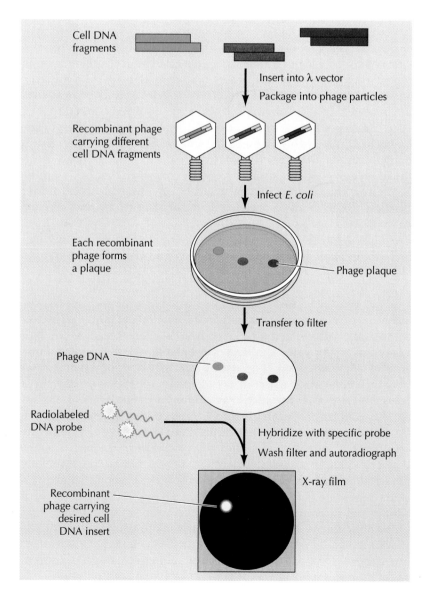

Figure 3.33
Screening a recombinant library by hybridization Fragments of cell DNA are cloned in a bacteriophage λ vector and packaged into phage particles, yielding a collection of recombinant phage carrying different cell inserts. The phages are used to infect bacteria, and the culture is overlaid with a filter. Some of the phages in each plaque are transferred to the filter, which is then hybridized with a radiolabeled probe to identify the phage plaque containing the desired gene. The appropriate phage plaque can then be isolated from the original culture plate.

hybridized with a radiolabeled probe to identify the phage plaques that contain the gene of interest. The appropriate plaque can then be isolated from the original plate in order to propagate the recombinant phage that carries the desired cell insert. Similar procedures can be used to screen bacterial colonies carrying plasmid DNA clones, so specific clones can be isolated by hybridization from either phage or plasmid libraries.

A variety of probes can be used for screening recombinant libraries. For example, a cDNA clone can be used as a probe to isolate the corresponding genomic clone, or a gene cloned from one species (e.g., mouse) can be used to isolate a related gene from a different species (e.g., human). In addition to isolated DNA fragments, synthetic oligonucleotides can be used as probes, enabling the isolation of genes on the basis of partial amino acid sequences of their encoded proteins. For example, oligonucleotides consisting of 15 to 20 bases can be synthesized on the basis of the partial amino acid sequence of a protein of interest. These oligonucleotides can then be used as probes to isolate cDNA clones, which (as already discussed) are much easier to sequence and to characterize than is the protein itself. It is thus possible to proceed experimentally from a partial peptide sequence of a protein to the isolation of a cloned gene.

An alternative approach to isolating a gene on the basis of its protein product is the use of antibodies as probes to screen expression libraries. In this case a **cDNA library** is generated in an expression vector that drives protein synthesis in *E. coli*. Phage plaques or bacterial colonies are then transferred to a filter as already described, but the filter is then reacted with an antibody (as in Western blotting) to identify clones that are producing the protein of interest.

These procedures for identifying molecular clones and detecting genes and gene products in cells illustrate the flexibility of recombinant DNA technology. Starting with any cloned gene, it is possible not only to determine the nucleotide sequence of that gene and use it as a probe for studies of gene organization and transcription, but also to express and raise antisera against its encoded protein. Conversely, genes can be cloned on the basis of limited characterization of a protein of interest, using either oligonucleotide or antibody probes. Thus, recombinant DNA has allowed experimental analyses to proceed either from DNA to protein or from protein to DNA, providing great versatility to the strategies currently employed for studies of eukaryotic genes and their encoded proteins.

Gene Function in Eukaryotes

The recombinant DNA techniques discussed in the preceding sections provide powerful approaches to the isolation and detailed characterization of the genes of eukaryotic cells. Understanding the function of a gene, however, requires analysis of the gene within cells or intact organisms—not simply as a molecular clone in bacteria. In classical genetics, the function of genes has generally been revealed by the altered phenotypes of mutant organisms. The advent of recombinant DNA has added a new dimension to studies of gene function. Namely, it has become possible to investigate the function of a cloned gene directly by reintroducing the cloned DNA into eukaryotic cells. In simpler eukaryotes, such as yeasts, this technique has made possible the isolation of molecular clones corresponding to virtually any mutant gene. In addition, there are several methods by which cloned genes can be introduced into cultured animal and plant cells, as well as intact organisms, for functional analysis. These approaches can be coupled with the ability to introduce mutations in cloned DNA *in vitro*, extending the power of recombinant DNA to allow functional studies of the genes of more complex eukaryotes.

Genetic Analysis in Yeasts

Yeasts are particularly advantageous for studies of eukaryotic molecular biology (see Chapter 1). The genome of *Saccharomyces cerevisiae*, which consists of approximately 1.2×10^7 base pairs, is 200 times smaller than the human genome. Moreover, yeasts can easily be grown in culture, reproducing with a division time of about 2 hours. Thus, yeasts offer the same basic advantages—a small genome and rapid reproduction—that are afforded by bacteria.

Mutations in yeasts can be identified as readily as in *E. coli*. For example, yeast mutants that require a particular amino acid or other nutrient for growth can easily be isolated. In addition, yeasts with defects in genes required for fundamental cell processes (in contrast to metabolic defects) can be isolated as **temperature-sensitive mutants**. Such mutants encode proteins that are functional at one temperature (the permissive temperature) but not another (the nonpermissive temperature), whereas normal proteins are functional at both. A yeast with a temperature-sensitive mutation in an essential gene can be identified by its ability to grow only at the permissive temperature. The ability to isolate such temperature-sensitive mutants has allowed the identification of yeast genes controlling many fundamental cell processes, such as RNA synthesis and processing, progression through the cell cycle, and transport of proteins between cellular compartments.

The relatively simple genetics of yeast also enables a gene corresponding to any yeast mutation to be cloned, simply on the basis of its functional activity (Figure 3.34). First, a genomic library of normal yeast DNA is prepared in vectors that replicate as plasmids in yeasts as well as in *E. coli*. The small size of the yeast genome means that a complete library consists of only a few thousand plasmids. A mixture of such plasmids is then used to transform a temperature-sensitive yeast mutant, and transformants that are

Figure 3.34
Cloning of yeast genes (A) A yeast vector. The vector contains a bacterial origin of replication (*ori*) and an ampicillin resistance gene (*Amp^r*), allowing it to be propagated as a plasmid in *E. coli*. In addition, the vector contains a yeast origin of replication and a marker gene (*LEU2*), allowing the selection of transformed yeast. The *LEU2* gene encodes an enzyme required for synthesis of the amino acid leucine, so transformation of yeast strains lacking this enzyme can be selected for by growth on medium lacking leucine. (B) Isolation of a yeast gene. A gene of interest is identified by a temperature-sensitive mutation, which allows yeast to grow at 25°C but not at 37°C. To isolate a clone of the gene, the temperature-sensitive yeasts are transformed with a plasmid library containing a collection of genes encompassing the entire yeast genome. All yeasts transformed by plasmid DNAs are able to grow on media lacking leucine at 25°C, but only those yeasts transformed by a plasmid carrying a normal copy of the gene of interest are able to grow at 37°C. The desired plasmid can be isolated from transformed yeasts that form colonies at the nonpermissive temperature.

(A)

LEU2
Yeast *ori*
Yeast DNA insert
Amp^r
E. coli ori

(B)

Temperature-sensitive yeast

Yeast plasmid library

LEU2

Different yeast DNA inserts

Transformation

Selection for transformed yeast by growth on media lacking leucine

All cells containing plasmids grow

25°C

Only cells with plasmid carrying normal copy of temperature-sensitive gene grow

37°C

Isolate plasmid carrying gene of interest

able to grow at the nonpermissive temperature are selected. Such transformants have acquired a normal copy of the gene of interest on plasmid DNA, which can then be easily isolated from the transformed yeast cells for further characterization.

Yeast genes encoding a wide variety of essential proteins have been identified in this manner. In many cases, such genes isolated from yeasts have also been useful in identifying and cloning related genes from mammalian cells. Thus, the simple genetics of yeast has not only provided an important model for eukaryotic cells, but has also led directly to the cloning of related genes from more complex eukaryotes.

Gene Transfer in Plants and Animals

Although the cells of complex eukaryotes are not amenable to the simple genetic manipulations possible in yeasts, gene function can still be assayed by the introduction of cloned DNA into plant and animal cells. Such experiments (generally called **gene transfer**) have proven critical to addressing a wide variety of questions, including studies of the mechanisms that regulate gene expression and protein processing. In addition, as discussed later in the book, gene transfer has enabled the identification and characterization of genes that control animal cell growth and differentiation, including a variety of genes responsible for the abnormal growth of human cancer cells.

DNA is usually introduced into animal cells in culture as a coprecipitate with calcium phosphate (Figure 3.35). The methodology was initially developed for the introduction of infectious viral DNAs into animal cells and is therefore frequently called **transfection** (a word derived from *trans*formation + in*fection*). Cells exposed to such a DNA precipitate take up the DNA and transport it to the nucleus, where it can be transcribed for several days—a phenomenon called **transient expression**. In a smaller fraction of cells (0.1%

Figure 3.35
Introduction of DNA into animal cells A eukaryotic gene of interest is cloned in a plasmid containing a drug resistance marker that can be selected for in cultured animal cells. The plasmid DNA is introduced into cultured cells as a calcium phosphate coprecipitate, which is taken up and expressed by a fraction of the cells for a few days (transient expression). Stably transformed cells, in which the plasmid DNA becomes integrated into chromosomal DNA, can be selected for by their ability to grow in drug-containing medium.

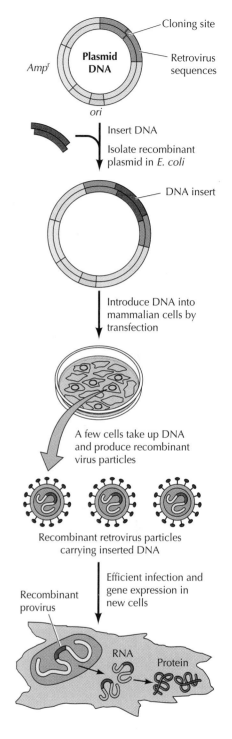

Figure 3.36
Retroviral vectors   The vector consists of retroviral sequences cloned in a plasmid that can be propagated in *E. coli*. Foreign DNA is inserted into the viral sequences, and recombinant plasmids are isolated in bacteria. Animal cells in culture are then transfected with the recombinant DNA. The DNA is taken up by a small fraction of the cells, which produce recombinant retroviruses carrying the inserted DNA. These recombinant retroviruses can be used to efficiently infect new cells, where the viral genome carrying the inserted gene integrates into chromosomal DNA as a provirus.

or less), the foreign DNA becomes stably integrated into the cell genome and is transferred to progeny cells at cell division just as any other cell gene is. These stably transformed cells can be isolated if the transfected DNA contains a selectable marker, such as resistance to a drug that inhibits the growth of normal cells. Thus, any cloned gene can be introduced into mammalian cells by being transferred together with a drug resistance marker that can be used to isolate stable transformants. The effects of such cloned genes on cell behavior—for example, cell growth or differentiation—can then be analyzed.

Various other methods, depending on the purpose of the experiment, can also be used to introduce DNA into mammalian cells. These methods include direct microinjection of DNA into the cell nucleus, incorporation of DNA into lipid vesicles (**liposomes**) that fuse with the plasma membrane, and exposure of cells to a brief electric pulse that transiently opens pores in the plasma membrane (**electroporation**). In addition, animal viruses can be used as vectors for more efficient introduction of cloned DNAs into cells. Retroviruses are particularly useful in this respect, since their life cycle involves the stable integration of viral DNA into the genome of infected cells (Figure 3.36). Consequently, retroviral vectors can be used to efficiently introduce cloned genes into a wide variety of cell types, making them an important vehicle for a broad range of applications, including gene therapy.

Cloned genes can also be introduced into the germ line of multicellular organisms, allowing them to be studied in the context of the intact animal rather than in cultured cells. Mice that carry such foreign genes (**transgenic mice**) are usually produced by microinjection of cloned DNA into the pronucleus of a fertilized egg (Figure 3.37). The injected eggs are then transferred

Figure 3.37
Production of transgenic mice   DNA is microinjected into a pronucleus of a fertilized mouse egg (fertilized eggs contain two pronuclei, one from the egg and one from the sperm). The microinjected eggs are then transferred to foster mothers and allowed to develop. Some of the offspring (transgenic) have incorporated the injected DNA into their genome.

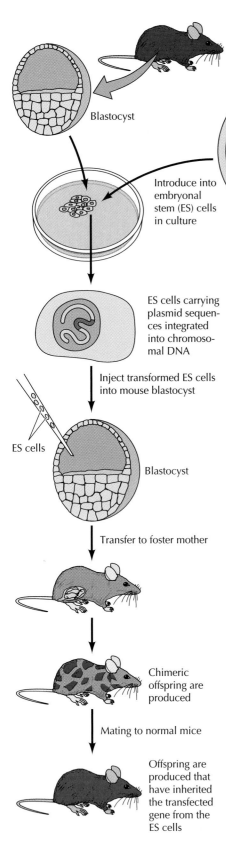

Blastocyst

Introduce into embryonal stem (ES) cells in culture

Drug-resistance marker

Plasmid DNA

Gene of interest

ES cells carrying plasmid sequences integrated into chromosomal DNA

Inject transformed ES cells into mouse blastocyst

ES cells

Blastocyst

Transfer to foster mother

Chimeric offspring are produced

Mating to normal mice

Offspring are produced that have inherited the transfected gene from the ES cells

Figure 3.38
Introduction of genes into mice via embryonal stem cells Embryonal stem (ES) cells are cultured cells derived from early mouse embryos (blastocysts). DNA can be introduced into these cells in culture, and stably transformed ES cells can be isolated. These transformed ES cells can then be injected into a recipient blastocyst, where they are able to participate in normal development of the embryo. Some of the progeny mice that develop after transfer of injected embryos to foster mothers therefore contain cells derived from the transformed ES cells, as well as from the normal cells of the blastocyst. Since these mice are a mixture of two different cell types, they are referred to as chimeric. Offspring carrying the transfected gene can then be produced by the breeding of chimeric mice in which descendants of the transformed ES cells have been incorporated into the germ line.

to foster mothers and allowed to develop to term. In a fraction of the progeny mice (often about 10%), the foreign DNA will have integrated into the genome of the fertilized egg and is therefore present in all cells of the animal. Since the foreign DNA is present in germ cells as well as in somatic cells, it is transferred by breeding to new progeny mice just as any other cell gene would be.

The properties of **embryonal stem (ES) cells** provide an alternative means of introducing cloned genes into mice (Figure 3.38). ES cells can be established in culture from early mouse embryos. They can also be reintroduced into early embryos, where they participate normally in development and can give rise to cells in all tissues of the mouse—including germ cells. It is thus possible to introduce cloned DNA into ES cells in culture, select stably transformed cells, and then introduce these cells back into mouse embryos. Such embryos give rise to chimeric offspring in which some cells are derived from the normal embryo cells and some from the transfected ES cells. In some such mice the transfected ES cells are incorporated into the germ line. Breeding these mice therefore leads to the direct inheritance of the transfected gene by their progeny.

Cloned DNAs can also be introduced into plant cells. One approach is to remove the plant cell wall, forming protoplasts that are surrounded only by a plasma membrane. DNA can then be introduced into such protoplasts by electroporation, as was described for animal cells. Alternatively, DNA can be introduced into intact plant cells by bombardment of the cells with DNA-coated microprojectiles, such as small particles of tungsten. The DNA-coated particles are shot directly into the plant cells; some of the cells are killed, but others survive and become stably transformed.

Vectors for more efficient introduction of recombinant DNA into plant cells have been developed from plant viruses. In addition, a plasmid from the bacterium *Agrobacterium tumefaciens* (the **Ti plasmid**) provides a novel vehicle for the introduction of cloned DNAs into various plants (Figure 3.39). In nature, *Agrobacterium* attaches to the leaves of plants and the Ti plasmid is transferred into plant cells, where it becomes incorporated into chromosomal DNA. Vectors developed from the Ti plasmid therefore provide an efficient means of introducing recombinant DNA into sensitive plant cells. Since many plants can be regenerated from single cultured cells (see Chapter 1), transgenic plants can be established directly from cells into which recombinant DNA has been introduced in culture—a much simpler procedure than the production of transgenic animals.

Figure 3.39
Introduction of genes into plant cells via the Ti plasmid   The Ti plasmid contains the T region, which is transferred to infected plant cells, and virulence (*vir*) genes, which function in T DNA transfer. In Ti plasmid vectors, foreign DNA is inserted into the T region. The recombinant plasmid is introduced into *Agrobacterium tumefaciens*, which is then used to infect cultured cells. The T region of the plasmid (carrying the inserted DNA) is transferred to the plant cells and integrates into chromosomal DNA. A transgenic plant can then be generated from the transformed cells.

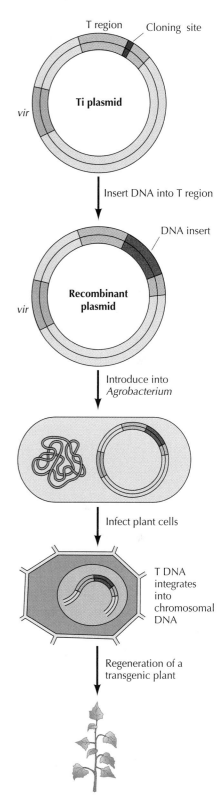

Mutagenesis of Cloned DNAs

In classical genetic studies (e.g., in bacteria or yeasts), mutants are the key to identifying genes and understanding their function by observing the altered phenotype of mutant organisms. In such studies, mutant genes are detected because they result in observable phenotypic changes—for example, temperature-sensitive growth or a specific nutritional requirement. The isolation of genes by recombinant DNA, however, has opened a different approach to mutagenesis. It is now possible to introduce any desired alteration into a cloned gene and to determine the effect of the mutation on gene function. Such procedures have been called **reverse genetics**, since a mutation is introduced into a gene first and its functional consequence is determined second. The ability to introduce specific mutations into cloned DNAs (***in vitro* mutagenesis**) has proven to be a powerful tool in studying the expression and function of eukaryotic genes.

Cloned genes can be altered by many *in vitro* mutagenesis procedures, which can lead to the introduction of deletions, insertions, or single nucleotide alterations. Deletions, for example, are frequently generated by the digestion of a plasmid DNA with a restriction endonuclease that cleaves at a single site in the molecule (Figure 3.40). The linear DNA is then treated with an exonuclease, which digests DNA from the ends of the molecule. The digested DNA can be ligated again to yield a series of circular plasmids with deletions around the restriction site that was used initially to make the molecule linear. These deleted DNAs can be cloned and sequenced to determine the extent of the deletion in each of a series of individual plasmids. The functional consequence of the deletions can then be determined by introduction of the characterized plasmids into cultured cells.

Deletion mutagenesis enables a relatively crude assessment of the functional role of a region of a cloned gene. More detailed information can be provided by procedures that alter smaller regions of the DNA molecule. One example of such a procedure is the use of synthetic oligonucleotides to generate mutations of single bases in a DNA sequence (Figure 3.41). In this procedure a synthetic oligonucleotide bearing the mutant base is used as a primer for DNA synthesis. Newly synthesized DNA molecules containing the mutation can then be isolated and characterized. For example, specific amino acids of a protein can be altered in order to characterize their role in protein function.

These examples only begin to illustrate the wide variety of procedures for producing a desired alteration in a cloned gene. The effects of such mutations on gene expression and function can then be determined by introduction of the gene into an appropriate cell type. *In vitro* mutagenesis has thus allowed detailed characterization of the functional roles of both the regulatory and protein-coding sequences of cloned genes.

Introducing Mutations into Cellular Genes

Although the transfer of cloned genes into cells (particularly in combination with *in vitro* mutagenesis) provides a powerful approach to studying gene

Figure 3.40
Deletion mutagenesis of a cloned gene Plasmid DNA is made linear by digestion with a restriction endonuclease that cleaves at a site within the gene of interest. The linear DNA is then treated with an exonuclease, which digests DNA starting from the ends of each molecule. This process yields a collection of digested DNAs, in which different extents of the gene have been deleted. These molecules can then be recircularized by treatment with DNA ligase, yielding a set of plasmids with different deletions around the initial site of restriction endonuclease digestion.

Figure 3.41
Mutagenesis with synthetic oligonucleotides An oligonucleotide carrying the desired base alteration is used to prime DNA synthesis from plasmid DNA, and the DNA is circularized by incubation with DNA ligase. This process yields plasmids in which one strand contains the normal base and the other strand the mutant base. Replication of the plasmid DNAs after transformation of *E. coli* therefore yields a mixture of both starting and mutant plasmids.

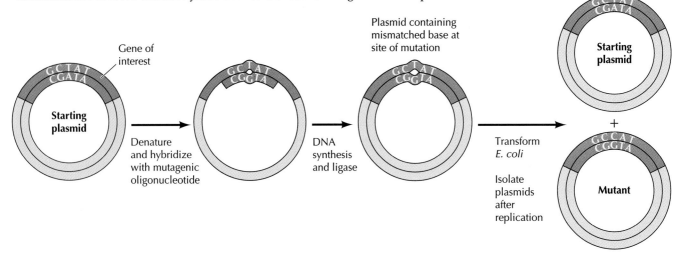

structure and function, such experiments fall short of defining the role of an unknown gene in a cell or intact organism. The cells used as recipients for transfer of cloned genes usually already have normal copies of the gene in their chromosomal DNAs, and these normal copies continue to perform their roles in the cell. Determining the biological role of a cloned gene therefore requires that the activity of the normal cellular gene copies be eliminated. As discussed in the following section, this can be readily accomplished in yeasts. In mammalian cells, this task is considerably more difficult, although several approaches to either inactivating the chromosomal copies of a cloned gene or inhibiting normal gene function are now available.

Mutating the chromosomal genes in yeasts is relatively easy because DNA introduced into yeast cells frequently undergoes **homologous recombination** with its chromosomal copy (Figure 3.42). In homologous recombination, the cloned yeast gene replaces the normal allele, so mutations introduced into the cloned gene *in vitro* become incorporated into the chromosomal copy of the yeast gene. In the simplest case, mutations that inactivate the cloned gene can be introduced in place of the normal gene copy in order to determine its role in cellular processes. Since yeasts have both haploid and diploid stages of their life cycle, even genes that are required for cell growth can be inactivated and studied. An inactive gene copy is introduced into diploid cells, which then contain one functional and one inactive copy of the target gene. The cells are induced to undergo meiosis, and the effect of gene inactivation on the progeny haploid cells can be observed.

Recombination between transferred DNA and the homologous chromosomal gene is a rare event in mammalian cells, so gene inactivation by this approach is more difficult than it is in yeasts. Possibly because the genomes of mammalian cells are so much larger than that of yeasts, most transfected

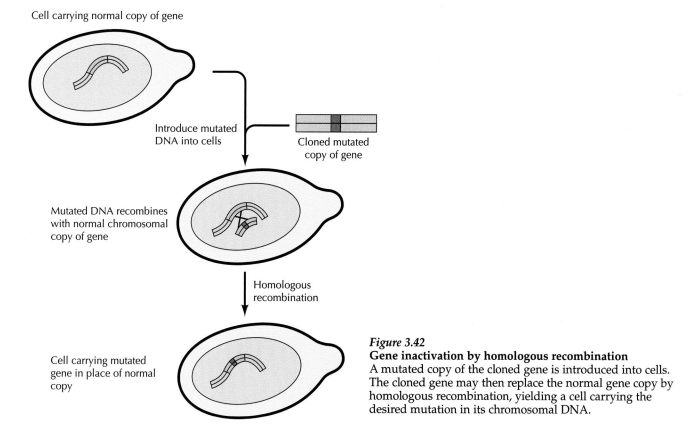

Cell carrying normal copy of gene

Introduce mutated DNA into cells

Cloned mutated copy of gene

Mutated DNA recombines with normal chromosomal copy of gene

Homologous recombination

Cell carrying mutated gene in place of normal copy

Figure 3.42
Gene inactivation by homologous recombination
A mutated copy of the cloned gene is introduced into cells. The cloned gene may then replace the normal gene copy by homologous recombination, yielding a cell carrying the desired mutation in its chromosomal DNA.

Figure 3.43
Inhibition of gene expression by antisense RNA or DNA Antisense RNA or single-stranded DNA is complementary to the mRNA of a gene of interest. Antisense nucleic acids therefore form hybrids with their target mRNAs, blocking the translation of mRNA into protein.

DNA that integrates into the recipient cell genome does so at random sites by recombination with unrelated sequences. However, various procedures have been developed to select and isolate the transformed cells in which homologous recombination has occurred, so genes in mammalian cells can be inactivated by this approach. Importantly, genes can be inactivated or disrupted not only in somatic cell lines but also in embryonal stem cells in culture. As already described, ES cells can be used to generate transgenic mice, so the effects of inactivation of a gene can be investigated in the context of the intact animal. The functions of hundreds of mouse genes have been investigated in this way, and such studies have been particularly important in revealing the roles of specific genes in mouse development.

An alternative to gene inactivation by homologous recombination is the use of **antisense nucleic acids** to block gene expression (Figure 3.43). In this approach, RNA or single-stranded DNA complementary to the mRNA of the gene of interest (antisense) is introduced into a cell. The antisense RNA or DNA hybridizes with the mRNA and blocks its translation into protein. Moreover, the RNA-DNA hybrids resulting from the introduction

Figure 3.44
Direct inhibition of protein function Microinjected antibodies can bind to proteins within cells, thereby inhibiting their normal function. In addition, some mutant proteins interfere with the function of normal proteins—for example, by competing with the normal protein for interaction with target molecules.

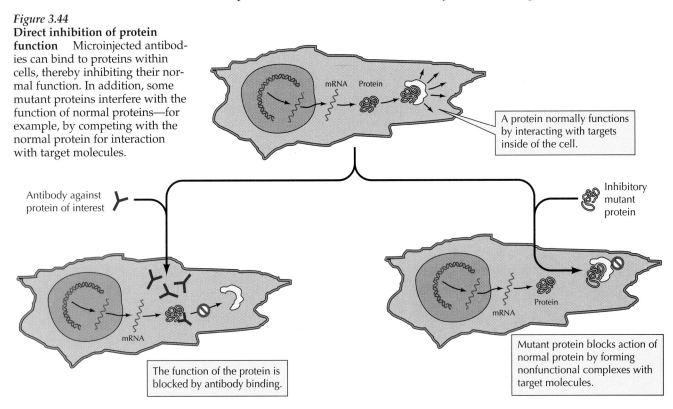

of antisense DNA molecules are usually degraded within the cell. Antisense RNAs can be introduced into cells directly by microinjection. Alternatively, cells can be transfected with vectors that have been engineered to express antisense RNA. Antisense DNA is usually in the form of short oligonucleotides (about 20 bases long) that are microinjected into cells. Alternatively, because cells are able to take up such oligonucleotides from the culture medium, antisense oligonucleotides can simply be added to the cell culture.

In addition to inactivating a gene or blocking mRNA translation, it is sometimes possible to interfere directly with the function of proteins within cells (Figure 3.44). One approach is to microinject antibodies that block the activity of the protein against which they are directed. Alternatively, some mutant proteins interfere with the function of their normal counterparts when they are expressed within the same cell—for example, by competing with the normal protein for binding to its target molecule. Cloned DNAs encoding such mutant proteins (called **dominant inhibitory mutants**) can be introduced into cells by gene transfer and used to study the effects of blocking normal gene function.

SUMMARY

HEREDITY, GENES, AND DNA

Genes and Chromosomes: Chromosomes are the carriers of genes.

Genes and Enzymes: A gene specifies the amino acid sequence of a polypeptide chain.

Identification of DNA as the Genetic Material: DNA was identified as the genetic material by bacterial transformation experiments.

The Structure of DNA: DNA is a double helix in which hydrogen bonds form between purines and pyrimidines on opposite strands. Because of specific base pairing—A with T and G with C—the two strands of a DNA molecule are complementary in sequence.

Replication of DNA: DNA replicates by semiconservative replication, in which the two strands separate and each serves as a template for synthesis of a new progeny strand.

EXPRESSION OF GENETIC INFORMATION

Colinearity of Genes and Proteins: The order of nucleotides in DNA specifies the order of amino acids in proteins.

The Role of Messenger RNA: Messenger RNA functions as an intermediate to convey information from DNA to the ribosomes, where it serves as a template for protein synthesis.

The Genetic Code: Transfer RNAs serve as adaptors between amino acids and mRNA during translation. Each amino acid is specified by a codon consisting of three nucleotides.

KEY TERMS

gene, allele, dominant, recessive, genotype, phenotype, chromosome, diploid, meiosis, haploid, mutation, recombination, genetic map

one gene–one enzyme hypothesis

transformation

semiconservative replication, DNA polymerase

central dogma, transcription, translation, messenger RNA (mRNA), RNA polymerase, ribosomal RNA (rRNA), transfer RNA (tRNA)

genetic code, *in vitro* translation, codon

retrovirus, reverse transcription,
reverse transcriptase

RNA Viruses and Reverse Transcription: DNA can be synthesized from RNA templates, as first discovered in retroviruses.

RECOMBINANT DNA

restriction endonuclease, gel
electrophoresis, restriction map

Restriction Endonucleases: Restriction endonucleases cleave specific DNA sequences, yielding defined fragments of DNA molecules.

molecular cloning, vector,
recombinant molecule, molecular
clone, DNA ligase, cDNA

Generation of Recombinant DNA Molecules: Recombinant DNA molecules consist of a DNA fragment of interest ligated to a vector that is able to replicate independently in an appropriate host cell.

plasmid, origin of replication,
cosmid, yeast artificial
chromosome (YAC)

Vectors for Recombinant DNA: A variety of vectors are used to clone different sizes of DNA fragments.

dideoxynucleotide,
autoradiography

DNA Sequencing: The nucleotide sequences of cloned DNA fragments can be readily determined.

expression vector, baculovirus

Expression of Cloned Genes: The proteins encoded by cloned genes can be expressed at high levels in either bacteria or eukaryotic cells.

polymerase chain reaction (PCR)

Amplification of DNA by the Polymerase Chain Reaction: PCR allows the amplification and isolation of specific fragments of DNA *in vitro*.

DETECTION OF NUCLEIC ACIDS AND PROTEINS

nucleic acid hybridization, probe,
Southern blotting, Northern
blotting, *in situ* hybridization

Nucleic Acid Hybridization: Nucleic acid hybridization allows the detection of specific DNA or RNA sequences.

Detection of Small Amounts of DNA or RNA by PCR: PCR provides a sensitive method for detecting small amounts of specific DNA or RNA molecules.

antibody, antigen, monoclonal
antibody, immunoblotting,
Western blotting, immunoprecipitation, SDS-polyacrylamide gel
electrophoresis (SDS-PAGE)

Antibodies as Probes for Proteins: Antibodies are used to detect specific proteins in cells or cell extracts.

recombinant DNA library,
genomic library, cDNA library

Probes for Screening Recombinant DNA Libraries: Specific DNA inserts can be detected in recombinant DNA libraries by the use of either nucleic acid hybridization or antibody probes.

GENE FUNCTION IN EUKARYOTES

temperature-sensitive mutant

Genetic Analysis in Yeasts: The simple genetics and rapid replication of yeasts facilitate the molecular cloning of a gene corresponding to any yeast mutation.

gene transfer, transfection,
transient expression, liposome,
electroporation, transgenic mouse,
embryonal stem (ES) cell,
Ti plasmid

Gene Transfer in Plants and Animals: Cloned genes can be introduced into complex eukaryotic cells and multicellular organisms for functional analysis.

reverse genetics,
in vitro mutagenesis

Mutagenesis of Cloned DNAs: *In vitro* mutagenesis of cloned DNAs is used to study the effect of engineered mutations on gene function.

***Introducing Mutations into Cellular Genes*:** Mutations can be introduced into chromosomal gene copies by homologous recombination with cloned DNA sequences. In addition, the expression or function of specific gene products can be blocked by antisense nucleic acids or dominant inhibitory mutants.

homologous recombination, antisense nucleic acids, dominant inhibitory mutant

Questions

1. What is the genetic map of a locus containing genes *x*, *y*, and *z*? The frequencies of recombination are 0.5% between *x* and *z*, 0.2% between *x* and *y*, and 0.7% between *y* and *z*.

2. You are studying an enzyme in which an active-site cysteine residue is encoded by the triplet UGU. How would mutating the third base to a C affect enzyme function? How about mutating it to an A?

3. Digestion of a 4-kb DNA molecule with *Eco*RI yields two fragments of 1 kb and 3 kb each. Digestion of the same molecule with *Hin*dIII yields fragments of 1.5 kb and 2.5 kb. Finally, digestion with *Eco*RI and *Hin*dIII in combination yields fragments of 0.5 kb, 1 kb, and 2.5 kb. Draw a restriction map indicating the positions of the *Eco*RI and *Hin*dIII cleavage sites.

4. Starting with DNA from a single sperm, how many copies of a specific gene sequence will be obtained after 10 cycles of PCR amplification? After 30 cycles?

5. You have cloned a cDNA of unknown function. How could you experimentally determine the subcellular localization of the protein it encodes?

6. You are interested in identifying the amino acid residues that are important for the catalytic activity of an enzyme. Assuming you have a cDNA clone available, what experimental strategies could you use?

References and Further Reading

General References

Watson, J. D., M. Gilman, J. Witkowski and M. Zoller. 1992. *Recombinant DNA*. 2nd ed. New York: W. H. Freeman.

Watson, J. D., N. H. Hopkins, J. W. Roberts, J. A. Steitz and A. M. Weiner. 1987. *Molecular Biology of the Gene*. 4th ed. Menlo Park, Calif: Benjamin Cummings.

Weaver, R. F. 1999. *Molecular Biology*. New York: McGraw-Hill.

Heredity, Genes, and DNA

Avery, O. T., C. M. MacLeod and M. McCarty. 1944. Studies on the chemical nature of the substance inducing transformation of pneumococcal types. *J. Exp. Med.* 79: 137–158. [P]

Franklin, R. E. and R. G. Gosling. 1953. Molecular configuration in sodium thymonucleate. *Nature* 171: 740–741. [P]

Kornberg, A. 1960. Biologic synthesis of deoxyribonucleic acid. *Science* 131: 1503–1508. [P]

Meselson, M. and F. W. Stahl. 1958. The replication of DNA in *Escherichia coli*. *Proc. Natl. Acad. Sci. USA* 44: 671–682. [P]

Watson, J.D. and F. H. C. Crick. 1953. Genetical implications of the structure of deoxyribonucleic acid. *Nature* 171: 964–967. [P]

Watson, J.D. and F. H. C. Crick. 1953. Molecular structure of nucleic acids: A structure for deoxyribose nucleic acid. *Nature* 171: 737–738. [P]

Wilkins, M. H. F., A. R. Stokes and H. R. Wilson. 1953. Molecular structure of deoxypentose nucleic acids. *Nature* 171: 738–740. [P]

Expression of Genetic Information

Baltimore, D. 1970. RNA-dependent DNA polymerase in virions of RNA tumour viruses. *Nature* 226: 1209–1211. [P]

Brenner, S., F. Jacob and M. Meselson. 1961. An unstable intermediate carrying information from genes to ribosomes for protein synthesis. *Nature* 190: 576–581. [P]

Crick, F. H. C., L. Barnett, S. Brenner and R. J. Watts-Tobin. 1961. General nature of the genetic code for proteins. *Nature* 192: 1227–1232. [P]

Ingram, V. M. 1957. Gene mutations in human hemoglobin: The chemical difference between normal and sickle cell hemoglobin. *Nature* 180: 326–328. [P]

Nirenberg, M. and P. Leder. 1964. RNA codewords and protein synthesis. *Science* 145: 1399–1407. [P]

Nirenberg, M. W. and J. H. Matthaei. 1961. The dependence of cell-free protein synthesis in *E. coli* upon naturally occurring or synthetic polyribonucleotides. *Proc. Natl. Acad. Sci. USA* 47: 1588–1602. [P]

Temin, H. M. and S. Mizutani. 1970. RNA-dependent DNA polymerase in virions of Rous sarcoma virus. *Nature* 226: 1211–1213. [P]

Yanofsky, C., B. C. Carlton, J. R. Guest, D. R. Helinski and U. Henning. 1964. On the colinearity of gene structure and protein structure. *Proc. Natl. Acad. Sci. USA* 51: 266–272. [P]

Recombinant DNA

Arnheim, N. and H. Erlich. 1992. Polymerase chain reaction strategy. *Ann. Rev. Biochem.* 61: 131–156. [R]

Ausubel, F. M., R. Brent, R. E. Kingston, D. D. Moore, J. G. Seidman, J. A. Smith and K. Struhl. eds. 1989. *Current Protocols in Molecular Biology*. New York: Greene Publishing and Wiley Interscience. [R]

Burke, D. T., G. F. Carle and M. V. Olson. 1987. Cloning of large segments of exogenous DNA into yeast by means of artificial chromosome vectors. *Science* 236: 806–812. [P]

Cohen, S. N., A. C. Y. Chang, H. W. Boyer and R. B. Helling. 1973. Construction of biologically functional bacterial plasmids in vitro. *Proc. Natl. Acad. Sci. USA* 70: 3240–3244. [P]

Glick, B. R. and J. J. Pasternak. 1994. *Molecular Biotechnology: Principles and Applications of Recombinant DNA*. Washington, D. C.: ASM Press.

Hames, B. D. and D. Glover. eds. 1995. *DNA Cloning. A Practical Approach*. Oxford, England: IRL Press.

McPherson, M. J., B. D. Hames, and G. Taylor. eds. 1995. *PCR. A Practical Approach*. Oxford, England: IRL Press.

Nathans, D. and H. O. Smith. 1975. Restriction endonucleases in the analysis and restructuring of DNA molecules. *Ann. Rev. Biochem.* 44: 273–293. [R]

Saiki, R. K., D. H. Gelfand, S. Stoffel, S. J. Scharf, R. Higuchi, G. T. Horn, K. B. Mullis and H. A. Erlich. 1988. Primer-directed enzymatic amplification of DNA with a thermostable DNA polymerase. *Science* 239: 487–491. [P]

Sambrook, J., E. F. Fritsch and T. Maniatis. 1989. *Molecular Cloning: A Laboratory Manual*. 2nd ed. Plainview, N.Y.: Cold Spring Harbor Laboratory Press.

Sanger, F., S. Nicklen and A. R. Coulson. 1977. DNA sequencing with chain-terminating inhibitors. *Proc. Natl. Acad. Sci. USA* 74: 5463–5467. [P]

Detection of Nucleic Acids and Proteins

Ausubel, F. M., R. Brent, R. E. Kingston, D. D. Moore, J. G. Seidman, J. A. Smith and K. Struhl. eds. 1989. *Current Protocols in Molecular Biology*. New York: Greene Publishing and Wiley Interscience.

Benton, W. D. and R. W. Davis. 1977. Screening λgt recombinant clones by hybridization to single plaques *in situ*. *Science* 196: 180–182. [P]

Broome, S. and W. Gilbert. 1978. Immunological screening method to detect specific translation products. *Proc. Natl. Acad. Sci. USA* 75: 2746–2749. [P]

Caruthers, M. H. 1985. Gene synthesis machines: DNA chemistry and its uses. *Science* 230: 281–285. [R]

Grunstein, M. and D. S. Hogness. 1975. Colony hybridization: A method for the isolation of cloned DNAs that contain a specific gene. *Proc. Natl. Acad. Sci. USA* 72: 3961–3965. [P]

Harlow, E. and D. Lane. 1999. *Using Antibodies: A Laboratory Manual*. Cold Spring Harbor, N.Y.: Cold Spring Harbor Laboratory Press.

Kohler, G. and C. Milstein. 1975. Continuous cultures of fused cells secreting antibody of predefined specificity. *Nature* 256: 495–497. [P]

Laemmli, U. K. 1970. Cleavage of structural proteins during the assembly of the head of bacteriophage T4. *Nature* 227: 680–685. [P]

Maniatis, T., R. C. Hardison, E. Lacy, J. Lauer, C. O'Connell, D. Quon, G. K. Sim and A. Efstratiadis. 1978. The isolation of structural genes from libraries of eucaryotic DNA. *Cell* 15: 687–701. [P]

Sambrook, J., E. F. Fritsch and T. Maniatis. 1989. *Molecular Cloning: A Laboratory Manual*. 2nd ed. Plainview, N.Y.: Cold Spring Harbor Laboratory Press.

Schildkraut, C. L., J. Marmur and P. Doty. 1961. The formation of hybrid DNA molecules, and their use in studies of DNA homologies. *J. Mol. Biol.* 3: 595–617. [P]

Southern, E. M. 1975. Detection of specific sequences among DNA fragments separated by gel electrophoresis. *J. Mol. Biol.* 98: 503–517. [P]

Gene Function in Eukaryotes

Botstein, D. and D. Shortle. 1985. Strategies and applications of *in vitro* mutagenesis. *Science* 229: 1193–1201. [R]

Bronson, S. K. and O. Smithies. 1994. Altering mice by homologous recombination using embryonic stem cells. *J. Biol. Chem.* 269: 27155–27158. [R]

Capecchi, M. R. 1989. Altering the genome by homologous recombination. *Science* 244: 1288–1292. [R]

Gasser, C. S. and R. T. Fraley. 1989. Genetically engineering plants for crop improvement. *Science* 244: 1293–1299. [R]

Gordon, J. W., G. A. Scangos, D. J. Plotkin, J. A. Barbosa and F. H. Ruddle. 1980. Genetic transformation of mouse embryos by microinjection of purified DNA. *Proc. Natl. Acad. Sci. USA* 77: 7380–7384. [P]

Herskowitz, I. 1987. Functional inactivation of genes by dominant negative mutations. *Nature* 329: 219–222. [R]

Horsch, R. B., J. E. Fry, N. L. Hoffmann, D. Eichholtz, S. G. Rogers and R. T. Fraley. 1985. A simple and general method for transferring genes into plants. *Science* 227: 1229–1231. [P]

Izant, J. G. and H. Weintraub. 1984. Inhibition of thymidine kinase gene expression by antisense RNA: A molecular approach to genetic analysis. *Cell* 36: 1007–1015. [P]

Jaenisch, R. 1988. Transgenic animals. *Science* 240: 1468–1474. [R]

Joyner, A. L., ed. 1993. *Gene Targeting. A Practical Approach*. Oxford, England: IRL Press.

Kaiser, C., S. Michaelis and A. Mitchell. 1994. *Methods in Yeast Genetics: A Laboratory Course Manual*. Plainview, N.Y.: Cold Spring Harbor Laboratory Press.

Kuhn, R., F. Schwenk, M. Aguet and K. Rajewsky. 1995. Inducible gene targeting in mice. *Science* 269: 1427–1429. [P]

Maliga, P., D. F. Klessig, A. R. Cashmore, W. Gruissem and J. E. Varner, eds. 1994. *Methods in Plant Molecular Biology: A Laboratory Course Manual*. Plainview, N.Y.: Cold Spring Harbor Laboratory Press.

Palmiter, R. D. and R. L. Brinster. 1986. Germ-line transformation of mice. *Ann. Rev. Genet.* 20: 465–499. [R]

Robertson, E., A. Bradley, M. Kuehn and M. Evans. 1986. Germline transmission of genes introduced into cultured pluripotential cells by retroviral vector. *Nature* 323: 445–448. [P]

Sedivy, J. M. 1999. Gene targeting and somatic cell genetics: A rebirth or a coming of age? *Trends Genet.* 15: 88–90. [R]

Smith, M. 1985. *In vitro* mutagenesis. *Ann. Rev. Genet.* 19: 423–462. [R]

Struhl, K. 1983. The new yeast genetics. *Nature* 305: 391–397. [R]

Thomas, K. R. and M. R. Capecchi, 1987. Site-directed mutagenesis by gene targeting in mouse embryo-derived stem cells. *Cell* 51: 503–512. [P]

Wagner, R. W. 1994. Gene inhibition using antisense oligonucleotides. *Nature* 372: 333–335. [R]

Wigler, M., R. Sweet, G. K. Sim, B. Wold, A. Pellicer, E. Lacy, T. Maniatis, S. Silverstein and R. Axel. 1979. Transformation of mammalian cells with genes from prokaryotes and eukaryotes. *Cell* 16: 777–785. [P]

PART II

The Flow of Genetic Information

4

The Organization of Cellular Genomes

AS THE GENETIC MATERIAL, DNA PROVIDES A BLUEPRINT that directs all cellular activities and specifies the developmental plan of multicellular organisms. An understanding of gene structure and function is therefore fundamental to an appreciation of the molecular biology of cells. The development of gene cloning represented a major step toward this goal, enabling scientists to dissect complex eukaryotic genomes and probe the functions of eukaryotic genes. Continuing advances in recombinant DNA technology have now brought us to the exciting point of determining the sequences of entire genomes, providing a new approach to deciphering the genetic basis of cell behavior.

As reviewed in Chapter 3, the initial applications of recombinant DNA were directed toward the isolation and analysis of individual genes. Within the last few years, the more global approach of sequencing entire genomes has also become feasible—not only for bacteria and yeasts, but also for the larger genomes of complex plants and animals, including humans. Complete genome sequences are now known for several bacteria, for yeast, and for prototypical animal genomes—those of the nematode *C. elegans* and the fruit fly *Drosophila melanogaster*. Substantial progress has also been made on determination of the human genome sequence, completion of which is expected in the near future. The sequences of these complete cellular genomes provide a rich harvest of information, including the identification of many hitherto unknown genes. The results of these genome sequencing projects can be expected to stimulate many years of future research in molecular and cellular biology, and to have a profound impact on our understanding of many inherited human diseases.

The Complexity of Eukaryotic Genomes

The genomes of most eukaryotes are larger and more complex than those of prokaryotes (Figure 4.1). This larger size of eukaryotic genomes is not inherently surprising, since one would expect to find more genes in organisms that are more complex. However, the genome size of many eukaryotes does not appear to be related to genetic complexity. For example, the

Figure 4.1
Genome size   The range of sizes of the genomes of representative groups of organisms are shown on a logarithmic scale.

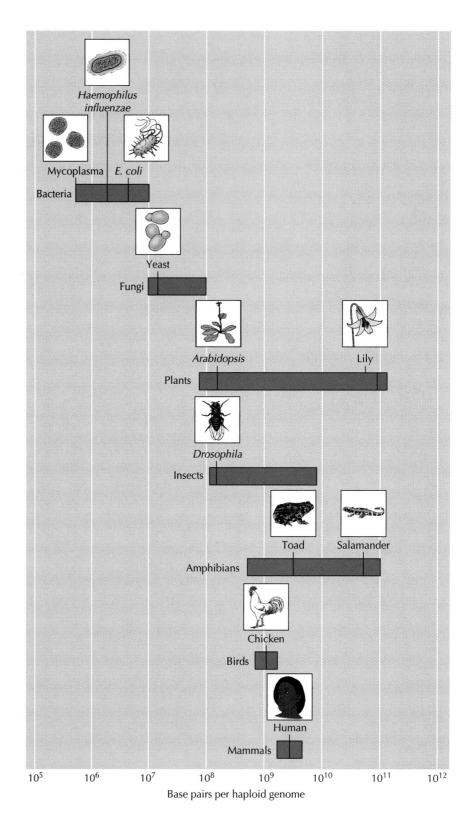

genomes of salamanders and lilies contain more than ten times the amount of DNA that is in the human genome, yet these organisms are clearly not ten times more complex than humans.

This apparent paradox was resolved by the discovery that the genomes of most eukaryotic cells contain not only functional genes but also large amounts of DNA sequences that do not code for proteins. The difference in the sizes of the salamander and human genomes thus reflects larger amounts of noncoding DNA, rather than more genes, in the genome of the salamander. The presence of large amounts of noncoding sequences is a general property of the genomes of complex eukaryotes. Thus, the thousandfold greater size of the human genome compared to that of *E. coli* is not due solely to a larger number of human genes. The human genome is thought to contain approximately 100,000 genes—only about 25 times more than *E. coli* has. Much of the complexity of eukaryotic genomes thus results from the abundance of several different types of noncoding sequences, which constitute most of the DNA of higher eukaryotic cells.

Introns and Exons

In molecular terms, a **gene** can be defined as a segment of DNA that is expressed to yield a functional product, which may be either an RNA (e.g., ribosomal and transfer RNAs) or a polypeptide. Some of the noncoding DNA in eukaryotes is accounted for by long DNA sequences that lie between genes (**spacer sequences**). However, large amounts of noncoding DNA are also found within most eukaryotic genes. Such genes have a split structure in which segments of coding sequence (called **exons**) are separated by noncoding sequences (intervening sequences, or **introns**) (Figure 4.2). The entire gene is transcribed to yield a long RNA molecule and the introns are then removed by splicing, so only exons are included in the mRNA. Although most introns have no known function, they account for a substantial fraction of DNA in the genomes of higher eukaryotes.

Introns were first discovered in 1977, independently in the laboratories of Phillip Sharp and Richard Roberts, during studies of the replication of adenovirus in cultured human cells. Adenovirus is a useful model for studies of gene expression, both because the viral genome is only about 3.5×10^4 base pairs long and because adenovirus mRNAs are produced at high levels in infected cells. One approach used to characterize the adenovirus mRNAs was to determine the locations of the corresponding viral genes by examination of RNA-DNA hybrids in the electron microscope. Because RNA-DNA hybrids are distinguishable from single-stranded DNA, the positions of RNA transcripts on a DNA molecule can be determined. Surprisingly, such experiments revealed that adenovirus mRNAs do not hybridize to only a single region of viral DNA (Figure 4.3). Instead, a single mRNA molecule hybridizes to several separated regions of the viral genome. Thus, the adenovirus mRNA does not correspond to an uninterrupted transcript of the template DNA; rather the mRNA is assembled from several distinct blocks of sequences that originated from different parts of the viral DNA. This was subsequently shown to occur by **RNA splicing**, which will be discussed in detail in Chapter 6.

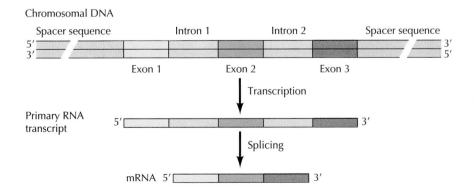

Figure 4.2
The structure of eukaryotic genes
Most eukaryotic genes contain segments of coding sequences (exons) interrupted by noncoding sequences (introns). Both exons and introns are transcribed to yield a long primary RNA transcript. The introns are then removed by splicing to form the mature mRNA.

Key Experiment

The Discovery of Introns

Spliced Segments at the 5′ Terminus of Adenovirus 2 Late mRNA

Susan M. Berget, Claire Moore, and Phillip A. Sharp
Massachusetts Institute of Technology, Cambridge, Massachusetts
Proceedings of the National Academy of Sciences USA, Volume 74, 1977, pages 3171–3175

The Context

Prior to molecular cloning, little was known about mRNA synthesis in eukaryotic cells. However, it was clear that this process is more complex in eukaryotes than in bacteria. The synthesis of eukaryotic mRNAs appeared to require not only transcription, but also processing reactions that modify the structure of primary transcripts. Most notably, eukaryotic mRNAs appeared to be synthesized as long primary transcripts, found in the nucleus, which were then cleaved to yield much shorter mRNA molecules that were exported to the cytoplasm.

These processing steps were generally assumed to involve the removal of sequences from the 5′ and 3′ ends of the primary transcripts. In this model, mRNAs embedded within long primary transcripts would be encoded by uninterrupted DNA sequences. This view of eukaryotic mRNA was changed radically by the discovery of splicing, made independently by Berget, Moore, and Sharp, and by Louise Chow, Richard Gelinas, Tom Broker, and Richard Roberts (An amazing sequence arrangement at the 5′ ends of adenovirus 2 messenger RNA, *Cell* 12: 1–8, 1977).

The Experiments

Both of the research groups that discovered splicing used adenovirus 2 to investigate mRNA synthesis in human cells. The major advantage of the virus is that it provides a model that is much simpler than the host cell. Viral DNA can be isolated directly from virus particles, and

mRNAs encoding the viral structural proteins are present in such high amounts that they can be purified directly from infected cells. Berget, Moore, and Sharp focused their experiments on an abundant mRNA that encodes a viral structural polypeptide known as the hexon.

To map the hexon mRNA on the viral genome, purified mRNA was hybridized to adenovirus DNA and the hybrid molecules were examined by electron microscopy. As expected, the body of the hexon mRNA formed hybrids with restriction fragments of adenovirus DNA that had previously been shown to contain the hexon gene. Surprisingly, however, sequences at the 5′ end of hexon mRNA failed to hybridize to DNA sequences adjacent to those encoding the body of the message, suggesting that the 5′ end of the mRNA had arisen from

sequences located elsewhere in the viral genome.

This possibility was tested by hybridization of hexon mRNA to a restriction fragment extending upstream of the hexon gene. The mRNA-DNA hybrids formed in this experiment displayed a complex loop structure (see figure). The body of the mRNA formed a long hybrid region with the previously identified hexon DNA sequences. Strikingly, the 5′ end of the hexon mRNA hybridized to three short upstream regions of DNA, which were separated from each other and from the body of the message by large single-stranded DNA loops. The sequences at the 5′ end of hexon mRNA thus appeared to be transcribed from three separate regions of the viral genome, which were spliced to the body of the mRNA during the processing of a long primary transcript.

The Impact

The discovery of splicing in adenovirus mRNA was quickly followed by similar experiments with cellular mRNAs, demonstrating that eukary-

An electron micrograph and tracing of hexon mRNA hybridized to adenovirus DNA. The single-stranded loops designated A, B, and C correspond to introns.

The Discovery of Introns *(continued)*

otic genes had a previously unexpected structure. Rather than being continuous, their coding sequences were interrupted by introns, which were removed from primary transcripts by splicing. Introns are now known to account for much of the DNA in eukaryotic genomes, and the roles of introns in the evolution and regulation of gene expression continue to be active areas of investigation. The discovery of splicing also stimulated intense interest in the mechanism of this unexpected RNA

processing reaction. As discussed in Chapter 6, these studies have not only illuminated new mechanisms of regulating gene expression; they have also revealed novel catalytic activities of RNA and provided critical evidence supporting the hypothesis that early evolution was based on self-replicating RNA molecules. The unexpected structure of adenovirus mRNAs has thus had a major impact on diverse areas of cellular and molecular biology.

Phillip Sharp

Richard Roberts

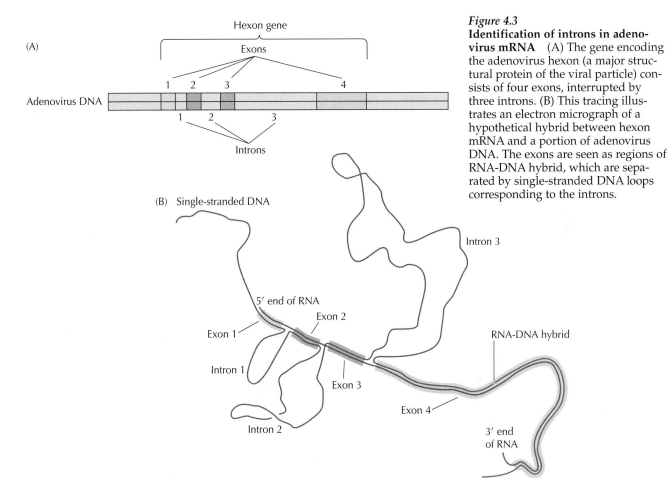

Figure 4.3
Identification of introns in adenovirus mRNA (A) The gene encoding the adenovirus hexon (a major structural protein of the viral particle) consists of four exons, interrupted by three introns. (B) This tracing illustrates an electron micrograph of a hypothetical hybrid between hexon mRNA and a portion of adenovirus DNA. The exons are seen as regions of RNA-DNA hybrid, which are separated by single-stranded DNA loops corresponding to the introns.

Soon after the discovery of introns in adenovirus, similar observations were made on cloned genes of eukaryotic cells. For example, electron microscopic analysis of RNA-DNA hybrids and subsequent nucleotide sequencing of cloned genomic DNAs and cDNAs indicated that the coding region of the mouse *β*-globin gene (which encodes the *β* subunit of hemoglobin) is interrupted by two introns that are removed from the mRNA by splicing (Figure 4.4). The intron-exon structure of many eukaryotic genes is

Figure 4.4
The mouse β-globin gene   This gene contains two introns, which divide the coding region among three exons. Exon 1 encodes amino acids 1 to 30, exon 2 encodes amino acids 31 to 104, and exon 3 encodes amino acids 105 to 146. Exons 1 and 3 also contain untranslated regions (UTRs) at the 5′ and 3′ ends of the mRNA, respectively.

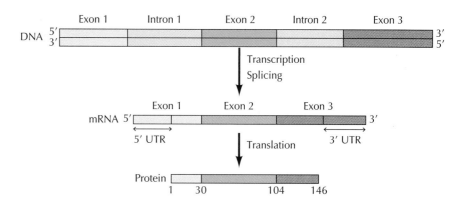

quite complicated, and the amount of DNA in the intron sequences is often greater than that in the exons. The chicken ovalbumin gene, for example, contains eight exons and seven introns distributed over approximately 7700 base pairs (7.7 **kilobases**, or **kb**) of genomic DNA. The exons total only about 1.9 kb, so approximately 75% of the gene consists of introns. An extreme example is the human gene that encodes the blood clotting protein factor VIII. This gene spans approximately 186 kb of DNA and is divided into 26 exons. The mRNA is only about 9 kb long, so the gene contains introns totaling more than 175 kb. On average, introns are estimated to account for about ten times more DNA than exons in the genes of higher eukaryotes.

Introns are present in most genes of complex eukaryotes, although they are not universal. Almost all histone genes, for example, lack introns, so introns are clearly not required for gene function in eukaryotic cells. In addition, introns are not found in most genes of simple eukaryotes, such as yeasts. Conversely, introns *are* present in rare genes of prokaryotes. The presence or absence of introns is therefore not an absolute distinction between prokaryotic and eukaryotic genes, although introns are much more prevalent in higher eukaryotes (both plants and animals), where they account for a substantial amount of total genomic DNA.

Most introns have no known cellular function, although a few have been found to encode functional RNAs or proteins. Introns are generally thought to represent remnants of sequences that were important earlier in evolution. In particular, introns may have helped accelerate evolution by facilitating recombination between protein-coding regions (exons) of different genes—a process known as exon shuffling. Exons frequently encode functionally distinct protein domains, so recombination between introns of different genes would result in new genes containing novel combinations of protein-coding sequences. As predicted by this hypothesis, DNA sequencing studies have demonstrated that some genes are chimeras of exons derived from several other genes, providing direct evidence that new genes can be formed by recombination between intron sequences.

It appears most likely that introns were present early in evolution, prior to the divergence of prokaryotic and eukaryotic cells. According to this hypothesis, introns played an important role in the initial assembly of protein-coding sequences in the ancient ancestors of present-day cells. Introns were subsequently lost from most genes of prokaryotes and simpler eukaryotes (e.g., yeasts) in response to evolutionary selection for rapid replication, which led to streamlining the genomes of these organisms. However, since rapid cell division is not an advantage to higher eukaryotes, introns have been retained in their genomes. Alternatively, introns may have arisen later in evolution as a result of the insertion of DNA sequences into genes that had already been formed as continuous protein-coding sequences. Exon shuffling would then have played an important role in the further evolution of genes in higher eukaryotes but would not account for the initial assembly of protein-coding sequences prior to the evolutionary divergence of prokaryotic and eukaryotic cells.

Gene Families and Pseudogenes

Another factor contributing to the large size of eukaryotic genomes is that some genes are repeated many times. Whereas most prokaryotic genes are represented only once in the genome, many eukaryotic genes are present in multiple copies, called **gene families**. In some cases, multiple copies of genes are needed to produce RNAs or proteins required in large quantities, such as ribosomal RNAs or histones. In other cases, distinct members of a gene family may be transcribed in different tissues or at different stages of development. For example, the α and β subunits of hemoglobin are both encoded by gene families in the human genome, with different members of these families being expressed in embryonic, fetal, and adult tissues (Figure 4.5). Members of many gene families (e.g., the globin genes) are clustered within a region of DNA; members of other gene families are dispersed to different chromosomes.

Gene families are thought to have arisen by duplication of an original ancestral gene, with different members of the family then diverging as a consequence of mutations during evolution. Such divergence can lead to the evolution of related proteins that are optimized to function in different tissues or at different stages of development. For example, fetal globins have a higher affinity for O_2 than do adult globins—a difference that allows the fetus to obtain O_2 from the maternal circulation.

As might be expected, however, not all mutations enhance gene function. Some gene copies have instead sustained mutations that result in their loss of ability to produce a functional gene product. For example, the human α- and β-globin gene families each contain two genes that have been inactivated by mutations. Such nonfunctional gene copies (called **pseudogenes**) represent evolutionary relics that significantly increase the size of eukaryotic genomes without making a functional genetic contribution.

Repetitive DNA Sequences

A substantial portion of eukaryotic genomes consists of highly repeated noncoding DNA sequences. These sequences, sometimes present in hundreds of thousands of copies per genome, were first demonstrated by Roy Britten and David Kohne during studies of the rates of reassociation of denatured frag-

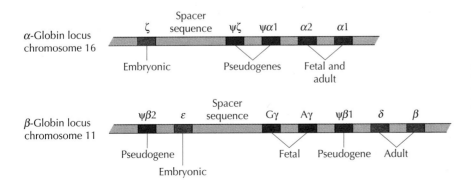

Figure 4.5
Globin gene families Members of the human α- and β-globin gene families are clustered on chromosomes 16 and 11, respectively. Each family contains genes that are specifically expressed in embryonic, fetal, and adult tissues, in addition to nonfunctional gene copies (pseudogenes).

Figure 4.6
Identification of repetitive sequences by DNA reassociation   The kinetics of the reassociation of fragments of *E. coli* and bovine DNAs are illustrated as a function of C_0t, which is the initial concentration of DNA multiplied by the time of incubation. The *E. coli* DNA reassociates at a uniform rate, consistent with each fragment of DNA being represented once in a genome of 4.7×10^6 base pairs. In contrast, the bovine DNA fragments exhibit two distinct steps in their reassociation. About 60% of the DNA fragments (the nonrepeated sequences) reassociate more slowly than *E. coli* DNA, as expected for sequences represented as single copies in the larger bovine genome (3×10^9 base pairs). However, the other 40% of the bovine DNA fragments (the repeated sequences) reassociate more rapidly than *E. coli* DNA, indicating that multiple copies of these sequences are present.

ments of cellular DNAs (Figure 4.6). Denatured strands of DNA hybridize to each other (reassociate), re-forming double-stranded molecules (see Figure 3.28). Since DNA reassociation is a bimolecular reaction (two separated strands of denatured DNA must collide with each other in order to hybridize), the rate of reassociation depends on the concentration of DNA strands. When fragments of *E. coli* DNA were denatured and allowed to hybridize with each other, all of the DNA reassociated at the same rate, as expected if each DNA sequence were represented once per genome. However, reassociation of fragments of DNA extracted from mammalian cells showed a very different pattern. Approximately 60% of the DNA fragments reassociated at the rate expected for sequences present once per genome, but the remainder reassociated much more rapidly than expected. The interpretation of these results was that some sequences were present in multiple copies and therefore reassociated more rapidly than those sequences that were represented only once per genome. In particular, these experiments indicated that approximately 40% of mammalian DNA consists of highly repetitive sequences, some of which are repeated 10^5 to 10^6 times.

Further analysis has identified several types of these highly repeated sequences. One class (called **simple-sequence DNA**) contains tandem arrays of thousands of copies of short sequences, ranging from 5 to 200 nucleotides. For example, one type of simple-sequence DNA in *Drosophila* consists of tandem repeats of the seven nucleotide unit ACAAACT. Because of their distinct base compositions, many simple-sequence DNAs can be separated from the rest of the genomic DNA by equilibrium centrifugation in CsCl density gradients. The density of DNA is determined by its base composition, with AT-rich sequences being less dense than GC-rich sequences. Therefore, an AT-rich simple-sequence DNA bands in CsCl gradients at a lower density than the bulk of *Drosophila* genomic DNA (Figure 4.7). Since such repeat-sequence DNAs band as "satellites" separate from the main band of DNA, they are frequently referred to as **satellite DNAs**. These sequences are repeated millions of times per genome, accounting for 10 to 20% of the DNA of most higher eukaryotes. Simple-sequence DNAs are not transcribed and do not convey functional genetic information. Some, however, may play important roles in chromosome structure.

Other repetitive DNA sequences are scattered throughout the genome rather than being clustered as tandem repeats. These sequences are classified as **SINEs** (short interspersed elements) or **LINEs** (long interspersed elements). The major SINEs in mammalian genomes are ***Alu* sequences**, so-called because they usually contain a single site for the restriction endonuclease *Alu*I. *Alu* sequences are approximately 300 base pairs long, and about a million such sequences are dispersed throughout the genome, accounting for nearly 10% of the total cellular DNA. Although *Alu* sequences are transcribed into RNA, they do not encode proteins and their function is unknown. The major human LINEs (which belong to the LINE 1, or L1, family) are about 6000 base pairs long and repeat approximately 50,000 times in the genome. L1 sequences are transcribed and at least some encode proteins, but like *Alu* sequences, they have no known function in cell physiology.

Both *Alu* and L1 sequences are examples of transposable elements, which are capable of moving to different sites in genomic DNA (see Chapter 5). Some of these sequences may help regulate gene expression, but most *Alu* and L1 sequences appear not to make a useful contribution to the cell. They may, however, have played important evolutionary roles by contributing to the generation of genetic diversity.

The Number of Genes in Eukaryotic Cells

Having discussed several kinds of noncoding DNA that contribute to the genomic complexity of higher eukaryotes, it is of interest to consider the total number of genes in eukaryotic genomes (Table 4.1). Assuming that the average polypeptide is approximately 400 amino acids long, the average size of the coding sequence of a gene is 1200 base pairs. In bacterial genomes, most of the DNA encodes proteins. For example, the genome of *E. coli* is approximately 4.6×10^6 base pairs long and contains 4288 genes, with nearly 90% of the DNA used as protein-coding sequence.

The yeast genome, which consists of 12×10^6 base pairs, is about 2.5 times the size of the genome of *E. coli*, but is still extremely compact. Only 4% of the genes of *Saccharomyces cerevisiae* contain introns, and these usually have only a single small intron near the start of the coding sequence. The average gene in yeast spans about 2000 base pairs, and approximately 70% of the yeast genome is used as protein-coding sequence, specifying a total of about 6000 proteins.

The genome of the nematode *C. elegans*, a relatively simple animal genome, is intermediate in size and complexity between the genomes of yeast and mammals. The *C. elegans* genome is 97×10^6 base pairs and contains approximately 19,000 protein-coding genes. Thus, while the genome of *C. elegans* is 8 times larger than that of yeast, it contains only about three times the number of genes. This correlates with the presence of a substan-

Figure 4.7
Satellite DNA Equilibrium centrifugation of *Drosophila* DNA in a CsCl gradient separates satellite DNAs (designated I–IV) with buoyant densities (in g/cm³) of 1.672, 1.687, and 1.705 from the main band of genomic DNA (buoyant density 1.701).

TABLE 4.1 The Numbers of Genes in Cellular Genomes

Organism	Genome size (Mb)[a]	Protein-coding sequence (percent)[b]	Number of genes[b]
E. coli	4.6	90	4,288
S. cerevisiae	12	70	5,885
C. elegans	97	25	19,099
Drosophila	180	13	13,600
Human	3,000	3	100,000

[a] Mb = millions of base pairs.
[b] Determined from the *E. coli*, *S. cerevisiae*, *C. elegans*, and *Drosophila* genomic sequences and estimated for the human genome.

tial number of introns in *C. elegans*. Each gene in *C. elegans* contains an average of five introns and spans an average of 5000 bases. Consistent with this, only about 25% of the *C. elegans* sequence corresponds to exons, versus 70% protein-coding sequence in the yeast genome. Although the genome of *Drosophila* is 180 x 10⁶ base pairs, *Drosophila* contains fewer genes than *C. elegans* (about 13,600). Protein-coding sequence thus corresponds to only about 13% of the *Drosophila* genome.

The genomes of higher animals (such as humans) are still more complex and contain large amounts of noncoding DNA. Thus, only a small fraction of the 3×10^9 base pairs of the human genome is expected to correspond to protein-coding sequence. Approximately one-third of the genome corresponds to highly repetitive sequences, leaving an estimated 2×10^9 base pairs for functional genes, pseudogenes, and nonrepetitive spacer sequences. If the average gene spans 10,000–20,000 base pairs (including introns), one might expect the human genome to consist of about 100,000 genes, with protein-coding sequences corresponding to only about 3% of human DNA. Although this estimate is generally accepted as plausible, it remains to be verified or corrected by the final results of human genome sequencing.

Chromosomes and Chromatin

Not only are the genomes of most eukaryotes much more complex than those of prokaryotes, but the DNA of eukaryotic cells is also organized differently from that of prokaryotic cells. The genomes of prokaryotes are contained in single chromosomes, which are usually circular DNA molecules. In contrast, the genomes of eukaryotes are composed of multiple chromosomes, each containing a linear molecule of DNA. Although the numbers and sizes of chromosomes vary considerably between different species (Table 4.2), their basic structure is the same in all eukaryotes. The DNA of eukaryotic cells is tightly bound to small basic proteins (histones) that package the DNA in an orderly way in the cell nucleus. This task is substantial, given the DNA content of most eukaryotes. For example, the total extended length of DNA in a human cell is nearly 2 m, but this DNA must fit into a nucleus with a diameter of only 5 to 10 μm. Although DNA packaging is also a problem in bacteria, the mechanism by which prokaryotic DNAs are packaged in the cell appears distinct from that of eukaryotes and is not well understood.

Chromatin

The complexes between eukaryotic DNA and proteins are called **chromatin**, which typically contains about twice as much protein as DNA. The major proteins of chromatin are the **histones**—small proteins containing a high proportion of basic amino acids (arginine and lysine) that facilitate binding to the negatively charged DNA molecule. There are five major types of histones—called H1, H2A, H2B, H3, and H4—which are very similar among different species of eukaryotes (Table 4.3). The histones are extremely abundant proteins in eukaryotic cells; together, their mass is approximately equal to that of the cell's DNA. In addition, chromatin contains an approximately equal mass of a wide variety of nonhistone chromo-

TABLE 4.2   Chromosome Numbers of Eukaryotic Cells

Organism	Genome size (Mb)[a]	Chromosome number[a]
Yeast (*Saccharomyces cerevisiae*)	12	16
Slime mold (*Dictyostelium*)	70	7
Arabidopsis thaliana	130	5
Corn	5,000	10
Onion	15,000	8
Lily	50,000	12
Nematode (*Caenorhabditis elegans*)	97	6
Fruit fly (*Drosophila*)	180	4
Toad (*Xenopus laevis*)	3,000	18
Lungfish	50,000	17
Chicken	1,200	39
Mouse	3,000	20
Cow	3,000	30
Dog	3,000	39
Human	3,000	23

[a] Both genome size and chromosome number are for haploid cells.
Mb = millions of base pairs.

somal proteins. There are more than a thousand different types of these proteins, which are involved in a range of activities, including DNA replication and gene expression. Histones are not found in eubacteria (e.g., *E. coli*), although the DNA of these bacteria is associated with other proteins that presumably function like histones to package the DNA within the bacterial cell. Archaebacteria, however, do contain histones that package their DNAs in structures similar to eukaryotic chromatin.

The basic structural unit of chromatin, the **nucleosome**, was described by Roger Kornberg in 1974 (Figure 4.8). Two types of experiments led to Kornberg's proposal of the nucleosome model. First, partial digestion of chromatin with micrococcal nuclease (an enzyme that degrades DNA) was found to yield DNA fragments approximately 200 base pairs long. In contrast, a similar digestion of naked DNA (not associated with proteins) yielded a continuous smear of randomly sized fragments. These results suggested that the binding of proteins to DNA in chromatin protects regions of the DNA from nuclease digestion, so that the enzyme can attack DNA only at sites separated by approximately 200 base pairs. Consistent with this notion, electron microscopy revealed that chromatin fibers have a beaded appearance, with the beads spaced at intervals of approximately 200 base pairs. Thus, both the nuclease digestion and the electron microscopic studies suggested that chromatin is composed of repeating 200-base-pair units, which were called nucleosomes.

TABLE 4.3   The Major Histone Proteins

Histone[a]	Molecular weight	Number of amino acids	Percentage Lysine + Arginine
H1	22,500	244	30.8
H2A	13,960	129	20.2
H2B	13,774	125	22.4
H3	15,273	135	22.9
H4	11,236	102	24.5

[a] Data are for rabbit (H1) and bovine histones.

Figure 4.8
The organization of chromatin in nucleosomes   (A) The DNA is wrapped around histones in nucleosome core particles and sealed by histone H1. Nonhistone proteins bind to the linker DNA between nucleosome core particles. (B) Gel electrophoresis of DNA fragments obtained by partial digestion of chromatin with micrococcal nuclease. The linker DNA between the nucleosome core particles is preferentially sensitive, so limited digestion of chromatin yields fragments corresponding to multiples of 200 base pairs. (C) An electron micrograph of an extended chromatin fiber, illustrating its beaded appearance. (B, courtesy of Roger Kornberg, Stanford University; C, courtesy of Ada L. Olins and Donald E. Olins, Oak Ridge National Laboratory.)

(A)

2 molecules each of
H2A, H2B, H3, and H4

DNA

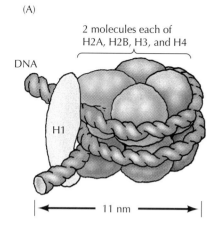

H1

|← 11 nm →|

(B)

Figure 4.9
Structure of a chromatosome   (A) The nucleosome core particle consists of 146 base pairs of DNA wrapped 1.65 turns around a histone octamer consisting of two molecules each of H2A, H2B, H3, and H4. A chromatosome contains two full turns of DNA (166 base pairs) locked in place by one molecule of H1. (B) Model of the nucleosome core particle. The DNA backbones are shown in brown and turquoise. The histones are shown in blue (H3), green (H4), yellow (H2A), and red (H2B). (B, from K. Luger et al., 1997. *Nature* 389: 251.)

More extensive digestion of chromatin with micrococcal nuclease was found to yield particles (called **nucleosome core particles**) that correspond to the beads visible by electron microscopy. Detailed analysis of these particles has shown that they contain 146 base pairs of DNA wrapped 1.65 times around a histone core consisting of two molecules each of H2A, H2B, H3, and H4 (the core histones) (Figure 4.9). One molecule of the fifth histone, H1, is bound to the DNA as it enters each nucleosome core particle. This forms a chromatin subunit known as a **chromatosome**, which consists of 166 base pairs of DNA wrapped around the histone core and held in place by H1 (a linker histone).

The packaging of DNA with histones yields a chromatin fiber approximately 10 nm in diameter that is composed of chromatosomes separated by linker DNA segments averaging about 80 base pairs in length (Figure 4.10). In the electron microscope, this 10-nm fiber has the beaded appearance that suggested the nucleosome model. Packaging of DNA into such a 10-nm chromatin fiber shortens its length approximately sixfold. The chromatin can then be further condensed by coiling into 30-nm fibers, the structure of which still remains to be determined. Interactions between histone H1 molecules appear to play an important role in this stage of chromatin condensation.

The extent of chromatin condensation varies during the life cycle of the cell. In interphase (nondividing) cells, most of the chromatin (called **euchro-**

Figure 4.10
Chromatin fibers   The packaging of DNA into nucleosomes yields a chromatin fiber approximately 10 nm in diameter. The chromatin is further condensed by coiling into a 30-nm fiber, containing about six nucleosomes per turn. (Photographs courtesy of Ada L. Olins and Donald E. Olins, Oak Ridge National Laboratory.)

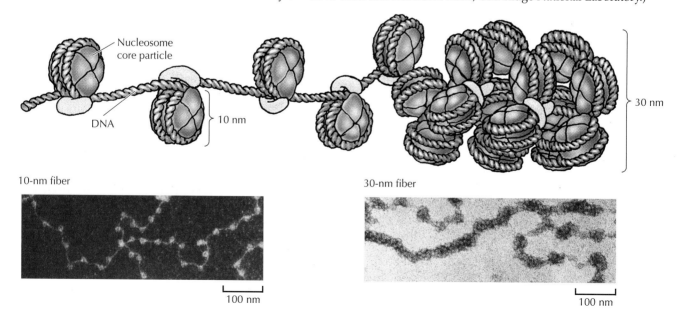

Nucleosome
core particle

DNA

10 nm

30 nm

10-nm fiber

30-nm fiber

|— 100 nm —|

|— 100 nm —|

1 µm

Figure 4.11
Interphase chromatin   Electron micrograph of an interphase nucleus. The euchromatin is distributed throughout the nucleus. The hetero-chromatin is indicated by arrowheads, and the nucleolus by an arrow. (Courtesy of Ada L. Olins and Donald E. Olins, Oak Ridge National Laboratory.)

matin) is relatively decondensed and distributed throughout the nucleus (Figure 4.11). During this period of the cell cycle, genes are transcribed and the DNA is replicated in preparation for cell division. Most of the euchromatin in interphase nuclei appears to be in the form of 30-nm fibers, organized into large loops containing approximately 50 to 100 kb of DNA. About 10% of the euchromatin, containing the genes that are actively transcribed, is in a more decondensed state (the 10-nm conformation) that allows transcription. Chromatin structure is thus intimately linked to the control of gene expression in eukaryotes, as will be discussed in Chapter 6.

In contrast to euchromatin, about 10% of interphase chromatin (called **heterochromatin**) is in a very highly condensed state that resembles the chromatin of cells undergoing mitosis. Heterochromatin is transcriptionally inactive and contains highly repeated DNA sequences, such as those present at centromeres and telomeres.

As cells enter mitosis, their chromosomes become highly condensed so that they can be distributed to daughter cells. The loops of 30-nm chromatin fibers are thought to fold upon themselves further to form the compact metaphase chromosomes of mitotic cells, in which the DNA has been condensed nearly 10,000-fold (Figure 4.12). Such condensed chromatin can no longer be used as a template for RNA synthesis, so transcription ceases during mitosis. Electron micrographs indicate that the DNA in metaphase chromosomes is organized into large loops attached to a protein scaffold (Figure 4.13), but we currently understand neither the detailed structure of this highly condensed chromatin nor the mechanism of chromatin condensation.

Figure 4.12
Chromatin condensation during mitosis   Scanning electron micrograph of metaphase chromosomes. Artificial color has been added. (Biophoto Associates/Photo Researchers Inc.)

10 µm

Protein scaffold   DNA loops

Figure 4.13
Structure of metaphase chromosomes   An electron micrograph of DNA loops attached to the protein scaffold of metaphase chromosomes that have been depleted of histones. (From J. R. Paulson and U. K. Laemmli, 1977. *Cell* 12: 817.)

Metaphase chromosomes are so highly condensed that their morphology can be studied using the light microscope (Figure 4.14). Several staining techniques yield characteristic patterns of alternating light and dark chromosome bands, which result from the preferential binding of stains or fluorescent dyes to AT-rich versus GC-rich DNA sequences. These bands are specific for each chromosome and appear to represent distinct chromosome regions. Genes can be localized to specific chromosome bands by *in situ* hybridization, indicating that the packaging of DNA into metaphase chromosomes is a highly ordered and reproducible process.

Centromeres

The **centromere** is a specialized region of the chromosome that plays a critical role in ensuring the correct distribution of duplicated chromosomes to daughter cells during mitosis (Figure 4.15). The cellular DNA is replicated during interphase, resulting in the formation of two copies of each chromosome prior to the beginning of mitosis. As the cell enters mitosis, chromatin condensation leads to the formation of metaphase chromosomes consisting of two identical sister chromatids. These sister chromatids are held together at the centromere, which is seen as a constricted chromosomal region. As mitosis proceeds, microtubules of the mitotic spindle attach to the centromere, and the two sister chromatids separate and move to opposite poles of the spindle. At the end of mitosis, nuclear membranes re-

Figure 4.14
Human metaphase chromosomes   (A) A light micrograph of human chromosomes spread from a metaphase cell. (B) Human chromosomes are arranged in pairs numbered from the largest (chromosome 1) to the smallest. The chromosomes shown are from a female, so there are 22 pairs of autosomes and two X chromosomes. (Craig Holmes/Biological Photo Service.)

(A)

(B)

10 μm

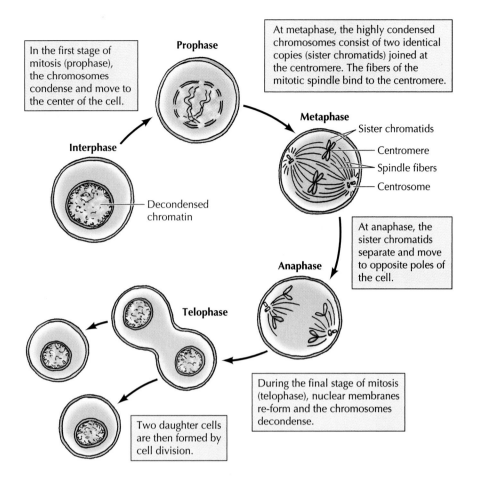

In the first stage of mitosis (prophase), the chromosomes condense and move to the center of the cell.

At metaphase, the highly condensed chromosomes consist of two identical copies (sister chromatids) joined at the centromere. The fibers of the mitotic spindle bind to the centromere.

At anaphase, the sister chromatids separate and move to opposite poles of the cell.

During the final stage of mitosis (telophase), nuclear membranes re-form and the chromosomes decondense.

Two daughter cells are then formed by cell division.

Interphase — Decondensed chromatin

Prophase

Metaphase — Sister chromatids, Centromere, Spindle fibers, Centrosome

Anaphase

Telophase

Figure 4.15
Chromosomes during mitosis Since DNA replicates during interphase, the cell contains two identical duplicated copies of each chromosome prior to entering mitosis.

form and the chromosomes decondense, resulting in the formation of daughter nuclei containing one copy of each parental chromosome.

The centromeres thus serve both as the sites of association of sister chromatids and as the attachment sites for microtubules of the mitotic spindle. They consist of specific DNA sequences to which a number of centromere-associated proteins bind, forming a specialized structure called the **kinetochore** (Figure 4.16). The binding of microtubules to kinetochore proteins mediates the attachment of chromosomes to the mitotic spindle. Proteins associated with the kinetochore then act as "molecular motors" that drive the movement of chromosomes along the spindle fibers, segregating the chromosomes to daughter nuclei.

Centromeric DNA sequences have been defined best in yeasts, where their function can be assayed by following the segregation of plasmids at mitosis (Figure 4.17). Plasmids that contain functional centromeres segregate like chromosomes and are equally distributed to daughter cells following mitosis. In the absence of a functional centromere, however, the plasmid does not segregate properly, and many daughter cells fail to inherit plasmid DNA. Assays of this type have enabled determination of

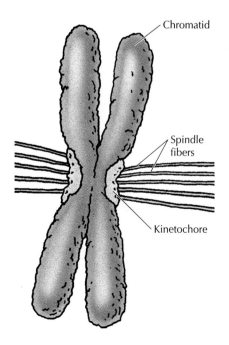

Chromatid

Spindle fibers

Kinetochore

Figure 4.16
The centromere of a metaphase chromosome The centromere is the region at which the two sister chromatids remain attached at metaphase. Specific proteins bind to centromeric DNA, forming the kinetochore, which is the site of spindle fiber attachment.

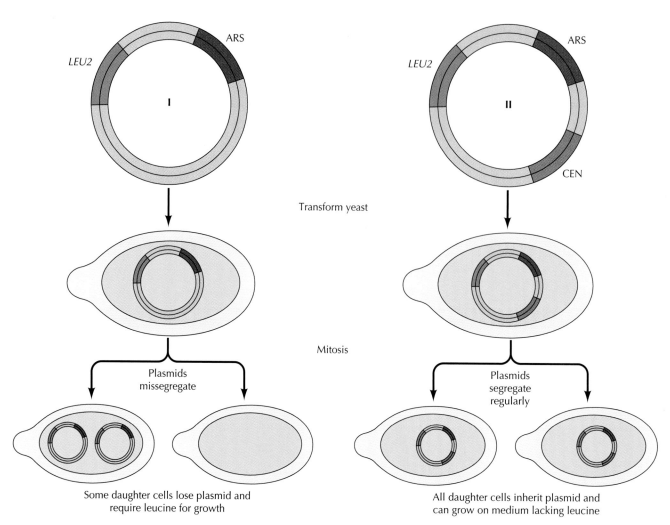

Transform yeast

Mitosis

Plasmids
missegregate

Plasmids
segregate
regularly

Some daughter cells lose plasmid and
require leucine for growth

All daughter cells inherit plasmid and
can grow on medium lacking leucine

Figure 4.17
Assay of a centromere in yeast   Both plasmids shown contain a selectable marker (*LEU2*) and DNA sequences that serve as origins of replication in yeast (ARS,
which stands for autonomously replicating sequence). However, plasmid I lacks a
centromere and is therefore frequently lost as a result of missegregation during
mitosis. In contrast, the presence of a centromere (CEN) in plasmid II ensures its
regular transmission to daughter cells.

the sequences required for centromere function. Such experiments first
showed that the centromere sequences of the well-studied yeast *Saccharomyces cerevisiae* are contained in approximately 125 base pairs consisting
of three sequence elements: two short sequences of 8 and 25 base pairs separated by 78 to 86 base pairs of very AT-rich DNA (Figure 4.18A).

The short centromere sequences defined in *S. cerevisiae*, however, do not
appear to reflect the situation in other eukaryotes. More recent studies have
defined the centromeres of the fission yeast *Schizosaccharomyces pombe* by a
similar functional approach. Although *S. cerevisiae* and *S. pombe* are both
yeasts, they appear to be as divergent from each other as either is from
humans and are quite different in many aspects of their cell biology. These
two yeast species thus provide complementary models for simple and easily studied eukaryotic cells. The centromeres of *S. pombe* span 40 to 100 kb
of DNA; they are approximately a thousand times larger than those of *S.*

(A) *S. cerevisiae*

(B) *S. pombe*

(C) *Drosophila melanogaster*

■ Transposons   ■ AAGAG Satellite
■ AATAT Satellite   □ Nonrepetitive DNA

Figure 4.18
Centromeres of *S. cerevisiae, S. pombe,* and *Drosophila melanogaster*
(A) The *S. cerevisiae* centromere (CEN) sequences consist of two short conserved sequences (CDEI and CDEIII) separated by 78 to 86 base pairs (bp) of AT-rich DNA (CDEII). The sequences shown are consensus sequences derived from analysis of the centromere sequences of individual yeast chromosomes. Pu = A or G; x = A or T; y = any base. (B) The arrangement of sequences at the centromere of *S. pombe* chromosome II is illustrated. The centromere consists of a central core (CC) of unique-sequence DNA, flanked by tandem repeats of three repetitive sequence elements (B, K, and L). (C) The *Drosophila* centromere consists of two satellite sequences, transposable elements, and nonrepetitive DNA.

cerevisiae. They consist of a central core of 4 to 7 kb of single-copy DNA flanked by repetitive sequences (Figure 4.18B). Not only the central core but also the flanking repeated sequences are required for centromere function, so the centromeres of *S. pombe* appear to be considerably more complex than those of *S. cerevisiae.*

Studies of a *Drosophila* chromosome have provided the first characterization of a centromere in higher eukaryotes (Figure 4.18C). The *Drosophila* centromere spans 420 kb, most of which (more than 85%) consists of two highly repeated satellite DNAs with the sequences AATAT and AAGAG. The remainder of the centromere consists of interspersed transposable elements, which are also found at other sites in the *Drosophila* genome, in addition to a nonrepetitive region of AT-rich DNA. Deletion of the satellite sequences and transposable elements, as well as the nonrepetitive DNA, reduced the activity of the centromere in functional assays. Thus, both repetitive and nonrepetitive sequences appear to contribute to kinetochore formation and centromere function.

Centromeres of humans and other mammals have not yet been defined by functional studies, but they have been identified by the binding of centromere-associated proteins. Mammalian centromeres are characterized by extensive regions of heterochromatin consisting of highly repetitive satellite DNA sequences. In humans and other primates the primary centromeric sequence is α satellite DNA, which is a 171-base-pair sequence arranged in tandem repeats spanning up to millions of base pairs. The α satellite DNA appears to play a role in centromere structure and function, since it has been found to bind centromere-associated proteins. However, the precise function of α satellite DNA, as well as the potential roles of other sequences in mammalian centromeres, remains to be established. Consistent with their large size, mammalian centromeres form large kinetochores that bind 30 to 40 microtubules, whereas only single microtubules bind to the centromeres of *S. cerevisiae.*

TABLE 4.4 Telomeric DNAs

Organism	Telomeric repeat sequence
Yeasts	
Saccharomyces cerevisiae	$G_{1-3}T$
Schizosaccharomyces pombe	$G_{2-5}TTAC$
Protozoans	
Tetrahymena	GGGGTT
Dictyostelium	$G_{1-8}A$
Plant	
Arabidopsis	AGGGTTT
Mammal	
Human	AGGGTT

Telomeres

The sequences at the ends of eukaryotic chromosomes, called **telomeres**, play critical roles in chromosome replication and maintenance. Telomeres were initially recognized as distinct structures because broken chromosomes were highly unstable in eukaryotic cells, implying that specific sequences are required at normal chromosomal termini. This was subsequently demonstrated by experiments in which telomeres from the protozoan *Tetrahymena* were added to the ends of linear molecules of yeast plasmid DNA. The addition of these telomeric DNA sequences allowed these plasmids to replicate as linear chromosome-like molecules in yeasts, demonstrating directly that telomeres are required for the replication of linear DNA molecules.

The telomere DNA sequences of a variety of eukaryotes are similar, consisting of repeats of a simple-sequence DNA containing clusters of G residues on one strand (Table 4.4). For example, the sequence of telomere repeats in humans and other mammals is AGGGTT, and the telomere repeat in *Tetrahymena* is GGGGTT. These sequences are repeated hundreds or thousands of times, thus spanning up to several kilobases, and terminate with an overhang of single-stranded DNA. Recent results suggest that the repeated sequences of telomere DNA form loops at the ends of chromosomes, thereby protecting the chromosome termini from degradation (Figure 4.19).

Telomeres play a critical role in replication of the ends of linear DNA molecules (see Chapter 5). DNA polymerase is able to extend a growing DNA chain but cannot initiate synthesis of a new chain at the terminus of a linear DNA molecule. Consequently, the ends of linear chromosomes cannot be replicated by the normal action of DNA polymerase. This problem has been solved by the evolution of a special mechanism, involving reverse transcriptase activity, to replicate telomeric DNA sequences. Maintenance of telomeres appears to be an important factor in determining the lifespan and reproductive capacity of cells, so studies of telomeres and telomerase have the promise of providing new insights into conditions such as aging and cancer.

The Sequences of Complete Genomes

Some of the most exciting endeavors in molecular biology are the current efforts directed at obtaining and analyzing the complete nucleotide sequences of both the human genome and the genomes of several model organisms, including *E. coli*, *Saccharomyces cerevisiae*, *Caenorhabditis elegans*, *Drosophila*, *Arabidopsis*, and the mouse. Recombinant DNA has had an enormous impact on our understanding of the molecular basis of cell biology by allowing the isolation and sequencing of a wide variety of important genes.

Until the last few years, scientists had generally focused on the cloning, sequencing, and further characterization of specific genes. Now, however, recombinant DNA technology allows the alternative approach of cloning and sequencing entire genomes. In principle, this approach has the potential of identifying all the genes in an organism, which then become accessible for investigations of their structure and function. Genome sequencing is thus providing scientists with a unique database, consisting of the nucleotide sequences of complete sets of genes. Since many of these genes have not been previously identified, determination of their functions will form the basis of many future studies in cell biology.

Prokaryotic Genomes

We now know the complete genome sequences of nearly two dozen different bacteria, and many others are in the process of being determined. The first complete sequence of a cellular genome, reported in 1995 by a team of researchers led by J. Craig Venter, was that of the bacterium *Haemophilus influenzae*, a common inhabitant of the human respiratory tract. The genome of *H. influenzae* is approximately 1.8×10^6 base pairs (1.8 **megabases, or Mb**), slightly less than half the size of the *E. coli* genome. The complete nucleotide sequence indicated that the *H. influenzae* genome is a circular molecule containing 1,830,137 base pairs of DNA. The sequence was then analyzed to identify the genes encoding rRNAs, tRNAs, and proteins. Potential protein-coding regions were identified by computer analysis of the DNA sequence to detect **open-reading frames**—long stretches of nucleotide sequence that can encode polypeptides because they contain none of the three chain-terminating codons (UAA, UAG, and UGA). Since these chain-terminating codons occur randomly once in every 21 codons (3 chain-terminating codons out of 64 total), open-reading frames that extend for more than a hundred codons usually represent functional genes.

This analysis identified six copies of rRNA genes, 54 different tRNA genes, and 1743 potential protein-coding regions in the *H. influenzae* genome (Figure 4.20). More than a thousand of these could be assigned a biological role (e.g., an enzyme of the citric acid cycle) on the basis of their relationships to known protein sequences, but the others represent genes of currently unknown function. The predicted coding sequences have an average size of approximately 900 base pairs, so they cover about 1.6 Mb of DNA. Protein-coding sequences thus correspond to more than 85% of the genome of *H. influenzae*, consistent with previous estimates that protein-coding sequences account for most of the DNA in bacterial genomes.

The complete sequence of the genome of *Mycoplasma genitalium* is of particular interest because mycoplasmas are the simplest present-day bacteria and contain the smallest genomes of all known cells. The genome of *M. genitalium* is only 580 kb (0.58 Mb) long and may represent the minimal set of genes required to maintain a self-replicating organism. Analysis of its DNA sequence indicates that *M. genitalium* contains only 470 predicted protein-coding sequences, which correspond to approximately 88% of genomic DNA (Figure 4.21). Many of these sequences were identified as genes encoding proteins involved in DNA replication, transcription, translation, membrane transport, and energy metabolism. However, *M. genitalium* contains many fewer genes for metabolic enzymes than does *H. influenzae*, reflecting its more limited metabolism. For example, many

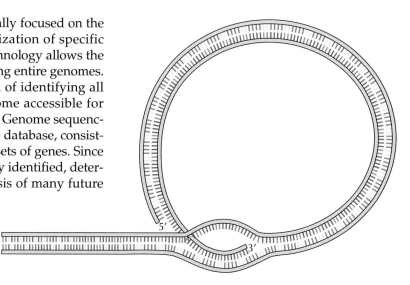

Figure 4.19
Structure of a telomere   Telomere DNA loops back on itself to form a circular structure that protects the ends of chromosomes.

Figure 4.20
The genome of *Haemophilus influenzae*
Predicted protein-coding regions are designated by colored bars. Numbers indicate base pairs of DNA. (From R. D. Fleischmann et al., 1995. *Science* 269: 496.)

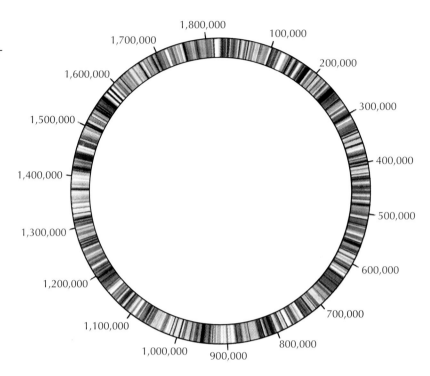

genes known to encode components of biosynthetic pathways are lacking in the genome of *M. genitalium*, consistent with its need to obtain amino acids and nucleotide precursors from a host organism. Interestingly, the *Mycoplasma* genome also includes approximately 150 genes of currently unknown function. Thus, even in the simplest of cells, the biological roles of many genes remain to be determined.

The sequence of the genome of the archaebacterium *Methanococcus jannaschii*, reported in 1996, provided major insights into the evolutionary rela-

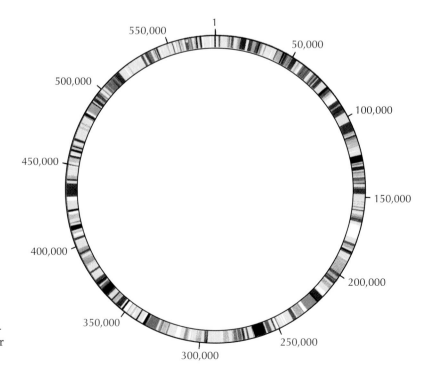

Figure 4.21
The genome of *Mycoplasma genitalium*
Predicted protein-coding sequences are designated by colored bars. Numbers indicate base pairs of DNA. (From C. M. Fraser et al., 1995. *Science* 270: 397.)

tionships between the archaebacteria, eubacteria, and eukaryotes. The genome of *M. jannaschii* is 1.7 Mb and contains 1738 predicted protein-coding sequences—similar in size to the genome of *H. influenzae*. However, only about one-third of the protein-coding sequences identified in *M. jannaschii* were related to known genes of either eubacteria or eukaryotes, indicating the distinct genetic composition of the archaebacteria. The genes of *M. jannaschii* encoding proteins involved in energy production and biosynthesis of cell constituents are related to those of eubacteria, suggesting that basic metabolic processes evolved in a common ancestor of both the archaebacteria and the eubacteria. Importantly, however, the *M. jannaschii* genes encoding proteins involved in DNA replication, transcription, and translation are more closely related to those of eukaryotes than to those of eubacteria. Genomic sequencing of this archaebacterium thus indicates that the archaebacteria and eukaryotes share a common line of evolutionary descent and are more closely related to each other than either is to the eubacteria (see Figure 1.8).

Although the relative simplicity and facile genetics of *E. coli* have made it a favored organism of molecular biologists, the 4.6-Mb *E. coli* genome was not completely sequenced until 1997. Analysis of the *E. coli* sequence revealed a total of 4288 genes, with protein-coding sequences accounting for 88% of the *E. coli* genome. Of the 4288 genes revealed by sequencing, 1835 had been previously identified and the functions of an additional 821 could be deduced by comparisons to the sequences of characterized genes of other organisms. However, the functions of 1632 *E. coli* genes (nearly 40% of the genome) could not be determined. Thus, even for an organism as thoroughly studied as *E. coli*, genomic sequencing demonstrates that a great deal remains to be learned about prokaryotic cell biology.

Sequencing the Yeast Genome

As noted already, the simplest eukaryotic genome (1.2×10^7 base pairs of DNA) is found in the yeast *Saccharomyces cerevisiae*. Moreover, yeasts grow rapidly and are subject to simple genetic manipulations. Thus, in many ways yeasts are model eukaryotic cells that can be studied much more readily than the cells of mammals or other higher eukaryotes. Consequently, the complete sequencing of an entire yeast chromosome in 1992 (Figure 4.22), followed by determination of the sequence of the complete yeast genome in 1996, were major steps in understanding the molecular biology of eukaryotic cells.

The *S. cerevisiae* genome contains about 6000 genes, including 5885 predicted protein-coding sequences, 140 ribosomal RNA genes, 275 transfer RNA genes, and 40 genes encoding small nuclear RNAs involved in RNA processing (discussed in Chapter 6). Yeast thus have a high density of protein-coding sequences, similar to bacterial genomes, with protein-coding sequences accounting for approximately 70% of total yeast DNA. Consistent with this, only 4% of yeast genes were found to contain introns. Moreover, those *S. cerevisiae* genes that do contain introns usually have only a single small intron near the beginning of the gene.

Computer analysis was able to assign a predicted function to approximately 3000 of the *S. cerevisiae* protein-coding sequences based on similarities to the sequences of known genes. Based on analysis of these genes, it appears that approximately 11% of yeast proteins function in metabolism, 3% in the production and storage of metabolic energy, 3% in DNA replication, repair, and recombination, 7% in transcription, 6% in translation, and 14% in protein sorting and transport. However, the functions of many of these genes are only known in general terms (such as "transcription factor"), so their precise roles within the cell still need to be determined. Moreover, since half of the proteins encoded by the yeast genome were unrelated

Figure 4.22
Yeast chromosome III The upper blue bars designate the clones used for DNA sequencing. Open-reading frames are indicated by arrows. (From S. G. Oliver et al., 1992. *Nature* 357: 38.)

to previously described genes, the functions of an additional 3000 unknown proteins remain to be elucidated by genetic and biochemical analyses.

Now that the sequence of the yeast genome has been completed, determination of the functions of these many new genes has become a major goal. Fortunately, yeasts are particularly amenable to functional analyses of unknown genes because of the facility with which normal chromosomal loci can be inactivated by homologous recombination with cloned sequences (see Figure 3.42). Therefore, direct functional analysis of yeast genes that were initially identified only on the basis of their nucleotide sequence can be systematically undertaken. Sequencing the yeast genome has thus opened the door to studying many new areas of the biology of this simple eukaryotic cell. Such studies are expected to reveal the functions of many new genes that are not restricted to yeasts but are common to all eukaryotes, including humans.

The Genomes of Caenorhabditis elegans, Drosophila melanogaster, and Arabidopsis thaliana

The genomes of *C. elegans*, *Drosophila*, and *Arabidopsis* are intermediate in size and complexity between those of yeasts and humans. Distinctive features of each of these organisms make them important models for genome analysis: *Arabidopsis* provides a simple model for a plant genome, *C. elegans* is widely used for studies of animal development, and *Drosophila* has been especially well analyzed genetically. The genomes of these organisms, however, are about tenfold larger than those of yeasts, introducing a new order of difficulty in genome mapping and sequencing. Nonetheless, the complete genome sequences of *C. elegans* and *Drosophila* have been determined, and sequencing of *Arabidopsis* is rapidly progressing.

Determination of the sequence of *C. elegans* in 1998 extended genome sequencing from unicellular organisms (bacteria and yeast) to a multicellular organism recognized as an important model for animal development. The initial phases of analysis of the *C. elegans* genome used DNA fragments cloned in cosmids, which accommodate DNA inserts of approximately 40 kb (see Figure 3.23). This approach, however, was unable to cover the complete genome, which was accomplished by the cloning of much larger pieces of DNA in **yeast artificial chromosome (YAC)** vectors (Figure 4.23). The unique feature of YACs is that they contain centromeres and telomeres, allowing them to replicate as linear chromosome-like molecules in yeasts.

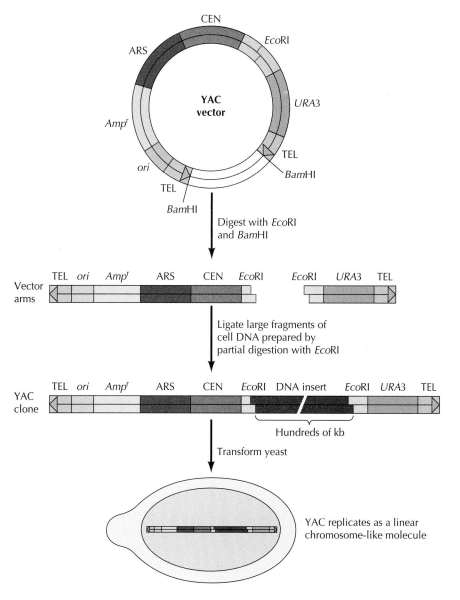

Figure 4.23
Cloning large DNA fragments in YACs   The vector contains a bacterial origin of replication (*ori*) and a gene conferring ampicillin resistance (*Amp*r), allowing it to be propagated in *E. coli*. In addition, it contains a yeast origin of replication (ARS), a centromere (CEN), a selectable marker allowing growth in the absence of uracil (*URA3*), and two telomeres (TEL). The vector is digested with *Eco*RI and *Bam*HI, yielding two vector arms terminating with telomeres and *Eco*RI sites. Large fragments of cell DNA prepared by partial digestion with *Eco*RI are ligated to the vector arms. Yeast cells are then transformed with the recombinant YACs, which replicate as linear chromosome-like molecules.

They can therefore be used to clone DNA fragments the size of yeast chromosomal DNAs—up to thousands of kilobases in length. The large DNA inserts that can be cloned in YACs are critically important for mapping complex genomes, as will be even more apparent in our following discussion of mapping genomes of still greater complexity, such as that of humans.

The *C. elegans* genome is 97×10^6 base pairs and contains 19,099 predicted protein-coding sequences—approximately three times the number of genes in yeast and one-fifth the number of genes predicted in humans. In contrast to the compact genome organization of yeast, genes in *C. elegans* span about 5 kilobases and contain an average of five introns. Protein-coding sequences thus account for only about 25% of the *C. elegans* genome, as compared to 70% of yeast and nearly 90% of bacterial genomes.

Approximately 40% of the predicted *C. elegans* proteins displayed significant similarity to known proteins of other organisms. As expected, there are substantially more similarities between the proteins of *C. elegans* and humans than between *C. elegans* and either yeast or bacteria. Proteins that are common between *C. elegans* and yeast may function in the basic cellular processes shared by these organisms, such as metabolism, DNA replication, transcription, translation, and protein sorting. These core biological processes appear to be carried out by a similar number of genes in both organisms, and it is likely that these genes will be shared by all eukaryotic cells. In contrast, the majority of *C. elegans* genes are not found in yeast and are likely to function in the more intricate regulatory activities required for the development of multicellular organisms. Elucidating the functions of these genes is likely to be particularly exciting in terms of understanding animal development. Although adult *C. elegans* are about 1 mm long and contain only 959 somatic cells in the entire body, they have all of the specialized cell types found in more complicated animals. Moreover, the complete pattern of cell divisions leading to *C. elegans* development has been described, including analysis of the connections made by all 302 neurons in the adult animal. Many of the genes involved in *C. elegans* development and differentiation have already been found to be related to genes involved in controlling the proliferation and differentiation of mammalian cells, substantiating the validity of *C. elegans* as a model for more complex animals. With little doubt, many more critical developmental control genes will be uncovered from studies of the *C. elegans* genomic sequence.

Drosophila is another key model for animal development, which has been particularly well-characterized genetically. The advantages of *Drosophila* for genetic analysis include its relatively simple genome and the fact that it can be easily maintained and bred in the laboratory. In addition, a special tool for genetic analysis in *Drosophila* is provided by the giant **polytene chromosomes** that are found in some tissues, such as the salivary glands of larvae. These chromosomes arise in nondividing cells as a consequence of repeated replication of DNA strands that fail to separate from each other. Thus, each polytene chromosome contains hundreds of identical DNA molecules aligned in parallel. Because of their size, these polytene chromosomes are visible in the light microscope, and appropriate staining procedures reveal a distinct banding pattern (Figure 4.24). The banding of polytene chromosomes provides a much greater degree of resolution than that achieved with metaphase chromosomes (e.g., see Figure 4.14). The polytene chromosomes are decondensed interphase chromosomes that contain actively expressed genes. More than 5000 bands are visible, each corresponding to an average length of approximately 20 kb of DNA. In contrast, the bands identified in human metaphase chromosomes contain several megabases of DNA.

Figure 4.24
Polytene chromosomes of *Drosophila*
A light micrograph of stained salivary gland chromosomes. The four chromosomes (X, 2, 3, and 4) are joined at their centromeres. (Peter J. Bryant/ Biological Photo Service.)

The banding pattern of polytene chromosomes thus provides a high-resolution physical map of the *Drosophila* genome. Gene deletions can often be correlated with the loss of a specific chromosomal band, thereby defining the physical location of the gene on the chromosome. In addition, cloned DNAs can be mapped by *in situ* hybridization to polytene chromosomes, often with sufficient resolution to localize cloned genes to specific bands (Figure 4.25). Thus, the map positions of cosmid or YAC clones (which span many bands) can readily be determined, providing the base for genomic sequence analysis.

Because of the power of *Drosophila* genetics, the complete sequencing of the *Drosophila* genome early in 2000 was a major milestone in genomic analysis. The genome of *Drosophila* consists of approximately 180×10^6 base pairs, of which about one-third is heterochromatin. The remaining 120×10^6 base pairs of the *Drosophila* genome contain approximately 13,600 genes, somewhat fewer than the number of genes in *C. elegans*. It is especially striking that a complex animal like *Drosophila* has only about twice the number of genes found in yeast, and the functional analysis of new genes that have been uncovered by sequencing the *Drosophila* genome will undoubtedly play a major role in understanding the molecular basis of animal development.

The complete sequences of *Arabidopsis thaliana* chromosomes 2 and 4 were reported in 1999. Together, these two chromosomes represent about 32% of the genome and contain approximately 7,800 genes. Some of these genes are related to genes of yeast and *C. elegans*, whereas others are novel and may represent genes that are specific for plants. The emerging sequence of the *Arabidopsis* genome is thus likely to reveal a total of about 20,000 genes, analysis of which can be expected to provide a new foundation for studies of plant biology.

Figure 4.25
***In situ* hybridization to a *Drosophila* polytene chromosome**   Hybridization of a YAC clone to a polytene chromosome is illustrated. The region of hybridization is indicated by an arrow. (Courtesy of Daniel L. Hartl, Harvard University.)

The Human Genome

The ultimate goal of genome analysis is to determine the complete nucleotide sequence of the human genome: 3×10^9 base pairs of DNA. To understand the magnitude of this undertaking (called the Human Genome Project), recall that the human genome is more than ten times larger than that of *Drosophila*; that the smallest human chromosome is several times larger than the entire yeast genome; and that the extended length of DNA that makes up the human genome is about 1 m long, whereas that of the fully sequenced bacterium *Haemophilus influenzae* is less than 1 mm long. Nonetheless, the sequencing of human DNA is well underway and its completion is expected in the near future. Identification of the estimated 100,000 human genes will not only open new horizons for research but will also have an immense impact on medicine by illuminating the genetic basis of many human diseases.

Human Gene Mapping

The human genome is distributed among 24 chromosomes (22 autosomes and the 2 sex chromosomes), each containing between 5×10^4 and 26×10^4 kb of DNA (Figure 4.26). The first stage of human gene mapping is the assignment of known genes to a chromosomal locus, usually defined as a metaphase chromosome band. Of the estimated 100,000 human genes, about 6000 genes of known function have been identified and positioned on the human chromosome map, using the basic approaches described in this section.

The first general method developed for localizing human genes to individual chromosomes or subchromosomal regions was **somatic cell hybridization** (Figure 4.27). Cells in culture can be induced to fuse with each other by treatment with certain viruses or chemical agents, such as polyethylene glycol. The nuclei in such cells also fuse, yielding hybrid cells that contain chromosomes derived from both parents. Such hybrids can be made not only between cells of the same species but also between cells of different species, such as human and mouse. Importantly, the human chromosomes are unstable in such human-mouse hybrids and are progressively lost during growth of the hybrid cells. It is therefore possible to derive hybrid cell lines that contain only one or a few human chromosomes. In some cases, hybrid cell lines containing only fragments of human chromosomes can also be obtained. The chromosomal locus of a human gene can then be determined by screening of a panel of somatic cell hybrids containing different sets of human chromosomes. For example, hybridization of a cloned human gene to DNAs extracted from such a panel of somatic cell hybrids can be used to map the cloned gene to a human chromosome (see Figure 4.27).

Higher-resolution chromosomal localization of cloned genes is possible with *in situ* hybridization to chromosomes, usually using fluorescent probes—a method generally referred to as **fluorescence *in situ* hybridization,** or **FISH**. *In situ* hybridization to metaphase chromosomes allows the mapping of a cloned gene to a locus defined by a chromosome band (Figure 4.28). Although FISH provides greater resolution than analysis of somatic cell hybrids, each band of metaphase chromosomes still contains thousands of kilobases of DNA. *In situ* hybridization to metaphase chromosomes therefore does not provide the detailed mapping information obtained by hybridization to the polytene chromosomes of *Drosophila*, which allows the localization of genes to interphase chromosome bands containing only 10 to 20 kb of DNA. Higher resolution can be obtained,

Figure 4.26
The human chromosomes   A schematic of human metaphase chromosomes
showing the banding pattern obtained after cytogenetic staining.

Figure 4.27
Somatic cell hybridization Human and mouse cells in culture are fused by treatment with a virus or chemical agent, yielding a hybrid cell containing both human and mouse chromosomes. (In this example only three human and two mouse chromosomes are shown.) Human chromosomes are unstable in such hybrids and are gradually lost during the outgrowth of individual clones of the hybrid cells. Therefore, a panel of hybrid cell lines is obtained, with each line containing a different complement of human chromosomes. Hybridization of a cloned gene to DNAs extracted from such a panel of hybrid cell lines can be used to map the gene to a specific human chromosome. In this illustration, a human probe hybridizes to DNAs of cell lines A and D, but not B and C. Since only hybrid cell lines A and D have retained chromosome 1, these results map the cloned gene to this human chromosome.

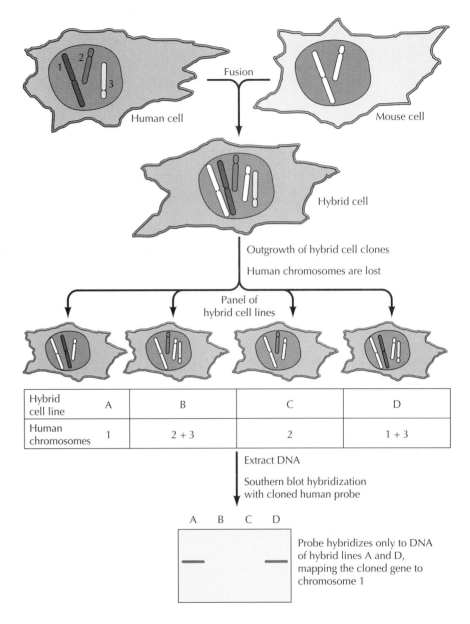

Hybrid cell line	A	B	C	D
Human chromosomes	1	2 + 3	2	1 + 3

Extract DNA

Southern blot hybridization with cloned human probe

Probe hybridizes only to DNA of hybrid lines A and D, mapping the cloned gene to chromosome 1

Figure 4.28
Fluorescence *in situ* hybridization A fluorescent probe for the gene encoding lamin B receptor is hybridized to stained human metaphase chromosomes (blue). Single gene hybridization signals are detected as red fluorescence. (Courtesy of K. L. Wydner and J. B. Lawrence, University of Massachusetts Medical Center.)

however, by hybridization to more extended human chromosomes from prometaphase or interphase cells, allowing the use of fluorescence *in situ* hybridization to map cloned genes to regions of about 100 kb.

Human genes can also be mapped by genetic linkage analysis, in which the distance between genes is estimated by recombination frequencies (see Figure 3.5). Genetic linkage was first used to map genes to the four *Drosophila* chromosomes and has subsequently been used to generate genetic maps of various other organisms, including *E. coli*, *S. cerevisiae*, *C. elegans*, and the mouse. However, construction of extensive genetic linkage maps generally requires the isolation of hundreds of single gene mutations, followed by extensive analysis of their inheritance in controlled breeding experiments. Although this approach applies to a variety of model organisms, it clearly cannot be applied directly to humans. This problem was overcome, however, by the realization that not only gene mutations but also any other detectable differences in DNA sequences between individuals can be used as genetic markers in human DNA. This approach greatly extended the range of available markers in the human genome and allowed analysis of recombination frequencies in small groups of families, leading to the development of a complete genetic map of the human genome.

The first sequences used as genetic markers were based on inherited differences in cleavage sites for restriction endonucleases. A single base change within a restriction endonuclease site is a readily detectable genetic marker because the mutated site is no longer cleaved by the enzyme in question. Two chromosomes that differ by such a mutation are then distinguishable on the basis of a **restriction fragment length polymorphism (RFLP)**, which arises because a particular cleavage site is present in only one of the two DNA molecules (Figure 4.29). A mutation that gives rise to an RFLP thus represents a genetic marker that can be detected by Southern blot hybridization of restriction endonuclease-digested DNA with an appropriate probe.

Many RFLPs have been identified simply by hybridization of cloned probes to Southern blots of restriction endonuclease-digested DNAs of a series of individuals. These RFLPs provide a large set of genetic markers that can be used to construct linkage maps by studies of a limited number

Figure 4.29
A restriction fragment length polymorphism (RFLP)   The region of DNA shown contains three cleavage sites for *Bam*HI in molecule I. In molecule II one of these sites has mutated from A to G, so it is no longer recognized by the restriction endonuclease. This mutation generates an RFLP, which can be detected by Southern blot hybridization of *Bam*HI-digested DNAs with a probe derived from the polymorphic region. In DNA I, the probe hybridizes to a 4-kb *Bam*HI fragment. In DNA II, however, the probe hybridizes to a 9-kb *Bam*HI fragment instead, because one of the *Bam*HI sites has been lost.

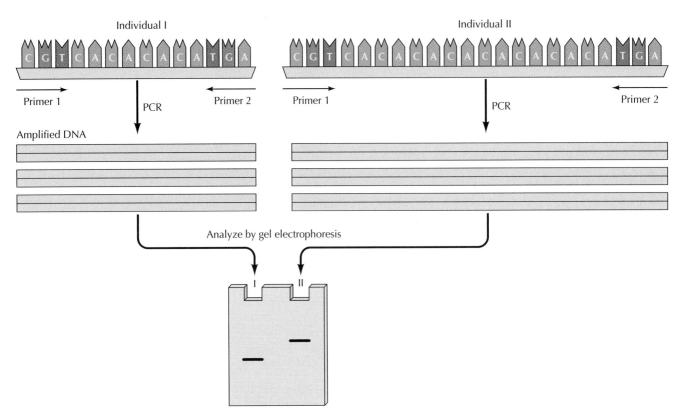

Figure 4.30
Detection of microsatellite markers by PCR   This microsatellite consists of tandem repeats of the dinucleotide CA. The number of repeats in such microsatellites varies with the individual (e.g., four repeats in individual I and eight repeats in individual II). These variations can be detected by PCR amplification with primers that flank the repeat region, since this process yields amplified DNA products of different lengths.

of families. For example, the first complete linkage map of the human genome was generated in 1987 by studies of the inheritance of 393 RFLPs in 21 families, each of which consisted of grandparents, parents, and children.

In addition to RFLPs, particularly useful genetic markers are tandem repeats of short nucleotide sequences (called **microsatellites**) distributed throughout the genome (Figure 4.30). For example, the human genome contains about 100,000 blocks of tandem repeats of the dinucleotide CA. Because the number of CA repeats at any given locus varies between individuals, differences in the number of tandem repeats can be used as genetic markers. Such variations can be easily scored by PCR using flanking-sequence primers, facilitating the use of microsatellites as genetic markers. The current linkage map of the human genome consists of more than 8000 loci (defined primarily by short tandem-repeat polymorphisms), which are spaced at an average distance of less than 1 Mb of DNA (Figure 4.31).

Once markers such as RFLPs and short tandem-repeat polymorphisms (STRPs) have been mapped, they become available for use as reference markers for mapping other genes. Thus, the map position of any human gene can be determined by analysis of its linkage to these polymorphic DNA sequences. The most important application of this approach has been the genetic mapping and eventual cloning of genes responsible for inherited human diseases. Such isolation of human disease genes provides

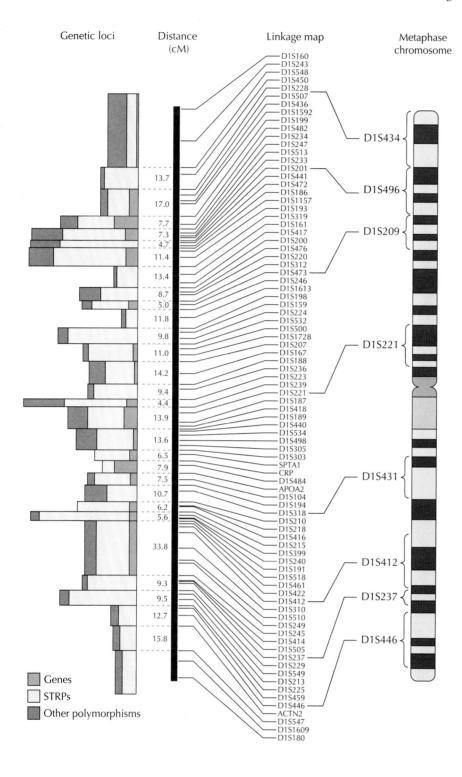

Genetic loci | Distance (cM) | Linkage map | Metaphase chromosome

Genes
STRPs
Other polymorphisms

Figure 4.31
Linkage map of human chromosome 1
The genetic linkage map is composed of loci containing short tandem-repeat polymorphisms (STRPs), with distances between markers indicated by frequencies of recombination (in centimorgans, or cM). One centimorgan corresponds to approximately 1 Mb of human DNA. Relationships between the genetic linkage map and metaphase chromosome bands are indicated to the right. The histogram to the left indicates the number of genes, STRPs, and other polymorphisms assigned to each map interval. (From Cooperative Human Linkage Center et al., 1994. Human Genetic Map: Genome Maps V. *Science* 265: 2055.)

much of the impetus for mapping and sequencing the human genome. The basic approach to gene mapping is to analyze a series of polymorphic markers (e.g., RFLPs) in families carrying the disease gene of interest (Figure 4.32). Genetic linkage between an RFLP and the disease gene is then indicated by coinheritance of the disease and the RFLP marker in affected family members. Given that the RFLP has been mapped to a chromosomal locus, this linkage analysis determines the map position of the disease gene on a human chromosome.

Figure 4.32
Linkage of a human disease gene to an RFLP In this example, the allele leading to development of the disease (*D*) is dominant over the normal wild-type (*WT*) allele, so individuals with one copy of *D* are affected (pink symbols). The disease gene is linked to an RFLP defined by two alleles (*A1* and *A2*) that are distinguishable by Southern blot hybridization. In the family shown, the affected parent is heterozygous for the RFLP (*A1/A2*), while the normal parent is homozygous at this locus (*A2/A2*). Two children inherit the mutant gene; one inherits the wild-type allele. RFLP analysis of DNAs then indicates that both the affected children, but not the normal child, have also inherited *A1* from their affected parent, demonstrating genetic linkage between *A1* and *D*.

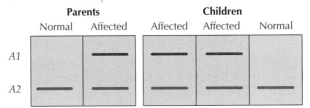

Genomic DNAs analyzed for *A1/A2* RFLP

Once the map position of a gene has been determined, that information can be used to isolate the gene as a molecular clone—a strategy known as **positional cloning**. This approach has been used to clone the genes for several human diseases; the gene for cystic fibrosis was one of the first examples. Cystic fibrosis is an inherited disease that affects approximately one in 2500 newborn Caucasians, although it is rare in other races. The disease is diagnosed in childhood and is inevitably fatal by about age 30. Most patients die from respiratory complications. The cystic fibrosis gene was first mapped by linkage to an RFLP, which was localized to chromosome 7 (Figure 4.33). Further mapping studies established that the cystic fibrosis gene is located between two RFLPs separated by approximately 1 Mb. Molecular clones from this region of DNA were isolated by the use of RFLPs as probes. These clones were then used as probes to isolate clones containing adjacent DNA sequences, leading to the isolation of a series of clones covering the region containing the cystic fibrosis gene. Analysis of these ge-

Figure 4.33
The cystic fibrosis gene Linkage to the MET and J3.11 RFLPs mapped the gene to the indicated region of chromosome 7, which was further delineated by an additional RFLP designated KM.19. The cystic fibrosis gene, which spans approximately 250 kb, was then isolated by positional cloning.

nomic clones, followed by cDNA cloning and sequencing, eventually identified a candidate gene that spans more than 250 kb of DNA and is transcribed to yield a 6-kb cDNA containing 24 exons. DNA sequencing confirmed that this gene is the cystic fibrosis gene by demonstrating that it is mutated in affected individuals.

The isolation of human disease genes has important implications not only for understanding the molecular and cellular basis of the disease, but also for prevention and treatment. An immediate benefit of cloning the gene responsible for an inherited disease is the potential for developing sensitive diagnostic tests for mutant alleles of the gene, such as the detection of mutations by PCR. These tests can identify individuals at risk for the disease and, at least for some diseases, enable appropriate preventive measures to be instituted before the disease develops. In addition, the ability to detect mutant alleles of disease genes opens the possibility of prenatal diagnosis to prevent transmission of the disease to future generations. Couples carrying mutant alleles of the gene can be identified, and their pregnancies monitored to determine if the mutant gene has been transmitted to the fetus. Because of the sensitivity of PCR analysis, mutations can be detected even in early embryos that have been derived by *in vitro* fertilization procedures prior to their return to the uterus.

In addition to these diagnostic measures, the cloning of disease genes suggests the potential use of **gene therapy** to correct the defect. The general strategy of gene therapy is to introduce a normal cDNA allele into the affected cells of patients. In some cases, genes can be introduced into target cells (e.g., lymphocytes) that can be maintained in culture and then returned to the patient. Alternatively, appropriate vectors can be used to introduce genes into some types of cells *in vivo*. For example, gene therapy trials for cystic fibrosis are evaluating the use of aerosols to deliver viral vectors to the affected epithelial cells lining the airways of the respiratory tract. Gene therapy is still in its early, experimental stages of development. However, the progress thus far clearly illustrates the general feasibility of the approach, suggesting that gene therapy will eventually become an important method of treating at least some human diseases.

Physical Mapping of the Human Genome

A physical map of cloned DNA segments is a necessary intermediate between the genetic map and the nucleotide sequence of the human genome. Because of the large size of the genome, YACs provide the obvious vector for physical mapping of cloned human DNA. The most useful markers for analyzing these clones appear to be **sequence-tagged sites (STSs)**, which are readily detected by PCR analysis. STSs are short segments (usually 200 to 500 base pairs) of nonrepetitive DNA whose nucleotide sequences have been determined. Many STSs, each located at a unique site in the genome, can thus be defined. Since an STS corresponds to a segment of sequenced DNA, it is detectable in any sample of DNA by PCR amplification using the appropriate primers (Figure 4.34). This ease of detection provides a considerable advantage over other methods, such as nucleic acid hybridization, for large-scale analysis of cloned DNAs. Moreover, since STSs are obtained directly from sequence data, they are available for any investigator to use; laboratories need not exchange cloned DNAs. STSs thus provide an accessible common language for genome mapping and sequencing.

An alternative approach to analyzing the entire genome has been to focus on protein-coding genes by sequencing cDNAs. Assuming that only about 3% of the genome corresponds to protein-coding sequences, this task is obviously much more modest than sequencing the entire genome. The

Gene Therapy for Adenosine Deaminase Deficiency

The Disease

Adenosine deaminase deficiency is one of the causes of severe combined immunodeficiency (SCID), a group of hereditary diseases in which lymphocytes fail to develop normally. Affected individuals are deficient in all functions of the immune system. SCID is rare, with a frequency of about one in a million. Because of the profound defect in immune function, infants with this disease are susceptible to recurrent infections by various organisms, including viruses, bacteria, fungi, and protozoans. Unless treated, these infections are usually fatal within two years.

Molecular and Cellular Basis

About 20% of SCID cases result from genetic deficiencies in the enzyme adenosine deaminase, which catalyzes the deamination of adenosine and deoxyadenosine. In the absence of adenosine deaminase, deoxyadenosine accumulates, resulting in corresponding increases in the concentration of deoxyadenosine triphosphate (dATP). High levels of dATP are toxic to proliferating cells because dATP inhibits the enzyme ribonucleotide reductase, which is required for synthesis of all four deoxyribonucleoside triphosphates. Consequently, increased concentrations of dATP block DNA synthesis. Although all cells have adenosine deaminase, a deficiency in its activity is particularly toxic to lymphocytes, because they lack what other types of cells have: additional enzymes that degrade dAMP, thereby preventing dATP accumulation. Thus, a deficiency in adenosine deaminase specifically blocks lymphocyte replication and prevents the development of a functional immune system.

Deoxyadenosine

Deoxyinosine

Adenosine deaminase

dAMP

dATP

Inhibition of ribonucleotide reductase and DNA synthesis

Metabolism of deoxyadenosine.

Prevention and Treatment

Some infants with adenosine deaminase deficiency can be cured by bone marrow transplantation, which is successful when compatible donor marrow can be obtained from an antigenically matched sibling. In other cases, the disease can be treated by enzyme replacement therapy, in which functional adenosine deaminase is administered to the patient by intravenous injection. Although this treatment is beneficial, it is not curative. Thus, adenosine deaminase deficiency represents a prime target for gene therapy.

In 1990, following preliminary studies in animals, the first clinical trial of gene therapy was initiated with the treatment of two children suffering from this disease. Lymphocytes obtained from the blood of these children were stimulated to proliferate in culture. A functional adenosine deaminase cDNA in a retroviral vector was then introduced into these lymphocytes, which were subsequently returned to the patients. These treatments were repeated 11 or 12 times over a two-year period. The results of this trial demonstrated that the vector was successfully transferred to the lymphocytes and continued to be maintained after treatment. Moreover, one of the patients showed substantial

Gene Therapy for Adenosine Deaminase Deficiency (continued)

improvement and makes about 25% of the normal level of adenosine deaminase in her lymphocytes. In the other patient, however, the infected lymphocytes do not produce enough adenosine deaminase to be clinically significant. The results of this initial trial thus support the general efficacy of gene therapy as a treatment for this disease, but they also demonstrate that more efficient vectors are needed to increase the numbers of infected cells and the levels of production of the introduced gene.

References

Blaese, R. M. and 18 others. 1995. T lymphocyte-directed gene therapy for ADA⁻ SCID: Initial results after 4 years. *Science* 270: 475–480.

Verma, I. M. and N. Somia. 1997. Gene therapy—promises, problems and prospects. *Nature* 389: 239–242.

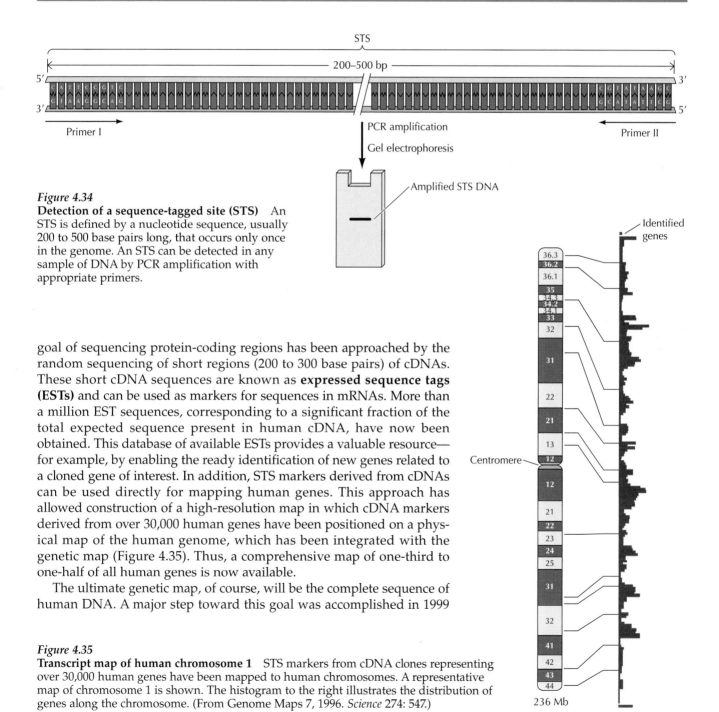

Figure 4.34
Detection of a sequence-tagged site (STS) An STS is defined by a nucleotide sequence, usually 200 to 500 base pairs long, that occurs only once in the genome. An STS can be detected in any sample of DNA by PCR amplification with appropriate primers.

goal of sequencing protein-coding regions has been approached by the random sequencing of short regions (200 to 300 base pairs) of cDNAs. These short cDNA sequences are known as **expressed sequence tags (ESTs)** and can be used as markers for sequences in mRNAs. More than a million EST sequences, corresponding to a significant fraction of the total expected sequence present in human cDNA, have now been obtained. This database of available ESTs provides a valuable resource— for example, by enabling the ready identification of new genes related to a cloned gene of interest. In addition, STS markers derived from cDNAs can be used directly for mapping human genes. This approach has allowed construction of a high-resolution map in which cDNA markers derived from over 30,000 human genes have been positioned on a physical map of the human genome, which has been integrated with the genetic map (Figure 4.35). Thus, a comprehensive map of one-third to one-half of all human genes is now available.

The ultimate genetic map, of course, will be the complete sequence of human DNA. A major step toward this goal was accomplished in 1999

Figure 4.35
Transcript map of human chromosome 1 STS markers from cDNA clones representing over 30,000 human genes have been mapped to human chromosomes. A representative map of chromosome 1 is shown. The histogram to the right illustrates the distribution of genes along the chromosome. (From Genome Maps 7, 1996. *Science* 274: 547.)

Figure 4.36
DNA sequence of human chromosome 22 The DNA sequence of the entire long arm of chromosome 22 has been determined. The positions of genes that are transcribed in either the rightwards (+) or leftwards (−) directions are indicated for a portion of the chromosome. (From I. Dunham et al., 1999. *Nature* 402: 489.)

with the sequencing of human chromosome 22. Chromosome 22 is one of the smallest human chromosomes, containing approximately 33.4×10^6 base pairs of DNA. Analysis of the sequence indicates that chromosome 22 contains at least 545 genes (Figure 4.36), although this is likely to be an underestimate because the presence of extensive introns makes it more difficult to identify genes in human DNA than in the DNA of lower eukaryotes. Thus, the sequence of chromosome 22 is consistent with the estimate of approximately 100,000 genes in the human genome, the complete sequence of which will be available in the near future.

KEY TERMS

gene, spacer sequence, exon, intron, RNA splicing, kilobase (kb)

gene family, pseudogene

simple-sequence DNA, satellite DNA, SINE, LINE, *Alu* sequence

chromatin, histone, nucleosome, nucleosome core particle, chromatosome, euchromatin, heterochromatin

SUMMARY

THE COMPLEXITY OF EUKARYOTIC GENOMES

Introns and Exons: Most eukaryotic genes have a split structure in which segments of coding sequence (exons) are interrupted by noncoding sequences (introns). In complex eukaryotes, introns account for about ten times more DNA than do exons.

Gene Families and Pseudogenes: Many eukaryotic genes are present in multiple copies, called gene families. Some members of gene families function in different tissues or at different stages of development. Other members of gene families (pseudogenes) have been inactivated by mutations and no longer represent functional genes.

Repetitive DNA Sequences: Approximately 40% of mammalian DNA consists of highly repetitive DNA sequences, some of which are present in 10^5 to 10^6 copies per genome.

The Number of Genes in Eukaryotic Cells: Only a small fraction of the genome in complex eukaryotes corresponds to protein-coding sequences. The human genome is estimated to contain about 100,000 genes.

CHROMOSOMES AND CHROMATIN

Chromatin: The DNA of eukaryotic cells is wrapped around histones to form nucleosomes. Chromatin can be further compacted by the folding of nucleosomes into higher-order structures, including the highly condensed metaphase chromosomes of cells undergoing mitosis.

Centromeres: Centromeres are specialized regions of eukaryotic chromosomes that serve as the sites of association between sister chromatids and the sites of spindle fiber attachment during mitosis.

centromere, kinetochore

Telomeres: Telomeres are specialized sequences required to maintain the ends of eukaryotic chromosomes.

telomere

THE SEQUENCES OF COMPLETE GENOMES

Prokaryotic Genomes: The genomes of nearly two dozen different bacteria, including *E. coli*, have been completely sequenced. The *E. coli* genome contains 4288 genes, with protein-coding sequences accounting for nearly 90% of the DNA.

megabase (Mb), open-reading frame

Sequencing the Yeast Genome: The first eukaryotic genome to be sequenced was that of *S. cerevisiae*. The yeast genome contains about 6000 genes, and protein-coding sequences account for approximately 70% of the genome.

The Genomes of **Caenorhabditis elegans, Drosophila melanogaster,** *and* **Arabidopsis thaliana:** The genome of *C. elegans* is the first sequenced genome of a multicellular organism. The *C. elegans* genome contains about 19,000 protein-coding sequences, which account for only about 25% of the genome. The majority of *C. elegans* genes are not found in yeast and are likely to function in the regulation and development of multicellular organisms. The genome of *Drosophila* contains approximately 13,600 genes, with protein-coding sequences accounting for about 13% of the genome. Genomic sequencing of *Arabidopsis* is actively ongoing and will yield the sequence of a key model for studies of plant biology.

yeast artificial chromosome (YAC), polytene chromosome

THE HUMAN GENOME

Human Gene Mapping: Human genes can be mapped by somatic cell hybridization, *in situ* hybridization, and genetic linkage analysis. The mapping of some human disease genes has allowed their isolation by positional cloning.

somatic cell hybridization, fluorescence *in situ* hybridization (FISH), restriction fragment length polymorphism (RFLP), microsatellite, positional cloning, gene therapy

Physical Mapping of the Human Genome: YAC clones have been used to construct maps of the human genome. The random sequencing of cDNA clones has provided markers for sequences present in mRNAs. Markers of cDNAs from over 30,000 human genes have been used to construct an integrated physical and genetic map of the human genome. This map provides a key resource for genomic analysis until the human DNA sequence is completed.

sequence-tagged site (STS), expressed sequence tag (EST)

Questions

1. Repetitive DNA sequences were first identified by studies of rates of DNA reassociation. What relative rates of reassociation are expected for sequences repeated 1000 times in the genome compared to genes with only a single copy?

2. How many histone molecules are associated with the chromosomes of a human cell? Assuming that a typical cell contains about 1 ng of protein, what fraction of total cellular protein corresponds to histones?

3. Why are YAC vectors useful for analysis of complex genomes? What is the role of telomeres in these vectors?

4. Assuming that yeast and *C. elegans* utilize a similar number of proteins to carry out basic cellular functions, what fraction of *C. elegans* genes are expected to encode proteins involved in aspects of cell regulation specific to multicellular organisms?

5. Approximately 30,000 cDNAs have been localized on the human genome map. What is the average distance between these markers?

References and Further Reading

General References

Hartl, D. L. and E. W. Jones. 1998. *Genetics: Principles and Analysis.* 4th ed. Sudbury, MA: Jones and Bartlett Publishers.

Pasternak, J. J. 1999. *An Introduction to Human Genetics.* Bethesda, MD: Fitzgerald Science Press.

Watson, J. D., M. Gilman, J. Witkowski, and M. Zoller. 1992. *Recombinant DNA.* 2nd ed. New York: Scientific American Books.

The Complexity of Eukaryotic Genomes

Berget, S. M., C. Moore and P. A. Sharp. 1977. Spliced segments at the 5′ terminus of adenovirus 2 late mRNA. *Proc. Natl. Acad. Sci. USA* 74: 3171–3175. [P]

Breathnach, R., J. L. Mandel and P. Chambon. 1977. Ovalbumin gene is split in chicken DNA. *Nature* 270: 314–319. [P]

Britten, R. J. and D. E. Kohne. 1968. Repeated sequences in DNA. *Science* 161: 529–540. [P]

Charlesworth, B., P. Sniegowski and W. Stephan. 1994. The evolutionary dynamics of repetitive DNA in eukaryotes. *Nature* 371: 215–220. [R]

Chow, L. T., R. E. Gelinas, T. R. Broker and R. J. Roberts. 1977. An amazing sequence arrangement at the 5′ ends of adenovirus 2 messenger RNA. *Cell* 12: 1–8. [P]

Fritsch, E. F., R. M. Lawn and T. Maniatis. 1980. Molecular cloning and characterization of the human β-like globin gene cluster. *Cell* 19: 959–972. [P]

Gilbert, W. 1985. Genes-in-pieces revisited. *Science* 228: 823–824. [R]

Gilbert, W., S. J. de Souza and M. Long. 1997. Origin of genes. *Proc. Natl. Acad. Sci. USA* 94: 7698–7703. [R].

Henikoff, S., E. A. Greene, S. Pietrokovski, P. Bork, T. K. Attwood and L. Hood. 1997. Gene families: The taxonomy of protein paralogs and chimeras. *Science* 278: 609–614. [R]

Hentschel, C. C. and M. L. Birnsteil. 1981. The organization and expression of histone gene families. *Cell* 25: 301–313. [R]

Kidwell, M. G. and D. Lisch. 1997. Transposable elements as sources of variation in animals and plants. *Proc. Natl. Acad. Sci. USA* 94: 7704–7711. [R]

Little, P. F. R. 1982. Globin pseudogenes. *Cell* 28: 683–684. [R]

Moran, J. V., R. J. DeBerardinis and H. H. Kazazian Jr. 1999. Exon shuffling by L1 retrotransposition. *Science* 283: 1530–1534. [P]

Schmid, C. W. and W. R. Jelinek. 1982. The *Alu* family of dispersed repetitive sequences. *Science* 216: 1065–1070. [R]

Singer, M. F. 1982. SINEs and LINEs: Highly repeated short and long interspersed sequences in mammalian genomes. *Cell* 28: 433–434. [R]

Singer, M. F. and J. Skowronski. 1985. Making sense out of LINEs: Long interspersed repeat sequences in mammalian genomes. *Trends Biochem. Sci.* 10: 119–122. [R]

Stoltzfus, A., D. F. Spencer, M. Zuker, J. M. Logsdon, Jr. and W. F. Doolittle. 1994. Testing the exon theory of genes: The evidence from protein structure. *Science* 265: 202–207. [R]

Tilghman, S. M., P. J. Curtis, D. C. Tiemeier, P. Leder and C. Weissmann. 1978. The intervening sequence of a mouse β-globin gene is transcribed within the 15S β-globin mRNA precursor. *Proc. Natl. Acad. Sci. USA* 75: 1309–1313. [P]

Chromosomes and Chromatin

Baum, M., V. K. Ngan and L. Clarke. 1994. The centromeric K-type repeat and the central core are together sufficient to establish a functional *Schizosaccharomyces pombe* centromere. *Mol. Biol. Cell* 5: 747–761. [P]

Blackburn, E. H. 1991. Structure and function of telomeres. *Nature* 350: 569–573. [R]

Bloom, K. 1993. The centromere frontier: Kinetochore components, microtubule-based motility, and the CEN-value paradox. *Cell* 73: 621–624. [R]

Carbon, J. 1984. Yeast centromeres: Structure and function. *Cell* 37: 351–353. [R]

Clarke, L. 1990. Centromeres of budding and fission yeasts. *Trends Genet.* 6: 150–154. [R]

Clarke, L. and J. Carbon. 1980. Isolation of a yeast centromere and construction of functional small circular chromosomes. *Nature* 287: 504–509. [P]

Comings, D. E. 1978. Mechanisms of chromosome banding and implications for chromosome structure. *Ann. Rev. Genet.* 12: 25–46. [R]

Greider, C. W. 1996. Telomere length regulation. *Ann. Rev. Biochem.* 65: 337–365. [R]

Greider, C. W. 1999. Telomeres do D-loop-T-loop. *Cell* 97: 419–422. [R]

Grunstein, M. 1992. Histones as regulators of genes. *Sci. Am.* 267(4): 68–74B. [R]

Heck, M. M. S. 1997. Condensins, cohesins, and chromosome architecture: How to make and break a mitotic chromosome. *Cell* 91: 5–8. [R]

Kornberg, R. D. 1974. Chromatin structure: A repeating unit of histones and DNA. *Science* 184: 868–871. [P]

Koshland, D. and A. Strunnikov. 1996. Mitotic chromosome condensation. *Ann. Rev. Cell Biol.* 12: 305–333. [R]

Luger, K., A. W. Mader, R. K. Richmond, D. F. Sargent and T. J. Richmond. 1997. Crystal structure of the nucleosome core particle at 2.8 Å resolution. *Nature* 389: 251–260. [P]

Murphy, T. D. and G. H. Karpen. 1998. Centromeres take flight: Alpha satellite and the quest for the human centromere. *Cell* 93: 317–320. [R]

Pardue, M. L. and J. G. Gall. 1970. Chromosomal localization of mouse satellite DNA. *Science* 168: 1356–1358. [P]

Paulson, J. R. and U. K. Laemmli. 1977. The structure of histone-depleted metaphase chromosomes. *Cell* 12: 817–828. [P]

Pluta, A. F., A. M. MacKay, A. M. Ainsztein, A. G. Goldberg and W. C. Earnshaw. 1995. The centromere: Hub of chromosomal activities. *Science* 270: 1591–1594. [R]

Reeve, J. N., K. Sandman and C. J. Daniels. 1997. Archael histones, nucleosomes, and transcription initiation. *Cell* 89: 999–1002. [R]

Richmond, T. J., J. T. Finch, B. Rushton, D. Rhodes and A. Klug. 1984. Structure of the nucleosome core particle at 7 Å resolution. *Nature* 311: 532–537. [R]

Saitoh, Y. and U. K. Laemmli. 1994. Metaphase chromosome structure: Bands arise from a differential folding path of the highly AT-rich scaffold. *Cell* 76: 609–622. [P]

Schulman, I. and K. S. Bloom. 1991. Centromeres: An integrated protein/DNA complex required for chromosome movement. *Ann. Rev. Cell Biol.* 7: 311–336. [R]

Sun, X., J. Wahlstrom and G. Karpen. 1997. Molecular structure of a functional *Drosophila* centromere. *Cell* 91: 1007–1019. [P]

Szostak, J. W. and E. H. Blackburn. 1982. Cloning yeast telomeres on linear plasmid vectors. *Cell* 29: 245–255. [P]

Van Holde, K. E. and J. Zlatanovai. 1995. Chromatin higher order structure: Chasing a mirage. *J. Biol. Chem.* 270: 8373–8376. [R]

Wiens, G. R. and P. K. Sorger. 1998. Centromeric chromatin and epigenetic effects in kinetochore assembly. *Cell* 93: 313–316. [R]

Willard, H. F. 1990. Centromeres of mammalian chromosomes. *Trends Genet.* 6: 410–416. [R]

Zakian, V. A. 1995 Telomeres: Beginning to understand the end. *Science* 270: 1601–1607. [R]

The Sequences of Complete Genomes

Adams, M. D. and 194 others. 2000. The genome sequence of *Drosophila melanogaster*. *Science* 287: 2185–2195. [P]

Blattner, F.R., G. Plunkett III., C. A. Bloch, N. T. Perna, V. Burland, M. Riley, J. Collado-Vides, J. D. Glasner, C. K. Rode, G. F. Mayhew, J. Gregor, N. W. Davis, H. A. Kirkpatrick, M. A. Goeden, D. J. Rose, B. Mau and Y. Shao. 1997. The complete genome sequence of *Escherichia coli* K-12. *Science* 277: 1453–1462. [P]

Bult, C. J. and 39 others. 1996. Complete genome sequence of the methanogenic Archaeon, *Methanococcus jannaschii*. *Science* 273: 1058–1073. [P]

Burke, D. T., G. F. Carle and M. V. Olson. 1987. Cloning of large segments of exogenous DNA into yeast by means of artificial chromosome vectors. *Science* 236: 806–812. [P]

Chervitz, S. A., L. Aravind, G. Sherlock, C. A. Ball, E. V. Koonin, S. S. Dwight, M. A. Harris, K. Dolinski, S. Mohr, T. Smith, S. Weng, J. M. Cherry and D. Botstein. 1998. Comparison of the complete protein sets of worm and yeast: Orthology and divergence. *Science* 282: 2022–2028. [R]

Fleischmann, R. D., and 39 others. 1995. Whole-genome random sequencing and assembly of *Haemophilus influenzae* Rd. *Science* 269: 496–512. [P]

Fraser, C. M. and 28 others. 1995. The minimal gene complement of *Mycoplasma genitalium*. *Science* 270: 397–403. [P]

Garza, D., J. W. Ajioka, D. T. Burke and D. L. Hartl. 1989. Mapping the *Drosophila* genome with yeast artificial chromosomes. *Science* 246: 641–646. [R]

Goffeau, A. and 15 others. 1996. Life with 6000 genes. *Science* 274: 546–567. [P]

Lin, X. and 36 others. 1999. Sequence and analysis of chromosome 2 of the plant *Arabidopsis thaliana*. *Nature* 402: 761–768. [P]

Meinke, D. W., J. M. Cherry, C. Dean, S. D. Rounsley and M. Koornneef. 1998. *Arabidopsis thaliana*: A model plant for genome analysis. *Science* 282: 662–682. [R]

Miklos, G. L. G. and G. M. Rubin. 1996. The role of the genome project in determining gene function: Insights from model organisms. *Cell* 86: 521–529. [R]

Oliver, S. G. and 146 others. 1992. The complete DNA sequence of yeast chromosome III. *Nature* 357: 38–46. [P]

Rubin, G. M. and 54 others. 2000. Comparative genomics of the eukaryotes. *Science* 287: 2204–2215. [R]

The *C. elegans* Sequencing Consortium. 1998. Genome sequence of the nematode *C. elegans*: A platform for investigating biology. *Science* 282: 2012–2018. [P]

The European Union *Arabidopsis* Genome Sequencing Consortium and The Cold Spring Harbor, Washington University in St. Louis and PE Biosystems *Arabidopsis* Sequencing Consortium. 1999. Sequence and analysis of chromosome 4 of the plant *Arabidopsis thaliana*. *Nature* 402: 769–776. [P]

The Human Genome

Adams, M. D., J. M. Kelley, J. D. Gocayne, M. Dubnick, M. H. Polymeropoulos, H. Xiao, C. R. Merril, A. Wu, B. Olde, R. F. Moreno, A. R. Kerlavage, W. R. McCombie and J. C. Venter. 1991. Complementary DNA sequencing: Expressed sequence tags and human genome project. *Science* 252: 1651–1656. [R]

Adams, M. D. and 85 others. 1995. Initial assessment of human gene diversity and expression patterns based upon 83 million nucleotides of cDNA sequence. *Nature* 377 [Suppl]: 3–174. [P]

Botstein, D., R. L. White, M. Skolnick and R. W. Davis. 1980. Construction of a genetic linkage map in man using restriction fragment length polymorphism. *Am. J. Hum. Genet.* 32: 314–331. [P]

Collins, F. S. 1992. Cystic fibrosis: Molecular biology and therapeutic implications. *Science* 256: 774–779. [R]

Collins, F. S., A. Patrinos, E. Jordan, A. Chakravarti, R. Gesteland, L. Walters and the members of the DOE and NIH planning groups. 1998. New goals for the U.S. human genome project: 1998–2003. *Science* 282: 682–689. [R]

Cooperative Human Linkage Center. 1994. A comprehensive human linkage map with centimorgan density. *Science* 265: 2049–2054. [P]

Crystal, R. G. 1995. Transfer of genes to humans: Early lessons and obstacles to success. *Science* 270: 404–410. [R]

Deloukas, P. and 64 others. 1998. A physical map of 30,000 human genes. *Science* 282: 744–746. [P]

Donis-Keller, H. and 32 others. 1987. A genetic linkage map of the human genome. *Cell* 51: 319–337. [P]

Dunham, I. and 216 others. 1999. The DNA sequence of human chromosome 22. *Nature* 402: 489–495. [P]

Nakamura, Y., M. Leppert, P. O'Connell, R. Wolff, T. Holm, M. Culver, C. Martin, E. Fujimoto, M. Hoff, E. Kumlin and R. White. 1987. Variable number of tandem repeat (VNTR) markers for human gene mapping. *Science* 235: 1616–1622. [P]

Olson, M., L. Hood, C. Cantor and D. Botstein. 1989. A common language for physical mapping of the human genome. *Science* 245: 1434–1435. [R]

Rommens, J. M., M. C. Iannuzzi, B.-S. Kerem, M. L. Drumm, G. Melmer, M. Dean, R. Rozmahel, J. L. Cole, D. Kennedy, N. Hidaka, M. Zsiga, M. Buchwald, J. R. Riordan, L.-C. Tsui and F. S. Collins. 1989. Identification of the cystic fibrosis gene: Chromosome walking and jumping. *Science* 245: 1059–1065. [P]

Ruddle, F. H. 1981. A new era in mammalian gene mapping: Somatic cell genetics and recombinant DNA methodologies. *Nature* 294: 115–120. [R]

Verma, I. M. and N. Somia. 1997. Gene therapy—promises, problems and prospects. *Nature* 389: 239–242. [R]

Weiss, M. C. and H. Green. 1967. Humanmouse hybrid cell lines containing partial complements of human chromosomes and functioning human genes. *Proc. Natl. Acad. Sci. USA* 58: 1104–1111. [P]

5

Replication, Maintenance, and Rearrangements of Genomic DNA

T HE FUNDAMENTAL BIOLOGICAL PROCESS OF REPRODUCTION requires the faithful transmission of genetic information from parent to offspring. Thus, the accurate replication of genomic DNA is essential to the lives of all cells and organisms. Each time a cell divides, its entire genome must be duplicated, and complex enzymatic machinery is required to copy the large DNA molecules that make up both prokaryotic and eukaryotic chromosomes. In addition, cells have evolved mechanisms to correct mistakes that sometimes occur during DNA replication and to repair DNA damage that can result from the action of environmental agents, such as radiation. Abnormalities in these processes result in a failure of accurate replication and maintenance of genomic DNA—a failure that can have disastrous consequences, such as the development of cancer.

Despite the importance of accurate DNA replication and maintenance, cell genomes are far from static. In order for species to evolve, mutations and gene rearrangements are needed to maintain genetic variation between individuals. Recombination between homologous chromosomes during meiosis plays an important role in this process by allowing parental genes to be rearranged into new combinations in the next generation. Rearrangements of DNA sequences within the genome are also thought to contribute to evolution by creating novel combinations of genetic information. In addition, some DNA rearrangements are programmed to regulate gene expression during the differentiation and development of individual cells and organisms. In humans, a prominent example is the rearrangement of antibody genes during development of the immune system. A careful balance between maintenance and variation of genetic information is thus critical to the development of individual organisms as well as to evolution of the species.

DNA Replication

As discussed in Chapter 3, DNA replication is a semiconservative process in which each parental strand serves as a template for the synthesis of a new complementary daughter strand. The central enzyme involved is DNA polymerase, which catalyzes the joining of deoxyribonucleoside

5'-triphosphates (dNTPs) to form the growing DNA chain. However, DNA replication is much more complex than a single enzymatic reaction. Other proteins are involved, and proofreading mechanisms are required to ensure that the accuracy of replication is compatible with the low frequency of errors that is needed for cell reproduction. Additional proteins and specific DNA sequences are also needed both to initiate replication and to copy the ends of eukaryotic chromosomes.

DNA Polymerases

DNA polymerase was first identified in lysates of *E. coli* by Arthur Kornberg in 1956. The ability of this enzyme to accurately copy a DNA template provided a biochemical basis for the mode of DNA replication that was initially proposed by Watson and Crick, so its isolation represented a landmark discovery in molecular biology. Ironically, however, this first DNA polymerase to be identified (now called DNA polymerase I) is not the major enzyme responsible for *E. coli* DNA replication. Instead, it is now clear that both prokaryotic and eukaryotic cells contain several different DNA polymerases that play distinct roles in the replication and repair of DNA.

The multiplicity of DNA polymerases was first revealed by the isolation of a mutant strain of *E. coli* that was deficient in polymerase I (Figure 5.1). Cultures of *E. coli* were treated with a chemical (a **mutagen**) that induces a high frequency of mutations, and individual bacterial colonies were isolated and screened to identify a mutant strain lacking polymerase I. Analysis of a few thousand colonies led to the isolation of the desired mutant, which was almost totally defective in polymerase I activity. Surprisingly, the mutant bacteria grew normally, leading to the conclusion that polymerase I is not required for DNA replication. On the other hand, the mutant bacteria were extremely sensitive to agents that damage DNA (e.g., ultraviolet light), suggesting that polymerase I is involved primarily in the repair of DNA damage rather than in DNA replication per se.

The conclusion that polymerase I is not required for replication implied that *E. coli* must contain other DNA polymerases, and subsequent experiments led to the identification of two such enzymes, now called DNA polymerases II and III. The potential roles of these enzymes were investigated by the isolation of appropriate mutants. Strains of *E. coli* with mutations in polymerase II were found to grow and otherwise behave normally, so the role of this enzyme in the cell is unknown. Temperature-sensitive polymerase III mutants, however, were unable to replicate their DNA at high temperature, and subsequent studies have confirmed that polymerase III is the major replicative enzyme in *E. coli*.

It is now known that, in addition to polymerase III, polymerase I is also required for replication of *E. coli* DNA. The original polymerase I mutant was not completely defective in that enzyme, and later experiments showed that the residual polymerase I activity in this strain plays a key role in the replication process. The replication of *E. coli* DNA thus involves two distinct DNA polymerases, the specific roles of which are discussed below.

Eukaryotic cells contain five DNA polymerases: α, β, γ, δ, and ε. Polymerase γ is located in mitochondria and is responsible for replication of mitochondrial DNA. The other four enzymes are located in the nucleus and are therefore candidates for involvement in nuclear DNA replication. Polymerases α, δ, and ε are most active in dividing cells, suggesting that they function in replication. In contrast, polymerase β is active in nondividing and dividing cells, suggesting that it may function primarily in the repair of DNA damage.

Figure 5.1
Isolation of a mutant deficient in polymerase I   A culture of *E. coli* was treated with a chemical mutagen, and individual bacterial colonies were isolated by growth on semisolid medium. Several thousand colonies were then cultured and screened to identify a mutant lacking polymerase I.

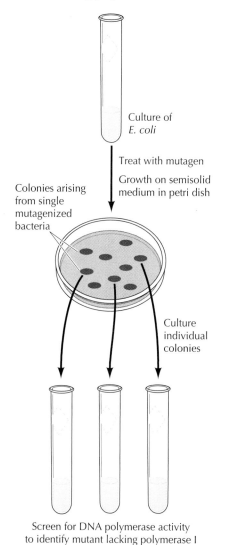

Culture of *E. coli*

Treat with mutagen
Growth on semisolid medium in petri dish

Colonies arising from single mutagenized bacteria

Culture individual colonies

Screen for DNA polymerase activity to identify mutant lacking polymerase I

Two types of experiments have provided further evidence addressing the roles of polymerases α, δ, and ε in DNA replication. First, replication of the DNAs of some animal viruses, such as SV40, can be studied in cell-free extracts. The ability to study replication *in vitro* has allowed direct identification of the enzymes involved, and analysis of such cell-free systems has shown that polymerases α and δ are required for SV40 DNA replication. Second, polymerases α, δ, and ε are found in yeasts as well as in mammalian cells, enabling the use of the powerful approaches of yeast genetics (see Chapter 3) to test their biological roles directly. Such studies indicate that yeast mutants lacking any of these three DNA polymerases are unable to proliferate, implying a critical role for polymerase ε as well as for α and δ. However, further studies have shown that the essential function of polymerase ε in yeast does not require its activity as a replicative DNA polymerase. Thus, polymerases α and δ appear to be sufficient for DNA replication both in cell-free systems and in yeast, so the role of polymerase ε remains to be established.

All known DNA polymerases share two fundamental properties that carry critical implications for DNA replication (Figure 5.2). First, all polymerases synthesize DNA only in the 5' to 3' direction, adding a dNTP to the 3' hydroxyl group of a growing chain. Second, DNA polymerases can add a new deoxyribonucleotide only to a preformed primer strand that is hydrogen-bonded to the template; they are not able to initiate DNA synthesis *de novo* by catalyzing the polymerization of free dNTPs. In this respect, DNA polymerases differ from RNA polymerases, which can initiate the synthesis of a new strand of RNA in the absence of a primer. As discussed later in this

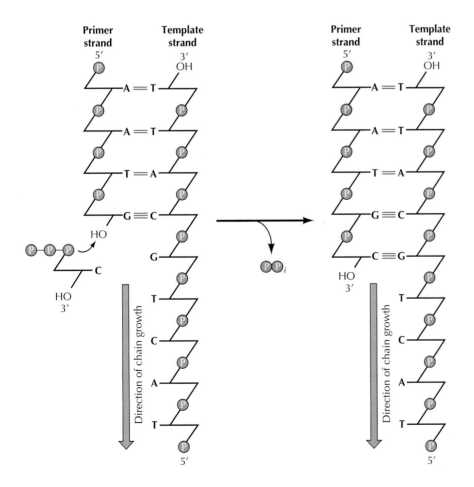

Figure 5.2
The reaction catalyzed by DNA polymerase   All known DNA polymerases add a deoxyribonucleoside 5'-triphosphate to the 3' hydroxyl group of a growing DNA chain (the primer strand).

chapter, these properties of DNA polymerases appear critical for maintaining the high fidelity of DNA replication that is required for cell reproduction.

The Replication Fork

DNA molecules in the process of replication were first analyzed by John Cairns in experiments in which *E. coli* were grown in the presence of radioactive thymidine, which allowed subsequent visualization of newly replicated DNA by autoradiography (Figure 5.3). In some cases, complete circular molecules in the process of replicating could be observed. These DNA molecules contained two **replication forks**, representing the regions of active DNA synthesis. At each fork the parental strands of DNA separated and two new daughter strands were synthesized.

The synthesis of new DNA strands complementary to both strands of the parental molecule posed an important problem to understanding the biochemistry of DNA replication. Since the two strands of double-helical DNA run in opposite (antiparallel) directions, continuous synthesis of two new strands at the replication fork would require that one strand be synthesized in the 5′ to 3′ direction while the other is synthesized in the opposite (3′ to 5′) direction. But DNA polymerase catalyzes the polymerization of dNTPs only in the 5′ to 3′ direction. How, then, can the other progeny strand of DNA be synthesized?

This enigma was resolved by experiments showing that only one strand of DNA is synthesized in a continuous manner in the direction of overall DNA replication; the other is formed from small, discontinuous pieces of DNA that are synthesized backward with respect to the direction of movement of the replication fork (Figure 5.4). These small pieces of newly synthesized DNA (called **Okazaki fragments** after their discoverer) are joined by the action of **DNA ligase**, forming an intact new DNA strand. The continuously synthesized strand is called the **leading strand**, since its elongation in the direction of replication fork movement exposes the template used for the synthesis of Okazaki fragments (the **lagging strand**).

Although the discovery of discontinuous synthesis of the lagging strand provided a mechanism for the elongation of both strands of DNA at the replication fork, it raised another question: Since DNA polymerase requires a primer and cannot initiate synthesis *de novo*, how is the synthesis of

Figure 5.3
Replication of *E. coli* DNA   (A) An autoradiograph showing bacteria that were grown in [³H]thymidine for two generations to label the DNA, which was then extracted and visualized by exposure to photographic film. (B) This schematic illustrates the two replication forks shown in (A). (From J. Cairns, 1963. *Cold Spring Harbor Symp. Quant. Biol.* 28: 43.)

(A)

100 μm

(B)

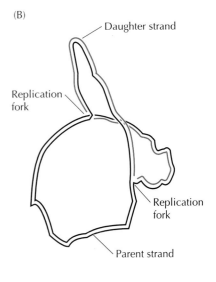

Daughter strand

Replication fork

Replication fork

Parent strand

Figure 5.4
Synthesis of leading and lagging strands of DNA The leading strand is synthesized continuously in the direction of replication fork movement. The lagging strand is synthesized in small pieces (Okazaki fragments) backward from the overall direction of replication. The Okazaki fragments are then joined by the action of DNA ligase.

Okazaki fragments initiated? The answer is that short fragments of RNA serve as primers for DNA replication (Figure 5.5). In contrast to DNA synthesis, the synthesis of RNA can initiate *de novo*, and an enzyme called **primase** synthesizes short fragments of RNA (e.g., three to ten nucleotides long) complementary to the lagging strand template at the replication fork. Okazaki fragments are then synthesized via extension of these RNA primers by DNA polymerase. An important consequence of such RNA priming is that newly synthesized Okazaki fragments contain an RNA-DNA joint, the discovery of which provided critical evidence for the role of RNA primers in DNA replication.

To form a continuous lagging strand of DNA, the RNA primers must eventually be removed from the Okazaki fragments and replaced with DNA.

Figure 5.5
Initiation of Okazaki fragments with RNA primers Short fragments of RNA serve as primers that can be extended by DNA polymerase.

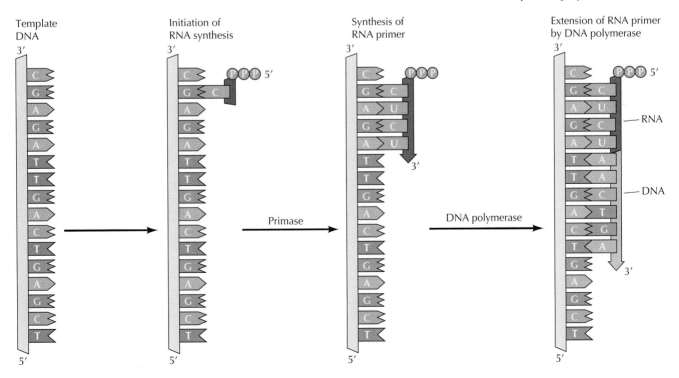

In *E. coli*, RNA primers are removed by the combined action of **RNase H**, an enzyme that degrades the RNA strand of RNA-DNA hybrids, and polymerase I. This is the aspect of *E. coli* DNA replication in which polymerase I plays a critical role. In addition to its DNA polymerase activity, polymerase I acts as an **exonuclease** that can hydrolyze DNA (or RNA) in either the 3' to 5' or 5' to 3' direction. The action of polymerase I as a 5' to 3' exonuclease removes ribonucleotides from the 5' ends of Okazaki fragments, allowing them to be replaced with deoxyribonucleotides to yield fragments consisting entirely of DNA (Figure 5.6). In eukaryotic cells, other exonucleases take the place of *E. coli* polymerase I in removing primers, and the gaps between Okazaki fragments are filled by the action of polymerase δ. As in prokaryotes, these DNA fragments can then be joined by DNA ligase.

The different DNA polymerases thus play distinct roles at the replication fork (Figure 5.7). In prokaryotic cells, polymerase III is the major replicative polymerase, functioning in the synthesis both of the leading strand of DNA and of Okazaki fragments by the extension of RNA primers. Polymerase I then removes RNA primers and fills the gaps between Okazaki fragments. In eukaryotic cells, however, two DNA polymerases are required to do what in *E. coli* is accomplished by polymerase III alone. Polymerase α is found in a complex with primase, and it appears to function in conjunction with primase to synthesize short RNA-DNA fragments during lagging strand synthesis. Polymerase δ can then synthesize both the leading and lagging strands, acting to extend the RNA-DNA primers initially synthesized by the polymerase α-primase complex. In addition, polymerase δ can take the place of *E. coli* polymerase I in filling the gaps between Okazaki fragments following primer removal.

Not only polymerases and primase but also a number of other proteins act at the replication fork. These additional proteins have been identified both by the analysis of *E. coli* mutants defective in DNA replication and by

Figure 5.6
Removal of RNA primers and joining of Okazaki fragments   Because of its 5' to 3' exonuclease activity, DNA polymerase I removes RNA primers and fills the gaps between Okazaki fragments with DNA. The resultant DNA fragments can then be joined by DNA ligase.

Figure 5.7
Roles of DNA polymerases in *E. coli* and mammalian cells The leading strand is synthesized by polymerase III (pol III) in *E. coli* and by polymerase δ (pol δ) in mammalian cells. In *E. coli*, lagging strand synthesis is initiated by primase, and RNA primers are extended by polymerase III. In mammalian cells, lagging strand synthesis is initiated by a complex of primase with polymerase α (pol α). The short RNA-DNA fragments synthesized by this complex are then extended by polymerase δ.

the purification of the mammalian proteins required for *in vitro* replication of SV40 DNA. One class of proteins required for replication binds to DNA polymerases, increasing the activity of the polymerases and causing them to remain bound to the template DNA so that they continue synthesis of a new DNA strand. Both *E. coli* polymerase III and eukaryotic polymerase δ are associated with two types of accessory proteins (sliding-clamp proteins and clamp-loading proteins) that load the polymerase onto the primer and maintain its stable association with the template (Figure 5.8). The clamp-

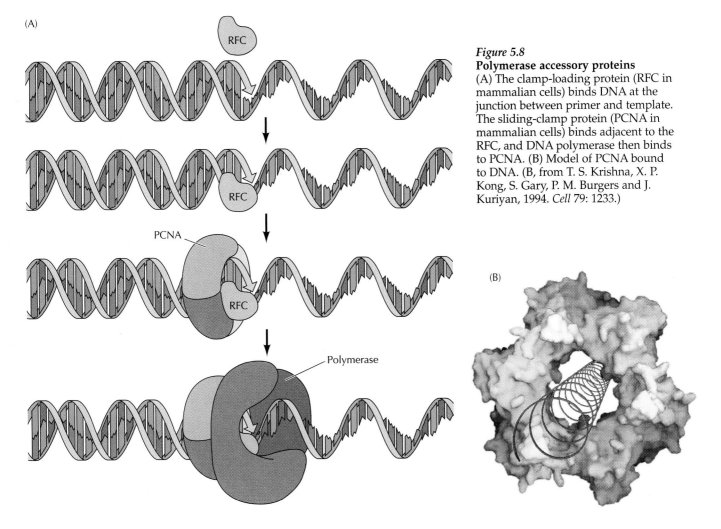

Figure 5.8
Polymerase accessory proteins
(A) The clamp-loading protein (RFC in mammalian cells) binds DNA at the junction between primer and template. The sliding-clamp protein (PCNA in mammalian cells) binds adjacent to the RFC, and DNA polymerase then binds to PCNA. (B) Model of PCNA bound to DNA. (B, from T. S. Krishna, X. P. Kong, S. Gary, P. M. Burgers and J. Kuriyan, 1994. *Cell* 79: 1233.)

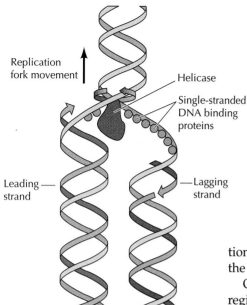

Replication
fork movement

Helicase

Single-stranded
DNA binding
proteins

Leading
strand

Lagging
strand

Figure 5.9
Action of helicases and single-stranded DNA-binding proteins
Helicases unwind the two strands of parental DNA ahead of the replication fork. The unwound DNA strands are then stabilized by single-stranded DNA-binding proteins so that they can serve as templates for new DNA synthesis.

loading proteins (called the γ complex in *E. coli* and replication factor C [RFC] in eukaryotes) specifically recognize and bind DNA at the junction between the primer and template. The sliding-clamp proteins (β protein in *E. coli* and proliferating cell nuclear antigen [PCNA] in eukaryotes) bind adjacent to the clamp-loading proteins, forming a ring around the template DNA. The clamp proteins then load the DNA polymerase onto DNA at the primer-template junction. The ring formed by the sliding clamp maintains the association of the polymerase with its template as replication proceeds, allowing the uninterrupted synthesis of many thousands of nucleotides of DNA.

Other proteins unwind the template DNA and stabilize single-stranded regions (Figure 5.9). **Helicases** are enzymes that catalyze the unwinding of parental DNA, coupled to the hydrolysis of ATP, ahead of the replication fork. **Single-stranded DNA-binding proteins** (e.g., eukaryotic replication factor A [RFA]) then stabilize the unwound template DNA, keeping it in an extended single-stranded state so that it can be copied by the polymerase.

As the strands of parental DNA unwind, the DNA ahead of the replication fork is forced to rotate. Unchecked, this rotation would cause circular DNA molecules (such as SV40 DNA or the *E. coli* chromosome) to become twisted around themselves, eventually blocking replication (Figure 5.10). This problem is solved by **topoisomerases,** enzymes that catalyze the reversible breakage and rejoining of DNA strands. There are two types of these enzymes: Type I topoisomerases break just one strand of DNA; type II topoisomerases introduce simultaneous breaks in both strands. The breaks introduced by type I and type II topoisomerases serve as "swivels" that allow the two strands of template DNA to rotate freely around each other so that replication can proceed without twisting the DNA ahead of

Figure 5.10
Action of topoisomerases during DNA replication   (A) As the two strands of template DNA unwind, the DNA ahead of the replication fork is forced to rotate in the opposite direction, causing circular molecules to become twisted around themselves. (B) This problem is solved by topoisomerases, which catalyze the reversible breakage and joining of DNA strands. The transient breaks introduced by these enzymes serve as swivels that allow the two strands of DNA to rotate freely around each other.

(A)

Unwinding of parental DNA

Twisting of DNA strands ahead of replication fork

(B)

Topoisomerase

Transient break serves as a swivel to allow free rotation of DNA strands the fork (see Figure 5.10). Although eukaryotic chromosomes are composed of linear rather than circular DNA molecules, their replication also requires topoisomerases; otherwise, the complete chromosomes would have to rotate continually during DNA synthesis.

Type II topoisomerase is needed not only to unwind DNA but also to unravel newly replicated circular DNA molecules that become interwined with each other. In eukaryotic cells, topoisomerase II appears to be involved in mitotic chromosome condensation. In addition, studies of yeast mutants, as well as experiments in *Drosophila* and mammalian cells, indicate that topoisomerase II is required for the separation of daughter chromatids at mitosis, suggesting that it functions to untangle newly replicated loops of DNA in the chromosomes of eukaryotes.

The enzymes involved in DNA replication act in a coordinated manner to synthesize both leading and lagging strands of DNA simultaneously at the replication fork (Figure 5.11). This task is accomplished by the formation of dimers of the replicative DNA polymerases (polymerase III in *E. coli* or polymerase δ in eukaryotes), each with its appropriate accessory proteins. One molecule of polymerase then acts in synthesis of the leading strand while the other acts in synthesis of the lagging strand. The lagging strand template is thought to form a loop at the replication fork so that the polymerase subunit engaged in lagging strand synthesis moves in the same overall direction as the other subunit, which is synthesizing the leading strand.

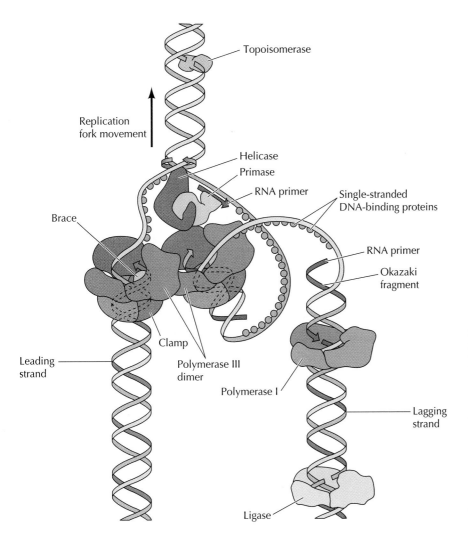

Figure 5.11
Model of the *E. coli* replication fork Helicase, primase, and two molecules of DNA polymerase III carry out coordinated synthesis of both the leading and lagging strands of DNA. The lagging strand template is folded so that the polymerase responsible for lagging strand synthesis moves in the same direction as overall movement of the fork. Topoisomerase acts as a swivel ahead of the fork, and DNA polymerase I and ligase remove RNA primers and join Okazaki fragments behind the fork.

The Fidelity of Replication

The accuracy of DNA replication is critical to cell reproduction, and estimates of mutation rates for a variety of genes indicate that the frequency of errors during replication corresponds to only one incorrect base per 10^9 to 10^{10} nucleotides incorporated. This error frequency is much lower than would be predicted simply on the basis of complementary base pairing. In particular, the standard configurations of nucleic acid bases are in equilibrium with rare alternative conformations (tautomeric forms) that hydrogen-bond with the wrong partner (e.g., G with T) with a frequency of about one incorrect base per 10^4 (Figure 5.12). The much higher degree of fidelity actually achieved results largely from the activities of DNA polymerase.

One mechanism by which DNA polymerase increases the fidelity of replication is by helping to select the correct base for insertion into newly synthesized DNA. The polymerase does not simply catalyze incorporation of whatever nucleotide is hydrogen-bonded to the template strand. Instead, it actively discriminates against incorporation of a mismatched base, presumably by adapting to the conformation of a correct base pair. The molecular mechanisms responsible for the ability of DNA polymerases to select against incorrect bases are not yet entirely understood, but this selectivity appears to increase the accuracy of replication about a hundredfold, reducing the expected error frequency from 10^{-4} to approximately 10^{-6}.

The other major mechanism responsible for the accuracy of DNA replication is the **proofreading** activity of DNA polymerase. As already noted, *E. coli* polymerase I has 3′ to 5′ as well as 5′ to 3′ exonuclease activity. The 5′ to 3′ exonuclease operates in the direction of DNA synthesis and helps remove RNA primers from Okazaki fragments. The 3′ to 5′ exonuclease operates in the reverse direction of DNA synthesis, and participates in proofreading newly synthesized DNA (Figure 5.13). Proofreading is effective because DNA polymerase requires a primer and is not able to initiate synthesis *de novo*. Primers that are hydrogen-bonded to the template are preferentially used, so when an incorrect base is incorporated, it is likely to be removed by the 3′ to 5′ exonuclease activity rather than being used to continue synthesis. Such 3′ to 5′ exonuclease activities are also associated with *E. coli* polymerase III and eukaryotic polymerases δ and ε. The 3′ to 5′ exonucleases of these polymerases selectively excise mismatched bases that have been incorporated at the end of a growing DNA chain, thereby increasing the accuracy of replication by a hundred- to a thousandfold.

The importance of proofreading may explain the fact that DNA polymerases require primers and catalyze the growth of DNA strands only in the 5′ to 3′ direction. When DNA is synthesized in the 5′ to 3′ direction, the energy required for polymerization is derived from hydrolysis of the 5′ triphosphate group of a free dNTP as it is added to the 3′ hydroxyl group of the growing chain (see Figure 5.2). If DNA were to be extended in the 3′ to 5′ direction, the energy of polymerization would instead have to be

Figure 5.12
Mismatching between rare configurations of nucleic acid bases In its normal configuration, guanine (G) specifically forms complementary hydrogen bonds (dashed lines) with cytosine (C). However, G occasionally assumes a rare configuration (tautomeric form) that instead forms hydrogen bonds with thymine (T).

Normal G—C pairing

Rare tautomeric form of G pairs with T

Figure 5.13
Proofreading by DNA polymerase A rare tautomeric form of G (G*) is incorporated in place of A as a result of mispairing with T on the template strand. The subsequent shift of G back to its normal form disrupts this base pairing, so the 3′ terminal G is no longer hydrogen-bonded to the template strand. This mismatch at the 3′ terminus of the growing chain is recognized and excised by the 3′ to 5′ exonuclease activity of DNA polymerase, which requires a primer hydrogen-bonded to the template strand in order to continue synthesis. Following excision of the mismatched G, DNA synthesis can proceed with incorporation of the correct nucleotide (A).

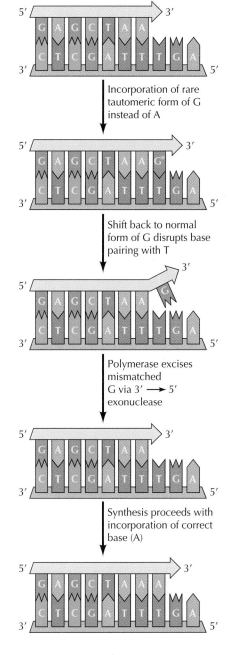

derived from hydrolysis of the 5′ triphosphate group of the terminal nucleotide already incorporated into DNA. This would eliminate the possibility of proofreading, because removal of a mismatched terminal nucleotide would also remove the 5′ triphosphate group needed as an energy source for further chain elongation. Thus, although the ability of DNA polymerase to extend a primer only in the 5′ to 3′ direction appears to make replication a complicated process, it is necessary for ensuring accurate duplication of the genetic material.

Combined with the ability to discriminate against the insertion of mismatched bases, the proofreading activity of DNA polymerases is sufficient to reduce the error frequency of replication to about one mismatched base per 10^9. Additional mechanisms (discussed in the section "DNA Repair") act to remove mismatched bases that have been incorporated into newly synthesized DNA, further ensuring correct replication of the genetic information.

Origins and the Initiation of Replication

The replication of both prokaryotic and eukaryotic DNAs starts at a unique sequence called the **origin of replication**, which serves as a specific binding site for proteins that initiate the replication process. The first origin to be defined was that of *E. coli*, in which genetic analysis indicated that replication always begins at a unique site on the bacterial chromosome. The *E. coli* origin has since been studied in detail and found to consist of 245 base pairs of DNA, elements of which serve as binding sites for proteins required to initiate DNA replication (Figure 5.14). The key step is the bind-

Figure 5.14
Origin of replication in *E. coli* Replication initiates at a unique site on the *E. coli* chromosome, designated the origin (*ori*). The first event is the binding of an initiator protein to *ori* DNA, which leads to partial unwinding of the template. The DNA continues to unwind by the actions of helicase and single-stranded DNA-binding proteins, and RNA primers are synthesized by primase. The two replication forks formed at the origin then move in opposite directions along the circular DNA molecule.

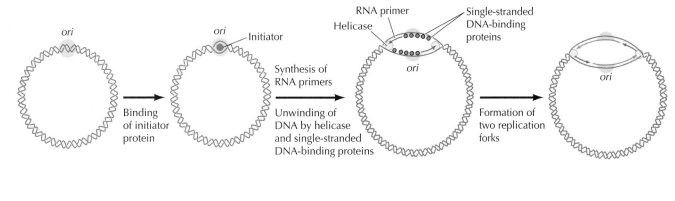

ing of an initiator protein to specific DNA sequences within the origin. The initiator protein begins to unwind the origin DNA and recruits the other proteins involved in DNA synthesis. Helicase and single-stranded DNA-binding proteins then act to continue unwinding and exposing the template DNA, and primase initiates the synthesis of leading strands. Two replication forks are formed and move in opposite directions along the circular *E. coli* chromosome.

The origins of replication of several animal viruses, such as SV40, have been studied as models for the initiation of DNA synthesis in eukaryotes. SV40 has a single origin of replication (consisting of 64 base pairs) that functions both in infected cells and in cell-free systems. Replication is initiated by a virus-encoded protein (called T antigen) that binds to the origin and also acts as a helicase. A single-stranded DNA-binding protein is required to stabilize the unwound template, and the DNA polymerase α-primase complex then initiates DNA synthesis.

Although single origins are sufficient to direct the replication of bacterial and viral genomes, multiple origins are needed to replicate the much larger genomes of eukaryotic cells within a reasonable period of time. For example, the entire genome of *E. coli* (4×10^6 base pairs) is replicated from a single origin in approximately 30 minutes. If mammalian genomes (3×10^9 base pairs) were replicated from a single origin at the same rate, DNA replication would require about 3 weeks (30,000 minutes). The problem is further exacerbated by the fact that the rate of DNA replication in mammalian cells is actually about tenfold lower than in *E. coli*, possibly as a result of the packaging of eukaryotic DNA in chromatin. Nonetheless, the genomes of mammalian cells are typically replicated within a few hours, necessitating the use of thousands of replication origins.

The presence of multiple replication origins in eukaryotic cells was first demonstrated by the exposure of cultured mammalian cells to radioactive thymidine for different time intervals, followed by autoradiography to detect newly synthesized DNA. The results of such studies indicated that DNA synthesis is initiated at multiple sites, from which it then proceeds in both directions along the chromosome (Figure 5.15). The replication origins in mammalian cells are spaced at intervals of approximately 50 to 300 kb; thus the human genome has about 30,000 origins of replication. The genomes of simpler eukaryotes also have multiple origins; for example, replication in yeasts initiates at origins separated by intervals of approximately 40 kb.

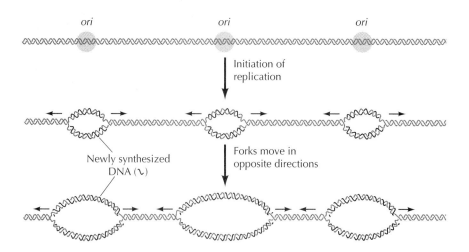

Figure 5.15
Replication origins in eukaryotic chromosomes   Replication initiates at multiple origins (*ori*), each of which produces two replication forks.

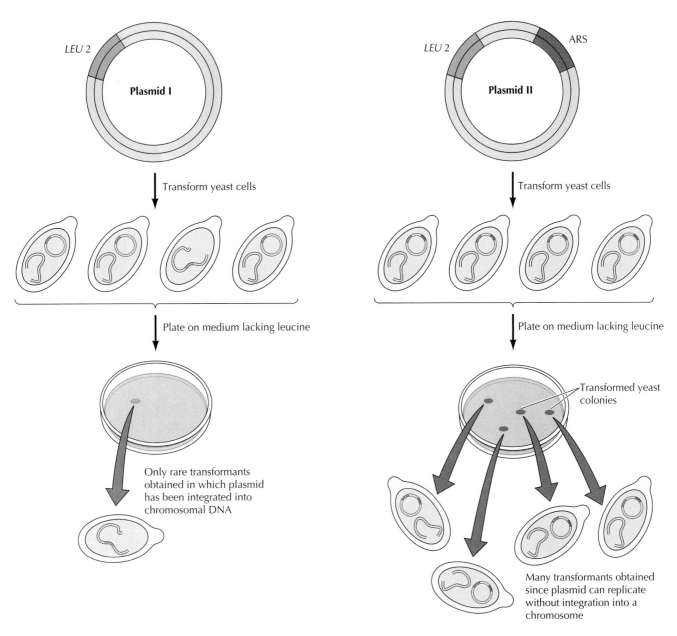

Figure 5.16
Identification of origins of replication in yeast   Both plasmids I and II contain a selectable marker gene (*LEU* 2) that allows transformed cells to grow on medium lacking leucine. Only plasmid II, however, contains an origin of replication (ARS). The transformation of yeasts with plasmid I yields only rare transformants in which the plasmid has integrated into chromosomal DNA. Plasmid II, however, is able to replicate without integration into a yeast chromosome (autonomous replication), so many more transformants result from its introduction into yeast cells.

The origins of replication of eukaryotic chromosomes have been studied best in yeasts, in which they have been identified as sequences that can support the replication of plasmids in transformed cells (Figure 5.16). This has provided a functional assay for these sequences, and several such elements (called **autonomously replicating sequences,** or **ARSs**) have been isolated. Their role as origins of replication has been verified by direct biochemical analysis, not only in plasmids but also in yeast chromosomal DNA.

Functional ARS elements span about 100 base pairs, including an 11-base-pair core sequence common to many different ARSs (Figure 5.17). This core sequence is essential for ARS function and has been found to be the binding site of a protein complex (called the **origin replication complex,** or **ORC**) that is required for initiation of DNA replication at yeast origins. The ORC complex appears to recruit other proteins (including DNA helicases) to the origin, leading to the initiation of replication. The mechanism of initiation of DNA replication in yeasts thus appears similar to that in prokaryotes and eukaryotic viruses; that is, an initiator protein specifically binds to origin sequences.

Figure 5.17
A yeast ARS element   The element
contains an 11-base-pair ARS consensus sequence (ACS), which is the specific binding site of the origin replication complex (ORC). Three additional
elements (B1, B2, and B3) are individu-
ally not essential but together contribute to ARS function.

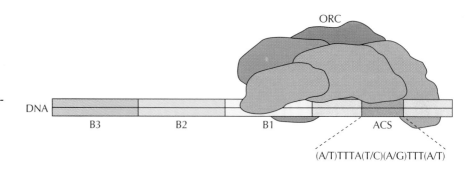

In contrast to the well-defined ARS elements in yeasts, much less is
known about the nature of replication origins in more complex eukaryotes.
However, recent experiments have shown that specific origin sequences are
required for initiation of DNA replication in mammalian cells. In addition,
proteins related to the yeast ORC proteins have been identified in a variety
of eukaryotes, including *Drosophila, C. elegans, Arabidopsis*, and humans,
and shown to be essential for DNA replication. It thus appears likely that
the basic mechanism used to initiate DNA replication is conserved in
eukaryotic cells.

Telomeres and Telomerase: Replicating the Ends of Chromosomes

Because DNA polymerases extend primers only in the 5′ to 3′ direction,
they are unable to copy the extreme 5′ ends of linear DNA molecules. Consequently, special mechanisms are required to replicate the terminal
sequences of the linear chromosomes of eukaryotic cells. These sequences
(**telomeres**) consist of tandem repeats of simple-sequence DNA (see Chapter 4). They are replicated by the action of a unique enzyme called **telomerase**, which is able to maintain telomeres by catalyzing their synthesis in
the absence of a DNA template.

Telomerase is a **reverse transcriptase**, one of a class of DNA polymerases, first discovered in retroviruses (see Chapter 3), that synthesize
DNA from an RNA template. Importantly, telomerase carries its own template RNA, which is complementary to the telomere repeat sequences, as
part of the enzyme complex. The use of this RNA as a template allows telomerase to generate multiple copies of the telomeric repeat sequences,
thereby maintaining telomeres in the absence of a conventional DNA template to direct their synthesis.

The mechanism of telomerase action was initially elucidated in 1985 by
Carol Greider and Elizabeth Blackburn during studies of the protozoan
Tetrahymena (Figure 5.18). The *Tetrahymena* telomerase is complexed to a
159-nucleotide-long RNA that includes the sequence 3′–AACCCCAAC–5′.
This sequence is complementary to the *Tetrahymena* telomeric repeat
(5′–TTGGGG–3′) and serves as the template for the synthesis of telomeric
DNA. The use of this RNA as a template allows telomerase to extend the 3′
end of chromosomal DNA by one repeat unit beyond its original length.
The complementary strand can then be synthesized by the polymerase α-
primase complex using conventional RNA priming. Removal of the RNA
primer leaves an overhanging 3′ end of chromosomal DNA, which can
form loops at the ends of eukaryotic chromosomes (see Figure 4.19).

Telomerase has been identified in a variety of eukaryotes, and genes
encoding telomerase RNAs have been cloned from *Tetrahymena*, yeasts,
mice, and humans. In each case, the telomerase RNA contains sequences
complementary to the telomeric repeat sequence of that organism (see

Telomeric DNA with overhanging 3' end

Newly synthesized lagging strand

Binding to telomerase RNA

Telomerase RNA

Telomerase

Telomerase reverse transcriptase activity

3' end of leading strand elongated by one repeat unit

Extension of lagging strand by primase and polymerase

Removal of RNA primer

Telomeric DNA extended by one repeat unit

Figure 5.18
Action of telomerase Telomeric DNA is a simple repeat sequence with an overhanging 3' end on the newly synthesized leading strand. Telomerase carries its own RNA molecule, which is complementary to telomeric DNA, as part of the enzyme complex. The overhanging end of telomeric DNA binds to the telomerase RNA, which then serves as a template for extension of the leading strand by one repeat unit. The lagging strand of telomeric DNA can then be elongated by conventional RNA priming and DNA polymerase activity.

Table 4.3). Moreover, the introduction of mutant telomerase RNA genes into yeasts has been shown to result in corresponding alterations of the chromosomal telomeric repeat sequences, directly demonstrating the function of telomerase in maintaining the termini of eukaryotic chromosomes.

DNA Repair

DNA, like any other molecule, can undergo a variety of chemical reactions. Because DNA uniquely serves as a permanent copy of the cell genome, however, changes in its structure are of much greater consequence than are alterations in other cell components, such as RNAs or proteins. Mutations can result from the incorporation of incorrect bases during DNA replication. In addition, various chemical changes occur in DNA either spontaneously (Figure 5.19) or as a result of exposure to chemicals or radiation (Figure 5.20). Such damage to DNA can block replication or transcription, and can result in a high frequency of mutations—consequences that are unacceptable from the standpoint of cell reproduction. To maintain the integrity of their genomes, cells have therefore had to evolve mechanisms to repair damaged DNA. These mechanisms of DNA repair can be divided into two general classes: (1) direct reversal of the chemical reaction responsible for DNA damage, and (2) removal of the damaged bases followed by their replacement with newly synthesized DNA. Where DNA repair fails, additional mechanisms have evolved to enable cells to cope with the damage.

(A) **Deamination**

Cytosine → Uracil

Adenine → Hypoxanthine

(B) **Depurination**

DNA chain

dGMP → AP site

Figure 5.19
Spontaneous damage to DNA
There are two major forms of spontaneous DNA damage: (A) deamination of adenine, cytosine, and guanine, and (B) depurination (loss of purine bases) resulting from cleavage of the bond between the purine bases and deoxyribose, leaving an apurinic (AP) site in DNA. dGMP = deoxyguanosine monophosphate.

(A) **Exposure to UV light**

Adjacent thymines in DNA

Thymine dimer

Cyclobutane ring

(B) **Alkylation**

Guanine

O^6-methylguanine

(C) **Reaction with carcinogen**

Guanine

Bulky group addition

Direct Reversal of DNA Damage

Most damage to DNA is repaired by removal of the damaged bases followed by resynthesis of the excised region. Some lesions in DNA, however, can be repaired by direct reversal of the damage, which may be a more efficient way of dealing with specific types of DNA damage that occur frequently. Only a few types of DNA damage are repaired in this way, particularly pyrimidine dimers resulting from exposure to ultraviolet (UV) light and alkylated guanine residues that have been modified by the addition of methyl or ethyl groups at the O^6 position of the purine ring.

UV light is one of the major sources of damage to DNA and is also the most thoroughly studied form of DNA damage in terms of repair mechanisms. Its importance is illustrated by the fact that exposure to solar UV

Figure 5.20
Examples of DNA damage induced by radiation and chemicals (A) UV light induces the formation of pyrimidine dimers, in which two adjacent pyrimidines (e.g., thymines) are joined by a cyclobutane ring structure. (B) Alkylation is the addition of methyl or ethyl groups to various positions on the DNA bases. In this example, alkylation of the O^6 position of guanine results in formation of O^6-methylguanine. (C) Many carcinogens (e.g., benzo-(*a*)pyrene) react with DNA bases, resulting in the addition of large bulky chemical groups to the DNA molecule.

Figure 5.21
Direct repair of thymine dimers
UV-induced thymine dimers can be repaired by photoreactivation, in which energy from visible light is used to split the bonds forming the cyclobutane ring.

irradiation is the cause of almost all skin cancer in humans. The major type of damage induced by UV light is the formation of **pyrimidine dimers,** in which adjacent pyrimidines on the same strand of DNA are joined by the formation of a cyclobutane ring resulting from saturation of the double bonds between carbons 5 and 6 (see Figure 5.20A). The formation of such dimers distorts the structure of the DNA chain and blocks transcription or replication past the site of damage, so their repair is closely correlated with the ability of cells to survive UV irradiation. One mechanism of repairing UV-induced pyrimidine dimers is direct reversal of the dimerization reaction. The process is called **photoreactivation** because energy derived from visible light is utilized to break the cyclobutane ring structure (Figure 5.21). The original pyrimidine bases remain in DNA, now restored to their normal state. As might be expected from the fact that solar UV irradiation is a major source of DNA damage for diverse cell types, the repair of pyrimidine dimers by photoreactivation is common to a variety of prokaryotic and eukaryotic cells, including *E. coli*, yeasts, and some species of plants and animals. Curiously, however, photoreactivation is not universal; many species (including humans) lack this mechanism of DNA repair.

Another form of direct repair deals with damage resulting from the reaction between alkylating agents and DNA. Alkylating agents are reactive compounds that can transfer methyl or ethyl groups to a DNA base, thereby chemically modifying the base (see Figure 5.20B). A particularly important type of damage is methylation of the O^6 position of guanine,

Figure 5.22
Repair of O^6-methylguanine O^6-methylguanine methyltransferase transfers the methyl group from O^6-methylguanine to a cysteine residue in the enzyme's active site.

because the product, O⁶-methylguanine, forms complementary base pairs with thymine instead of cytosine. This lesion can be repaired by an enzyme (called O⁶-methylguanine methyltransferase) that transfers the methyl group from O⁶-methylguanine to a cysteine residue in its active site (Figure 5.22). The potentially mutagenic chemical modification is thus removed, and the original guanine is restored. Enzymes that catalyze this direct repair reaction are widespread in both prokaryotes and eukaryotes, including humans.

Excision Repair

Although direct repair is an efficient way of dealing with particular types of DNA damage, excision repair is a more general means of repairing a wide variety of chemical alterations to DNA. Consequently, the various types of excision repair are the most important DNA repair mechanisms in both prokaryotic and eukaryotic cells. In excision repair, the damaged DNA is recognized and removed, either as free bases or as nucleotides. The resulting gap is then filled in by synthesis of a new DNA strand, using the undamaged complementary strand as a template. Three types of excision repair—base-excision repair, nucleotide-excision repair, and mismatch repair—enable cells to cope with a variety of different kinds of DNA damage.

The repair of uracil-containing DNA is a good example of **base-excision repair**, in which single damaged bases are recognized and removed from the DNA molecule (Figure 5.23). Uracil can arise in DNA by two mechanisms: (1) Uracil (as dUTP [deoxyuridine triphosphate]) is occasionally incorporated in place of thymine during DNA synthesis, and (2) uracil can be formed in DNA by the deamination of cytosine (see Figure 5.19A). The second mechanism is of much greater biological significance because it alters the normal pattern of complementary base pairing and thus represents a mutagenic event. The excision of uracil in DNA is catalyzed by **DNA glycosylase**, an enzyme that cleaves the bond linking the base (uracil) to the deoxyribose of the DNA backbone. This reaction yields free uracil and an apyrimidinic site—a sugar with no base attached. DNA glycosylases also recognize and remove other abnormal bases, including hypoxanthine formed by the deamination of adenine, pyrimidine dimers, alkylated purines other than O⁶-alkylguanine, and bases damaged by oxidation or ionizing radiation.

The result of DNA glycosylase action is the formation of an apyridiminic or apurinic site (generally called an AP site) in DNA. Similar AP sites are formed as the result of the spontaneous loss of purine bases (see Figure 5.19B), which occurs at a significant rate under normal cellular conditions. For example, each cell in the human body is estimated to lose several thousand purine bases daily. These sites are repaired by **AP endonuclease**, which cleaves adjacent to the AP site (see Figure 5.23). The remaining deoxyribose moiety is then removed, and the resulting single-base gap is filled by DNA polymerase and ligase.

Figure 5.23
Base-excision repair   In this example, uracil (U) has been formed by deamination of cytosine (C) and is therefore opposite a guanine (G) in the complementary strand of DNA. The bond between uracil and the deoxyribose is cleaved by a DNA glycosylase, leaving a sugar with no base attached in the DNA (an AP site). This site is recognized by AP endonuclease, which cleaves the DNA chain. The remaining deoxyribose is removed by deoxyribosephosphodiesterase. The resulting gap is then filled by DNA polymerase and sealed by ligase, leading to incorporation of the correct base (C) opposite the G.

Whereas DNA glycosylases recognize only specific forms of damaged bases, other excision repair systems recognize a wide variety of damaged bases that distort the DNA molecule, including UV-induced pyrimidine dimers and bulky groups added to DNA bases as a result of the reaction of many carcinogens with DNA (see Figure 5.20C). This widespread form of DNA repair is known as **nucleotide-excision repair**, because the damaged bases (e.g., a thymine dimer) are removed as part of an oligonucleotide containing the lesion (Figure 5.24).

In *E. coli*, nucleotide-excision repair is catalyzed by the products of three genes (*uvrA*, *B*, and *C*) that were identified because mutations at these loci result in extreme sensitivity to UV light. The protein UvrA recognizes damaged DNA and recruits UvrB and UvrC to the site of the lesion. UvrB and UvrC then cleave on the 3′ and 5′ sides of the damaged site, respectively, thus excising an oligonucleotide consisting of 12 or 13 bases. The UvrABC complex is frequently called an **excinuclease**, a name that reflects its ability to directly *exci*se an oligonucleotide. The action of a helicase is then required to remove the damage-containing oligonucleotide from the dou-

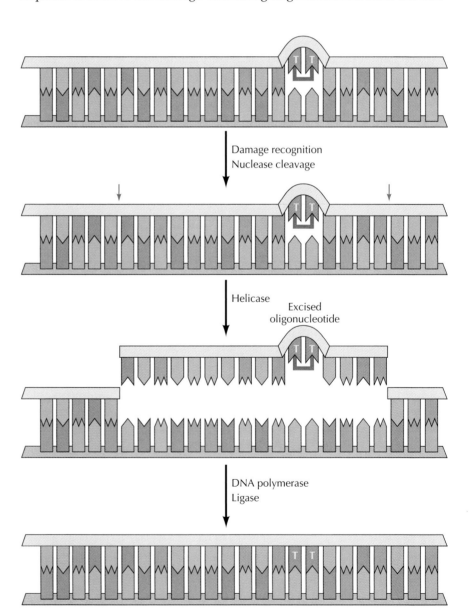

Figure 5.24
Nucleotide-excision repair of thymine dimers Damaged DNA is recognized and then cleaved on both sides of a thymine dimer by 3′ and 5′ nucleases. Unwinding by a helicase results in excision of an oligonucleotide containing the damaged bases. The resulting gap is then filled by DNA polymerase and sealed by ligase.

Damage recognition
Nuclease cleavage

Helicase

Excised
oligonucleotide

DNA polymerase
Ligase

ble-stranded DNA molecule, and the resulting gap is filled by DNA polymerase I and sealed by ligase.

Nucleotide-excision repair systems have also been studied extensively in eukaryotes, particularly in yeasts and in humans. In yeasts, as in *E. coli*, several genes involved in DNA repair (called *RAD* genes for *rad*iation sensitivity) have been identified by the isolation of mutants with increased sensitivity to UV light. In humans, DNA repair genes have been identified largely by studies of individuals suffering from inherited diseases resulting from deficiencies in the ability to repair DNA damage. The most extensively studied of these diseases is xeroderma pigmentosum (XP), a rare genetic disorder that affects approximately one in 250,000 people. Individuals with this disease are extremely sensitive to UV light and develop multiple skin cancers on the regions of their bodies that are exposed to sunlight. In 1968 James Cleaver made the key discovery that cultured cells from XP patients were deficient in the ability to carry out nucleotide-excision repair. This observation not only provided the first link between DNA repair and cancer, but also suggested the use of XP cells as an experimental system to identify human DNA repair genes. The identification of human DNA repair genes has been accomplished by studies not only of XP cells, but also of two other human diseases resulting from DNA repair defects (Cockayne's syndrome and trichothiodystrophy) and of UV-sensitive mutants of rodent cell lines. The availability of mammalian cells with defects in DNA repair has allowed the cloning of repair genes based on the ability of wild-type alleles to restore normal UV sensitivity to mutant cells in gene transfer assays, thereby opening the door to experimental analysis of nucleotide-excision repair in mammalian cells.

Molecular cloning has now identified seven different repair genes (designated *XPA* through *XPG*) that are mutated in cases of xeroderma pigmentosum, as well as in some cases of Cockayne's syndrome, trichothiodystrophy, and UV-sensitive mutants of rodent cells. Table 5.1 lists the enzymes encoded by these genes. Some UV-sensitive rodent cells have mutations in yet another repair gene, called *ERCC1* (for *e*xcision *r*epair *c*ross *c*omplementing), which has not been found to be mutated in known human diseases. It is notable that the proteins encoded by these human DNA repair genes are closely related to proteins encoded by yeast *RAD* genes, indicating that nucleotide-excision repair is highly conserved throughout eukaryotes.

With cloned yeast and human repair genes available, it has been possible to purify their encoded proteins and develop *in vitro* systems to study the repair process. Although some steps remain to be fully elucidated, these studies have led to the development of a basic model for nucleotide-excision repair in eukaryotic cells. In mammalian cells, the XPA protein (and possibly also XPC) initiates repair by recognizing damaged DNA and forming complexes with other proteins involved in the repair process. These include the XPB and XPD proteins, which act as helicases that unwind the damaged DNA. In addition, the binding of XPA to damaged DNA leads to the recruitment of XPF (as a heterodimer with ERCC1) and XPG to the repair complex. XPF/ERCC1 and XPG are endonucleases, which cleave DNA on the 5' and 3' sides of the damaged site, respectively. This cleavage excises an oligonucleotide consisting of approximately 30 bases. The resulting gap then appears to be filled in by DNA polymerase δ or ε (in association with RFC and PCNA) and sealed by ligase.

An intriguing feature of nucleotide-excision repair is its relationship to transcription. A connection between transcription and repair was first suggested by experiments showing that transcribed strands of DNA are

TABLE 5.1 Enzymes Involved in Nucleotide-Excision Repair

Human	Yeast	Function
XPA	RAD14	Damage recognition
XPB	RAD25	Helicase
XPC	RAD4	DNA binding
XPD	RAD3	Helicase
XPF	RAD1	5' nuclease
XPG	RAD2	3' nuclease
ERCC1	RAD10	Dimer with XPF

Molecular Medicine

Colon Cancer and DNA Repair

The Disease

Cancers of the colon and rectum (colorectal cancers) are one of the most common types of cancer in Western countries, accounting for nearly 140,000 cancer cases per year in the United States (approximately 10% of the total cancer incidence). Most colon cancers (like other types of cancer) are not inherited diseases; that is, they are not transmitted directly from parent to offspring. However, two inherited forms of colon cancer have been described. In both of these syndromes, the inheritance of a cancer susceptibility gene results in a very high likelihood of cancer development. One inherited form of colon cancer (familial adenomatous polyposis) is extremely rare, accounting for less than 1% of total colon cancer incidence. The second inherited form of colon cancer (hereditary nonpolyposis colorectal cancer, or HNPCC) is much more common and accounts for up to 15% of all colon cancer cases. Indeed, HNPCC is one of the most common inherited diseases, affecting as many as one in 200 people. Although colon cancers are the most common manifestation of this disease, affected individuals also suffer an increased incidence of other types of cancer, including cancers of the ovary and endometrium.

Molecular and Cellular Basis

Like other cancers, colorectal cancer results from mutations in genes that regulate cell proliferation, leading to the uncontrolled growth of cancer cells. In most cases these mutations occur sporadically in somatic cells. In hereditary cancers, however, inherited germ-line mutations predispose the individual to cancer development.

A striking advance was made in 1993 with the discovery that a gene responsible for approximately 50% of HNPCC cases encodes an enzyme involved in mismatch repair of DNA; this gene is a human homolog of the *E. coli MutS* gene. Subsequent studies have shown that three other genes, responsible for most remaining cases of HNPCC, are homologs of *MutL* and thus are also involved in the mismatch repair pathway. Defects in these genes appear to result in a high frequency of mutations in other cell genes, with a correspondingly high likelihood that some of these mutations will eventually lead to the development of cancer by affecting genes that regulate cell proliferation.

Prevention and Treatment

As with other inherited diseases, identification of the genes responsible for HNPCC allows individuals at risk for this inherited cancer to be identified by genetic testing. Moreover, prenatal genetic diagnosis may be of great importance to carriers of HNPCC mutations who are planning a family. However, the potential benefits of detecting these mutations are not limited to preventing the transmission of mutant genes to the next generation; their detection may also help prevent the development of cancer in affected individuals.

In terms of disease prevention, a key characteristic of colon cancer is that it develops gradually over several years. Early diagnosis of the disease substantially improves the chances for patient survival. The initial stage of colon cancer development is the outgrowth of small benign polyps, which eventually become malignant and invade the surrounding connective tissue. Prior to the development of malignancy, however, polyps can be easily removed surgically, effectively preventing the outgrowth of a malignant tumor. Polyps and early stages of colon cancer can be detected by examination of the colon with a thin lighted tube (colonoscopy), so frequent colonoscopy of HNPCC patients may allow polyps to be removed before cancer develops. In addition, several drugs are being tested as potential inhibitors of colon cancer development, and these drugs may be of significant benefit to HNPCC patients. By allowing the timely application of such preventive measures, the identification of mutations responsible for HNPCC may make a significant contribution to disease prevention.

A colon polyp visualized by colonoscopy. (Custom Medical Stock Photo.)

repaired more rapidly than nontranscribed strands in both *E. coli* and mammalian cells. Since DNA damage blocks transcription, this transcription-repair coupling is thought to be advantageous by allowing the cell to preferentially repair damage to actively expressed genes. In *E. coli*, the mechanism of transcription-repair coupling involves recognition of RNA poly-

Figure 5.25
Mismatch repair in *E. coli* The mismatch repair system detects and excises mismatched bases in newly replicated DNA, which is distinguished from the parental strand because it has not yet been methylated. MutS binds to the mismatched base, followed by MutL. The binding of MutL activates MutH, which cleaves the unmodified strand opposite a site of methylation. MutS and MutL, together with a helicase and an exonuclease, then excise the portion of the unmodified strand that contains the mismatch. The gap is then filled by DNA polymerase and sealed by ligase.

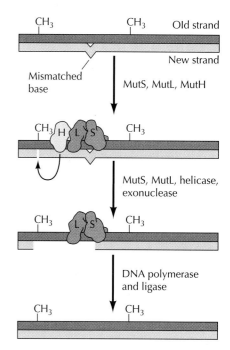

merase stalled at a lesion in the DNA strand being transcribed. The stalled RNA polymerase is recognized by a protein called transcription-repair coupling factor, which displaces RNA polymerase and recruits the UvrABC excinuclease to the site of damage.

Although the molecular mechanism of transcription-repair coupling in mammalian cells is not yet known, it is noteworthy that the XPB and XPD helicases are components of a multisubunit transcription factor (called TFIIH) that is required to initiate the transcription of eukaryotic genes (see Chapter 6). Thus, these helicases appear to be required for the unwinding of DNA during both transcription and nucleotide-excision repair, providing a direct biochemical link between these two processes. Patients suffering from Cockayne's syndrome are also characterized from a failure to preferentially repair transcribed DNA strands, suggesting that the proteins encoded by the two genes known to be responsible for this disease (*CSA* and *CSB*) function in transcription-coupled repair. In addition, one of the genes responsible for inherited breast cancer in humans (*BRCA1*) appears to encode a protein specifically involved in transcription-coupled repair of oxidative DNA damage, suggesting that defects in this type of DNA repair can lead to the development of one of the most common cancers in women.

A third excision repair system recognizes mismatched bases that are incorporated during DNA replication. Many such mismatched bases are removed by the proofreading activity of DNA polymerase. The ones that are missed are subject to later correction by the **mismatch repair** system, which scans newly replicated DNA. If a mismatch is found, the enzymes of this repair system are able to identify and excise the mismatched base specifically from the newly replicated DNA strand, allowing the error to be corrected and the original sequence restored.

In *E. coli*, the ability of the mismatch repair system to distinguish between parental DNA and newly synthesized DNA is based on the fact that DNA of this bacterium is modified by the methylation of adenine residues within the sequence GATC to form 6-methyladenine (Figure 5.25). Since methylation occurs after replication, newly synthesized DNA strands are not methylated and thus can be specifically recognized by the mismatch repair enzymes. Mismatch repair is initiated by the protein MutS, which recognizes the mismatch and forms a complex with two other proteins called MutL and MutH. The MutH endonuclease then cleaves the unmethylated DNA strand at a GATC sequence. MutL and MutS then act together with an exonuclease and a helicase to excise the DNA between the strand break and the mismatch, with the resulting gap being filled by DNA polymerase and ligase.

Eukaryotes have a similar mismatch repair system, although the mechanism by which eukaryotic cells identify newly replicated DNA differs from that used by *E. coli*. In mammalian cells, it appears that the strand-specificity of mismatch repair is determined by the presence of single-strand breaks (which would be present in newly replicated DNA) in the strand to be repaired (Figure 5.26). The eukaryotic homologs of MutS and MutL then bind to the mismatched base and direct excision of the DNA between the strand break and the mismatch, as in *E. coli*. The importance of this repair

Figure 5.26
Mismatch repair in mammalian cells Mismatch repair in mammalian cells is similar to *E. coli*, except that the newly replicated strand is distinguished from the parental strand because it contains strand breaks. MutS and MutL bind to the mismatched base and direct excision of the DNA between the strand break and the mismatch.

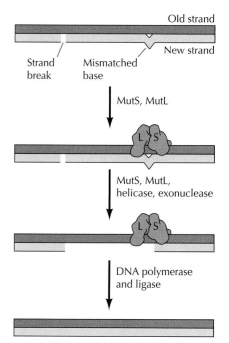

system is dramatically illustrated by the fact that mutations in the human homologs of *MutS* and *MutL* are responsible for a common type of inherited colon cancer (hereditary nonpolyposis colorectal cancer, or HNPCC). HNPCC is one of the most common inherited diseases; it affects as many as one in 200 people and is responsible for about 15% of all colorectal cancers in this country. The relationship between HNPCC and defects in mismatch repair was discovered in 1993, when two groups of researchers cloned the human homolog of *MutS* and found that mutations in this gene were responsible for about half of all HNPCC cases. Subsequent studies have shown that most of the remaining cases of HNPCC are caused by mutations in one of three human genes that are homologs of *MutL*.

Postreplication Repair

The direct reversal and excision repair systems act to correct DNA damage before replication, so that replicative DNA synthesis can proceed using an undamaged DNA strand as a template. Should these systems fail, however, the cell has alternative mechanisms for dealing with damaged DNA at the replication fork. Pyrimidine dimers and many other types of lesions cannot be copied by the normal action of DNA polymerases, so replication is

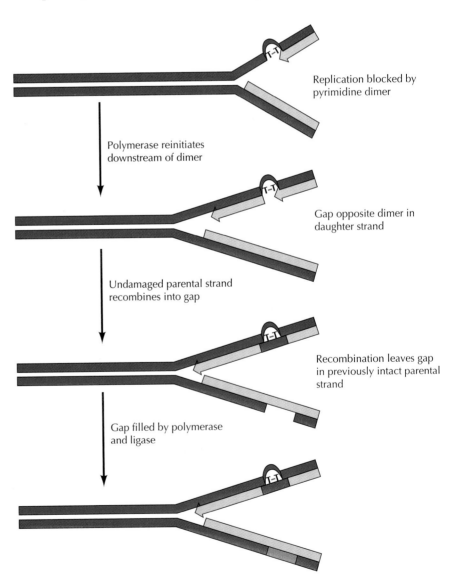

Replication blocked by pyrimidine dimer

Polymerase reinitiates downstream of dimer

Gap opposite dimer in daughter strand

Undamaged parental strand recombines into gap

Recombination leaves gap in previously intact parental strand

Gap filled by polymerase and ligase

Figure 5.27
Postreplication repair   The presence of a thymine dimer blocks replication, but DNA polymerase can bypass the lesion and reinitiate replication at a new site downstream of the dimer. The result is a gap opposite the dimer in the newly synthesized DNA strand. In recombinational repair, this gap is filled by recombination with the undamaged parental strand. Although this leaves a gap in the previously intact parental strand, the gap can be filled by the actions of polymerase and ligase, using the intact daughter strand as a template. Two intact DNA molecules are thus formed, and the remaining thymine dimer eventually can be removed by excision repair.

blocked at the sites of such damage. Downstream of the damaged site, however, replication can be initiated again by the synthesis of an Okazaki fragment and can proceed along the damaged template strand (Figure 5.27). The result is a daughter strand that has a gap opposite the site of damage to the parental strand. One of two types of mechanisms may be used to repair such gaps in newly synthesized DNA: recombinational repair or error-prone repair.

Recombinational repair depends on the fact that one strand of the parental DNA was undamaged and therefore was copied during replication to yield a normal daughter molecule (see Figure 5.27). The undamaged parental strand can be used to fill the gap opposite the site of damage in the other daughter molecule by recombination between homologous DNA sequences (see the next section). Because the resulting gap in the previously intact parental strand is opposite an undamaged strand, it can be filled in by DNA polymerase. Although the other parent molecule still retains the original damage (e.g., a pyrimidine dimer), the damage now lies opposite a normal strand and can be dealt with later by excision repair. By a similar mechanism, recombination with an intact DNA molecule can be used to repair double strand breaks, which are frequently introduced into DNA by radiation and other damaging agents.

In **error-prone repair**, a gap opposite a site of DNA damage is filled by newly synthesized DNA. Since the new DNA is synthesized from a damaged template strand, this form of DNA synthesis is very inaccurate and leads to frequent mutations. It is used only in bacteria that have been subjected to potentially lethal conditions, such as extensive UV irradiation. Such treatments induce the SOS response, which may be viewed as a mechanism for dealing with extreme environmental stress. The SOS response includes inhibition of cell division and induction of repair systems to cope with a high level of DNA damage. Under these conditions, error-prone repair mechanisms are used, presumably as a way of dealing with damage so extensive that cell death is the only alternative.

Recombination Between Homologous DNA Sequences

Accurate DNA replication and repair of DNA damage are essential to maintaining genetic information and ensuring its accurate transmission from parent to offspring. From the standpoint of evolution, however, it is also important to generate genetic diversity. Genetic differences between individuals provide the essential starting material of natural selection, which allows species to evolve and adapt to changing environmental conditions. Recombination plays an important role in this process by allowing genes to be reassorted into different combinations. For example, genetic recombination results in the exchange of genes between paired homologous chromosomes during meiosis (see Figure 3.4). However, increasing genetic diversity is not the only role of recombination. As discussed in the previous section, recombination is also an important mechanism for repairing damaged DNA. In addition, recombination is involved in rearrangements of specific DNA sequences that alter the expression and function of some genes during development and differentiation. Thus, recombination plays important roles in the lives of individual cells and organisms, as well as contributing to the genetic diversity of the species.

This section discusses the molecular mechanisms of recombination between DNA molecules that share extensive sequence homology. Examples include recombination between paired eukaryotic chromosomes during meiosis and recombination between bacterial chromosomes during mating. Since this type of recombination involves the exchange of genetic information between two homologous DNA molecules, it does not alter the overall arrangement of the genes on a chromosome. Other types of recombination, however, do not require extensive sequence homology and therefore can occur between unrelated DNAs. Recombination events of this type lead to gene rearrangements, which are discussed later in the chapter.

DNA Molecules Recombine by Breaking and Rejoining

Genetic recombination was first defined by studies of *Drosophila*, on the basis of the observation that genes on different copies of homologous chromosomes can reassort during meiosis. With the subsequent discovery that genes consist of DNA, two alternative models to explain recombination at the molecular level were considered (Figure 5.28). The "copy choice" model proposed that the recombinant molecule is generated during DNA synthesis, as a result of copying first one parental DNA and then switching to copy a different template. The alternative proposal was that recombination results from the breakage and rejoining of two parental DNA molecules rather than by synthesis of new DNA.

These alternatives were first distinguished in 1961 by studies of recombination between the genomes of bacterial viruses (Figure 5.29). Infection of *E. coli* with viruses carrying different genetic markers was known to yield recombinant progeny. To determine if this recombination involved breakage and rejoining of the parental DNAs, one of the parental viruses was grown in medium containing the heavy isotopes of carbon (^{13}C) and nitrogen (^{15}N), while the other was grown in medium containing the normal light isotopes (^{12}C and ^{14}N). The result was parental viruses having different densities, so they could be separated by equilibrium density centrifugation in a CsCl gradient. *E. coli* were then infected with these differentially labeled parental viruses under conditions in which replication was inhibited, and the progeny viruses produced were analyzed for both their density and their genetic characteristics. The important result was that genetic recombinant viruses were obtained that had intermediate densities, indicating that they had acquired DNA from both parents, as predicted by the breakage-and-rejoining, but not the copy choice, model.

Figure 5.28
Models of recombination In copy choice, recombination occurs during the synthesis of daughter DNA molecules. DNA replication starts with one parental DNA template and then switches to a second parental molecule, resulting in the synthesis of recombinant daughter DNAs containing sequences homologous to both parents. In breakage and rejoining, recombination occurs as a result of breakage and crosswise rejoining of parental DNA molecules.

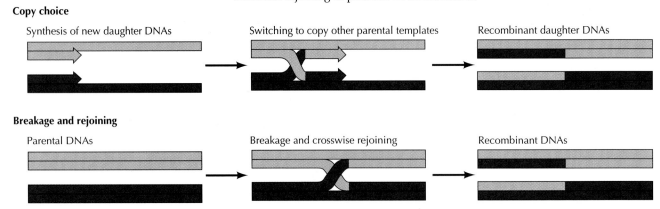

Copy choice

Synthesis of new daughter DNAs   Switching to copy other parental templates   Recombinant daughter DNAs

Breakage and rejoining

Parental DNAs   Breakage and crosswise rejoining   Recombinant DNAs

Figure 5.29
Experimental demonstration of recombination by breakage and rejoining Genetically distinct parental viruses were grown in medium containing either light or heavy isotopes of carbon (^{12}C or ^{13}C) and nitrogen (^{14}N or ^{15}N) to density-label their DNAs. *E. coli* were infected under conditions in which replication was inhibited, and the progeny viruses were harvested and analyzed by equilibrium centrifugation in a CsCl gradient to determine the density of genetic recombinants. The recombinant viruses were found to have intermediate densities, indicating that they had acquired DNA from both parents by a breakage and rejoining mechanism.

Models of Homologous Recombination

The finding that recombination occurs by breakage and rejoining raises a critical question: How can two parental DNA molecules be broken at precisely the same point, so that they can rejoin without mutations resulting from the gain or loss of nucleotides at the break point? During recombination between homologous DNA molecules (**general homologous recombination**), this alignment is provided, not surprisingly, by base pairing between complementary DNA strands (Figure 5.30). Overlapping single strands are exchanged between homologous DNA molecules, leading to the formation of a heteroduplex region, in which the two strands of the recombinant double helix are derived from different parents. If the heteroduplex region contains a genetic difference, the result is a single progeny DNA molecule that contains two genetic markers. In some cases, mispaired bases in a heteroduplex may be recognized and corrected by

Figure 5.30
Homologous recombination by complementary base pairing Parental DNAs are broken at staggered sites, and overlapping single-stranded regions are exchanged via base pairing with homologous sequences. The result is a heteroduplex region, in which the two DNA strands are derived from different parental molecules.

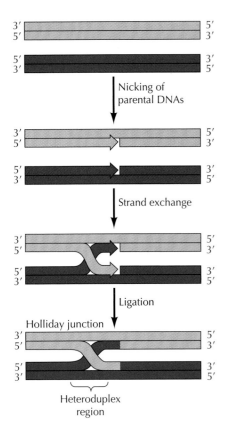

Figure 5.31
The Holliday model for homologous recombination   Single-strand nicks are introduced at the same position on both parental molecules. The nicked strands then exchange by complementary base pairing, and ligation produces a crossed-strand intermediate called a Holliday junction.

Figure 5.32
Identification of Holliday junctions by electron microscopy   Electron micrograph of a Holliday junction that was detected during recombination of plasmid DNAs in *E. coli*. An interpretive drawing of the structure is shown below. The molecule illustrates a Holliday junction in the open configuration resulting from rotation of the crossed-strand intermediate (see Figure 5.33). (Courtesy of Huntington Potter, University of South Florida, and David Dressler, University of Oxford.)

mismatch repair systems, as discussed in preceding sections of this chapter. Genetic evidence for the formation and repair of such heteroduplex regions, obtained in studies of recombination in fungi as well as in bacteria, led to the development of a molecular model for recombination in 1964. This model, known as the **Holliday model** (after Robin Holliday), has continued to provide the basis for current thinking about recombination mechanisms, although it has been modified as new data have been obtained.

The original version of the Holliday model proposed that recombination is initiated by the introduction of nicks at the same position on the two parental DNA molecules (Figure 5.31). The nicked DNA strands partially unwind, and each invades the other molecule by pairing with the complementary unbroken strand. Ligation of the broken strands then produces a crossed-strand intermediate, known as a **Holliday junction**, that is the central intermediate in recombination. The direct demonstration of Holliday junctions by electron microscopy has provided clear support for this model of recombination (Figure 5.32).

Once a Holliday junction is formed, it can be resolved by cutting and rejoining of the crossed strands to yield recombinant molecules (Figure 5.33). This can occur in two different ways, depending on the orientation of the Holliday junction, which can readily form two different isomers. In the isomer resulting from the initial strand exchange, the crossed strands are those that were nicked at the start of the recombination process. However, simple rotation of this structure yields a different isomer in which the unbroken parental strands are crossed. Resolution of these different isomers has distinct genetic consequences. In the first case, the progeny molecules have heteroduplex regions but are nonrecombinant for DNA that flanks these regions. If isomerization occurs, however, cutting and rejoining of the crossed strands results in progeny molecules that are recombinant for DNA that flanks the heteroduplex regions. The structure of the Holliday junction thus provides the possibility of generating both recombinant and nonrecombinant heteroduplexes, consistent with the genetic data upon which the Holliday model was based.

One modification of the Holliday model, proposed in 1975, eliminates a potential difficulty with the initial proposal—namely, how can both parental molecules be nicked simultaneously at the same position to initiate recombination? In this modified version, recombination is initiated by a nick in only one of the parental molecules (Figure 5.34). The nicked strand is then displaced, and the resulting single strand invades the other parental molecule by homologous base pairing. This process produces a displaced loop of DNA, which can then be cleaved and joined to the other parental molecule. The result is a crossed-strand Holliday junction, which can be resolved into recombinant or nonrecombinant heteroduplex molecules as already described.

Still another modification of the Holliday model suggests that recombination is initiated by a double-strand break, rather than a single-strand nick. Exonucleases at the site of the break then generate single-stranded tails, which can invade a homologous double-stranded molecule. Again the result is a Holliday junction, which can be resolved to yield either recombinant or nonrecombinant heteroduplexes. This double-strand break repair model appears to be particularly applicable to meiotic recombination in yeasts.

Multiple alternatives may thus account for the initial stages of recombination between two DNA molecules, and the details of recombination

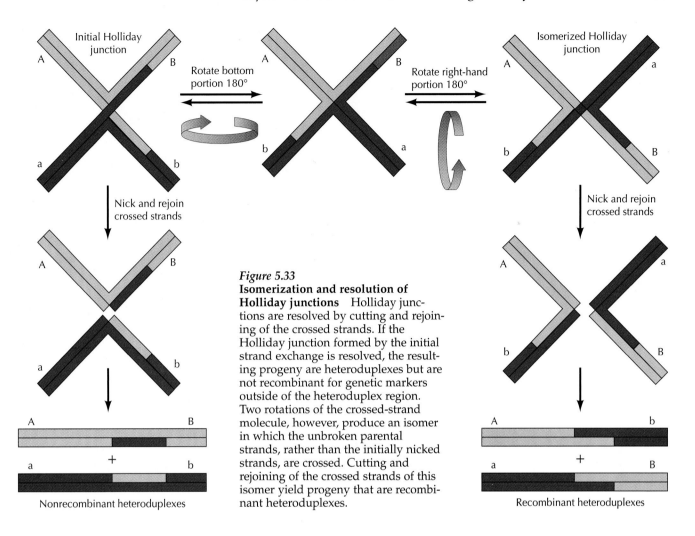

Figure 5.33
Isomerization and resolution of Holliday junctions Holliday junctions are resolved by cutting and rejoining of the crossed strands. If the Holliday junction formed by the initial strand exchange is resolved, the resulting progeny are heteroduplexes but are not recombinant for genetic markers outside of the heteroduplex region. Two rotations of the crossed-strand molecule, however, produce an isomer in which the unbroken parental strands, rather than the initially nicked strands, are crossed. Cutting and rejoining of the crossed strands of this isomer yield progeny that are recombinant heteroduplexes.

Figure 5.34
Initiation of recombination by nicking only one parental molecule The nicked DNA strand invades the other parental molecule by homologous base pairing, thereby displacing a single-stranded loop of DNA. This loop is then cleaved and pairs with the first parental molecule, yielding a crossed-strand Holliday junction.

mechanisms, particularly in eukaryotic cells, have not been fully eluci-dated. But the crossed-strand Holliday junction, generated by strand exchange leading to the formation of a heteroduplex region, remains the central intermediate in consideration of the recombination process.

Enzymes Involved in Homologous Recombination

Most of the enzymes currently known to be involved in recombination have been identified by analysis of recombination-defective mutants of *E. coli*. Such genetic analysis has established that recombination requires specific enzymes, in addition to proteins (such as DNA polymerase, ligase, and single-stranded DNA-binding proteins) that function in multiple aspects of DNA metabolism. The identification of genes required for effi-cient recombination in *E. coli* led to the isolation of their encoded proteins, which have been characterized by biochemical analysis in cell-free systems. These studies have elucidated the action of several enzymes in catalyzing the formation and resolution of Holliday junctions.

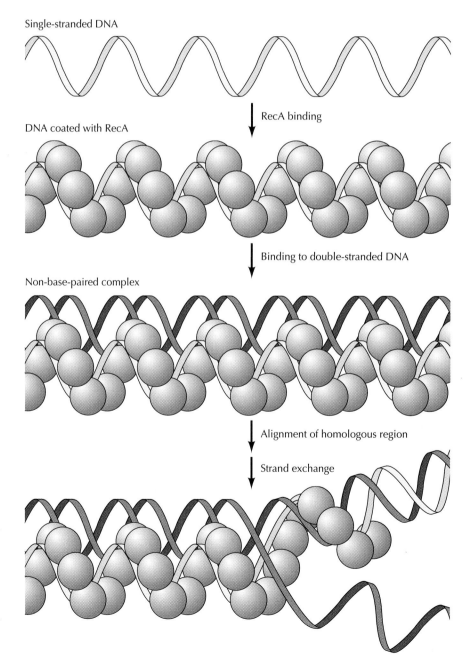

Single-stranded DNA

RecA binding

DNA coated with RecA

Binding to double-stranded DNA

Non-base-paired complex

Alignment of homologous region

Strand exchange

Figure 5.35
Function of the RecA protein RecA initially binds to single-stranded DNA to form a protein-DNA filament. The RecA protein that coats the single-stranded DNA then binds to a second, double-stranded DNA molecule to form a non-base-paired complex. Complementary base pairing and strand exchange follow, forming a het-eroduplex region.

The central protein involved in homologous recombination is **RecA**, which promotes the exchange of strands between homologous DNAs that causes heteroduplexes to form (Figure 5.35). The action of RecA can be considered in three stages. First, the RecA protein binds to single-stranded DNA, coating the DNA to form a protein-DNA filament. Because RecA has two DNA binding sites, the RecA protein bound to single-stranded DNA is able to bind a second, double-stranded DNA molecule, forming a complex between the two DNAs. This nonspecific RecA-mediated association is followed by specific base pairing between the single-stranded DNA and its complement. The RecA protein then catalyzes strand exchange, with the single strand originally coated with RecA displacing its homologous strand to form a heteroduplex. Thus, the RecA protein is capable of catalyzing, by itself, the strand exchange reactions that are central to the formation of Holliday junctions.

In yeast, a RecA-related protein, designated RAD51, is required for genetic recombination as well as for the repair of double-strand breaks. RAD51 is not only structurally similar to RecA; like RecA, it is also able to catalyze strand exchange reactions *in vitro*. Proteins related to RAD51 have been identified in complex eukaryotes, including humans, indicating that proteins related to RecA play key roles in homologous recombination in both prokaryotic and eukaryotic cells.

Most recombination events in *E. coli* also require the RecBCD enzyme, which is a complex of three proteins (RecB, C, and D). The properties of RecBCD are consistent with the hypothesis that it initiates recombination by providing the single-stranded DNA to which RecA binds. RecBCD accomplishes this task by unwinding and nicking double-stranded DNA (Figure 5.36). The RecBCD complex binds to the end of a DNA molecule and then

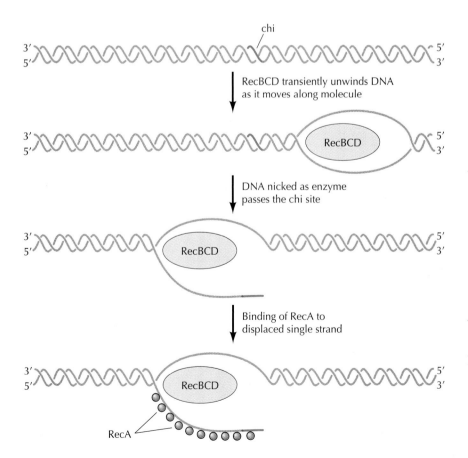

Figure 5.36
Initiation of recombination by RecBCD The *E. coli* RecBCD complex binds to the end of a DNA molecule and unwinds the DNA as it travels along the molecule. When it encounters a specific sequence (called a chi site), it nicks the DNA strand. Continued unwinding then forms a displaced single strand to which RecA can bind.

Holliday junction

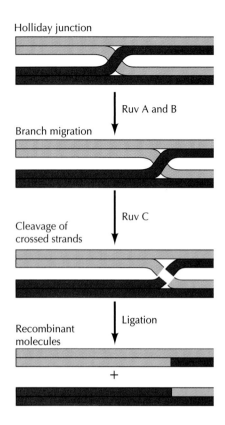

Branch migration

Cleavage of
crossed strands

Recombinant
molecules

Figure 5.37
Branch migration and resolution of Holliday junctions   Two *E. coli* proteins
(RuvA and RuvB) together catalyze the movement of the crossed-strand site in
Holliday junctions (branch migration). RuvC resolves the Holliday junctions by
cleaving the crossed strands, which are then joined by ligase.

acts as a helicase, transiently unwinding the DNA as it travels along the
molecule. When it encounters a specific nucleotide sequence (GCTGGTGG,
called a chi site), RecBCD acts as a nuclease to introduce a single-strand
nick. It then continues to unwind the double helix, forming a displaced single strand to which RecA can bind to initiate strand exchange.

Once a Holliday junction is formed, three other *E. coli* proteins (RuvA, B,
and C) become involved in recombination (Figure 5.37). RuvA and RuvB
act as a complex to drive the migration of the site at which the strands cross
in the Holliday junction, thereby varying the extent of the heteroduplex
region and the position at which the crossed strands will be cut and
rejoined. RuvC then resolves Holliday junctions by cleaving the crossed
DNA strands. Rejoining of the cleaved strands by ligation completes the
process, yielding two recombinant molecules.

In yeasts, Holliday junctions are resolved by a complex of RAD1 and
RAD10, with RAD1 cleaving single-stranded DNA at the crossover junction. RAD1 and RAD10 are homologs of the mammalian XPF and ERCC1
DNA repair proteins and also cleave damaged DNA during nucleotide-
excision repair (see Table 5.1). RAD1 and RAD10 homologs thus may have
a conserved function in recombination in eukaryotic cells.

DNA Rearrangements

Homologous recombination results in the reassortment of genes between
chromosome pairs without altering the arrangement of genes within the
genome. In contrast, other types of recombinational events lead to rearrangements of genomic DNA. Some of these DNA rearrangements are
important in controlling gene expression in specific cell types; others may
play an evolutionary role by contributing to genetic diversity.

The discovery that genes can move to different chromosomal locations
came from Barbara McClintock's studies of corn in the 1940s. Purely on the
basis of genetic analysis, McClintock described novel genetic elements that
could move to different locations in the genome and alter the expression of
adjacent genes. Nearly three decades elapsed, however, before the physical
basis of McClintock's work was elucidated by the discovery of transposable
elements in bacteria and the notion of movable genetic elements became
widely accepted by scientists. Several types of DNA rearrangements,
including the transposition of elements initially described by McClintock,
are now recognized in both prokaryotic and eukaryotic cells.

Site-Specific Recombination

In contrast to general homologous recombination, which occurs at any
extensive region of sequence homology, **site-specific recombination** occurs
between specific DNA sequences, which are usually homologous over only
a short stretch of DNA. The principal interaction in this process is mediated
by proteins that recognize the specific DNA target sequences rather than by
complementary base pairing.

The prototype of site-specific recombination has been provided by studies
of the bacteriophage λ. When λ infects *E. coli*, it can either replicate to cause

cell lysis or it can integrate into the bacterial chromosome, forming a prophage that is then maintained as part of the *E. coli* genome (a process called **lysogeny**) (Figure 5.38). Under appropriate conditions, DNA integration can be reversed, resulting in excision of the λ DNA and initiation of lytic viral replication. Both the integration and the excision of λ DNA involve site-specific recombination between viral and host cell DNA sequences.

 E. coli DNA and λ DNA recombine at specific sites, called attachment (*att*) sites. Thus, integration of λ DNA involves recombination between *att*

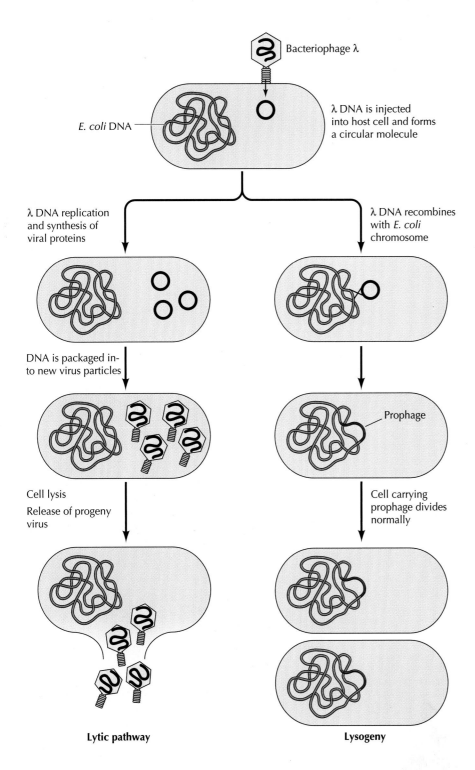

Figure 5.38
Lytic and lysogenic pathways of bacteriophage λ Infection of *E. coli* is initiated by the injection of λ DNA, which then becomes circular within the host cell. In lytic infection, the λ DNA replicates and directs the synthesis of viral proteins. The viral DNA is then packaged into progeny virus particles, which are released upon cell lysis. In lysogenic infection, the λ DNA recombines with the host genome to form a prophage that is integrated into the *E. coli* chromosome. The integrated λ DNA does not direct the synthesis of progeny viruses, but is instead replicated along with the rest of the bacterial genome.

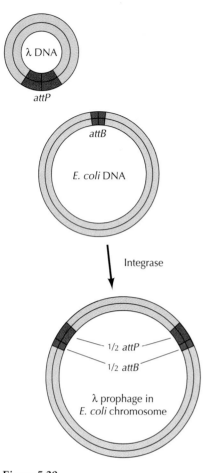

Figure 5.39
Integration of λ DNA by site-specific recombination Integration results from recombination between specific sequences in the λ and *E. coli* genomes, called *attP* and *attB*, respectively. The process is catalyzed by a virus-encoded enzyme (integrase), which recognizes both *attP* and *attB* sequences.

sites of the phage (*attP*) and the bacterium (*attB*), which are about 240 and 25 nucleotides long, respectively (Figure 5.39). The process is mediated by a λ protein called integrase (Int), which specifically binds to both *attP* and *attB* sequences. Int initially binds to *attP*, forming a complex in which the *attP* DNA is wrapped around multiple copies of the Int protein. The Int-*attP* complex binds to *attB*, aligning the phage and bacterial *att* sites. The phage and bacterium then exchange strands within a 15-nucleotide core sequence shared by *attB* and *attP* (Figure 5.40). The Int protein introduces staggered cuts within the core homology region of *attB* and *attP*, catalyzes

Figure 5.40
Mechanism of λ site-specific recombination Site-specific recombination occurs within a 15-nucleotide homologous core sequence shared by *attP* and *attB*. Integrase (Int) cleaves at specific sites within this sequence to generate staggered single-stranded DNA tails. It then catalyzes strand exchange and ligation, resulting in recombination between *attP* and *attB* and integration of λ DNA.

strand exchange, and then ligates the broken strands, integrating the λ DNA into the *E. coli* chromosome. The Int protein also acts in excision of the λ prophage, which is essentially the reverse of integration.

Site-specific recombination is important not only in the interaction of viruses such as λ with their host cells, but also in programmed gene rearrangements within cell genomes. In vertebrates, site-specific recombination is critical to the development of the immune system, which recognizes foreign substances (**antigens**) and provides protection against infectious agents. There are two major classes of immune responses, which are mediated by B and T lymphocytes. B lymphocytes secrete antibodies (**immunoglobulins**) that react with soluble antigens; T lymphocytes express cell surface proteins (called **T cell receptors**) that react with antigens expressed on the surfaces of other cells. The key feature of both immunoglobulins and T cell receptors is their enormous diversity, which enables different antibody or T cell receptor molecules to recognize a vast array of foreign antigens. For example, each individual is capable of producing more than 10^{11} different antibody molecules, which is far in excess of the total number of genes in the human genome (approximately 10^5). Rather than being encoded in germ-line DNA, these diverse antibodies (and T cell receptors) are encoded by unique lymphocyte genes that are formed during development of the immune system as a result of site-specific recombination between distinct segments of immunoglobulin and T cell receptor genes.

The role of site-specific recombination in the formation of immunoglobulin genes was first demonstrated by Susumu Tonegawa in 1976. Immunoglobulins consist of pairs of identical heavy and light polypeptide chains (Figure 5.41). Both the heavy and light chains are composed of C-terminal constant regions and N-terminal variable regions. The variable regions, which have different amino acid sequences in different immunoglobulin molecules, are responsible for antigen binding, and it is the diversity of variable region amino acid sequences that allows different individual antibodies to recognize unique antigens. Although every individual is capable of producing a vast spectrum of different antibodies, each B lymphocyte produces only a single type of antibody. Tonegawa's key discovery was that each antibody is encoded by unique genes formed by site-specific recombination during B lymphocyte development. These gene rearrangements create different immunoglobulin genes in different individual B lymphocytes, so the population of approximately 10^{12} B lymphocytes in the human body includes cells capable of producing antibodies against a diverse array of foreign antigens.

The genes that encode immunoglobulin light chains consist of three regions: a V region that encodes the 95 to 96 N-terminal amino acids of the polypeptide variable region; a joining (J) region that encodes the 12 to 14 C-terminal amino acids of the polypeptide variable region; and a C region that encodes the polypeptide constant region (Figure 5.42). The major class of light-chain genes in the mouse are formed from combinations of approximately 250 V regions and four J regions with a single C region. Site-specific recombination during lymphocyte development leads to a gene rearrangement in which a single V region recombines with a single J region to generate a functional light-chain gene. Different V and J regions are rearranged in different B lymphocytes, so the possible combinations of 250 V regions with 4 J regions can generate approximately 1000 (4 × 250) unique light chains.

The heavy-chain genes include a fourth region (known as the diversity, or D, region), which encodes amino acids lying between V and J (Figure 5.43).

Figure 5.41
Structure of an immunoglobulin
Immunoglobulins are composed of two heavy chains and two light chains, joined by disulfide bonds. Both the heavy and the light chains consist of variable and constant regions.

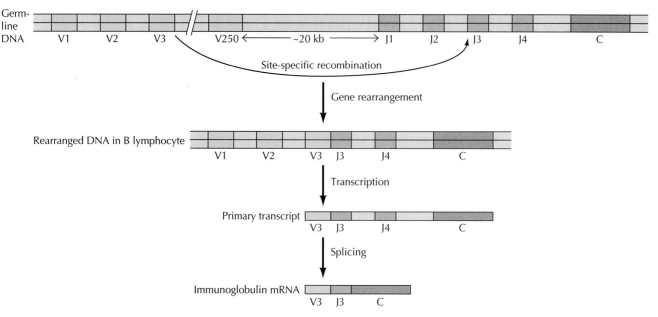

Figure 5.42
Rearrangement of immunoglobulin light-chain genes   Each light-chain gene
(mouse κ light chains are illustrated) consists of a constant region (C), a joining
region (J), and a variable region (V). There are approximately 250 different
V regions, which are separated from J and C by about 20 kb in germ-line DNA.
During the development of B lymphocytes, site-specific recombination joins one of
the V regions to one of the four J regions. This rearrangement activates transcription, resulting in the formation of a primary transcript containing the rearranged
VJ region together with the remaining J regions and C. The remaining unused
J regions and the introns between J and C are then removed by splicing, yielding
a functional mRNA.

Figure 5.43
**Rearrangement of immuno-
globulin heavy-chain genes**
The heavy-chain genes contain D regions in addition to
V, J, and C regions. First the
D and J segments join. Then
a V segment is joined to the
rearranged DJ region. The
introns between J and C are
removed by splicing to yield
heavy-chain mRNA.

Figure 5.44
Structure of a T cell receptor   T cell receptors consist of two polypeptide chains
(α and β) that span the plasma membrane and are joined by disulfide bonds. Both
the α and β chains are composed of variable and constant regions.

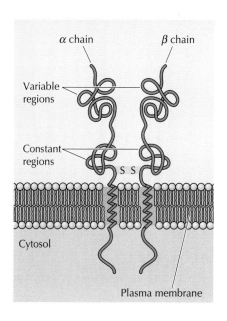

Assembly of a functional heavy-chain gene requires two recombination
events: A D region first recombines with a J region, and a V region then
recombines with the rearranged DJ segment. In the mouse, there are about
500 heavy-chain V regions, 12 D regions, and 4 J regions, so the total number of heavy chains that can be generated by the recombination events is
24,000 ($500 \times 12 \times 4$).

Combinations between the 1000 different light chains and 24,000 different heavy chains formed by site-specific recombination can generate
approximately 2×10^7 different immunoglobulin molecules. This diversity
is further increased because the joining of immunoglobulin gene segments
is often imprecise, with one to several nucleotides frequently being lost or
gained at the sites of joining. The mutations resulting from these deletions
and insertions increase the diversity of immunoglobulin variable regions
approximately a hundredfold, corresponding to the formation of about 10^5
different light chains and 2×10^6 heavy chains, which can then combine to
form more than 10^{11} distinct antibodies. Still further antibody diversity is
generated after the formation of rearranged immunoglobulin genes by a
process known as somatic hypermutation, which results in the introduction of frequent mutations into the variable regions of both heavy-chain
and light-chain genes.

T cell receptors similarly consist of two chains (called α and β), each of
which contains variable and constant regions (Figure 5.44). The genes
encoding these polypeptides are generated by recombination between V
and J segments (the α chain) or between V, D, and J segments (the β chain),
analogous to the formation of immunoglobulin genes. Site-specific recombination between these distinct segments of DNA, in combination with
mutations introduced during recombination, generates a degree of diversity in T cell receptors that is similar to that in immunoglobulins. However,
T cell receptors differ from immunoglobulins in that they are not subject to
the introduction of further diversity by somatic hypermutation.

VDJ recombination is mediated by signal sequences adjacent to the coding sequences of each gene segment (Figure 5.45). These recombination signal sequences are recognized and cleaved by a complex of two proteins,
called RAG1 and RAG2, which are specifically expressed in lymphocytes.
Subsequent joining of the coding ends of the gene segments then yields a
rearranged immunoglobulin or T cell receptor gene. Interestingly, RAG1
and RAG2 are closely related to the enzymes that catalyze DNA transposition in bacteria, as discussed in the next section of this chapter.

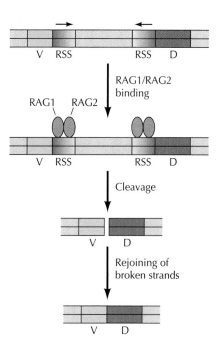

Figure 5.45
VDJ recombination   The coding segments of immunoglobulin and T cell receptor genes (e.g., a V and D segment) are flanked by short recombination signal
sequences (RSS), which are in opposite orientations at the 5' and 3' ends of the
coding sequences. The RSS are recognized by a complex of the lymphocyte-
specific recombination proteins RAG1 and RAG2, which cleave the DNA between
the coding sequence and the RSS. The broken coding strands are then rejoined to
yield a rearranged gene segment.

Key Experiment

Rearrangement of Immunoglobulin Genes

Evidence for Somatic Rearrangement of Immunoglobulin Genes Coding for Variable and Constant Regions

Nobumichi Hozumi and Susumu Tonegawa
Basel Institute for Immunology, Basel, Switzerland

Proceedings of the National Academy of Sciences, USA, Volume 73, 1976, pages 3628–3632

The Context

The ability of the vertebrate immune system to recognize a seemingly infinite variety of foreign molecules implies that lymphocytes can produce a correspondingly vast array of antibodies. Since this antibody diversity is key to immune recognition, understanding the mechanism by which an apparently unlimited number of distinct immunoglobulins are encoded in genomic DNA is a central issue in immunology.

Prior to the experiments of Hozumi and Tonegawa, protein sequencing of multiple immunoglobulins had demonstrated that both heavy and light chains consist of distinct variable and constant regions. Genetic studies further indicated that mice inherit only single copies of the constant-region genes. These observations first led to the proposal that immunoglobulins are encoded by multiple variable-region genes that can associate with a single constant-region gene. The discovery of immunoglobulin gene rearrangements by Hozumi and Tonegawa provided the first direct experimental support for this hypothesis and laid the groundwork for understanding the molecular basis of antibody diversity.

The Experiments

Hozumi and Tonegawa tested the possibility that the genes encoding immunoglobulin variable and constant regions were joined at the DNA level during lymphocyte development. Their experimental approach was to use restriction endonuclease digestion to compare the organization of variable-region and constant-region sequences in DNAs extracted from mouse embryos and from cells of a mouse plasmacytoma (a B lymphocyte tumor that produces a single species of immunoglobulin).

Embryo and plasmacytoma DNAs were digested with the restriction endonuclease *Bam*HI, and DNA fragments of different sizes were separated by electrophoresis in agarose gels. The gel was then cut into slices, and DNA extracted from each slice was hybridized with radiolabeled probes that had been prepared from immunoglobulin mRNA isolated from the plasmacytoma cells. Two probes were used, corresponding either to the complete immunoglobulin mRNA or to the 3′ half of the mRNA, consisting only of constant-region sequences.

The critical result was that completely different patterns of variable-region and constant-region sequences were detected in embryo versus plasmacytoma DNAs (see figure). In embryo DNA, the complete probe hybridized to two *Bam*HI fragments

Gel electrophoresis of embryo and plasmacytoma DNAs digested with *Bam*HI and hybridized to probes corresponding to either the whole or the 3′ half of the plasmacytoma mRNA. Data are presented as the radioactivity detected in hybrid molecules with DNA from each gel slice.

Rearrangement of Immunoglobulin Genes *(continued)*

of approximately 8.6 and 5.6 kb, respectively. Only the 8.6-kb fragment hybridized to the 3′ probe, suggesting that the 8.6-kb fragment contained constant-region sequences and the 5.6-kb fragment contained variable-region sequences. In striking contrast, both probes hybridized to only a single 3.4-kb fragment in plasmacytoma DNA. The interpretation of these results was that the variable- and constant-region sequences were separated in embryo DNA but rearranged to form a single immunoglobulin gene during lymphocyte development.

The Impact

The initial results of Hozumi and Tonegawa, based on the relatively indirect approach of restriction endonuclease mapping, were con-

firmed and extended by the molecular cloning and sequencing of immunoglobulin genes. Such studies have now unambiguously established that these genes are generated by site-specific recombination between distinct segments of DNA in B lymphocytes. In T lymphocytes, similar DNA rearrangements are responsible for formation of the genes encoding T cell receptors. Thus, site-specific recombination and programmed gene rearrangements are central to the development of the immune system.

Further studies have shown that the variable regions of immunoglobulins and T cell receptors are generated by rearrangements of two or three distinct segments of DNA. The ability of these segments to recombine, together with a high frequency of mutations

Susumu Tonegawa

introduced at the recombination sites, is largely responsible for immunoglobulin and T cell receptor diversity. The discovery of immunoglobulin gene rearrangements thus provided the basis for understanding how the immune system can recognize and respond to a virtually unlimited range of foreign substances.

Transposition via DNA Intermediates

Site-specific recombination occurs between two specific sequences that contain at least a small core of homology. In contrast, **transposition** involves the movement of sequences throughout the genome and has no requirement for sequence homology. Elements that move by transposition, such as those first described by McClintock, are called **transposable elements,** or **transposons**. They are divided into two general classes, depending on whether they transpose via DNA intermediates or via RNA intermediates. The first class of transposable elements is discussed here; transposition via RNA intermediates is considered in the next section.

The best-studied transposons are those of bacteria, all of which move via DNA intermediates (Figure 5.46). The simplest of these elements are the insertion sequences, ranging in size from about 800 to 2000 nucleotides. Insertion sequences consist only of a gene for the enzyme involved in transposition (transposase) flanked by short inverted repeats, which are the

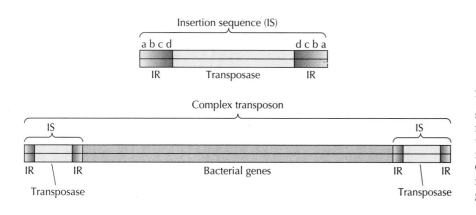

Figure 5.46
Bacterial transposons Insertion sequences (IS) range from 800 to 2000 nucleotides and contain a gene for transposase flanked by inverted repeats (IR) of about 20 nucleotides. Complex transposons consist of two insertion sequences flanking other genes and are typically 5 to 20 kb long.

Figure 5.47
Transposition of insertion sequences
Simple transposition does not involve replication of the transposon DNA. Transposase cleaves at both ends of the transposon and introduces a staggered cut in the target DNA. The overhanging ends of target DNA are then joined to the transposon, and gaps resulting from the staggered cuts at the target site are repaired. The result is the formation of short direct repeats of target-site DNA (5 to 10 nucleotides long) flanking the integrated transposon.

sites at which transposase acts. Complex transposons consist of two insertion sequences flanking other genes, which move as a unit.

Insertion sequences move from one chromosomal site to another without replicating their DNA (Figure 5.47). Transposase introduces a staggered break in the target DNA and cleaves at the ends of the transposon inverted-repeat sequences. Although transposase acts specifically at the transposon inverted repeats, it is usually less specific with respect to the sequence of the target DNA, so it catalyzes the movement of transposons throughout the genome. Following the cleavage of transposon and target site DNAs, transposase joins the overhanging ends of the target DNA to the transposable element. The resulting gap in the target-site DNA is repaired by DNA synthesis, followed by ligation to the other strand of the transposon. The result of this process is a short direct repeat of the target-site DNA on both sides of the transposable element—a hallmark of transposon integration.

This transposition mechanism causes the transposon to move from one chromosomal site to another. Other types of transposons move by a more complex mechanism, in which the transposon is replicated in concert with its integration into a new target site. This mechanism results in the integration of one copy of the transposon into a new position in the genome, while another copy remains at its original location.

Transposons that move via DNA intermediates are present in eukaryotes as well as in bacteria. The original transposable elements described by McClintock in corn move by a nonreplicative mechanism, as do some transposable elements in *Drosophila* and other insects. Like most bacterial transposons, these elements move to many different target sites throughout the genome. Although the movement of these transposons itself is not likely to be useful to the cells in which it occurs, the mutations induced by transposon movement may play an important role in evolution.

In yeasts and protozoans, however, transposition by a replicative mechanism is responsible for programmed DNA rearrangements that regulate gene expression. In these cases transposition is initiated by the action of a site-specific nuclease that cleaves a specific target site, at which a copy of the transposable element is then inserted. Transposable elements are thus capable not only of moving to nonspecific sites throughout the genome, but also of participating in specific gene rearrangements that result in programmed changes in gene expression.

Transposition via RNA Intermediates

Most transposons in eukaryotic cells move via RNA intermediates rather than DNA intermediates. The mechanism of transposition of these elements is similar to the replication of retroviruses, which have provided the prototype system for studying this class of movable DNA sequences.

Retroviruses contain RNA genomes in their virus particles but replicate via the synthesis of a DNA provirus, which is integrated into the chromosomal DNA of infected cells (see Figure 3.15). A DNA copy of the viral RNA is synthesized by the viral enzyme **reverse transcriptase**. The mechanism by which this occurs results in the synthesis of a DNA molecule that contains direct repeats of several hundred nucleotides at both ends (Figure 5.48). These repeated sequences, called **long terminal repeats**, or **LTRs**, arise from duplication of the sites on viral RNA at which primers bind to initiate DNA synthesis. The LTR sequences thus play central roles in reverse transcription, in addition to being involved in the integration and subsequent transcription of proviral DNA.

Like all DNA polymerases, reverse transcriptase requires a primer, which in the case of retroviruses is a tRNA molecule bound at a specific site (the primer-binding site) close to the 5′ terminus of the viral RNA (Figure 5.49). Since DNA synthesis proceeds in the 5′ to 3′ direction, only a short piece of DNA is synthesized before reverse transcriptase reaches the end of its template. Continuation of DNA synthesis then depends on the ability of reverse transcriptase to "jump" to the 3′ end of the template RNA molecule. This is accomplished via an RNase H activity of reverse transcriptase, which degrades the RNA strand of DNA-RNA hybrids. As a result, the newly synthesized DNA is converted to a single-stranded molecule, which can hybridize to a short repeated sequence present at both the 5′ and the 3′ ends of the viral RNA. DNA synthesis can then continue, yielding a single-stranded DNA complementary to viral RNA. Synthesis of the opposite strand of DNA is initiated by a fragment of viral RNA that acts as a primer, at a site near the 3′ end of the template DNA strand. Again the result is a short piece of DNA, which includes the primer-binding site copied from the tRNA used as the initial primer for reverse transcription. The primer-binding sequence of the tRNA is then degraded by RNase H, leaving an overhanging DNA strand that again "jumps" to pair with its complementary sequence at the other end of the template. DNA

Figure 5.48
The organization of retroviral DNA
The integrated proviral DNA is flanked by long terminal repeats (LTRs), which are direct repeats of several hundred nucleotides. Viral genes, including genes for reverse transcriptase and for structural proteins of the virus particle, are located between the LTRs. The integrated provirus is flanked by short direct repeats of host DNA.

Figure 5.49
Generation of LTRs during reverse transcription LTRs consist of three sequence elements: a short repeat sequence (R) of about 20 nucleotides that is present at both ends of the viral RNA; a sequence unique to the 5′ end of viral RNA (U5); and a sequence unique to the 3′ end of viral RNA (U3). Repeats of these sequences are generated during DNA synthesis as reverse transcriptase jumps twice between the ends of its template. Synthesis is initiated using a tRNA primer bound to a primer-binding site (PBS) adjacent to U5 at the 5′ end of the viral RNA. The polymerase copies R, and the RNA strand of the RNA-DNA hybrid is then degraded by RNase H. The polymerase then jumps to the 3′ end of the viral RNA in order to synthesize a complete DNA strand complementary to the RNA template. The polymerase jumps again during synthesis of the second strand of DNA, which is also initiated by a primer bound close to the 5′ end of its template. The result of these jumps is the formation of LTRs that contain U3-R-U5 sequences.

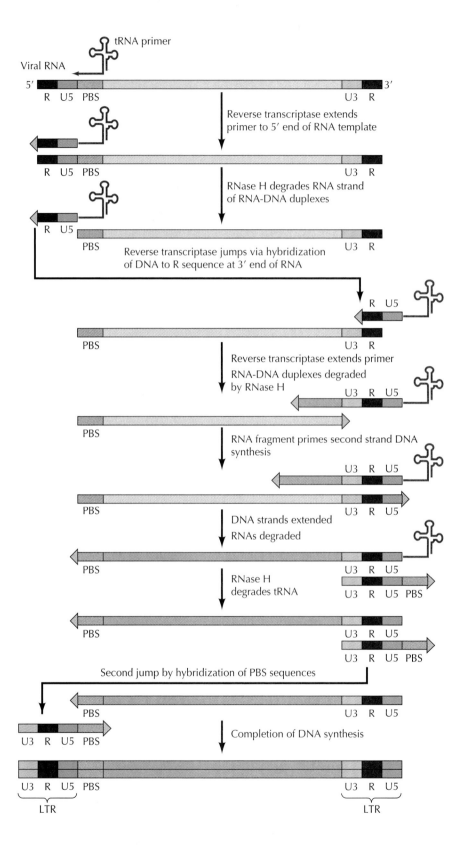

synthesis can then continue once more, finally yielding a linear DNA with LTRs at both ends.

The linear viral DNA integrates into the host cell chromosome by a process that resembles the integration of DNA transposable elements. Integration is catalyzed by a viral integration protein and occurs at many dif-

Figure 5.50
Structure of a class I retrotransposon
The yeast Ty1 transposable element displays the same organization as a retrovirus. Protein-coding sequences, including genes for reverse transcriptase and integration proteins, are flanked by LTRs (called δ elements) of about 330 base pairs (bp). The integrated transposon is flanked by short direct repeats of target-site DNA.

ferent target sequences in cellular DNA. The integration protein cleaves two bases from the ends of viral DNA and introduces a staggered cut at the target site in cellular DNA. The overhanging ends of cellular DNA are then joined to the termini of viral DNA, and the gap is filled by DNA synthesis. The integrated provirus is therefore flanked by a direct repeat of cell sequences, similar to the repeats that flank DNA transposons.

The viral life cycle continues with transcription of the integrated provirus, which yields viral genomic RNA as well as mRNAs that direct the synthesis of viral proteins (including reverse transcriptase and integration protein). The genomic RNA is then packaged into viral particles, which are released from the host cell. These progeny viruses can infect a new cell, initiating another round of DNA synthesis and integration. The net effect can be viewed as the movement of the provirus from one chromosomal site to another, via the synthesis and reverse transcription of an RNA intermediate.

The retroviruses can be considered one type of **retrotransposon**—an element that moves via RNA intermediates. Other retrotransposons differ from retroviruses in that they are not packaged into infectious particles and therefore cannot spread from one cell to another. However, these retrotransposons can move to new chromosomal sites within the same cell, via mechanisms fundamentally similar to those involved in retrovirus replication.

Some retrotransposons (called class I retrotransposons) are structurally similar to retroviruses (Figure 5.50). Retrotransposons of this type have been studied in yeasts, *Drosophila*, mice, and plants. They have LTR sequences at both ends; they encode reverse transcriptase and integration proteins; and they transpose (like retroviruses) via transcription into RNA, synthesis of a new DNA copy by reverse transcriptase, and integration into cellular DNA. Class II retrotransposons differ from retroviruses in that they do not contain LTR sequences, although they do encode their own reverse transcriptase. Retrotransposons of this type are found in mammals, insects (including *Drosophila*), plants, and trypanosomes. In mammals, the major class of these retrotransposons consists of the highly repetitive long interspersed elements (**LINEs**), which are repeated approximately 50,000 times in the genome and account for about 10% of genomic DNA (see Chapter 4). A full-length LINE element is 6 to 7 kb long, although many members of the family are truncated at their 5′ end (Figure 5.51). At their 3′ end, LINEs have tracts of A-rich sequences thought to be derived by reverse transcription of the poly-A tails that are added to mRNAs following transcription (see Chapter 6). Like other transposable elements, LINEs are flanked by short direct repeats of the target-site DNA, indicating that integration involves staggered cuts and repair synthesis.

Figure 5.51
Structure of human LINEs LINEs lack LTRs, but they do encode reverse transcriptase. They have tracts of A-rich sequences (designated A_n) at their 3′ ends, which are thought to arise from reverse transcription of poly-A tails added to the 3′ end of mRNAs. Like other transposable elements, LINEs are flanked by short direct repeats of target-site DNA.

LINE

Figure 5.52
Model for reverse transcription and integration of non-LTR-containing retrotransposons   Reverse transcription, primed by a broken end of DNA at the target site, initiates within the poly-A tail at the 3′ end of retrotransposon RNA. Synthesis of the opposite strand of retrotransposon DNA is similarly primed by the other strand of DNA at the target site.

Since LINEs do not contain LTR sequences, the mechanism of their reverse transcription and subsequent integration into chromosomal DNA must differ from that of retroviruses and class I (LTR-containing) retrotransposons. In particular, reverse transcription is primed by a broken end of chromosomal DNA at the integration target site, resulting from cleavage of the target site DNA by a nuclease encoded by the retrotransposon (Figure 5.52). Reverse transcription then initiates within the poly-A tract at the 3′ end of the transposon RNA and continues along the molecule. The opposite strand of DNA is synthesized using the other broken end of target-site DNA as primer, resulting in simultaneous synthesis and integration of the retrotransposon DNA.

Other sequence elements, which do not encode their own reverse transcriptase, also transpose via RNA intermediates. These elements include the highly repetitive short interspersed elements (**SINEs**), of which there are nearly a million copies in mammalian genomes (see Chapter 4). The major family of these elements consists of the *Alu* sequences, which are about 300 bases long. These sequences have A-rich tracts at their 3′ end and are flanked by short duplications of target-site DNA sequences, a structure similar to that of retrotransposons that lack LTR sequences (e.g., LINEs). SINEs arose by reverse transcription of small RNAs, including tRNAs and small cytoplasmic RNAs involved in protein transport. Since SINEs no longer encode functional RNA products, they represent pseudogenes that arose via RNA-mediated transposition. Pseudogenes of many protein-coding genes (called **processed pseudogenes**) have similarly arisen by reverse transcription of mRNAs (Figure 5.53). Such processed pseudogenes are readily recognized not only because they terminate in an A-rich tract but also because the introns present in the corresponding normal gene have been removed during mRNA processing. The transposition of SINEs and of other processed pseudogenes is thought to proceed similarly to the transposition of LINEs. However, since these elements do not include genes for reverse transcriptase or a nuclease, their transposition presumably involves the action of reverse transcriptases and nucleases that are encoded elsewhere in the genome—probably by class I or II retrotransposons, such as LINEs.

Although the highly repetitive SINEs and LINEs account for a significant fraction of genomic DNA, their transpositions to random sites in the genome are not likely to be useful for the cell in which they are located. These transposons induce mutations when they integrate at a new target

Figure 5.53
Formation of a processed pseudogene
The gene illustrated contains three exons, separated by two introns. The introns are removed from the primary transcript by splicing, and a poly-A tail is added to the 3′ end of the mRNA. Reverse transcription and integration then yield a processed pseudogene, which does not contain introns and has an A-rich tract at its 3′ end. The processed pseudogene is flanked by short direct repeats of target-site DNA that were generated during its integration.

site, and like mutations induced by other agents, most mutations resulting from transposon integration are expected to be harmful to the cell. Indeed, mutations resulting from the transposition of both LINEs and *Alu* elements have been associated with some cases of hemophilia, muscular dystrophy, and colon cancer. On the other hand, some mutations resulting from the movement of transposable elements may be beneficial, contributing in a positive way to evolution of the species. For example, some retrotransposons in mammalian genomes have been found to contain regulatory sequences that control the expression of adjacent genes.

In addition to their role as mutagens, retrotranspons may contribute to genetic diversity by stimulating DNA rearrangements. Sequences of cellular DNA adjacent to LINEs are frequently carried along during the process of transposition. Consequently, the transposition of LINEs can result in the movement of cellular DNA sequences to new genomic sites. Since LINEs can integrate into active genes, the associated transposition of cellular DNA sequences can lead to the formation of new combinations of regulatory and/or coding sequences and contribute directly to the evolution of new genes.

Gene Amplification

The DNA rearrangements that have been discussed so far alter the position of a DNA sequence within the genome. **Gene amplification** may be viewed as a different type of alteration in genome structure; it increases the number of copies of a gene within a cell. Gene amplification results from repeated rounds of DNA replication, yielding multiple copies of a particular region (Figure 5.54). The amplified DNA sequences can be found either as free extrachromosomal molecules or as tandem arrays of sequences within a chromosome. In either case, the result is increased expression of the amplified gene, simply because more copies of the gene are available to be transcribed.

In some cases, gene amplification is responsible for developmentally programmed increases in gene expression. The prototypical example is amplification of the ribosomal RNA genes in amphibian oocytes (eggs). Eggs are extremely large cells, with correspondingly high requirements for protein synthesis. Amphibian oocytes in particular are about a million times larger in volume than typical somatic cells and must support large amounts of protein synthesis during early development. This requires

Figure 5.54
DNA amplification   Repeated rounds of DNA replication yield multiple copies of a particular chromosomal region.

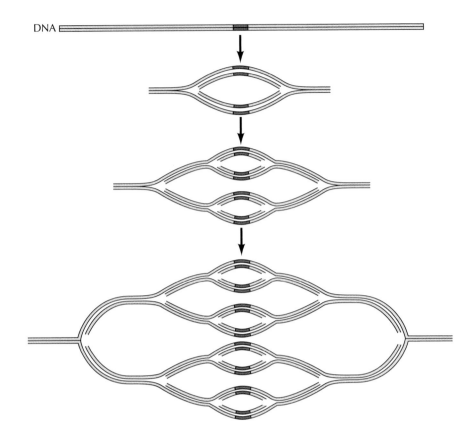

increased synthesis of ribosomal RNAs, which is accomplished in part by amplification of the ribosomal RNA genes. As discussed in Chapter 4, there are already several hundred copies of ribosomal RNA genes per genome, so that enough ribosomal RNA can be produced to meet the needs of somatic cells. In amphibian eggs, these genes are amplified an additional 2000-fold, to approximately 1 million copies per oocyte. Another example of programmed gene amplification occurs in *Drosophila*, where the genes that encode eggshell proteins (chorion genes) are amplified in ovarian cells to support the requirement for large amounts of these proteins. Like other programmed gene rearrangements, however, gene amplification is a relatively infrequent event that occurs in highly specialized cell types; it is not a common mechanism of gene regulation.

Gene amplification also occurs as an abnormal event in cancer cells, where it results in the increased expression of genes that contribute to uncontrolled cell growth. Such gene amplification was first recognized in cancer cells that had become resistant to methotrexate, a drug commonly used in cancer chemotherapy. Methotrexate inhibits the enzyme dihydrofolate reductase, which is involved in the synthesis of dNTPs and is therefore required for DNA synthesis. Resistance to methotrexate frequently develops by amplification of the dihydrofolate reductase gene, leading to increased production of the enzyme and consequently the loss of effective inhibition by methotrexate. In addition, gene amplification in cancer cells frequently results in the increased expression of genes that drive cell proliferation (oncogenes) and thereby directly contributes to tumor development (see Chapter 15). For example, amplification of the oncogene *erb*B-2 is frequently involved in human breast cancers. Thus, as with other types of DNA rearrangements, gene amplification can have either beneficial or deleterious consequences for the cell or organism in which it occurs.

SUMMARY	**KEY TERMS**

DNA REPLICATION

DNA Polymerases: Different DNA polymerases play distinct roles in DNA replication and repair in both prokaryotic and eukaryotic cells. All known DNA polymerases synthesize DNA only in the 5′ to 3′ direction by the addition of dNTPs to a preformed primer strand of DNA.

DNA polymerase, mutagen

The Replication Fork: Parental strands of DNA separate and serve as templates for the synthesis of two new strands at the replication fork. One new DNA strand (the leading strand) is synthesized in a continuous manner; the other strand (the lagging strand) is formed by the joining of small fragments of DNA that are synthesized backward with respect to the overall direction of replication. DNA polymerases and various other proteins act in a coordinated manner to synthesize both leading and lagging strands of DNA.

replication fork, Okazaki fragment, DNA ligase, leading strand, lagging strand, primase, RNase H, exonuclease, helicase, single-stranded DNA-binding protein, topoisomerase

The Fidelity of Replication: DNA polymerases increase the accuracy of replication both by selecting the correct base for insertion and by proofreading newly synthesized DNA to eliminate mismatched bases.

proofreading

Origins and the Initiation of Replication: DNA replication starts at specific origins of replication, which contain binding sites for proteins that initiate the process.

origin of replication, autonomously replicating sequence (ARS), origin replication complex (ORC)

Telomeres and Telomerase: Replicating the Ends of Chromosomes: Telomeric repeat sequences at the ends of chromosomes are replicated by the action of a reverse transcriptase (telomerase) that carries its own template RNA.

telomere, telomerase, reverse transcriptase

DNA REPAIR

Direct Reversal of DNA Damage: A few types of common DNA lesions, such as pyrimidine dimers and alkylated guanine residues, are repaired by direct reversal of the damage.

pyrimidine dimer, photoreactivation

Excision Repair: Most types of DNA damage are repaired by excision of the damaged DNA. The resulting gap is filled by newly synthesized DNA, using the undamaged complementary strand as a template. In base-excision repair, specific types of single damaged bases are removed from the DNA molecule. In contrast, nucleotide excision repair systems recognize a wide variety of lesions that distort the structure of DNA and remove the damaged bases as part of an oligonucleotide. A third excision repair system specifically removes mismatched bases from newly synthesized DNA strands.

base-excision repair, DNA glycosylase, AP endonuclease, nucleotide-excision repair, excinuclease, mismatch repair

Postreplication Repair: Replication of damaged DNA can result in the synthesis of a daughter strand containing a gap opposite the site of damage in the parental template strand. Such gaps in newly synthesized DNA can be filled by recombination with the undamaged parental strand or by error-prone repair synthesis using the damaged parental strand as a template.

recombinational repair, error-prone repair

RECOMBINATION BETWEEN HOMOLOGOUS DNA SEQUENCES

DNA Molecules Recombine by Breaking and Rejoining: The molecular mechanism of recombination involves the breaking and rejoining of parental DNA molecules.

general homologous recombination, Holliday model, Holliday junction

Models of Homologous Recombination: Alignment between homologous DNA molecules is provided by complementary base pairing. Nicked strands of parental DNA invade the other parental molecule, yielding a crossed-strand intermediate known as a Holliday junction. Recombinant molecules are then formed by cleavage and rejoining of the crossed strands.

RecA

Enzymes Involved in Homologous Recombination: The central enzyme of homologous recombination is RecA, which catalyzes the exchange of strands between homologous DNAs. Other enzymes nick and unwind parental DNAs and resolve Holliday junctions.

DNA REARRANGEMENTS

site-specific recombination, lysogeny, antigen, immunoglobulin, T cell receptor

Site-Specific Recombination: Site-specific recombination takes place between specific DNA sequences that are recognized by proteins that mediate the process. In vertebrates, site-specific recombination plays a critical role in generating immunoglobulin and T cell receptor genes during development of the immune system.

transposition, transposable element, transposon

Transposition via DNA Intermediates: Most DNA transposons move throughout the genome with no requirement for specific DNA sequences at their sites of insertion. In yeasts and protozoans, however, the transposition of some DNA sequences to specific target sites results in programmed DNA rearrangements that regulate gene expression.

retrovirus, reverse transcriptase, long terminal repeat (LTR), retrotransposon, LINE, SINE, processed pseudogene

Transposition via RNA Intermediates: Most transposons in eukaryotic cells move by reverse transcription of RNA intermediates, similar to the replication of retroviruses. These retrotransposons include the highly repeated LINE and SINE sequences of mammalian genomes.

gene amplification

Gene Amplification: Gene amplification results from repeated replication of a chromosomal region. In some cases, gene amplification provides a mechanism for increasing gene expression during development. Gene amplification also frequently occurs in cancer cells, where it can result in the elevated expression of genes that contribute to uncontrolled cell proliferation.

Questions

1. How is the fidelity of DNA replication affected by the fact that DNA polymerase adds nucleotides only to a primer strand that is hydrogen-bonded to the template?

2. Would you expect the RNA fragments synthesized by primase to be accurate copies of the template DNA? How does this affect the overall accuracy of DNA replication?

3. Patients with xeroderma pigmentosum suffer an extremely high incidence of skin cancer but have not been found to have correspondingly high incidences of cancers of internal organs (e.g., colon cancer). What might this suggest about the kinds of DNA damage responsible for most internal cancers?

4. *RecA* mutants of *E. coli* are sensitive to UV irradiation in addition to being recombination-deficient. Why?

5. What phenotype would you predict for a mutant mouse lacking one of the genes required for site-specific recombination in lymphocytes?

6. Many of the drugs in clinical use and under evaluation for the treatment of AIDS are inhibitors of the HIV reverse transcriptase. What reverse transcriptases in human cells might also be inhibited by these drugs? What would be the consequences of inhibiting these enzymes?

References and Further Reading

DNA Replication

Aladjem, M. I., L. W. Rodewald, J. L. Kolman and G. M. Wahl. 1998. Genetic dissection of a mammalian replicator in the human β-globin locus. *Science* 281: 1005–1009. [P]

Baker, T. A. and S. P. Bell. 1998. Polymerases and the replisome: Machines within machines. *Cell* 92: 296–305. [R]

Bambara, R. A., R. S. Murante and L. A. Henricksen. 1997. Enzymes and reactions at the eukaryotic DNA replication fork. *J. Biol. Chem.* 272: 4647–4650. [R]

Blackburn, E. H. 1992. Telomerases. *Ann. Rev. Biochem.* 61: 113–129. [R]

Cairns, J. 1963. The chromosome of *Escherichia coli*. *Cold Spring Harbor Symp. Quant. Biol.* 28: 43–46. [P]

Dutta, A. and S. P. Bell. 1997. Initiation of DNA replication in eukaryotic cells. *Ann. Rev. Cell Dev. Biol.* 13: 293–332. [R]

Echols, H. and M. F. Goodman. 1991. Fidelity mechanisms in DNA replication. *Ann. Rev. Biochem.* 60: 477–511. [R]

Gavin, K. A., M. Hidaka and B. Stillman. 1995. Conserved initiator proteins in eukaryotes. *Science* 270: 1667–1671. [P]

Greider, C. W. 1996. Telomere length regulation. *Ann. Rev. Biochem.* 65: 337–365. [R]

Heintz, N. H., L. Dailey, P. Held and N. Heintz. 1992. Eukaryotic replication origins as promoters of bidirectional DNA synthesis. *Trends Genet.* 8: 376–381. [R]

Huberman, J. A. 1998. Choosing a place to begin. *Science* 281: 929–930. [R]

Huberman, J. A. and A. D. Riggs. 1968. On the mechanism of DNA replication in mammalian chromosomes. *J. Mol. Biol.* 32: 327–341. [P]

Kesti, T., K. Flick, S. Keranen, J. E. Syvaoja and C. Wittenberg. 1999. DNA polymerase ε catalytic domains are dispensable for DNA replication, DNA repair, and cell viability. *Mol. Cell* 3: 679–685. [P]

Kornberg, A. and T. A. Baker. 1992. *DNA Replication*. 2nd ed. New York: W. H. Freeman.

Kornberg, A., I. R. Lehman, M. J. Bessman and E. S. Simms. 1956. Enzymic synthesis of deoxyribonucleic acid. *Biochim. Biophys. Acta* 21: 197–198. [P]

Kunkel, T. A. 1992. DNA replication fidelity. *J. Biol. Chem.* 267: 18251–18254. [R]

Lohman, T. M. and K. P. Bjornson. 1996. Mechanisms of helicase-catalyzed DNA unwinding. *Ann. Rev. Biochem.* 65: 169–214. [R]

Nakamura, T. M. and T. R. Cech. 1998. Reversing time: Origin of telomerase. *Cell* 92: 587–590. [R]

Newlon, C. S. 1997. Putting it all together: Building a prereplicative complex. *Cell* 91: 717–720. [R]

Ogawa, T. and T. Okazaki. 1980. Discontinuous DNA replication. *Ann. Rev. Biochem.* 49: 421–457. [R]

Stinthcomb, D. T., K. Struhl and R. W. Davis. 1979. Isolation and characterization of a yeast chromosomal replicator. *Nature* 282: 39–43. [P]

Waga, S. and B. Stillman. 1994. Anatomy of a DNA replication fork revealed by reconstitution of SV40 DNA replication in vitro. *Nature* 369: 207–212. [P]

Waga, S. and B. Stillman. 1998. The DNA replication fork in eukaryotic cells. *Ann. Rev. Biochem.* 67: 721–751. [R]

Wang, J. C. 1996. DNA topoisomerases. *Ann. Rev. Biochem.* 65: 635–692. [R]

West, S. C. 1996. DNA helicases: New breeds of translocating motors and molecular pumps. *Cell* 86: 177–180. [R]

Wold, M. S. 1997. Replication protein A: A heterotrimeric, single-stranded DNA-binding protein required for eukaryotic DNA metabolism. *Ann. Rev. Biochem.* 66: 61–92. [R]

Zakian, V. A. 1995. Telomeres: Beginning to understand the end. *Science* 270: 1601–1607. [R]

DNA Repair

Cleaver, J. E. 1968. Defective repair replication of DNA in xeroderma pigmentosum. *Nature* 218: 652–656. [P]

Fishel, R., M. K. Lescoe, M. R. S. Rao, N. G. Copeland, N. A. Jenkins, J. Garber, M. Kane and R. Kolodner. 1993. The human mutator gene homolog *MSH2* and its association with hereditary nonpolyposis colon cancer. *Cell* 75: 1027–1038. [P]

Friedberg, E. C. 1996. Relationships between DNA repair and transcription. *Ann. Rev. Biochem.* 65: 15–42. [R]

Friedberg, E. C., G. C. Walker and W. Siede. 1995. *DNA Repair and Mutagenesis*. Washington, D.C.: ASM Press.

Gowen, L. C., A. V. Avrutskaya, A. M. Latour, B. H. Koller and S. A. Leadon. 1998. BRCA1 required for transcription-coupled repair of oxidative DNA damage. *Science* 281: 1009–1012. [P]

Hanawalt, P. C. 1994. Transcription-coupled repair and human disease. *Science* 266: 1957–1958. [R]

Leach, F. S. and 34 others. 1993. Mutations of a *mutS* homolog in hereditary nonpolyposis colorectal cancer. *Cell* 75: 1215–1225. [P]

Lehmann, A. R. 1995. Nucleotide excision repair and the link with transcription. *Trends Biochem. Sci.* 20: 402–405. [R]

Modrich, P. and R. Lahue. 1996. Mismatch repair in replication fidelity, genetic recombination, and cancer biology. *Ann. Rev. Biochem.* 65: 101–133. [R]

Modrich, P. 1997. Strand-specific mismatch repair in mammalian cells. *J. Biol. Chem.* 272: 24727–24730. [R]

Sancar, A. 1996. DNA excision repair. *Ann. Rev. Biochem.* 65: 43–81. [R]

Seeberg, E., L. Eide and M. Bjoras. 1995. The base excision repair pathway. *Trends Biochem. Sci.* 20: 391–397. [R]

Wood, R. D. 1996. DNA repair in eukaryotes. *Ann. Rev. Biochem.* 65: 135–167. [R]

Wood, R. D. 1997. Nucleotide excision repair in mammalian cells. *J. Biol. Chem.* 272: 23465–23468. [R]

Recombination Between Homologous DNA Sequences

Baumann, P. and S. C. West. 1998. Role of the human RAD51 protein in homologous recombination and double-stranded-break repair. *Trends Biochem. Sci.* 23: 247–251. [R]

DasGupta, C., A. M. Wu, R. Kahn, R. P. Cunningham and C. M. Radding. 1981. Concerted strand exchange and formation of Holliday structures by *E. coli* RecA protein. *Cell* 25: 507–516. [P]

Holliday, R. 1964. A mechanism for gene conversion in fungi. *Genet. Res.* 5: 282–304. [P]

Kowalczykowski, S. C. and A. K. Eggleston. 1994. Homologous pairing and DNA strand-exchange proteins. *Ann. Rev. Biochem.* 63: 991–1043. [R]

Meselson, M. and C. M. Radding. 1975. A general model for genetic recombination. *Proc. Natl. Acad. Sci. USA* 72: 358–361. [P]

Meselson, M. and J. J. Weigle. 1961. Chromosome breakage accompanying genetic recombination in bacteriophage. *Proc. Natl. Acad. Sci. USA* 47: 857–868. [P]

Potter, H. and D. Dressler. 1976. On the mechanism of genetic recombination: Electron microscopic observation of recombination intermediates. *Proc. Natl. Acad. Sci. USA* 73: 3000–3004. [P]

Radding, C. M. 1991. Helical interactions in homologous pairing and strand exchange driven by RecA protein. *J. Biol. Chem.* 266: 5355–5358. [R]

Shinohara, A. and T. Ogawa. 1995. Homologous recombination and the roles of double-strand breaks. *Trends Biochem. Sci.* 20: 387–391. [R]

Stahl, F. 1996. Meiotic recombination in yeast: Coronation of the double-strand-break repair model. *Cell* 87: 965–968. [R]

Szostak, J. W., T. L. Orr-Weaver, R. J. Rothstein and F. W. Stahl. 1983. The double-strand-break repair model for recombination. *Cell* 33: 25–35. [P]

Taylor, A. F. 1992. Movement and resolution of Holliday junctions by enzymes from *E. coli. Cell* 69: 1063–1065. [R]

West, S. C. 1992. Enzymes and molecular mechanisms of genetic recombination. *Ann. Rev. Biochem.* 61: 603–640. [R]

West, S. C. 1994. The processing of recombination intermediates: Mechanistic insights from studies of bacterial proteins. *Cell* 76: 9–15. [R]

DNA Rearrangements

Boeke, J. D., D. J. Garfinkel, C. A. Styles and G. R. Fink. 1985. Ty elements transpose through an RNA intermediate. *Cell* 40: 491–500. [P]

Borst, P. and D. R. Greaves. 1987. Programmed gene rearrangements altering gene expression. *Science* 235: 658–667. [R]

Craig, N. L. 1995. Unity in transposition reactions. *Science* 270: 253–254. [R]

Craig, N. L. 1997. Target site selection in transposition. *Ann. Rev. Biochem.* 66: 437–474. [R]

Davis, M. M. 1990. T cell receptor gene diversity and selection. *Ann. Rev. Biochem.* 59: 475–496. [R]

Fedoroff, N. V. 1989. About maize transposable elements and development. *Cell* 56: 181–191. [R]

Fedoroff, N. and D. Botstein. 1992. *The Dynamic Genome: Barbara McClintock's Ideas in the Century of Genetics.* Plainview, N.Y.: Cold Spring Harbor Laboratory Press.

Finnegan, D. J. 1989. Eukaryotic transposable elements and genome evolution. *Trends Genet.* 5: 103–107. [R]

Gilboa, E., S. W. Mitra, S. Goff and D. Baltimore. 1979. A detailed model of reverse transcription and tests of crucial aspects. *Cell* 18: 93–100. [P]

Grandbastien, M.-A. 1992. Retroelements in higher plants. *Trends Genet.* 8: 103–108. [R]

Hozumi, N. and S. Tonegawa. 1976. Evidence for somatic rearrangement of immunoglobulin genes coding for variable and constant regions. *Proc. Natl. Acad. Sci. USA* 73: 3628–3632. [P]

Kazazian, H. H. Jr. and J. V. Moran. 1998. The impact of L1 retrotransposons on the human genome. *Nature Genet.* 19: 19–24. [R]

Kidwell, M. G. and D. Lisch. 1997. Transposable elements as sources of variation in animals and plants. *Proc. Natl. Acad. Sci. USA* 94: 7704–7711. [R]

Landy, A. 1989. Dynamic, structural, and regulatory aspects of λ site-specific recombination. *Ann. Rev. Biochem.* 58: 913–949. [R]

Levin, H. L. 1997. It's prime time for reverse transcriptase. *Cell* 88: 5–8. [R]

Lewis, S. and M. Gellert. 1989. The mechanism of antigen receptor gene assembly. *Cell* 59: 585–588. [R]

McClintock, B. 1956. Controlling elements and the gene. *Cold Spring Harbor Symp. Quant. Biol.* 21: 197–216. [P]

Moran, J. V., R. J. DeBerardinis and H. H. Kazazian Jr. 1999. Exon shuffling by L1 retrotransposition. *Science* 283: 1530–1534. [P]

Roth, D. B. and N. L. Craig. 1998. VDJ recombination: A transposase goes to work. *Cell* 94: 411–414. [R]

Singer, M. F. 1995. Unusual reverse transcriptases. *J. Biol. Chem.* 270: 24623–24626. [R]

Stark, G. R., M. Debatisse, E. Giulotto and G. M. Wahl. 1989. Recent progress in understanding mechanisms of mammalian DNA amplification. *Cell* 57: 901–908. [R]

Stark, G. R. and G. M. Wahl. 1984. Gene amplification. *Ann. Rev. Biochem.* 53: 447–491. [R]

Temin, H. M. 1989. Retrons in bacteria. *Nature* 339: 254–255. [R]

Tonegawa, S. 1983. Somatic generation of antibody diversity. *Nature* 302: 575–581. [R]

Weiner, A. M., P. L. Deininger and A. Efstratiadis. 1986. Nonviral retroposons: Genes, pseudogenes, and transposable elements generated by the reverse flow of genetic information. *Ann. Rev. Biochem.* 55: 631–661. [R]

Whitcomb, J. M. and S. H. Hughes. 1992. Retroviral reverse transcription and integration: Progress and problems. *Ann. Rev. Cell Biol.* 8: 275–306. [R]

6

RNA Synthesis and Processing

CHAPTERS 4 AND 5 DISCUSSED THE ORGANIZATION and maintenance of genomic DNA, which can be viewed as the set of genetic instructions governing all cellular activities. These instructions are implemented via the synthesis of RNAs and proteins. Importantly, the behavior of a cell is determined not only by what genes it inherits, but also by which of those genes are expressed at any given time. Regulation of gene expression allows cells to adapt to changes in their environments and is responsible for the distinct activities of the multiple differentiated cell types that make up complex plants and animals. Muscle cells and liver cells, for example, contain the same genes; the functions of these cells are determined not by differences in their genomes, but by regulated patterns of gene expression that govern development and differentiation.

The first step in expression of a gene, the transcription of DNA into RNA, is the primary level at which gene expression is regulated in both prokaryotic and eukaryotic cells. RNAs in eukaryotic cells are then modified in various ways—for example, introns are removed by splicing—to convert the primary transcript into its functional form. Different types of RNA play distinct roles in cells: Messenger RNAs (mRNAs) serve as templates for protein synthesis; ribosomal RNAs (rRNAs) and transfer RNAs (tRNAs) function in mRNA translation. Still other small RNAs function in splicing and protein sorting in eukaryotes. Transcription and RNA processing are discussed in this chapter. The final step in gene expression, the translation of mRNA to protein, is the subject of Chapter 7.

Transcription in Prokaryotes

As in most areas of molecular biology, studies of *E. coli* have provided the model for subsequent investigations of transcription in eukaryotic cells. As reviewed in Chapter 3, mRNA was discovered first in *E. coli*. *E. coli* was also the first organism from which RNA polymerase was purified and studied. The basic mechanisms by which transcription is regulated were likewise elucidated by pioneering experiments in *E. coli*, in which regulated gene expression allows the cell to respond to variations in the environment, such

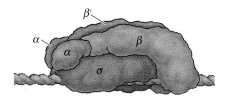

Figure 6.1
***E. coli* RNA polymerase** The complete enzyme consists of five subunits: two α, one β, one β′, and one σ. The σ subunit is relatively weakly bound and can be dissociated from the other four subunits, which constitute the core polymerase.

as changes in the availability of nutrients. An understanding of transcription in *E. coli* has thus provided the foundation for studies of the far more complex mechanisms that regulate gene expression in eukaryotic cells.

RNA Polymerase and Transcription

The principal enzyme responsible for RNA synthesis is **RNA polymerase**, which catalyzes the polymerization of ribonucleoside 5′-triphosphates (NTPs) as directed by a DNA template. The synthesis of RNA is similar to that of DNA, and like DNA polymerase, RNA polymerase catalyzes the growth of RNA chains always in the 5′ to 3′ direction. Unlike DNA polymerase, however, RNA polymerase does not require a preformed primer to initiate the synthesis of RNA. Instead, transcription initiates *de novo* at specific sites at the beginning of genes. The initiation process is particularly important because this is the primary step at which transcription is regulated.

E. coli RNA polymerase, like DNA polymerase, is a complex enzyme made up of multiple polypeptide chains. The intact enzyme consists of four different types of subunits, called α, β, β′, and σ (Figure 6.1). The σ subunit is relatively weakly bound and can be separated from the other subunits, yielding a core polymerase consisting of two α, one β, and one β′ subunits. The core polymerase is fully capable of catalyzing the polymerization of NTPs into RNA, indicating that σ is not required for the basic catalytic activity of the enzyme. However, the core polymerase does not bind specifically to the DNA sequences that signal the normal initiation of transcription; therefore, the σ subunit is required to identify the correct sites for transcription initiation. The selection of these sites is a critical element of transcription because synthesis of a functional RNA must start at the beginning of a gene.

The DNA sequence to which RNA polymerase binds to initiate transcription of a gene is called the **promoter**. The DNA sequences involved in promoter function were first identified by comparisons of the nucleotide sequences of a series of different genes isolated from *E. coli*. These comparisons revealed that the region upstream of the transcription initiation site contains two sets of sequences that are similar in a variety of genes. These common sequences encompass six nucleotides each, and are located approximately 10 and 35 base pairs upstream of the transcription start site (Figure 6.2). They are called the –10 and –35 elements, denoting their position relative to the transcription initiation site, which is defined as the +1 position. The sequences at the –10 and –35 positions in different promoters are not identical, but they are all similar enough to establish consensus sequences—the bases most frequently found at each position.

Several types of experimental evidence support the functional importance of the –10 and –35 promoter elements. First, genes with promoters that differ from the consensus sequences are transcribed less efficiently than genes whose promoters match the consensus sequences more closely. Second, mutations introduced in either the –35 or –10 consensus sequences have strong effects on promoter function. Third, the sites at which RNA polymerase binds to promoters have been directly identified by **footprinting** experiments, which are widely used to determine the sites at which proteins bind to DNA (Figure 6.3). In experiments of this type, a DNA fragment is radiolabeled at one end. The labeled DNA is incubated with the protein

Figure 6.2
Sequences of *E. coli* promoters *E. coli* promoters are characterized by two sets of sequences located 10 and 35 base pairs upstream of the transcription start site (+1). The consensus sequences shown correspond to the bases most frequently found in different promoters.

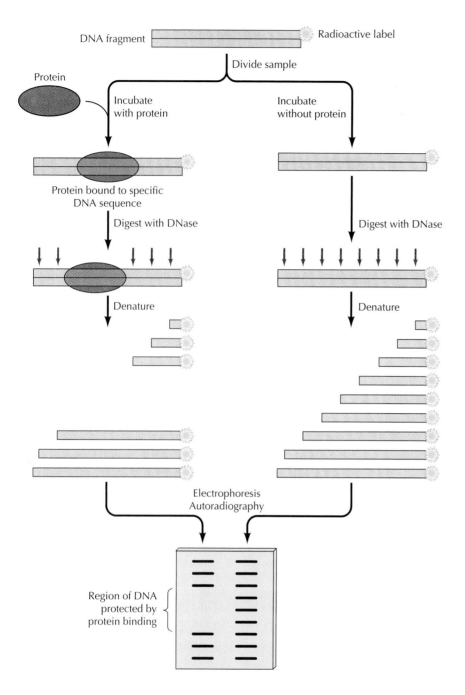

Figure 6.3
DNA footprinting   A sample containing fragments of DNA radiolabeled at one end is divided into two, and one half of the sample is incubated with a protein that binds to a specific DNA sequence within the fragment. Both samples are then digested with DNase, under conditions such that the DNase introduces an average of one cut per molecule. The region of DNA bound to the protein is protected from DNase digestion. The DNA-protein complexes are then denatured, and the sizes of the radiolabeled DNA fragments produced by DNase digestion are analyzed by electrophoresis (as for DNA sequencing). Fragments of DNA resulting from DNase cleavage within the region protected by protein binding are missing from the sample of DNA that was incubated with protein.

of interest (e.g., RNA polymerase) and then subjected to partial digestion with DNase. The principle of the method is that the regions of DNA to which the protein binds are protected from DNase digestion. These regions can therefore be identified by comparison of the digestion products of the protein-bound DNA with those resulting from identical DNase treatment of a parallel sample of DNA that was not incubated with protein. Variations of this basic method, which employ chemical reagents to modify and cleave DNA at particular nucleotides, can be used to identify the specific DNA bases that are in contact with protein. Such footprinting analysis has shown that RNA polymerase generally binds to promoters over approximately a 60-base-pair region, extending from –40 to +20 (i.e., from 40 nucleotides upstream to 20 nucleotides downstream of the transcription start site). The σ subunit binds specifically to sequences in both the –35 and –10 promoter regions, substantiating the importance of these sequences in promoter function. In addition, some *E. coli* promoters have a third sequence, located upstream of the –35 region, that serves as a specific binding site for the RNA polymerase α subunit.

In the absence of σ, RNA polymerase binds nonspecifically to DNA with low affinity. The role of σ is to direct the polymerase to promoters by binding specifically to both the –35 and –10 sequences, leading to the initiation of transcription at the beginning of a gene (Figure 6.4). The initial binding between the polymerase and a promoter is referred to as a closed-promoter complex because the DNA is not unwound. The polymerase then unwinds approximately 15 bases of DNA around the initiation site to form an open-promoter complex in which single-stranded DNA is available as a template for transcription. Transcription is initiated by the joining of two free NTPs. After addition of about the first 10 nucleotides, σ is released from the poly-

Figure 6.4
Transcription by *E. coli* RNA polymerase The polymerase initially binds nonspecifically to DNA and migrates along the molecule until the σ subunit binds to the –35 and –10 promoter elements, forming a closed-promoter complex. The polymerase then unwinds DNA around the initiation site, and transcription is initiated by the polymerization of free NTPs. The σ subunit then dissociates from the core polymerase, which migrates along the DNA and elongates the growing RNA chain.

Figure 6.5
Transcription termination   The termination of transcription is signaled by a GC-rich inverted repeat followed by four A residues. The inverted repeat forms a stable stem-loop structure in the RNA, causing the RNA to dissociate from the DNA template.

merase, which then leaves the promoter and moves along the template DNA to continue elongation of the growing RNA chain. As it travels, the polymerase unwinds the template DNA ahead of it and rewinds the DNA behind it, maintaining an unwound region of about 17 base pairs in the region of transcription.

RNA synthesis continues until the polymerase encounters a termination signal, at which point transcription stops, the RNA is released from the polymerase, and the enzyme dissociates from its DNA template. The simplest and most common type of termination signal in *E. coli* consists of a symmetrical inverted repeat of a GC-rich sequence followed by four or more A residues (Figure 6.5). Transcription of the GC-rich inverted repeat results in the formation of a segment of RNA that can form a stable stem-loop structure by complementary base pairing. The formation of such a self-complementary structure in the RNA disrupts its association with the DNA template and terminates transcription. Because hydrogen bonding between A and U is weaker than that between G and C, the presence of A residues downstream of the inverted repeat sequences is thought to facilitate the dissociation of the RNA from its template. Other types of transcription termination signals, in both prokaryotic and eukaryotic cells, depend on the binding of proteins that terminate transcription to specific DNA sequences, rather than on the formation of a stem-loop structure in the RNA.

Repressors and Negative Control of Transcription

The pioneering studies of gene regulation in *E. coli* were carried out by François Jacob and Jacques Monod in the 1950s. These investigators and their colleagues analyzed the expression of enzymes involved in the metabolism of lactose, which can be used as a source of carbon and energy via cleavage to glucose and galactose (Figure 6.6). The enzyme that catalyzes the cleavage of lactose (β-galactosidase) and other enzymes involved in lac-

Figure 6.6
Metabolism of lactose β-galactosidase catalyzes the hydrolysis of lactose to glucose and galactose.

tose metabolism are expressed only when lactose is available for use by the bacteria. Otherwise, the cell is able to economize by not investing energy in the synthesis of unnecessary RNAs and proteins. Thus, lactose induces the synthesis of enzymes involved in its own metabolism. In addition to requiring β-galactosidase, lactose metabolism involves the products of two other closely linked genes: lactose permease, which transports lactose into the cell, and a transacetylase, whose function in lactose metabolism is still unknown. On the basis of purely genetic experiments, Jacob and Monod deduced the mechanism by which the expression of these genes was regulated, thereby formulating a model that remains fundamental to our understanding of transcriptional regulation.

The starting point in this analysis was the isolation of mutants that were defective in regulation of the genes involved in lactose utilization. These mutants were of two types: constitutive mutants, which expressed all three genes even when lactose was not available, and noninducible mutants, which failed to express the genes even in the presence of lactose. Genetic mapping localized these regulatory mutants to two distinct loci, called o and i, with o located immediately upstream of the structural gene for β-galactosidase. Mutations affecting o resulted in constitutive expression; mutants of i were either constitutive or noninducible.

The function of these regulatory genes was probed by experiments in which two strains of bacteria were mated, resulting in diploid cells containing genes derived from both parents (Figure 6.7). Analysis of gene expression in such diploid bacteria provided critical insights by defining which alleles of these regulatory genes are dominant and which recessive. For example, when bacteria containing a normal i gene (i^+) were mated with bacteria carrying an i gene mutation resulting in constitutive expression (an i^- mutation), the resulting diploid bacteria displayed normal inducibility; therefore, the normal i^+ gene was dominant over the i^- mutant. In contrast, matings between normal bacteria and bacteria with an o^c mutation (constitutive expression) yielded diploids with the constitutive expression phenotype, indicating that o^c is dominant over o^+. Additional experiments in which mutations in o and i were combined with different mutations in the structural genes showed that o affects the expression of only the genes to which it is physically linked, whereas i affects the expression of genes on both chromosome copies in diploid bacteria. Thus, in an o^c/o^+ cell, only the structural genes that are linked to o^c are constitutively expressed. In contrast, in an i^+/i^- cell, structural genes on both chromosomes are regulated normally. These results led to the conclusion that o represents a region of DNA

Figure 6.7
Regulation of β-galactosidase in diploid *E. coli* The mating of two bacterial strains results in diploid cells that contain genes from both parents. In these examples, it is assumed that the genes encoding β-galactosidase (the z genes) can be distinguished on the basis of structural gene mutations, designated z^1 and z^2. In an i^+/i^- diploid (left), both structural genes are inducible; therefore, i^+ is dominant over i^- and affects expression of z genes on both chromosomes. In contrast, in an o^c/o^+ diploid (right), the z gene linked to o^c is constitutively expressed, whereas that linked to o^+ is inducible. Therefore, o affects expression of only the adjacent z gene on the same chromosome.

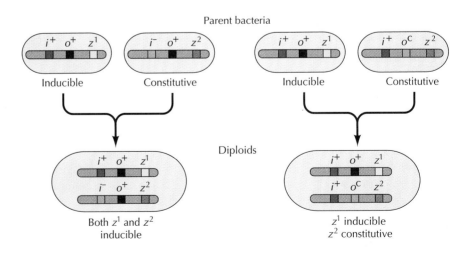

Parent bacteria

i^+ o^+ z^1 — Inducible i^- o^+ z^2 — Constitutive i^+ o^+ z^1 — Inducible i^+ o^c z^2 — Constitutive

Diploids

i^+ o^+ z^1 / i^- o^+ z^2 i^+ o^+ z^1 / i^+ o^c z^2

Both z^1 and z^2 inducible z^1 inducible z^2 constitutive

Figure 6.8
Negative control of the *lac* operon
The *i* gene encodes a repressor which, in the absence of lactose (top), binds to the operator (*o*) and blocks transcription of the three structural genes (*z*, β-galactosidase; *y*, permease; and *a*, transacetylase). Lactose induces expression of the operon by binding to the repressor (bottom), which prevents the repressor from binding to the operator. P = promoter; Pol = polymerase.

that controls the transcription of adjacent genes, whereas the *i* gene encodes a regulatory factor (e.g., a protein) that can diffuse throughout the cell and control genes on both chromosomes.

The model of gene regulation developed on the basis of these experiments is illustrated in Figure 6.8. The genes encoding β-galactosidase, permease, and transacetylase are expressed as a single unit, called an **operon**. Transcription of the operon is controlled by *o* (the **operator**), which is adjacent to the transcription initiation site. The *i* gene encodes a protein that regulates transcription by binding to the operator. Since *i*⁻ mutants (which result in constitutive gene expression) are recessive, it was concluded that these mutants failed to make a functional gene product. This result implies that the normal *i* gene product is a **repressor**, which blocks transcription when bound to *o*. The addition of lactose leads to induction of the operon because lactose binds to the repressor, thereby preventing it from binding to the operator DNA. In noninducible *i* mutants (which are dominant over *i*⁺), the repressor fails to bind lactose, so expression of the operon cannot be induced.

The model neatly fits the results of the genetic experiments from which it was derived. In *i*⁻ cells, the repressor is not made, so the *lac* operon is constitutively expressed. Diploid *i*⁺/*i*⁻ cells are normally inducible, since functional repressor is encoded by the *i*⁺ allele. Finally, in *o*ᶜ mutants a functional operator has been lost and repressor cannot be bound. Consequently, *o*ᶜ mutants are dominant but affect the expression only of linked structural genes.

Confirmation of this basic model has since come from a variety of experiments, including Walter Gilbert's isolation, in the 1960s, of the *lac* repressor and analysis of its binding to operator DNA. Molecular analysis has defined the operator as approximately 30 base pairs of DNA, starting a few bases before the transcription initiation site. Footprinting analysis has identified this region as the site to which the repressor binds, blocking transcription. As predicted, lactose binds to the repressor, which then no longer binds to operator DNA. Also as predicted, *o*ᶜ mutations alter sequences within the operator, thereby preventing repressor binding and resulting in constitutive gene expression.

The central principle of gene regulation exemplified by the lactose operon is that control of transcription is mediated by the interaction of regulatory proteins with specific DNA sequences. This general mode of regulation is broadly applicable to both prokaryotic and eukaryotic cells. Regulatory sequences like the operator are called ***cis*-acting control elements**, because they affect the expression of only linked genes on the same DNA

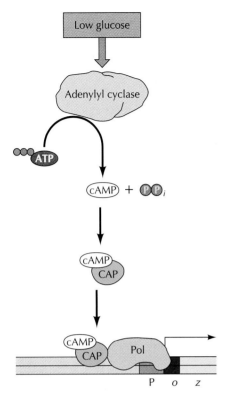

Figure 6.9
Positive control of the *lac* **operon by glucose** Low levels of glucose activate adenylyl cyclase, which converts ATP to cyclic AMP (cAMP). Cyclic AMP then binds to the catabolite activator protein (CAP) and stimulates its binding to regulatory sequences of various operons concerned with the metabolism of alternative sugars, such as lactose. CAP interacts with the α subunit of RNA polymerase to activate transcription.

molecule. On the other hand, proteins like the repressor are called ***trans-acting factors*** because they can affect the expression of genes located on other chromosomes within the cell. The *lac* operon is an example of negative control because binding of the repressor blocks transcription. This, however, is not always the case; many *trans*-acting factors are activators rather than inhibitors of transcription.

Positive Control of Transcription

The best-studied example of positive control in *E. coli* is the effect of glucose on the expression of genes that encode enzymes involved in the breakdown (catabolism) of other sugars (including lactose) that provide alternative sources of carbon and energy. Glucose is preferentially utilized, so as long as glucose is available, enzymes involved in catabolism of alternative energy sources are not expressed. For example, if *E. coli* are grown in medium containing both glucose and lactose, the *lac* operon is not induced and only glucose is used by the bacteria. Thus, glucose represses the *lac* operon even in the presence of the normal inducer (lactose).

Glucose repression (generally called catabolite repression) is now known to be mediated by a positive control system, which is coupled to levels of cyclic AMP (cAMP) (Figure 6.9). In bacteria, the enzyme adenylyl cyclase, which converts ATP to cAMP, is regulated such that levels of cAMP increase when glucose levels drop. cAMP then binds to a transcriptional regulatory protein called catabolite activator protein (CAP). The binding of cAMP stimulates the binding of CAP to its target DNA sequences, which in the *lac* operon are located approximately 60 bases upstream of the transcription start site. CAP then interacts with the α subunit of RNA polymerase, facilitating the binding of polymerase to the promoter and activating transcription.

Transcriptional Attenuation

Both the positive and negative control mechanisms that we have discussed act at the level of initiation of transcription. An additional mechanism, **transcriptional attenuation**, regulates the expression of some genes by controlling the ability of RNA polymerase to continue elongation past specific sites. This mode of regulation has been described best in the *E. coli trp* operon, which encodes five enzymes involved in biosynthesis of the amino acid tryptophan. These genes are expressed only when tryptophan is not available to the cell in its environment, since otherwise the synthesis of additional tryptophan is unnecessary.

The *trp* operon is regulated in part by a repressor that, when bound to tryptophan, blocks transcription (Figure 6.10). However, transcriptional attenuation provides an additional level of control that results in more stringent regulation than could be achieved by repression of initiation alone. The site of attenuation is located 162 nucleotides downstream of the transcription start site. If tryptophan is abundant, most transcription terminates at this site; only if tryptophan is scarce does transcription continue to yield functional Trp mRNA.

The mechanism of attenuation depends on the fact that translation in bacteria is coupled with transcription, so ribosomes begin translating the 5'end of an mRNA while it is still being synthesized. Thus, the rate of translation can affect the structure of the growing RNA chain, which in turn determines whether further transcription can continue. Transcription termination is signaled by a stem-loop structure that forms by complementary base pairing between two specific sequences of the growing Trp mRNA chain (Figure 6.11). This structure forms if translation of the growing chain is proceeding at a normal rate, as it does when tryptophan is present in adequate supply. If tryptophan is scarce, however, protein synthesis stalls at a

Figure 6.10
Regulation of the tryptophan operon   The operon contains five structural genes involved in the biosynthesis of tryptophan: *trpE*, *D*, *C*, *B*, and *A*. Expression of these genes is controlled at two levels. The *trpR* gene encodes a repressor that, in the presence of tryptophan, binds to the operator (*o*) to block transcription. In addition, expression is mediated by an attenuator sequence that prematurely terminates transcription when high levels of tryptophan are present. In this case, the attenuated RNA consists of only a short leader sequence (L). P = promoter.

Figure 6.11
Mechanism of transcriptional attenuation   The *trp* mRNA is translated while still being synthesized. In the presence of high levels of tryptophan, the ribosomes proceed along the message slightly behind the site of transcription. Under these conditions, the mRNA regions designated 3 and 4 hybridize to form a stem-loop structure that signals the termination of transcription. In the presence of low levels of tryptophan, however, the ribosomes stall at region 1 of the mRNA, which contains two adjacent codons for tryptophan. In this case, since region 2 is not bound to a ribosome, it is free to form an alternative stem-loop structure by hybridizing to region 3. This hybridization prevents formation of the 3-4 stem loop, and transcription is able to continue past the attenuator sequence.

critical region of the message. If this occurs, the ribosomes bound to the mRNA block formation of the transcription-terminating stem loop, allowing Trp mRNA synthesis to continue.

The critical region of Trp mRNA contains two adjacent tryptophan codons, so the rate of translation is highly dependent on tryptophan levels; this is the link between transcriptional attenuation and the availability of tryptophan. If tryptophan levels in the cell are low, the ribosome stalls at this point and transcription of Trp mRNA continues. If tryptophan is abundant, translation continues and transcription is terminated.

Eukaryotic RNA Polymerases and General Transcription Factors

Although transcription proceeds by the same fundamental mechanisms in all cells, it is considerably more complex in eukaryotic cells than in bacteria. This is reflected in two distinct differences between the prokaryotic and eukaryotic systems. First, whereas all genes are transcribed by a single RNA polymerase in bacteria, eukaryotic cells contain multiple different RNA polymerases that transcribe distinct classes of genes. Second, rather than binding directly to promoter sequences, eukaryotic RNA polymerases need to interact with a variety of additional proteins to specifically initiate transcription. This increased complexity of eukaryotic transcription presumably facilitates the sophisticated regulation of gene expression needed to direct the activities of the many different cell types of multicellular organisms.

Eukaryotic RNA Polymerases

Eukaryotic cells contain three distinct nuclear RNA polymerases that transcribe different classes of genes (Table 6.1). Protein-coding genes are transcribed by RNA polymerase II to yield mRNAs; ribosomal RNAs (rRNAs) and transfer RNAs (tRNAs) are transcribed by RNA polymerases I and III. RNA polymerase I is specifically devoted to transcription of the three largest species of rRNAs, which are designated 28S, 18S, and 5.8S according to their rates of sedimentation during velocity centrifugation. RNA polymerase III transcribes the genes for tRNAs and for the smallest species of ribosomal RNA (5S rRNA). Some of the small RNAs involved in splicing and protein transport (snRNAs and scRNAs) are also transcribed by RNA polymerase III, while others are polymerase II transcripts. In addition, separate RNA polymerases (which are similar to bacterial RNA polymerases) are found in chloroplasts and mitochondria, where they specifically transcribe the DNAs of those organelles.

All three of the nuclear RNA polymerases are complex enzymes, consisting of 8 to 14 different subunits each. Although they recognize different promoters and transcribe distinct classes of genes, they share several common features. The two largest subunits of all three eukaryotic RNA polymerases are related to the β and β' subunits of the single *E. coli* RNA polymerase. In addition, five subunits of the eukaryotic RNA polymerases are common to all three different enzymes. Consistent with these structural similarities, the different eukaryotic polymerases share several functional properties, including the need to interact with other proteins to appropriately initiate transcription.

General Transcription Factors and Initiation of Transcription by RNA Polymerase II

Because RNA polymerase II is responsible for the synthesis of mRNA from protein-coding genes, it has been the focus of most studies of transcription in eukaryotes. Early attempts at studying this enzyme indicated that its activity

TABLE 6.1 Classes of Genes Transcribed by Eukaryotic RNA Polymerases

Type of RNA synthesized	RNA polymerase
Nuclear genes	
mRNA	II
tRNA	III
rRNA	
5.8S, 18S, 28S	I
5S	III
snRNA and scRNA	II and III[a]
Mitochondrial genes	Mitochondrial[b]
Chloroplast genes	Chloroplast[b]

[a] Some small nuclear (sn) and small cytoplasmic (sc) RNAs are transcribed by polymerase II and others by polymerase III.

[b] The mitochondrial and chloroplast RNA polymerases are similar to bacterial enzymes.

is different from that of prokaryotic RNA polymerase. The accurate transcription of bacterial genes that can be accomplished *in vitro* simply by the addition of purified RNA polymerase to DNA containing a promoter is not possible in eukaryotic systems. The basis of this difference was elucidated in 1979, when Robert Roeder and his colleagues discovered that RNA polymerase II is able to initiate transcription only if additional proteins are added to the reaction. Thus, transcription in the eukaryotic system appeared to require distinct initiation factors that (in contrast to bacterial σ factors) were not associated with the polymerase.

Biochemical fractionation of nuclear extracts has now led to the identification of specific proteins (called **transcription factors**) that are required for RNA polymerase II to initiate transcription. Indeed, the identification and characterization of these factors represents a major part of ongoing efforts to understand transcription in eukaryotic cells. Two general types of transcription factors have been defined. **General transcription factors** are involved in transcription from all polymerase II promoters and therefore constitute part of the basic transcription machinery. Additional transcription factors (discussed later in the chapter) bind to DNA sequences that control the expression of individual genes and are thus responsible for regulating gene expression.

Five general transcription factors are required for initiation of transcription by RNA polymerase II in reconstituted *in vitro* systems (Figure 6.12). The promoters of many genes transcribed by polymerase II contain a sequence similar to TATAA 25 to 30 nucleotides upstream of the transcription start site. This sequence (called the **TATA box**) resembles the –10 sequence element of bacterial promoters, and the results of introducing mutations into TATAA sequences have demonstrated their role in the initiation of transcription. The first step in

Figure 6.12
Formation of a polymerase II transcription complex Many polymerase II promoters have a TATA box (consensus sequence TATAA) 25 to 30 nucleotides upstream of the transcription start site. This sequence is recognized by transcription factor TFIID, which consists of the TATA-binding protein (TBP) and TBP-associated factors (TAFs). TFIIB(B) then binds to TBP, followed by binding of the polymerase in association with TFIIF(F). Finally, TFIIE(E) and TFIIH(H) associate with the complex.

Figure 6.13
Model of the TBP-TFIIB complex bound to DNA The DNA is shown as a stick figure consisting of yellow and green strands, with the site of transcription initiation designated +1. TBP consists of two repeats, colored light blue and dark blue. TFIIB repeats are colored orange and magenta. Note that TBP binding bends the DNA by approximately 110°. (From D. B. Nikolov et al., 1995. *Nature* 377: 119.)

formation of a transcription complex is the binding of a general transcription factor called TFIID to the TATA box (*TF* indicates *t*ranscription *f*actor; *II* indicates polymerase *II*). TFIID is itself composed of multiple subunits, including the **TATA-binding protein (TBP)**, which binds specifically to the TATAA consensus sequence, and 10-12 other polypeptides, called **TBP-associated factors (TAFs)**. TBP then binds a second general transcription factor (TFIIB) forming a TBP-TFIIB complex at the promoter (Figure 6.13). TFIIB in turn serves as a bridge to RNA polymerase, which binds to the TBP-TFIIB complex in association with a third factor, TFIIF.

Following recruitment of RNA polymerase II to the promoter, the binding of two additional factors (TFIIE and TFIIH) is required for initiation of transcription. TFIIH is a multisubunit factor that appears to play at least two important roles. First, two subunits of TFIIH are helicases, which may unwind DNA around the initiation site. (These subunits of TFIIH are also required for nucleotide excision repair, as discussed in Chapter 5.) Another subunit of TFIIH is a protein kinase that phosphorylates repeated sequences present in the C-terminal domain of the largest subunit of RNA polymerase II. Phosphorylation of these sequences is thought to release the polymerase from its association with the initiation complex, allowing it to proceed along the template as it elongates the growing RNA chain.

In addition to a TATA box, the promoters of many genes transcribed by RNA polymerase II contain a second important sequence element (an initiator, or Inr, sequence) that spans the transcription start site. Moreover, some RNA polymerase II promoters contain only an Inr element, with no TATA box. Initiation at these promoters still requires TFIID (and TBP), even though TBP obviously does not recognize these promoters by binding directly to the TATA sequence. Instead, other subunits of TFIID (TAFs) appear to bind to the Inr sequences. This binding recruits TBP to the promoter, and TFIIB, polymerase II, and additional transcription factors then assemble as already described. TBP thus plays a central role in initiating polymerase II transcription, even on promoters that lack a TATA box.

Despite the development of *in vitro* systems and the characterization of several general transcription factors, much remains to be learned concerning

Polymerase II
holoenzyme

TATAA

Figure 6.14
RNA polymerase II holoenzyme   The holoenzyme consists of a preformed complex of RNA polymerase II, the general transcription factors TFIIB, TFIIE, TFIIF, and TFIIH, and several other proteins that activate transcription. This complex can be recruited directly to a promoter via interaction with TFIID (TBP + TAFs).

the mechanism of polymerase II transcription in eukaryotic cells. The sequential recruitment of transcription factors described here represents the minimal system required for transcription *in vitro*; additional factors may be needed within the cell. Furthermore, RNA polymerase II appears to be able to associate with some transcription factors *in vivo* prior to the assembly of a transcription complex on DNA. In particular, preformed complexes of RNA polymerase II with TFIIB, TFIIE, TFIIF, TFIIH, and other transcriptional regulatory proteins have been detected in both yeast and mammalian cells. These large complexes (called polymerase II holoenzymes) can be recruited to a promoter via direct interaction with TFIID (Figure 6.14). The relative contributions of stepwise assembly of individual factors versus recruitment of the RNA polymerase II holoenzyme to promoters within the cell thus remain to be determined.

Transcription by RNA Polymerases I and III

As previously discussed, distinct RNA polymerases are responsible for the transcription of genes encoding ribosomal and transfer RNAs in eukaryotic cells. All three RNA polymerases, however, require additional transcription factors to associate with appropriate promoter sequences. Furthermore, although the three different polymerases in eukaryotic cells recognize distinct types of promoters, a common transcription factor—the TATA-binding protein (TBP)—appears to be required for initiation of transcription by all three enzymes.

RNA polymerase I is devoted solely to the transcription of ribosomal RNA genes, which are present in tandem repeats. Transcription of these genes yields a large 45S pre-rRNA, which is then processed to yield the 28S, 18S, and 5.8S rRNAs (Figure 6.15). The promoter of ribosomal RNA genes spans about 150 base pairs just upstream of the transcription initiation site. These promoter sequences are recognized by two transcription

Figure 6.15
The ribosomal RNA gene The ribosomal DNA (rDNA) is transcribed to yield a large RNA molecule (45S pre-rRNA), which is then cleaved into 28S, 18S, and 5.8S rRNAs.

factors, UBF (upstream binding factor) and SL1 (selectivity factor 1), which bind cooperatively to the promoter and then recruit polymerase I to form an initiation complex (Figure 6.16). The SL1 transcription factor is composed of four protein subunits, one of which, surprisingly, is TBP. The role of TBP has been demonstrated directly by the finding that yeasts carrying mutations in TBP are defective not only for transcription by polymerase II, but also for transcription by polymerases I and III. Thus, TBP is a common transcription factor required by all three classes of eukaryotic RNA polymerases. Since the promoter for ribosomal RNA genes does not contain a TATA box, TBP does not bind to specific promoter sequences. Instead, the association of TBP with ribosomal RNA genes is mediated by the binding of other proteins in the SL1 complex to the promoter, a situation similar to the association of TBP with the Inr sequences of polymerase II genes that lack TATA boxes.

The genes for tRNAs, 5S rRNA, and some of the small RNAs involved in splicing and protein transport are transcribed by polymerase III. These genes are characterized by promoters that lie within, rather than upstream of, the transcribed sequence (Figure 6.17). The most thoroughly studied of the genes transcribed by polymerase III are the 5S rRNA genes of *Xenopus*. TFIIIA (which is the first transcription factor to have been purified) initiates assembly of a transcription complex by binding to specific DNA sequences in the 5S rRNA promoter. This binding is followed by the sequential binding of TFIIIC, TFIIIB, and the polymerase. The promoters for the tRNA genes differ from the 5S rRNA promoter in that they do not contain the DNA sequence recognized by TFIIIA. Instead, TFIIIC binds directly to the promoter of tRNA genes, serving to recruit TFIIIB and polymerase to form a transcription complex. TFIIIB is composed of multiple subunits, one of which (once again) is the TATA-binding protein, TBP. Thus, although the three RNA polymerases of eukaryotic cells recognize different promoters, TBP appears to be a common element that links promoter recognition with polymerase recruitment to the transcription complex.

Figure 6.16
Initiation of rDNA transcription Two transcription factors, UBF and SL1, bind cooperatively to the rDNA promoter and recruit RNA polymerase I to form an initiation complex. One subunit of SL1 is the TATA-binding protein (TBP).

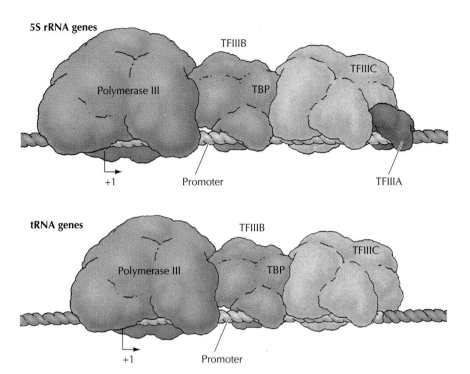

5S rRNA genes

TFIIIB

TFIIIC

Polymerase III

TBP

+1   Promoter   TFIIIA

tRNA genes

TFIIIB

TFIIIC

Polymerase III

TBP

+1   Promoter

Figure 6.17
Transcription of polymerase III genes
The promoters of 5S rRNA and tRNA genes are downstream of the transcription initiation site. Transcription of the 5S rRNA gene is initiated by the binding of TFIIIA, followed by the binding of TFIIIC, TFIIIB, and the polymerase. The tRNA promoters do not contain a binding site for TFIIIA, and TFIIIA is not required for their transcription. Instead, TFIIIC initiates the transcription of tRNA genes by binding to promoter sequences, followed by the association of TFIIIB and polymerase. The TATA-binding protein (TBP) is a subunit of TFIIIB.

Regulation of Transcription in Eukaryotes

Although the control of gene expression is far more complex in eukaryotes than in bacteria, the same basic principles apply. The expression of eukaryotic genes is controlled primarily at the level of initiation of transcription, although in some cases transcription may be attenuated and regulated at subsequent steps. As in bacteria, transcription in eukaryotic cells is controlled by proteins that bind to specific regulatory sequences and modulate the activity of RNA polymerase. The intricate task of regulating gene expression in the many differentiated cell types of multicellular organisms is accomplished primarily by the combined actions of multiple different transcriptional regulatory proteins. In addition, the packaging of DNA into chromatin and its modification by methylation impart further levels of complexity to the control of eukaryotic gene expression.

cis-*Acting Regulatory Sequences: Promoters and Enhancers*

As already discussed, transcription in bacteria is regulated by the binding of proteins to *cis*-acting sequences (e.g., the *lac* operator) that control the transcription of adjacent genes. Similar *cis*-acting sequences regulate the expression of eukaryotic genes. These sequences have been identified in mammalian cells largely by the use of gene transfer assays to study the activity of suspected regulatory regions of cloned genes (Figure 6.18). The eukaryotic regulatory sequences are usually ligated to a reporter gene that encodes an easily detectable enzyme. The expression of the reporter gene following its transfer into cultured cells then provides a sensitive assay for the ability of the cloned regulatory sequences to direct transcription. Biologically active regulatory regions can thus be identified, and *in vitro* mutagenesis can be used to determine the roles of specific sequences within the region.

Genes transcribed by RNA polymerase II have two core promoter elements, the TATA box and the Inr sequence, that serve as specific binding sites for general transcription factors. Other *cis*-acting sequences serve as binding sites for a wide variety of regulatory factors that control the expres-

Figure 6.18
Identification of eukaryotic regulatory sequences   The regulatory sequence of a cloned eukaryotic gene is ligated to a reporter gene that encodes an easily detectable enzyme. The resulting plasmid is then introduced into cultured recipient cells by transfection. An active regulatory sequence directs transcription of the reporter gene, expression of which is then detected in the transfected cells.

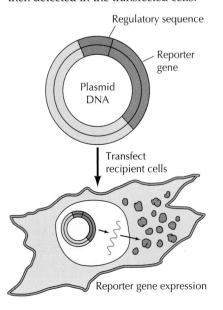

Regulatory sequence

Reporter gene

Plasmid DNA

Transfect recipient cells

Reporter gene expression

Figure 6.19
A eukaryotic promoter   The promoter of the thymidine kinase gene of herpes simplex virus contains three sequence elements upstream of the TATA box that are required for efficient transcription: a CCAAT box and two GC boxes (consensus sequence GGGCGG).

sion of individual genes. These *cis*-acting regulatory sequences are frequently, though not always, located upstream of the TATA box. For example, two regulatory sequences that are found in many eukaryotic genes were identified by studies of the promoter of the herpes simplex virus gene that encodes thymidine kinase (Figure 6.19). Both of these sequences are located within 100 base pairs upstream of the TATA box: Their consensus sequences are CCAAT and GGGCGG (called a GC box). Specific proteins that bind to these sequences and stimulate transcription have since been identified.

In contrast to the relatively simple organization of CCAAT and GC boxes in the herpes thymidine kinase promoter, many genes in mammalian cells are controlled by regulatory sequences located farther away (sometimes more than 10 kilobases) from the transcription start site. These sequences, called **enhancers**, were first identified by Walter Schaffner in 1981 during studies of the promoter of another virus, SV40 (Figure 6.20). In addition to a TATA box and a set of six GC boxes, two 72-base-pair repeats located farther upstream are required for efficient transcription from this promoter. These sequences were found to stimulate transcription from other promoters as well as from that of SV40, and, surprisingly, their activity depended on neither their distance nor their orientation with respect to the transcription initiation site (Figure 6.21). They could stimulate transcription when placed either upstream or downstream of the promoter, in either a forward or backward orientation.

The ability of enhancers to function even when separated by long distances from transcription initiation sites at first suggested that they work by mechanisms different from those of promoters. However, this has turned out not to be the case: Enhancers, like promoters, function by binding transcription factors that then regulate RNA polymerase. This is possible because of DNA looping, which allows a transcription factor bound to a distant enhancer to interact with RNA polymerase or general transcription factors at the promoter (Figure 6.22). Transcription factors bound to distant enhancers can thus work by the same mechanisms as those bound adjacent to promoters, so there is no fundamental difference between the actions of enhancers and those of *cis*-acting regulatory sequences adjacent to transcription start sites. Interestingly, although enhancers were first identified in mammalian cells, they have subsequently been found in bacteria—an unusual instance in which studies of eukaryotes served as a model for the simpler prokaryotic systems.

The binding of specific transcriptional regulatory proteins to enhancers is responsible for the control of gene expression during development and differentiation, as well as during the response of cells to hormones and growth factors. One of the most thoroughly studied mammalian enhancers controls the transcription of immunoglobulin genes in B lymphocytes. Gene transfer experiments have established that the immunoglobulin enhancer is active in

Figure 6.20
The SV40 enhancer   The SV40 promoter for early gene expression contains a TATA box and six GC boxes arranged in three sets of repeated sequences. In addition, efficient transcription requires an upstream enhancer consisting of two 72-base-pair (bp) repeats.

Figure 6.21
Action of enhancers Without an enhancer, the gene is transcribed at a low basal level (A). Addition of an enhancer, E—for example, the SV40 72-base-pair repeats—stimulates transcription. The enhancer is active not only when placed just upstream of the promoter (B), but also when inserted up to several kilobases either upstream or downstream from the transcription start site (C and D). In addition, enhancers are active in either the forward or backward orientation (E).

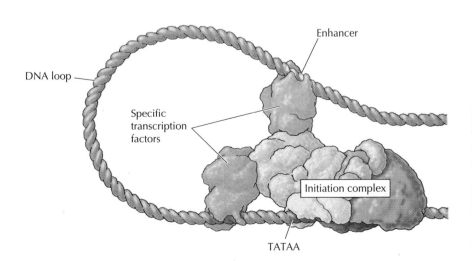

Figure 6.22
DNA looping Transcription factors bound at distant enhancers are able to interact with general transcription factors at the promoter because the intervening DNA can form loops. There is therefore no fundamental difference between the action of transcription factors bound to DNA just upstream of the promoter and to distant enhancers.

Figure 6.23
The immunoglobulin enhancer
The immunoglobulin heavy-chain enhancer spans about 200 bases and contains nine functional sequence elements (E, μE1–5, π, μB, and OCT), which together stimulate transcription in B lymphocytes.

lymphocytes, but not in other types of cells. Thus, this regulatory sequence is at least partly responsible for tissue-specific expression of the immunoglobulin genes in the appropriate differentiated cell type.

An important aspect of enhancers is that they usually contain multiple functional sequence elements that bind different transcriptional regulatory proteins. These proteins work together to regulate gene expression. The immunoglobulin heavy-chain enhancer, for example, spans approximately 200 base pairs and contains at least nine distinct sequence elements that serve as protein-binding sites (Figure 6.23). Mutation of any one of these sequences reduces but does not abolish enhancer activity, indicating that the functions of individual proteins that bind to the enhancer are at least partly redundant. Many of the individual sequence elements of the immunoglobulin enhancer by themselves stimulate transcription in nonlymphoid cells. The restricted activity of the intact enhancer in B lymphocytes therefore does not result from the tissue-specific function of each of its components. Instead, tissue-specific expression results from the combination of the individual sequence elements that make up the complete enhancer. These elements include some *cis*-acting regulatory sequences that bind transcriptional activators that are expressed specifically in B lymphocytes, as well as other regulatory sequences that bind repressors in nonlymphoid cells. Thus, the immunoglobulin enhancer contains negative regulatory elements that inhibit transcription in inappropriate cell types, as well as positive regulatory elements that activate transcription in B lymphocytes. The overall activity of the enhancer is greater than the sum of its parts, reflecting the combined action of the proteins associated with each of its individual sequence elements.

Transcriptional Regulatory Proteins

The isolation of a variety of transcriptional regulatory proteins has been based on their specific binding to promoter or enhancer sequences. Protein binding to these DNA sequences is commonly analyzed by two types of experiments. The first, footprinting, was described earlier in connection with the binding of RNA polymerase to prokaryotic promoters (see Figure 6.3). The second approach is the **electrophoretic-mobility shift assay**, in which a radiolabeled DNA fragment is incubated with a protein preparation and then subjected to electrophoresis through a nondenaturing gel (Figure 6.24). Protein binding is detected as a decrease in the electrophoretic mobility of the DNA fragment, since its migration through the gel is slowed by the bound protein. The combined use of footprinting and electrophoretic-mobility shift assays has led to the correlation of protein-binding sites with the regulatory elements of enhancers and promoters, indicating that these sequences generally constitute the recognition sites of specific DNA-binding proteins.

One of the prototypes of eukaryotic transcription factors was initially identified by Robert Tjian and his colleagues during studies of the transcription of SV40 DNA. This factor (called Sp1, for specificity protein 1) was found to stimulate transcription from the SV40 promoter, but not from several other promoters, in cell-free extracts. Then, stimulation of transcription by Sp1 was found to depend on the presence of the GC boxes in the SV40 promoter: If these sequences were deleted, stimulation by Sp1 was abol-

Figure 6.24
Electrophoretic-mobility shift assay
A sample containing radiolabeled fragments of DNA is divided into two, and one half of the sample is incubated with a protein that binds to a specific DNA sequence. Samples are then analyzed by electrophoresis in a nondenaturing gel so that the protein remains bound to DNA. Protein binding is detected by the slower migration of DNA-protein complexes compared to that of free DNA. Only a fraction of the DNA in the sample is actually bound to protein, so both DNA-protein complexes and free DNA are detected following incubation of the DNA with protein.

ished. Moreover, footprinting experiments established that Sp1 binds specifically to the GC box sequences. Taken together, these results indicate that the GC box represents a specific binding site for a transcriptional activator—Sp1. Similar experiments have established that many other transcriptional regulatory sequences, including the CCAAT sequence and the various sequence elements of the immunoglobulin enhancer, also represent recognition sites for sequence-specific DNA-binding proteins (Table 6.2).

The specific binding of Sp1 to the GC box not only established the action of Sp1 as a sequence-specific transcription factor; it also suggested a general approach to the purification of transcription factors. The isolation of these proteins initially presented a formidable challenge because they are present in very small quantities (e.g., only 0.001% of total cell protein) that are difficult to purify by conventional biochemical techniques. This problem was overcome in the purification of Sp1 by **DNA-affinity chromatography** (Figure 6.25). Multiple copies of oligonucleotides corresponding to the GC box sequence were bound to a solid support, and cell extracts were passed through the oligonucleotide column. Because Sp1 bound to the GC box with high affinity, it was specifically retained on the column while other proteins were not. Highly purified Sp1 could thus be obtained and used for further studies, including partial determination of its amino acid sequence, which in turn led to cloning of the gene for Sp1.

TABLE 6.2 Examples of Transcription Factors and Their DNA-Binding Sites

Transcription factor	Consensus binding site
Specificity protein 1 (Sp1)	GGGCGG
CCAAT/Enhancer binding protein (C/EBP)	CCAAT
Activator protein 1 (AP1)	TGACTCA
Octamer binding proteins (OCT-1 and OCT-2)	ATGCAAAT
E-box binding proteins (E12, E47, E2-2)	CANNTG[a]

[a] N stands for any nucleotide.

Figure 6.25
Purification of Sp1 by DNA-affinity chromatography A double-stranded oligonucleotide containing repeated GC box sequences is bound to agarose beads, which are poured into a column. A mixture of cell proteins containing Sp1 is then applied to the column; because Sp1 specifically binds to the GC box oligonucleotide, it is retained on the column while other proteins flow through. Washing the column with high salt buffer then dissociates Sp1 from the GC box DNA, yielding purified Sp1.

The general approach of DNA-affinity chromatography, first optimized for the purification of Sp1, has been used successfully to isolate a wide variety of sequence-specific DNA-binding proteins from eukaryotic cells. Protein purification has been followed by gene cloning and nucleotide sequencing, leading to the accumulation of a great deal of information on the structure and function of these critical regulatory proteins.

Structure and Function of Transcriptional Activators

Because transcription factors are central to the regulation of gene expression, understanding the mechanisms of their action is a major area of ongoing research in cell and molecular biology. The most thoroughly studied of these proteins are **transcriptional activators**, which, like Sp1, bind to regulatory DNA sequences and stimulate transcription. In general, these factors have been found to consist of two domains: One region of the protein specifically binds DNA; the other activates transcription by interacting with other components of the transcriptional machinery (Figure 6.26). Transcriptional activators appear to be modular proteins, in the sense that the DNA binding and activation domains of different factors can frequently be interchanged using recombinant DNA techniques. Such manipulations result in hybrid transcription factors, which activate transcription by binding to promoter or enhancer sequences determined by the specificity of their DNA-binding domains. It therefore appears that the basic function of the DNA-binding domain is to anchor the transcription factor to the proper site on DNA; the activation domain then independently stimulates transcription by interacting with other proteins.

Key Experiment

Isolation of a Eukaryotic Transcription Factor

Affinity Purification of Sequence-Specific DNA-Binding Proteins

James T. Kadonaga and Robert Tjian
University of California, Berkeley
Proceedings of the National Academy of Sciences, USA, 1986, Volume 83, pages 5889–5893

The Context

Starting with studies of the *lac* operon by Jacob and Monod, it became clear that transcription is regulated by proteins that bind to specific DNA sequences. One of the prototype systems for studies of gene expression in eukaryotic cells was the monkey virus SV40, in which several regulatory DNA sequences were identified in the early 1980s. In 1983 William Dynan and Robert Tjian first demonstrated that one of these sequence elements (the GC box) is the specific binding site of a protein detectable in nuclear extracts of human cells. This protein (called Sp1 for specificity protein 1) not only binds to the GC box sequence; it also stimulates transcription *in vitro*, demonstrating that it is a sequence-specific transcriptional activator.

To study the mechanism of Sp1 action, it then became necessary to obtain the transcription factor in pure form and eventually to clone the *Sp1* gene. The isolation of pure Sp1 thus became a high priority, but it also posed a daunting technical challenge.

Sp1 and other transcription factors appeared to represent only about 0.001% of total cell protein, so they could not be purified by conventional biochemical techniques. James Kadonaga and Robert Tjian solved this problem by developing a method of DNA-affinity chromatography that led to the purification not only of Sp1 but also of many other eukaryotic transcription factors, thereby opening the door to molecular analysis of transcriptional regulation in eukaryotic cells.

The Experiments

The DNA-affinity chromatography method developed by Kadonaga and Tjian exploited the specific high-affinity binding of Sp1 to the GC box sequence, GGGCGG. Synthetic oligonucleotides containing multiple copies of this sequence were coupled to solid beads, and a crude nuclear extract was passed through a column consisting of beads linked to GC box DNA. The beads were then washed to remove proteins that had failed to bind specifically to the oligonucleotides. Finally, the beads were washed with a high salt buffer (0.5 *M* KCl), which disrupted the binding of Sp1 to DNA, thereby releasing Sp1 from the column.

Gel electrophoresis demonstrated that the crude nuclear extract

Robert Tjian

initially applied to the column was a complex mixture of proteins (see figure). In contrast, approximately 90% of the protein recovered after two cycles of DNA-affinity chromatography corresponded to only two polypeptides, which were identified as Sp1 by DNA binding and by their activity in *in vitro* transcription assays. Thus, Sp1 had been successfully purified by DNA-affinity chromatography.

The Impact

In their 1986 paper, Kadonaga and Tjian stated that the DNA-affinity chromatography technique "should be generally applicable for the purification of other sequence-specific DNA binding proteins." This prediction has been amply verified; many eukaryotic transcription factors have been purified by this method. The genes that encode still other transcription factors have been isolated by an alternative approach (developed independently in 1988 in the laboratories of Phillip Sharp and Steven McKnight) in which cDNA expression libraries are screened with oligonucleotide probes to detect recombinant proteins that bind specifically to the desired DNA sequences. The ability to isolate sequence-specific DNA-binding proteins by these methods has led to detailed characterization of the structure and function of a wide variety of transcriptional regulatory proteins, providing the basis for our current understanding of gene expression in eukaryotic cells.

Gel figure (markers in kilodaltons):
205.0
116.0
97.4
66.0
45.0
29.0

1 2 3

Purification of Sp1. Gel electrophoresis of proteins initially present in the crude nuclear extract (lane 1) and of proteins obtained after either one or two sequential cycles of DNA-affinity chromatography (lanes 2 and 3, respectively). The sizes of marker proteins (in kilodaltons) are indicated to the left of the gel, and the Sp1 polypeptides are indicated by arrows.

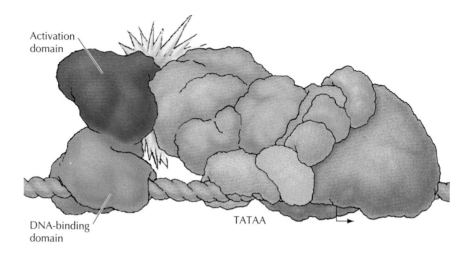

Many different transcription factors have now been identified in eukaryotic cells, as might be expected, given the intricacies of tissue-specific and
inducible gene expression in complex multicellular organisms. Molecular
characterization has revealed that the DNA-binding domains of many of
these proteins are related to one another (Figure 6.27). **Zinc finger
domains** contain repeats of cysteine and histidine residues that bind zinc
ions and fold into looped structures ("fingers") that bind DNA. These
domains were initially identified in the polymerase III transcription factor
TFIIIA but are also common among transcription factors that regulate
polymerase II promoters, including Sp1. Other examples of transcription
factors that contain zinc finger domains are the **steroid hormone receptors**, which regulate gene transcription in response to hormones such as
estrogen and testosterone.

The **helix-turn-helix** motif was first recognized in prokaryotic DNA-
binding proteins, including the *E. coli* catabolite activator protein (CAP). In
these proteins, one helix makes most of the contacts with DNA, while the
other helices lie across the complex to stabilize the interaction. In eukaryotic cells, helix-turn-helix proteins include the **homeodomain** proteins,
which play critical roles in the regulation of gene expression during
embryonic development. The genes encoding these proteins were first discovered as developmental mutants in *Drosophila*. Some of the earliest recognized *Drosophila* mutants (termed homeotic mutants in 1894) resulted in
the development of flies in which one body part was transformed into
another. For example, in the homeotic mutant called *Antennapedia*, legs
rather than antennae grow out of the head of the fly (Figure 6.28). Genetic
analysis of these mutants, pioneered by Ed Lewis in the 1940s, has shown
that *Drosophila* contains nine homeotic genes, each of which specifies the
identity of a different body segment. Molecular cloning and analysis of
these genes then indicated that they contain conserved sequences of 180
base pairs (called **homeoboxes**) that encode the DNA-binding domains
(homeodomains) of transcription factors. A wide variety of additional
homeodomain proteins have since been identified in fungi, plants, and
other animals, including humans. Vertebrate homeobox genes are strikingly similar to their *Drosophila* counterparts in both structure and function, demonstrating the highly conserved roles of these transcription factors in animal development.

Two other families of DNA-binding proteins, **leucine zipper** and **helix-
loop-helix** proteins, contain DNA-binding domains formed by dimerization

(A) Zinc fingers

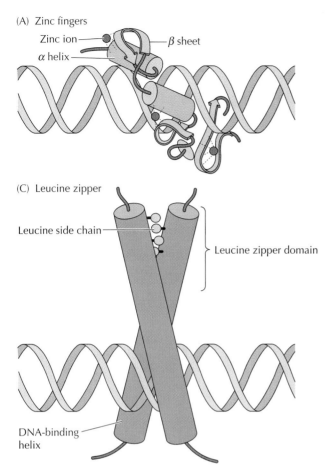

Zinc ion

α helix

β sheet

(B) Helix-turn-helix

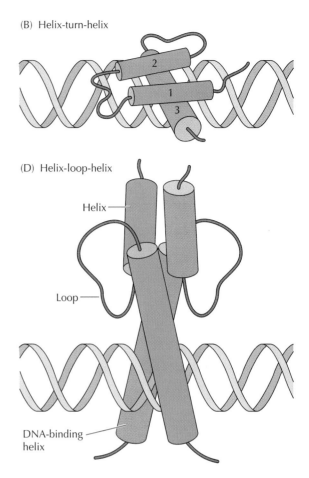

(C) Leucine zipper

Leucine side chain

Leucine zipper domain

DNA-binding helix

(D) Helix-loop-helix

Helix

Loop

DNA-binding helix

Figure 6.27
Families of DNA-binding domains (A) Zinc finger domains consist of loops in which an α helix and a β sheet coordinately bind a zinc ion. (B) Helix-turn-helix domains consist of three (or in some cases four) helical regions. One helix (helix 3) makes most of the contacts with DNA, while helices 1 and 2 lie on top and stabilize the interaction. (C) The DNA-binding domains of leucine zipper proteins are formed from two distinct polypeptide chains. Interactions between the hydrophobic side chains of leucine residues exposed on one side of a helical region (the leucine zipper) are responsible for dimerization. Immediately following the leucine zipper is a DNA-binding helix, which is rich in basic amino acids. (D) Helix-loop-helix domains are similar to leucine zippers, except that the dimerization domains of these proteins each consist of two helical regions separated by a loop.

(A)

(B)

Figure 6.28
The *Antennapedia* mutation *Antennapedia* mutant flies have legs growing out of their heads in place of antennae. (A) Head of a normal fly. (B) Head of an *Antennapedia* mutant. (Courtesy of F. Rudolf Turner, Indiana University.)

Molecular Medicine

The Pit-1 Transcription Factor and Growth Hormone Deficiency

The Disease

Growth hormone is a protein produced by the pituitary gland that is required for normal growth and development. An estimated one out of every 5,000 to 10,000 newborn infants is deficient in growth hormone production. These infants develop into children with short stature—pituitary dwarfs. Such cases of growth hormone deficiency can arise from a variety of causes, including inherited mutations in genes affecting growth hormone production. In some cases, affected children are deficient not only in growth hormone but also in one or more of the other hormones produced by the pituitary gland, such as thyroid-stimulating hormone (TSH), adrenocorticotrophic hormone (ACTH), gonadotrophins, and prolactin. A well-characterized subset of these patients are deficient in production of growth hormone, thyroid-stimulating hormone, and prolactin but produce the other pituitary hormones at normal levels.

Molecular and Cellular Basis

The combined human pituitary hormone deficiencies involving growth hormone, thyroid-stimulating hormone, and prolactin closely resemble the phenotypes of certain dwarf strains of mice. These dwarf mice were known to lack the specific pituitary cell types responsible for growth hormone, thyroid-stimulating hormone, and prolactin secretion (called somatotrophs, thyrotrophs, and lactotrophs, respectively). In 1990 these defects were found to result from mutations in the gene encoding a transcription factor called Pit-1. Two groups of researchers extended these findings to humans with the identification of *Pit-1* mutations in patients with combined deficiencies in growth hormone, thyroid-stimulating hormone, and prolactin. Molecular analysis has demonstrated that Pit-1 is a homeodomain-containing protein required for transcription of the growth hormone and prolactin genes, as well as for the establishment of somatotrophs, thyrotrophs, and lactotrophs during pituitary development.

Prevention and Treatment

Children with growth hormone deficiency can be treated effectively by human growth hormone injections. Until recently, this hormone was available only from human brains and consequently was in extremely short supply. In 1979, however, human growth hormone cDNA was successfully expressed in *E. coli*. This recombinant hormone was approved for clinical use in 1985, providing a readily available source of human growth hormone for the treatment of pituitary dwarfs. When growth hormone deficiency results from inherited *Pit-1* mutations, genetic screen-

An individual of normal height (circus producer P. T. Barnum) and a pituitary dwarf who appeared in the circus as General Tom Thumb. (Bettmann Archive, Inc.)

ing provides the further potential for presymptomatic diagnosis and treatment of affected children.

Reference

Parks, J. S, H. Abdul-Latif, E. Kinoshita, L. R. Meacham., R. W. Pfaffle and M. R. Brown. 1993. Genetics of growth hormone gene expression. *Hormone Res.* 40: 54–61.

of two polypeptide chains. The leucine zipper contains four or five leucine residues spaced at intervals of seven amino acids, resulting in their hydrophobic side chains being exposed at one side of a helical region. This region serves as the dimerization domain for the two protein subunits, which are held together by hydrophobic interactions between the leucine side chains. Immediately following the leucine zipper is a region rich in positively charged amino acids (lysine and arginine) that binds DNA. The helix-loop-helix proteins are similar in structure, except that their dimerization

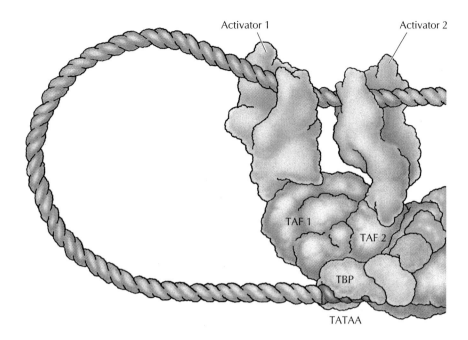

Activator 1   Activator 2

TAF 1

TAF 2

TBP

TATAA

Figure 6.29
Synergistic action of transcriptional activators   Different transcriptional activators can interact with the general transcription factor TFIID by binding to different TAFs.

domains are each formed by two helical regions separated by a loop. An important feature of both leucine zipper and helix-loop-helix transcription factors is that different members of these families can dimerize with each other. Thus, the combination of distinct protein subunits can form an expanded array of factors that can differ both in DNA sequence recognition and in transcription-stimulating activities. Both leucine zipper and helix-loop-helix proteins play important roles in regulating tissue-specific and inducible gene expression, and the formation of dimers between different members of these families is a critical aspect of the control of their function.

The activation domains of transcription factors are not as well characterized as their DNA-binding domains. Some, called acidic activation domains, are rich in negatively charged residues (aspartate and glutamate); others are rich in proline or glutamine residues. These activation domains are thought to stimulate transcription by interacting with general transcription factors, such as TFIIB or TFIID, thereby facilitating the assembly of a transcription complex on the promoter. For example, the activation domains of several transcription factors (including Sp1) have been shown to interact with TFIID by binding to TBP-associated factors (TAFs) (Figure 6.29). An important feature of these interactions is that different activators can bind to different general transcription factors or TAFs, providing a mechanism by which the combined action of multiple factors can synergistically stimulate transcription—a key feature of transcriptional regulation in eukaryotic cells.

Eukaryotic Repressors

Gene expression in eukaryotic cells is regulated by repressors as well as by transcriptional activators. Like their prokaryotic counterparts, eukaryotic repressors bind to specific DNA sequences and inhibit transcription. In some cases, eukaryotic repressors simply interfere with the binding of other transcription factors to DNA (Figure 6.30A). For example, the binding of a repressor near the transcription start site can block the interaction of RNA polymerase or general transcription factors with the promoter, which is similar to the action of repressors in bacteria. Other repressors compete with activators for binding to specific regulatory sequences. Some such repressors contain the same DNA-binding domain as the activator but lack its

Figure 6.30
Action of eukaryotic repressors
(A) Some repressors block the binding of activators to regulatory sequences. (B) Other repressors have active repression domains that inhibit transcription by interactions with general transcription factors.

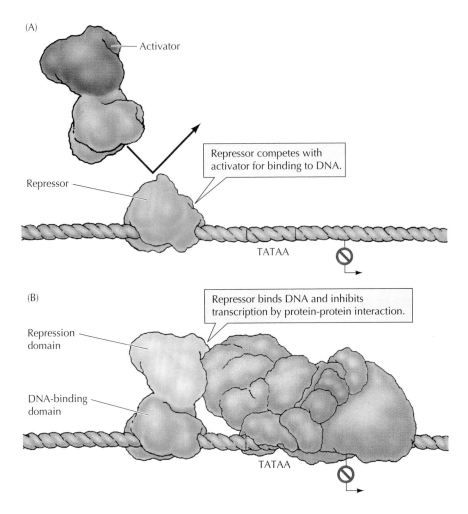

activation domain. As a result, their binding to a promoter or enhancer blocks the binding of the activator, thereby inhibiting transcription.

In contrast to repressors that simply interfere with activator binding, many repressors (called active repressors) contain specific functional domains that inhibit transcription via protein-protein interactions (Figure 6.30B). The first such active repressor was described in 1990 during studies of a gene called *Krüppel*, which is involved in embryonic development in *Drosophila*. Molecular analysis of the Krüppel protein demonstrated that it contains a discrete repression domain, which is linked to a zinc finger DNA-binding domain. The Krüppel repression domain could be interchanged with distinct DNA-binding domains of other transcription factors. These hybrid molecules also repressed transcription, indicating that the Krüppel repression domain inhibits transcription via protein-protein interactions, irrespective of its site of binding to DNA.

Many active repressors have since been found to play key roles in the regulation of transcription in animal cells, in many cases serving as critical regulators of cell growth and differentiation. As with transcriptional activators, several distinct types of repression domains have been identified. For example, the repression domain of Krüppel is rich in alanine residues, whereas other repression domains are rich in proline or acidic residues. The functional targets of repressors are also diverse. Some repressors inhibit transcription by interacting with general transcription factors, such as TFIID; others are thought to interact with specific activator proteins.

The regulation of transcription by repressors as well as by activators considerably extends the range of mechanisms that control the expression of eukaryotic genes. One important role of repressors may be to inhibit the expression of tissue-specific genes in inappropriate cell types. For example, as noted earlier, a repressor-binding site in the immunoglobulin enhancer is thought to contribute to its tissue-specific expression by suppressing transcription in nonlymphoid cell types. Other repressors play key roles in the control of cell proliferation and differentiation in response to hormones and growth factors (see Chapters 13 and 14).

Relationship of Chromatin Structure to Transcription

In the preceding discussion, the transcription of eukaryotic genes was considered as if they were present within the nucleus as naked DNA. However, this is not the case. The DNA of all eukaryotic cells is tightly bound to histones, forming chromatin. The basic structural unit of chromatin is the nucleosome, which consists of 146 base pairs of DNA wrapped around two molecules each of histones H2A, H2B, H3, and H4, with one molecule of histone H1 bound to the DNA as it enters the nucleosome core particle (see Figure 4.9). The chromatin is then further condensed by being coiled into higher-order structures organized into large loops of DNA. This packaging of eukaryotic DNA in chromatin clearly has important consequences in terms of its availability as a template for transcription, so chromatin structure is a critical aspect of gene expression in eukaryotic cells. Indeed, both activators and repressors regulate transcription in eukaryotes not only by interacting with general transcription factors and other components of the transcriptional machinery, but also by inducing changes in the structure of chromatin.

The relationship between chromatin structure and transcription is evident at several levels. First, actively transcribed genes are found in decondensed chromatin, corresponding to the extended 10-nm chromatin fibers discussed in Chapter 4 (see Figure 4.10). For example, microscopic visualization of the polytene chromosomes of *Drosophila* indicates that regions of the genome that are actively engaged in RNA synthesis correspond to decondensed chromosome regions (Figure 6.31). Similarly, actively transcribed genes in vertebrate cells are present in a decondensed fraction of chromatin that is more accessible to transcription factors than is the rest of the genome.

10 μm

Figure 6.31
Decondensed chromosome regions in *Drosophila*   A light micrograph showing decondensed regions of polytene chromosomes (arrows), which are active in RNA synthesis. (Courtesy of Joseph Gall, Carnegie Institute.)

Decondensation of chromatin, however, is not sufficient to make the DNA an accessible template for transcription. Even in decondensed chromatin, actively transcribed genes remain bound to histones and packaged in nucleosomes, so transcription factors and RNA polymerase are still faced with the problem of interacting with chromatin rather than with naked DNA. The tight winding of DNA around the nucleosome core particle is a major obstacle to transcription, affecting both the ability of transcription factors to bind DNA and the ability of RNA polymerase to transcribe through a chromatin template. This inhibitory effect of nucleosomes is relieved by acetylation of histones and by the binding of two nonhistone chromosomal proteins (called **HMG-14** and **HMG-17**) to nucleosomes of actively transcribed genes. (HMG stands for high-mobility group proteins; these proteins migrate rapidly during gel electrophoresis.) Additional proteins called **nucleosome remodeling factors** facilitate the binding of transcription factors to chromatin by altering nucleosome structure.

Acetylation of histones has been correlated with transcriptionally active chromatin in a wide variety of cell types (Figure 6.32). The core histones (H2A, H2B, H3 and H4) have two domains: a histone fold domain, which is involved in interactions with other histones and in wrapping DNA around the nucleosome core particle, and an amino-terminal tail domain, which extends outside of the nucleosome. The amino-terminal tail domains are rich in lysine and can be modified by acetylation at specific lysine residues. Acetylation reduces the net positive charge of the histones, and may weaken their binding to DNA as well as altering their interactions with other proteins. Importantly, recent experiments have provided direct evidence that histone acetylation facilitates the binding of transcription factors to nucleosomal DNA, indicating that histone acetylation increases the accessibility of chromatin to DNA-binding proteins. In addition, direct links between histone acetylation and transcriptional regulation have come from experiments showing that transcriptional activators and repressors are associated with histone acetyltransferases and deacetylases, respectively. This association was first revealed by cloning a gene encoding a histone acetyltransferase

Figure 6.32
Histone acetylation   (A) The core histones have histone-fold domains, which interact with other histones and with DNA in the nucleosome, and N-terminal tails, which extend outside of the nucleosome. The N-terminal tails of the core histones (e.g., H3) are modified by the addition of acetyl groups (Ac) to the side chains of specific lysine residues. (B) Transcriptional activators and repressors are associated with coactivators and corepressors, which have histone acetyltransferase (HAT) and histone deacetylase (HDAC) activities, respectively. Histone acetylation is characteristic of actively transcribed chromatin and may weaken the binding of histones to DNA or alter their interactions with other proteins.

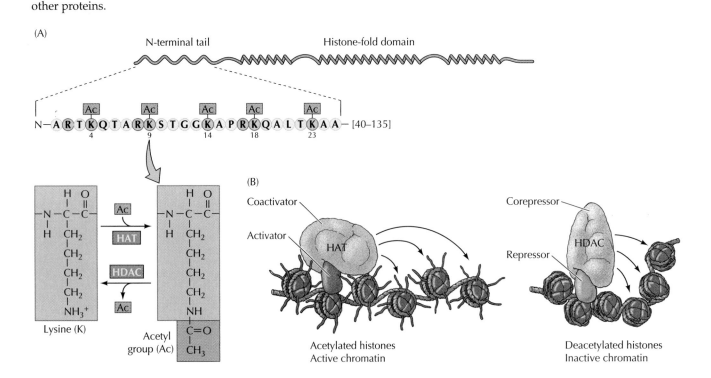

from *Tetrahymena*. Unexpectedly, the sequence of this histone acetyltransferase revealed that it was closely related to a previously known yeast transcriptional coactivator called Gcn5p, which stimulates transcription in association with several different sequence-specific transcriptional activators. Further experiments revealed that Gcn5p itself has histone acetyltransferase activity, suggesting that transcriptional activation results directly from histone acetylation. These results have been extended by the finding that several mammalian transcriptional coactivators are also histone acetyltransferases, as is a general transcription factor (TAFII250, a component of TFIID). Conversely, histone deacetylases (which remove the acetyl groups from histone tails) are associated with transcriptional repressors in both yeast and mammalian cells. Histone acetylation is thus regulated by both transcriptional activators and repressors, indicating that it plays a key role in eukaryotic gene expression.

Nucleosome remodeling factors are protein complexes that facilitate the binding of transcription factors by altering nucleosome structure (Figure 6.33). The mechanism of action of nucleosome remodeling factors is not yet clear, but they appear to increase the accessibility of nucleosomal DNA to other proteins (such as transcription factors) without removing the histones. One possibility is that they catalyze the sliding of histone octamers along the DNA molecule, thereby repositioning nucleosomes to facilitate transcription factor binding. The mechanisms by which nucleosome remodeling factors are targeted to actively transcribed genes also remain to be established, although some studies suggest that they can be brought to enhancer or promoter sites in association with transcriptional activators or as components of the RNA polymerase II holoenzyme (see Figure 6.14).

Perhaps surprisingly, the packaging of DNA in nucleosomes does not present an impassable barrier to transcriptional elongation by RNA polymerase, which is able to transcribe through a nucleosome core by disrupting histone-DNA contacts. The ability of RNA polymerase to transcribe chromatin templates is facilitated by acetylation of histones and by the association of the nonhistone chromosomal proteins HMG-14 and HMG-17 with the nucleosomes of actively transcribed genes. The binding sites of these proteins on nucleosomes overlap the binding site of histone H1, and HMG-14 and HMG-17 appear to stimulate transcription by altering the interaction of histone H1 with nucleosomes to maintain a decondensed chromatin structure that facilitates transcription through a nucleosome template. As with nucleosome remodeling factors, the signals that target HMG-14 and HMG-17 to actively transcribed genes remain to be elucidated by future research.

DNA Methylation

The methylation of DNA is another general mechanism by which control of transcription in vertebrates is linked to chromatin structure. Cytosine residues in vertebrate DNA can be modified by the addition of methyl groups at the 5-carbon position (Figure 6.34). DNA is methylated specifically at the C's that precede G's in the DNA chain (CpG dinucleotides). This methylation is correlated with reduced transcriptional activity of genes that contain high frequencies of CpG dinucleotides in the vicinity of their promoters. Methylation inhibits transcription of these genes via the action of a protein, MeCP2, that specifically binds to methylated DNA and represses transcription. Interestingly, MeCP2 functions as a complex with histone deacetylase, linking DNA methylation to alterations in histone acetylation and nucleosome structure.

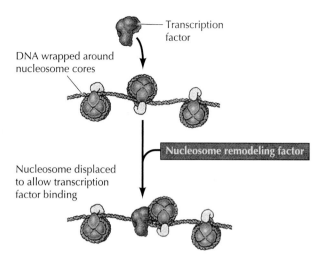

Figure 6.33
Nucleosome remodeling factors
Nucleosome remodeling factors facilitate the binding of transcription factors to chromatin by repositioning nucleosomes on the DNA.

Figure 6.34
DNA methylation A methyl group is added to the 5-carbon position of cytosine residues in DNA.

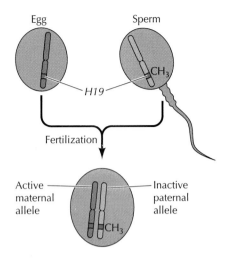

Figure 6.35
Genomic imprinting The *H19* gene is specifically methylated during development of male germ cells. Therefore, sperm contain a methylated *H19* allele and eggs contain an unmethylated allele. Following fertilization, the methylated paternal allele remains transcriptionally inactive, and only the unmethylated maternal allele is expressed in the embryo.

Although DNA methylation is capable of inhibiting transcription, its general significance in gene regulation is unclear. In many cases, methylation of inactive genes is thought to be a consequence, rather than the primary cause, of their lack of transcriptional activity. However, an important regulatory role of DNA methylation has been established in the phenomenon known as **genomic imprinting**, which controls the expression of some genes involved in the development of mammalian embryos. In most cases, both the paternal and maternal alleles of a gene are expressed in diploid cells. However, there are a few imprinted genes (over two dozen have been described in mice and humans) whose expression depends on whether they are inherited from the mother or from the father. In some cases, only the paternal allele of an imprinted gene is expressed, and the maternal allele is transcriptionally inactive. For other imprinted genes, the maternal allele is expressed and the paternal allele is inactive.

Although the biological role of genomic imprinting is uncertain, DNA methylation appears to distinguish between the paternal and maternal alleles of imprinted genes. A good example is the gene *H19*, which is transcribed only from the maternal copy (Figure 6.35). The *H19* gene is specifically methylated during the development of male, but not female, germ cells. The union of sperm and egg at fertilization therefore yields an embryo containing a methylated paternal allele and an unmethylated maternal allele of the gene. These differences in methylation are maintained following DNA replication by an enzyme that specifically methylates CpG sequences of a daughter strand that is hydrogen-bonded to a methylated parental strand (Figure 6.36). The paternal *H19* allele therefore remains methylated, and transcriptionally inactive, in embryonic cells and somatic tissues. However, the paternal *H19* allele becomes demethylated in the germ line, allowing a new pattern of methylation to be established for transmittal to the next generation.

Figure 6.36
Maintenance of methylation patterns In parental DNA, both strands are methylated at complementary CpG sequences. Following replication, only the parental strand of each daughter molecule is methylated. The newly synthesized daughter strands are then methylated by an enzyme that specifically recognizes CpG sequences opposite a methylation site.

RNA Processing and Turnover

Although transcription is the first and most highly regulated step in gene expression, it is usually only the beginning of the series of events required to produce a functional RNA. Most newly synthesized RNAs must be modified in various ways to be converted to their functional forms. Bacterial mRNAs are an exception; as discussed earlier in this chapter, they are used immediately as templates for protein synthesis while still being transcribed. However, the primary transcripts of both rRNAs and tRNAs must undergo a series of processing steps in prokaryotic as well as eukaryotic cells. Primary transcripts of eukaryotic mRNAs similarly undergo extensive modifications, including the removal of introns by splicing, before they are transported from the nucleus to the cytoplasm to serve as templates for protein synthesis. Regulation of these processing steps provides an additional level of control of gene expression, as does regulation of the rates at which different mRNAs are subsequently degraded within the cell.

Processing of Ribosomal and Transfer RNAs

The basic processing of ribosomal and transfer RNAs in prokaryotic and eukaryotic cells is similar, as might be expected given the fundamental roles of these RNAs in protein synthesis. As discussed previously, eukaryotes have four species of ribosomal RNAs (see Table 6.1), three of which (the 28S, 18S, and 5.8S rRNAs) are derived by cleavage of a single long precursor transcript, called a **pre-rRNA** (Figure 6.37). Prokaryotes have three ribosomal RNAs (23S, 16S, and 5S), which are equivalent to the 28S, 18S, and 5S rRNAs of eukaryotic cells and are also formed by the processing of a single pre-rRNA transcript. The only rRNA that is not processed extensively is the 5S rRNA in eukaryotes, which is transcribed from a separate gene.

Prokaryotic and eukaryotic pre-rRNAs are processed in several steps. Initial cleavages of bacterial pre-rRNA yield separate precursors for the three individual rRNAs; these are then further processed by secondary

Figure 6.37
Processing of ribosomal RNAs
Prokaryotic cells contain three rRNAs (16S, 23S, and 5S), which are formed by cleavage of a pre-rRNA transcript. Eukaryotic cells (e.g., human cells) contain four rRNAs. One of these (5S rRNA) is transcribed from a separate gene; the other three (18S, 28S, and 5.8S) are derived from a common pre-rRNA. Following cleavage, the 5.8S rRNA (which is unique to eukaryotes) becomes hydrogen-bonded to 28S rRNA.

cleavages to the final products. In eukaryotic cells, pre-rRNA is first cleaved at a site adjacent to the 5.8S rRNA on its 5′ side, yielding two separate precursors that contain the 18S and the 28S + 5.8S rRNAs, respectively. Further cleavages then convert these to their final products, with the 5.8S rRNA becoming hydrogen-bonded to the 28S molecule. In addition to these cleavages, rRNA processing involves the addition of methyl groups to the bases and sugar moieties of specific nucleotides, although the function of these modifications is not known.

Like rRNAs, tRNAs in both bacteria and eukaryotes are synthesized as longer precursor molecules (**pre-tRNAs**), some of which contain several individual tRNA sequences (Figure 6.38). In bacteria, some tRNAs are included in the pre-rRNA transcripts. The processing of the 5′ end of pre-tRNAs involves cleavage by an enzyme called **RNase P**, which is of special interest because it is a prototypical model of a reaction catalyzed by an RNA enzyme. RNase P consists of RNA and protein molecules, both of which are required for maximal activity. In 1983 Sidney Altman and his colleagues demonstrated that the isolated RNA component of RNase P is itself capable of catalyzing pre-tRNA cleavage. These experiments established that RNase P is a **ribozyme**—an enzyme in which RNA rather than protein is responsible for catalytic activity.

The 3′ end of tRNAs is generated by the action of a conventional protein RNase, but the processing of this end of the tRNA molecule also involves an unusual activity: the addition of a CCA terminus. All tRNAs have the sequence CCA at their 3′ ends. This sequence is the site of amino acid attach-

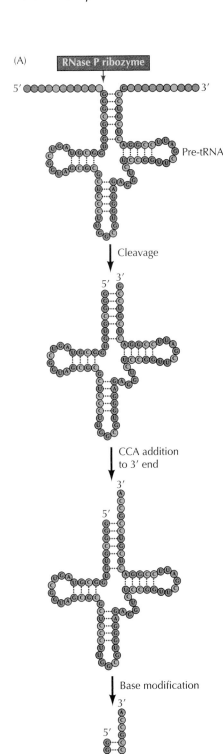

Figure 6.38
Processing of transfer RNAs (A) Transfer RNAs are derived from pre-tRNAs, some of which contain several individual tRNA molecules. Cleavage at the 5′ end of the tRNA is catalyzed by the RNase P ribozyme; cleavage at the 3′ end is catalyzed by a conventional protein RNase. A CCA terminus is then added to the 3′ end of many tRNAs in a posttranscriptional processing step. Finally, some bases are modified at characteristic positions in the tRNA molecule. In this example, these modified nucleosides include dihydrouridine (DHU), methylguanosine (mG), inosine (I), ribothymidine (T), and pseudouridine (ψ). (B) Structure of modified bases. Ribothymidine, dihydrouridine, and pseudouridine are formed by modification of uridines in tRNA. Inosine and methylguanosine are formed by the modification of guanosines.

(B) Modified bases

ment, so it is required for tRNA function during protein synthesis. The CCA terminus is encoded in the DNA of some tRNA genes, but in others it is not, instead being added as an RNA processing step by an enzyme that recognizes and adds CCA to the 3' end of all tRNAs that lack this sequence.

Another unusual aspect of tRNA processing is the extensive modification of bases in tRNA molecules. Approximately 10% of the bases in tRNAs are altered to yield a variety of modified nucleotides at specific positions in tRNA molecules (see Figure 6.38). The functions of most of these modified bases are unknown, but some play important roles in protein synthesis by altering the base-pairing properties of the tRNA molecule (see Chapter 7).

Some pre-tRNAs, as well as pre-rRNAs in a few organisms, contain introns that are removed by splicing. These processing steps are discussed in the next section, together with other splicing reactions.

Processing of mRNA in Eukaryotes

In contrast to the processing of ribosomal and transfer RNAs, the processing of messenger RNAs represents a major difference between prokaryotic and eukaryotic cells. In bacteria, ribosomes have immediate access to mRNA and translation begins on the nascent mRNA chain while transcription is still in progress. In eukaryotes, mRNA synthesized in the nucleus must first be transported to the cytoplasm before it can be used as a template for protein synthesis. Moreover, the initial products of transcription in eukaryotic cells (**pre-mRNAs**) are extensively modified before export from the nucleus. The processing of pre-mRNA includes modification of both ends of the molecule, as well as the removal of introns from its middle (Figure 6.39).

Figure 6.39
Processing of eukaryotic messenger RNAs   The processing of mRNA involves modification of the 5' terminus by capping with 7-methylguanosine (m^7G), modification of the 3' terminus by polyadenylation, and removal of introns by splicing. The 5' cap is formed by the addition of a GTP in reverse orientation to the 5' end of the mRNA, forming a 5'-to-5' linkage. The added G is then methylated at the N-7 position, and methyl groups are added to the riboses of the first one or two nucleotides in the mRNA.

Figure 6.40
Formation of the 3′ ends of eukaryotic mRNAs Polyadenyl-ation signals in mammalian cells consist of the hexanucleotide AAUAAA in addition to upstream and downstream (G-U rich) elements. An endonuclease cleaves the pre-mRNA 10 to 30 nucleotides downstream of the AAUAAA, usually at a CA sequence. Poly-A polymerase then adds a poly-A tail consisting of about 200 A's to the 3′ end of the RNA.

The 5′ end of pre-mRNAs is modified soon after its synthesis by the addition of a structure called a **7-methylguanosine cap**. Capping is initiated by the addition of a GTP in reverse orientation to the 5′ terminal nucleotide of the pre-mRNA. Then methyl groups are added to this G residue and to the ribose moieties of one or two 5′ nucleotides of the RNA chain. The 5′ cap aligns eukaryotic mRNAs on the ribosome during translation (see Chapter 7).

The 3′ end of most eukaryotic mRNAs is defined not by termination of transcription, but by cleavage of the primary transcript and addition of a **poly-A tail**—a processing reaction called **polyadenylation** (Figure 6.40). Signals for polyadenylation include several sequence elements. The most conserved of these is the hexanucleotide AAUAAA in mammalian cells, which is located 10 to 30 nucleotides upstream of the site of polyadenylation. Less conserved sequences that also contribute to signaling polyadenylation are found both upstream and downstream of the AAUAAA. These sequences are recognized by a complex of proteins, including an endonuclease that cleaves the RNA chain and a separate poly-A polymerase that adds a poly-A tail of about 200 nucleotides to the transcript. The act of polyadenylation signals the termination of transcription, which usually occurs several hundred nucleotides downstream of the site of poly-A addition. The proteins that catalyze polyadenylation are associated with RNA polymerase II, providing a link between transcription and formation of the 3′ end of mRNA.

Most mRNAs in eukaryotes are polyadenylated, and poly-A tails have been shown to regulate both translation and mRNA stability. In addition, polyadenylation plays an important regulatory role in early development, where changes in the length of poly-A tails control mRNA translation. For example, many mRNAs are stored in unfertilized eggs in an untranslated form with short poly-A tails (usually 30 to 50 nucleotides long). Fertilization stimulates the lengthening of the poly-A tails of these stored mRNAs, which in turn activates their translation and the synthesis of protiens required for early embryonic development.

The most striking modification of pre-mRNAs is the removal of introns by splicing. As discussed in Chapter 4, the coding sequences of most eukaryotic genes are interrupted by noncoding sequences (introns) that are precisely excised from the mature mRNA. Most genes contain multiple introns, which typically account for about ten times more pre-mRNA sequences than the exons do. The unexpected discovery of introns in 1977 generated an active research effort directed toward understanding the mechanism of splicing, which had to be highly specific to yield functional mRNAs. Further studies of splicing have not only illuminated new mechanisms of gene regulation; they have also revealed novel catalytic activities of RNA molecules.

Splicing Mechanisms

The key to understanding pre-mRNA splicing was the development of *in vitro* systems that efficiently carried out the splicing reaction (Figure 6.41). Pre-mRNAs were synthesized *in vitro* by the cloning of structural genes (with their introns) adjacent to promoters for bacteriophage RNA polymerases, which could readily be isolated in large quantities. Transcription of these plasmids could then be used to prepare large amounts of pre-mRNAs that, when added to nuclear extracts of mammalian cells, were found to be correctly spliced. As with transcription, the use of such *in vitro* systems has allowed splicing to be analyzed in much greater detail than would have been possible in intact cells.

Analysis of the reaction products and intermediates formed *in vitro* revealed that pre-mRNA splicing proceeds in two steps (Figure 6.42). First, the pre-mRNA is cleaved at the 5′ splice site, and the 5′ end of the

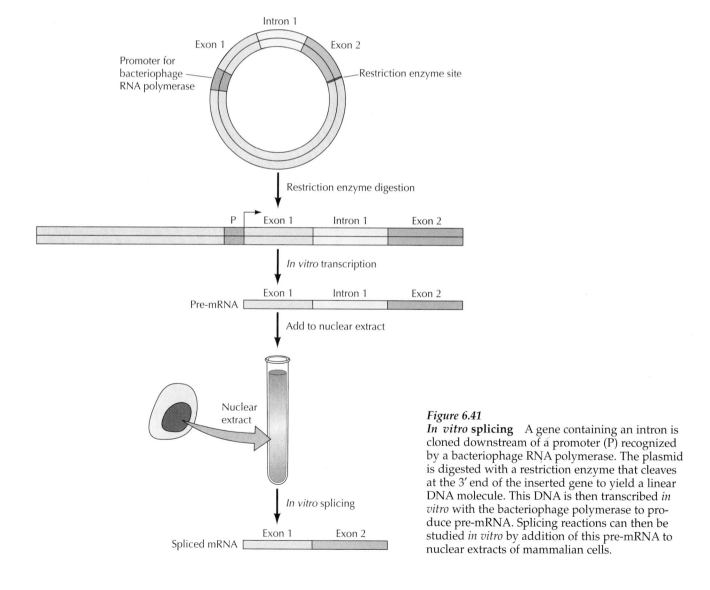

Figure 6.41
In vitro splicing   A gene containing an intron is cloned downstream of a promoter (P) recognized by a bacteriophage RNA polymerase. The plasmid is digested with a restriction enzyme that cleaves at the 3′ end of the inserted gene to yield a linear DNA molecule. This DNA is then transcribed *in vitro* with the bacteriophage polymerase to produce pre-mRNA. Splicing reactions can then be studied *in vitro* by addition of this pre-mRNA to nuclear extracts of mammalian cells.

Figure 6.42
Splicing of pre-mRNA The splicing reaction proceeds in two steps. The first step involves cleavage at the 5′ splice site (SS) and joining of the 5′ end of the intron to an A within the intron (the branch point). This reaction yields a lariat-like intermediate, in which the intron forms a loop. The second step is cleavage at the 3′ splice site and simultaneous ligation of the exons, resulting in excision of the intron as a lariat-like structure.

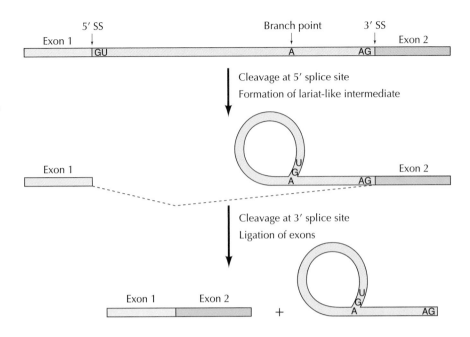

intron is joined to an adenine nucleotide within the intron (near its 3′ end). In this step an unusual bond forms between the 5′ end of the intron and the 2′ hydroxyl group of the adenine nucleotide. The resulting intermediate is a lariat-like structure, in which the intron forms a loop. The second step in splicing then proceeds with simultaneous cleavage at the 3′ splice site and ligation of the two exons. The intron is thus excised as a lariat-like structure, which is then linearized and degraded within the nucleus of intact cells.

These reactions define three critical sequence elements of pre-mRNAs: sequences at the 5′ splice site, sequences at the 3′ splice site, and sequences within the intron at the branch point (the point at which the 5′ end of the intron becomes ligated to form the lariat-like structure) (see Figure 6.42). Pre-mRNAs contain similar consensus sequences at each of these positions, allowing the splicing apparatus to recognize pre-mRNAs and carry out the cleavage and ligation reactions involved in the splicing process.

Biochemical analysis of nuclear extracts has revealed that splicing takes place in large complexes, called **spliceosomes**, composed of proteins and RNAs. The RNA components of the spliceosome are five types of **small nuclear RNAs (snRNAs)** called U1, U2, U4, U5, and U6. These snRNAs, which range in size from approximately 50 to nearly 200 nucleotides, are complexed with six to ten protein molecules to form small nuclear ribonucleoprotein particles (**snRNPs**), which play central roles in the splicing process. The U1, U2, and U5 snRNPs each contain a single snRNA molecule, whereas U4 and U6 snRNAs are complexed to each other in a single snRNP.

The first step in spliceosome assembly is the binding of U1 snRNP to the 5′ splice site of pre-mRNA (Figure 6.43). This recognition of 5′ splice sites involves base pairing between the 5′ splice site consensus sequence and a complementary sequence at the 5′ end of U1 snRNA (Figure 6.44). U2 snRNP then binds to the branch point, by similar complementary base pairing between U2 snRNA and branch point sequences. A preformed complex consisting of U4/U6 and U5 snRNPs is then incorporated into the spliceosome, with U5 binding to sequences upstream of the 5′ splice site.

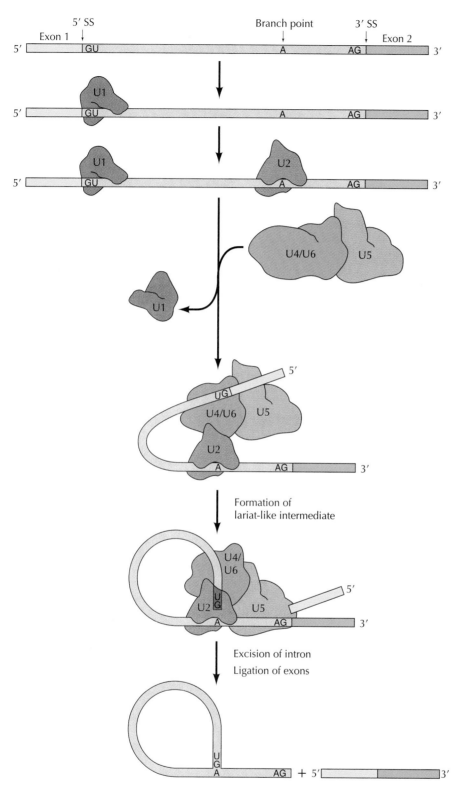

Figure 6.43
Assembly of the spliceosome The first step in spliceosome assembly is the binding of U1 snRNP to the 5′ splice site (SS), followed by the binding of U2 snRNP to the branch point. A preformed complex consisting of U4/U6 and U5 snRNPs then enters the spliceosome. U5 binds to sequences upstream of the 5′ splice site, and U6 dissociates from U4 and displaces U1 prior to formation of the lariat-like intermediate. U5 then binds to the 3′ splice site, followed by excision of the intron and ligation of the exons.

The splicing reaction is then accompanied by rearrangements of the snRNAs. Prior to the first reaction step (formation of the lariat-like intermediate, see Figure 6.42), U6 dissociates from U4 and displaces U1 at the 5′ splice site. U5 then binds to sequences at the 3′ splice site, followed by excision of the intron and ligation of the exons.

Figure 6.44
Binding of U1 snRNA to the 5′ splice site The 5′ terminus of U1 snRNA binds to consensus sequences at 5′ splice sites by complementary base pairing.

Not only do the snRNAs recognize consensus sequences at the branch points and splice sites of pre-mRNAs; they also catalyze the splicing reaction directly. The catalytic role of RNAs in splicing was demonstrated by the discovery that some RNAs are capable of **self-splicing**; that is, they can catalyze the removal of their own introns in the absence of other protein or RNA factors. Self-splicing was first described by Tom Cech and his colleagues during studies of the 28S rRNA of the protozoan *Tetrahymena*. This RNA contains an intron of approximately 400 bases that is precisely removed following incubation of the pre-rRNA in the absence of added proteins. Further studies have revealed that splicing is catalyzed by the intron, which acts as a ribozyme to direct its own excision from the pre-rRNA molecule. The discovery of self-splicing of *Tetrahymena* rRNA, together with the studies of RNase P already discussed, provided the first demonstrations of the catalytic activity of RNA.

Additional studies have revealed self-splicing RNAs in mitochondria, chloroplasts, and bacteria. These self-splicing RNAs are divided into two classes on the basis of their reaction mechanisms (Figure 6.45). The first step in splicing for group I introns (e.g., *Tetrahymena* pre-rRNA) is cleavage at the 5′ splice site mediated by a guanosine cofactor. The 3′ end of the free exon then reacts with the 3′ splice site to excise the intron as a linear RNA. In contrast, the self-splicing reactions of group II introns (e.g., some mitochondrial pre-mRNAs) closely resemble those characteristic of nuclear pre-mRNA splicing, in which cleavage of the 5′ splice site results from attack by an adenosine nucleotide in the intron. As with pre-mRNA splicing, the result is a lariat-like intermediate, which is then excised.

The similarity between spliceosome-mediated pre-mRNA splicing and self-splicing of group II introns strongly suggested that the active catalytic components of the spliceosome were RNAs rather than proteins. In particular, these similarities suggested that pre-mRNA splicing was catalyzed by the snRNAs of the spliceosome. Continuing studies of pre-mRNA splicing have provided clear support for this view; U2, U5, and U6 snRNAs have been identified as catalytic components of the spliceosome. Pre-mRNA splicing is thus considered to be an RNA-based reaction, catalyzed by spliceosome snRNAs acting analogously to group II self-splicing introns. Protein components of the snRNPs are also required, however, and participate in both assembly of the spliceosome and the splicing reaction. In addition, a number of proteins that are not snRNP components play auxiliary roles in splicing and spliceosome assembly. One important role of both snRNP and non-snRNP proteins is to identify and select the splice sites that are recognized by the snRNAs. For example, protein splicing factors are required for the binding of both U1 and U2 snRNPs to the appropriate sites on pre-mRNA. Because many pre-mRNAs contain multiple introns, the splicing machinery must be able to identify and join the appropriate 5′ and 3′ splice sites to produce a functional mRNA. The mechanism of such splice site selection is not understood, but protein splicing factors clearly play an important role.

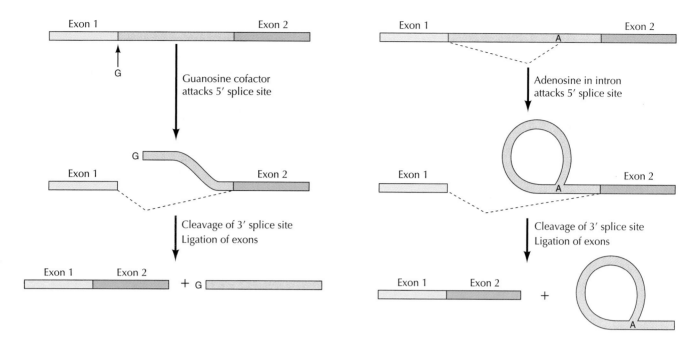

Figure 6.45
Self-splicing introns Group I and group II self-splicing introns are distinguished by their reaction mechanisms. In group I introns, the first step in splicing is cleavage of the 5′ splice site by reaction with a guanosine cofactor. The result is a linear intermediate with a G added to the 5′ end of the intron. In group II introns (as in pre-mRNA splicing), the first step is cleavage of the 5′ splice site by reaction with an A within the intron, forming a lariat-like intermediate. In both cases, the second step is simultaneous cleavage of the 3′ splice site and ligation of the exons.

Finally, it should be noted that the mechanism of pre-tRNA splicing differs from both types of self-splicing reactions and from the splicing of nuclear pre-mRNA. Pre-tRNA splicing involves conventional protein enzymes with no known role for RNA catalysis. Instead, a protein endonuclease cleaves at the splice sites, excising the intron as a linear RNA. This step is followed by ligation of the exons, yielding the mature tRNA molecule.

Alternative Splicing

The central role of splicing in the processing of pre-mRNA opens the possibility of regulation of gene expression by control of the activity of the splicing machinery. Moreover, since most pre-mRNAs contain multiple introns, different mRNAs can be produced from the same gene by different combinations of 5′ and 3′ splice sites. The possibility of joining exons in varied combinations provides a novel means of controlling gene expression by generating multiple mRNAs (and therefore multiple proteins) from the same pre-mRNA. This process, called **alternative splicing**, occurs frequently in genes of complex eukaryotes and provides an important mechanism for tissue-specific and developmental regulation of gene expression.

One interesting example of alternative splicing is provided by some genes that encode transcriptional regulatory proteins. In several cases, alternative splicing of these pre-mRNAs yields products with dramatically different functions—namely, the ability to act as either activators or repressors of transcription (Figure 6.46). As discussed earlier, transcriptional activators consist of two distinct domains: a DNA-binding domain and an activation domain. These domains are generally encoded in separate exons, so alternative splicing allows them to be reassorted into different combinations, thereby enabling the production of activators and repressors from the same gene. Splicing that yields an mRNA that contains exons encoding both DNA-binding and activation domains will result in synthesis of an activator protein. Alternative splicing, however, may result in the formation of an mRNA encoding a DNA-binding domain but lacking an activation domain. Translation of such an alternatively spliced mRNA will result in synthesis of a repressor, which will suppress gene expression by competing with the activator for binding to target DNA sequences (see Figure 6.30).

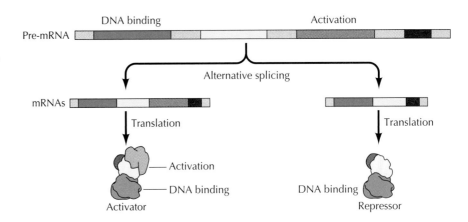

Figure 6.46
Alternative splicing of pre-mRNA that encodes a transcription factor
In this example, a transcriptional regulatory protein is encoded by four exons: the first encodes the DNA-binding domain; the third encodes the activation domain. The pre-mRNA is subject to two alternative-splicing pathways. The first results in ligation of all four exons, yielding an mRNA that encodes a transcriptional activator. In the second splicing pathway, however, exon 2 is joined directly to exon 4, yielding an mRNA that does not include the third exon. This mRNA is then translated to yield a protein lacking the activation domain, which therefore functions as a repressor rather than an activator of transcription.

Because patterns of alternative splicing can vary in different tissues, regulation of splicing provides an important means of regulating tissue-specific gene expression. Although the mechanism by which the correct splice sites are selected in a pre-mRNA is not known, several protein factors have been identified that contribute to splice site selection and can affect the use of alternative splice sites in a pre-mRNA molecule. Variations in the expression of such splicing factors in different cell types may result in tissue-specific patterns of alternative splicing, thereby contributing to the regulation of gene expression during development and differentiation. A particularly notable example is provided by sex determination in *Drosophila*, where alternative splicing of the same pre-mRNA determines whether a fly is male or female.

RNA Editing

RNA editing refers to RNA processing events (other than splicing) that alter the protein-coding sequences of some mRNAs. This unexpected form of RNA processing was first discovered in mitochondrial mRNAs of trypanosomes, in which U residues are added and deleted at multiple sites along the molecule. More recently, editing has also been described in mitochondrial mRNAs of other organisms, chloroplast mRNAs of higher plants, and nuclear mRNAs of some mammalian genes.

Editing in mammalian nuclear mRNAs, as well as in mitochondrial and chloroplast RNAs of higher plants, involves single base changes as a result of base modification reactions, similar to those involved in tRNA processing. In mammalian cells, RNA editing reactions include the deamination of cytosine to uridine and of adenosine to inosine. One of the best-studied examples is editing of the mRNA for apolipoprotein B, which transports lipids in the blood. In this case, tissue-specific RNA editing results in two different forms of apolipoprotein B (Figure 6.47). In humans, Apo-B100 (4536 amino acids) is synthesized in the liver by translation of the unedited mRNA. However, a shorter protein (Apo-B48, 2152 amino acids) is synthesized in the intestine as a result of translation of an edited mRNA in which a C has been changed to a U by deamination. This alteration changes the codon for glutamine (CAA) in the unedited mRNA to a translation termination codon (UAA) in the edited mRNA, resulting in synthesis of the shorter Apo-B protein. Tissue-specific editing of Apo-B mRNA thus results in the expression of structurally and functionally different proteins in liver and intestine. The full-length Apo-B100 produced by the liver transports lipids in the circulation; Apo-B48 functions in the absorption of dietary lipids by the intestine.

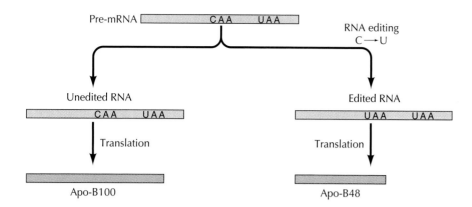

Figure 6.47
Editing of apolipoprotein B mRNA
In human liver, unedited mRNA is translated to yield a 4536-amino-acid protein called Apo-B100. In human intestine, however, the mRNA is edited by a base modification that changes a specific C to a U. This modification changes the codon for glutamine (CAA) to a termination codon (UAA), resulting in synthesis of a shorter protein (Apo-B48, consisting of only 2152 amino acids).

RNA Degradation

The processing steps discussed in the previous section result in the formation of mature mRNAs, which then direct protein synthesis. However, what may be considered the final aspect of the processing of an RNA molecule is its eventual degradation within the cell. Since the intracellular level of any RNA is determined by a balance between synthesis and degradation, the rate at which individual RNAs are degraded is another level at which gene expression can be controlled. Both ribosomal and transfer RNAs are very stable, and this stability largely accounts for the high levels of these RNAs (greater than 90% of all RNA) in both prokaryotic and eukaryotic cells. In contrast, bacterial mRNAs are rapidly degraded, usually having half-lives of only 2 to 3 minutes. This rapid turnover of bacterial mRNAs allows the cell to respond quickly to alterations in its environment, such as changes in the availability of nutrients required for growth. In eukaryotic cells, however, different mRNAs are degraded at different rates, providing an additional parameter to the regulation of eukaryotic gene expression.

The degradation of most eukaryotic mRNAs is initiated by shortening of their poly-A tails. Then follows removal of the 5' cap and degradation of the RNA by nucleases acting from both ends. The half-lives of mRNAs in mammalian cells vary from less than 30 minutes to approximately 20 hours. The unstable mRNAs frequently code for regulatory proteins, including certain transcription factors, whose levels within the cell vary rapidly in response to environmental stimuli. These mRNAs often contain specific AU-rich sequences near their 3' ends that appear to signal rapid degradation by promoting deadenylation.

The stability of some mRNAs can also be regulated in response to extracellular signals. A good example is provided by the mRNA that encodes transferrin receptor—a cell surface protein involved in the uptake of iron by mammalian cells. The amount of transferrin receptor within cells is regulated by the availability of iron, largely as a result of modulation of the stability of its mRNA (Figure 6.48). In the presence of adequate amounts of iron, transferrin receptor mRNA is rapidly degraded as a result of specific nuclease cleavage at a sequence near its 3' end. If an adequate supply of iron is not available, however, the mRNA is stabilized, resulting in increased synthesis of transferrin receptor and more iron uptake by the cell. This regulation is mediated by a protein that binds to specific sequences (called the iron response element, or IRE) near the 3' end of transferrin receptor mRNA and protects the mRNA from cleavage. The binding of this regulatory protein to the IRE is in turn controlled by the levels of iron within the cell: If iron is scarce, the protein binds to the IRE and protects

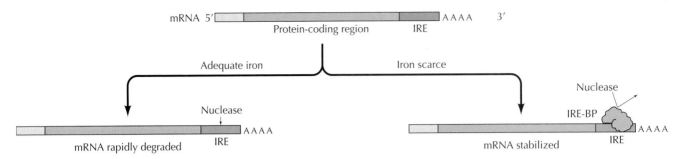

Figure 6.48
Regulation of transferrin receptor mRNA stability The levels of transferrin receptor mRNA are regulated by the availability of iron. If the supply of iron is adequate, the mRNA is rapidly degraded as a result of nuclease cleavage near the 3′ end. If iron is scarce, however, a regulatory protein (called the iron response element–binding protein, or IRE-BP) binds to a sequence near the 3′ end of the mRNA (the iron response element, or IRE), protecting the mRNA from nuclease cleavage.

transferrin receptor mRNA from degradation. Similar changes in the stability of other mRNAs are involved in the regulation of gene expression by certain hormones. Thus, although transcription remains the primary level at which gene expression is regulated, variations in the rate of mRNA degradation also play an important role in controlling steady-state levels of mRNAs within the cell.

KEY TERMS

RNA polymerase, promoter, footprinting

operon, operator, repressor, *cis*-acting control element, *trans*-acting factor

transcriptional attenuation

transcription factor, general transcription factor, TATA box, TATA-binding protein (TBP), TBP-associated factor (TAF)

SUMMARY

TRANSCRIPTION IN PROKARYOTES

RNA Polymerase and Transcription: E. coli RNA polymerase consists of α, β, β′, and σ subunits. Transcription is initiated by the binding of σ to promoter sequences. After synthesis of the first few nucleotides of RNA, the core polymerase dissociates from σ and travels along the template DNA as it elongates the RNA chain. Transcription then continues until the polymerase encounters a termination signal.

Repressors and Negative Control of Transcription: The prototype model for gene regulation in bacteria is the *lac* operon, which is regulated by the binding of a repressor to specific DNA sequences near the promoter.

Positive Control of Transcription: Some bacterial genes are regulated by transcriptional activators rather than repressors.

Transcriptional Attenuation: Transcription can also be regulated by control of the ability of RNA polymerase to elongate past specific sites in the template DNA.

EUKARYOTIC RNA POLYMERASES AND GENERAL TRANSCRIPTION FACTORS

Eukaryotic RNA Polymerases: Eukaryotic cells contain three distinct nuclear RNA polymerases that transcribe genes encoding mRNAs (polymerase II), rRNAs (polymerase I), and tRNAs (polymerase III).

General Transcription Factors and Initiation of Transcription by RNA Polymerase II: Eukaryotic RNA polymerases do not bind directly to promoter sequences; they require additional proteins (general transcription factors) to initiate transcription. The promoter sequences of many polymerase II genes are recognized by the TATA-binding protein, which recruits additional transcription factors and RNA polymerase to the promoter.

Transcription by RNA Polymerases I and III: RNA polymerases I and III also require additional transcription factors to bind to the promoters of rRNA and tRNA genes.

REGULATION OF TRANSCRIPTION IN EUKARYOTES

cis-Acting Regulatory Sequences: Promoters and Enhancers: Transcription of eukaryotic genes is controlled by proteins that bind to regulatory sequences, which can be located up to several kilobases away from the transcription start site. Enhancers typically contain binding sites for multiple proteins that work together to regulate gene expression.

enhancer

Transcriptional Regulatory Proteins: Many eukaryotic transcription factors have been isolated on the basis of their binding to specific DNA sequences.

electrophoretic-mobility shift assay, DNA-affinity chromatography

Structure and Function of Transcriptional Activators: Transcriptional activators are modular proteins, consisting of distinct DNA-binding and activation domains. DNA-binding domains mediate association with specific regulatory sequences; activation domains stimulate transcription by interacting with general transcription factors.

transcriptional activator, zinc finger domain, steroid hormone receptor, helix-turn-helix, homeodomain, homeobox, leucine zipper, helix-loop-helix

Eukaryotic Repressors: Gene expression in eukaryotic cells is regulated by repressors as well as by activators. Some repressors interfere with the binding of activators or general transcription factors to DNA. Other repressors contain discrete repression domains that inhibit transcription by interacting with either general transcription factors or transcriptional activators.

Relationship of Chromatin Structure to Transcription: The packaging of DNA in nucleosomes presents an impediment to transcription in eukaryotic cells. Modification of histones by acetylation increases the accessibility of nucleosomal DNA to transcription factors, and this modification of chromatin is tightly linked to transcriptional regulation. Enzymes that catalyze histone acetylation are associated with transcriptional activators, whereas histone deacetylases are associated with repressors. In addition, nucleosome remodeling factors facilitate the binding of transcription factors to DNA by altering nucleosome structure. RNA polymerase is then able to transcribe through nucleosomes by disrupting histone-DNA contacts. This transcription is facilitated by the nonhistone chromosomal proteins HMG-14 and HMG-17, which are specifically associated with actively transcribed genes.

HMG-14, HMG-17, nucleosome remodeling factor

DNA Methylation: Methylation of cytosine residues can inhibit the transcription of vertebrate genes. Regulation of gene expression by methylation plays an important role in genomic imprinting, which controls the transcription of some genes involved in mammalian development.

genomic imprinting

RNA PROCESSING AND TURNOVER

Processing of Ribosomal and Transfer RNAs: Ribosomal and transfer RNAs are derived by cleavage of long primary transcripts in both prokaryotic and eukaryotic cells. Methyl groups are added to rRNAs, and various bases are modified in tRNAs.

pre-rRNA, pre-tRNA, RNase P, ribozyme

pre-mRNA, 7-methylguanosine cap, poly-A tail, polyadenylation

Processing of mRNA in Eukaryotes: Eukaryotic pre-mRNAs are modified by the addition of 7–methylguanosine caps and 3' poly-A tails, in addition to the removal of introns by splicing.

spliceosome, small nuclear RNA (snRNA), snRNP, self-splicing

Splicing Mechanisms: Splicing of nuclear pre-mRNAs takes place in large complexes, called spliceosomes, composed of proteins and small nuclear RNAs (snRNAs). The snRNAs recognize sequences at the splice sites of pre-mRNAs and catalyze the splicing reaction. Some mitochondrial, chloroplast, and bacterial RNAs undergo self-splicing, in which the splicing reaction is catalyzed by intron sequences.

alternative splicing

Alternative Splicing: Exons can be joined in various combinations as a result of alternative splicing, which provides an important mechanism for tissue-specific control of gene expression in complex eukaryotes.

RNA editing

RNA Editing: Some mRNAs are modified by processing events that alter their protein-coding sequences. Editing of mitochondrial mRNAs in some protozoans involves the addition and deletion of U residues at multiple sites in the molecule. Other forms of RNA editing in plant and mammalian cells involve the modification of specific bases.

RNA Degradation: mRNAs in eukaryotic cells are degraded at different rates, providing an additional mechanism for control of gene expression. In some cases, rates of mRNA degradation are regulated by extracellular signals.

Questions

1. The consensus sequence of the *E. coli* –10 promoter element is TATAAT. You are comparing two promoters that have –10 element sequences of TATGAT and CATGAT, respectively. Which would you expect to be transcribed more efficiently?

2. You are working with two strains of *E. coli*. One contains a wild-type β-galactosidase gene and an i^- mutation; the other contains a temperature-sensitive β-galactosidase gene and an o^c mutation. After mating these strains, you assay for the production of β-galactosidase at both permissive and non-permissive temperatures in the absence of lactose. What do you expect to find?

3. You are comparing the requirements for *in vitro* basal transcription of two polymerase II genes, one containing a TATA box and the other containing only an Inr sequence. Does transcription from these promoters require TBP or TFIID?

4. You are studying the enhancer of a gene that normally is expressed only in neurons. Constructs in which this enhancer is linked to a reporter gene are expressed in neuronal cells but not in fibroblasts. However, if you mutate a specific sequence element within the enhancer, you find expression in both fibroblasts and neuronal cells. What type of regulatory protein would you expect to bind to that enhancer element?

5. A transcription factor is found to activate transcription by binding to different DNA sequences in muscle cells and liver cells. How might alternative splicing be involved in determining this tissue specificity of activator function?

References and Further Reading

Transcription in Prokaryotes

Burgess, R. R., A. A. Travers, J. J. Dunn and E. K. F. Bautz. 1969. Factor stimulating transcription by RNA polymerase. *Nature* 221: 43–44. [P]

Busby, S. and R. H. Ebright. 1994. Promoter structure, promoter recognition, and transcription activation in prokaryotes. *Cell* 79: 743–746. [R]

Gilbert, W. and B. Muller-Hill. 1966. Isolation of the *lac* repressor. *Proc. Natl. Acad. Sci. USA* 56: 1891–1899. [P]

Hochschild, A. and S. L. Dove. 1998. Protein-protein contacts that activate and repress prokaryotic transcription. *Cell* 92: 597–600. [R]

Jacob, F. and J. Monod. 1961. Genetic and regulatory mechanisms in the synthesis of proteins. *J. Mol. Biol.* 3: 318–356. [P]

Lewis, M., G. Chang, N. C. Horton, M. A. Kirchner, H. C. Pace, M. A. Schumacher, R. G. Brennan and P. Lu. 1996. Crystal structure of the lactost operon repressor and its complexes with DNA and inducer. *Science* 271: 1247–1254. [P]

Pardee, A. B., F. Jacob and J. Monod. 1959. The genetic control and cytoplasmic expression of "inducibility" in the synthesis of β-galactosidase by *E. coli. J. Mol. Biol.* 1: 165–178. [P]

Uptain, S. M., C. M. Kane and M. J. Chamberlin. 1997. Basic mechanisms of transcript elongation and its regulation. *Ann. Rev. Biochem.* 66: 117–172. [R]

Yanofsky, C. 1981. Attenuation in the control of expression of bacterial operons. *Nature* 289: 751–758. [R]

Eukaryotic RNA Polymerases and General Transcription Factors

Buratowski, S. 1994. The basics of basal transcription by RNA polymerase II. *Cell* 77: 1–3. [R]

Burley, S. K. and R. G. Roeder. 1996. Biochemistry and structural biology of transcription factor IID (TFIID). *Ann. Rev. Biochem.* 65: 769–799. [R]

Geiduschek, E. P. and G. P. Tocchini-Valentini. 1988. Transcription by RNA polymerase III. *Ann. Rev. Biochem.* 57: 873–914. [R]

Hahn, S. 1998. The role of TAFs in RNA polymerase II transcription. *Cell* 95: 579–582. [R]

Hernandez, N. 1993. TBP, a universal eukaryotic transcription factor? *Genes Dev.* 7: 1291–1308. [R]

Matsui, T., J. Segall, P. A. Weil and R. G. Roeder. 1980. Multiple factors are required for accurate initiation of transcription by purified RNA polymerase II. *J. Biol. Chem.* 255: 11992–11996. [P]

Myer, V. E. and R. A. Young. 1998. RNA polymerase II holoenzymes and subcomplexes. *J. Biol. Chem.* 273: 27757–27760. [R]

Nikolov, D. B. and S. K. Burley. 1997. RNA polymerase II transcription initiation: a structural view. *Proc. Natl. Acad. Sci. USA* 94: 15–22. [R]

Nikolov, D. B., H. Chen, E. D. Halay, A. A. Usheva, K. Hisatake, D. K. Lee, R. G. Roeder and S. K. Burley. 1995. Crystal structure of a TFIIB-TBP-TATA-element ternary complex. *Nature* 377: 119–128. [P]

Orphanides, G., T. Lagrange and D. Reinberg. 1996. The general transcription factors of RNA polymerase II. *Genes Dev.* 10: 2657–2683. [R]

Weil, P. A., D. S. Luse, J. Segall and R. G. Roeder. 1979. Selective and accurate transcription at the Ad2 major late promoter in a soluble system dependent on purified RNA polymerase II and DNA. *Cell* 18: 469–484. [P]

Zawel, L. and D. Reinberg. 1995. Common themes in assembly and function of eukaryotic transcription complexes. *Ann. Rev. Biochem.* 64: 533–561. [R]

Regulation of Transcription in Eukaryotes

Atchison, M. L. 1988. Enhancers: Mechanisms of action and cell specificity. *Ann. Rev. Cell. Biol.* 4: 127–153. [R]

Banerji, J., S. Rusconi and W. Schaffner. 1981. Expression of a β-globin gene is enhanced by remote SV40 DNA sequences. *Cell* 27: 299–308. [P]

Barlow, D. P. 1995. DNA methylation and genomic imprinting. *Cell* 77: 473–476. [R]

Bird, A. 1992. The essentials of DNA methylation. *Cell* 70: 5–8. [R]

Brent, R. and M. Ptashne. 1985. A eukaryotic transcriptional activator bearing the DNA specificity of a prokaryotic repressor. *Cell* 43: 729–736. [P]

Brownell, J. E., J. Zhou, T. Ranalli, R. Kobayashi, D. G. Edmondson, S. Y. Roth and C. D. Allis. 1996. *Tetrahymena* histone acetyltransferase A: A homolog to yeast Gcn5p linking histone acetylation to gene activation. *Cell* 84: 843–851. [P]

Buratowski, S. 1995. Mechanisms of gene activation. *Science* 270: 1773–1774. [R]

Dynan, W. S. and R. Tjian. 1983. The promoter-specific transcription factor Sp1 binds to upstream sequences in the SV40 early promoter. *Cell* 35: 79–87. [P]

Felsenfeld, G. 1996. Chromatin unfolds. *Cell* 86: 13–19. [R]

Gehring, W. J., Y. Q. Qian, M. Billeter, K. Furukubo-Tokunaga, A. F. Schier, D. Resendez-Perez, M. Affolter, G. Otting and K. Wuthrich. 1994. Homeodomain-DNA recognition. *Cell* 78: 211–223. [R]

Goodrich, J. A., G. Cutler, and R. Tjian. 1996. Contacts in context: Promoter specificity and macromolecular interactions in transcription. *Cell* 84: 825–830. [R]

Grunstein, M. 1992. Histones as regulators of genes. *Sci. Amer.* 267(4): 68–74B. [R]

Hamiche, A., R. Sandaltzopoulos, D. A. Gdula and C. Wu. 1999. ATP-dependent histone octamer sliding mediated by the chromatin remodeling complex NURF. *Cell* 97: 833–842. [P]

Hanna-Rose, W. and U. Hansen. 1996. Active repression mechanisms of eukaryotic transcription repressors. *Trends Genet.* 12: 229–234. [R]

Johnson, P. F. and S. L. McKnight. 1989. Eukaryotic transcriptional regulatory proteins. *Ann. Rev. Biochem.* 58: 799–839. [R]

Jones, P. L., G. J. Veenstra, P. A. Wade, D. Vermaak, S. U. Kass, N. Landsberger, J. Strouboulis and A. P. Wolffe. 1998. Methylated DNA and MeCP2 recruit histone deacetylase to repress transcription. *Nature Genet.* 19: 187–191. [P]

Kadonaga, J. T. 1998. Eukaryotic transcription: An interlaced network of transcription factors and chromatin-modifying machines. *Cell* 92: 307–313. [R]

Kadonaga, J. T. and R. Tjian. 1986. Affinity purification of sequence-specific DNA binding proteins. *Proc. Natl. Acad. Sci. USA* 83: 5889–5893. [P]

Kornberg, R. D. and Y. Lorch. 1992. Chromatin structure and transcription. *Ann. Rev. Cell Biol.* 8: 563–587. [R]

Langst, G., E. J. Bonte, D. F. V. Corona and P. B. Becker. 1999. Nucleosome movement by CHRAC and ISWI without disruption or trans-displacement of the histone octamer. *Cell* 97: 843–852. [P]

Licht, J. D., M. J. Grossel, J. Figge and U. M. Hansen. 1990. *Drosophila Krüppel* protein is a transcriptional repressor. *Nature* 346: 76–79. [P]

Maniatis, T., S. Goodbourn and J. A. Fischer. 1987. Regulation of inducible and tissue-specific gene expression. *Science* 236: 1237–1244. [R]

McGinnis, W. and R. Krumlauf. 1992. Homeobox genes and axial patterning. *Cell* 68: 283–302. [R]

McKnight, S. L. and R. Kingsbury. 1982. Transcriptional control signals of a eukaryotic protein-coding gene. *Science* 217: 316–324. [P]

Mitchell, P. and R. Tjian. 1989. Transcriptional regulation in mammalian cells by sequence-specific DNA binding proteins. *Science* 245: 371–378. [R]

Nan, X., H. H. Ng, C. A. Johnson, C. D. Laherty, B. M. Turner, R. N. Eisenman and A. Bird. 1998. Transcriptional repression by the methyl-CpG-binding protein MeCP2 involves a histone deacetylase complex. *Nature* 393: 386–389. [R]

Pabo, C. O. and R. T. Sauer. 1992. Transcription factors: Structural families and principles of DNA recognition. *Ann. Rev. Biochem.* 61: 1053–1095. [R]

Ptashne, M. and A. Gann. 1997. Transcriptional activation by recruitment. *Nature* 386: 569–577. [R]

Razin, A. and H. Cedar. 1994. DNA methylation and genomic imprinting. *Cell* 77: 473–476. [R]

Roberts, S. G. E. and M. R. Green. 1994. Activator-induced conformational change in general transcription factor TFIIB. *Nature* 371: 717–720. [P]

Roth, S. Y. and C. D. Allis. 1996. The subunit-exchange model of histone acetylation. *Trends Cell Biol.* 6: 371–375. [R]

Staudt, L. M. and M. J. Lenardo. 1991. Immunoglobulin gene transcription. *Ann. Rev. Immunol.* 9: 373–398. [R]

Taunton, J., C. A. Hassig and S. L. Schreiber. 1996. A mammalian histone deacetylase related to the yeast transcriptional regulator Rpd3p. *Science* 272: 408–411. [P]

Tilghman, S. M. 1999. The sins of the fathers and mothers: Genomic imprinting in mammalian development. *Cell* 96: 185–193. [R]

Tjian, R. and T. Maniatis. 1994. Transcriptional activation: A complex puzzle with few easy pieces. *Cell* 77: 5–8. [R]

Wade, P. A., D. Pruss and A. P. Wolffe. 1997. Histone acetylation: Chromatin in action. *Trends Biochem. Sci.* 22: 128–132. [R]

Wolffe, A. 1998. *Chromatin: Structure and Function.* 3rd. ed. New York: Academic Press.

Workman, J. L. and R. E. Kingston. 1998. Alteration of nucleosome structure as a mechanism of transcriptional regulation. *Ann. Rev. Biochem.* 67: 545–579. [R]

Wu, C. 1997. Chromatin remodeling and the control of gene expression. *J. Biol. Chem.* 272: 28171–28174. [R]

RNA Processing and Turnover

Abelson, J., C. R. Trotta and H. Li. 1998. tRNA splicing. *J. Biol. Chem.* 273:12685–12688. [R]

Bass, B. L. 1997. RNA editing and hypermutation by adenosine deamination. *Trends Biochem. Sci.* 22: 157–162. [R]

Beelman, C. A. and R. Parker. 1995. Degradation of mRNA in eukaryotes. *Cell* 81: 179–183. [R]

Berget, S. M. 1995. Exon recognition in vertebrate splicing. *J. Biol. Chem.* 270: 2411–2414. [R]

Cech, T. R. 1990. Self-splicing of group I introns. *Ann. Rev. Biochem.* 59: 543–568. [R]

Curtis, D., R. Lehmann and P. D. Zamore. 1995. Translational regulation in development. *Cell* 81: 171–178. [R]

Foulkes, N. S. and P. Sassone-Corsi. 1992. More is better: Activators and repressors from the same gene. *Cell* 68: 411–414. [R]

Frank, D. N. and N. R. Pace. 1998. Ribonuclease P: Unity and diversity in a tRNA processing ribozyme. *Ann. Rev. Biochem.* 67: 153–180. [R]

Green, M. R. 1991. Biochemical mechanisms of constitutive and regulated pre-mRNA splicing. *Ann. Rev. Cell Biol.* 7: 559–599. [R]

Guerrier-Takada, C., K. Gardiner, T. Marsh, N. Pace and S. Altman. 1983. The RNA moiety of ribonuclease P is the catalytic subunit of the enzyme. *Cell* 35: 849–857. [P]

Jacobson, A. and S. W. Peltz. 1996. Interrelationships of the pathways of mRNA decay and translation in eukaryotic cells. *Ann. Rev. Biochem.* 65: 693–739. [R]

Keller, W. 1995. No end yet to messenger RNA 3′ processing! *Cell* 81: 829–832. [R]

Klausner, R. D., T. A. Rouault and J. B. Harford. 1993. Regulating the fate of mRNA: The control of cellular iron metabolism. *Cell* 72: 19–28. [R]

Kramer, A. 1996. The structure and function of proteins involved in mammalian premRNA splicing. *Ann. Rev. Biochem.* 65: 367–409. [R]

Kruger, K., P. J. Grabowski, A. Zaug, A. J. Sands, D. E. Gottschling and T. R. Cech. 1982. Self-splicing RNA: Autoexcision and autocyclization of the ribosomal RNA intervening sequence of *Tetrahymena. Cell* 31: 147–157. [P]

Maniatis, T. 1991. Mechanisms of alternative pre-mRNA splicing. *Science* 251: 33–34. [R]

McKeown, M. 1992. Alternative mRNA splicing. *Ann. Rev. Cell Biol.* 8: 133–155. [R]

Newman, A. J. 1997. The role of U5 snRNP in pre-mRNA splicing. *EMBO J.* 16: 5797–5800. [R]

Nilsen, T. W. 1994. RNA–RNA interactions in the spliceosome: Unraveling the ties that bind. *Cell* 78: 1–4. [R]

Padgett, R. A., M. M. Konarska, P. J. Grabowski, S. F. Hardy and P. A. Sharp. 1984. Lariat RNAs as intermediates and products in the splicing of messenger RNA precursors. *Science* 225: 898–903. [P]

Padgett, R. A., S. M. Mount, J. A. Steitz and P. A. Sharp. 1983. Splicing of messenger RNA precursors is inhibited by antisera to small nuclear ribonucleoprotein. *Cell* 35: 101–107. [P]

Proudfoot, N. 1996. Ending the message is not so simple. *Cell* 87: 779–781. [R]

Ruskin, B., A. R. Krainer, T. Maniatis and M. R. Green. 1984. Excision of an intact intron as a novel lariat structure during pre-mRNA splicing *in vitro. Cell* 38: 317–331. [P]

Sachs, A. B. 1993. Messenger RNA degradation in eukaryotes. *Cell* 74: 413–421. [R]

Scott, J. 1995. A place in the world for RNA editing. *Cell* 81: 833–836. [R]

Sharp, P. A. 1994. Split genes and RNA splicing. *Cell* 77: 805–815. [R]

Staley, J. P. and C. Guthrie. 1998. Mechanical devices of the spliceosome: Motors, clocks, springs, and things. *Cell* 92: 315–326. [R]

Steinmetz, E. J. 1997. Pre-mRNA processing and the CTD of RNA polymerase II: The tail that wags the dog? *Cell* 89: 491–494. [R]

Valcarcel, J. and M. R. Green. 1996. The SR protein family: Pleiotropic functions in pre-mRNA splicing. *Trends Biochem. Sci.* 21: 296–301. [R]

7

Protein Synthesis, Processing, and Regulation

TRANSCRIPTION AND RNA PROCESSING ARE FOLLOWED BY TRANSLATION, the synthesis of proteins as directed by mRNA templates. Proteins are the active players in most cell processes, implementing the myriad tasks that are directed by the information encoded in genomic DNA. Protein synthesis is thus the final stage of gene expression. However, the translation of mRNA is only the first step in the formation of a functional protein. The polypeptide chain must then fold into the appropriate three-dimensional conformation and, frequently, undergo various processing steps before being converted to its active form. These processing steps, particularly in eukaryotes, are intimately related to the sorting and transport of different proteins to their appropriate destinations within the cell.

Although the expression of most genes is regulated primarily at the level of transcription (see Chapter 6), gene expression can also be controlled at the level of translation, and this control is an important element of gene regulation in both prokaryotic and eukaryotic cells. Of even broader significance, however, are the mechanisms that control the activities of proteins within cells. Once synthesized, most proteins can be regulated in response to extracellular signals by either covalent modifications or by association with other molecules. In addition, the levels of proteins within cells can be controlled by differential rates of protein degradation. These multiple controls of both the amounts and activities of intracellular proteins ultimately regulate all aspects of cell behavior.

Translation of mRNA

Proteins are synthesized from mRNA templates by a process that has been highly conserved throughout evolution (reviewed in Chapter 3). All mRNAs are read in the 5′ to 3′ direction, and polypeptide chains are synthesized from the amino to the carboxy terminus. Each amino acid is specified by three bases (a codon) in the mRNA, according to a nearly universal genetic code. The basic mechanics of protein synthesis are also the same in all cells: Translation is carried out on ribosomes, with tRNAs serving as adaptors between the mRNA template and the amino acids being incorporated into protein. Protein synthesis thus involves interactions between three types of RNA molecules (mRNA templates, tRNAs, and rRNAs), as well as various proteins that are required for translation.

Transfer RNAs

During translation, each of the 20 amino acids must be aligned with their corresponding codons on the mRNA template. All cells contain a variety of **tRNAs** that serve as adaptors for this process. As might be expected, given their common function in protein synthesis, different tRNAs share similar overall structures. However, they also possess unique identifying sequences that allow the correct amino acid to be attached and aligned with the appropriate codon in mRNA.

Transfer RNAs are approximately 70 to 80 nucleotides long and have characteristic cloverleaf structures that result from complementary base pairing between different regions of the molecule (Figure 7.1). X-ray crystallography studies have further shown that all tRNAs fold into similar compact L shapes, which are likely required for the tRNAs to fit onto ribosomes during the translation process. The adaptor function of the tRNAs involves two separated regions of the molecule. All tRNAs have the sequence CCA at their 3′ terminus, and amino acids are covalently attached to the ribose of the terminal adenosine. The mRNA template is then recognized by the **anticodon** loop, located at the other end of the folded tRNA, which binds to the appropriate codon by complementary base pairing.

The incorporation of the correctly encoded amino acids into proteins depends on the attachment of each amino acid to an appropriate tRNA, as well as on the specificity of codon-anticodon base pairing. The attachment of amino acids to specific tRNAs is mediated by a group of enzymes called **aminoacyl tRNA synthetases**, which were discovered by Paul Zamecnik and Mahlon Hoagland in 1957. Each of these enzymes recognizes a single amino acid, as well as the correct tRNA (or tRNAs) to which that amino acid should be attached. The reaction proceeds in two steps (Figure 7.2). First, the amino acid is activated by reaction with ATP to form an aminoacyl AMP synthetase intermediate. The activated amino acid is then joined to the

Figure 7.1
Structure of tRNAs The structure of yeast phenylalanyl tRNA is illustrated in open "cloverleaf" form (A) to show complementary base pairing. Modified bases are indicated as mG, methylguanosine; mC, methylcytosine; DHU, dihydrouridine; T, ribothymidine; Y, a modified purine (usually adenosine); and ψ, pseudouridine. The folded form of the molecule is shown in (B) and a space-filling model in (C). (C, courtesy of Dan Richardson.)

Figure 7.2
Attachment of amino acids to tRNAs In the first reaction step, the amino acid is joined to AMP, forming an aminoacyl AMP intermediate. In the second step, the amino acid is transferred to the 3′ CCA terminus of the acceptor tRNA and AMP is released. Both steps of the reaction are catalyzed by aminoacyl tRNA synthetases.

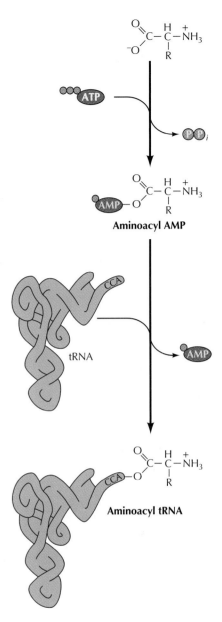

3′ terminus of the tRNA. The aminoacyl tRNA synthetases must be highly selective enzymes that recognize both individual amino acids and specific base sequences that identify the correct acceptor tRNAs. In some cases, the high fidelity of amino acid recognition results in part from a proofreading function by which incorrect aminoacyl AMPs are hydrolyzed rather than being joined to tRNA during the second step of the reaction. Recognition of the correct tRNA by the aminoacyl tRNA synthetase is also highly selective; the synthetase recognizes specific nucleotide sequences (in most cases including the anticodon) that uniquely identify each species of tRNA.

After being attached to tRNA, an amino acid is aligned on the mRNA template by complementary base pairing between the mRNA codon and the anticodon of the tRNA. Codon-anticodon base pairing is somewhat less stringent than the standard A-U and G-C base pairing discussed in preceding chapters. The significance of this unusual base pairing in codon-anticodon recognition relates to the redundancy of the genetic code. Of the 64 possible codons, three are stop codons that signal the termination of translation; the other 61 encode amino acids (see Table 3.1). Thus, most of the amino acids are specified by more than one codon. In part, this redundancy results from the attachment of many amino acids to more than one species of tRNA. *E. coli*, for example, contain about 40 different tRNAs that serve as acceptors for the 20 different amino acids. In addition, some tRNAs are able to recognize more than one codon in mRNA, as a result of nonstandard base pairing (called wobble) between the tRNA anticodon and the third position of some complementary codons (Figure 7.3). Relaxed base pairing at this position results partly from the formation of G-U base pairs and partly from the modification of guanosine to inosine in the anticodons of several tRNAs during processing (see Figure 6.38). Inosine can base-pair with either C, U, or A in the third position, so its inclusion in the anticodon allows a single tRNA to recognize three different codons in mRNA templates.

The Ribosome

Ribosomes are the sites of protein synthesis in both prokaryotic and eukaryotic cells. First characterized as particles detected by ultracentrifugation of cell lysates, ribosomes are usually designated according to their rates of sedimentation: 70S for bacterial ribosomes and 80S for the somewhat larger ribosomes of eukaryotic cells. Both prokaryotic and eukaryotic ribosomes are composed of two distinct subunits, each containing characteristic proteins and **rRNAs**. The fact that cells typically contain many ribosomes reflects the central importance of protein synthesis in cell metabolism. *E. coli*, for example, contain about 20,000 ribosomes, which account for approximately 25% of the dry weight of the cell, and rapidly growing mammalian cells contain about 10 million ribosomes.

The general structures of prokaryotic and eukaryotic ribosomes are similar, although they differ in some details (Figure 7.4). The small subunit (designated 30S) of *E. coli* ribosomes consists of the 16S rRNA and 21 proteins; the large subunit (50S) is composed of the 23S and 5S rRNAs and 34 proteins. Each ribosome contains one copy of the rRNAs and one copy of each of the ribosomal proteins, with one exception: One protein of the 50S

Figure 7.3
Nonstandard codon-anticodon base pairing   Base pairing at the third codon position is relaxed, allowing G to pair with U, and inosine (I) in the anticodon to pair with U, C, or A. Two examples of abnormal base pairing, allowing phenylalanyl (Phe) tRNA to recognize either UUC or UUU codons and alanyl (Ala) tRNA to recognize GCU, GCC, or GCA, are illustrated.

Phenylalanyl tRNA pairing

Alanyl tRNA pairing

subunit is present in four copies. The subunits of eukaryotic ribosomes are larger and contain more proteins than their prokaryotic counterparts have. The small subunit (40S) of eukaryotic ribosomes is composed of the 18S rRNA and approximately 30 proteins; the large subunit (60S) contains the 28S, 5.8S, and 5S rRNAs and about 45 proteins.

A noteworthy feature of ribosomes is that they can be formed *in vitro* by self-assembly of their RNA and protein constituents. As first described in

(A)

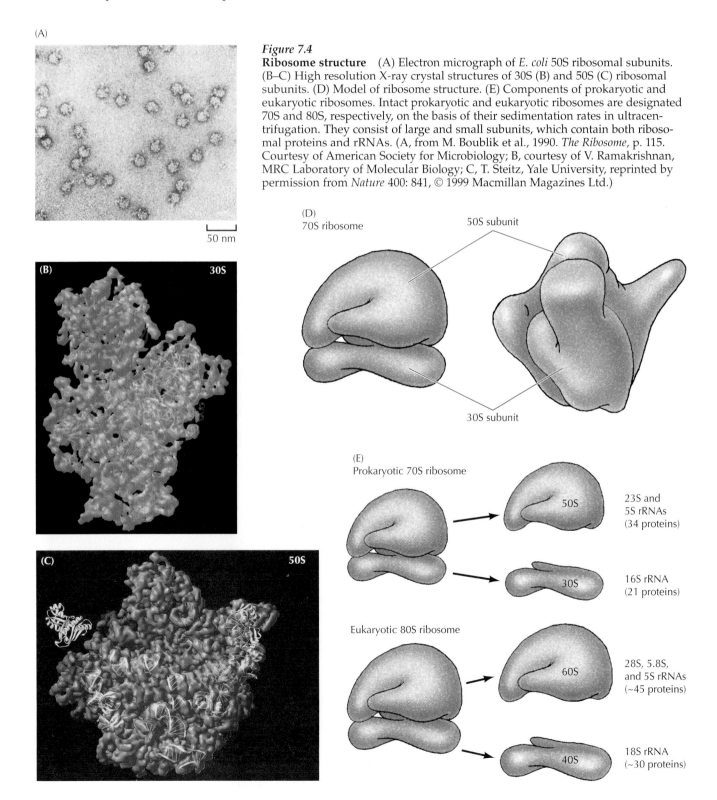

Figure 7.4
Ribosome structure   (A) Electron micrograph of *E. coli* 50S ribosomal subunits. (B–C) High resolution X-ray crystal structures of 30S (B) and 50S (C) ribosomal subunits. (D) Model of ribosome structure. (E) Components of prokaryotic and eukaryotic ribosomes. Intact prokaryotic and eukaryotic ribosomes are designated 70S and 80S, respectively, on the basis of their sedimentation rates in ultracentrifugation. They consist of large and small subunits, which contain both ribosomal proteins and rRNAs. (A, from M. Boublik et al., 1990. *The Ribosome*, p. 115. Courtesy of American Society for Microbiology; B, courtesy of V. Ramakrishnan, MRC Laboratory of Molecular Biology; C, T. Steitz, Yale University, reprinted by permission from *Nature* 400: 841, © 1999 Macmillan Magazines Ltd.)

50 nm

(B)   30S

(C)   50S

(D)
70S ribosome

50S subunit

30S subunit

(E)
Prokaryotic 70S ribosome

50S   23S and 5S rRNAs (34 proteins)

30S   16S rRNA (21 proteins)

Eukaryotic 80S ribosome

60S   28S, 5.8S, and 5S rRNAs (~45 proteins)

40S   18S rRNA (~30 proteins)

1968 by Masayasu Nomura, purified ribosomal proteins and rRNAs can be mixed together and, under appropriate conditions, will re-form a functional ribosome. Although ribosome assembly *in vivo* (particularly in eukaryotic cells) is considerably more complicated, the ability of ribosomes to self-assemble *in vitro* has provided an important experimental tool, allowing analysis of the roles of individual proteins and rRNAs.

Like tRNAs, rRNAs form characteristic secondary structures by complementary base pairing (Figure 7.5). In association with ribosomal proteins the rRNAs fold further, into distinct three-dimensional structures. Initially, rRNAs were thought to play a structural role, providing a scaffold upon which ribosomal proteins assemble. However, with the discovery of the catalytic activity of other RNA molecules (e.g., RNase P and the self-splicing introns discussed in Chapter 6), the possible catalytic role of rRNA became widely considered. Consistent with this hypothesis, rRNAs were found to be absolutely required for the *in vitro* assembly of functional ribosomes. On

Figure 7.5
The structure of 16S rRNA
Complementary base pairing results in the formation of a distinct secondary structure. (From S. Stern, T. Powers, L.-I. Changchien and H. F. Noller, 1989. *Science* 244: 783.)

the other hand, the omission of many ribosomal proteins resulted in a decrease, but not a complete loss, of ribosome activity.

Direct evidence for the catalytic activity of rRNA first came from experiments of Harry Noller and his colleagues in 1992. These investigators demonstrated that the large ribosomal subunit is able to catalyze the formation of peptide bonds (the peptidyl transferase reaction) even after approximately 95% of the ribosomal proteins have been removed by standard protein extraction procedures. In contrast, treatment with RNase completely abolishes peptide bond formation, providing strong support for the hypothesis that the formation of a peptide bond is an RNA-catalyzed reaction. Further studies have confirmed and extended these results by demonstrating that the peptidyl transferase reaction can be catalyzed by synthetic fragments of 23S rRNA in the total absence of any ribosomal protein. Thus, the fundamental reaction of protein synthesis is catalyzed by ribosomal RNA. Rather than being the primary catalytic constituents of ribosomes, ribosomal proteins are now thought to facilitate proper folding of the rRNA and to enhance ribosome function by properly positioning the tRNAs.

The direct involvement of rRNA in the peptidyl transferase reaction has important evolutionary implications. RNAs are thought to have been the first self-replicating macromolecules (see Chapter 1). This notion is strongly supported by the fact that ribozymes, such as RNase P and self-splicing introns, can catalyze reactions that involve RNA substrates. The role of rRNA in the formation of peptide bonds extends the catalytic activities of RNA beyond self-replication to direct involvement in protein synthesis. Additional studies indicate that the *Tetrahymena* rRNA ribozyme can catalyze the attachment of amino acids to RNA, lending credence to the possibility that the original aminoacyl tRNA synthetases were RNAs rather than proteins. The ability of RNA molecules to catalyze the reactions required for protein synthesis as well as for self-replication may provide an important link for understanding the early evolution of cells.

The Organization of mRNAs and the Initiation of Translation

Although the mechanisms of protein synthesis in prokaryotic and eukaryotic cells are similar, there are also differences, particularly in the signals that determine the positions at which synthesis of a polypeptide chain is initiated on an mRNA template (Figure 7.6). Translation does not simply begin at the

Prokaryotic mRNA

Eukaryotic mRNA

Figure 7.6
Prokaryotic and eukaryotic mRNAs Both prokaryotic and eukaryotic mRNAs contain untranslated regions (UTRs) at their 5' and 3' ends. Eukaryotic mRNAs also contain 5' 7-methylguanosine (m^7G) caps and 3' poly-A tails. Prokaryotic mRNAs are frequently polycistronic: They encode multiple proteins, each of which is translated from an independent start site. Eukaryotic mRNAs are usually monocistronic, encoding only a single protein.

Key Experiment

Catalytic Role of Ribosomal RNA

Unusual Resistance of Peptidyl Transferase to Protein Extraction Procedures

Harre F. Noller, Vernita Hoffarth, and Ludwika Zimniak

University of California at Santa Cruz

Science, Volume 256, 1992, pages 1416–1419

The Context

The role of ribosomes in protein synthesis was elucidated in the 1960s. During this period, ribosomes were characterized as particles consisting of both proteins and RNAs, and the reconstitution of functional ribosomes from purified components was accomplished. At that time peptide bond formation (the peptidyl transferase reaction) was generally assumed to be catalyzed by ribosomal proteins, with the rRNAs delegated to a supporting role in ribosome structure. As early as the 1970s, however, evidence began to suggest that the rRNAs might play a more direct role in protein synthesis. For example, many ribosomal proteins were found to be nonessential for ribosome function. Conversely, the sequences of some parts of rRNA were shown to be extremely well conserved in evolution, suggesting a critical functional role for these portions of the rRNA molecule.

In the early 1980s, the catalytic activity of RNA molecules was established by Tom Cech's studies of the *Tetrahymena* ribozyme and Sidney Altman's studies of RNase P. These discoveries of RNA catalysis provided a precedent for the hypothesis that rRNA is directly involved in catalyzing peptide bond formation. Compelling evidence in support of such a role for rRNA was provided in this 1992 paper by Harry Noller and his colleagues.

The Experiments

To study the catalytic activity of rRNA, Noller and colleagues used a simplified model reaction to assay peptidyl transferase activity. This reaction measures the transfer of radioactively labeled *N*-formylmethionine from a fragment of tRNA to the amino group of puromycin, an antibiotic that resembles an aminoacyl tRNA and can form peptide bonds with a growing peptide chain. The advantage of this peptidyl transferase model reaction is that it can be carried out by isolated 50S ribosomal subunits; small ribosomal subunits, other protein factors, and mRNA are not required.

The investigators then tested the role of rRNA by assaying the peptidyl transferase activity of 50S subunits from which the ribosomal proteins had been removed by standard protein extraction procedures. An important aspect of the experiments was the use of ribosomes from the

bacterium *Thermus aquaticus*. Because these bacteria live at high temperatures, the structure of their rRNA was thought to be more stable than that of *E. coli* rRNA. The critical result was that the peptidyl transferase activity of *T. aquaticus* ribosomes was completely resistant to a vigorous extraction procedure involving treatment with detergents, proteases, and phenol (see figure). Most strikingly, full peptidyl transferase activity was retained even after repeated extractions that removed 95% of the ribosomal protein. In contrast, peptidyl transferase of either intact or extracted ribosomes was extremely sensitive to even a brief treatment with RNase. Although these experiments could not exclude a possible role for the remaining ribosomal pro-

					f-Met- puro
Protein extraction	−	−	+	+	
RNase	−	+	−	+	

The peptidyl transferase reaction is assayed by the formation of radioactive *N*-formylmethionine-puromycin (f-Met-puro), which is detected by electrophoresis and autoradiography. Ribosomes from *T. aquaticus* were subjected to protein extraction or treatment with RNase, as indicated.

5′ end of the mRNA; it starts at specific initiation sites. The 5′ terminal portions of both prokaryotic and eukaryotic mRNAs are therefore noncoding sequences, referred to as **5′ untranslated regions**. Eukaryotic mRNAs usually encode only a single polypeptide chain, but many prokaryotic mRNAs encode multiple polypeptides that are synthesized independently from distinct initiation sites. For example, the *E. coli lac* operon consists of three genes that are translated from the same mRNA (see Figure 6.8). Messenger RNAs

Catalytic Role of Ribosomal RNA (continued)

teins, they provided strong support for the direct participation of 23S rRNA in the peptidyl transferase reaction.

The Impact

The results of Noller's experiments have been confirmed and extended by studies of Kimitsuna Watanabe and colleagues (Nitta et al., *Science* 281: 666–669, 1998) showing that synthetic fragments of 23S ribosomal RNA, in the total absence of any ribosomal protein, can catalyze the formation of peptide bonds. These results remove any doubts as to the catalytic role of RNA in this reaction and demonstrate that the fundamen-

tal reaction of protein synthesis is catalyzed by ribosomal RNA. These findings not only have a striking impact on our understanding of ribosome function but also profoundly extend the previously described catalytic activities of RNA molecules and provide substantial new support for the hypothesis that an early period of evolution was an RNA world populated by self-replicating RNA molecules. This hypothesis had previously been based on the ability of RNA molecules to catalyze the reactions required for their own replication. The discovery that RNA could catalyze reactions involved in protein synthesis provided a clear link be-

Harry F. Noller

tween the RNA world and the flow of genetic information in present-day cells, with rRNA continuing to function in the key reaction of peptide bond formation.

---

that encode multiple polypeptides are called **polycistronic**, whereas **monocistronic** mRNAs encode a single polypeptide chain. Finally, both prokaryotic and eukaryotic mRNAs end in noncoding **3′ untranslated regions**.

In both prokaryotic and eukaryotic cells, translation always initiates with the amino acid methionine, usually encoded by AUG. Alternative initiation codons, such as GUG, are used occasionally in bacteria, but when they occur at the beginning of a polypeptide chain, these codons direct the incorporation of methionine rather than of the amino acid they normally encode (GUG normally encodes valine). In most bacteria, protein synthesis is initiated with a modified methionine residue (*N*-formylmethionine), whereas unmodified methionines initiate protein synthesis in eukaryotes (except in mitochondria and chloroplasts, whose ribosomes resemble those of bacteria).

The signals that identify initiation codons are different in prokaryotic and eukaryotic cells, consistent with the distinct functions of polycistronic and monocistronic mRNAs (Figure 7.7). Initiation codons in bacterial

Figure 7.7
Signals for translation initiation Initiation sites in prokaryotic mRNAs are characterized by a Shine-Delgarno sequence that precedes the AUG initiation codon. Base pairing between the Shine-Delgarno sequence and a complementary sequence near the 3′ terminus of 16S rRNA aligns the mRNA on the ribosome. In contrast, eukaryotic mRNAs are bound to the 40S ribosomal subunit by their 5′ 7-methylguanosine caps. The ribosome then scans along the mRNA until it encounters an AUG initiation codon.

Initiation

Ribosome binds mRNA at start codon

Elongation

Direction of ribosome movement

Polypeptide chain elongates by successively adding amino acids

Termination

When a stop codon is encountered, polypeptide is released and ribosome dissociates

Figure 7.8
Overview of translation

mRNAs are preceded by a specific sequence (called a **Shine-Delgarno sequence**, after its discoverers) that aligns the mRNA on the ribosome for translation by base-pairing with a complementary sequence near the 3′ terminus of 16S rRNA. This base-pairing interaction enables bacterial ribosomes to initiate translation not only at the 5′ end of an mRNA but also at the internal initiation sites of polycistronic messages. In contrast, ribosomes recognize most eukaryotic mRNAs by binding to the 7-methylguanosine cap at their 5′ terminus (see Figure 6.39). The ribosomes then scan downstream of the 5′ cap until they encounter an AUG initiation codon. Sequences that surround AUGs affect the efficiency of initiation, so in many cases the first AUG in the mRNA is bypassed and translation initiates at an AUG farther downstream. However, eukaryotic mRNAs have no sequence equivalent to the Shine-Delgarno sequence of prokaryotic mRNAs. Translation of eukaryotic mRNAs is instead initiated at a site determined by scanning from the 5′ terminus, consistent with their functions as monocistronic messages that encode only single polypeptides.

The Process of Translation

Translation is generally divided into three stages: initiation, elongation, and termination (Figure 7.8). In both prokaryotes and eukaryotes the first step of the initiation stage is the binding of a specific initiator methionyl tRNA and the mRNA to the small ribosomal subunit. The large ribosomal subunit then joins the complex, forming a functional ribosome on which elongation of the polypeptide chain proceeds. A number of specific nonribosomal proteins are also required for the various stages of the translation process (Table 7.1).

TABLE 7.1 Translation Factors

| Role | Translation factors | |
	Prokaryotes	**Eukaryotes**
Initiation	IF-1, IF-2, IF-3	eIF-1, eIF-1A, eIF-2, eIF-2B, eIF-3, eIF-4A, eIF-4B, eIF-4E, eIF-4G, eIF-5
Elongation	EF-Tu, EF-Ts, EF-G	eEF-1α, eEF-1$\beta\gamma$, eEF-2
Termination	RF-1, RF-2, RF-3	eRF-1, eRF-3

Molecular Medicine

Antibiotics and Protein Synthesis

The Disease

Bacteria are responsible for a wide variety of potentially lethal infectious diseases, including tuberculosis, bacterial pneumonia, childhood meningitis, infections of wounds and burns, syphilis, and gonorrhea. Prior to the 1940s, physicians had no effective treatments for these bacterial infections. At that time, the first antibiotics became available for clinical use. Previously untreatable infections became curable, and a significant increase in average life span has been attributed to the introduction of antibiotics into clinical practice. More than a hundred different antibiotics are now in use, providing effective treatment for bacterial infections.

Molecular and Cellular Basis

To be effective clinically, an antibiotic must kill or inhibit the growth of bac-

teria without being toxic to humans. Thus, most clinically useful antibiotics are directed against targets that are present in bacteria but not in human cells. Penicillin, for example, inhibits synthesis of the bacterial cell wall (see Chapter 12). Many other commonly used antibiotics, however, inhibit different steps in protein synthesis (see table). Some of these antibiotics—including streptomycin, tetracycline, chloramphenicol, and erythromycin—are specific for prokaryotic ribosomes and therefore are effective agents for the treatment of bacterial infections. Other antibiotic inhibitors of protein synthesis act against both prokaryotes and eukaryotes (e.g., puromycin) or against eukaryotes only (e.g., cycloheximide). Although these antibiotics are obviously not useful clinically, they have provided important experimental tools for studies of protein

synthesis in both prokaryotic and eukaryotic cells.

Prevention and Treatment

The use of antibiotics has had a major impact on modern medicine by enabling physicians to cure otherwise life-threatening bacterial infections. Unfortunately, however, mutations can lead to the development of antibiotic-resistant strains of bacteria. Many strains of bacteria are now resistant to one or more antibiotics, so physicians sometimes have to try several different antibiotics before they find one that is effective. Moreover, the spread of resistant bacterial strains has made some antibiotics obsolete. Most seriously, strains of bacteria that are resistant to multiple antibiotics have emerged, and a few strains of bacteria, frequently found in hospitals, are now resistant to all but one or a few known antibiotics. The emergence of these strains with multiple resistance raises the specter of untreatable infections caused by the spread of antibiotic-resistant bacteria—a scenario that would herald a return to the pre-antibiotic era of infectious diseases. Controlling antibiotic-resistant bacteria, both by optimizing the use of currently available antibiotics and by developing new antibiotics, thus represents an important concern in current medical practice.

Antibiotic Inhibitors of Protein Synthesis

Antibiotic	Target cells	Effect
Streptomycin	Prokaryotic	Inhibits initiation and causes misreading
Tetracycline	Prokaryotic	Inhibits binding of aminoacyl tRNAs
Chloramphenicol	Prokaryotic	Inhibits peptidyl transferase activity
Erythromycin	Prokaryotic	Inhibits translocation
Puromycin	Prokaryotic and eukaryotic	Causes premature chain termination
Cycloheximide	Eukaryotic	Inhibits peptidyl transferase activity

The first translation step in bacteria is the binding of three **initiation factors** (IF-1, IF-2, and IF-3) to the 30S ribosomal subunit (Figure 7.9). The mRNA and initiator *N*-formylmethionyl tRNA then join the complex, with IF-2 (which is bound to GTP) specifically recognizing the initiator tRNA. IF-3 is then released, allowing a 50S ribosomal subunit to associate with the complex. This association triggers the hydrolysis of GTP bound to IF-2, which leads to the release of IF-1 and IF-2 (bound to GDP). The result is the formation of a 70S initiation complex (with mRNA and initiator tRNA bound to the ribosome) that is ready to begin peptide bond formation during the elongation stage of translation.

Initiation in eukaryotes is more complicated and requires at least ten proteins (each consisting of multiple polypeptide chains), which are designated eIFs (*e*ukaryotic *i*nitiation *f*actors; see Table 7.1). The factors eIF-1,

Figure 7.9
Initiation of translation in bacteria Three initiation factors (IF-1, IF-2, and IF-3) first bind to the 30S ribosomal subunit. This step is followed by binding of the mRNA and the initiator *N*-formylmethionyl (fMet) tRNA, which is recognized by IF-2 bound to GTP. IF-3 is then released, and a 50S subunit binds to the complex, triggering the hydrolysis of bound GTP, followed by the release of IF-1 and IF-2 bound to GDP.

eIF-1A, and eIF-3 bind to the 40S ribosomal subunit, and eIF-2 (in a complex with GTP) associates with the initiator methionyl tRNA (Figure 7.10). The mRNA is recognized and brought to the ribosome by the eIF-4 group of factors. The 5′ cap of the mRNA is recognized by eIF-4E. Another factor, eIF-4G, binds to both eIF-4E and to a protein (poly-A binding protein or PABP) associated with the poly-A tail at the 3′ end of the mRNA. Eukaryotic initiation factors thus recognize both the 5′ and 3′ ends of mRNAs, accounting for the stimulatory effect of polyadenylation on translation. The initiation factors eIF-4E and eIF-4G, in association with eIF-4A and eIF-4B, then bring the mRNA to the 40S ribosomal subunit, with eIF-4G interacting with eIF-3. The 40S ribosomal subunit, in association with the bound methionyl tRNA and eIFs, then scans the mRNA to identify the AUG initiation codon. When the AUG codon is reached, eIF-5 triggers the hydrolysis of GTP bound to eIF-2. Initiation factors (including eIF-2 bound to GDP) are then released, and a 60S subunit binds to the 40S subunit to form the 80S initiation complex of eukaryotic cells.

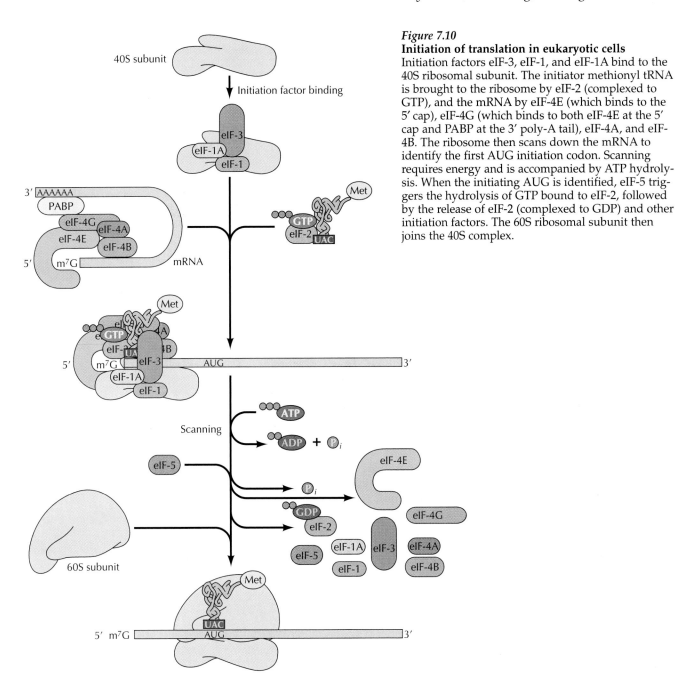

Figure 7.10
Initiation of translation in eukaryotic cells
Initiation factors eIF-3, eIF-1, and eIF-1A bind to the 40S ribosomal subunit. The initiator methionyl tRNA is brought to the ribosome by eIF-2 (complexed to GTP), and the mRNA by eIF-4E (which binds to the 5′ cap), eIF-4G (which binds to both eIF-4E at the 5′ cap and PABP at the 3′ poly-A tail), eIF-4A, and eIF-4B. The ribosome then scans down the mRNA to identify the first AUG initiation codon. Scanning requires energy and is accompanied by ATP hydrolysis. When the initiating AUG is identified, eIF-5 triggers the hydrolysis of GTP bound to eIF-2, followed by the release of eIF-2 (complexed to GDP) and other initiation factors. The 60S ribosomal subunit then joins the 40S complex.

After the initiation complex has formed, translation proceeds by elongation of the polypeptide chain. The mechanism of elongation in prokaryotic and eukaryotic cells is very similar (Figure 7.11). The ribosome has three sites for tRNA binding, designated the P (peptidyl), A (aminoacyl), and E (exit) sites. The initiator methionyl tRNA is bound at the P site. The first step in elongation is the binding of the next aminoacyl tRNA to the A site by pairing with the second codon of the mRNA. The aminoacyl tRNA is escorted to the ribosome by an **elongation factor** (EF-Tu in prokaryotes, eEF-1α in eukaryotes), which is complexed to GTP. The GTP is hydrolyzed to GDP as the correct aminoacyl tRNA is inserted into the A site of the ribosome and the elongation factor bound to GDP is released. The requirement for hydrolysis of GTP before EF-Tu or eEF-1α is released

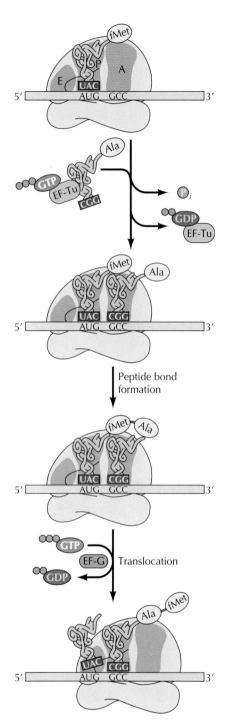

Figure 7.11
Elongation stage of translation   The ribosome has three tRNA-binding sites, designated P (peptidyl), A (aminoacyl), and E (exit). The initiating *N*-formylmethionyl tRNA is positioned in the P site, leaving an empty A site. The second aminoacyl tRNA (e.g., alanyl tRNA) is then brought to the A site by EF-Tu (complexed with GTP). Following GTP hydrolysis, EF-Tu (complexed with GDP) leaves the ribosome, with alanyl tRNA inserted into the A site. A peptide bond is then formed, resulting in the transfer of methionine to the aminoacyl tRNA at the A site. The ribosome then moves three nucleotides along the mRNA. This movement translocates the peptidyl (Met-Ala) tRNA to the P site and the uncharged tRNA to the E site, leaving an empty A site ready for addition of the next amino acid. Translocation is mediated by EF-G, coupled to GTP hydrolysis. The process, illustrated here for prokaryotic cells, is very similar in eukaryotes. (Table 7.1 gives the names of the eukaryotic elongation factors.)

from the ribosome is the rate-limiting step in elongation and provides a time interval during which an incorrect aminoacyl tRNA, which would bind less strongly to the mRNA codon, can dissociate from the ribosome rather than being used for protein synthesis. Thus, the expenditure of a high-energy GTP at this step is an important contribution to accurate protein synthesis; it allows time for proofreading of the codon-anticodon pairing before the peptide bond forms.

Once EF-Tu (or eEF-1α) has left the ribosome, a peptide bond can be formed between the initiator methionyl tRNA at the P site and the second aminoacyl tRNA at the A site. This reaction is catalyzed by the large ribosomal subunit, with the rRNA playing a critical role (as already discussed). The result is the transfer of methionine to the aminoacyl tRNA at the A site of the ribosome, forming a peptidyl tRNA at this position and leaving the uncharged initiator tRNA at the P site. The next step in elongation is translocation, which requires another elongation factor (EF-G in prokaryotes, eEF-2 in eukaryotes) and is again coupled to GTP hydrolysis. During translocation, the ribosome moves three nucleotides along the mRNA, positioning the next codon in an empty A site. This step translocates the peptidyl tRNA from the A site to the P site, and the uncharged tRNA from the P site to the E site. The ribosome is then left with a peptidyl tRNA bound at the P site, and an empty A site. The binding of a new aminoacyl tRNA to the A site then induces the release of the uncharged tRNA from the E site, leaving the ribosome ready for insertion of the next amino acid in the growing polypeptide chain.

As elongation continues, the EF-Tu (or eEF-1α) that is released from the ribosome bound to GDP must be reconverted to its GTP form (Figure 7.12). This conversion requires a third elongation factor, EF-Ts (eEF-1$\beta\gamma$ in eukaryotes), which binds to the EF-Tu/GDP complex and promotes the exchange of bound GDP for GTP. This exchange results in the regeneration of EF-Tu/GTP, which is now ready to escort a new aminoacyl tRNA to the A site of the ribosome, beginning a new cycle of elongation. The regulation of EF-Tu by GTP binding and hydrolysis illustrates a common means of the regulation of protein activities. As will be discussed in later chapters, similar mechanisms control the activities of a wide variety of proteins involved in the regulation of cell growth and differentiation, as well as in protein transport and secretion.

Elongation of the polypeptide chain continues until a stop codon (UAA, UAG, or UGA) is translocated into the A site of the ribosome. Cells do not contain tRNAs with anticodons complementary to these termination signals; instead, they have **release factors** that recognize the signals and ter-

Figure 7.12
Regeneration of EF-Tu/GTP   EF-Tu complexed to GTP escorts the aminoacyl tRNA to the ribosome. The bound GTP is hydrolyzed as the correct tRNA is inserted, so EF-Tu complexed to GDP is released. The EF-Tu/GDP complex is inactive and unable to bind another tRNA. In order for translation to continue, the active EF-Tu/GTP complex must be regenerated by another factor, EF-Ts, which stimulates the exchange of the bound GDP for free GTP.

minate protein synthesis (Figure 7.13). Prokaryotic cells contain two release factors that recognize termination codons: RF-1 recognizes UAA or UAG, and RF-2 recognizes UAA or UGA (see Table 7.1). In eukaryotic cells a single release factor (eRF-1) recognizes all three termination codons. Both prokaryotic and eukaryotic cells also contain release factors (RF-3 and eRF-3, respectively) that do not recognize specific termination codons but act together with RF-1 (or eRF-1) and RF-2. The release factors bind to a termination codon at the A site and stimulate hydrolysis of the bond between the tRNA and the polypeptide chain at the P site, resulting in release of the completed polypeptide from the ribosome. The tRNA is then released, and the ribosomal subunits and the mRNA template dissociate.

Messenger RNAs can be translated simultaneously by several ribosomes in both prokaryotic and eukaryotic cells. Once one ribosome has moved away from the initiation site, another can bind to the mRNA and begin synthesis of a new polypeptide chain. Thus, mRNAs are usually translated by a series of ribosomes, spaced at intervals of about 100 to 200 nucleotides (Figure 7.14). The group of ribosomes bound to an mRNA molecule is called a polyribosome, or **polysome**. Each ribosome within the group functions independently to synthesize a separate polypeptide chain.

Figure 7.13
Termination of translation   A termination codon (e.g., UAA) at the A site is recognized by a release factor rather than by a tRNA. The result is the release of the completed polypeptide chain, followed by the dissociation of tRNA and mRNA from the ribosome.

(A)

Figure 7.14

50 nm

Polysomes   Messenger RNAs are translated by a series of multiple ribosomes (a polysome). (A) Electron micrograph of a eukaryotic polysome. (B) Schematic of a generalized polysome. Note that the ribosomes closer to the 3′ end of the mRNA have longer polypeptide chains. (A, from M. Boublik et al., 1990. *The Ribosome*, p. 117. Courtesy of American Society for Microbiology.)

(B)

Growing polypeptide chain

5′

3′

Direction of ribosome movement

Regulation of Translation

Although transcription is the primary level at which gene expression is controlled, the translation of mRNAs is also regulated in both prokaryotic and eukaryotic cells. One mechanism of translational regulation is the binding of repressor proteins, which block translation, to specific mRNA sequences. The best understood example of this mechanism in eukaryotic cells is regulation of the synthesis of ferritin, a protein that stores iron within the cell. The translation of ferritin mRNA is regulated by the supply of iron: More ferritin is synthesized if iron is abundant (Figure 7.15). This regulation is mediated by a protein which (in the absence of iron) binds to a sequence (the iron response element, or IRE) in the 5′ untranslated region of ferritin mRNA, blocking its translation. In the presence of iron, the repressor no longer binds to the IRE and ferritin translation is able to proceed.

It is interesting to note that the regulation of translation of ferritin mRNA by iron is similar to the regulation of transferrin receptor mRNA stability, which was discussed in the previous chapter (see Figure 6.48). Namely, the stability of transferrin receptor mRNA is regulated by protein binding to an IRE in its 3′ untranslated region. The same protein binds to the IREs of both ferritin and transferrin receptor mRNAs. However, the consequences of protein binding to the two IREs are quite different. Protein bound to the transferrin receptor IRE protects the mRNA from degradation rather than inhibiting its translation. These distinct effects presumably result from the different locations of the IRE in the two mRNAs. To function as a repressor-binding site, the IRE must be located within 70 nucleotides of the 5′ cap of ferritin mRNA, suggesting that protein binding to the IRE blocks translation by interfering with cap recognition and binding of the 40S ribosomal subunit. Rather than inhibiting translation, protein binding to the same sequence in the 3′ untranslated region of transferrin receptor mRNA protects the mRNA from nuclease degradation. Binding of the same regulatory protein to different sites on mRNA molecules can thus have distinct effects on gene

Figure 7.15

Translational regulation of ferritin The mRNA contains an iron response element (IRE) near its 5′ cap. In the presence of adequate supplies of iron, translation of the mRNA proceeds normally. If iron is scarce, however, a protein (called the iron response element–binding protein, or IRE-BP) binds to the IRE, blocking translation of the mRNA.

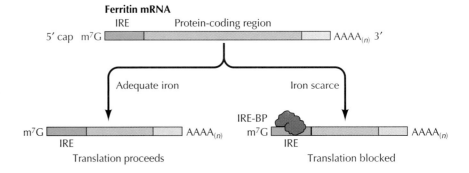

Ferritin mRNA

IRE   Protein-coding region

5′ cap   m⁷G   AAAA₍n₎ 3′

Adequate iron   Iron scarce

m⁷G   AAAA₍n₎   IRE-BP   m⁷G   AAAA₍n₎

IRE   IRE

Translation proceeds   Translation blocked

expression, in one case inhibiting translation and in the other stabilizing the mRNA to increase protein synthesis.

Another mechanism of translational regulation in eukaryotic cells, resulting in global effects on overall translational activity rather than on the translation of specific mRNAs, involves modulation of the activity of initiation factors, particularly eIF-2. As already discussed, eIF-2 (complexed with GTP) binds to the initiator methionyl tRNA, bringing it to the ribosome. The subsequent release of eIF-2 is accompanied by GTP hydrolysis, leaving eIF-2 as an inactive GDP complex. To participate in another cycle of initiation, the eIF-2/GTP complex must be regenerated by the exchange of bound GDP for GTP. This exchange is mediated by another factor, eIF-2B. The control of eIF-2 activity by GTP binding and hydrolysis is thus similar to that of EF-Tu (see Figure 7.12). However, the regulation of eIF-2 provides a critical control point in a variety of eukaryotic cells. In particular, eIF-2 can be phosphorylated by regulatory protein kinases. This phosphorylation blocks the exchange of bound GDP for GTP, thereby inhibiting initiation of translation. One type of cell in which such phosphorylation occurs is the reticulocyte, which is devoted to the synthesis of hemoglobin (Figure 7.16). The translation of globin mRNA is controlled by the availability of heme: The mRNA is translated

Figure 7.16
Regulation of translation by phosphorylation of eIF-2   Translation in reticulocytes (which is devoted to synthesis of hemoglobin) is controlled by the supply of heme, which regulates the activity of eIF-2. The active form of eIF-2 (complexed with GTP) escorts initiator methionyl tRNA to the ribosome (see Figure 7.10). The eIF-2 is then released from the ribosome in an inactive GDP-bound form, which must be reactivated by exchange of GTP for the bound GDP. If adequate heme is available, this exchange occurs and translation is able to proceed. If heme supplies are inadequate, however, a protein kinase that phosphorylates eIF-2 is activated. Phosphorylation of eIF-2 blocks the exchange of GTP for GDP, so eIF-2/GTP cannot be regenerated and translation is inhibited.

only if adequate heme is available to form functional hemoglobin molecules. In the absence of heme, a protein kinase that phosphorylates eIF-2 is activated, and further translation is inhibited. Similar mechanisms have been found to control protein synthesis in other cell types, particularly virus-infected cells in which viral protein synthesis is inhibited by interferon.

Other studies have implicated eIF-4E, which binds to the 5′ cap of mRNAs, as a translational regulatory protein. For example, the hormone insulin stimulates protein synthesis in adipocytes and muscle cells. This effect of insulin is mediated, at least in part, by phosphorylation of proteins associated with eIF-4E, resulting in stimulation of eIF-4E activity and increased rates of translational initiation.

Translational regulation is particularly important during early development. As discussed in Chapter 6, a variety of mRNAs are stored in oocytes in an untranslated form; the translation of these stored mRNAs is activated at fertilization or later stages of development. One mechanism of such translational regulation is the controlled polyadenylation of oocyte mRNAs. Many untranslated mRNAs are stored in oocytes with short poly-A tails (approximately 20 nucleotides). These stored mRNAs are subsequently recruited for translation at the appropriate stage of development by the lengthening of their poly-A tails to several hundred nucleotides. In addition, the translation of some mRNAs during development appears to be regulated by repressor proteins that bind to specific sequences in their 3′ untranslated regions. These regulatory proteins may also direct mRNAs to specific regions of eggs or embryos, allowing localized synthesis of the encoded proteins during embryonic development.

Protein Folding and Processing

Translation completes the flow of genetic information within the cell. The sequence of nucleotides in DNA has now been converted to the sequence of amino acids in a polypeptide chain. The synthesis of a polypeptide, however, is not equivalent to the production of a functional protein. To be useful, polypeptides must fold into distinct three-dimensional conformations, and in many cases multiple polypeptide chains must assemble into a functional complex. In addition, many proteins undergo further modifications, including cleavage and the covalent attachment of carbohydrates and lipids, that are critical for the function and correct localization of proteins within the cell.

Chaperones and Protein Folding

The three-dimensional conformations of proteins result from interactions between the side chains of their constituent amino acids, as reviewed in Chapter 2. The classic principle of protein folding is that all the information required for a protein to adopt the correct three-dimensional conformation is provided by its amino acid sequence. This was initially established by Christian Anfinsen's experiments demonstrating that denatured RNase can spontaneously refold *in vitro* to its active conformation (see Figure 2.17). Protein folding thus appeared to be a self-assembly process that did not require additional cellular factors. More recent studies, however, have shown that this is not an adequate description of protein folding within the cell. The proper folding of proteins within cells is mediated by the activities of other proteins.

Proteins that facilitate the folding of other proteins are called molecular **chaperones**. The term "chaperone" was first used by Ron Laskey and his colleagues to describe a protein (nucleoplasmin) that is required for the assembly of nucleosomes from histones and DNA. Nucleoplasmin binds to histones and mediates their assembly into nucleosomes, but nucleoplasmin itself is not incorporated into the final nucleosome structure. Chaperones thus act as catalysts that facilitate assembly without being part of the assembled complex. Subsequent studies have extended the concept to include proteins that mediate a variety of other assembly processes, particularly protein folding.

It is important to note that chaperones do not convey additional information required for the folding of polypeptides into their correct three-dimensional conformations; the folded conformation of a protein is determined solely by its amino acid sequence. Rather, chaperones catalyze protein folding by assisting the self-assembly process. They appear to function by binding to and stabilizing unfolded or partially folded polypeptides that are intermediates along the pathway leading to the final correctly folded state. In the absence of chaperones, unfolded or partially folded polypeptide chains would be unstable within the cell, frequently folding incorrectly or aggregating into insoluble complexes. The binding of chaperones stabilizes these unfolded polypeptides, thereby preventing incorrect folding or aggregation and allowing the polypeptide chain to fold into its correct conformation.

A good example is provided by chaperones that bind to nascent polypeptide chains that are still being translated on ribosomes, thereby preventing incorrect folding or aggregation of the amino-terminal portion of the polypeptide before synthesis of the chain is finished (Figure 7.17). Presumably, this interaction is particularly important for proteins in which the carboxy terminus (the last to be synthesized) is required for correct folding of the amino terminus. In such cases, chaperone binding stabilizes the amino-terminal portion in an unfolded conformation until the rest of the polypeptide chain is synthesized and the completed protein can fold correctly. Chaperones also stabilize unfolded polypeptide chains during their transport into subcellular organelles—for example, during the transfer of proteins into mitochondria from the cytosol (Figure 7.18). Proteins are transported across the mitochondrial membrane in partially unfolded conformations that are stabilized by chaperones in the cytosol. Chaperones within the mitochondrion then facilitate transfer of the polypeptide chain across the membrane and its subsequent folding within the organelle. In addition, chaperones are involved in the assembly of proteins that consist of multiple polypeptide chains, in the assembly of macromolecular struc-

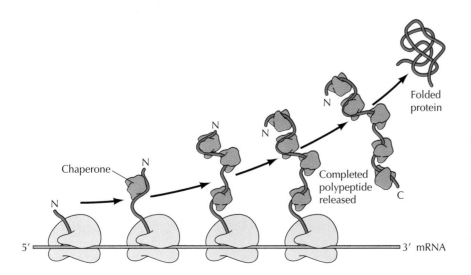

Figure 7.17
Action of chaperones during translation Chaperones bind to the amino (N) terminus of the growing polypeptide chain, stabilizing it in an unfolded configuration until synthesis of the polypeptide is completed. The completed protein is then released from the ribosome and is able to fold into its correct three-dimensional conformation.

Figure 7.18
Action of chaperones during protein transport   A partially unfolded polypeptide is transported from the cytosol to a mitochondrion. Cytosolic chaperones stabilize the unfolded configuration. Mitochondrial chaperones facilitate transport and subsequent folding of the polypeptide chain within the organelle.

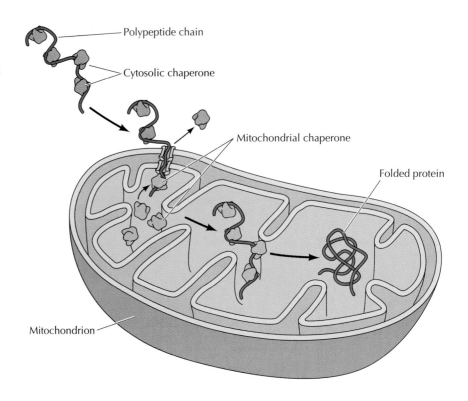

Polypeptide chain

Cytosolic chaperone

Mitochondrial chaperone

Folded protein

Mitochondrion

tures (e.g., nucleoplasmin), and (as discussed later in this chapter) in the regulation of protein degradation.

Many of the proteins now known to function as molecular chaperones (Table 7.2) were initially identified as **heat-shock proteins**, a group of proteins expressed in cells that have been subjected to elevated temperatures or other forms of environmental stress. The heat-shock proteins (abbreviated Hsp), which are highly conserved in both prokaryotic and eukaryotic cells, are thought to stabilize and facilitate the refolding of proteins that have been partially denatured as a result of exposure to elevated temperature. However, many members of the heat-shock protein family are expressed and have essential cellular functions under normal growth conditions. These proteins serve as molecular chaperones, which are needed for polypeptide folding and transport under normal conditions as well as in cells subjected to environmental stress.

The Hsp70 and Hsp60 families of heat-shock proteins appear to be particularly important in the general pathways of protein folding in both prokaryotic and eukaryotic cells. The proteins of both families function by binding to unfolded regions of polypeptide chains. Members of the Hsp70 family stabilize unfolded polypeptide chains during translation (see, for example, Figure 7.17) as well as during the transport of polypeptides into a variety of subcellular compartments, such as mitochondria and the endoplasmic reticulum. These proteins bind to short segments (seven or eight amino acid residues) of unfolded polypeptides, maintaining the polypeptide chain in an unfolded configuration and preventing aggregation.

Members of the Hsp60 family (also called **chaperonins**) facilitate the folding of proteins into their

TABLE 7.2 Molecular Chaperones

| Protein family | Chaperone proteins | |
	Prokaryotes	Eukaryotes
Hsp70	DnaK	Hsc73 (cytosol)
		BiP (endoplasmic reticulum)
		SSC1 (mitochondria)
		ctHsp70 (chloroplasts)
Hsp60	GroEL	TriC (cytosol)
		Hsp60 (mitochondria)
		Cpn60 (chloroplasts)
Hsp90	HtpG	Hsp90 (cytosol)
		Grp94 (endoplasmic reticulum)

native conformations. Each chaperonin consists of 14 subunits of approximately 60 kilodaltons (kd) each, arranged in two stacked rings to form a "double doughnut" structure (Figure 7.19). Unfolded polypeptide chains are shielded from the cytosol by being bound within the central cavity of the chaperonin cylinder. In this isolated environment protein folding can proceed while aggregation of unfolded segments of the polypeptide chain is prevented by their binding to the chaperonin. The binding of unfolded polypeptides to the chaperonin is a reversible reaction that is coupled to the hydrolysis of ATP as a source of energy. ATP hydrolysis thus drives multiple rounds of release and rebinding of unfolded regions of the polypeptide chain to the chaperonin, allowing the polypeptide to fold gradually into the correct conformation.

In some cases, members of the Hsp70 and Hsp60 families have been found to act together in a sequential fashion. For example, Hsp70 and Hsp60 family members act sequentially during the transport of proteins into mitochondria and during the folding of newly synthesized proteins in *E. coli* (Figure 7.20). First, an Hsp70 chaperone stabilizes nascent polypeptide chains until protein synthesis is completed. The unfolded polypeptide chain is then transferred to an Hsp60 chaperonin, within which protein folding takes place, yielding a protein correctly folded into its functional three-dimensional conformation. Members of the Hsp70 and Hsp60 families are found in the cytosol and in subcellular organelles (e.g., mitochondria) of eukaryotic cells, as well as in bacteria (see Table 7.2), so the sequential action of Hsp70 and Hsp60 appears to represent a general pathway of protein folding. An alternative pathway for the folding of some proteins in the cytosol and endoplasmic reticulum may involve the sequential actions of Hsp70 and Hsp90 family members, although the function of Hsp90 is not yet well understood.

Figure 7.19
Structure of a chaperonin GroEL, a member of the Hsp60 family, is a porous cylinder composed of two stacked rings. Each ring consists of seven subunits. (Courtesy of Paul B. Sigler, Yale University.)

Partially folded intermediate

Hsp70

Hsp60

Transfer to Hsp60

Folded protein

Figure 7.20
Sequential actions of Hsp70 and Hsp60 chaperones Chaperones of the Hsp70 family bind to and stabilize unfolded polypeptide chains during translation. The unfolded polypeptide is then transferred to chaperones of the Hsp60 family, within which protein folding takes place. ATP hydrolysis is required for release of the unfolded polypeptide from Hsp70 as well as for folding within Hsp60.

Incorrect disulfide bonds   Correct disulfide bonds

Figure 7.21
**The action of protein disulfide iso-
merase**   Protein disulfide isomerase
(PDI) catalyzes the breakage and
rejoining of disulfide bonds, resulting
in exchanges between paired disulfides in a polypeptide chain. The
enzyme forms a disulfide bond with a
cysteine residue of the polypeptide
and then exchanges its paired disulfide with another cysteine residue. In
this example, PDI catalyzes the conversion of two incorrect disulfide
bonds (1-2 and 3-4) to the correct
pairing (1-3 and 2-4).

Enzymes and Protein Folding

In addition to chaperones, which facilitate protein folding by binding to and
stabilizing partially folded intermediates, cells contain at least two types of
enzymes that catalyze protein folding by breaking and re-forming covalent
bonds. The formation of disulfide bonds between cysteine residues is
important in stabilizing the folded structures of many proteins (see Figure
2.16). **Protein disulfide isomerase**, which was discovered by Christian
Anfinsen in 1963, catalyzes the breakage and re-formation of these bonds
(Figure 7.21). For proteins that contain multiple cysteine residues, protein
disulfide isomerase (PDI) plays an important role by promoting rapid
exchanges between paired disulfides, thereby allowing the protein to attain
the pattern of disulfide bonds that is compatible with its stably folded conformation. Disulfide bonds are generally restricted to secreted proteins and
some membrane proteins because the cytosol contains reducing agents that
maintain cysteine residues in their reduced (—SH form), thereby preventing
the formation of disulfide (S—S) linkages. In eukaryotic cells, disulfide
bonds form in the endoplasmic reticulum, in which an oxidizing environment is maintained. Consistent with the role of disulfide bonds in stabilizing secreted proteins, the activity of PDI in the endoplasmic reticulum is
correlated with the level of protein secretion in different types of cells.

The second enzyme that plays a role in protein folding catalyzes the isomerization of peptide bonds that involve proline residues (Figure 7.22).
Proline is an unusual amino acid in that the equilibrium between the *cis*
and *trans* conformations of peptide bonds that precede proline residues is
only slightly in favor of the *trans* form. In contrast, peptide bonds between
other amino acids are almost always in the *trans* form. Isomerization
between the *cis* and *trans* configurations of prolyl peptide bonds, which
could otherwise represent a rate-limiting step in protein folding, is catalyzed by the enzyme **peptidyl prolyl isomerase**. This enzyme is widely
distributed in both prokaryotic and eukaryotic cells and can catalyze the
refolding of at least some proteins. However, its physiologically important
substrates and role within cells have not yet been determined.

Protein Cleavage

Cleavage of the polypeptide chain (**proteolysis**) is an important step in the
maturation of many proteins. A simple example is removal of the initiator
methionine from the amino terminus of many polypeptides, which occurs
soon after the amino terminus of the growing polypeptide chain emerges
from the ribosome. Additional chemical groups, such as acetyl groups or
fatty acid chains (discussed shortly), are then frequently added to the
amino-terminal residues.

Proteolytic modifications of the amino terminus also play a part in the
translocation of many proteins across membranes, including secreted proteins
in both bacteria and eukaryotes as well as proteins destined for incorporation

Figure 7.22
**The action of peptidyl prolyl iso-
merase**   Peptidyl prolyl isomerase
catalyzes the isomerization of peptide
bonds that involve proline between
the *cis* and *trans* conformations.

trans   *cis*

ER membrane

Signal sequence

mRNA 5'

Signal peptidase

Cleavage of signal sequence

Translocated protein

3'

Direction of ribosome movement

Figure 7.23
The role of signal sequences in membrane translocation   Signal sequences target the translocation of polypeptide chains across the plasma membrane of bacteria or into the endoplasmic reticulum of eukaryotic cells (shown here). The signal sequence, a stretch of hydrophobic amino acids at the amino terminus of the polypeptide chain, inserts into a membrane channel as it emerges from the ribosome. The rest of the polypeptide is then translocated through the channel and the signal sequence is cleaved by the action of signal peptidase, releasing the mature translocated protein.

into the plasma membrane, lysosomes, mitochondria, and chloroplasts of eukaryotic cells. These proteins are targeted for transport to their destinations by amino-terminal sequences that are removed by proteolytic cleavage as the protein crosses the membrane. For example, amino-terminal **signal sequences**, usually about 20 amino acids long, target secreted proteins to the plasma membrane of bacteria or to the endoplasmic reticulum of eukaryotic cells while translation is still in progress (Figure 7.23). The signal sequence, which consists predominantly of hydrophobic amino acids, is inserted into the membrane as it emerges from the ribosome. The remainder of the polypeptide chain passes through a channel in the membrane as translation proceeds. The signal sequence is then cleaved by a specific membrane protease (**signal peptidase**), and the mature protein is released. In eukaryotic cells, the translocation of growing polypeptide chains into the endoplasmic reticulum is the first step in targeting proteins for secretion, incorporation into the plasma membrane, or incorporation into lysosomes. The mechanisms that direct the transport of proteins to these destinations, as well as the role of other targeting sequences in directing the import of proteins into mitochondria and chloroplasts, will be discussed in detail in Chapters 9 and 10.

In other important instances of proteolytic processing, active enzymes or hormones form via cleavage of larger precursors. Insulin, which is synthesized as a longer precursor polypeptide, is a good example. Insulin forms by two cleavages. The initial precursor (preproinsulin) contains an amino-terminal signal sequence that targets the polypeptide chain to the endoplasmic reticulum (Figure 7.24). Removal of the signal sequence during transfer to the endoplasmic reticulum yields a second precursor, called proinsulin. This precursor is then converted to insulin, which consists of two chains held together by disulfide bonds, by proteolytic removal of an internal peptide. Other proteins activated by similar cleavage processes include digestive enzymes and the proteins involved in blood clotting.

It is interesting to note that the proteins of many animal viruses are derived from the cleavage of larger precursors. One particularly important example of the role of proteolysis in virus replication is provided by HIV. In the replication of HIV, a virus-encoded protease cleaves precursor polypeptides to form the viral structural proteins. Because of its central role in virus replication, the HIV protease (in addition to reverse transcriptase) is an important target for the development of drugs used for treating AIDS. Indeed, such protease inhibitors are now among the most effective agents available for combating this disease.

Preproinsulin

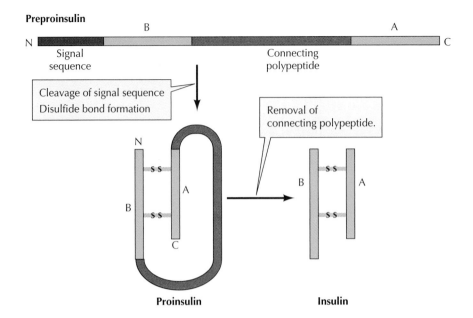

Figure 7.24
Proteolytic processing of insulin
The mature insulin molecule consists of two polypeptide chains (A and B) joined by disulfide bonds. It is synthesized as a precursor polypeptide (preproinsulin) containing an amino-terminal signal sequence that is cleaved during transfer of the growing polypeptide chain to the endoplasmic reticulum. This cleavage yields a second precursor (proinsulin), which is converted to insulin by further proteolysis, removing the internal connecting polypeptide.

Glycosylation

Many proteins, particularly in eukaryotic cells, are modified by the addition of carbohydrates, a process called **glycosylation**. The proteins to which carbohydrate chains have been added (called **glycoproteins**) are usually secreted or localized to the cell surface, although some nuclear and cytosolic proteins are also glycosylated. The carbohydrate moieties of glycoproteins play important roles in protein folding in the endoplasmic reticulum, in the targeting of proteins for delivery to the appropriate intracellular compartments, and as recognition sites in cell-cell interactions.

Glycoproteins are classified as either *N*-linked or *O*-linked, depending on the site of attachment of the carbohydrate side chain (Figure 7.25). In *N*-linked glycoproteins, the carbohydrate is attached to the nitrogen atom in the side chain of asparagine. In *O*-linked glycoproteins, the oxygen atom in the side chain of serine or threonine is the site of carbohydrate attachment. The sugars directly attached to these positions are usually either *N*-acetylglucosamine or *N*-acetylgalactosamine, respectively.

Most glycoproteins in eukaryotic cells are destined either for secretion or for incorporation into the plasma membrane. These proteins are usually transferred into the endoplasmic reticulum (with the cleavage of a signal sequence) while their translation is still in progress. Glycosylation is also initiated in the endoplasmic reticulum before translation is complete. The first step is the transfer of a common oligosaccharide consisting of 14 sugar residues (2 *N*-acetylglucosamine, 3 glucose, and 9 mannose) to an asparagine residue of the growing polypeptide chain (Figure 7.26). The oligosaccharide is assembled within the endoplasmic reticulum on a lipid carrier (**dolichol phosphate**). It is then transferred as an intact unit to an acceptor

(N-linkage)

(O-linkage)

Figure 7.25
Linkage of carbohydrate side chains to glycoproteins The carbohydrate chains of *N*-linked glycoproteins are attached to asparagine; those of *O*-linked glycoproteins are attached to either serine (shown) or threonine. The sugars joined to the amino acids are usually either *N*-acetylglucosamine (*N*-linked) or *N*-acetylgalactosamine (*O*-linked).

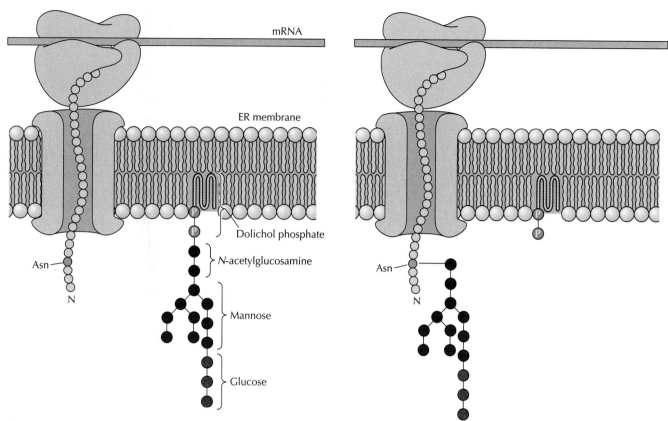

Figure 7.26
Synthesis of N-linked glycoproteins   The first step in glycosylation is the addition of an oligosaccharide consisting of 14 sugar residues to a growing polypeptide chain in the endoplasmic reticulum (ER). The oligosaccharide (which consists of two *N*-acetylglucosamine, nine mannose, and three glucose residues) is assembled on a lipid carrier (dolichol phosphate) in the ER membrane. It is then transferred as a unit to an acceptor asparagine residue of the polypeptide.

asparagine (Asn) residue within the sequence Asn-X-Ser or Asn-X-Thr (where X is any amino acid other than proline).

In further processing, the common *N*-linked oligosaccharide is modified. Three glucose residues and one mannose are removed while the glycoprotein is in the endoplasmic reticulum. The oligosaccharide is then further modified in the Golgi apparatus, to which glycoproteins are transferred from the endoplasmic reticulum. These modifications (which will be discussed in Chapter 9) include both the removal and addition of carbohydrate residues as the glycoprotein is transported through the compartments of the Golgi (Figure 7.27). The *N*-linked oligosaccharides of different glycoproteins are processed to different extents, depending on both the enzymes present in different cells and on the accessibility of the oligosaccharide to the enzymes that catalyze its modification. Glycoproteins with inaccessible oligosaccharides do not have new sugars added to them in the Golgi. The relatively simple oligosaccharides of these glycoproteins are called high-mannose oligosaccharides because they contain a high proportion of mannose residues, similar to the common oligosaccharide originally added in the endoplasmic reticulum. In contrast, glycoproteins with accessible oligosaccharides are processed more extensively, resulting in the formation of a variety of complex oligosaccharides.

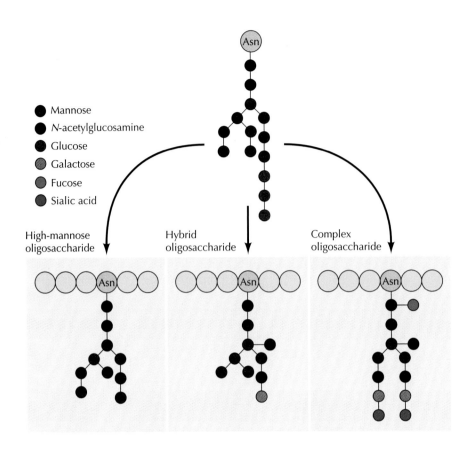

Figure 7.27
Examples of N-linked oligosaccharides Various oligosaccharides form from further modifications of the common 14-sugar unit initially added in the endoplasmic reticulum (see Figure 7.26). In high-mannose oligosaccharides, the glucose residues and some mannose residues are removed, but no other sugars are added. In the synthesis of complex oligosaccharides, more mannose residues are removed and other sugars are added. Hybrid oligosaccharides are intermediate between high-mannose and complex oligosaccharides. The structures shown are representative examples.

O-linked oligosaccharides are also added within the Golgi apparatus. In contrast to the *N*-linked oligosaccharides, *O*-linked oligosaccharides are formed by the addition of one sugar at a time and usually consist of only a few residues (Figure 7.28). Many cytoplasmic and nuclear proteins, including a variety of transcription factors, are also modified by the addition of single *O*-linked *N*-acetylglucosamine residues, catalyzed by a different enzyme system. However, the roles of carbohydrates in the function of these cytoplasmic and nuclear glycoproteins are not yet understood.

Attachment of Lipids

Some proteins in eukaryotic cells are modified by the attachment of lipids to the polypeptide chain. Such modifications frequently target and anchor these proteins to the plasma membrane, with which the hydrophobic lipid is able to interact (see Figure 2.48). Three general types of lipid additions—N-myristoylation, prenylation, and palmitoylation—are common in eukaryotic proteins associated with the cytosolic face of the plasma membrane. A fourth type of modification, the addition of glycolipids, plays an important role in anchoring some cell surface proteins to the extracellular face of the plasma membrane.

In some proteins, a fatty acid is attached to the amino terminus of the growing polypeptide chain during translation. In this process, called **N-myristoylation**, myristic acid (a 14-carbon fatty acid) is attached to an

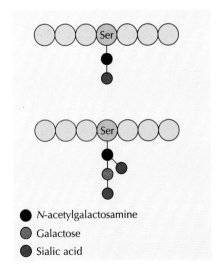

Figure 7.28
Examples of O-linked oligosaccharides *O*-linked oligosaccharides usually consist of only a few carbohydrate residues, which are added one sugar at a time.

Figure 7.29
Addition of a fatty acid by N-myristoylation The initiating methionine is removed, leaving glycine at the N terminus of the polypeptide chain. Myristic acid (a 14-carbon fatty acid) is then added.

N-terminal glycine residue (Figure 7.29). The glycine is usually the second amino acid incorporated into the polypeptide chain; the initiator methionine is removed by proteolysis before fatty acid addition. Many proteins that are modified by N-myristoylation are associated with the inner face of the plasma membrane, and the role of the fatty acid in this association has been clearly demonstrated by analysis of mutant proteins in which the N-terminal glycine is changed to an alanine. This substitution prevents myristoylation and blocks the function of the mutant proteins by inhibiting their membrane association.

Lipids can also be attached to the side chains of cysteine, serine, and threonine residues. One important example of this type of modification is **prenylation**, in which specific types of lipids (prenyl groups) are attached to the sulfur atoms in the side chains of cysteine residues located near the C terminus of the polypeptide chain (Figure 7.30). Many plasma mem-

Figure 7.30
Prenylation of a C-terminal cysteine residue The type of prenylation shown affects Ras proteins and proteins of the nuclear envelope (nuclear lamins). These proteins terminate with a cysteine residue (Cys) followed by two aliphatic amino acids (A) and any other amino acid (X) at the C terminus. The first step in their modification is addition of the 15-carbon farnesyl group to the side chain of cysteine (farnesylation). This step is followed by proteolytic removal of the three C-terminal amino acids and methylation of the cysteine, which is now at the C terminus.

Palmitoylation

Figure 7.31
Palmitoylation   Palmitate (a 16-carbon fatty acid) is added to the side chain of an internal cysteine residue.

brane–associated proteins involved in the control of cell growth and differentiation are modified in this way, including the Ras oncogene proteins, which are responsible for the uncontrolled growth of many human cancers (see Chapter 15). Prenylation of these proteins proceeds by three steps. First, the prenyl group is added to a cysteine located three amino acids from the carboxy terminus of the polypeptide chain. The prenyl groups added in this reaction are either farnesyl (15 carbons, as shown in Figure 7.30) or geranylgeranyl (20 carbons). The amino acids following the cysteine residue are then removed, leaving cysteine at the carboxy terminus. Finally, a methyl group is added to the carboxyl group of the C-terminal cysteine residue.

The biological significance of prenylation is indicated by the fact that mutations of the critical cysteine block the membrane association and function of Ras proteins. Because farnesylation is a relatively rare modification of cellular proteins, interest in this reaction has been stimulated by the possibility that inhibitors of the key enzyme (farnesyl transferase) might prove useful as drugs for the treatment of cancers that involve Ras proteins. Such inhibitors of farnesylation have been found to interfere with the growth of cancer cells in experimental models and are undergoing evaluation of their efficacy against human tumors in clinical trials.

In the third type of fatty acid modification, **palmitoylation**, palmitic acid (a 16-carbon fatty acid) is added to sulfur atoms of the side chains of internal cysteine residues (Figure 7.31). Like N-myristoylation and prenylation, palmitoylation plays an important role in the association of some proteins with the cytosolic face of the plasma membrane.

Finally, lipids linked to oligosaccharides (**glycolipids**) are added to the C-terminal carboxyl groups of some proteins, where they serve as anchors that attach the proteins to the external face of the plasma membrane. Because the glycolipids attached to these proteins contain phosphatidylinositol, they are usually called **glycosylphosphatidylinositol**, or **GPI**, **anchors** (Figure 7.32). The oligosaccharide portions of GPI anchors are attached to the terminal carboxyl group of polypeptide chains. The inositol head group of phosphatidylinositol is in turn attached to the oligosaccharide, so the carbohydrate serves as a bridge between the protein and the fatty acid chains of the phospholipid. The GPI anchors are synthesized and added to proteins as a preassembled unit within the endoplasmic reticulum. Their addition is accompanied by cleavage of a peptide consisting of about 20 amino acids from the C terminus of the polypeptide chain. The modified protein is then transported to the cell surface, where the fatty acid chains of the GPI anchor mediate its attachment to the plasma membrane.

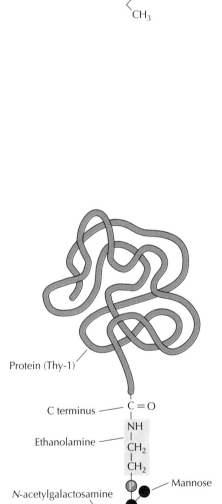

Figure 7.32
Structure of a GPI anchor   The GPI anchor, attached to the C terminus, anchors the protein in the plasma membrane. The anchor is joined to the C-terminal amino acid by an ethanolamine, which is linked to an oligosaccharide that consists of mannose, N-acetylgalactosamine, and glucosamine residues. The oligosaccharide is in turn joined to the inositol head group of phosphatidylinositol. The two fatty acid chains of the lipid are embedded in the plasma membrane. The GPI anchor shown here is that of a rat protein, Thy-1.

Regulation of Protein Function

A critical function of proteins is their activity as enzymes, which are needed to catalyze almost all biological reactions. Regulation of enzyme activity thus plays a key role in governing cell behavior. This is accomplished in part at the level of gene expression, which determines the amount of any enzyme (protein) synthesized by the cell. A further level of control is then obtained by regulation of protein function, which allows the cell to regulate not only the amounts but also the activities of its protein constituents. Regulation of the activities of some of the proteins involved in transcription and translation has already been discussed in this and the preceding chapter, and many further examples of regulated protein function in the control of cell behavior will be evident throughout the remainder of this book. This section discusses the three general mechanisms by which the activities of cellular proteins are controlled.

Regulation by Small Molecules

Most enzymes are controlled by changes in their conformation, which in turn alter catalytic activity. In many cases such conformational changes result from the binding of small molecules, such as amino acids or nucleotides, that regulate enzyme activity. This type of regulation commonly is responsible for controlling metabolic pathways by feedback inhibition. For example, the end products of many biosynthetic pathways (e.g., amino acids) inhibit the enzymes that catalyze the first step in their synthesis, thus ensuring an adequate supply of the product while preventing the synthesis of excess amounts (Figure 7.33).

Feedback inhibition is an example of **allosteric regulation**, in which a regulatory molecule binds to a site on an enzyme that is distinct from the catalytic site (*allo* = "other"; *steric* = "site"). The binding of such a regulatory molecule alters the conformation of the protein, thereby changing the shape of the catalytic site and affecting catalytic activity (see Figure 2.29). One of the best-studied allosteric enzymes is aspartate transcarbamylase, which catalyzes the first step in the synthesis of pyrimidine nucleotides and is regulated by feedback inhibition by cytidine triphosphate (CTP). Aspartate transcarbamylase consists of 12 distinct polypeptide chains: six catalytic subunits and six regulatory subunits. The binding of CTP to the regulatory subunits induces a major rearrangement of subunit positions, thereby inhibiting enzymatic activity (Figure 7.34).

Many transcription factors (discussed in Chapter 6) are also regulated by the binding of small molecules. For example, the binding of lactose or a metabolite to the *E. coli lac* repressor induces a conformational change that prevents the repressor from binding DNA. In eukaryotic cells, steroid hormones similarly control gene expression by binding to transcriptional regulatory proteins.

The regulation of translation factors such as EF-Tu by GTP binding (see Figure 7.12) illustrates another common mechanism by which the activities of intracellular proteins are controlled. In this case, the GTP-bound form of the protein is its active conformation, while the GDP-bound form is inactive. Many cellular proteins are similarly regulated by GTP or GDP binding. These proteins include the Ras oncogene proteins, which have been studied intensively because of their roles in the control of cell proliferation and in human cancers. X-ray crystallography analysis of these proteins has been particularly interesting, revealing subtle but functionally very important conformational differences between the inactive GDP-bound and active GTP-bound forms (Figure 7.35). This small difference in protein con-

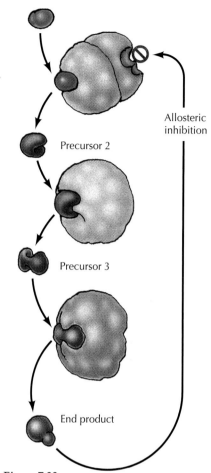

Figure 7.33
Feedback inhibition The end product of a biochemical pathway acts as an allosteric inhibitor of the enzyme that catalyzes the first step in its synthesis.

Active conformation

Inactive conformation

Figure 7.34
Allosteric regulation of aspartate transcarbamylase The arrangement of two catalytic (green) and two regulatory (orange) subunits of aspartate transcarbamylase is shown in the active and inactive conformations. The complete enzyme consists of six catalytic and six regulatory subunits. (Courtesy of Irving Geis. © I. Geis and E. Gouaux, from coordinates by W. N. Lipscomb.)

formation determines whether Ras (in the active GTP-bound form) can interact with its target molecule, which signals the cell to divide. The importance of such subtle differences in protein conformation is dramatically illustrated by the fact that mutations in *ras* genes contribute to the development of about 15% of human cancers. Such mutations alter the structure of the Ras proteins so that they are locked in the active GTP-bound conformation and continually signal cell division, thereby driving the uncontrolled growth of cancer cells. In contrast, normal Ras proteins alternate between the GTP- and GDP-bound conformations, such that they are active only following stimulation by the hormones and growth factors that normally control cell proliferation in multicellular organisms.

Protein Phosphorylation

The examples discussed in the previous section involve noncovalent associations of proteins with small-molecule inhibitors or activators. Since no covalent bonds form, the binding of these regulatory molecules to the protein is readily reversible, allowing the cell to respond rapidly to environmental changes. The activity of many proteins, however, is also regulated

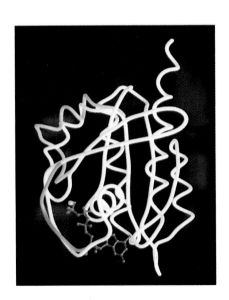

Figure 7.35
Conformational differences between active and inactive Ras proteins The Ras proteins alternate between active GTP-bound and inactive GDP-bound forms. The major effect of GTP binding versus GDP binding is alteration of the conformation of two regions of the molecule, designated the switch I and II regions. The backbone of the GTP complex is shown here in white; the backbone of the GDP complex of switch I and II are in blue and yellow, respectively. The guanine nucleotide is in red and Mg^{2+} is yellow. (Courtesy of Sung-Hou Kim, University of California, Berkeley.)

by covalent modifications. One example of this type of regulation is the activation of some enzymes by proteolytic cleavage of inactive precursors. As noted previously in this chapter, digestive enzymes and proteins involved in blood clotting are regulated by this mechanism. Since proteolysis is irreversible, however, it provides a means of controlling enzyme activation rather than of turning proteins on and off in response to changes in the environment. In contrast, other covalent modifications—particularly phosphorylation—are readily reversible within the cell and function, as allosteric regulation does, to reversibly activate or inhibit a wide variety of cellular proteins in response to environmental signals.

Protein phosphorylation is catalyzed by **protein kinases**, most of which transfer phosphate groups from ATP to the hydroxyl groups of the side chains of serine, threonine, or tyrosine residues (Figure 7.36). Most protein kinases phosphorylate either serine and threonine or tyrosine residues: These enzymes are called **protein-serine/threonine kinases** or **protein-tyrosine kinases**, respectively. Protein phosphorylation is reversed by **protein phosphatases**, which catalyze the hydrolysis of phosphorylated amino acid residues. Like protein kinases, most protein phosphatases are specific either for serine and threonine or for tyrosine residues, although some protein phosphatases recognize all three phosphoamino acids.

The combined action of protein kinases and protein phosphatases mediates the reversible phosphorylation of many cellular proteins. Frequently, protein kinases function as components of signal transduction pathways in which one kinase activates a second kinase, which may act on yet another kinase. The sequential action of a series of protein kinases can transmit a signal received at the cell surface to target proteins within the cell, resulting in changes in cell behavior in response to environmental stimuli.

The prototype of the action of protein kinases came from studies of glycogen metabolism by Ed Fischer and Ed Krebs in 1955. In muscle cells the hormone epinephrine (adrenaline) signals the breakdown of glycogen

Figure 7.36
Protein kinases and phosphatases
Protein kinases catalyze the transfer of a phosphate group from ATP to the side chains of serine and threonine (protein-serine/threonine kinases) or tyrosine (protein-tyrosine kinases) residues. Protein phosphatases catalyze the removal of phosphate groups from the same amino acids by hydrolysis.

Figure 7.37
Regulation of glycogen breakdown by protein phosphorylation   The binding of epinephrine (adrenaline) to its cell surface receptor triggers the production of cyclic AMP (cAMP), which activates cAMP-dependent protein kinase. cAMP-dependent protein kinase phosphorylates and activates phosphorylase kinase, which in turn phosphorylates and activates glycogen phosphorylase. Glycogen phosphorylase then catalyzes the breakdown of glycogen to glucose-1-phosphate.

to glucose-1-phosphate, providing an available source of energy for increased muscular activity. Glycogen breakdown is catalyzed by the enzyme glycogen phosphorylase, which is regulated by a protein kinase (Figure 7.37). Epinephrine binds to a cell surface receptor that triggers the conversion of ATP to cyclic AMP (cAMP), which then binds to and activates a protein kinase, called cAMP-dependent protein kinase. This kinase phosphorylates and activates a second protein kinase, called phosphorylase kinase. Phosphorylase kinase in turn phosphorylates and activates glycogen phosphorylase, leading to glucose production. The activating phosphorylations of both phosphorylase kinase and glycogen phosphorylase can be reversed by specific phosphatases, so removal of the initial stimulus (epinephrine) inhibits further glycogen breakdown.

The signaling pathway that leads to activation of glycogen phosphorylase is initiated by the binding of small molecules at the cell surface—epinephrine binding to its receptor and cAMP binding to cAMP-dependent protein kinase. The signal is then transmitted to its intracellular target by the sequential action of protein kinases. Similar signaling pathways, in which protein kinases and phosphatases play central roles, are involved in regulating almost all aspects of the behavior of eukaryotic cells (see Chapters 13 and 14). Aberrations in these pathways, frequently involving abnormalities of protein kinases, are also responsible for many diseases associated with improper regulation of cell growth and differentiation, particularly the development of cancer.

Protein-Protein Interactions

Many proteins consist of multiple subunits, each of which is an independent polypeptide chain. In some proteins the subunits are identical; other proteins are composed of two or more distinct polypeptides. In either case, interactions between the polypeptide chains are important in regulation of protein activity. The importance of these interactions is evident in many allosteric enzymes, such as aspartate transcarbamylase, in which the binding of a regulatory molecule alters protein conformation by changing the interactions between subunits.

Many other enzymes are similarly regulated by protein-protein interactions. A good example is cAMP-dependent protein kinase, which is composed of two regulatory and two catalytic subunits (Figure 7.38). In this state, the enzyme is inactive; the regulatory subunits inhibit the enzymatic activity of the catalytic subunits. The enzyme is activated by cAMP, which binds to the regulatory subunits and induces a conformational change leading to dissociation of the complex; the free catalytic subunits are then enzymatically active protein kinases. Cyclic AMP thus acts as an allosteric regulator by altering protein-protein interactions.

The transcriptional regulatory proteins discussed in Chapter 6 provide another important example of protein-protein interactions. Many eukaryotic transcription factors function as activators or repressors via protein-protein interactions with components of the general transcription machin-

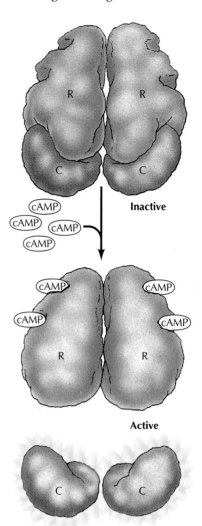

Figure 7.38
Regulation of cAMP-dependent protein kinase   In the inactive state, the enzyme consists of two regulatory (R) and two catalytic (C) subunits. Cyclic AMP binds to the regulatory subunits, inducing a conformational change that leads to their dissociation from the catalytic subunits. The free catalytic subunits are enzymatically active.

ery. As discussed in later chapters, similar protein-protein interactions, which can themselves be regulated by the binding of small molecules and phosphorylation, play critical roles in the control of many different aspects of cell behavior.

Protein Degradation

The levels of proteins within cells are determined not only by rates of synthesis, but also by rates of degradation. The half-lives of proteins within cells vary widely, from minutes to several days, and differential rates of protein degradation are an important aspect of cell regulation. Many rapidly degraded proteins function as regulatory molecules, such as transcription factors. The rapid turnover of these proteins is necessary to allow their levels to change quickly in response to external stimuli. Other proteins are rapidly degraded in response to specific signals, providing another mechanism for the regulation of intracellular enzyme activity. In addition, faulty or damaged proteins are recognized and rapidly degraded within cells, thereby eliminating the consequences of mistakes made during protein synthesis. In eukaryotic cells, two major pathways—the ubiquitin-proteasome pathway and lysosomal proteolysis—mediate protein degradation.

The Ubiquitin-Proteasome Pathway

The major pathway of selective protein degradation in eukaryotic cells uses **ubiquitin** as a marker that targets cytosolic and nuclear proteins for rapid proteolysis (Figure 7.39). Ubiquitin is a 76-amino-acid polypeptide that is highly conserved in all eukaryotes (yeasts, animals, and plants). Proteins are marked for degradation by the attachment of ubiquitin to the amino group of the side chain of a lysine residue. Additional ubiquitins are then added to form a multiubiquitin chain. Such polyubiquinated proteins are recognized and degraded by a large, multisubunit protease complex, called the **proteasome**. Ubiquitin is released in the process, so it can be reused in another cycle. It is noteworthy that both the attachment of ubiquitin and the degradation of marked proteins require energy in the form of ATP.

Since the attachment of ubiquitin marks proteins for rapid degradation, the stability of many proteins is determined by whether they become ubiquitinated. Ubiquitination is a multistep process. First, ubiquitin is activated by being attached to the ubiquitin-activating enzyme, E1. The ubiquitin is then transferred to a second enzyme, called ubiquitin-conjugating enzyme (E2). The final transfer of ubiquitin to the target protein is then mediated by a third enzyme, called ubiquitin ligase or E3, which is responsible for the selective recognition of appropriate substrate proteins. In some cases, the ubiquitin is first transferred from E2 to E3 and then to the target protein (see Figure 7.39). In other cases, the ubiquitin may be transferred directly from E2 to the target protein in a complex with E3. Most cells contain a single E1, but have many E2s and multiple families of E3 enzymes. Different members of the E2 and E3 families recognize different substrate proteins, and the specificity of these enzymes is what selectively targets cellular

Figure 7.39
The ubiquitin-proteasome pathway
Proteins are marked for rapid degradation by the covalent attachment of several molecules of ubiquitin. Ubiquitin is first activated by the enzyme E1. Activated ubiquitin is then transferred to one of several different ubiquitin-conjugating enzymes (E2). In most cases, the ubiquitin is then transferred to a ubiquitin ligase (E3) and then to a specific target protein. Multiple ubiquitins are then added, and the polyubiquinated proteins are degraded by a protease complex (the proteasome).

proteins for degradation by the ubiquitin-proteasome pathway.

A number of proteins that control fundamental cellular processes, such as gene expression and cell proliferation, are targets for regulated ubiquitination and proteolysis. An interesting example of such controlled degradation is provided by proteins (known as cyclins) that regulate progression through the division cycle of eukaryotic cells (Figure 7.40). The entry of all eukaryotic cells into mitosis is controlled in part by cyclin B, which is a regulatory subunit of a protein kinase called Cdc2 (see Chapter 14). The association of cyclin B with Cdc2 is required for activation of the Cdc2 kinase, which initiates the events of mitosis (including chromosome condensation and nuclear envelope breakdown) by phosphorylating various cellular proteins. Cdc2 also activates a ubiquitin-mediated proteolysis system that degrades cyclin B toward the end of mitosis. This degradation of cyclin B inactivates Cdc2, allowing the cell to exit mitosis and progress to interphase of the next cell cycle. The ubiquitination of cyclin B is a highly selective reaction, targeted by a 9-amino-acid cyclin B sequence called the destruction box. Mutations of this sequence prevent cyclin B proteolysis and lead to the arrest of dividing cells in mitosis, demonstrating the importance of regulated protein degradation in controlling the fundamental process of cell division.

Lysosomal Proteolysis

The other major pathway of protein degradation in eukaryotic cells involves the uptake of proteins by **lysosomes**. Lysosomes are membrane-enclosed organelles that contain an array of digestive enzymes, including several proteases (see Chapter 9). They have several roles in cell metabolism, including the digestion of extracellular proteins taken up by endocytosis as well as the gradual turnover of cytoplasmic organelles and cytosolic proteins.

The containment of proteases and other digestive enzymes within lysosomes prevents uncontrolled degradation of the contents of the cell. Therefore, in order to be degraded by lysosomal proteolysis, cellular proteins must first be taken up by lysosomes. One pathway for this uptake of cellular proteins, **autophagy**, involves the formation of vesicles (autophagosomes) in which small areas of cytoplasm or cytoplasmic organelles are enclosed in membranes derived from the endoplasmic reticulum (Figure 7.41). These vesicles then fuse with lysosomes, and the degradative lysoso-

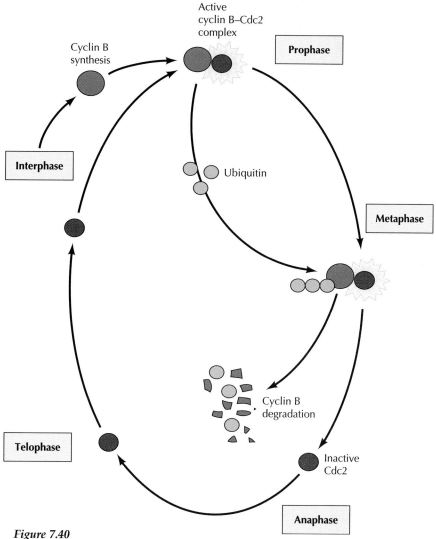

Cyclin B synthesis

Active cyclin B–Cdc2 complex

Prophase

Interphase

Ubiquitin

Metaphase

Cyclin B degradation

Inactive Cdc2

Telophase

Anaphase

Figure 7.40
Cyclin degradation during the cell cycle   The progression of eukaryotic cells through the division cycle is controlled in part by the synthesis and degradation of cyclin B, which is a regulatory subunit of the Cdc2 protein kinase. Synthesis of cyclin B during interphase leads to the formation of an active cyclin B–Cdc2 complex, which induces entry into mitosis. Rapid degradation of cyclin B then leads to inactivation of the Cdc2 kinase, allowing the cell to exit mitosis and return to interphase of the next cell cycle.

mal enzymes digest their contents. The uptake of proteins into autophagosomes appears to be nonselective, so it results in the eventual slow degradation of long-lived cytoplasmic proteins.

However, not all protein uptake by lysosomes is nonselective. For example, lysosomes are able to take up and degrade certain cytosolic proteins in a selective manner as a response to cellular starvation. The proteins degraded by lysosomal proteases under these conditions contain amino acid sequences similar to the broad consensus sequence Lys-Phe-Glu-Arg-Gln, which presumably targets them to lysosomes. A member of the Hsp70 family of molecular chaperones is also required for the lysosomal degradation of these proteins, presumably acting to unfold the polypeptide chains during their transport across the lysosomal membrane. The proteins susceptible to degradation by this pathway are thought to be normally long-

Figure 7.41
The lysosome system Lysosomes contain various digestive enzymes, including proteases. Lysosomes take up cellular proteins by fusion with autophagosomes, which are formed by the enclosure of areas of cytoplasm or organelles (e.g., a mitochondrion) in fragments of the endoplasmic reticulum. This fusion yields a phagolysosome, which digests the contents of the autophagosome.

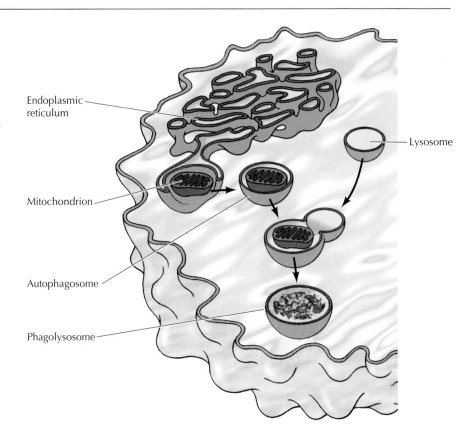

lived but dispensable proteins. Under starvation conditions, these proteins are sacrificed to provide amino acids and energy, allowing some basic metabolic processes to continue.

KEY TERMS	SUMMARY
	TRANSLATION OF mRNA
tRNA, anticodon, aminoacyl tRNA synthetase	*Transfer RNAs:* Transfer RNAs serve as adaptors that align amino acids on the mRNA template. Aminoacyl tRNA synthetases attach amino acids to the appropriate tRNAs, which then bind to mRNA codons by complementary base pairing.
ribosome, rRNA	*The Ribosome:* Ribosomes consist of two subunits, which are composed of proteins and ribosomal RNAs. The 23S rRNA is the primary catalyst of peptide bond formation.
5' untranslated region, polycistronic, monocistronic, 3' untranslated region, Shine-Delgarno sequence	*The Organization of mRNAs and the Initiation of Translation:* Translation of both prokaryotic and eukaryotic mRNAs initiates with a methionine residue. In bacteria, initiation codons are preceded by a sequence that aligns the mRNA on the ribosome by base pairing with 16S rRNA. In eukaryotes, initiation codons are identified by scanning from the 5' end of the mRNA, which is recognized by its 7-methylguanosine cap.
initiation factor, elongation factor, release factor, polysome	*The Process of Translation:* Translation is initiated by the binding of methionyl tRNA and mRNA to the small ribosomal subunit. The large ribosomal subunit then joins the complex, and the polypeptide chain elongates until the ribosome reaches a termination codon in the mRNA. A variety of nonribosomal factors are required for initiation, elongation, and termination of translation in both prokaryotic and eukaryotic cells.

Regulation of Translation: Translation of specific mRNAs can be regulated by the binding of repressor proteins. General translational activity can be regulated by modification of initiation factors. Polyadenylation of mRNA is also an important mechanism for the regulation of translation during early development.

PROTEIN FOLDING AND PROCESSING

Chaperones and Protein Folding: Molecular chaperones facilitate the intracellular folding of polypeptide chains into their correct three-dimensional conformations by binding to and stabilizing unfolded or partially folded polypeptide chains.

chaperone, heat-shock protein, chaperonin

Enzymes and Protein Folding: At least two types of enzymes, protein disulfide isomerase and peptidyl prolyl isomerase, catalyze protein folding by breaking and re-forming covalent bonds.

protein disulfide isomerase, peptidyl prolyl isomerase

Protein Cleavage: Proteolysis is an important step in the processing of many proteins. For example, secreted proteins and proteins incorporated into most eukaryotic organelles are targeted to their destinations by amino-terminal sequences that are removed by proteolytic cleavage as the polypeptide chain crosses the membrane.

proteolysis, signal sequence, signal peptidase

Glycosylation: Many eukaryotic proteins, particularly secreted and plasma membrane proteins, are modified by the addition of carbohydrates in the endoplasmic reticulum and Golgi apparatus.

glycosylation, glycoprotein, dolichol phosphate

Attachment of Lipids: Covalently attached lipids frequently target and anchor proteins to the plasma membrane.

N-myristoylation, prenylation, palmitoylation, glycolipid, glycosylphosphatidylinositol (GPI) anchor

REGULATION OF PROTEIN FUNCTION

Regulation by Small Molecules: Many proteins are regulated by the binding of small molecules, such as amino acids and nucleotides, that induce changes in protein conformation and activity.

allosteric regulation

Protein Phosphorylation: Reversible phosphorylation, which controls the activities of a wide variety of cellular proteins, results from the action of protein kinases and phosphatases.

protein kinase, protein-serine/ threonine kinase, protein-tyrosine kinase, protein phosphatase

Protein-Protein Interactions: Interactions between polypeptide chains are important in the regulation of allosteric enzymes and other cellular proteins.

PROTEIN DEGRADATION

The Ubiquitin-Proteasome Pathway: The major pathway of selective protein degradation in eukaryotic cells uses ubiquitin as a marker that targets proteins for rapid proteolysis by the proteasome.

ubiquitin, proteasome

Lysosomal Proteolysis: Lysosomal proteases degrade extracellular proteins taken up by endocytosis and are responsible for the slow degradation of cytoplasmic organelles and long-lived cytosolic proteins. Some proteins are targeted for selective degradation in lysosomes as a response to cell starvation.

lysosome, autophagy

Questions

1. You wish to express a cloned eukaryotic cDNA in bacteria. What type of sequence must you add in order for the mRNA to be translated on prokaryotic ribosomes?

2. What effect would an inhibitor of polyadenylation be expected to have on protein synthesis in fertilized eggs?

3. Why is a member of the Hsp70 family required for selective lysosomal degradation of proteins in starved cells, but not for the degradation of proteins taken up by autophagy?

4. You are interested in studying a protein expressed on the surface of liver cells. How could treatment of these cells with a phospholipase enable you to determine whether your protein is a transmembrane protein or is attached to the cell surface by a GPI anchor?

5. You are studying an enzyme that is activated by phosphorylation of a serine residue at position 59. How would mutation of this serine to threonine be expected to affect the enzyme's activity? What effect would you predict following mutation of the serine to alanine?

References and Further Reading

Translation of mRNA

Arnez, J. G. and D. Moras. 1997. Structural and functional considerations of the aminoacetylation reaction. *Trends Biochem. Sci.* 22: 211–216. [R]

Ban, N., P. Nissen, J. Hansen, M. Capel, P. B. Moore and T. A. Steitz. 1999. Placement of protein and RNA structures into a 5 Å-resolution map of the 50S ribosomal subunit. *Nature* 400: 841–847. [P]

Chen, J.-J. and I. M. London. 1995. Regulation of protein synthesis by heme-regulated eIF-2α kinase. *Trends Biochem. Sci.* 20: 105–108. [R]

Clemons, W. M. Jr., J. L. C. May, B. T. Wimberly, J. P. McCutcheon, M. S. Capel and V. Ramakrishnan. 1999. Structure of a bacterial 30S ribosomal subunit at 5.5 Å resolution. *Nature* 400: 833–840. [P]

Crick, F. H. C. 1966. Codon-anticodon pairing: The wobble hypothesis. *J. Mol. Biol.* 19: 548–555. [P]

Gray, N. K. and M. Wickens. 1998. Control of translation initiation in animals. *Ann. Rev. Cell Dev. Biol.* 14: 399–458. [R]

Green, R. and H. F. Noller. 1997. Ribosomes and translation. *Ann. Rev. Biochem.* 66: 679–716. [R]

Illangeskare, M., G. Sanchez, T. Nickles and M. Yarus. 1995. Aminoacyl-RNA synthesis catalyzed by an RNA. *Science* 267: 643–647. [P]

Klausner, R. D., T. A. Rouault and J. B. Harford. 1993. Regulating the fate of mRNA: The control of cellular iron metabolism. *Cell* 72: 19–28. [R]

Nakamura, Y., K. Ito and L. A. Isaksson. 1996. Emerging understanding of translation termination. *Cell* 87: 147–150. [R]

Nitta, I., Y. Kamada, H. Noda, T. Ueda and K. Watanabe. 1998. Reconstitution of peptide bond formation with *Escherichia coli* 23S ribosomal RNA domains. *Science* 281: 666–669. [P]

Noller, H. F., V. Hoffarth and L. Zimniak. 1992. Unusual resistance of peptidyl transferase to protein extraction procedures. *Science* 256: 1416–1419. [P]

Nomura, M. 1997. Reflections on the days of ribosome reconstitution research. *Trends Biochem. Sci.* 22: 275–279. [R]

Pestova, T. V. and C. U. T. Hellen. 1999. Ribosome recruitment and scanning: What's new? *Trends Biochem. Sci.* 24: 85–87. [R]

Piccirilli, J. A., T. S. McConnell, A. J. Zaug, H. F. Noller and T. R. Cech. 1992. Aminoacyl esterase activity of the *Tetrahymena* ribozyme. *Science* 256: 1420–1424. [P]

Ramakrishnan, V. and S. W. White. 1998. Ribosomal protein structures: Insights into the architecture, machinery and evolution of the ribosome. *Trends Biochem. Sci.* 23: 208–212. [R]

Sachs, A. B., P. Sarnow and M. W. Hentze. 1997. Starting at the beginning, middle, and end: Translation initiation in eukaryotes. *Cell* 89: 831–838. [R]

Saks, M. E., J. R. Sampson and J. N. Abelson. 1994. The transfer RNA identity problem: A search for rules. *Science* 263: 191–197. [R]

Wilson, K. S. and H. F. Noller. 1998. Molecular movement inside the translational engine. *Cell* 92: 337–349. [R]

Protein Folding and Processing

Braig, K., Z. Otwinowski, R. Hegde, D. C. Bolsvert, A. Joachimiak, A. L. Horwich and P. B. Sigler. 1994. The crystal structure of the bacterial chaperonin GroEL at 2.8 Å. *Nature* 371: 578–586. [P]

Buchner, J. 1999. Hsp90 & Co.—a holding for folding. *Trends Biochem. Sci.* 24: 136–141. [R]

Bukau, B. and A. L. Horwich. 1998. The Hsp70 and Hsp60 chaperone machines. *Cell* 92: 351–366. [R]

Caplan, A. J. 1999. Hsp90's secrets unfold: New insights from structural and functional studies. *Trends Cell Biol.* 9: 262–268. [R]

Dalbey, R. E. and G. von Heijne. 1992. Signal peptidases in prokaryotes and eukaryotes—a new protease family. *Trends Biochem. Sci.* 17: 474–478. [R]

Englund, P. T. 1993. The structure and biosynthesis of glycosyl phosphatidylinositol protein anchors. *Ann. Rev. Biochem.* 62: 121–138. [R]

Freedman, R. B., T. R. Hirst and M. F. Tuite. 1994. Protein disulphide isomerase: Building bridges in protein folding. *Trends Biochem. Sci.* 19: 331–336. [R]

Gahmberg, C. G. and M. Tolvanen. 1996. Why mammalian cell surface proteins are glycoproteins. *Trends Biochem. Sci.* 21: 308–311. [R]

Gething, M.-J. and J. Sambrook. 1992. Protein folding in the cell. *Nature* 355: 33–45. [R]

Gierasch, L. M. 1989. Signal sequences. *Biochemistry* 28: 923–930. [R]

Gilbert, H. F. 1997. Protein disulfide isomerase and assisted protein folding. *J. Biol. Chem.* 272: 29399–29402. [R]

Hart, G. W. 1997. Dynamic *O*-linked glycosylation of nuclear and cytoskeletal proteins. *Ann. Rev. Biochem.* 66: 315–335. [R]

Hartl, F. U., R. Hlodan and T. Langer. 1994. Molecular chaperones in protein folding: The art of avoiding sticky situations. *Trends Biochem. Sci.* 19: 20–25. [R]

Hirschberg, C. B. and M. D. Snider. 1987. Topography of glycosylation in the rough endoplasmic reticulum and Golgi apparatus. *Ann. Rev. Biochem.* 56: 63–87. [R]

Johnson, D. R., R. S. Bhatnagar, L. J. Knoll and J. I. Gordon. 1994. Genetic and biochemical studies of protein N-myristoylation. *Ann. Rev. Biochem.* 63: 869–914. [R]

Johnson, J. L. and E. A. Craig. 1997. Protein folding *in vivo*: Unraveling complex pathways. *Cell* 90: 201–204. [R]

Kornfield, R. and S. Kornfield. 1985. Assembly of asparagine-linked oligosaccharides. *Ann. Rev. Biochem.* 45: 631–664. [R]

Sigler, P. B., Z. Xu, H. S. Rye, S. G. Burston, W. A. Fenton and A. L. Horwich. 1998. Structure and function in GroEL-mediated protein folding. *Ann. Rev. Biochem.* 67: 581–608. [R]

Udenfriend, S. and K. Kodukula. 1995. How glycosylphosphatidylinositol-anchored membrane proteins are made. *Ann. Rev. Biochem.* 64: 563–591. [R]

Zhang, F. L. and P. J. Casey. 1996. Protein prenylation: Molecular mechanisms and functional consequences. *Ann. Rev. Biochem.* 65: 241–269. [R]

Regulation of Protein Function

Barford, D. 1996. Molecular mechanisms of the protein serine/threonine phosphatases. *Trends Biochem. Sci.* 21: 407–412. [R]

Edelman, A. M., D. K. Blumenthal and E. G. Krebs. 1987. Protein serine/threonine kinases. *Ann. Rev. Biochem.* 56: 567–613. [R]

Fauman, E. B. and M. A. Saper. 1996. Structure and function of the protein tyrosine phosphatases. *Trends Biochem. Sci.* 21: 413–417. [R]

Fischer, E. H. and E. G. Krebs. 1989. Commentary on "The phosphorylase β to α converting enzyme of rabbit skeletal muscle." *Biochim. Biophys. Acta* 1000: 297–301. [R]

Hanks, S. K., A. M. Quinn and T. Hunter. 1988. The protein kinase family: Conserved features and deduced phylogeny of the catalytic domains. *Science* 241: 42–52. [R]

Hunter, T. 1995. Protein kinases and phosphatases: The yin and yang of protein phosphorylation and signaling. *Cell* 80: 225–236. [R]

Kantrowitz, E. R. and W. N. Lipscomb. 1990. *Escherichia coli* aspartate transcarbamoylase: The molecular basis for a concerted allosteric transition. *Trends. Biochem. Sci.* 15: 53–59. [R]

Milburn, M. V., L. Tong, A. M. DeVos, A. Brunger, Z. Yamaizumi, S. Nishimura and S.-H. Kim. 1990. Molecular switch for signal transduction: Structural differences between active and inactive forms of protooncogenic ras proteins. *Science* 247: 939–945. [P]

Monod, J., J.-P. Changeux and F. Jacob. 1963. Allosteric proteins and cellular control systems. *J. Mol. Biol.* 6: 306–329. [P]

Taylor, S. S., D. R. Knighton, J. Zheng, L. F. R. Eyck and J. M. Sowadski. 1992. Structural framework for the protein kinase family. *Ann. Rev. Cell Biol.* 8: 429–462. [R]

Protein Degradation

Coux, O., K. Tanaka and A. L. Goldberg. 1996. Structure and functions of the 20S and 26S proteasomes. *Ann. Rev. Biochem.* 65: 801–847. [R]

Deshaies, R. J. 1995. Make it or break it: The role of ubiquitin-dependent proteolysis in cellular regulation. *Trends Cell Biol.* 5: 428–434. [R]

Dice, J. F. 1990. Peptide sequences that target cytosolic proteins for lysosomal proteolysis. *Trends Biochem. Sci.* 15: 305–309. [R]

Glotzer, M., A. W. Murray and M. W. Kirschner. 1991. Cyclin is degraded by the ubiquitin pathway. *Nature* 349: 132–138. [P]

Hershko, A. and A. Ciechanover. 1998. The ubiquitin system. *Ann. Rev. Biochem.* 67: 425–479. [R]

Laney, J. D. and M. Hochstrasser. 1999. Substrate targeting in the ubiquitin system. *Cell* 97: 427–430. [R]

Varshavsky, A. 1997. The ubiquitin system. *Trends Biochem. Sci.* 22: 383–387. [R]

PART III

Cell Structure and Function

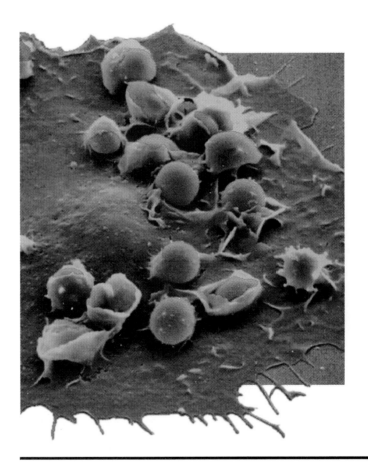

8

The Nucleus

THE PRESENCE OF A NUCLEUS IS THE PRINCIPAL FEATURE that distinguishes eukaryotic from prokaryotic cells. By housing the cell's genome, the nucleus serves both as the repository of genetic information and as the cell's control center. DNA replication, transcription, and RNA processing all take place within the nucleus, with only the final stage of gene expression (translation) localized to the cytoplasm.

By separating the genome from the cytoplasm, the nuclear envelope allows gene expression to be regulated by mechanisms that are unique to eukaryotes. Whereas prokaryotic mRNAs are translated while their transcription is still in process, eukaryotic mRNAs undergo posttranscriptional processing (e.g., splicing) before being transported from the nucleus to the cytoplasm. The presence of a nucleus thus allows gene expression to be regulated by posttranscriptional mechanisms, such as alternative splicing. By limiting the access of proteins to the genetic material, the nuclear envelope also provides novel opportunities for the control of gene expression at the level of transcription. For example, the expression of some eukaryotic genes is controlled by the regulated transport of transcription factors from the cytoplasm into the nucleus—a form of transcriptional regulation unavailable to prokaryotes. The separation of the genome from the site of mRNA translation thus plays a central role in eukaryotic gene expression.

The Nuclear Envelope and Traffic between the Nucleus and Cytoplasm

The nuclear envelope separates the contents of the nucleus from the cytoplasm and provides the structural framework of the nucleus. The nuclear membranes, acting as barriers that prevent the free passage of molecules between the nucleus and the cytoplasm, maintain the nucleus as a distinct biochemical compartment. The sole channels through the nuclear envelope are provided by the nuclear pore complexes, which allow the regulated exchange of molecules between the nucleus and cytoplasm. The selective traffic of proteins and RNAs through the nuclear pore complexes not only establishes the internal composition of the nucleus, but also plays a critical role in regulating eukaryotic gene expression.

(A)

(B)

Endoplasmic reticulum

Nuclear pore complex

Outer membrane

Chromatin

Nucleolus

Inner membrane

0.2 μm

(C)

Outer membrane

Perinuclear space

Inner membrane

Smooth endoplasmic reticulum

Ribosomes | Nuclear pore complex | Nuclear lamina | Nucleolus | Chromatin | Rough endoplasmic reticulum

Figure 8.1
The nuclear envelope
(A) An electron micrograph of a nucleus. The inner and outer nuclear membranes are joined at nuclear pore complexes (arrows). (B) An electron micrograph illustrating the continuity of the outer nuclear membrane with the endoplasmic reticulum. (C) Schematic of the nuclear envelope. The inner nuclear membrane is lined by the nuclear lamina, which serves as an attachment site for chromatin. (A, David M. Phillips/Photo Researchers, Inc.; B, courtesy of Dr. Werner W. Franke, German Cancer Research Center, Heidelberg.)

Structure of the Nuclear Envelope

The **nuclear envelope** has a complex structure, consisting of two nuclear membranes, an underlying nuclear lamina, and nuclear pore complexes (Figure 8.1). The nucleus is surrounded by a system of two concentric membranes, called the inner and outer **nuclear membranes**. The outer nuclear membrane is continuous with the endoplasmic reticulum, so the space between the inner and outer nuclear membranes is directly connected with the lumen of the endoplasmic reticulum. In addition, the outer nuclear membrane is functionally similar to the membranes of the endoplasmic reticulum (see Chapter 9) and has ribosomes bound to its cytoplasmic surface. In contrast, the inner nuclear membrane carries unique proteins that are specific to the nucleus.

Figure 8.2
Electron micrograph showing nuclear pores   Many nuclear pores (arrows) are visible in this freeze-fracture preparation of the nuclear envelope. (Photo Researchers, Inc.)

0.5 μm

The critical function of the nuclear membranes is to act as a barrier that separates the contents of the nucleus from the cytoplasm. Like other cell membranes, the nuclear membranes are phospholipid bilayers, which are permeable only to small nonpolar molecules (see Figure 2.49). Other molecules are unable to diffuse through the phospholipid bilayer. The inner and outer nuclear membranes are joined at nuclear pore complexes, the sole channels through which small polar molecules and macromolecules are able to travel through the nuclear envelope (Figure 8.2). As discussed in the next section, the nuclear pore complex is a complicated structure that is responsible for the selective traffic of proteins and RNAs between the nucleus and the cytoplasm.

Underlying the inner nuclear membrane is the **nuclear lamina**, a fibrous meshwork that provides structural support to the nucleus (Figure 8.3). The

Figure 8.3
Electron micrograph of the nuclear lamina   The lamina is a meshwork of filaments underlying the inner nuclear membrane. (From U. Aebi, J. Cohn, L. Buhle and L. Gerace, 1986. *Nature* 323: 560.)

0.5 μm

Lamin polypeptide

Dimer

Head-to-tail association of dimers

Polymer

Side-by-side association of polymers

Filament

Figure 8.4
Model of lamin assembly   The lamin polypeptides form dimers in which the central α-helical regions of two polypeptide chains are wound around each other. Further assembly may involve the head-to-tail association of dimers to form linear polymers and the side-by-side association of polymers to form filaments.

nuclear lamina is composed of one or more related proteins called **lamins**. Most mammalian cells, for example, contain four different lamins, designated A, B_1, B_2, and C. All the lamins are 60- to 80-kilodalton (kd) fibrous proteins that are related to the intermediate filament proteins of the cytoskeleton (see Chapter 11). Like other intermediate filament proteins, the lamins associate with each other to form filaments (Figure 8.4). The first stage of this association is the interaction of two lamins to form a dimer in which the α-helical regions of two polypeptide chains are wound around each other in a structure called a coiled coil. These lamin dimers then associate with each other to form the filaments that make up the nuclear lamina. The association of lamins with the inner nuclear membrane is facilitated by the posttranslational addition of lipid—in particular, prenylation of C-terminal cysteine residues (see Figure 7.30). In addition, the lamins bind to inner nuclear membrane proteins, which may help organize the lamin filaments into a meshwork and mediate their attachment to the membrane.

In addition to providing structural support to the nucleus, the nuclear lamina is thought to serve as a site of chromatin attachment. Chromatin within the nucleus is organized into large loops of DNA, some of which appear to be bound to the nuclear envelope. The lamins bind chromatin and may help mediate this interaction.

The Nuclear Pore Complex

The **nuclear pore complexes** are the only channels through which small polar molecules, ions, and macromolecules (proteins and RNAs) are able to travel between the nucleus and the cytoplasm. The nuclear pore complex is an extremely large structure with a diameter of about 120 nm and an estimated molecular mass of approximately 125 million daltons—about 30 times the size of a ribosome. In vertebrates, the nuclear pore complex is composed of 50 to 100 different proteins. By controlling the traffic of molecules between the nucleus and cytoplasm, the nuclear pore complex plays a fundamental role in the physiology of all eukaryotic cells. RNAs that are synthesized in the nucleus must be efficiently exported to the cytoplasm, where they function in protein synthesis. Conversely, proteins required for nuclear functions (e.g., transcription factors) must be transported into the nucleus from their sites of synthesis in the cytoplasm. In addition, many proteins shuttle continuously between the nucleus and the cytoplasm. The regulated traffic of proteins and RNAs through the nuclear pore complex thus determines the composition of the nucleus and plays a key role in gene expression.

Depending on their size, molecules can travel through the nuclear pore complex by one of two different mechanisms (Figure 8.5). Small molecules and some proteins with molecular mass less than approximately 50 kd pass freely across the nuclear envelope in either direction: cytoplasm to nucleus or nucleus to cytoplasm. These molecules diffuse passively through open aqueous channels, estimated to have diameters of approximately 9 nm, in the nuclear pore complex. Most proteins and RNAs, however, are unable to pass through these open channels. Instead, these macromolecules pass through the nuclear pore complex by an active process in which appropriate proteins and RNAs are recognized and selectively transported in only one direction (nucleus to cytoplasm or cytoplasm to nucleus). The traffic of these molecules occurs through regulated channels in the nuclear pore complex that, in response to appropriate signals, can open to a diameter of more than 25 nm—a size sufficient to accommodate large ribonucleoprotein complexes, such as ribosomal subunits. It is through these regulated channels that nuclear proteins are selectively imported from the cytoplasm to the nucleus while RNAs are exported from the nucleus to the cytoplasm.

Visualization of nuclear pore complexes by electron microscopy reveals a structure with eightfold symmetry organized around a large central channel (Figure 8.6), which is the route through which proteins and RNAs cross the nuclear envelope. Detailed structural studies, including computer-based image analysis, have led to the development of three-dimensional models of the nuclear pore complex (Figure 8.7). These studies indicate that the nuclear pore complex consists of an assembly of eight spokes arranged around a central channel. The spokes are connected to rings at the nuclear and cytoplasmic surfaces, and the spoke-ring assembly is anchored within the nuclear envelope at sites of fusion between the inner and outer nuclear membranes. Protein filaments extend from both the cytoplasmic and nuclear rings, forming a distinct basketlike structure on the nuclear side. The central channel is approximately 40 nm in diameter, which is wide

Figure 8.5
Molecular traffic through nuclear pore complexes   Small molecules are able to pass rapidly through open channels in the nuclear pore complex by passive diffusion. In contrast, macromolecules are transported by a selective, energy-dependent mechanism that acts predominantly to import proteins to the nucleus and export RNAs to the cytoplasm.

0.2 μm

Figure 8.6
Electron micrograph of nuclear pore complexes   In this face-on view, isolated nuclear pore complexes appear to consist of eight structural subunits surrounding a central channel. (Courtesy of Dr. Ron Milligan, Scripps Research Institute.)

Figure 8.7
Model of the nuclear pore complex
The complex consists of an assembly
of eight spokes attached to rings on
the cytoplasmic and nuclear sides of
the nuclear envelope. The spoke-ring
assembly surrounds a central channel
containing the central transporter.
Cytoplasmic filaments extend from the
cytoplasmic ring, and filaments forming the nuclear basket extend from the
nuclear ring.

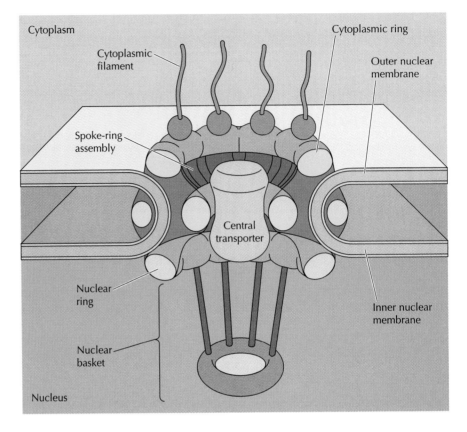

enough to accommodate the largest particles able to cross the nuclear envelope. It contains a structure called the central transporter, through which
the active transport of macromolecules is thought to occur.

Selective Transport of Proteins to and from the Nucleus

The basis for selective traffic across the nuclear envelope is best understood
for proteins that are imported from the cytoplasm to the nucleus. Such proteins are responsible for all aspects of genome structure and function; they
include histones, DNA polymerases, RNA polymerases, transcription factors, splicing factors, and many others. These proteins are targeted to the
nucleus by specific amino acid sequences, called **nuclear localization signals**, that direct their transport through the nuclear pore complex.

The first nuclear localization signal to be mapped in detail was characterized by Alan Smith and colleagues in 1984. These investigators studied simian
virus 40 (SV40) T antigen, a virus-encoded protein that initiates viral DNA
replication in infected cells (see Chapter 5). As expected for a replication protein, T antigen normally is localized to the nucleus. The signal responsible for
its nuclear localization was first identified by the finding that mutation of a
single lysine residue prevents nuclear import, resulting instead in the accumulation of T antigen in the cytoplasm. Subsequent studies defined the
T antigen nuclear localization signal as the seven-amino-acid sequence Pro-
Lys-Lys-Lys-Arg-Lys-Val. Not only was this sequence necessary for the
nuclear transport of T antigen, but its addition to other, normally cytoplasmic,
proteins was also sufficient to direct their accumulation in the nucleus.

Nuclear localization signals have since been identified in many other proteins. Most of these sequences, like that of T antigen, are short stretches rich
in basic amino acid residues (lysine and arginine). In many cases, however,
the amino acids that form the nuclear localization signal are close together but

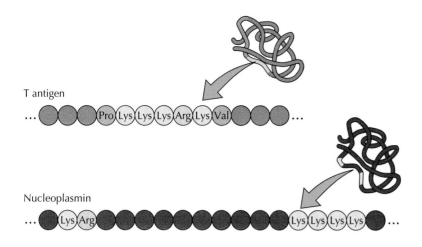

T antigen

Nucleoplasmin

Figure 8.8
Nuclear localization signals The T antigen nuclear localization signal is a single stretch of amino acids. In contrast, the nuclear localization signal of nucleoplasmin is bipartite, consisting of a Lys-Arg sequence, followed by a Lys-Lys-Lys-Lys sequence located ten amino acids farther downstream.

not immediately adjacent to each other. For example, the nuclear localization signal of nucleoplasmin (a protein involved in chromatin assembly) consists of two parts: a Lys-Arg pair followed by four lysines located ten amino acids farther downstream (Figure 8.8). Both the Lys-Arg and Lys-Lys-Lys-Lys sequences are required for nuclear targeting, but the ten amino acids between these sequences can be mutated without affecting nuclear localization. Because this nuclear localization sequence is composed of two separated elements, it is referred to as bipartite. Similar bipartite motifs appear to function as the localization signals of many nuclear proteins; thus they may be more common than the simpler nuclear localization signal of T antigen. In addition, some proteins, such as ribosomal proteins, contain distinct nuclear localization signals which are unrelated to the basic amino acid-rich nuclear localization signals of either nucleoplasmin or T antigen.

Protein import through the nuclear pore complex can be operationally divided into two steps, distinguished by whether they require energy (Figure 8.9). In the first step, which does not require energy, proteins that

Figure 8.9
Protein import through the nuclear pore complex Proteins are transported through the nuclear pore complex in two steps. In the example shown, a protein with a classical basic amino acid–rich nuclear localization sequence (NLS) is recognized by importin α, which forms a complex with importin β. Importin β binds to the cytoplasmic filaments of the nuclear pore complex, bringing the target protein to the nuclear pore. The protein and importin α are then translocated through the nuclear pore complex in a second, energy-requiring step, which requires GTP hydrolysis by the Ran protein.

Identification of Nuclear Localization Signals

A Short Amino Acid Sequence Able to Specify Nuclear Location

Daniel Kalderon, Bruce L. Roberts, William D. Richardson, and Alan E. Smith
National Institute for Medical Research, Mill Hill, London
Cell, Volume 39, 1984, 499–509

The Context

Maintaining the nucleus as a distinct biochemical compartment requires a mechanism by which proteins are segregated between the nucleus and the cytoplasm. Studies in the 1970s established that small molecules diffuse rapidly across the nuclear envelope, but that most proteins are unable to do so. It therefore appeared likely that nuclear proteins are specifically recognized and selectively imported to the nucleus from their sites of synthesis on cytoplasmic ribosomes.

Earlier experiments of Günter Blobel and his colleagues had established that proteins are targeted to the endoplasmic reticulum by signal sequences consisting of short stretches of amino acids (see Chapter 9). In this 1984 paper, Alan Smith and his colleagues extended this principle to the targeting of nuclear proteins by identifying a short amino acid sequence that serves as a nuclear localization signal.

The Experiments

The viral protein SV40 T antigen was used as a model for studies of nuclear localization in animal cells. T antigen is a 94-kd protein that is required for SV40 DNA replication and is nor-mally localized to the nucleus of SV40-infected cells. Previous studies in both Alan Smith's laboratory and in the laboratory of Janet Butel (Lanford and Butel, 1984, *Cell* 37: 801–813) had shown that mutation of Lys-128 to either Thr or Asn prevented the normal nuclear accumulation of T antigen in both rodent and monkey cells. Rather than being transported to the nucleus, these mutant T antigens remained in the cytoplasm, suggesting that Lys-128 was part of a nuclear localization signal. Kalderon and col-leagues tested this hypothesis using two distinct experimental approaches.

First, they determined the effects of different deletions on the subcellular localization of T antigen. Mutant T antigens bearing deletions that eliminated amino acids either between residues 1 and 126 or between residue 136 and the C terminus were found to accumulate normally in the nucleus. In contrast, a mutant with a deletion of amino acids 127 to 132 remained in the cytoplasm. Thus, the amino acid sequence extending from residue 127 to 132 appeared to be responsible for nuclear localization of T antigen.

To determine whether this amino acid sequence was able to target other proteins to the nucleus, the investigators constructed chimeras in which the T antigen amino acid sequence was fused to proteins that were nor-

(A)   (B)

Cells were microinjected with plasmid DNAs encoding chimeric proteins in which SV40 amino acids were fused to pyruvate kinase. Cellular localization of the fusion proteins was then determined by immunofluorescence microscopy. (A) The fusion protein contains an intact SV40 nuclear localization signal (amino acids 126 to 132). (B) The nuclear localization signal has been inactivated by deletion of amino acids 131 and 132.

contain nuclear localization signals bind to the nuclear pore complex but do not pass through the pore. In this initial step, nuclear localization signals are recognized by a cytosolic receptor protein, and the receptor-substrate complex binds to the nuclear pore. The prototype receptor, called **importin**, consists of two subunits. One subunit (importin α) binds to the basic amino acid–rich nuclear localization signals of proteins such as T antigen and nucleoplasmin. The second subunit (importin β) binds to

Identification of Nuclear Localization Signals *(continued)*

mally cytoplasmic. These experiments established that the addition of T antigen amino acids 126 to 132 to either β-galactosidase or pyruvate kinase is sufficient to specify the nuclear accumulation of these otherwise cytoplasmic proteins (see figure). This short amino acid sequence of SV40 T antigen thus functions as a nuclear localization signal, which is both necessary and sufficient to target proteins for nuclear import.

The Impact

As Kalderon and colleagues suggested in their 1984 paper, the nuclear

localization signal of SV40 T antigen has proved to "represent a prototype of similar sequences in other nuclear proteins." By targeting proteins for nuclear import, these signals are key to establishing the biochemical identity of the nucleus and maintaining the fundamental division of eukaryotic cells into nuclear and cytoplasmic compartments. Nuclear localization signals are now known to be recognized by cytoplasmic receptors that transport their substrate proteins to the nuclear pore complex. Although the mechanism of transport through the nuclear pore complex remains to

Alan Smith

be elucidated, the identification of nuclear localization signals was a key advance in understanding nuclear protein import.

---

the cytoplasmic filaments of the nuclear pore complex, bringing the target protein to the nuclear pore. Other types of nuclear localization signals, such as those of ribosomal proteins, are recognized by distinct receptors which are related to importin β and function similarly to importin β during the transport of their target proteins into the nucleus.

The second step in nuclear import, translocation through the nuclear pore complex, is an energy-dependent process that requires GTP hydrolysis. A key player in the translocation process is a small GTP-binding protein called **Ran**, which is related to the Ras proteins (Figure 8.10). The conformation and activity of Ran is regulated by GTP binding and hydrolysis, like Ras or several of the translation factors involved in protein synthesis (see Figure 7.12). Enzymes that stimulate GTP binding to Ran are localized to the nuclear side of the nuclear envelope whereas

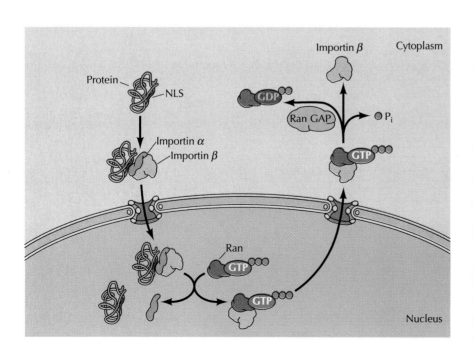

Figure 8.10
Role of the Ran protein in nuclear import Transport through the nuclear pore complex is driven by a gradient of Ran/GTP, with a high concentration of Ran/GDP in the cytoplasm and a high concentration of Ran/GTP in the nucleus. Complexes form between target proteins containing nuclear localization signals (NLS), importin α, and importin β in the cytoplasm where Ran is in the GDP-bound form. Following transport through the nuclear pore complex, Ran/GTP binds to importin β, releasing importin α and the target protein in the nucleus. The Ran/GTP-importin β complex is then transported back to the cytoplasm, where the Ran GTPase-activating protein (Ran GAP) stimulates hydrolysis of the bound GTP to form Ran/GDP. This conversion of Ran/GTP to Ran/GDP is accompanied by release of importin β.

enzymes that stimulate GTP hydrolysis are localized to the cytoplasmic side. Consequently, there is a gradient of Ran/GTP across the nuclear envelope, with a high concentration of Ran/GTP in the nucleus and a high concentration of Ran/GDP in the cytoplasm. This gradient of Ran/GTP is thought to determine the directionality of nuclear transport, and GTP hydrolysis by Ran appears to account for most (if not all) of the energy required for nuclear import. Importin β forms a complex with importin α and its associated target protein on the cytoplasmic side of the nuclear pore complex, in the presence of a high concentration of Ran/GDP. This complex is then transported through the nuclear pore to the nucleus, where a high concentration of Ran/GTP is present. At the nuclear side of the pore, Ran/GTP binds to importin β, displacing importin α and the target protein. As a result, the target protein is released within the nucleus. The Ran/GTP-importin β complex is then exported to the cytosol, where the bound GTP is hydrolyzed to GDP, releasing importin β to participate in another cycle of nuclear import.

Some proteins remain within the nucleus following their import from the cytoplasm, but many others shuttle back and forth between the nucleus and the cytoplasm. Some of these proteins act as carriers in the transport of other molecules, such as RNAs; others coordinate nuclear and cytoplasmic functions (e.g., by regulating the activities of transcription factors). Proteins are targeted for export from the nucleus by specific amino acid sequences, called **nuclear export signals**. Like nuclear localization signals, nuclear export signals are recognized by receptors within the nucleus that direct protein transport through the nuclear pore complex to the cytoplasm. Interestingly, the nuclear export receptors (called **exportins**) are related to importin β. Like importin β, the exportins bind to Ran, which is required for nuclear export as well as for nuclear import (Figure 8.11). Strikingly, however, Ran/GTP promotes the formation of stable complexes between exportins and their target proteins, whereas it dissociates the complexes between importins and their targets. This effect of Ran/GTP binding on exportins dictates the movement of proteins containing nuclear export signals from the nucleus to the cytoplasm. Thus,

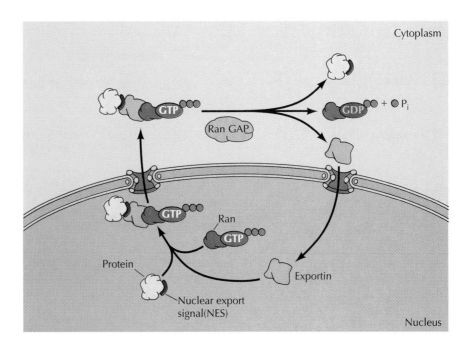

Figure 8.11
Nuclear export Complexes between target proteins bearing nuclear export signals (NES), exportins, and Ran/GTP form in the nucleus. Following transport through the nuclear pore complex, Ran GAP stimulates the hydrolysis of bound GTP, leading to formation of Ran/GDP and release of the target protein and exportin in the cytoplasm. Exportin is then transported back to the nucleus.

exportins form stable complexes with their target proteins in association with Ran/GTP within the nucleus. Following transport to the cytosolic side of the nuclear envelope, GTP hydrolysis leads to dissociation of the target protein, which is released into the cytoplasm.

Regulation of Nuclear Protein Import

An intriguing aspect of the transport of proteins into the nucleus is that it is another level at which the activities of nuclear proteins can be controlled. Transcription factors, for example, are functional only when they are present in the nucleus, so regulation of their import to the nucleus is a novel means of controlling gene expression. As will be discussed in Chapter 13, the regulated nuclear import of both transcription factors and protein kinases plays an important role in controlling the behavior of cells in response to changes in the environment, because it provides a mechanism by which signals received at the cell surface can be transmitted to the nucleus.

In one mechanism of regulation, transcription factors (or other proteins) associate with cytoplasmic proteins that mask their nuclear localization signals; because their signals are no longer recognizable, these proteins remain in the cytoplasm. A good example is provided by the transcription factor NF-κB, which activates transcription of immunoglobulin-κ light chains in B lymphocytes (Figure 8.12). In unstimulated cells, NF-κB is found as an inactive complex with an inhibitory protein (IκB) in the cytoplasm. Binding to IκB appears to mask the NF-κB nuclear localization signal, thus preventing NF-κB from being transported into the nucleus. In stimulated cells, IκB is phosphorylated and degraded by ubiquitin-mediated proteolysis, allowing NF-κB to enter the nucleus and activate transcription of its target genes.

The nuclear import of other transcription factors is regulated directly by their phosphorylation, rather than by association with inhibitory proteins. For example, the yeast transcription factor SWI5 is imported into the nucleus only at a specific stage of the cell cycle (see Figure 8.12). Otherwise, SWI5 is retained in the cytoplasm as a result of phosphorylation at serine residues adjacent to its nuclear localization signal, preventing nuclear import. Regulated dephosphorylation of these sites activates SWI5 at the appropriate stage of the cell cycle by permitting its translocation to the nucleus.

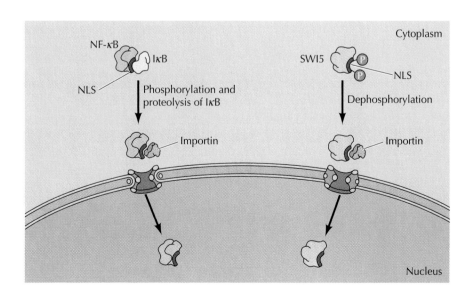

Figure 8.12
Regulation of nuclear import of transcription factors   The transcription factor NF-κB is maintained as an inactive complex with IκB, which masks its nuclear localization sequence (NLS), in the cytoplasm. In response to appropriate extracellular signals, IκB is phosphorylated and degraded by proteolysis, allowing the import of NF-κB to the nucleus. In contrast, the yeast transcription factor SWI5 is maintained in the cytoplasm by phosphorylation in the vicinity of its nuclear localization sequence. Regulated dephosphorylation exposes the NLS and allows SWI5 to be transported to the nucleus at the appropriate stage of the cell cycle.

(A)   (B)   (C)   (D)

0.1 μm

Figure 8.13
Transport of a ribonucleoprotein complex   Insect salivary gland cells produce large ribonucleoprotein complexes (RNPs), which contain 35 to 40 kb of RNA and have a total mass of approximately 30 million daltons. This series of electron micrographs shows the attachment of such an RNP to a nuclear pore complex (A) and the unfolding of the RNA during its translocation to the cytoplasm (B–D). (From H. Mehlin et al., 1992. *Cell* 69: 605.)

Transport of RNAs

Whereas many proteins are selectively transported from the cytoplasm into the nucleus, most RNAs are exported from the nucleus to the cytoplasm. Since proteins are synthesized in the cytoplasm, the export of mRNAs, rRNAs, and tRNAs is a critical step in gene expression in eukaryotic cells. Like protein import, the export of RNAs through nuclear pore complexes is an active, energy-dependent process that requires the Ran GTP-binding protein.

RNAs are transported across the nuclear envelope as RNA-protein complexes, which in some cases are large enough to visualize by electron microscopy (Figure 8.13). The substrates for transport are ribonucleoprotein complexes rather than naked RNAs, and RNAs are targeted for transport from the nucleus by nuclear export signals on the proteins bound to them. These proteins are recognized by exportins and transported from the nucleus to the cytoplasm as described earlier (see Figure 8.11). Pre-mRNAs and mRNAs are associated with a set of at least 20 proteins (forming **heterogeneous nuclear ribonucleoproteins**, or **hnRNPs**) throughout their processing in the nucleus and eventual transport to the cytoplasm. At least two of these hnRNP proteins contain nuclear export signals and are thought to function as carriers of mRNAs during their export to the cytoplasm. As discussed in a later section of this chapter, ribosomal RNAs are

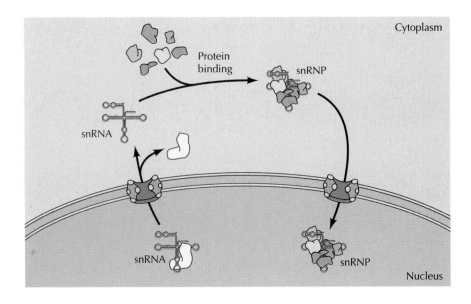

Figure 8.14
Transport of snRNAs between nucleus and cytoplasm   Small nuclear RNAs are initially exported from the nucleus to the cytoplasm, where they associate with proteins to form snRNPs. The assembled snRNPs are then transported back into the nucleus.

assembled with ribosomal proteins in the nucleolus, and intact ribosomal subunits are then transported to the cytoplasm. Their export from the nucleus appears to be mediated by nuclear export signals present on ribosomal proteins. For tRNAs, the specific proteins that mediate nuclear export remain to be identified.

In contrast to mRNAs, tRNAs, and rRNAs, which function in the cytoplasm, the snRNAs function within the nucleus as components of the RNA processing machinery. Perhaps surprisingly, these RNAs are initially transported from the nucleus to the cytoplasm, where they associate with proteins to form functional snRNPs and then return to the nucleus (Figure 8.14). Proteins that bind to the 5′ caps of snRNAs appear to be involved in the export of the snRNAs to the cytoplasm, whereas sequences present on the snRNP proteins are responsible for the transport of snRNPs from the cytoplasm to the nucleus.

Internal Organization of the Nucleus

The nucleus is more than a container in which chromatin, RNAs, and nuclear proteins move freely in aqueous solution. Instead, the nucleus appears to have an internal structure that organizes the genetic material and localizes some nuclear functions to discrete sites. The most obvious aspect of the internal organization of the nucleus is the nucleolus, which, as discussed in the following section, is the site at which the rRNA genes are transcribed and ribosomal subunits are assembled. Additional elements of internal nuclear structure are suggested by the organization of chromosomes and by the potential localization of functions such as DNA replication and pre-mRNA processing to distinct nuclear domains.

Chromosomes and Higher-Order Chromatin Structure

Chromatin becomes highly condensed during mitosis to form the compact metaphase chromosomes that are distributed to daughter nuclei (see Figure 4.12). During interphase, some of the chromatin (**heterochromatin**) remains highly condensed and is transcriptionally inactive; the remainder of the chromatin (**euchromatin**) is decondensed and distributed throughout the nucleus (Figure 8.15). Cells contain two types of heterochromatin. Constitutive heterochromatin contains DNA sequences that are never transcribed, such as the satellite sequences present at centromeres. Facultative heterochromatin contains sequences that are not transcribed in the cell being examined, but are transcribed in other cell types. Consequently, the amount of facultative heterochromatin varies depending on the transcriptional activity of the cell. Much of the heterochromatin is localized to the periphery of the nucleus, possibly because one of the principal proteins associated with heterochromatin binds to a protein of the inner nuclear membrane.

The phenomenon of **X chromosome inactivation** provides an example of the role of heterochromatin in gene expression. In many animals, including humans, females have two X chromosomes, and males have one X and one Y chromosome. The X chromosome contains thousands of genes that are not present on the much smaller Y chromosome (see Figure 4.26). Thus, females have twice as many X chromosome genes as males have. Despite this difference, female and male cells contain equal amounts of the proteins encoded by X chromosome genes. This results from a dosage compensation mechanism in which one of the two X chromosomes in female cells is inactivated by being converted to heterochromatin early in development. Consequently, only one copy of the X chromosome is available for transcription in either female or male cells. The mechanism of X chromosome inactiva-

Figure 8.15
Heterochromatin in interphase nuclei
The euchromatin is distributed throughout the nucleus. The heterochromatin is indicated by arrowheads, and the nucleolus by an arrow. (Courtesy of Ada L. Olins and Donald E. Olins, Oak Ridge National Laboratory.)

1 μm

tion is fascinating though not yet fully understood; it appears to involve the action of a regulatory RNA that coats the inactive X chromosome and induces its conversion to heterochromatin.

Although interphase chromatin appears to be uniformly distributed, the chromosomes are actually arranged in an organized fashion and divided into discrete functional domains that play an important role in regulating gene expression. The nonrandom distribution of chromatin within the interphase nucleus was first suggested in 1885 by C. Rabl, who proposed that each chromosome occupies a distinct territory, with centromeres and telomeres attached to opposite sides of the nuclear envelope (Figure 8.16). This basic model of chromosome organization was confirmed nearly a hundred years later (in 1984) by detailed studies of polytene chromosomes in *Drosophila* salivary glands. Rather than randomly winding around one another, each chromosome was found to occupy a discrete region of the

(A)   (B)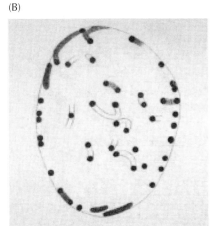

Figure 8.16
Chromosome organization
Reproduction of hand-drawn sketches of chromosomes in salamander cells. (A) Complete chromosomes. (B) Telomeres only (located at the nuclear membrane). (From C. Rabl, 1885. *Morphologisches Jahrbuch* 10: 214.)

Figure 8.17
Organization of *Drosophila* chromosomes   (A) A model of the nucleus, showing the five chromosome arms in different colors. The positions of telomeres and centromeres are indicated. (B) The two arms of chromosome 3 are shown to illustrate the topological separation between chromosomes. (From D. Mathog et al., 1984. *Nature* 308: 414.)

nucleus (Figure 8.17). The chromosomes are closely associated with the nuclear envelope at many sites, with their centromeres and telomeres clustered at opposite poles.

Individual chromosomes also occupy distinct territories within the nuclei of mammalian cells (Figure 8.18). Actively transcribed genes appear to be localized to the periphery of these territories, adjacent to channels separating the chromosomes. Newly transcribed RNAs are thought to be released into these channels between chromosomes, where RNA processing takes place.

Like the DNA in metaphase chromosomes (see Figure 4.13), the chromatin in interphase nuclei appears to be organized into looped domains containing approximately 50 to 100 kb of DNA. A good example of this looped-domain organization is provided by the highly transcribed chromosomes of amphibian oocytes, in which actively transcribed regions of DNA can be visualized as extended loops of decondensed chromatin (Figure 8.19). These chromatin domains appear to represent discrete functional units, which independently regulate gene expression.

The effects of chromosome organization on gene expression have been demonstrated by a variety of experiments showing that the position of a gene in chromosomal DNA affects the level at which the gene is expressed. For example, the transcriptional activity of genes introduced into transgenic mice depends on their sites of integration in the mouse genome. This effect of chromosomal position on gene expression can be alleviated by sequences known as **locus control regions**, which result in a high level of expression of the introduced genes irrespective of their site of integration. In contrast to transcriptional enhancers (see Chapter 6), locus control regions stimulate only transfected genes that are integrated into chromosomal DNA; they do not affect the expression of unintegrated plasmid DNAs in transient assays. In addition, rather than affecting individual promoters, locus control regions appear to activate large chromosome domains, presumably by inducing long-range alterations in chromatin structure.

(A)

Copies of chromosome 4

(B)

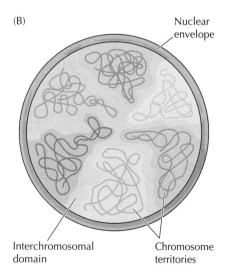

Nuclear envelope

Interchromosomal domain

Chromosome territories

Figure 8.18
Organization of chromosomes in the mammalian nucleus   (A) Probes to repeated sequences on chromosome 4 were hybridized to a human cell. The two copies of chromosome 4, identified by yellow fluorescence, occupy distinct territories in the nucleus. (B) A model of chromosome organization. The chromosomes occupy discrete territories, separated by intrachromosomal domains in which RNA processing and transport is thought to occur. (A, courtesy of Thomas Cremer, Ludwig Maximilians University, from A. I. Lamond and W. C. Earnshaw, 1998. *Science* 280: 547.)

Figure 8.19
Looped chromatin domains   Light micrograph of a chromosome of amphibian oocytes, showing decondensed loops of actively transcribed chromatin extending from an axis of highly condensed nontranscribed chromatin. (Courtesy of Joseph Gall, Carnegie Institute.)

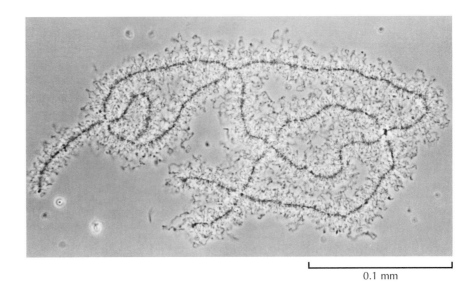

0.1 mm

The separation between chromosomal domains is maintained by boundary sequences or **insulator elements**, which prevent the chromatin structure of one domain from spreading to its neighbors. In addition, insulators act as barriers that prevent enhancers in one domain from acting on promoters located in an adjacent domain. Like locus control regions, insulators function only in the context of chromosomal DNA, suggesting that they regulate higher-order chromatin structure. Although the mechanisms of action of locus control regions and insulators remain to be elucidated, their functions clearly indicate the importance of higher-order chromatin organization in the control of eukaryotic gene expression.

Functional Domains within the Nucleus

An internal organization of the nucleus is indicated also by the localization of some nuclear processes to distinct regions of the nucleus. Rather than taking place throughout the nucleus, activities such as DNA replication and pre-mRNA processing may be localized to discrete subnuclear structures or domains. The nature and function of these nuclear substructures are not yet clear, however, and understanding the organization of functional domains within the nucleus is an incompletely explored area of cell biology.

The nuclei of mammalian cells appear to contain clustered sites of DNA replication within which the replication of multiple DNA molecules takes place. These discrete sites of DNA replication have been defined by experiments in which newly synthesized DNA was visualized within cell nuclei (Figure 8.20). This was accomplished by labeling cells with bromodeoxyuridine—an analog of thymidine that can be incorporated into DNA and then detected by staining with fluorescent antibodies. In such experiments, the newly replicated DNA was detected in approximately 200 discrete clusters distributed throughout the nucleus. Since approximately 4000

Fluorescent foci
of DNA replication

Figure 8.20
Clustered sites of DNA replication   Newly replicated DNA was labeled by a brief exposure of cells to bromodeoxyuridine, which is incorporated into DNA in place of thymidine. This substitution allows detection of newly synthesized DNA by immunofluorescence following staining of the nuclei with an antibody against bromodeoxyuridine. Note that the newly replicated DNA is present in discrete clusters distributed throughout the nucleus. (Courtesy of Ronald Berezney, SUNY/Buffalo.)

Molecular Medicine

Systemic Lupus Erythematosus

The Disease

Systemic lupus erythematosus is a chronic autoimmune disease in which patients produce antibodies against their own normal cell constituents, particularly nuclear antigens. The disease affects women about ten times more frequently than it affects men and has an incidence of about one in 700 among women between the ages of 20 and 60. The disease is characterized by the formation of large amounts of antibody-antigen complexes (immune complexes) that are deposited in tissues and blood vessels throughout the body. The tissue damage and inflammation resulting from deposition of these immune complexes produces a variety of clinical manifestations, including skin rashes, arthritis, and kidney disease. Damage to the kidney is particularly common and can be severe enough to cause kidney failure and death.

Molecular and Cellular Basis

The fundamental abnormality in systemic lupus erythematosus is the production of autoantibodies that react with constituents of normal cells. These antibodies are frequently directed against components of the nucleus, and the presence of antinuclear antibodies is the principal diagnostic test for the disease. The auto-

antibodies detected in systemic lupus erythematosus patients are directed against many different nuclear targets, including DNA, histones, nonhistone chromosomal proteins, and snRNPs. Indeed, the snRNPs involved in pre-mRNA splicing were first characterized by the use of antibodies obtained from systemic lupus erythematosus patients.

Because so many different autoantibodies are produced, this disease is thought to result from a defect in immune regulation that leads to a general stimulation of B lymphocytes. Genetic factors contribute to the risk of disease, but the basic cause of the breakdown in immune regulation that leads to systemic lupus erythematosus remains unknown.

Prevention and Treatment

In the absence of understanding the fundamental defects that cause the production of autoantibodies, there are no currently available approaches for either preventing or curing systemic lupus erythematosus. Instead, current treatments are restricted to arresting the pathological developments of the disease. Anti-inflammatory drugs help reduce the inflammation and tissue damage resulting from the deposition of immune com-

Immunofluorescent staining of nuclei with sera from a patient with systemic lupus erythematosus. (Courtesy of Joan Steitz, Yale University.)

plexes. In life-threatening cases, patients are also treated with immunosuppressive drugs that inhibit the proliferation of lymphocytes and blunt the immune response. It is hoped that continuing research will lead to an understanding of the defects that result in autoimmunity, ultimately allowing the development of more effective therapies based on restoring normal immune function.

Reference

Tan, E. M. 1989. Antinuclear antibodies: Diagnostic markers for autoimmune diseases and probes for cell biology. *Adv. Immunol.* 44: 93–151.

---

origins of replication are active in a diploid mammalian cell at any given time, each of these clustered sites of DNA replication must contain approximately 40 replication forks. Thus, DNA replication appears to take place in large structures that contain multiple replication complexes organized into distinct functional domains, which have been called replication factories.

Actively transcribed genes appear to be distributed throughout the nucleus, but components of the splicing machinery are concentrated in discrete subnuclear structural domains. The localization of splicing components to discrete domains within the nucleus has been demonstrated by immunofluorescent staining with antibodies against snRNPs and splicing factors (Figure 8.21). Rather than being distributed uniformly throughout the nucleus, these components of the splicing apparatus are concentrated in

Figure 8.21
Localization of splicing components   Staining with immunofluorescent antibodies indicates that splicing factors are concentrated in discrete domains within the nucleus. (Courtesy of David L. Spector, Cold Spring Harbor Laboratory.)

20 to 50 discrete structures termed nuclear speckles. It is thought that speckles are storage sites of splicing components, which are then recruited from the speckles to actively transcribed genes where pre-mRNA processing occurs.

In addition to speckles, nuclei contain several other types of morphologically distinct structures, collectively called nuclear bodies. The two major types of these nuclear structures are coiled bodies and PML bodies. Coiled bodies are enriched in snRNPs and are believed to function as sites of snRNP assembly. The function of PML bodies is unknown; they are not enriched in snRNPs and do not appear to be major sites of transcription or DNA replication. Thus, although nuclear bodies indicate the presence of substructural domains within the nucleus, their functions remain to be fully elucidated.

Some scientists believe that a central component of the internal organization of the nucleus is the nuclear matrix, which is defined as the structural skeleton of the nucleus. The concept of the nuclear matrix was proposed in 1975 on the basis of experiments in which nuclei were treated with DNase to digest most of the DNA and extracted with high salt buffers to remove histones and other nuclear proteins. Such treatments left a residual network of fibers (the nuclear matrix), which was suggested to provide an internal structural framework for the nucleus, analogous to the role of the cytoskeleton as the structural framework of the cell. In addition, it has been proposed that the nuclear matrix serves to organize and anchor functional domains of the nucleus, including chromatin loops, DNA replication factories, splicing domains, and structures involved in mRNA transport.

Many scientists, however, are not convinced that the nuclear matrix exists in the living cell. They believe that the matrix observed after extraction of nuclei may be the result of artificial aggregation of proteins and nucleic acids during preparation. Although the nuclear matrix is visualized after extraction of nuclei by several different methods, its molecular composition has not been clearly defined. In the absence of definitive characterization of its structural components, the nuclear matrix remains an area of dispute among cell biologists.

The Nucleolus

The most prominent substructure within the nucleus is the **nucleolus** (see Figure 8.1), which is the site of rRNA transcription and processing, and of ribosome assembly. As discussed in the preceding chapter, cells require large numbers of ribosomes to meet their needs for protein synthesis. Actively growing mammalian cells, for example, contain 5 million to 10 million ribosomes that must be synthesized each time the cell divides. The nucleolus is a ribosome production factory, designed to fulfill the need for large-scale production of rRNAs and assembly of the ribosomal subunits.

Ribosomal RNA Genes and the Organization of the Nucleolus

The nucleolus, which is not surrounded by a membrane, is organized around the chromosomal regions that contain the genes for the 5.8S, 18S, and 28S rRNAs. Eukaryotic ribosomes contain four types of RNA, designated the

Figure 8.22
Ribosomal RNA genes.   Each rRNA gene is a single transcription unit containing the 18S, 5.8S, and 28S rRNAs and transcribed spacer sequences. The rRNA genes are organized in tandem arrays, separated by nontranscribed spacer DNA.

5S, 5.8S, 18S, and 28S rRNAs (see Figure 7.4). The 5.8S, 18S, and 28S rRNAs are transcribed as a single unit within the nucleolus by RNA polymerase I, yielding a 45S ribosomal precursor RNA (Figure 8.22). The 45S pre-rRNA is processed to the 18S rRNA of the 40S (small) ribosomal subunit and to the 5.8S and 28S rRNAs of the 60S (large) ribosomal subunit. Transcription of the 5S rRNA, which is also found in the 60S ribosomal subunit, takes place outside of the nucleolus and is catalyzed by RNA polymerase III.

To meet the need for transcription of large numbers of rRNA molecules, all cells contain multiple copies of the rRNA genes. The human genome, for example, contains about 200 copies of the gene that encodes the 5.8S, 18S, and 28S rRNAs and approximately 2000 copies of the gene that encodes 5S rRNA. The genes for 5.8S, 18S, and 28S rRNAs are clustered in tandem arrays on five different human chromosomes (chromosomes 13, 14, 15, 21, and 22); the 5S rRNA genes are present in a single tandem array on chromosome 1.

The importance of ribosome production is particularly evident in oocytes, in which the rRNA genes are amplified to support the synthesis of the large numbers of ribosomes required for early embryonic development. In *Xenopus* oocytes, the rRNA genes are amplified approximately 2000-fold, resulting in about 1 million copies per cell. These amplified rRNA genes are distributed to thousands of nucleoli (Figure 8.23), which support the accumulation of nearly 10^{12} ribosomes per oocyte.

Morphologically, nucleoli consist of three distinguishable regions: the fibrillar center, dense fibrillar component, and granular component (Figure 8.24). These different regions are thought to represent the sites of progressive stages of rRNA transcription, processing, and ribosome assembly. The rRNA genes are located in the fibrillar centers, with transcription occurring primarily at the boundary of the fibrillar centers and dense fibrillar component. Processing of the pre-rRNA is initiated in the dense fibrillar component and continues in the granular component, where the rRNA is assembled with ribosomal proteins to form nearly completed preribosomal subunits, ready for export to the cytoplasm.

Following each cell division, nucleoli form around the chromosomal regions that contain the 5.8S, 18S, and 28S rRNA genes, which are therefore called **nucleolar organizing regions**. The formation of nucleoli requires the transcription of 45S pre-rRNA, which appears to lead to the fusion of small prenucleolar bodies that contain processing factors and other components of the nucleolus. In most cells, the initially separate nucleoli then fuse to form a single nucleolus. The size of the nucleolus depends on the metabolic activ-

Figure 8.23
Nucleoli in amphibian oocytes.   The amplified rRNA genes of *Xenopus* oocytes are clustered in multiple nucleoli (darkly stained spots). (From D. D. Brown and I. B. Dawid, 1968. *Science* 160: 272.)

Figure 8.24
Structure of the nucleolus   An electron micrograph illustrating the fibrillar center (FC), dense fibrillar component (DFC), and granular component (G) of a nucleolus. (Courtesy of David L. Spector, Cold Spring Harbor Laboratory.)

3 μm

ity of the cell, with large nucleoli found in cells that are actively engaged in protein synthesis. This variation is due primarily to differences in the size of the granular component, reflecting the levels of ribosome synthesis.

Transcription and Processing of rRNA

Each nucleolar organizing region contains a cluster of tandemly repeated rRNA genes that are separated from each other by nontranscribed spacer DNA. These genes are very actively transcribed by RNA polymerase I, allowing their transcription to be readily visualized by electron microscopy (Figure 8.25). In such electron micrographs, each of the tandemly arrayed rRNA genes is surrounded by densely packed growing RNA chains, forming a structure that looks like a Christmas tree. The high density of growing RNA chains reflects that of RNA polymerase molecules, which are present at a maximal density of approximately one polymerase per hundred base pairs of template DNA.

The primary transcript of the rRNA genes is the large 45S pre-rRNA, which contains the 18S, 5.8S, and 28S rRNAs as well as transcribed spacer regions (Figure 8.26). External transcribed spacers are present at both the 5′ and 3′ ends of the pre-rRNAs, and two internal transcribed spacers lie between the 18S, 5.8S, and 28S rRNA sequences. The initial processing step is a cleavage within the external transcribed spacer near the 5′ end of the pre-rRNA, which takes place during the early stages of transcription. This cleavage requires the U3 small nucleolar RNP (see below) that remains attached to the 5′ end of the pre-rRNA, forming the characteristic knobs seen in Figure 8.25. Once transcription is complete, the external transcribed spacer at the 3′ end of the molecule is removed. In human cells, this step is followed by a cleavage at the 5′ end of the 5.8S region, yielding separate

Figure 8.25
Transcription of rRNA genes   An electron micrograph of nucleolar chromatin, showing three rRNA genes separated by nontranscribed spacer DNA. Each rRNA gene is surrounded by an array of growing RNA chains, resulting in a Christmas tree appearance. (Courtesy of O. L. Miller, Jr.)

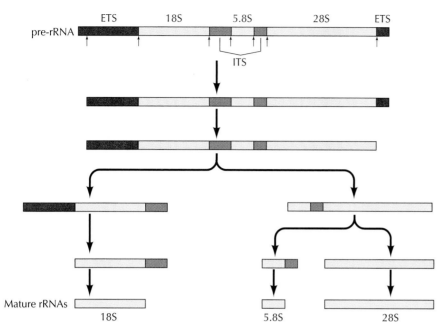

ETS 18S 5.8S 28S ETS
pre-rRNA
ITS

Mature rRNAs

18S   5.8S   28S

Figure 8.26
Processing of pre-rRNA The 45S pre-rRNA transcript contains external transcribed spacers (ETS) at both ends and internal transcribed spacers (ITS) between the sequences of 18S, 5.8S, and 28S rRNAs. The pre-rRNA is processed via a series of cleavages (illustrated for human pre-rRNA) to yield the mature rRNA species.

precursors to the 18S and 5.8S + 28S rRNAs. Additional cleavages then result in formation of the mature rRNAs. Processing follows a similar pattern in other species, although there are differences in the order of some of the cleavages.

In addition to cleavage, the processing of pre-rRNA involves a substantial amount of base modification resulting both from the addition of methyl groups to specific bases and ribose residues and from the conversion of uridine to pseudouridine (see Figure 6.38). In animal cells, pre-rRNA processing involves the methylation of approximately a hundred ribose residues and ten bases, in addition to the formation of about a hundred pseudouridines. Most of these modifications occur during or shortly after synthesis of the pre-rRNA, although a few take place at later stages of pre-rRNA processing.

The processing of pre-rRNA requires the action of both proteins and RNAs that are localized to the nucleolus. The involvement of small nuclear RNAs (snRNAs) in pre-mRNA splicing was discussed in Chapter 6. Nucleoli contain a large number (about 200) of **small nucleolar RNAs (snoRNAs)** that function in pre-rRNA processing. Like the spliceosomal snRNAs, the snoRNAs are complexed with proteins, forming snoRNPs. Individual snoRNPs consist of single snoRNAs associated with eight to ten proteins. The snoRNPs then assemble on the pre-rRNA to form processing complexes in a manner analogous to the formation of spliceosomes on pre-mRNA.

Some snoRNAs are responsible for the cleavages of pre-rRNA into 18S, 5.8S, and 28S products. For example, the most abundant nucleolar snoRNA is U3, which is present in about 200,000 copies per cell. As already noted, U3 is required for the initial cleavage of pre-rRNA within the 5′ external transcribed spacer sequences. Similarly, U8 snoRNA is responsible for cleavage of pre-rRNA to 5.8S and 28S rRNAs, and U22 snoRNA is responsible for cleavage of pre-rRNA to 18S rRNA.

The majority of snoRNAs, however, function to direct the specific base modifications of pre-rRNA, including the methylation of specific ribose residues and the formation of pseudouridines (Figure 8.27). Most of the

Figure 8.27
Role of snoRNAs in base modification of pre-rRNA The snoRNAs contain short sequences complementary to rRNA. Base pairing between snoRNAs and pre-rRNA targets the enzymes that catalyze base modification (e.g., methylation) to the appropriate sites on pre-rRNA.

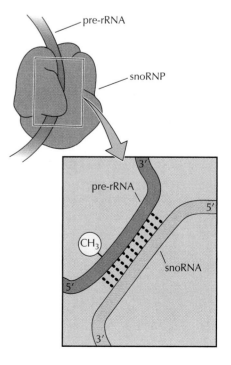

pre-rRNA
snoRNP
pre-rRNA
CH₃
snoRNA

snoRNAs contain short sequences of approximately 15 nucleotides that are complementary to 18S or 28S rRNA. Importantly, these regions of complementarity include the sites of base modification in the rRNA. By base pairing with specific regions of the pre-rRNA, the snoRNAs act as guide RNAs that target the enzymes responsible for ribose methylation or pseudouridylation to the correct site on the pre-rRNA molecule.

Ribosome Assembly

The formation of ribosomes involves the assembly of the ribosomal precursor RNA with both ribosomal proteins and 5S rRNA (Figure 8.28). The genes that encode ribosomal proteins are transcribed outside of the nucleolus by RNA polymerase II, yielding mRNAs that are translated on cytoplasmic ribosomes. The ribosomal proteins are then transported from the cytoplasm to the nucleolus, where they are assembled with rRNAs to form preribosomal particles. Although the genes for 5S rRNA are also transcribed outside of the nucleolus, in this case by RNA polymerase III, 5S rRNAs similarly are assembled into preribosomal particles within the nucleolus.

The association of ribosomal proteins with rRNA begins while the pre-rRNA is still being synthesized, and more than half of the ribosomal proteins are complexed with the pre-rRNA prior to its cleavage. The remaining ribosomal proteins and the 5S rRNA are incorporated into preribosomal particles as cleavage of the pre-rRNA proceeds. The smaller ribosomal subunit, which contains only the 18S rRNA, matures more rapidly than the larger subunit, which contains 28S, 5.8S, and 5S rRNAs. Consequently, most of the preribosomal particles in the nucleolus represent precursors to the large subunit. The final stages of ribosome maturation follow the export of preribosomal particles to the cytoplasm, forming the active 40S and 60S subunits of eukaryotic ribosomes.

Figure 8.28
Ribosome assembly Ribosomal proteins are imported to the nucleolus from the cytoplasm and begin to assemble on pre-rRNA prior to its cleavage. As the pre-rRNA is processed, additional ribosomal proteins and the 5S rRNA (which is synthesized elsewhere in the nucleus) assemble to form preribosomal particles. The final steps of maturation follow the export of preribosomal particles to the cytoplasm, yielding the 40S and 60S ribosomal subunits.

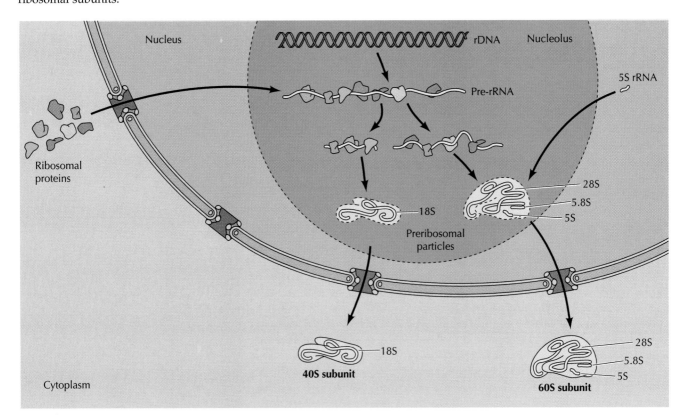

The Nucleus during Mitosis

A unique feature of the nucleus is that it disassembles and re-forms each time most cells divide. At the beginning of mitosis, the chromosomes condense, the nucleolus disappears, and the nuclear envelope breaks down, resulting in the release of most of the contents of the nucleus into the cytoplasm. At the end of mitosis, the process is reversed: The chromosomes decondense, and nuclear envelopes re-form around the separated sets of daughter chromosomes. Chapter 14 presents a comprehensive discussion of mitosis; in this section we will consider the mechanisms involved in the disassembly and re-formation of the nucleus. The process is controlled largely by reversible phosphorylation and dephosphorylation of nuclear proteins resulting from the action of the **Cdc2** protein kinase, which is a critical regulator of mitosis in all eukaryotic cells.

Dissolution of the Nuclear Envelope

In most cells, the disassembly of the nuclear envelope marks the end of the prophase of mitosis (Figure 8.29). However, this disassembly of the nucleus is not a universal feature of mitosis and does not occur in all cells. Some unicellular eukaryotes (e.g., yeasts) undergo so-called closed mitosis, in which the nuclear envelope remains intact (Figure 8.30). In closed mitosis, the daughter chromosomes migrate to opposite poles of the nucleus, which then divides in two. The cells of higher eukaryotes, however, usually undergo open mitosis, which is characterized by breakdown of the nuclear

Figure 8.29
The nucleus during mitosis
Micrographs illustrating the progressive stages of mitosis in a plant cell. During prophase, the chromosomes condense, the nucleolus disappears, and the nuclear envelope breaks down. At metaphase, the condensed chromosomes align on the center of the spindle. The daughter chromosomes then move to opposite poles of the spindle (anaphase), and during telophase the chromosomes decondense and the nuclei re-form. Chromosomes are stained blue and spindle microtubules are stained red. (Courtesy of Andrew S. Bajer, University of Oregon.)

Prophase

Prometaphase

Metaphase

Anaphase

Telophase

Figure 8.30
Closed and open mitosis In closed mitosis, the nuclear envelope remains intact and chromosomes migrate to opposite poles of a spindle within the nucleus. In open mitosis, the nuclear envelope breaks down and then reforms around the two sets of separated chromosomes.

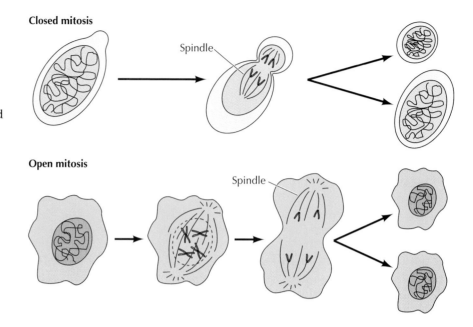

Closed mitosis

Spindle

Open mitosis

Spindle

Figure 8.31
Dissolution of the nuclear lamina The nuclear lamina consists of a meshwork of lamin filaments. At mitosis, Cdc2 and other protein kinases phosphorylate the lamins, causing the filaments to dissociate into free lamin dimers.

envelope. The daughter chromosomes then migrate to opposite poles of the mitotic spindle, and new nuclei reassemble around them.

Disassembly of the nuclear envelope, which parallels a similar breakdown of the endoplasmic reticulum, involves changes in all three of its components: The nuclear membranes are fragmented into vesicles, the nuclear pore complexes dissociate, and the nuclear lamina depolymerizes. The best understood of these events is depolymerization of the nuclear lamina—the meshwork of filaments underlying the nuclear membrane. The nuclear lamina is composed of fibrous proteins, lamins, which associate with each other to form filaments. Disassembly of the nuclear lamina results from phosphorylation of the lamins, which causes the filaments to break down into individual lamin dimers (Figure 8.31). Phosphorylation of the lamins is catalyzed by the Cdc2 protein kinase, which was introduced

Lamin filament

Phosphorylation

Lamin dimers

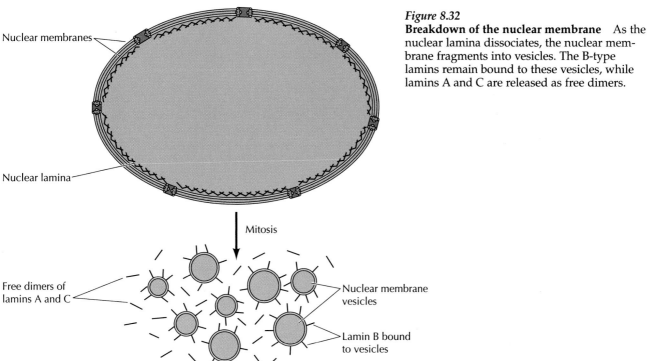

Figure 8.32
Breakdown of the nuclear membrane   As the nuclear lamina dissociates, the nuclear membrane fragments into vesicles. The B-type lamins remain bound to these vesicles, while lamins A and C are released as free dimers.

in Chapter 7 (see Figure 7.40) and will be discussed in detail in Chapter 14 as a central regulator of mitosis. Cdc2 (as well as other protein kinases activated in mitotic cells) phosphorylates all the different types of lamins, and treatment of isolated nuclei with Cdc2 has been shown to be sufficient to induce depolymerization of the nuclear lamina. Moreover, the requirement for lamin phosphorylation in the breakdown of the nuclear lamina has been demonstrated directly by the construction of mutant lamins that can no longer be phosphorylated. When genes encoding these mutant lamins were introduced into cells, their expression was found to block normal breakdown of the nuclear lamina as the cells entered mitosis.

In concert with dissolution of the nuclear lamina, the nuclear membrane fragments into vesicles (Figure 8.32). The B-type lamins remain associated with these vesicles, but lamins A and C dissociate from the nuclear membrane and are released as free dimers in the cytosol. This difference arises because the B-type lamins are permanently modified by the addition of lipid (prenyl groups), whereas the C-terminal prenyl groups of A- and C-type lamins are removed by proteolysis following their incorporation into the lamina. The nuclear pore complexes also dissociate into subunits as a result of phosphorylation of several nuclear pore proteins. Integral nuclear membrane proteins are also phosphorylated at mitosis, and phosphorylation of these proteins may be important in vesicle formation as well as in dissociation of the nuclear membrane from both chromosomes and the nuclear lamina.

Chromosome Condensation

The other major change in nuclear structure during mitosis is chromosome condensation. The interphase chromatin, which is already packaged into nucleosomes, condenses approximately a thousandfold further to form the compact chromosomes seen in mitotic cells (Figure 8.33). This condensation is needed to allow the chromosomes to move along the mitotic spindle without becoming tangled or broken during their distribution to daughter

2 μm

Figure 8.33
Chromosome condensation   Electron micrograph showing the condensation of individual chromosomes during the prophase of mitosis. (K. G. Murti/Visuals Unlimited.)

cells. DNA in this highly condensed state can no longer be transcribed, so all RNA synthesis stops during mitosis. As the chromosomes condense and transcription ceases, the nucleolus also disappears.

The condensed DNA in metaphase chromosomes appears to be organized into large loops, each encompassing about a hundred kilobases of DNA, which are attached to a protein scaffold (see Figure 4.13). Despite its fundamental importance, the mechanism of chromosome condensation during mitosis is not understood. The basic unit of chromatin structure is the nucleosome, which consists of 146 base pairs of DNA wrapped around a histone core containing two molecules each of histones H2A, H2B, H3, and H4 (see Figure 4.8). One molecule of histone H1 is bound to the DNA as it enters each nucleosome core particle, and interactions between these H1 molecules are involved in the folding of chromatin into higher-order, more compact structures. Histone H1 is a substrate for the Cdc2 protein kinase and is phosphorylated during mitosis of most cells, consistent with its phosphorylation playing a role in mitotic chromosome condensation. However, recent experiments have shown that phosphorylation of histone H1 is not required for chromosome condensation, so the potential role of H1 phosphorylation is unclear. In contrast, phosphorylation of histone H3 has been found to be required for condensation of mitotic chromosomes, although the mechanism by which H3 phosphorylation affects chromosome condensation remains to be elucidated.

Recent studies have also identified protein complexes called condensins that play a major role in chromosome condensation. Condensins are required for chromosome condensation in extracts of mitotic cells and appear to function by wrapping DNA around itself, thereby compacting chromosomes into the condensed mitotic structure. Condensins are phosphorylated and activated by the Cdc2 protein kinase, providing a direct link between activation of Cdc2 and mitotic chromosome condensation.

Re-formation of the Interphase Nucleus

During the completion of mitosis (telophase), two new nuclei form around the separated sets of daughter chromosomes (see Figure 8.29). Chromosome decondensation and reassembly of the nuclear envelope appear to be signaled by inactivation of Cdc2, which was responsible for initiating mitosis by phosphorylating cellular target proteins, including the lamins, histone H3, and condensins. The progression from metaphase to anaphase involves the activation of a ubiquitin-mediated proteolysis system that inactivates Cdc2 by degrading its regulatory subunit, cyclin B (see Figure 7.40). Inactivation of Cdc2 leads to the dephosphorylation of the proteins that were phosphorylated at the initiation of mitosis, resulting in exit from mitosis and the re-formation of interphase nuclei.

The initial step in re-formation of the nuclear envelope is the binding of the vesicles formed during nuclear membrane breakdown to the surface of chromosomes (Figure 8.34). This interaction of membrane vesicles with chromosomes may be mediated by both lamins and integral membrane

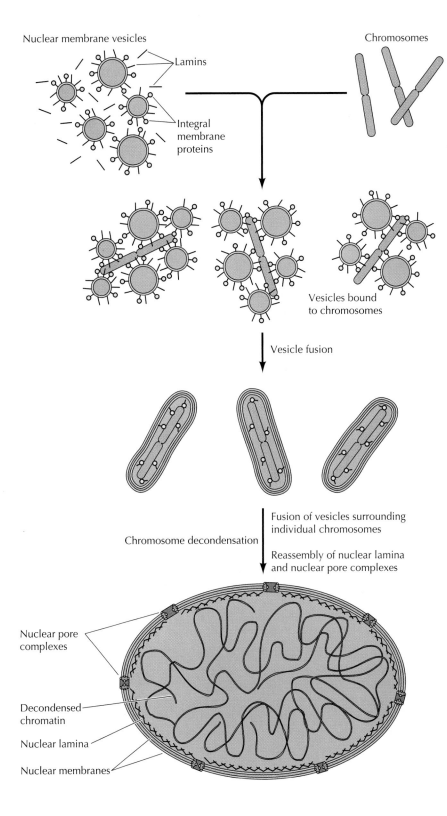

Figure 8.34
Re-formation of the nuclear envelope
The first step in reassembly of the nuclear envelope is the binding of membrane vesicles to chromosomes, which may be mediated by both integral membrane proteins and B-type lamins. The vesicles then fuse, the nuclear lamina reassembles, and the chromosomes decondense.

proteins of the inner nuclear membrane. The vesicles then fuse to form a double membrane around the chromosomes. This is followed by reassembly of the nuclear pore complexes, re-formation of the nuclear lamina, and chromosome decondensation. The vesicles first fuse to form membranes around individual chromosomes, which then fuse with each other to form a complete single nucleus.

The initial re-formation of the nuclear envelope around condensed chromosomes excludes cytoplasmic molecules from the newly assembled nucleus. The new nucleus is then able to expand via the selective import of nuclear proteins from the cytoplasm. Because nuclear localization signals are not cleaved from proteins that are imported to the nucleus, the same nuclear proteins that were released into the cytoplasm following disassembly of the nuclear envelope at the beginning of mitosis can be reimported into the new nuclei formed after mitosis. The nucleolus, too, re-forms as the chromosomes decondense and transcription of the rRNA genes begins, completing the return from mitosis to an interphase nucleus.

KEY TERMS

nuclear envelope, nuclear membrane, nuclear lamina, lamin

nuclear pore complex

nuclear localization signal, importin, Ran, nuclear export signal, exportin

heterogeneous nuclear ribonucleoproteins (hnRNPs)

heterochromatin, euchromatin, X chromosome inactivation, locus control region, insulator element

SUMMARY

THE NUCLEAR ENVELOPE AND TRAFFIC BETWEEN THE NUCLEUS AND CYTOPLASM

Structure of the Nuclear Envelope: The nuclear envelope separates the contents of the nucleus from the cytoplasm, maintaining the nucleus as a distinct biochemical compartment that houses the genetic material and serves as the site of transcription and RNA processing in eukaryotic cells. The nuclear envelope consists of the inner and outer nuclear membranes, which are joined at nuclear pore complexes, and an underlying nuclear lamina.

The Nuclear Pore Complex: Nuclear pore complexes are large structures that provide the only routes through which molecules can travel between the nucleus and cytoplasm. Small molecules are able to diffuse freely through open channels in the nuclear pore complex. Macromolecules are selectively transported in an energy-dependent process.

Selective Transport of Proteins to and from the Nucleus: Proteins destined for import to the nucleus contain nuclear localization signals that are recognized by receptors that direct transport through the nuclear pore complex. Proteins that shuttle back and forth between the nucleus and the cytoplasm contain nuclear export signals that target them for transport from the nucleus to the cytoplasm. The small GTP-binding protein Ran is required for translocation through the nuclear pore complex and determines the directionality of transport.

Regulation of Nuclear Protein Import: The activity of some proteins, such as transcription factors, is controlled by regulation of their import to the nucleus.

Transport of RNAs: RNAs are transported through the nuclear pore complex as ribonucleoprotein complexes. Messenger RNAs, ribosomal RNAs, and transfer RNAs are exported from the nucleus to function in protein synthesis. Small nuclear RNAs are initially transported from the nucleus to the cytoplasm, where they associate with proteins to form snRNPs; then they return to the nucleus.

INTERNAL ORGANIZATION OF THE NUCLEUS

Chromosomes and Higher-Order Chromatin Structure: The interphase nucleus contains transcriptionally inactive, highly condensed heterochromatin as well as decondensed euchromatin. Interphase chromosomes are organized within the nucleus and divided into large looped domains that function as independent units.

Functional Domains within the Nucleus: Some nuclear processes, such as DNA replication and pre-mRNA metabolism, may be localized to discrete subnuclear structures or domains.

THE NUCLEOLUS

Ribosomal RNA Genes and the Organization of the Nucleolus: The nucleolus is organized around the genes for ribosomal RNAs. It is the site of rRNA transcription and processing, and of ribosome assembly.

nucleolus, nucleolar organizing region

Transcription and Processing of rRNA: The primary transcript of the rRNA genes is 45S pre-rRNA, which is processed to yield 18S, 5.8S, and 28S rRNAs. Processing of pre-rRNA is mediated by small nucleolar RNAs (snoRNAs).

small nucleolar RNAs (snoRNAs)

Ribosome Assembly: Ribosomal subunits are assembled within the nucleolus from rRNAs and ribosomal proteins.

THE NUCLEUS DURING MITOSIS

Dissolution of the Nuclear Envelope: Entry into mitosis is signaled by activation of the Cdc2 protein kinase. In most cells, the nuclear envelope breaks down at the end of prophase. Depolymerization of the nuclear lamina results from phosphorylation of the lamins by Cdc2 and other protein kinases.

Cdc2

Chromosome Condensation: Phosphorylation of histones H1 and H3 is correlated with the condensation of mitotic chromosomes, and H3 phosphorylation is required for proper chromosome condensation. A complex of proteins called condensin is activated by Cdc2 phosphorylation and functions in chromosome condensation by wrapping the DNA into a compact structure.

Re-formation of the Interphase Nucleus: Inactivation of Cdc2 at the end of mitosis leads to re-formation of the nuclear envelope and chromosome decondensation. Nuclear proteins are then selectively imported through nuclear pore complexes.

Questions

1. You are studying a transcription factor that is regulated by phosphorylation of serine residues that inactivate its nuclear localization signal. How would mutating these serines to alanines affect subcellular localization of the transcription factor and expression of its target gene?

2. How would mutational inactivation of the nuclear export signal of a protein that normally shuttles back and forth between the nucleus and cytoplasm affect its subcellular distribution?

3. You have constructed a mutant importin β that no longer binds to Ran. What effect would expression of this mutant importin have on the traffic of proteins between the nucleus and the cytoplasm?

4. Consider the construct (diagrammed below) in which a single enhancer (E) acts on two separate promoters (P1 and P2). How would insertion of an insulator between E and P1 affect transcription from P1 and P2? What about insertion of an insulator between P1 and P2?

——E——— P1——— P2———

5. Treatment of cells with actinomycin D (a general inhibitor of transcription) prevents the formation of nucleoli following mitosis. Treatment with α-amanitin (a specific inhibitor of RNA polymerase II) does not have this effect. Why?

References and Further Reading

The Nuclear Envelope and Traffic between the Nucleus and Cytoplasm

Akey, C. W. and M. Radermacher. 1993. Architecture of the *Xenopus* nuclear pore complex revealed by three-dimensional cryo-electron microscopy. *J. Cell Biol.* 122: 1–19. [P]

Chook, Y. M. and G. Blobel. 1999. Structure of the nuclear transport complex karyopherin-β2-Ran·GppNHp. *Nature* 399: 230–237. [P]

Cingolani, G. C. Petosa, K. Weis and C. W. Muller. 1999. Structure of importin-β bound to the IBB domain of importin-α. *Nature* 399: 221–229. [P]

Daneholt, B. 1997. A look at messenger RNP moving through the nuclear pore. *Cell* 88: 585–588. [R]

Dingwall, C. and R. A. Laskey. 1991. Nuclear targeting sequences—a consensus? *Trends Biochem. Sci.* 16: 478–481. [R]

Franke, W. W. 1987. Nuclear lamins and cytoplasmic intermediate filament proteins: A growing multigene family. *Cell* 48: 3–4. [R]

Gant, T. M. and K. L. Wilson. 1997. Nuclear assembly. *Ann. Rev. Cell Biol.* 13: 669–695. [R]

Gerace, L. and B. Burke. 1988. Functional organization of the nuclear envelope. *Ann. Rev. Cell Biol.* 4: 353–374. [R]

Gerace, L. and R. Foisner. 1994. Integral membrane proteins and dynamic organization of the nuclear envelope. *Trends Cell Biol.* 4: 127–131. [R]

Hinshaw, J. E., B. O. Carragher and R. A. Milligan. 1992. Architecture and design of the nuclear pore complex. *Cell* 69: 1133–1141. [P]

Kalderon, D., B. L. Roberts, W. D. Richardson and A. E. Smith. 1984. A short amino acid sequence able to specify nuclear location. *Cell* 39: 499–509. [P]

Lanford, R. E. and J. S. Butel. 1984. Construction and characterization of an SV40 mutant defective in nuclear transport of T antigen. *Cell* 37: 801–813. [P]

Mattaj, I. W. and L. Englmeier. 1998. Nucleocytoplasmic transport: The soluble phase. *Ann. Rev. Biochem.* 67: 265–306. [R]

Moore, M. S. 1998. Ran and nuclear transport. *J. Biol. Chem.* 273: 22857–22860. [R]

Nigg, E. A. 1997. Nucleocytoplasmic transport: Signals, mechanisms and regulation. *Nature* 386: 779–786. [R]

Ohno, M., M. Fornerod and I. W. Mattaj. 1998. Nucleocytoplasmic transport: The last 200 nanometers. *Cell* 92: 327–336. [R]

Pante, N. and U. Aebi. 1993. The nuclear pore complex. *J. Cell Biol.* 122: 977–984. [R]

Talcott, B. and M. S. Moore. 1999. Getting across the nuclear pore complex. *Trends Cell Biol.* 9: 312–318. [R]

Ullman, K. S., M. A. Powers and D. J. Forbes. 1997. Nuclear export receptors: From importin to exportin. *Cell* 90: 967–970. [R]

Vandromme, M., C. Gauthier-Rouviere, N. Lamb and A. Fernandez. 1996. Regulation of transcription factor localization: Fine-tuning of gene expression. *Trends Biochem. Sci.* 21: 59–64. [R]

Vetter, I. R., A. Arndt, U. Kutay, D. Gorlich and A. Wittinghofer. 1999. Structural view of the Ran-importin β interaction at 2.3 Å resolution. *Cell* 67: 635–646. [P]

Weis, K. 1998. Importins and exportins: How to get in and out of the nucleus. *Trends Biochem. Sci.* 23: 185–189. [R]

Wozniak, R. W., M. P. Rout and J. D. Aitchison. 1998. Karyopherins and kissing cousins. *Trends Cell Biol.* 8: 184–188. [R]

Internal Organization of the Nucleus

Berezney, R. and D. S. Coffey. 1975. Nuclear protein matrix: Association with newly synthesized DNA. *Science* 189: 291–293. [P]

Corces, V. G. 1995. Chromatin insulators: Keeping enhancers under control. *Nature* 376: 462–463. [R]

Dillon, N. and F. Grosveld. 1993. Transcriptional regulation of multigene loci: Multilevel control. *Trends Genet.* 9: 134–137. [R]

Gasser, S. M. and U. K. Laemmli. 1987. A glimpse at chromosomal order. *Trends Genet.* 3: 16–22. [R]

Gilson, E., T. Laroche and S. M. Gasser. 1993. Telomeres and the functional architecture of the nucleus. *Trends Cell Biol.* 3: 128–134. [R]

Hozak, P. and P. R. Cook. 1994. Replication factories. *Trends Cell Biol.* 4: 48–49. [R]

Lamond, A. I. And W. C. Earnshaw. 1998. Structure and function in the nucleus. *Science* 280: 547–553. [R]

Lyon, M. F. 1992. Some milestones in the history of X-chromosome inactivation. *Ann. Rev. Genet.* 26: 17–28. [R]

Manuelidis, L. 1990. A view of interphase chromosomes. *Science* 250: 1533–1540. [R]

Matera, A. G. 1999. Nuclear bodies: Multifaceted subdomains of the interchromatin space. *Trends Cell Biol.* 9: 302–309. [R]

Mathog, D., M. Hochstrasser, Y. Gruenbaum, H. Saumweber and J. Sedat. 1984. Characteristic folding pattern of polytene chromosomes in *Drosophila* salivary gland nuclei. *Nature* 308: 414–421. [P]

Misteli, T., J. F. Caceres and D. L. Spector. 1997. The dynamics of a pre-mRNA splicing factor in living cells. *Nature* 387: 523–527. [P]

Misteli, T. and D. L. Spector. 1997. Protein phosphorylation and the nuclear organization of pre-mRNA splicing. *Trends Cell Biol.* 7: 135–138. [R]

Panning, B. and R. Jaenisch. 1998. RNA and the epigenetic regulation of X chromosome inactivation. *Cell* 93: 305–308. [R]

Pederson, T. 1998. Thinking about a nuclear matrix. *J. Mol. Biol.* 27: 147–159. [R]

Spector, D. L. 1993. Macromolecular domains within the cell nucleus. *Ann. Rev. Cell Biol.* 9: 265–315. [R]

The Nucleolus

Bachellerie, J.-P. and J. Cavaille. 1997. Guiding ribose methylation of rRNA. *Trends Biochem. Sci.* 22: 257–261. [R]

Maxwell, E. S. and M. J. Fournier. 1995. The small nucleolar RNAs. *Ann. Rev. Biochem.* 35: 897–934. [R]

Miller, O. L., Jr. and B. Beatty. 1969. Visualization of nucleolar genes. *Science* 164: 955–957. [P]

Scheer, U., M. Thiry and G. Goessens. 1993. Structure, function and assembly of the nucleolus. *Trends Cell Biol.* 3: 236–241. [R]

Smith, C. M. and J. A. Steitz. 1997. Sno storm in the nucleolus: New roles for myriad small RNPs. *Cell* 89: 669–672. [R]

Sollner-Webb, B. and E. B. Mougey. 1991. News from the nucleolus: rRNA gene expression. *Trends Biochem. Sci.* 16: 58–62. [R]

Sommerville, J. 1986. Nucleolar structure and ribosome biogenesis. *Trends Biochem. Sci.* 11: 438–442. [R]

The Nucleus during Mitosis

Gant, T. M. and K. L. Wilson. 1997. Nuclear assembly. *Ann. Rev. Cell Biol.* 13: 669–695. [R]

Heald, R. and F. McKeon. 1990. Mutations of phosphorylation sites in lamin A that prevent nuclear lamina disassembly in mitosis. *Cell* 61: 579–589. [P]

Kimura, K., M. Hirano, R. Kobayashi and T. Hirano. 1998. Phosphorylation and activation of 13S condensin by Cdc2 *in vitro*. *Science* 282: 487–490. [P]

Koshland, D. and A. Strunnikov. 1996. Mitotic chromosome condensation. *Ann. Rev. Cell Biol.* 12: 305–333. [R]

Marshall, I. C. B. and K. L. Wilson. 1997. Nuclear envelope assembly after mitosis. *Trends Cell Biol.* 7: 69–74. [R]

Murray, A. and T. Hunt. 1993. *The Cell Cycle: An Introduction.* New York: W. H. Freeman.

Murray, A. W. 1998. How to compact DNA. *Science* 282: 425–427. [R]

Peter, M., J. Nakagawa, M. Doreé, J. C. Labbé and E. A. Nigg 1990. *In vitro* disassembly of the nuclear lamina and M phase-specific phosphorylation of lamins by cdc2 kinase. *Cell* 61: 591–602. [P]

Ward, G. E. and M. W. Kirschner. 1990. Identification of cell cycle-regulated phosphorylation sites on nuclear lamin C. *Cell* 61: 561–577. [P]

Wei, Y., L. Yu, J. Brown, M. A. Gorovsky and C. D. Allis. 1999. Phosphorylation of histone H3 is required for proper chromosome condensation and segregation. *Cell* 97: 99–109. [P]

9

Protein Sorting and Transport

The Endoplasmic Reticulum, Golgi Apparatus, and Lysosomes

IN ADDITION TO THE PRESENCE OF A NUCLEUS, eukaryotic cells are distinguished from prokaryotic cells by the presence of membrane-enclosed organelles within their cytoplasm. These organelles provide discrete compartments in which specific cellular activities take place, and the resulting subdivision of the cytoplasm allows eukaryotic cells to function efficiently in spite of their large size (about a thousand times the volume of bacteria).

Because of the complex internal organization of eukaryotic cells, the sorting and targeting of proteins to their appropriate destinations are considerable tasks. The first step of protein sorting takes place while translation is still in progress. Many proteins destined for the endoplasmic reticulum, the Golgi apparatus, lysosomes, the plasma membrane, and secretion from the cell are synthesized on ribosomes that are bound to the membrane of the endoplasmic reticulum. As translation proceeds, the polypeptide chains are transported into the endoplasmic reticulum, where protein folding and processing take place. From the endoplasmic reticulum, proteins are transported in vesicles to the Golgi apparatus, where they are further processed and sorted for transport to lysosomes, the plasma membrane, or secretion from the cell. The endoplasmic reticulum, Golgi apparatus, and lysosomes are thus distinguished from other cytoplasmic organelles by their common involvement in protein processing and connection by vesicular transport.

The Endoplasmic Reticulum

The **endoplasmic reticulum** (**ER**) is a network of membrane-enclosed tubules and sacs (cisternae) that extends from the nuclear membrane throughout the cytoplasm (Figure 9.1). The entire endoplasmic reticulum is enclosed by a continuous membrane and is the largest organelle of most eukaryotic cells. Its membrane may account for about half of all cell membranes, and the space enclosed by the ER (the lumen, or cisternal space) may represent about 10% of the total cell volume. As discussed below, there are two distinct types of ER that perform different functions within the cell.

347

(A) Rough endoplasmic reticulum

(B) Smooth endoplasmic reticulum

Figure 9.1
The endoplasmic reticulum (ER)
(A) Electron micrograph of rough ER in rat liver cells. Ribosomes are attached to the cytosolic face of the ER membrane. (B) Electron micrograph of smooth ER in Leydig cells of the testis, which are active in steroid hormone synthesis. (A, Richard Rodewald, University of Virginia/Biological Photo Service; B, Dan Fawcett/Photo Researchers, Inc.)

The **rough ER**, which is covered by ribosomes on its outer surface, functions in protein processing. The **smooth ER** is not associated with ribosomes and is involved in lipid, rather than protein, metabolism.

The Endoplasmic Reticulum and Protein Secretion

The role of the endoplasmic reticulum in protein processing and sorting was first demonstrated by George Palade and his colleagues in the 1960s (Figure 9.2). These investigators studied the fate of newly synthesized proteins in specialized cells of the pancreas (pancreatic acinar cells) that secrete digestive enzymes into the small intestine. Because most proteins synthesized by these cells are secreted, Palade and coworkers were able to study

Figure 9.2
The secretory pathway   Pancreatic acinar cells, which secrete most of their newly synthesized proteins into the digestive tract, were labeled with radioactive amino acids to study the intracellular pathway taken by secreted proteins. After a short incubation with radioactive amino acids (3-minute label), autoradiography revealed that newly synthesized proteins were localized to the rough ER. Following further incubation with nonradioactive amino acids (a chase), proteins were found to move from the ER to the Golgi apparatus and then, within secretory vesicles, to the plasma membrane and cell exterior.

Radiolabeled protein

Secretory vesicles

Golgi apparatus

Rough endoplasmic reticulum

3-minute label

7-minute chase

120-minute chase

the pathway taken by secreted proteins simply by labeling newly synthesized proteins with radioactive amino acids. The location of the radiolabeled proteins within the cell was then determined by autoradiography, revealing the cellular sites involved in the events leading to protein secretion. After a brief exposure of pancreatic acinar cells to radioactive amino acids, newly synthesized proteins were detected in the rough ER, which was therefore identified as the site of synthesis of proteins destined for secretion. If the cells were then incubated for a short time in media containing nonradioactive amino acids (a process known as a chase), the radiolabeled proteins were detected in the Golgi apparatus. Following longer chase periods, the radiolabeled proteins traveled from the Golgi apparatus to the cell surface in **secretory vesicles**, which then fused with the plasma membrane to release their contents outside of the cell.

These experiments defined a pathway taken by secreted proteins, the **secretory pathway**: rough ER → Golgi → secretory vesicles → cell exterior. Further studies extended these results and demonstrated that this pathway is not restricted to proteins destined for secretion from the cell. Plasma membrane and lysosomal proteins also travel from the rough ER to the Golgi and then to their final destinations. Still other proteins travel through the initial steps of the secretory pathway but are then retained and function within either the ER or the Golgi apparatus.

The entrance of proteins into the ER thus represents a major branch point for the traffic of proteins within eukaryotic cells. Proteins destined for secretion or incorporation into the ER, Golgi apparatus, lysosomes, or plasma membrane are initially targeted to the ER. In mammalian cells, most proteins are transferred into the ER while they are being translated on membrane-bound ribosomes (Figure 9.3). In contrast, proteins destined to remain in the cytosol or to be incorporated into the nucleus, mitochondria, chloroplasts, or peroxisomes are synthesized on free ribosomes and released into the cytosol when their translation is complete.

Figure 9.3
Overview of protein sorting   In mammalian cells, the initial sorting of proteins to the ER takes place while translation is in progress. Proteins synthesized on free ribosomes either remain in the cytosol or are transported to the nucleus, mitochondria, chloroplasts, or peroxisomes. In contrast, proteins synthesized on membrane-bound ribosomes are translocated into the ER while their translation is in progress. They may be either retained within the ER or transported to the Golgi apparatus and, from there, to lysosomes, the plasma membrane, or the cell exterior via secretory vesicles.

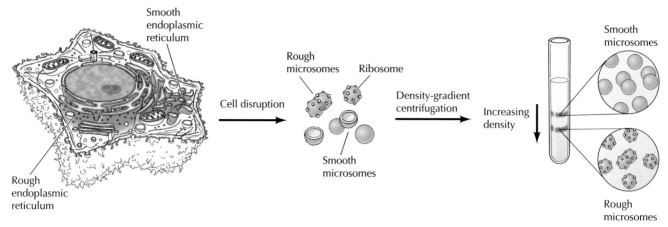

Figure 9.4
Isolation of rough ER When cells are disrupted, the ER fragments into small vesicles called microsomes. The microsomes derived from the rough ER (rough microsomes) are lined with ribosomes on their outer surface. Because ribosomes contain large amounts of RNA, the rough microsomes are denser than smooth microsomes and can be isolated by equilibrium density-gradient centrifugation.

Targeting Proteins to the Endoplasmic Reticulum

Proteins can be translocated into the ER either during their synthesis on membrane-bound ribosomes (cotranslational translocation) or after their translation has been completed on free ribosomes in the cytosol (posttranslational translocation). In mammalian cells, most proteins enter the ER cotranslationally, whereas both cotranslational and posttranslational pathways are used in yeast. The first step in the cotranslational pathway is the association of ribosomes with the ER. Ribosomes are targeted for binding to the ER membrane by the amino acid sequence of the polypeptide chain being synthesized, rather than by intrinsic properties of the ribosome itself. Free and membrane-bound ribosomes are functionally indistinguishable, and all protein synthesis initiates on ribosomes that are free in the cytosol. Ribosomes engaged in the synthesis of proteins that are destined for secretion are then targeted to the endoplasmic reticulum by a **signal sequence** at the amino terminus of the growing polypeptide chain. These signal sequences are short stretches of hydrophobic amino acids that are cleaved from the polypeptide chain during its transfer into the ER lumen.

The general role of signal sequences in targeting proteins to their appropriate locations within the cell was first elucidated by studies of the import of secretory proteins into the ER. These experiments used *in vitro* preparations of rough ER, which were isolated from cell extracts by density-gradient centrifugation (Figure 9.4). When cells are disrupted, the ER breaks up into small vesicles called **microsomes**. Because the vesicles derived from the rough ER are covered with ribosomes, they can be separated from similar vesicles derived from the smooth ER or from other membranes (e.g., the plasma membrane). In particular, the large amount of RNA within ribosomes increases the density of the membrane vesicles to which they are attached, allowing purification of vesicles derived from the rough ER (rough microsomes) by equilibrium centrifugation in density gradients.

David Sabatini and Günter Blobel first proposed in 1971 that the signal for ribosome attachment to the ER was an amino acid sequence near the amino terminus of the growing polypeptide chain. This hypothesis was supported by the results of *in vitro* translation of mRNAs encoding secreted proteins, such as immunoglobulins (Figure 9.5). If an mRNA encoding a secreted protein was translated on free ribosomes *in vitro*, it was found that the protein produced was slightly larger than the normal secreted protein. If microsomes were added to the system, however, the *in vitro*-translated protein was incorporated into the microsomes and cleaved to the correct size. These experiments led to a more detailed formulation of the signal hypothesis, which proposed that an amino-terminal leader sequence tar-

Translation on free ribosomes

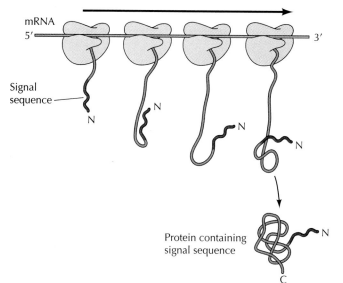

mRNA
5'
3'

Signal
sequence

N

N

N

N

Protein containing
signal sequence

C

N

Translation with microsomes present

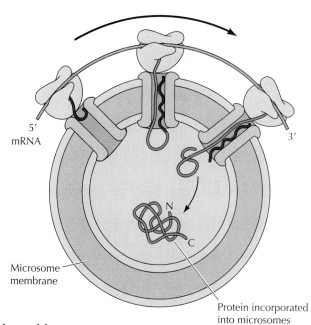

5'
mRNA

3'

N

C

Microsome
membrane

Protein incorporated
into microsomes

gets the polypeptide chain to the microsomes and is then cleaved by a microsomal protease. Many subsequent findings have substantiated this model, including recombinant DNA experiments demonstrating that addition of a signal sequence to a normally nonsecreted protein is sufficient to direct the incorporation of the recombinant protein into the rough ER.

The mechanism by which secretory proteins are targeted to the ER during their translation (the cotranslational pathway) is now well understood. The signal sequences span about 20 amino acids, including a stretch of hydrophobic residues, usually at the amino terminus of the polypeptide chain (Figure 9.6). As they emerge from the ribosome, signal sequences are recognized and bound by a **signal recognition particle (SRP)** consisting of six polypeptides and a small cytoplasmic RNA (**7SL RNA**). SRP binds the ribosome as well as the signal sequence, inhibiting further translation and targeting the entire complex (the SRP, ribosome, and growing polypeptide chain) to the rough ER by binding to the SRP receptor on the ER membrane (Figure 9.7). Binding to the receptor releases the SRP from both the ribosome and the signal sequence of the growing polypeptide chain. The ribosome then binds to a protein translocation complex in the ER membrane, and the signal sequence is inserted into a membrane channel. In both yeast and mammalian cells, the translocation channels through the ER membrane are complexes of three transmembrane proteins, called the Sec61 proteins. The yeast and mammalian Sec61 proteins are closely related to the plasma membrane proteins that translocate secreted polypeptides in bacteria, demonstrating a striking conservation of the protein secretion machinery in prokaryotic and eukaryotic cells. Transfer of the ribosome from the SRP to the Sec61 complex allows translation to resume, and the growing polypeptide chain is transferred directly into the Sec61 channel and across the ER membrane as translation proceeds. Thus, the process of protein synthesis directly drives the transfer of growing polypeptide chains through the Sec61 channel and into the ER. As translocation proceeds, the signal

Figure 9.5
Incorporation of secretory proteins into microsomes Secretory proteins are targeted to the ER by a signal sequence at their amino (N) terminus, which is removed during incorporation of the growing polypeptide chain into the ER. This was demonstrated by experiments showing that translation of secretory protein mRNAs on free ribosomes yielded proteins that retained their signal sequences and were therefore slightly larger than the normal secreted proteins. However, when microsomes were added to the system, the growing polypeptide chains were incorporated into the microsomes and the signal sequences were removed by proteolytic cleavage.

Figure 9.6
The signal sequence of growth hormone Most signal sequences contain a stretch of hydrophobic amino acids, preceded by basic residues (e.g., arginine).

Cleavage site of
signal peptidase

Met Ala Thr Gly Ser Arg Thr Ser Leu Leu Leu Ala Phe Gly Leu Leu Cys Leu Pro Trp Leu Gln Glu Gly Ser Ala Phe Pro Thr

Key Experiment

The Signal Hypothesis

Transfer of Proteins across Membranes. I. Presence of Proteolytically Processed and Unprocessed Nascent Immunoglobulin Light Chains on Membrane-Bound Ribosomes of Murine Myeloma

Günter Blobel and Bernhard Dobberstein
Rockefeller University, New York
Journal of Cell Biology, 1975, Volume 67, pages 835–851

The Context

How are specific polypeptide chains transferred across the appropriate membranes? Studies in the 1950s and 1960s indicated that secreted proteins are synthesized on membrane-bound ribosomes and transferred across the membrane during their synthesis. However, this did not explain why ribosomes engaged in the synthesis of secreted proteins attach to membranes while ribosomes synthesizing cytosolic proteins do not. A hypothesis to explain this difference was first suggested by Günter Blobel and David Sabatini in 1971. At that time, they proposed that (1) mRNAs to be translated on membrane-bound ribosomes contain a unique set of codons just 3' of the initiation site, (2) translation of these codons yields a unique sequence at the amino terminus of the growing polypeptide chain (the signal sequence), and (3) the signal sequence triggers attachment of the ribosome to the membrane. In 1975, Blobel and Dobberstein reported a series of experiments that provided critical support for this notion. In addition, they pro-

posed "a somewhat more detailed version of this hypothesis, henceforth referred to as the signal hypothesis."

The Experiments

Myelomas are cancers of B lymphocytes that actively secrete immunoglobulins, so they provide a good model for studies of secreted proteins. Previous studies in Cesar Milstein's laboratory had shown that the proteins produced by *in vitro* translation of immunoglobulin light-chain mRNA contain about 20 amino acids at their amino terminus that are not present in the secreted light chains. This result led to the suggestion that these amino acids direct binding of

the ribosome to the membrane. To test this idea, Blobel and Dobberstein investigated the synthesis of light chains by membrane-bound ribosomes from myeloma cells.

As expected from earlier work, *in vitro* translation of light-chain mRNA on free ribosomes yielded a protein that was larger than the secreted light chain (see figure). In contrast, *in vitro* translation of mRNA associated with membrane-bound ribosomes from myeloma cells yielded a protein that was the same size as the normally secreted light chain. Moreover, the light chains synthesized by ribosomes that remained bound to microsomes were resistant to digestion by added proteases, indicating that the light

In vitro translation of immunoglobulin light-chain mRNA on free ribosomes (lane 1) yields a product that migrates slower than secreted light chains (lane S) in gel electrophoresis. In contrast, light chains synthesized by *in vitro* translation on membrane-bound ribosomes (lane 2) are the same size as secreted light chains. In addition, the products of *in vitro* translation on membrane-bound ribosomes were unaffected by subsequent digestion with proteases (lane 3), indicating that they were protected from the proteases by insertion into microsomes.

sequence is cleaved by **signal peptidase** and the polypeptide is released into the lumen of the ER.

Many proteins in yeast, as well as a few proteins in mammalian cells, are targeted to the ER after their translation is complete (posttranslational translocation), rather than being transferred into the ER during synthesis on membrane-bound ribosomes. These proteins are synthesized on free cytosolic ribosomes, and their posttranslational incorporation into the ER does not require SRP. Instead, their signal sequences are recognized by distinct receptor proteins (the Sec62/63 complex) associated with the Sec61

The Signal Hypothesis (continued)

chains had been transferred into the microsomes.

These results indicated that an amino-terminal signal sequence is removed by a microsomal protease as growing polypeptide chains are transferred across the membrane. The results were interpreted in terms of a more detailed version of the signal hypothesis. As stated by Blobel and Dobberstein, "the essential feature of the signal hypothesis is the occurrence of a unique sequence of codons, located immediately to the right of the initiation codon, which is present only in those mRNAs whose translation products are to be transferred across a membrane."

The Impact

The selective transfer of proteins across membranes is critical to the maintenance of the membrane-enclosed organelles of eukaryotic cells. To maintain the identity of these organelles, proteins must be translocated specifically across the appropriate membranes. The signal hypothesis provided the conceptual basis for understanding this phenomenon. Not only has this basic model been firmly substantiated for the transfer of secreted proteins into the endoplasmic reticulum, but it also has provided the framework for understanding the targeting of proteins to all the other

Günter Blobel

membrane-enclosed compartments of the cell, thereby impacting virtually all areas of cell biology.

complex in the ER membrane (Figure 9.8). Cytosolic chaperones are required to maintain the polypeptide chains in an unfolded conformation so they can enter the Sec61 channel, and another chaperone within the ER (called BiP) is required to pull the polypeptide chain through the channel and into the ER. It appears that the binding of polypeptide chains to BiP is needed to drive the posttranslational translocation of proteins into the ER,

Figure 9.7
Cotranslational targeting of secretory proteins to the ER   Step 1: As the signal sequence emerges from the ribosome, it is recognized and bound by the signal recognition particle (SRP). Step 2: The SRP escorts the complex to the ER membrane, where it binds to the SRP receptor. Step 3: The SRP is released, the ribosome binds to a membrane translocation complex of Sec61 proteins, and the signal sequence is inserted into a membrane channel. Step 4: Translation resumes, and the growing polypeptide chain is translocated across the membrane. Step 5: Cleavage of the signal sequence by signal peptidase releases the polypeptide into the lumen of the ER.

3′ mRNA

5′

Signal sequence

3′

5′

SRP

Step 1

Step 2

5′   3′   **Step 3**   5′   3′   **Step 4**   5′   3′

SRP receptor

Endoplasmic reticulum lumen

Sec61 translocation complex

Signal peptidase

Step 5

Figure 9.8
Posttranslational translocation of proteins into the ER   Proteins destined for posttranslational import to the ER are synthesized on free ribosomes and maintained in an unfolded conformation by cytosolic chaperones. Their signal sequences are recognized by the Sec62/63 complex, which is associated with the Sec61 translocation channel in the ER membrane. The Sec63 protein is also associated with a chaperone protein (BiP), which acts as a molecular ratchet to drive protein translocation into the ER.

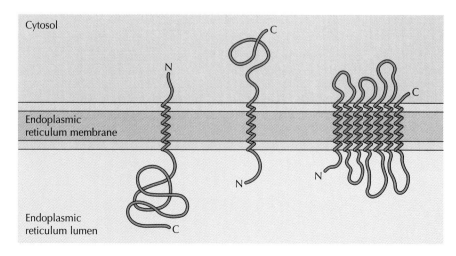

whereas the cotranslational translocation of growing polypeptide chains is driven directly by the process of protein synthesis.

Insertion of Proteins into the ER Membrane

Proteins destined for secretion or residence within the lumen of the ER, Golgi apparatus, or lysosomes are translocated across the ER membrane and released into the lumen of the ER as already described. However, proteins destined for incorporation into the plasma membrane or the membranes of the ER, Golgi, or lysosomes are initially inserted into the ER membrane

Figure 9.9
Orientations of membrane proteins Integral membrane proteins span the membrane via α-helical regions of 20 to 25 hydrophobic amino acids, which can be inserted in a variety of orientations. The two proteins at left and center each span the membrane once, but they differ in whether the amino (N) or carboxy (C) terminus is on the cytosolic side. On the right is an example of a protein that has multiple membrane-spanning regions.

instead of being released into the lumen. From the ER membrane, they proceed to their final destination along the same pathway as that of secretory proteins: ER→ Golgi→ plasma membrane or lysosomes. These proteins are transported along this pathway as membrane components, however, rather than as soluble proteins.

Integral membrane proteins are embedded in the membrane by hydrophobic regions that span the phospholipid bilayer (see Figure 2.48). The membrane-spanning portions of these proteins are usually α-helical regions consisting of 20 to 25 hydrophobic amino acids. The formation of an α helix maximizes hydrogen bonding between the peptide bonds, and the hydrophobic amino acid side chains interact with the fatty acid tails of the phospholipids. However, different integral membrane proteins differ in how they are inserted (Figure 9.9). For example, whereas some integral membrane proteins span the membrane only once, others have multiple membrane-spanning regions. In addition, some proteins are oriented in the membrane with their amino terminus on the cytosolic side; others have their carboxy terminus exposed to the cytosol. These orientations of proteins inserted into the ER, Golgi, lysosomal, and plasma membranes are established as the growing polypeptide chains are translocated into the ER. The lumen of the ER is topologically equivalent to the exterior of the cell, so the domains of plasma membrane proteins that are exposed on the cell surface correspond to the regions of polypeptide chains that are translocated into the ER (Figure 9.10).

The most straightforward mode of insertion into the ER membrane results in the synthesis of transmembrane proteins oriented with their carboxy termini exposed to the cytosol (Figure 9.11). These proteins have a normal amino-terminal signal sequence, which is cleaved by signal peptidase during translocation of the polypeptide chain across the ER membrane through the Sec61 channel. They are then anchored in the membrane by a second membrane-spanning α helix in the middle of the protein. This transmembrane sequence, called a stop-transfer sequence, signals closure of the Sec61 channel. Further translocation of the polypeptide

Figure 9.10
Topology of the secretory pathway The lumens of the endoplasmic reticulum and Golgi apparatus are topologically equivalent to the exterior of the cell. Consequently, those portions of polypeptide chains that are translocated into the ER are exposed on the cell surface following transport to the plasma membrane.

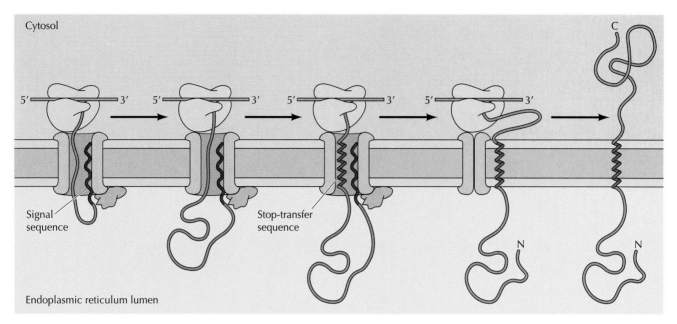

Figure 9.11
Insertion of a membrane protein with a cleavable signal sequence and a single stop-transfer sequence The signal sequence is cleaved as the polypeptide chain crosses the membrane, so the amino terminus of the polypeptide chain is exposed in the ER lumen. However, translocation of the polypeptide chain across the membrane is halted by a transmembrane stop-transfer sequence that closes the Sec61 translocation channel and exits the channel laterally to anchor the protein in the ER membrane. Continued translation results in a membrane-spanning protein with its carboxy terminus on the cytosolic side.

Figure 9.12
Insertion of membrane proteins with internal noncleavable signal sequences Internal noncleavable signal sequences can lead to the insertion of polypeptide chains in either orientation in the ER membrane. (A) The signal sequence directs insertion of the polypeptide such that its amino terminus is exposed on the cytosolic side. The remainder of the polypeptide chain is translocated into the ER as translation proceeds. The signal sequence is not cleaved, so it acts as a membrane-spanning sequence that anchors the protein in the membrane with its carboxy terminus in the lumen of the ER. (B) Other internal signal sequences are oriented to direct the transfer of the amino-terminal portion of the polypeptide across the membrane. Continued translation results in a protein that spans the ER membrane with its amino terminus in the lumen and its carboxy terminus in the cytosol. Note that this orientation is the same as that resulting from insertion of a protein that contains a cleavable signal sequence followed by a stop-transfer sequence (see Figure 9.11).

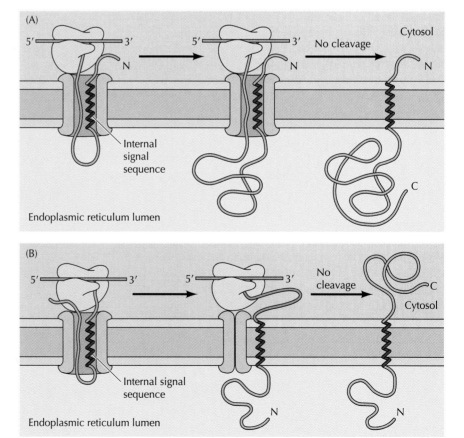

chain across the ER membrane is thus blocked, so the carboxy-terminal portion of the growing polypeptide chain is synthesized in the cytosol. The transmembrane domain then exits the translocation channel laterally to enter the lipid bilayer. The insertion of these proteins in the membrane thus involves the sequential action of two distinct elements: a cleavable amino-terminal signal sequence that initiates translocation across the membrane and a transmembrane stop-transfer sequence that anchors the protein in the membrane.

Proteins can also be anchored in the ER membrane by internal signal sequences that are not cleaved by signal peptidase (Figure 9.12). These internal signal sequences are recognized by the SRP and brought to the ER membrane as already discussed. Because they are not cleaved by signal peptidase, however, these signal sequences act as transmembrane α helices that exit the translocation channel and anchor proteins in the ER membrane. Importantly, internal signal sequences can be oriented so as to direct the translocation of either the amino or carboxy terminus of the polypeptide chain across the membrane. Therefore, depending on the orientation of the signal sequence, proteins inserted into the membrane by this mechanism can have either their amino or carboxy terminus exposed to the cytosol.

Proteins that span the membrane multiple times are thought to be inserted as a result of an alternating series of internal signal sequences and transmembrane stop-transfer sequences. For example, an internal signal sequence can result in membrane insertion of a polypeptide chain with its amino terminus on the cytosolic side (Figure 9.13). If a stop-transfer sequence

Figure 9.13
Insertion of a protein that spans the membrane multiple times   In this example, an internal signal sequence results in insertion of the polypeptide chain with its amino terminus on the cytosolic side of the membrane. A stop-transfer sequence then signals closure of the translocation channel, causing the polypeptide chain to form a loop within the lumen of the ER, and translation continues in the cytosol. A second internal signal sequence reopens the channel, triggering reinsertion of the polypeptide chain into the ER membrane and forming a loop in the cytosol. The process can be repeated many times, resulting in the insertion of proteins with multiple membrane-spanning regions.

is then encountered, the polypeptide will form a loop in the ER lumen, and protein synthesis will continue on the cytosolic side of the membrane. If a second signal sequence is encountered, the growing polypeptide chain will again be inserted into the ER, forming another looped domain on the cytosolic side of the membrane. This can be followed by yet another stop-transfer sequence and so forth, so that an alternating series of signal and stop-transfer sequences can result in the insertion of proteins that span the membrane multiple times, with looped domains exposed on both the lumenal and cytosolic sides.

Protein Folding and Processing in the ER

The folding of polypeptide chains into their correct three-dimensional conformations, the assembly of polypeptides into multisubunit proteins, and the covalent modifications involved in protein processing were discussed in Chapter 7. For proteins that enter the secretory pathway, many of these events occur either during translocation across the ER membrane or within the ER lumen. One such processing event is the proteolytic cleavage of the signal peptide as the polypeptide chain is translocated across the ER membrane. The ER is also the site of protein folding, assembly of multisubunit proteins, disulfide bond formation, the initial stages of glycosylation, and the addition of glycolipid anchors to some plasma membrane proteins. Indeed, the primary role of lumenal ER proteins is to catalyze the folding and assembly of newly translocated polypeptides.

As already discussed, proteins are translocated across the ER membrane as unfolded polypeptide chains while their translation is still in progress. These polypeptides, therefore, fold into their three-dimensional conformations within the ER, assisted by the molecular chaperones that facilitate the folding of polypeptide chains (see Chapter 7). For example, one of the major proteins within the ER lumen is a member of the Hsp70 family of chaperones called BiP. BiP is thought to bind to the unfolded polypeptide chain as it crosses the membrane and then mediates protein folding and the assembly of multisubunit proteins within the ER (Figure 9.14). Correctly assembled proteins are released from BiP and are available for transport to the Golgi apparatus. Abnormally folded or improperly assembled proteins, however, remain bound to BiP and are consequently retained within the ER or degraded, rather than being transported farther along the secretory pathway.

Figure 9.14
Protein folding in the ER The molecular chaperone BiP binds to polypeptide chains as they cross the ER membrane and facilitates protein folding and assembly within the ER.

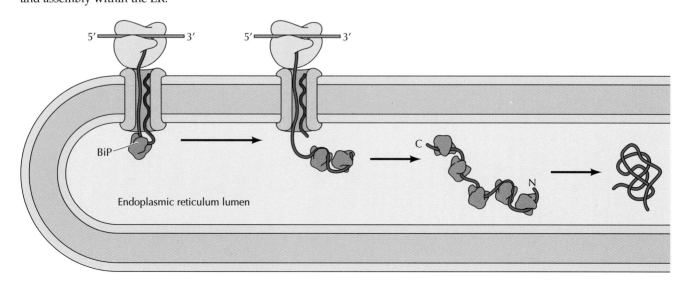

BiP

Endoplasmic reticulum lumen

The formation of disulfide bonds between the side chains of cysteine residues is an important aspect of protein folding and assembly within the ER. These bonds do not form in the cytosol, which is characterized by a reducing environment that maintains cysteine residues in their reduced (—SH) state. In the ER, however, an oxidizing environment promotes disulfide (S—S) bond formation, and disulfide bonds formed in the ER play important roles in the structure of secreted and cell surface proteins. Disulfide bond formation is facilitated by the enzyme **protein disulfide isomerase** (see Figure 7.21), which is located in the ER lumen.

Proteins are also glycosylated on specific asparagine residues (*N*-linked glycosylation) within the ER while their translation is still in process (Figure 9.15). As discussed in Chapter 7, oligosaccharide units consisting of 14 sugar residues are added to acceptor asparagine residues of growing polypeptide chains as they are translocated into the ER. The oligosaccharide is synthesized on a lipid (dolichol) carrier anchored in the ER membrane. It is then transferred as a unit to acceptor asparagine residues in the consensus sequence Asn-X-Ser/Thr by a membrane-bound enzyme called oligosaccharyl transferase. Four sugar residues (three glucose and one mannose) are removed while the protein is still within the ER, and the protein is modified further after being transported to the Golgi apparatus (discussed later in this chapter).

Some proteins are anchored in the plasma membrane by glycolipids rather than by membrane-spanning regions of the polypeptide chain. Because these membrane-anchoring glycolipids contain phosphatidylinositol, they are called **glycosylphosphatidylinositol (GPI) anchors**, the structure of which was illustrated in Figure 7.32. The GPI anchors are assembled in the ER membrane. They are then added immediately after completion of protein synthesis to the carboxy terminus of some proteins anchored in the membrane by a C-terminal membrane-spanning region (Figure 9.16). The transmembrane region of the protein is exchanged for the GPI anchor, so these proteins remain attached to the membrane only by their associated

Figure 9.15
Protein glycosylation in the ER

Oligosaccharide is transferred from a dolichol lipid carrier to polypeptide chains during their translocation across the ER membrane.

Three glucose residues are removed by two separate enzymes.

One mannose is removed.

Endoplasmic reticulum lumen

Figure 9.16
Addition of GPI anchors Glycosyl-phosphatidylinositol (GPI) anchors contain two fatty acid chains, an oligosaccharide portion consisting of inositol and other sugars, and ethanolamine (see Figure 7.32 for a more detailed structure). The GPI anchors are assembled in the ER and added to polypeptides anchored in the membrane by a carboxy-terminal membrane-spanning region. The membrane-spanning region is cleaved, and the new carboxy terminus is joined to the NH_2 group of ethanolamine immediately after translation is completed, leaving the protein attached to the membrane by the GPI anchor.

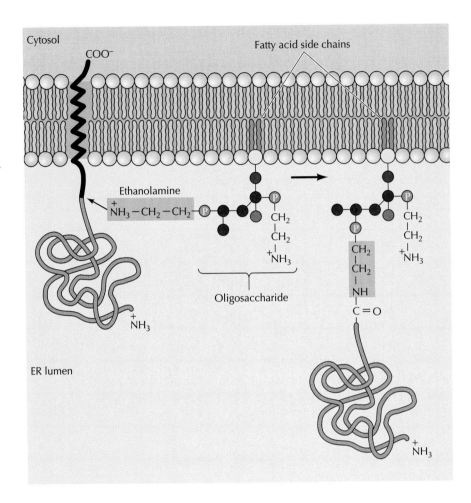

glycolipid. Like transmembrane proteins, they are transported to the cell surface as membrane components via the secretory pathway. Their orientation within the ER dictates that GPI-anchored proteins are exposed on the outside of the cell, with the GPI anchor mediating their attachment to the plasma membrane.

The Smooth ER and Lipid Synthesis

In addition to its activities in the processing of secreted and membrane proteins, the ER is the major site at which membrane lipids are synthesized in eukaryotic cells. Because they are extremely hydrophobic, lipids are synthesized in association with already existing cellular membranes rather than in the aqueous environment of the cytosol. Although some lipids are synthesized in association with other membranes, most are synthesized in the ER. They are then transported from the ER to their ultimate destinations either in vesicles or by carrier proteins, as discussed later in this chapter and in Chapter 10.

The membranes of eukaryotic cells are composed of three main types of lipids: phospholipids, glycolipids, and cholesterol. Most of the phospholipids, which are the basic structural components of the membrane, are derived from glycerol. They are synthesized on the cytosolic side of the ER membrane, from water-soluble cytosolic precursors (Figure 9.17). Fatty acids are first transferred from coenzyme A carriers to glycerol-3-phosphate by a membrane-bound enzyme, and the resulting phospholipid (phosphatidic acid) is inserted into the membrane. Enzymes on the cytoso-

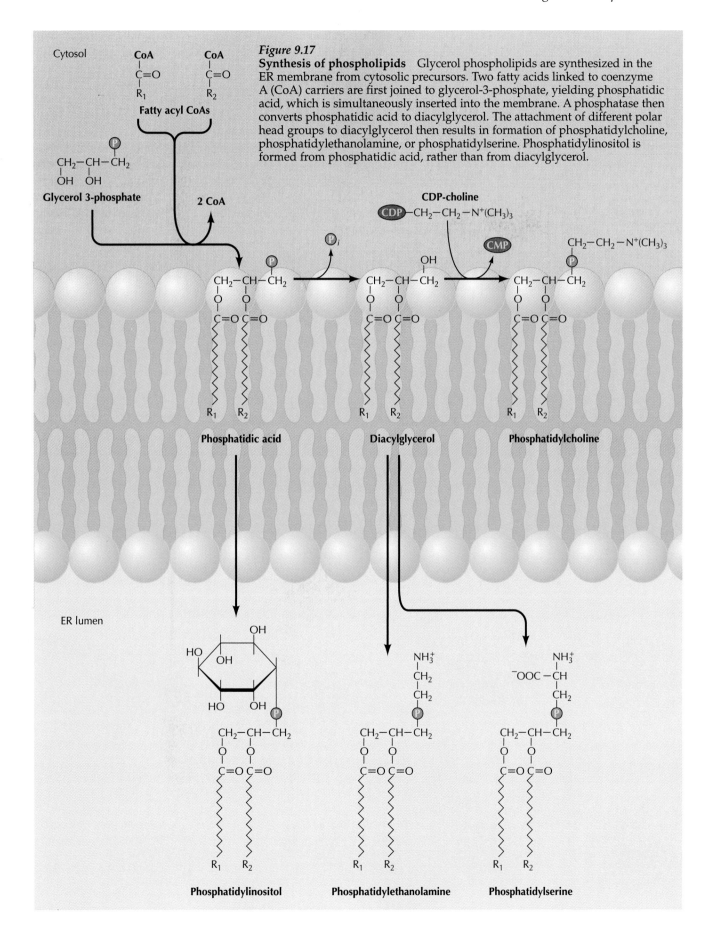

Figure 9.17
Synthesis of phospholipids Glycerol phospholipids are synthesized in the ER membrane from cytosolic precursors. Two fatty acids linked to coenzyme A (CoA) carriers are first joined to glycerol-3-phosphate, yielding phosphatidic acid, which is simultaneously inserted into the membrane. A phosphatase then converts phosphatidic acid to diacylglycerol. The attachment of different polar head groups to diacylglycerol then results in formation of phosphatidylcholine, phosphatidylethanolamine, or phosphatidylserine. Phosphatidylinositol is formed from phosphatidic acid, rather than from diacylglycerol.

lic face of the ER membrane then catalyze the addition of different polar head groups, resulting in formation of phosphatidylcholine, phosphatidylserine, phosphatidylethanolamine, or phosphatidylinositol.

The synthesis of these phospholipids on the cytosolic side of the ER membrane allows the hydrophobic fatty acid chains to remain buried in the membrane while membrane-bound enzymes catalyze their reactions with water-soluble precursors (e.g., CDP-choline) in the cytosol. Because of this topography, however, new phospholipids are added only to the cytosolic half of the ER membrane (Figure 9.18). To maintain a stable membrane, some of these newly synthesized phospholipids must therefore be transferred to the other (lumenal) half of the ER bilayer. This transfer does not occur spontaneously because it requires the passage of a polar head group through the membrane. Instead, membrane proteins called **flippases** catalyze the rapid translocation of phospholipids across the ER membrane, resulting in even growth of both halves of the bilayer.

In addition to its role in synthesis of the glycerol phospholipids, the ER also serves as the major site of synthesis of two other membrane lipids: cholesterol and ceramide (Figure 9.19). As discussed later, ceramide is converted to either glycolipids or sphingomyelin (the only membrane phospholipid not derived from glycerol) in the Golgi apparatus. The ER is thus responsible for synthesis of either the final products or the precursors of all the major lipids of eukaryotic membranes.

Smooth ER is abundant in cell types that are particularly active in lipid metabolism. For example, steroid hormones are synthesized (from cholesterol) in the ER, so large amounts of smooth ER are found in steroid-producing cells, such as those in the testis and ovary. In addition, smooth ER is abundant in the liver, where it contains enzymes that metabolize various lipid-soluble compounds. These detoxifying enzymes inactivate a number of potentially harmful drugs (e.g., phenobarbital) by converting them to water-soluble compounds that can be eliminated from the body in the urine. The smooth ER is thus involved in multiple aspects of the metabolism of lipids and lipid-soluble compounds.

Export of Proteins and Lipids from the ER

Both proteins and lipids travel along the secretory pathway in transport vesicles, which bud from the membrane of one organelle and then fuse with the membrane of another. Thus, molecules are exported from the ER in vesicles that bud from the ER and carry their cargo first to the ER-Golgi intermediate compartment and then to the Golgi apparatus (Figure 9.20). Subsequent steps in the secretory pathway involve vesicular transport between different compart-

Cytosol

ER membrane

Lumen

Phospholipid synthesis

Newly synthesized lipids added only to cytosolic half of bilayer

Flippase

Growth of both halves of phospholipid bilayer

Figure 9.18
Translocation of phospholipids across the ER membrane
Because phospholipids are synthesized on the cytosolic side of the ER membrane, they are added only to the cytosolic half of the bilayer. They are then translocated across the membrane by phospholipid flippases, resulting in even growth of both halves of the phospholipid bilayer.

Cholesterol

Ceramide

Figure 9.19
Structure of cholesterol and ceramide
The hydrogens attached to the ring carbons of cholesterol are not shown.

ments of the Golgi and from the Golgi to lysosomes or the plasma membrane. In each case, proteins within the lumen of one organelle are packaged into the budding transport vesicle and then released into the lumen of the recipient organelle following vesicle fusion. Membrane proteins and lipids are transported similarly, and it is noteworthy that their topological orientation is maintained as they travel from one membrane-enclosed organelle to

Rough endoplasmic reticulum

Vesicle budding

Transport vesicle

Vesicle fusion

ERGIC

Golgi

Figure 9.20
Vesicular transport from the ER to the Golgi Proteins and lipids are carried from the ER to the Golgi in transport vesicles that bud from the membrane of the ER and then fuse to form the vesicles and tubules of the ER-Golgi intermediate compartment (ERGIC). Lumenal ER proteins are taken up by the vesicles and released into the lumen of the Golgi. Membrane proteins maintain the same orientation in the Golgi as in the ER.

another. For example, the domains of a protein exposed on the cytosolic side of the ER membrane will also be exposed on the cytosolic side of the Golgi and plasma membranes, whereas protein domains exposed on the lumenal side of the ER membrane will be exposed on the lumenal side of the Golgi and on the exterior of the cell (see Figure 9.10).

While most proteins travel from the ER to the Golgi, some proteins must be retained within the ER rather than proceeding along the secretory pathway. In particular, proteins that function within the ER (including BiP, signal peptidase, protein disulfide isomerase, and other enzymes discussed earlier) must be retained within that organelle. Export to the Golgi versus retention in the ER is thus the first branch point encountered by proteins being sorted to their correct destinations in the secretory pathway. Similar branch points arise at each subsequent stage of transport, such as retention in the Golgi versus export to lysosomes or the plasma membrane. In each case, specific localization signals target proteins to their correct intracellular destinations.

The distinction between proteins exported from and those retained in the ER appears to be governed by two distinct types of targeting sequences that specifically mark proteins as either (1) destined for transport to the Golgi or (2) destined for retention in the ER. Many proteins are retained in the ER lumen as a result of the presence of the targeting sequence Lys-Asp-Glu-Leu (KDEL, in the single-letter code) at their carboxy terminus. If this sequence is deleted from a protein that is normally retained in the ER (e.g., BiP), the mutated protein is instead transported to the Golgi and secreted from the cell. Conversely, addition of the KDEL sequence to the carboxy terminus of proteins that are normally secreted causes them to be retained in the ER. The retention of some transmembrane proteins in the ER is similarly dictated by short C-terminal sequences that contain two lysine residues (KKXX sequences).

Interestingly, the KDEL and KKXX signals do not prevent soluble ER proteins from being packaged into vesicles and carried to the Golgi. Instead, these signals cause resident ER proteins to be selectively retrieved from the ER-Golgi intermediate compartment or the Golgi complex and returned to the ER via a recycling pathway (Figure 9.21). Proteins bearing the KDEL and KKXX sequences appear to bind to specific recycling receptors in the membranes of these compartments and are then selectively transported back to the ER.

The action of the KDEL and KKXX sequences as retention/retrieval signals indicates that there is a nonselective bulk flow of proteins through the secretory pathway leading from the ER to the cell surface. This bulk flow from the ER to the Golgi may be responsible for the export of many proteins from the ER. However, it also appears that some proteins destined for secretion are marked by signals that actively direct their export from the ER. Protein export from the ER can thus take place not only by bulk flow, but also by a regulated pathway that specifically recognizes targeting signals that mediate selective transport of proteins to the Golgi apparatus.

The Golgi Apparatus

The **Golgi apparatus**, or **Golgi complex**, functions as a factory in which proteins received from the ER are further processed and sorted for transport to their eventual destinations: lysosomes, the plasma membrane, or secretion. In addition, as noted earlier, glycolipids and sphingomyelin are synthesized within the Golgi. In plant cells, the Golgi apparatus further serves as the site at which the complex polysaccharides of the cell wall are synthesized. The Golgi apparatus is thus involved in processing the broad range of cellular constituents that travel along the secretory pathway.

Figure 9.21
Retrieval of resident ER proteins
Proteins destined to remain in the lumen of the ER are marked by the sequence Lys-Asp-Glu-Leu (KDEL) at their carboxy terminus. These proteins are exported from the ER to the Golgi in the nonselective bulk flow of proteins through the secretory pathway, but they are recognized by a receptor in the ER-Golgi intermediate compartment (ERGIC) or the Golgi apparatus and selectively returned to the ER.

Organization of the Golgi

Morphologically the Golgi is composed of flattened membrane-enclosed sacs (cisternae) and associated vesicles (Figure 9.22). A striking feature of the Golgi apparatus is its distinct polarity in both structure and function. Proteins from the ER enter at its *cis* face (entry face), which is convex and usually oriented toward the nucleus. They are then transported through the Golgi and exit from its concave *trans* face (exit face). As they pass through the Golgi, proteins are modified and sorted for transport to their eventual destinations within the cell.

Distinct processing and sorting events appear to take place in an ordered sequence within different regions of the Golgi complex, so the Golgi is usually considered to consist of multiple discrete compartments. Although the number of such compartments has not been established, the Golgi is most commonly viewed as consisting of four functionally distinct regions: the *cis* **Golgi network**, the **Golgi stack** (which is divided into the *medial* and *trans* subcompartments), and the *trans* **Golgi network** (Figure 9.23). Proteins from the ER are transported to the ER-Golgi intermediate compartment and then enter the Golgi apparatus at the *cis* Golgi network. They then progress to the *medial* and *trans* compartments of the Golgi stack, within which most metabolic activities of the Golgi apparatus take place. The modified proteins, lipids, and polysaccharides then move to the *trans* Golgi network, which acts as a sorting and distribution center, directing molecular traffic to lysosomes, the plasma membrane, or the cell exterior.

Figure 9.22
Electron micrograph of a Golgi apparatus   The Golgi apparatus consists of a stack of flattened cisternae and associated vesicles. Proteins and lipids from the ER enter the Golgi apparatus at its *cis* face and exit at its *trans* face. (Courtesy of Dr. L. Andrew Staehelin, University of Colorado at Boulder.)

Figure 9.23
Regions of the Golgi apparatus
Vesicles from the ER fuse to form the ER-Golgi intermediate compartment, and proteins from the ER are then transported to the *cis* Golgi network. Resident ER proteins are returned from the ER-Golgi intermediate compartment and the *cis* Golgi network via the recycling pathway. The *medial* and *trans* compartments of the Golgi stack correspond to the cisternae in the middle of the Golgi complex and are the sites of most protein modifications. Proteins are then carried to the *trans* Golgi network, where they are sorted for transport to the plasma membrane, secretion, or lysosomes.

Although the Golgi apparatus was first described over 100 years ago, the mechanism by which proteins move through the Golgi apparatus has still not been established and is an area of controversy among cell biologists. One possibility is that transport vesicles carry proteins between the cisternae of the Golgi compartments. However, there is considerable experimental support for an alternative model proposing that proteins are simply carried through compartments of the Golgi within the Golgi cisternae, which gradually mature and progressively move through the Golgi in the *cis* to *trans* direction.

Protein Glycosylation within the Golgi

Protein processing within the Golgi involves the modification and synthesis of the carbohydrate portions of glycoproteins. One of the major aspects of this processing is the modification of the *N*-linked oligosaccharides that were added to proteins in the ER. As discussed earlier in this chapter, proteins are modified within the ER by the addition of an oligosaccharide consisting of 14 sugar residues (see Figure 9.15). Three glucose residues and one mannose are then removed while the polypeptides are still in the ER. Following transport to the Golgi apparatus, the *N*-linked oligosaccharides of these glycoproteins are subject to extensive further modifications.

N-linked oligosaccharides are processed within the Golgi apparatus in an ordered sequence of reactions (Figure 9.24). The first modification of proteins destined for secretion or for the plasma membrane is the removal of three additional mannose residues. This is followed by the sequential addition of an *N*-acetylglucosamine, the removal of two more mannoses, and the addition of a fucose and two more *N*-acetylglucosamines. Finally, three galactose and three sialic acid residues are added. As noted in Chapter 7, different glycoproteins are modified to different extents during their passage through the Golgi, depending on both the structure of the protein and on the amount of processing enzymes that are present within the Golgi complexes of different types of cells. Consequently, proteins can emerge from the Golgi with a variety of different *N*-linked oligosaccharides.

Figure 9.24
Processing of *N*-linked oligosaccharides in the Golgi   The *N*-linked oligosaccharides of glycoproteins transported from the ER are further modified by an ordered sequence of reactions in the Golgi.

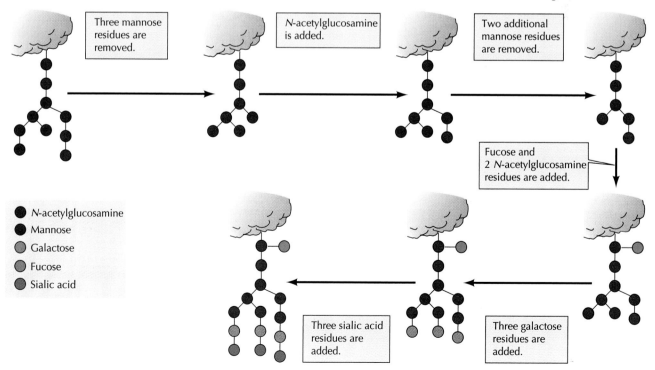

- N-acetylglucosamine
- Mannose
- Galactose
- Fucose
- Sialic acid

Three mannose residues are removed.

N-acetylglucosamine is added.

Two additional mannose residues are removed.

Fucose and 2 N-acetylglucosamine residues are added.

Three galactose residues are added.

Three sialic acid residues are added.

Figure 9.25
Targeting of lysosomal proteins by phosphorylation of mannose residues
Proteins destined for incorporation into lysosomes are specifically recognized and modified by the addition of phosphate groups to the 6 position of mannose residues. In the first step of the reaction, *N*-acetylglucosamine phosphates are transferred to mannose residues from UDP-*N*-acetylglucosamine. The *N*-acetylglucosamine groups are then removed, leaving mannose-6-phosphates.

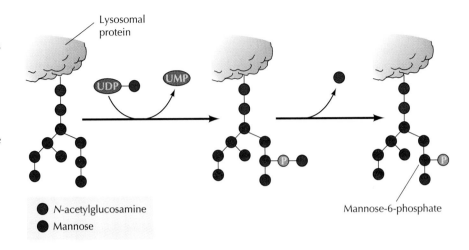

The processing of the *N*-linked oligosaccharide of lysosomal proteins differs from that of secreted and plasma membrane proteins. Rather than the initial removal of three mannose residues, proteins destined for incorporation into lysosomes are modified by mannose phosphorylation. In the first step of this reaction, *N*-acetylglucosamine phosphates are added to specific mannose residues, probably while the protein is still in the *cis* Golgi network (Figure 9.25). This is followed by removal of the *N*-acetylglucosamine group, leaving **mannose-6-phosphate** residues on the *N*-linked oligosaccharide. Because of this modification, these residues are not removed during further processing. Instead, these phosphorylated mannose residues are specifically recognized by a mannose-6-phosphate receptor in the *trans* Golgi network, which directs the transport of these proteins to lysosomes.

The phosphorylation of mannose residues is thus a critical step in sorting lysosomal proteins to their correct intracellular destination. The specificity of this process resides in the enzyme that catalyzes the first step in the reaction sequence—the selective addition of *N*-acetylglucosamine phosphates to lysosomal proteins. This enzyme recognizes a structural determinant that is present on lysosomal proteins but not on proteins destined for the plasma membrane or secretion. This recognition determinant is not a simple sequence of amino acids; rather, it is formed in the folded protein by the juxtaposition of amino acid sequences from different regions of the polypeptide chain. In contrast to the signal sequences that direct protein translocation to the ER, the recognition determinant that leads to mannose phosphorylation, and thus ultimately targets proteins to lysosomes, depends on the three-dimensional conformation of the folded protein. Such determinants are called **signal patches**, in contrast to the linear targeting signals discussed earlier in this chapter.

Proteins can also be modified by the addition of carbohydrates to the side chains of acceptor serine and threonine residues within specific sequences of amino acids (*O*-linked glycosylation) (see Figure 7.28). These modifications take place in the Golgi apparatus by the sequential addition of single sugar residues. The serine or threonine is usually linked directly to *N*-acetylgalactosamine, to which other sugars can then be added. In some cases, these sugars are further modified by the addition of sulfate groups.

Lipid and Polysaccharide Metabolism in the Golgi

In addition to its activities in processing and sorting glycoproteins, the Golgi apparatus functions in lipid metabolism—in particular, in the synthesis of glycolipids and sphingomyelin. As discussed earlier, the glycerol phospholipids,

Figure 9.26
Synthesis of sphingomyelin and glycolipids   Ceramide, which is synthesized in the ER, is converted either to sphingomyelin (a phospholipid) or to glycolipids in the Golgi apparatus. In the first reaction, a phosphorylcholine group is transferred from phosphatidylcholine to ceramide. Alternatively, a variety of different glycolipids can be synthesized by the addition of one or more sugar residues (e.g., glucose).

cholesterol, and ceramide are synthesized in the ER. Sphingomyelin and glycolipids are then synthesized from ceramide in the Golgi apparatus (Figure 9.26). Sphingomyelin (the only nonglycerol phospholipid in cell membranes) is synthesized by the transfer of a phosphorylcholine group from phosphatidylcholine to ceramide. Alternatively, the addition of carbohydrates to ceramide can yield a variety of different glycolipids.

   Sphingomyelin is synthesized on the lumenal surface of the Golgi, but glucose is added to ceramide on the cytosolic side. Glucosylceramide then apparently flips, however, and additional carbohydrates are added on the lumenal side of the membrane. Neither sphingomyelin nor the glycolipids are then able to translocate across the Golgi membrane, so they are found only in the lumenal half of the Golgi bilayer. Following vesicular transport, they are correspondingly localized to the exterior half of the plasma membrane, with their polar head groups exposed on the cell surface. As will be discussed in Chapter 12, the oligosaccharide portions of glycolipids are important surface markers in cell-cell recognition.

   In plant cells, the Golgi apparatus has the additional task of serving as the site where complex polysaccharides of the cell wall are synthesized. As discussed further in Chapter 12, the plant cell wall is composed of three major types of polysaccharides. Cellulose, the predominant constituent, is a simple linear polymer of glucose residues. It is synthesized at the cell surface by enzymes in the plasma membrane. The other cell wall polysaccharides (hemicelluloses and pectins), however, are complex, branched chain molecules that are synthesized in the Golgi apparatus and then transported in vesicles to the cell surface. The synthesis of these cell wall polysaccharides is a major cellular function, and as much as 80% of the metabolic activity of the Golgi apparatus in plant cells may be devoted to polysaccharide synthesis.

Protein Sorting and Export from the Golgi Apparatus

Proteins, as well as lipids and polysaccharides, are transported from the Golgi apparatus to their final destinations through the secretory pathway. This involves the sorting of proteins into different kinds of transport vesicles, which bud from the *trans* Golgi network and deliver their contents to the appropriate cellular locations (Figure 9.27). Some proteins are carried from the Golgi to the plasma membrane by a constitutive secretory pathway, which accounts for the incorporation of new proteins and lipids into the plasma membrane, as well as for the continuous secretion of proteins from the cell. Other proteins are transported to the cell surface by a distinct

Transport from the Golgi apparatus
Proteins are sorted in the *trans* Golgi network and transported in vesicles to their final destinations. In the absence of specific targeting signals, proteins are carried to the plasma membrane by constitutive secretion. Alternatively, proteins can be diverted from the constitutive secretion pathway and targeted to other destinations, such as lysosomes or regulated secretion from the cells.

pathway of regulated secretion or are specifically targeted to other intracellular destinations, such as lysosomes in animal cells or vacuoles in yeast.

Proteins that function within the Golgi apparatus must be retained within that organelle, rather than being transported along the secretory pathway. In contrast to the ER, all of the proteins retained within the Golgi complex are associated with the Golgi membrane rather than being soluble proteins within the lumen. The signals responsible for retention of some proteins within the Golgi have been localized to their transmembrane domains, which retain proteins within the Golgi apparatus by preventing them from being packaged in the transport vesicles that leave the *trans* Golgi network. In addition, like the KKXX sequences of resident ER membrane proteins, signals in the cytoplasmic tails of some Golgi proteins mediate the retrieval of these proteins from subsequent compartments along the secretory pathway.

The constitutive secretory pathway, which operates in all cells, leads to continual unregulated protein secretion. However, some cells also possess a distinct regulated secretory pathway in which specific proteins are secreted in response to environmental signals. Examples of regulated secretion include the release of hormones from endocrine cells, the release of neurotransmitters from neurons, and the release of digestive enzymes from the pancreatic acinar cells discussed at the beginning of this chapter (see Figure 9.2). Proteins are sorted into the regulated secretory pathway in the *trans* Golgi network, where they are packaged into specialized secretory vesicles. These secretory vesicles, which are larger than other transport vesicles, store their contents until specific signals direct their fusion with the plasma membrane. For example, the digestive enzymes produced by pancreatic acinar cells are stored in secretory vesicles until the presence of food in the stomach and small intestine triggers their secretion. The sorting of proteins into the regulated secretory pathway appears to involve the recognition of signal patches shared by multiple proteins that enter this pathway. These proteins selectively aggregate in the *trans* Golgi network and are then released by budding as secretory vesicles.

A further complication in the transport of proteins to the plasma membrane arises in many epithelial cells, which are polarized when they are organized into tissues. The plasma membrane of such cells is divided into two separate regions, the **apical domain** and the **basolateral domain**, that contain specific proteins related to their particular functions. For example, the apical membrane of intestinal epithelial cells faces the lumen of the intestine and is specialized for the efficient absorption of nutrients; the remainder of the cell is covered by the basolateral membrane (Figure 9.28). Distinct domains of the plasma membrane are present not only in epithelial cells, but also in other cell types. Thus, the constitutive secretory pathway must selectively transport proteins from the *trans* Golgi network to these distinct domains of the plasma membrane. This is accomplished by the selective packaging of proteins into at least two types of constitutive secretory vesicles that leave the *trans* Golgi network targeted specifically for either the apical or basolateral plasma membrane domains of the cell.

The best-characterized pathway of protein sorting in the Golgi is the selective transport of proteins to lysosomes. As already discussed, lumenal lysosomal proteins are marked by mannose-6-phosphates that are formed by modification of their *N*-linked oligosaccharides shortly after entry into the Golgi apparatus. A specific receptor in the membrane of the *trans*

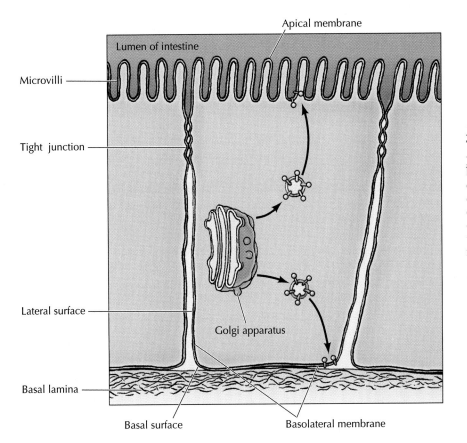

Figure 9.28
Transport to the plasma membrane of polarized cells   The plasma membranes of polarized epithelial cells are divided into apical and basolateral domains. In this example (intestinal epithelium), the apical surface of the cell faces the lumen of the intestine, the lateral surfaces are in contact with neighboring cells, and the basal surface rests on a sheet of extracellular matrix (the basal lamina). The apical membrane is characterized by the presence of microvilli, which facilitate the absorption of nutrients by increasing surface area. Specific proteins are targeted to either the apical or basolateral membranes in the *trans* Golgi network. Tight junctions between neighboring cells maintain the identity of the apical and basolateral membranes by preventing the diffusion of proteins between these domains.

Figure 9.29
A plant cell vacuole   The large central vacuole functions as a lysosome in addition to storing nutrients and maintaining osmotic balance. (E. H. Newcombe/Biological Photo Service.)

2 μm

Golgi network then recognizes these mannose-6-phosphate residues. The resulting complexes of receptor plus lysosomal enzyme are packaged into transport vesicles destined for lysosomes. Lysosomal membrane proteins are targeted by sequences in their cytoplasmic tails, rather than by mannose-6-phosphates.

In yeasts and plant cells, which lack lysosomes, proteins are transported from the Golgi apparatus to an additional destination: the **vacuole** (Figure 9.29). Vacuoles assume the functions of lysosomes in these cells as well as performing a variety of other tasks, such as the storage of nutrients and the maintenance of turgor pressure and osmotic balance. In contrast to lysosomal targeting, proteins are directed to vacuoles by short peptide sequences instead of carbohydrate markers.

The Mechanism of Vesicular Transport

As is evident from the preceding sections of this chapter, transport vesicles play a central role in the traffic of molecules between different membrane-enclosed compartments of the secretory pathway. As discussed in Chapter 12, vesicles are similarly involved in the transport of materials taken up at the cell surface. Vesicular transport is thus a major cellular activity, responsible for molecular traffic between a variety of specific membrane-enclosed compartments. The selectivity of such transport is therefore key to maintaining the functional organization of the cell. For example, lysosomal enzymes must be transported specifically from the Golgi apparatus to lysosomes—not to the plasma membrane or to the ER. Some of the signals that target proteins to specific organelles, such as lysosomes, were discussed earlier in this chapter. These proteins are transported within vesicles, so the specificity of transport is based on the selective packaging of the intended cargo into vesicles that recognize and fuse only with the appropriate target membrane. Because of the central importance of vesicular transport to the organization of eukaryotic cells, understanding the molecular mechanisms that control vesicle packaging, budding, and fusion is a major area of research in cell biology.

Experimental Approaches to Understanding Vesicular Transport

Progress toward elucidating the molecular mechanisms of vesicular transport has been made by three distinct experimental approaches: (1) isolation of yeast mutants that are defective in protein transport and sorting; (2) reconstitution of vesicular transport in cell-free systems; and (3) biochemical analysis of synaptic vesicles, which are responsible for the regulated secretion of neurotransmitters by neurons. Each of these experimental systems has distinct advantages for understanding particular aspects of the transport process. Most important, however, is the fact that results from all three of these avenues of investigation have converged, indicating that similar molecular mechanisms regulate secretion in cells as different as yeasts and mammalian neurons.

As in other areas of cell biology, yeasts have proved to be advantageous in studying the secretory pathway because they are readily amenable to genetic analysis. In particular, Randy Schekman and his colleagues have pioneered the isolation of yeast mutants defective in vesicular transport. These include mutants that are defective at various stages of protein secretion (*sec* mutants), mutants that are unable to transport proteins to the vacuole, and mutants that are unable to retain resident ER proteins. The isolation of such mutants in yeasts led directly to the molecular cloning and analysis of the corresponding genes, thereby identifying a number of proteins involved in various steps of the secretory pathway. For example, the role of Sec61 as a major component of the protein translocation channel in the endoplasmic reticulum was discussed earlier in this chapter.

Biochemical studies of vesicular transport using reconstituted systems have complemented these genetic studies and have enabled the direct isolation of transport proteins from mammalian cells. The first cell-free transport system was developed by James Rothman and colleagues, who analyzed protein transport between compartments of the Golgi apparatus (Figure 9.30). The experimental design exploited a mutant mammalian cell line that lacked the enzyme required to transfer *N*-acetylglucosamine residues to the *N*-linked oligosaccharide at an early stage of its modification in the Golgi apparatus (see Figure 9.24). Consequently, the glycoproteins produced by this mutant cell line lacked added *N*-acetylglucosamine units. However, if Golgi stacks isolated from the mutant cell line were incubated with stacks isolated from normal cells, *N*-acetylglucosamine residues were added to glycoproteins synthesized by the mutant cells. A variety of experiments established that this resulted from vesicular transport of proteins from the Golgi stacks of the mutant cell line to the Golgi stacks of normal cells, so the addition of *N*-acetylglucosamine provided a readily detectable marker for vesicular transport in this reconstituted system. Similar reconstituted systems have been developed to analyze transport between other compartments, including transport from the ER to the Golgi and transport from the Golgi to secretory vesicles, vacuoles, and the plasma membrane. The development of these *in vitro* systems has enabled biochemical studies of the transport process and functional analysis of proteins identified by mutations in yeasts, as well as direct isolation of some of the proteins involved in vesicle budding and fusion.

Critical insights into the molecular mechanisms of vesicular transport have also come from studies of synaptic transmission in neurons, which represents a highly specialized form of regulated secretion. A synapse is the junction of a neuron with another cell, which may be either another neuron or an effector, such as a muscle cell. Information is transmitted across the synapse by chemical neurotransmitters, such as acetylcholine, which are stored in **synaptic vesicles**. Stimulation of the transmitting neuron triggers the fusion of synaptic vesicles with the plasma membrane, causing neurotransmitters to be released and stimulating the postsynaptic neuron or effector cell. Synaptic vesicles are extremely abundant in the brain, allowing them to be purified in large amounts for biochemical analysis. Some of the proteins isolated from synaptic vesicles are closely related to proteins that have been shown to play critical roles in vesicular transport by yeast genetics and reconstitution experiments, so biochemical analysis of these proteins have provided important insights into the molecular mechanisms of vesicle fusion.

Coat Proteins and Vesicle Budding

The first step in vesicular transport is the formation of a vesicle by budding from the membrane. The cytoplasmic surfaces of transport vesicles are coated with proteins, and it appears to be the assembly of these protein coats that drives vesicle budding by distorting membrane conformation. Three kinds of coated vesicles, which appear to function in different types of vesicular transport, have been characterized. The first to be described were the **clathrin-coated vesicles**, which are responsible for the uptake of extracellular molecules from the plasma membrane by endocytosis (see Chapter 12) as well as the transport of molecules from the *trans* Golgi network to lysosomes. Two other types of coated vesicles have been identified as budding from the ER and Golgi complex. These vesicles are called non-clathrin-coated or **COP-coated vesicles** (COP stands for coat protein). One class of these vesicles (COPII-coated vesicles) bud from the ER and carry their cargo forward along the secretory pathway, to the Golgi apparatus. In contrast, COPI-coated vesicles bud from the ER-Golgi intermediate compartment or the Golgi apparatus and function in the retrieval pathways that serve to retain resident proteins in the Golgi and ER. For example, COPI-coated vesicles transport resident ER proteins marked by the KDEL or KKXX retrieval signals back to the ER from the ER-Golgi intermediate compartment or the *cis* Golgi network.

The coats of clathrin-coated vesicles are composed of two types of protein complexes, clathrin and adaptor proteins, which assemble on the cytosolic side of membranes (Figure 9.31). **Clathrin** plays a structural role by assembling into a basketlike lattice structure that distorts the membrane and drives vesicle budding. The binding of clathrin to membranes is mediated by a second class of proteins, called adaptor proteins. Different adaptor proteins are responsible for the assembly of clathrin-coated vesicles at the plasma membrane and at the *trans* Golgi network, and it is the adaptor proteins that are involved in selecting the specific molecules to be incorporated into the vesicles. For example, the AP-1 adaptor protein involved in budding from the *trans* Golgi network binds to the cytosolic portion of the mannose-6-phosphate receptor, thereby directing proteins destined for lysosomes into clathrin-coated vesicles.

The coats of COPI- and COPII-coated vesicles are composed of distinct protein complexes, which function analogously to clathrin and adaptor proteins in vesicle budding. Interestingly, components of the COPI coat interact with the KKXX motif responsible for the retrieval of ER proteins from the Golgi apparatus, consistent with a role for COPI-coated vesicles in recycling from the Golgi to the ER.

The assembly of vesicle coats also requires GTP-binding proteins, which appear to regulate the binding of coat proteins to the membrane. The budding of both clathrin-coated and COPI-coated vesicles from the Golgi complex requires a GTP-binding protein called ARF (ADP-ribosylation factor),

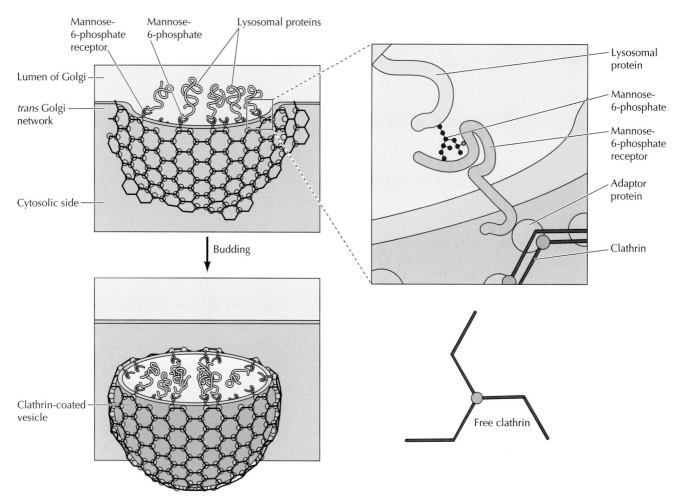

Figure 9.31
Incorporation of lysosomal proteins into clathrin-coated vesicles Proteins targeted for lysosomes are marked by mannose-6-phosphates, which bind to mannose-6-phosphate receptors in the *trans* Golgi network. The mannose-6-phosphate receptors span the Golgi membrane and serve as binding sites for cytosolic adaptor proteins, which in turn bind clathrin. Clathrins consist of three protein chains that associate with each other to form a basketlike lattice that distorts the membrane and drives vesicle budding.

while the budding of COPII-coated vesicles from the ER requires a distinct GTP-binding protein called Sar1. The role of these proteins is illustrated by the function of ARF in assembly of COPI-coated vesicles (Figure 9.32). The first step in vesicle formation is the association of ARF bound to GDP with the Golgi membrane. Proteins in the Golgi membrane then stimulate the exchange of the GDP bound to ARF for GTP, and the COPI coat proteins bind to the ARF/GTP complex. Assembly of the coat is then followed by deformation of the membrane and vesicle budding. ARF then hydrolyzes its bound GTP, leading to the conversion of ARF to the GDP-bound state and the dissociation of coat proteins from the vesicle membrane.

Vesicle Fusion

The fusion of a transport vesicle with its target involves two types of events. First, the transport vesicle must specifically recognize the correct target membrane; for example, a vesicle carrying lysosomal enzymes has to deliver its cargo only to lysosomes. Second, the vesicle and target membranes must fuse, thereby delivering the contents of the vesicle to the target organelle. Research over the last several years has led to development of a model of vesicle fusion in which specific recognition between a vesicle and its target is mediated by interactions between unique pairs of transmembrane proteins, followed by fusion between the phospholipid bilayers of the vesicle and target membranes.

Proteins involved in vesicle fusion were initially identified in James Rothman's laboratory by biochemical analysis of reconstituted vesicular transport

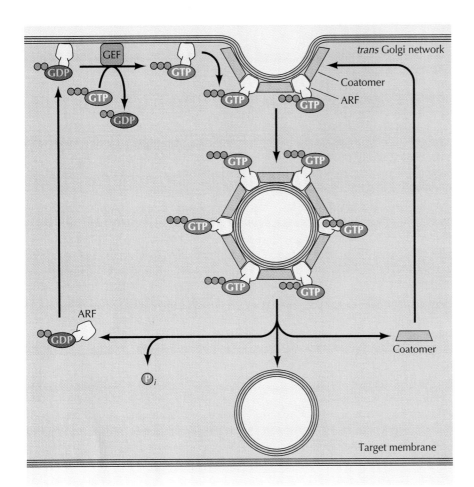

Figure 9.32
Role of ARF in the formation of COP-coated vesicles   ARF alternates between
GTP-bound and GDP-bound states. When bound to GDP, ARF associates with the
membrane of the *trans* Golgi network, where guanine nucleotide exchange factors
(GEF) promote the exchange of the bound GDP for GTP. In its GTP-bound state,
ARF promotes the binding of COPI coat protein (coatomer), leading to vesicle bud-
ding. Hydrolysis of the bound GTP then converts ARF to the GDP-bound state,
leading to disassembly of the vesicle coat prior to fusion with the target membrane.

systems from mammalian cells (see Figure 9.30). Analysis of the proteins
involved in vesicle fusion in these systems led Rothman and his colleagues
to propose a general model, called the **SNARE hypothesis**, in which vesicle
fusion is mediated by interactions between specific pairs of proteins, called
SNAREs, on the vesicle and target membranes (v-SNAREs and t-SNAREs,
respectively) (Figure 9.33). This hypothesis was supported by the identification of SNAREs that were present on synaptic vesicles and by the finding of
yeast secretion mutants that appeared to encode SNAREs required for a variety of vesicle transport events. For example, transport from the ER to the
Golgi in yeast requires specific SNAREs that are located on both the vesicle
and target membranes. The formation of complexes between v-SNAREs on
the vesicle and t-SNAREs on the target membranes then leads to membrane
fusion, by mechanisms which remain to be fully understood.

In addition to SNAREs, vesicle fusion requires at least two other types of
proteins. The **Rab** proteins are a family of small GTP-binding proteins that
are related to the Ras proteins, which were discussed in Chapter 7. More

Vesicle

v-SNARE

Rab

GTP

Rab

GTP

t-SNARE

Target

↓ Docking

Vesicle

Rab

GTP

Rab

GTP

↓ Fusion

GDP

GDP

Rab

Rab

NSF

NSF

SNAPs

SNAPs

↓ Disassembly of SNARE complexes

Figure 9.33
Vesicle fusion Vesicle fusion is mediated by binding between specific pairs of v-SNAREs and t-SNAREs on the vesicle and target membranes, respectively. Rab GTP-binding proteins are required to facilitate formation of v-SNARE/t-SNARE complexes. Following membrane fusion, the NSF/SNAP proteins disassemble the SNARE complex.

than 30 different Rab proteins have been identified and shown to function in specific vesicle transport processes. They may function in several steps of vesicle trafficking, including interacting with SNAREs to regulate and facilitate the formation of v-SNARE/t-SNARE complexes.

Following the formation of complexes between complementary SNAREs and membrane fusion, a complex of two additional proteins (the NSF/SNAP complex) is needed to complete the process of vesicle transport. The NSF/SNAP proteins are recruited to membranes following the formation of v-SNARE/t-SNARE complexes, and they are not required directly for either vesicle/target pairing or for the fusion of paired membranes. Instead, the NSF/SNAP proteins act after membrane fusion to disassemble the SNARE complex, thereby allowing the SNAREs to be reutilized for subsequent rounds of vesicle transport.

Lysosomes

Lysosomes are membrane-enclosed organelles that contain an array of enzymes capable of breaking down all types of biological polymers—proteins, nucleic acids, carbohydrates, and lipids. Lysosomes function as the digestive system of the cell, serving both to degrade material taken up from outside the cell and to digest obsolete components of the cell itself. In their simplest form, lysosomes are visualized as dense spherical vacuoles, but they can display considerable variation in size and shape as a result of differences in the materials that have been taken up for digestion (Figure 9.34). Lysosomes thus represent morphologically diverse organelles defined by the common function of degrading intracellular material.

Lysosomal Acid Hydrolases

Lysosomes contain about 50 different degradative enzymes that can hydrolyze proteins, DNA, RNA, polysaccharides, and lipids. Mutations in the genes that encode these enzymes are responsible for more than 30 different human genetic diseases, which are called **lysosomal storage diseases** because undegraded material accumulates within the lysosomes of affected individuals. Most of these diseases result from deficiencies in single lysosomal enzymes. For example, Gaucher's disease (the most common of these disorders) results from a mutation in the gene that encodes a lysosomal enzyme required for the breakdown of glycolipids. An intriguing exception is I-cell disease, which is caused by a deficiency in the enzyme that catalyzes the first step in the tagging of lysosomal enzymes with mannose-6-phosphate in the Golgi apparatus (see Figure 9.25). The result is a general failure of lysosomal enzymes to be incorporated into lysosomes.

All of the lysosomal enzymes are acid hydrolases, which are active at the acidic pH (about 5) that is maintained within lysosomes but not at the neutral pH (about 7.2) characteristic of the rest of the cytoplasm (Figure 9.35). The requirement of these lysosomal hydrolases for acidic pH provides double protection against uncontrolled digestion of the contents of the cytosol; even if the lysosomal membrane were to break down, the released acid hydrolases would be inactive at the neutral pH of the cytosol. To maintain their acidic internal pH, lysosomes must actively concentrate H^+ ions (pro-

Figure 9.34
Electron micrograph of lysosomes and mitochondria in a mammalian cell Lysosomes are indicated by arrows. (Visuals Unlimited/K. G. Murti.)

0.5 μm

tons). This is accomplished by a proton pump in the lysosomal membrane, which actively transports protons into the lysosome from the cytosol. This pumping requires expenditure of energy in the form of ATP hydrolysis, since it maintains approximately a hundredfold higher H⁺ concentration inside the lysosome.

Endocytosis and Lysosome Formation

One of the major functions of lysosomes is the digestion of material taken up from outside the cell by **endocytosis**, which is discussed in detail in Chapter 12. However, the role of lysosomes in the digestion of material taken up by endocytosis relates not only to the function of lysosomes but also to their formation. In particular, lysosomes are formed by the fusion of transport vesicles budded from the *trans* Golgi network with endosomes, which contain molecules taken up by endocytosis at the plasma membrane.

The formation of lysosomes thus represents an intersection between the secretory pathway, through which lysosomal proteins are processed, and the endocytic pathway, through which extracellular molecules are taken up at the cell surface (Figure 9.36). Material from outside the cell is taken up in clathrin-coated endocytic vesicles, which bud from the plasma membrane and then fuse with early **endosomes**. Membrane components are then recycled to the plasma membrane (discussed in detail in Chapter 12) and the early endosomes gradually mature into late endosomes, which are the precursors to lysosomes. One of the important changes during endosome maturation is the lowering of the internal pH to about 5.5, which plays a key role in the delivery of lysosomal acid hydrolases from the *trans* Golgi network.

As discussed earlier, acid hydrolases are targeted to lysosomes by mannose-6-phosphate residues, which are recognized by mannose-6-phosphate receptors in the *trans* Golgi network and packaged into clathrin-coated vesicles. Following removal of the clathrin coat, these transport vesicles fuse with late endosomes, and the acidic internal pH causes the hydrolases to dissociate from the mannose-6-phosphate receptor (see Figure 9.36). The hydrolases are thus released into the lumen of the endosome, while the receptors remain in the membrane and are eventually recycled to the Golgi. Late endosomes then mature into lysosomes as they acquire a full complement of acid hydrolases, which digest the molecules originally taken up by endocytosis.

Figure 9.35
Organization of the lysosome
Lysosomes contain a variety of acid hydrolases that are active at the acidic pH maintained within the lysosome, but not at the neutral pH of the cytosol. The acidic internal pH of lysosomes results from the action of a proton pump in the lysosomal membrane, which imports protons from the cytosol coupled to ATP hydrolysis.

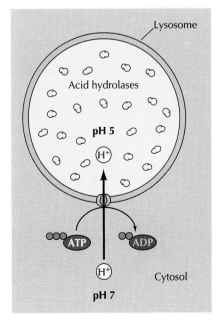

Figure 9.36
Endocytosis and lysosome formation
Molecules are taken up from outside the cell in endocytic vesicles, which fuse with early endosomes. Membrane components are recycled as the early endosomes mature into late endosomes. Transport vesicles carrying acid hydrolases from the Golgi apparatus then fuse with late endosomes, which mature into lysosomes as they acquire a full complement of lysosomal enzymes. The acid hydrolases dissociate from the mannose-6-phosphate receptor when the transport vesicles fuse with late endosomes, and the receptors are recycled to the Golgi apparatus.

Molecular Medicine

Gaucher's Disease

The Disease
Gaucher's disease is the most common of the lysosomal storage diseases, which are caused by a failure of lysosomes to degrade substances that they normally break down. The resulting accumulation of nondegraded compounds leads to an increase in the size and number of lysosomes within the cell, eventually resulting in cellular malfunction and pathological consequences to affected organs. Gaucher's disease is found primarily in the Jewish population, where it has a frequency of about 1 in 2500 individuals. There are three types of the disease, which differ in severity and nervous system involvement. In the most common form of the disease (type I), the nervous system is not involved; the disease is manifest as spleen and liver enlargement and development of bone lesions. Many patients with this form of the disease have no serious symptoms, and their life span is unaffected. The more severe forms of the disease (types II and III) are much rarer and found in both Jewish and non-Jewish populations. The most devastating is type II disease, in which extensive neurological involvement is evident in infancy and patients die early in life. Type III disease, intermediate in severity between types I and II, is characterized by the onset of neurological symptoms (including dementia and spasticity) by about age ten.

Molecular and Cellular Basis
Gaucher's disease is caused by deficiency of the lysosomal enzyme glucocerebrosidase, which catalyzes the hydrolysis of glucocerebroside to glucose and ceramide (see figure). This enzyme deficiency was demonstrated in 1965, and the responsible gene was cloned in 1985. Since then, more than 30 different mutations responsible for Gaucher's disease have been identified. Interestingly, the severity of the

disease can be largely predicted from the nature of these mutations. For example, patients with a mutation leading to the relatively conservative amino acid substitution of serine for asparagine have type I disease; patients with a mutation leading to substitution of proline for leucine have more severe enzyme deficiencies and develop type II or III disease.

Except for the very rare type II and III forms of the disease, the only cells affected in Gaucher's disease are macrophages. Because their function is to eliminate aged and damaged cells by phagocytosis, macrophages continually ingest large amounts of lipids, which are normally degraded in lysosomes. Deficiencies of glucocerebrosidase are therefore particularly evident in macrophages of these organs, consistent with spleen and liver abnormalities being the primary manifestation in most cases of Gaucher's disease.

Prevention and Treatment
Gaucher's disease is a prime example of a disease that can be treated by enzyme replacement therapy, in which exogenous administration of an enzyme is used to correct an enzyme defect. This approach to treatment of lysosomal storage diseases was suggested by Christian deDuve in the 1960s, based on the idea that exogenously administered enzymes

might be taken up by endocytosis and transported to lysosomes. In type I Gaucher's disease, this approach is particularly attractive because the single target cell is the macrophage. In the 1970s, it was discovered that macrophages express cell surface receptors that bind mannose residues on extracellular glycoproteins and then internalize these proteins by endocytosis. This finding suggested that exogenously administered glucocerebrosidase could be specifically targeted to macrophages by modifications that would expose mannose residues. Enzyme prepared from human placenta was appropriately modified, and clinical studies have clearly demonstrated its effectiveness in the treatment of Gaucher's disease.

Unfortunately, enzyme replacement therapy for Gaucher's disease is expensive: The cost of enzyme alone for treatment of a single patient at recommended doses is between $50,000 and $500,000 per year. The expense of this treatment puts it far beyond the resources of individual patients, and has raised a number of societal issues concerning the cost of expensive drugs for the treatment of rare disorders.

References
Beutler, E. 1995. Gaucher's disease. *Adv. Genet.* 32: 17–49.

Beutler, E. 1996. The cost of treating Gaucher's disease. *Nature Med.* 2: 523–524.

Glucocerebroside → Glucocerebrosidase → **Glucose** + **Ceramide**

The enzyme deficiency in Gaucher's disease prevents the hydrolysis of glucocerebroside to glucose and ceramide.

Phagocytosis and Autophagy

In addition to degrading molecules taken up by endocytosis, lysosomes digest material derived from two other routes: phagocytosis and autophagy (Figure 9.37). In **phagocytosis**, specialized cells, such as macrophages, take up and degrade large particles, including bacteria, cell debris, and aged cells that need to be eliminated from the body. Such large particles are taken up in phagocytic vacuoles (**phagosomes**), which then fuse with lysosomes, resulting in digestion of their contents. The lysosomes formed in this way (**phagolysosomes**) can be quite large and heterogeneous, since their size and shape is determined by the content of material that is being digested.

Lysosomes are also responsible for **autophagy**, the gradual turnover of the cell's own components. The first step of autophagy appears to be the

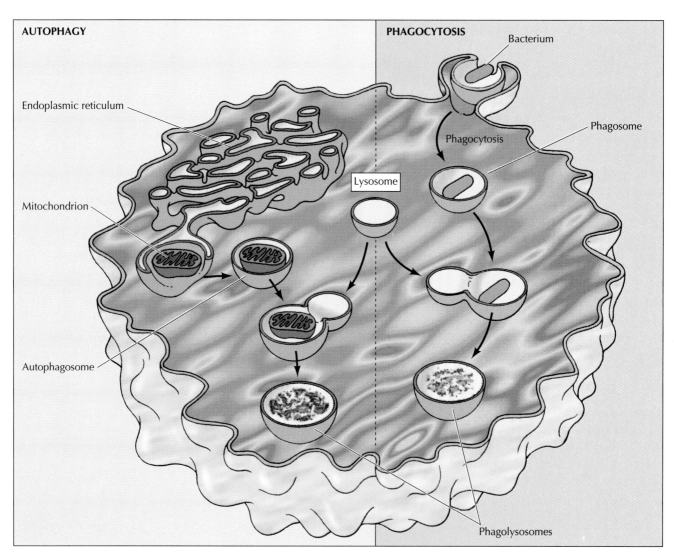

Figure 9.37
Lysosomes in phagocytosis and autophagy In phagocytosis, large particles (such as bacteria) are taken up into phagocytic vacuoles or phagosomes. In autophagy, internal organelles (such as mitochondria) are enclosed by membrane fragments from the ER, forming autophagosomes. Both phagosomes and autophagosomes fuse with lysosomes to form large phagolysosomes, in which their contents are digested.

enclosure of an organelle (e.g., a mitochondrion) in membrane derived from the ER. The resulting vesicle (an **autophagosome**) then fuses with a lysosome, and its contents are digested (see Figure 9.37). As discussed in Chapter 7, autophagy is responsible for the gradual turnover of cytoplasmic organelles.

SUMMARY

KEY TERMS

THE ENDOPLASMIC RETICULUM

The Endoplasmic Reticulum and Protein Secretion: The endoplasmic reticulum is the first branch point in protein sorting. In mammalian cells, proteins destined for secretion, lysosomes, or the plasma membrane are translated on membrane-bound ribosomes and transferred into the rough ER as their translation proceeds.

endoplasmic reticulum (ER), rough ER, smooth ER, secretory vesicle, secretory pathway

Targeting Proteins to the Endoplasmic Reticulum: Proteins can be targeted to the ER either while their translation is still in progress or following completion of translation in the cytosol. In mammalian cells, most proteins are translocated into the ER while they are being translated on membrane-bound ribosomes. Ribosomes engaged in the synthesis of secreted proteins are targeted to the endoplasmic reticulum by signal sequences at the amino terminus of the polypeptide chain. Growing polypeptide chains are then translocated into the ER through protein channels and released into the ER lumen by cleavage of the signal sequence.

signal sequence, microsome, signal recognition particle (SRP), 7SL RNA, signal peptidase

Insertion of Proteins into the ER Membrane: Integral membrane proteins of the plasma membrane or the membranes of the ER, Golgi apparatus, and lysosomes are initially inserted into the membrane of the ER. Rather than being translocated into the ER lumen, these proteins are anchored by membrane-spanning α helices that stop the transfer of the growing polypeptide chain across the membrane.

Protein Folding and Processing in the ER: Polypeptide chains are folded into their correct three-dimensional conformations within the ER. The ER is also the site of *N*-linked glycosylation and addition of GPI anchors.

protein disulfide isomerase, glycosylphosphatidylinositol (GPI) anchor

The Smooth ER and Lipid Synthesis: The ER is the major site of lipid synthesis in eukaryotic cells, and smooth ER is abundant in cells that are active in lipid metabolism and detoxification of lipid-soluble drugs.

flippase

Export of Proteins and Lipids from the ER: Proteins and lipids are transported in vesicles from the ER to the Golgi apparatus. Resident ER proteins are marked by sequences that signal their return from the Golgi to the ER by a recycling pathway. Other targeting sequences mediate the selective packaging of exported proteins into vesicles that transport them to the Golgi.

THE GOLGI APPARATUS

Organization of the Golgi: The Golgi apparatus functions in protein processing and sorting as well as in the synthesis of lipids and polysaccharides. Proteins are transported from the endoplasmic reticulum to the *cis* Golgi network. From there they are transported to the Golgi stack, which represents the site of most metabolic activities of the Golgi apparatus.

Golgi apparatus, Golgi complex, *cis* Golgi network, Golgi stack, *trans* Golgi network

Modified proteins are transported from the Golgi stack to the *trans* Golgi network, where they are sorted and packaged in vesicles for transport to lysosomes, the plasma membrane, or the exterior of the cell.

mannose-6-phosphate, signal patch

Protein Glycosylation within the Golgi: The *N*-linked oligosaccharides added to proteins in the ER are modified within the Golgi. Those proteins destined for lysosomes are specifically phosphorylated on mannose residues, and mannose-6-phosphate serves as a targeting signal that directs their transport to lysosomes from the *trans* Golgi network. *O*-linked glycosylation also takes place within the Golgi.

Lipid and Polysaccharide Metabolism in the Golgi: The Golgi apparatus is the site of synthesis of glycolipids, sphingomyelin, and the complex polysaccharides of plant cell walls.

apical domain, basolateral domain, vacuole

Protein Sorting and Export from the Golgi Apparatus: Proteins are sorted in the *trans* Golgi network for packaging into transport vesicles targeted for secretion, the plasma membrane, lysosomes, or yeast and plant vacuoles. In polarized cells, proteins are specifically targeted to the apical and basolateral domains of the plasma membrane.

THE MECHANISM OF VESICULAR TRANSPORT

synaptic vesicle

Experimental Approaches to Understanding Vesicular Transport: The mechanism of vesicular transport has been elucidated through studies of yeast mutants, reconstituted cell-free systems, and synaptic vesicles.

clathrin-coated vesicle, COP-coated vesicle, clathrin

Coat Proteins and Vesicle Budding: The cytoplasmic surfaces of vesicles are coated with proteins that drive vesicle budding and select the specific molecules to be transported.

SNARE hypothesis, Rab

Vesicle Fusion: Vesicle binding to the correct target membrane is mediated by interactions between pairs of transmembrane proteins, which leads to membrane fusion.

LYSOSOMES

lysosome, lysosomal storage disease

Lysosomal Acid Hydrolases: Lysosomes contain an array of acid hydrolases that degrade proteins, nucleic acids, polysaccharides, and lipids. These enzymes function specifically at the acidic pH maintained within lysosomes.

endocytosis, endosome

Endocytosis and Lysosome Formation: Extracellular molecules taken up by endocytosis are transported to endosomes, which mature to lysosomes as lysosomal acid hydrolases are delivered from the Golgi.

phagocytosis, phagosome, phagolysosome, autophagy, autophagosome

Phagocytosis and Autophagy: Lysosomes are responsible for the degradation of large particles taken up by phagocytosis and for the gradual digestion of the cell's own components by autophagy.

Questions

1. You have generated a recombinant cDNA in which a signal sequence has been added to the amino terminus of a nuclear protein. Where would you predict the protein encoded by this cDNA to be localized?

2. Sec61 is a critical component of the protein channel through the ER membrane. In Sec61 mutant cells, what is the fate of proteins that are normally localized to the Golgi apparatus?

3. Why are the carbohydrate groups of glycoproteins always exposed on the surface of the cell?

4. What effect would the addition of a lysosome-targeting signal have on the subcellular localization of a protein that is normally cytosolic? How would it affect localization of a protein that is normally secreted?

5. What is the predicted fate of lysosomal acid hydrolases in I-cell disease, in which cells are deficient in the enzyme required for formation of mannose-6-phosphate residues?

References and Further Reading

The Endoplasmic Reticulum

Abeijon, C. and C. B. Hirschberg. 1992. Topography of glycosylation reactions in the endoplasmic reticulum. *Trends. Biochem. Sci.* 17: 32–36. [R]

Beckmann, R. D. R. Bubeck, P. Grassuci, A. Penczek, G. Verschoor, G. Blobel and J. Frank. 1997. Alignment of conduits for the nascent polypeptide chain in the ribosome-Sec61 complex. *Science* 278: 2123–2130. [P]

Bishop, W. R. and R. M. Bell. 1988. Assembly of phospholipids into cellular membranes: Biosynthesis, transmembrane movement, and intracellular translocation. *Ann. Rev. Cell Biol.* 4: 580–610. [R]

Blobel, G. and B. Dobberstein. 1975. Transfer of proteins across the membrane. I. Presence of proteolytically processed and unprocessed nascent immunoglobulin light chains on membrane-bound ribosomes of murine myeloma. *J. Cell Biol.* 67: 835–851. [P]

Corsi, A. K. and R. Schekman. 1996. Mechanism of polypeptide translocation into the endoplasmic reticulum. *J. Biol. Chem.* 271: 30299–30302. [R]

Englund, P. T. 1993. The structure and biosynthesis of glycosyl phosphatidylinositol anchors. *Ann. Rev. Biochem.* 62: 121–138. [R]

Fiedler, K. and K. Simons. 1995. The role of N-glycans in the secretory pathway. *Cell* 81: 309–312. [R]

Hegde, R. S. and V. R. Lingappa. 1999. Regulation of protein biogenesis at the endoplasmic reticulum membrane. *Trends Cell Biol.* 9: 132–137. [R]

Kent, C. 1995. Eukaryotic phospholipid biosynthesis. *Ann. Rev. Biochem.* 64: 315–343. [R]

Matlack, K. E. S., W. Mothes and T. A. Rapoport. 1998. Protein translocation: Tunnel vision. *Cell* 92: 381–390. [R]

Matlack, K. E. S., B. Misselwitz, K. Plath and T. A. Rapoport. 1999. BiP acts as a molecular ratchet during posttranslational transport of prepro-α factor across the ER membrane. *Cell* 97: 553–564. [P]

Menon, A. K. 1995. Flippases. *Trends Cell Biol.* 5: 355–360. [R]

Palade, G. 1975. Intracellular aspects of the process of protein synthesis. *Science* 189: 347–358. [R]

Powers, T. and P. Walter. 1997. A ribosome at the end of the tunnel. *Science* 278: 2072–2073. [R]

Rapoport, T. A., B. Jungnickel and U. Kutay. 1996. Protein transport across the eukaryotic endoplasmic reticulum and bacterial inner membranes. *Ann. Rev. Biochem.* 65: 271–303. [R]

Singer, S. J. 1990. The structure and insertion of integral proteins in membranes. *Ann. Rev. Cell Biol.* 6: 247–296. [R]

Teasdale, R. D. and M. R. Jackson. 1996. Signal-mediated sorting of membrane proteins between the endoplasmic reticulum and the Golgi apparatus. *Ann. Rev. Cell Dev. Biol.* 12: 27–54. [R]

Udenfriend, S. and K. Kodukula. 1995. How glycosylphosphatidylinositol anchored membrane proteins are made. *Ann. Rev. Biochem.* 64: 563–591. [R]

Warren, G. and I Mellman. 1999. Bulk flow redux? *Cell* 98:125–127. [R]

The Golgi Apparatus

Allan, B. B. and W. E. Balch. 1999. Protein sorting by directed maturation of Golgi compartments. *Science* 285:63–66. [R]

Bannykh, S. I., N. Nishimura, and W. E. Balch. 1998. Getting into the Golgi. *Trends Cell Biol.* 8:21–25. [R]

Baranski, T. J., P. L. Faust and S. Kornfeld. 1990. Generation of a lysosomal enzyme targeting signal in the secretory protein pepsinogen. *Cell* 63: 281–291. [P]

Burgess, T. L. and R. B. Kelly. 1987. Constitutive and regulated secretion of proteins. *Ann. Rev. Cell Biol.* 3: 243–293. [R]

Chrispeels, M. J. and N. V. Raikhel. 1992. Short peptide domains target proteins to plant vacuoles. *Cell* 68: 613–616. [R]

Conibear, E. and T. H. Stevens. 1995. Vacuolar biogenesis in yeast: Sorting out the sorting proteins. *Cell* 83: 513–516. [R]

Driouich, A., L. Faye and L. A. Staehelin. 1993. The plant Golgi apparatus: A factory for complex polysaccharides and glycoproteins. *Trends Biochem. Sci.* 18: 210–214. [R]

Farquhar, M. G. and G. E. Palade. 1998. The Golgi apparatus: 100 years of progress and controversy. *Trends Cell Biol.* 8: 2–10. [R]

Folsch, H., H. Ohno, J. S. Bonifacino and I. Mellman. 1999. A novel clathrin adaptor complex mediates basolateral targeting in polarized epithelial cells. *Cell* 99: 189–196. [P]

Glick, B. S. and V. Malhotra. 1998. The curious state of the Golgi apparatus. *Cell* 95: 883–889. [R]

Hirschberg, C. B. and C. B. Snider. 1987. Topography of glycosylation in the rough endoplasmic reticulum and Golgi apparatus. *Ann. Rev. Biochem.* 56: 63–87. [R]

Kornfeld, R. and S. Kornfeld. 1985. Assembly of asparagine-linked oligosaccharides. *Ann. Rev. Biochem.* 54: 631–664. [R]

Kornfeld, S. and I. Mellman. 1989. The biogenesis of lysosomes. *Ann. Rev. Cell Biol.* 5: 483–525. [R]

Mostov, K., M. B. A. ter Beest and S. J. Chapin. 1999. Catch the μ1B train to the basolateral surface. *Cell* 99: 121–122. [R]

Munro, S. 1998. Localization of proteins to the Golgi apparatus. *Trends Cell Biol.* 8: 11–15. [R]

Nothwehr, S. F. and T. H. Stevens. 1994. Sorting of membrane proteins in the yeast secretory pathway. *J. Biol. Chem.* 269: 10185–10188. [R]

Pelham, H. R. B. 1998. Getting through the Golgi complex. *Trends Cell Biol.* 8: 45–49. [R]

Pelham, H. R. B. and S. Munro. 1993. Sorting of membrane proteins in the secretory pathway. *Cell* 75: 603–605. [R]

Rodriguez-Boulan, E. and S. K. Powell. 1992. Polarity of epithelial and neuronal cells. *Ann. Rev. Cell Biol.* 8: 395–427. [R]

Rothman, J. E. 1994. Mechanisms of intracellular protein transport. *Nature* 355: 409–415. [R]

van Meer, G. 1998. Lipids of the Golgi membrane. *Trends Cell Biol.* 8:29–33. [R]

Varki, A. 1998. Factors controlling the glycosylation potential of the Golgi apparatus. *Trends Cell Biol.* 8: 34–40. [R]

The Mechanism of Vesicular Transport

Balch, W. E., W. G. Dunphy, W. A. Braeli and J. E. Rothman. 1984. Reconstitution of the transport of protein between successive compartments of the Golgi measured by the coupled incorporation of *N*-acetylglucosamine. *Cell* 39: 405–416. [P]

Gonzales, L., Jr. and R. H. Scheller. 1999. Regulation of membrane trafficking: Structural insights from a Rab/effector complex. *Cell* 96: 755–758. [R]

Kirchhausen, T. 1999. Adaptors for clathrin-mediated traffic. *Ann. Rev. Cell Dev. Biol.* 15: 705–732. [R]

Novick, P. and P. Brennwald. 1993. Friends and family: The role of the Rab GTPases in vesicular traffic. *Cell* 75: 597–601. [R]

Novick, P., C. Field and R. Schekman. 1980. Identification of 23 complementation groups required for post-translational events in the yeast secretory pathway. *Cell* 21: 205–215. [P]

Nuoffer, C. and W. E. Balch. 1994. GTPases: Multifunctional molecular switches regulating vesicular traffic. *Ann. Rev. Biochem.* 63: 949–990. [R]

O'Connor, V., G. J. Augustine and H. Betz. 1994. Synaptic vesicle exocytosis: Molecules and models. *Cell* 76: 785–787. [R]

Odorizzi, G., C. R. Cowles and S. D. Emr. 1998. The AP-3 complex: A coat of many colors. *Trends Cell Biol.* 8: 282–288. [R]

Pishvaee, B. and G. S. Payne. 1998. Clathrin coats—threads laid bare. *Cell* 95: 443–446. [R]

Rothman, J. E. 1994. Mechanisms of intracellular protein transport. *Nature* 372: 55–62. [R]

Schekman, R. and L. Orci. 1996. Coat proteins and vesicle budding. *Science* 271: 1526–1533. [R]

Schimoller, F., I. Simon and S. R. Pfeffer. 1998. Rab GTPases, directors of vesicle docking. *J. Biol. Chem.* 273: 22161–22164. [R]

Schmid, S. L. 1997. Clathrin-coated vesicle formation and protein sorting: An integrated process. *Ann. Rev. Biochem.* 66: 511–548. [R]

Smith, C. J. and B. M. Pearse. 1999. Clathrin: Anatomy of a coat protein. *Trends Cell Biol.* 9: 335–338. [R]

Söllner, T., S. W. Whiteheart, M. Brunner, H. Erdjument-Bromage, S. Geromanos, P. Tempst and J. E. Rothman. 1993. SNAP receptors implicated in vesicle targeting and fusion. *Nature* 362: 318–324. [P]

Springer, S., A. Spang and R. Schekman. 1999. A primer on vesicle budding. *Cell* 97: 145–148. [R]

Lysosomes

Dunn, W. A., Jr. 1994. Autophagy and related mechanisms of lysosome-mediated protein degradation. *Trends Cell Biol.* 4: 139–143. [R]

Forgac, M. 1999. Structure and properties of the vacuolar (H$^+$)-ATPases. *J. Biol. Chem.* 274: 12951–12954. [R]

Fukuda, M. 1991. Lysosomal membrane glycoproteins. *J. Biol. Chem.* 266: 21327–21330. [R]

Gruenberg, J. and K. E. Howell. 1989. Membrane traffic in endocytosis: Insights from cell-free assays. *Ann. Rev. Cell Biol.* 5: 453–481. [R]

Kornfeld, S. 1992. Structure and function of the mannose-6-phosphate/insulinlike growth factor II receptors. *Ann. Rev. Biochem.* 61: 307–330. [R]

Kornfeld, S. and I. Mellman. 1989. The biogenesis of lysosomes. *Ann. Rev. Cell Biol.* 5: 483–525. [R]

Neufeld, E. F. 1991. Lysosomal storage diseases. *Ann. Rev. Biochem.* 60: 257–280. [R]

10

Bioenergetics and Metabolism

Mitochondria, Chloroplasts, and Peroxisomes

I N ADDITION TO BEING INVOLVED IN PROTEIN SORTING AND TRANSPORT, cytoplasmic organelles provide specialized compartments in which a variety of metabolic activities take place. The generation of metabolic energy is a major activity of all cells, and two cytoplasmic organelles are specifically devoted to energy metabolism and the production of ATP. Mitochondria are responsible for generating most of the useful energy derived from the breakdown of lipids and carbohydrates, and chloroplasts use energy captured from sunlight to generate both ATP and the reducing power needed to synthesize carbohydrates from CO_2 and H_2O. The third organelle discussed in this chapter, the peroxisome, contains enzymes involved in a variety of different metabolic pathways, including the breakdown of fatty acids and the metabolism of a by-product of photosynthesis.

Mitochondria, chloroplasts, and peroxisomes differ from the organelles discussed in the preceding chapter not only in their functions, but also in their mechanism of assembly. Rather than being synthesized on membrane-bound ribosomes and translocated into the endoplasmic reticulum, proteins destined for peroxisomes, mitochondria, and chloroplasts are synthesized on free ribosomes in the cytosol and imported into their target organelles as completed polypeptide chains. Mitochondria and chloroplasts also contain their own genomes, which include some genes that are transcribed and translated within the organelle. Protein sorting to the cytoplasmic organelles discussed in this chapter is thus distinct from the pathways of vesicular transport that connect the endoplasmic reticulum, Golgi apparatus, lysosomes, and plasma membrane.

Mitochondria

Mitochondria play a critical role in the generation of metabolic energy in eukaryotic cells. As reviewed in Chapter 2, they are responsible for most of the useful energy derived from the breakdown of carbohydrates and fatty acids, which is converted to ATP by the process of oxidative phosphorylation. Most mitochondrial proteins are translated on free cytosolic ribosomes and imported into the organelle by specific targeting signals. In addition,

387

mitochondria are unique among the cytoplasmic organelles already discussed in that they contain their own DNA, which encodes tRNAs, rRNAs, and some mitochondrial proteins. The assembly of mitochondria thus involves proteins encoded by their own genomes and translated within the organelle, as well as proteins encoded by the nuclear genome and imported from the cytosol.

Organization and Function of Mitochondria

Mitochondria are surrounded by a double-membrane system, consisting of inner and outer mitochondrial membranes separated by an intermembrane space (Figure 10.1). The inner membrane forms numerous folds (**cristae**), which extend into the interior (or **matrix**) of the organelle. Each of these components plays distinct functional roles, with the matrix and inner membrane representing the major working compartments of mitochondria.

The matrix contains the mitochondrial genetic system as well as the enzymes responsible for the central reactions of oxidative metabolism (Figure 10.2). As discussed in Chapter 2, the oxidative breakdown of glucose and fatty acids is the principal source of metabolic energy in animal cells. The initial stages of glucose metabolism (glycolysis) occur in the cytosol, where glucose is converted to pyruvate (see Figure 2.32). Pyruvate is then transported into mitochondria, where its complete oxidation to CO_2 yields the bulk of usable energy (ATP) obtained from glucose metabolism. This involves the initial oxidation of pyruvate to acetyl CoA, which is then broken down to CO_2 via the citric acid cycle (see Figures 2.33 and 2.34). The oxidation of fatty acids also yields acetyl CoA (see Figure 2.36), which is similarly metabolized by the citric acid cycle in mitochondria. The enzymes of the citric acid cycle (located in the matrix of mitochondria) thus are central players in the oxidative breakdown of both carbohydrates and fatty acids.

The oxidation of acetyl CoA to CO_2 is coupled to the reduction of NAD^+ and FAD to NADH and $FADH_2$, respectively. Most of the energy derived from oxidative metabolism is then produced by the process of oxidative phosphorylation (discussed in detail in the next section), which takes place

Figure 10.1
Structure of a mitochondrion
Mitochondria are bounded by a double-membrane system, consisting of inner and outer membranes. Folds of the inner membrane (cristae) extend into the matrix. (Micrograph by K. R. Porter/Photo Researchers, Inc.)

0.5 μm

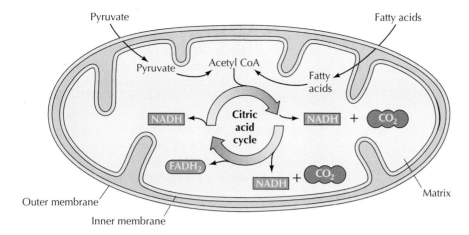

Figure 10.2
**Metabolism in the matrix of mito-
chondria**   Pyruvate and fatty acids
are imported from the cytosol and
converted to acetyl CoA in the mitochondrial matrix. Acetyl CoA is then
oxidized to CO_2 via the citric acid
cycle, the central pathway of oxidative
metabolism.

in the inner mitochondrial membrane. The high-energy electrons from
NADH and $FADH_2$ are transferred through a series of carriers in the membrane to molecular oxygen. The energy derived from these electron transfer reactions is converted to potential energy stored in a proton gradient
across the membrane, which is then used to drive ATP synthesis. The inner
mitochondrial membrane thus represents the principal site of ATP generation, and this critical role is reflected in its structure. First, its surface area is
substantially increased by its folding into cristae. In addition, the inner
mitochondrial membrane contains an unusually high percentage (greater
than 70%) of proteins, which are involved in oxidative phosphorylation as
well as in the transport of metabolites (e.g., pyruvate and fatty acids)
between the cytosol and mitochondria. Otherwise, the inner membrane is
impermeable to most ions and small molecules—a property critical to
maintaining the proton gradient that drives oxidative phosphorylation.

In contrast to the inner membrane, the outer mitochondrial membrane
is freely permeable to small molecules. This is because it contains proteins
called **porins**, which form channels that allow the free diffusion of molecules smaller than about 6000 daltons. The composition of the intermembrane space is therefore similar to the cytosol with respect to ions and
small molecules. Consequently, the inner mitochondrial membrane is the
functional barrier to the passage of small molecules between the cytosol
and the matrix and maintains the proton gradient that drives oxidative
phosphorylation.

The Genetic System of Mitochondria

Mitochondria contain their own genetic system, which is separate and distinct from the nuclear genome of the cell. As reviewed in Chapter 1, mitochondria are thought to have evolved from bacteria that developed a symbiotic relationship in which they lived within larger cells (**endosymbiosis**).
This hypothesis has recently been substantiated by the results of DNA
sequence analysis, which revealed striking similarities between the genomes
of mitochondria and of the bacterium *Rickettsia prowazekii*. *Rickettsia* are
intracellular parasites which, like mitochondria, are only able to reproduce
within eukaryotic cells. Consistent with their similar symbiotic lifestyles, the
genomic DNA sequences of *Rickettsia* and mitochondria suggest that they
share a common ancestor, from which the genetic system of present-day
mitochondria evolved.

Mitochondrial genomes are usually circular DNA molecules, like those
of bacteria, which are present in multiple copies per organelle. They vary

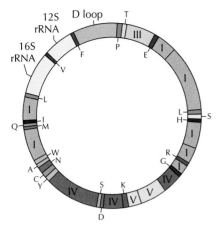

Figure 10.3
The human mitochondrial genome
The genome contains 13 protein-coding sequences, which are designated as components of respiratory complexes I, III, IV, or V. In addition, the genome contains genes for 12S and 16S rRNAs and for 22 tRNAs, which are designated by the one-letter code for the corresponding amino acid. The region of the genome designated "D loop" contains an origin of DNA replication and transcriptional promoter sequences.

considerably in size between different species. The genomes of human and most other animal mitochondria are only about 16 kb, but substantially larger mitochondrial genomes are found in yeasts (approximately 80 kb) and plants (more than 200 kb). However, these larger mitochondrial genomes are composed predominantly of noncoding sequences and do not appear to contain significantly more genetic information. For example, the largest sequenced mitochondrial genome is that of the plant *Arabidopsis thaliana*. Although *Arabidopsis* mitochondrial DNA is approximately 367 kb, it encodes only 32 proteins: just more than twice the number encoded by human mitochondrial DNA. The largest number of mitochondrial genes has been found in mitochondrial DNA of the protozoan *Reclinomonas americana*, which is 69 kb and contains 97 genes. The mitochondrial genome of *Reclinomonas* appears to more closely resemble the bacterial genome from which mitochondria evolved than most present-day mitochondrial genomes, which encode only a small number of proteins that are essential components of the oxidative phosphorylation system. In addition, mitochondrial genomes encode all of the ribosomal RNAs and most of the transfer RNAs needed for translation of these protein-coding sequences within mitochondria. Other mitochondrial proteins are encoded by nuclear genes, which are thought to have been transferred to the nucleus from the ancestral mitochondrial genome.

The human mitochondrial genome encodes 13 proteins involved in electron transport and oxidative phosphorylation (Figure 10.3). In addition, human mitochondrial DNA encodes 16S and 12S rRNAs and 22 tRNAs, which are required for translation of the proteins encoded by the organelle genome. The two rRNAs are the only RNA components of animal and yeast mitochondrial ribosomes, in contrast to the three rRNAs of bacterial ribosomes (23S, 16S, and 5S). Plant mitochondrial DNAs, however, also encode a third rRNA of 5S. The mitochondria of plants and protozoans also differ in importing and utilizing tRNAs encoded by the nuclear as well as the mitochondrial genome, whereas in animal mitochondria, all the tRNAs are encoded by the organelle.

The small number of tRNAs encoded by the mitochondrial genome highlights an important feature of the mitochondrial genetic system—the use of a slightly different genetic code, which is distinct from the "universal" genetic code used by both prokaryotic and eukaryotic cells (Table 10.1). As discussed in Chapter 3, there are 64 possible triplet codons, of which 61 encode the 20 different amino acids incorporated into proteins (see Table 3.1). Many tRNAs in both prokaryotic and eukaryotic cells are able to recognize more than a single codon in mRNA because of "wobble," which allows some mispairing between the tRNA anticodon and the third position of certain complementary codons (see Figure 7.3). However, at least 30 different tRNAs are required to translate the universal code according to the wobble rules. Yet human mitochondrial DNA encodes only 22 tRNA species, and these are the only tRNAs used for translation of mitochondrial mRNAs. This is accomplished by an extreme form of wobble in which U in the anticodon of the tRNA can pair with any of the four bases in the third codon position of mRNA, allowing four codons to be recognized by a single tRNA. In addition, some codons specify different amino acids in mitochondria than in the universal code.

Like the DNA of nuclear genomes, mitochondrial DNA can be altered by mutations, which are frequently deleterious to the organelle. Since almost all the mitochondria of fertilized eggs are contributed by the oocyte rather than by the sperm, germ-line mutations in mitochondrial DNA are transmitted to the next generation by the mother. Such mutations have been associated with a number of diseases. For example, Leber's hereditary optic neuropathy, a disease that leads to blindness, can be caused by mutations in mitochondrial genes that encode components of the electron transport chain. In addition, the progressive accumulation of mutations in mitochondrial DNA during the lifetime of individuals has been suggested to contribute to the process of aging.

Protein Import and Mitochondrial Assembly

In contrast to the RNA components of the mitochondrial translation apparatus (rRNAs and tRNAs), most mitochondrial genomes do not encode the proteins required for DNA replication, transcription, or translation. Instead, the genes that encode proteins required for the replication and expression of mitochondrial DNA are contained in the nucleus. In addition, the nucleus contains the genes that encode most of the mitochondrial proteins required for oxidative phosphorylation and all of the enzymes involved in mitochondrial metabolism (e.g., enzymes of the citric acid cycle). The proteins encoded by these genes (more than 95% of mitochondrial proteins) are synthesized on free cytosolic ribosomes and imported into mitochondria as completed polypeptide chains. Because of the double-membrane structure of mitochondria, the import of proteins is considerably more complicated than the transfer of a polypeptide across a single phospholipid bilayer. Proteins targeted to the matrix have to cross both the inner and outer mitochondrial membranes, while other proteins need to be sorted to distinct compartments within the organelle (e.g., the intermembrane space).

The import of proteins to the matrix is the best-understood aspect of mitochondrial protein sorting (Figure 10.4). Most proteins are targeted to mitochondria by amino-terminal sequences of 20 to 35 amino acids (called **presequences**) that are removed by proteolytic cleavage following their import into the organelle. The presequences of mitochondrial proteins, first characterized by Gottfried Schatz, contain multiple positively charged amino acid residues, usually in an amphipathic α helix. The first step in protein import is the binding of these presequences to receptors on the surface of mitochondria. The polypeptide chains are then inserted into a protein complex that directs translocation across the outer membrane (the translocase of the outer membrane or Tom complex). The proteins are then transferred to a second protein complex in the inner membrane (the translocase of the inner membrane or Tim complex). Continuing protein translocation requires the electrochemical potential established across the inner mitochondrial membrane during electron transport. As discussed in the next section of this chapter, the transfer of high-energy electrons from NADH and FADH$_2$ to molecular oxygen is coupled to the transfer of protons from the mitochondrial matrix to the intermembrane space. Since protons are charged particles, this transfer establishes an electric potential across the inner membrane, with the matrix being negative. During protein import, this electric potential drives translocation of the positively charged presequence.

To be translocated across the mitochondrial membrane, proteins must be at least partially unfolded. Consequently, protein import into mitochondria requires molecular chaperones in addition to the membrane proteins involved in translocation (see Figure 10.4). On the cytosolic side, members of the Hsp70 family of chaperones maintain proteins in a partially unfolded state so that they can be inserted into the mitochondrial membrane. As they cross the inner membrane, the unfolded polypeptide chains bind to another member of the Hsp70 family, which is associated with the Tim complex and acts as a motor that drives protein import. The polypeptide is then trans-

TABLE 10.1 Differences between the Universal and Mitochondrial Genetic Codes

Codon	Universal code	Human mitochondrial code
UGA	STOP	Trp
AGA	Arg	STOP
AGG	Arg	STOP
AUA	Ile	Met

Other codons vary from the universal code in yeast and plant mitochondria.

Figure 10.4
Import of proteins into mitochondria
Proteins are targeted for mitochondria by an amino-terminal presequence containing positively charged amino acids. Proteins are maintained in a partially unfolded state by association with a cytosolic Hsp70 and are recognized by a receptor on the surface of mitochondria. The unfolded polypeptide chains are then translocated through the Tom complex in the outer membrane and transferred to the Tim complex in the inner membrane. The voltage component of the electrochemical gradient is required for translocation across the inner membrane. The presequence is cleaved by a matrix protease, and a mitochondrial Hsp70 binds the polypeptide chain as it crosses the inner membrane, driving further protein translocation. A mitochondrial Hsp60 then facilitates folding of the imported polypeptide within the matrix.

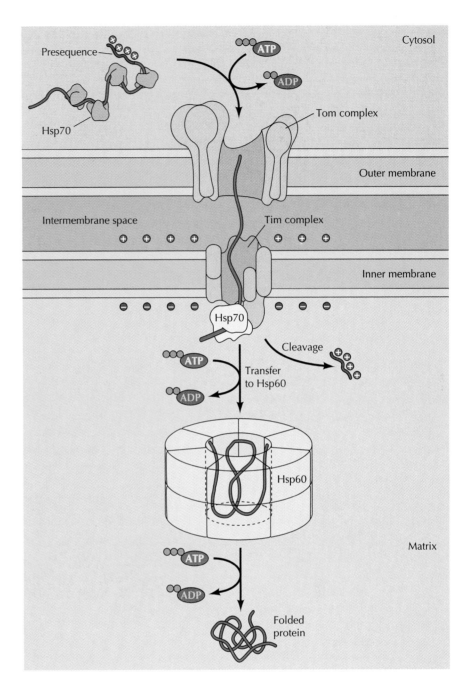

ferred to a chaperone of the Hsp60 family (a chaperonin), within which protein folding takes place. Since these interactions of polypeptide chains with molecular chaperones depend on ATP, protein import requires ATP both outside and inside the mitochondria, in addition to the electric potential across the inner membrane.

As noted above, some mitochondrial proteins are targeted to the outer membrane, inner membrane, or intermembrane space rather than to the matrix, so additional mechanisms are needed to direct these proteins to the correct submitochondrial compartment. These proteins are targeted to their destinations by a second sorting signal following the positively charged presequence that directs mitochondrial import. The targeting of proteins to the mitochondrial membranes appears to be mediated by hydrophobic stop-transfer sequences that halt translocation of the polypeptide chains through the Tim or Tom complexes, leading to their insertion into the inner or outer mitochondrial membranes, respectively (Figure 10.5). Proteins may be targeted to the intermembrane space by several different mechanisms (Figure 10.6). Some proteins are transferred across the outer membrane through the Tom complex but are then released within the intermembrane space instead of being transferred to the Tim complex. Other proteins are transferred to the Tim complex but are then released into the intermembrane space as a result of cleavage of hydrophobic stop-transfer sequences. Still other proteins may be completely imported into the mitochondrial matrix and then exported back across the inner membrane to the intermembrane space.

Figure 10.5
Insertion of mitochondrial membrane proteins Proteins targeted for the mitochondrial membranes contain hydrophobic stop-transfer sequences that halt their translocation through the Tom or Tim complexes and lead to their incorporation into the outer or inner membranes, respectively.

Figure 10.6
Sorting proteins to the intermembrane space Proteins can be targeted to the intermembrane space by several mechanisms. Some proteins (I) are translocated through the Tom complex and released into the intermembrane space. Other proteins (II) are transferred from the Tom complex to the Tim complex, but they contain hydrophobic stop-transfer sequences that halt translocation through the Tim complex. These stop-transfer sequences are then cleaved to release the proteins into the intermembrane space. Still other proteins (III) are imported to the matrix, as depicted in Figure 10.4. Removal of the presequence within the matrix then exposes a hydrophobic signal sequence, which targets the protein back across the inner membrane to the intermembrane space.

Molecular Medicine

Diseases of Mitochondria: Leber's Hereditary Optic Neuropathy

The Disease

Leber's hereditary optic neuropathy (LHON) is a rare inherited disease that results in blindness because of degeneration of the optic nerve. Vision loss usually occurs between the ages of 15 and 35, and is generally the only manifestation of the disease. Not all individuals who inherit the genetic defects responsible for LHON develop the disease, and females are affected less frequently than males. This propensity to affect males might suggest that LHON is an X-linked disease. This is not the case, however, because males never transmit LHON to their offspring. Instead, the inheritance of LHON is entirely by maternal transmission. This characteristic is consistent with cytoplasmic rather than nuclear inheritance of LHON, since the cytoplasm of fertilized eggs is derived almost entirely from the oocyte.

Molecular and Cellular Basis

In 1988 Douglas Wallace and his colleagues identified a mutation in the mitochondrial DNA of LHON patients. This mutation (at base pair 11778) affects one of the subunits of complex I of the electron transport chain (NADH dehydrogenase), resulting in the substitution of a histidine for an arginine. The 11778 mutation accounts for approximately half of all cases of LHON. Three other mutations of mitochondrial DNA have also been identified as primary causes of LHON. Two of these mutations affect other subunits of complex I, while the third affects cytochrome *b*, which is a component of complex III (see figure). Together, these four mutations account for more than 80% of LHON cases. A fifth mutation (at base pair 14459), affecting a complex I subunit, can cause either LHON or muscular disorders.

The mutations causing LHON reduce the capacity of mitochondria to carry out oxidative phosphorylation and generate ATP. This has the greatest effect on those tissues that are most dependent on oxidative phosphorylation, so defects in components of mitochondria can lead to clinical manifestations in specific organs, rather than to systemic disease. The central nervous system (including the brain and optic nerve) is most highly dependent on oxidative metabolism, consistent with blindness being the primary clinical manifestation resulting from the mitochondrial DNA mutations responsible for LHON.

As already noted, inheritance of LHON mutations does not always lead to development of the disease; only about 10% of females and 50% of males possessing a mutation suffer vision loss. One factor that may con-

LHON mutations in mitochondrial DNA.

- Complex I genes (NADH dehydrogenase)
- Complex III genes (Ubiquinol: cytochrome *c* oxidoreductase)
- Transfer RNA genes
- Complex IV genes (Cytochrome *c* oxidase)
- Complex V genes (ATP synthase)
- Ribosomal RNA genes

Diseases of Mitochondria: Leber's Hereditary Optic Neuropathy *(continued)*

tribute to this low incidence of disease among carriers of LHON mutations is that each cell contains thousands of copies of mitochondrial DNA, which can be present in mixtures of mutant and normal mitochondria. These mitochondria are randomly distributed to daughter cells at cell division, so the population of mitochondria can change as cells divide, leading to the formation of cells containing either greater or lesser proportions of mutant organelles. Importantly however, many individuals who bear predominantly mutant mitochondrial DNAs still fail to develop the disease. Thus, additional genetic or environmental factors, which have yet to be identified, appear to play a significant role in the development of LHON.

Prevention and Treatment

The identification of the mitochondrial DNA mutations responsible for

LHON allows molecular diagnosis of the disease, which can be important in establishing a definitive diagnosis of patients without a family history. However, the detection of mutations in mitochondrial DNA is of little value for screening members of affected families or for family planning. This contrasts to the utility of detecting inherited mutations of nuclear genes, where molecular analysis can determine whether a family member or embryo has inherited a mutant or wild-type allele. In LHON, however, mutant mitochondria are present in large numbers and are maternally transmitted to all offspring. As noted above, not all such offspring develop the disease, but this cannot be predicted by genetic analysis.

The finding that LHON is caused by mutations of mitochondrial DNA suggests the potential of new therapies. One approach is metabolic ther-

apy intended to enhance oxidative phosphorylation by administration of substrates or cofactors in the electron transport pathway, such as succinate or coenzyme Q. Another possibility that has been considered for treatment of LHON is gene therapy designed to relocate a normal gene allele to the nucleus. An appropriate targeting signal would be added to direct the gene product to mitochondria, where it could substitute for the defective mitochondrial-encoded protein.

References

Brown, M. D., D. S. Voljavec, M. T. Lott, I. MacDonald and D. C. Wallace. 1992. Leber's hereditary optic neuropathy: A model for mitochondrial neurodegenerative diseases. *FASEB J.* 6: 2791–2799.

Riordan-Eva, P. and A. E. Harding. 1995. Leber's hereditary optic neuropathy: The clinical relevance of different mitochondrial DNA mutations. *J. Med. Genet.* 32: 81–87.

Not only the proteins, but also the phospholipids of mitochondrial membranes are imported from the cytosol. In animal cells, phosphatidylcholine and phosphatidylethanolamine are synthesized in the ER and carried to mitochondria by **phospholipid transfer proteins**, which extract single phospholipid molecules from the membrane of the ER. The lipid can then be transported through the aqueous environment of the cytosol, buried in a hydrophobic binding site of the protein, and released when the complex reaches a new membrane, such as that of mitochondria. The mitochondria then synthesize phosphatidylserine from phosphatidylethanolamine, in addition to catalyzing the synthesis of the unusual phospholipid **cardiolipin**, which contains four fatty acid chains (Figure 10.7).

Figure 10.7
Structure of cardiolipin Cardiolipin is an unusual "double" phospholipid, containing four fatty acid chains, that is found primarily in the inner mitochondrial membrane.

The Mechanism of Oxidative Phosphorylation

Most of the usable energy obtained from the breakdown of carbohydrates or fats is derived by oxidative phosphorylation, which takes place within mitochondria. For example, the breakdown of glucose by glycolysis and the citric acid cycle yields a total of four molecules of ATP, ten molecules of NADH, and two molecules of FADH$_2$ (see Chapter 2). Electrons from NADH and FADH$_2$ are then transferred to molecular oxygen, coupled to the formation of an additional 32 to 34 ATP molecules by oxidative phosphorylation. Electron transport and oxidative phosphorylation are critical activities of protein complexes in the inner mitochondrial membrane, which ultimately serve as the major source of cellular energy.

The Electron Transport Chain

During **oxidative phosphorylation**, electrons derived from NADH and FADH$_2$ combine with O$_2$, and the energy released from these oxidation/reduction reactions is used to drive the synthesis of ATP from ADP. The transfer of electrons from NADH to O$_2$ is a very energy-yielding reaction,

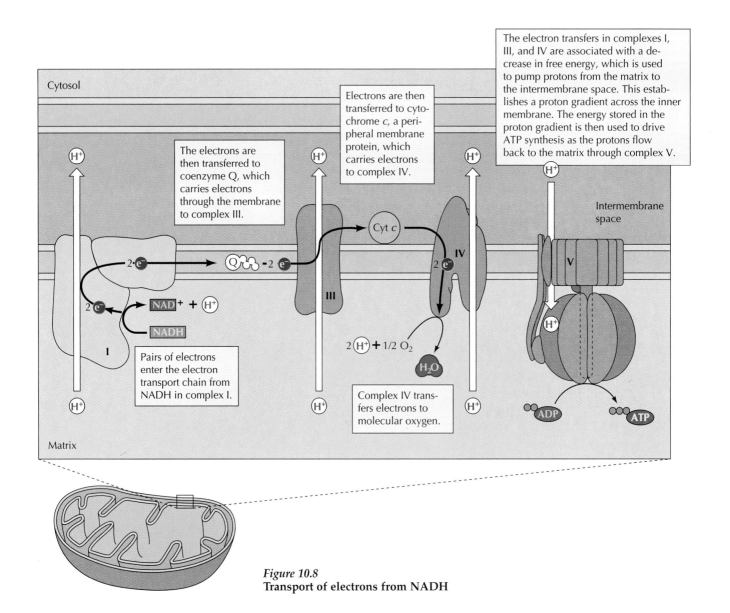

Figure 10.8
Transport of electrons from NADH

with $\Delta G^{\circ\prime} = -52.5$ kcal/mol for each pair of electrons transferred. To be harvested in usable form, this energy must be produced gradually, by the passage of electrons through a series of carriers, which constitute the **electron transport chain**. These carriers are organized into four complexes in the inner mitochondrial membrane. A fifth protein complex then serves to couple the energy-yielding reactions of electron transport to ATP synthesis.

Electrons from NADH enter the electron transport chain in complex I, which consists of nearly 40 polypeptide chains (Figure 10.8). These electrons are initially transferred from NADH to flavin mononucleotide and then, through an iron-sulfur carrier, to coenzyme Q—an energy-yielding process with $\Delta G^{\circ\prime} = -16.6$ kcal/mol. **Coenzyme Q** (also called **ubiquinone**) is a small, lipid-soluble molecule that carries electrons from complex I through the membrane to complex III, which consists of about ten polypeptides. In complex III, electrons are transferred from cytochrome *b* to cytochrome *c*—an energy-yielding reaction with $\Delta G^{\circ\prime} = -10.1$ kcal/mol. **Cytochrome *c***, a peripheral membrane protein bound to the outer face of the inner membrane, then carries electrons to complex IV (**cytochrome oxidase**), where they are finally transferred to O_2 ($\Delta G^{\circ\prime} = -25.8$ kcal/mol).

A distinct protein complex (complex II), which consists of four polypeptides, receives electrons from the citric acid cycle intermediate, succinate (Figure 10.9). These electrons are transferred to $FADH_2$, rather than to NADH,

Figure 10.9
Transport of electrons from FADH$_2$ Electrons from succinate enter the electron transport chain via FADH$_2$ in complex II. They are then transferred to coenzyme Q and carried through the rest of the electron transport chain as described in Figure 10.8. The transfer of electrons from FADH$_2$ to coenzyme Q is not associated with a significant decrease in free energy, so protons are not pumped across the membrane at complex II.

Key Experiment

The Chemiosmotic Theory

Coupling of Phosphorylation to Electron and Hydrogen Transfer by a Chemiosmotic Type of Mechanism

Peter Mitchell
University of Edinburgh, Edinburgh, Scotland
Nature, 1961, Volume 191, pages 144–148

The Context

By the 1950s it had been clearly established that oxidative phosphorylation involved the stepwise transfer of electrons through a series of carriers to molecular oxygen. But how the energy derived from these electron transfer reactions was converted to ATP remained a mystery. The natural assumption was that ADP was converted to ATP by direct transfer of high-energy phosphate groups from some other intermediate, as was known to occur during glycolysis. Thus, it was postulated that high-energy intermediates were produced as a result of electron transfer reactions, and that these intermediates

then drove ATP synthesis by phosphate group transfer.

The search for these postulated high-energy intermediates became a central goal of research during the 1950s and 1960s. But despite many false leads, no such intermediates were to be found. Moreover, several features of oxidative phosphorylation were difficult to reconcile with the orthodox hypothesis that ATP synthesis was driven by simple phosphate group transfer. In particular, phosphorylation was closely associated with membranes and was inhibited by a variety of compounds that disrupted membrane structure. These considerations led Peter Mitchell to propose a funda-

mentally different mechanism of energy coupling, in which ATP synthesis was driven by an electrochemical gradient across a membrane rather than by the elusive high-energy intermediates sought by other investigators.

The Experiments

The fundamental proposal of the chemiosmotic hypothesis was that the "intermediate" that coupled electron transport to ATP synthesis was a proton electrochemical gradient across the membrane. Mitchell postulated that such a gradient was produced by electron transport, and that the flow of protons back across the membrane in the energetically favorable direction was then coupled to ATP synthesis (see figure).

The hypothesis of chemiosmotic coupling clearly explained the lack of success in identifying a chemical high-energy intermediate, as well as the fact that intact membranes were required to synthesize ATP. Yet it was

---

and then to coenzyme Q. From coenzyme Q, electrons are transferred to complex III and then to complex IV as already described. In contrast to the transfer of electrons from NADH to coenzyme Q at complex I, the transfer of electrons from $FADH_2$ to coenzyme Q is not associated with a significant decrease in free energy and, therefore, is not coupled to ATP synthesis. Consequently, the passage of electrons derived from $FADH_2$ through the electron transport chain yields free energy only at complexes III and IV.

The free energy derived from the passage of electrons through complexes I, III, and IV is harvested by being coupled to the synthesis of ATP. Importantly, the mechanism by which the energy derived from these electron transport reactions is coupled to ATP synthesis is fundamentally different from the synthesis of ATP during glycolysis or the citric acid cycle. In the latter cases, a high-energy phosphate is transferred directly to ADP from the other substrate of an energy-yielding reaction. For example, in the final reaction of glycolysis, the high-energy phosphate of phosphoenolpyruvate is transferred to ADP, yielding pyruvate plus ATP (see Figure 2.32). Such direct transfer of high-energy phosphate groups does not occur during electron transport. Instead, the energy derived from electron transport is coupled to the generation of a proton gradient across the inner mitochondrial membrane. The potential energy stored in this gradient is then

The Chemiosmotic Theory (continued)

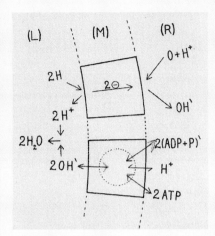

Mitchell's representation of chemiosmotic coupling between an electron transport system (top) and an ATP-generating system (bottom) in a membrane (M) enclosing aqueous phase L within aqueous phase R.

a radical concept that went against the biochemical dogma of the time. In a concluding paragraph of this 1961 paper, Mitchell took a philosophical view of his revolutionary proposal:

> In the exact sciences, cause and effect are no more than events linked in sequence. Biochemists now generally accept the idea that metabolism is the cause of membrane transport. The underlying thesis of the hypothesis put forward here is that if the processes that we call metabolism and transport represent events in a sequence, not only can metabolism be the cause of transport, but also transport can be the cause of metabolism.

The Impact

Mitchell's hypothesis was greeted with skepticism and remained the subject of acrimonious debate for more than a decade. However, the wealth of supporting evidence obtained by Mitchell and his colleagues, as well as by other investigators, eventually led to general acceptance of the chemiosmotic hypothesis—which became known instead as the chemiosmotic theory. It is now accepted not only as the basis for the generation of ATP during oxidative phosphorylation and photosynthesis in bacteria, mitochondria, and chloroplasts, but also for the energy-requiring transport of a variety of molecules across cell membranes.

Mitchell's work was recognized with a Nobel prize in 1978. The lecture he delivered on that occasion began as follows:

Peter Mitchell

> Although I had hoped that the chemiosmotic rationale of vectorial metabolism and biological energy transfer might one day come to be generally accepted, it would have been presumptuous of me to expect it to happen. Was it not Max Planck who remarked that a new scientific idea does not triumph by convincing its opponents, but rather because its opponents eventually die? The fact that what began as the chemiosmotic hypothesis has now been acclaimed as the chemiosmotic theory ... has therefore both astonished and delighted me, particularly because those who were formerly my most capable opponents are still in the prime of their scientific lives.

harvested by a fifth protein complex, which couples the energetically favorable flow of protons back across the membrane to the synthesis of ATP.

Chemiosmotic Coupling

The mechanism of coupling electron transport to ATP generation, **chemiosmotic coupling**, is a striking example of the relationship between structure and function in cell biology. The hypothesis of chemiosmotic coupling was first proposed in 1961 by Peter Mitchell, who suggested that ATP is generated by the use of energy stored in the form of proton gradients across biological membranes, rather than by direct chemical transfer of high-energy groups. Biochemists were initially highly skeptical of this proposal, and the chemiosmotic hypothesis took more than a decade to win general acceptance in the scientific community. Overwhelming evidence eventually accumulated in its favor, however, and chemiosmotic coupling is now recognized as a general mechanism of ATP generation, operating not only in mitochondria but also in chloroplasts and in bacteria, where ATP is generated via a proton gradient across the plasma membrane.

Electron transport through complexes I, III, and IV is coupled to the transport of protons out of the interior of the mitochondrion (see Figure

10.8). Thus, the energy-yielding reactions of electron transport are coupled to the transfer of protons from the matrix to the intermembrane space, which establishes a proton gradient across the inner membrane. Complexes I and IV appear to act as proton pumps that transfer protons across the membrane as a result of conformational changes induced by electron transport. In complex III, protons are carried across the membrane by coenzyme Q, which accepts protons from the matrix at complexes I or II and releases them into the intermembrane space at complex III. Complexes I and III each transfer four protons across the membrane per pair of electrons. In complex IV, two protons per pair of electrons are pumped across the membrane and another two protons per pair of electrons are combined with O_2 to form H_2O within the matrix. Thus, the equivalent of four protons per pair of electrons are transported out of the mitochondrial matrix at each of these three complexes. This transfer of protons from the matrix to the intermembrane space plays the critical role of converting the energy derived from the oxidation/reduction reactions of electron transport to the potential energy stored in a proton gradient.

Because protons are electrically charged particles, the potential energy stored in the proton gradient is electric as well as chemical in nature. The electric component corresponds to the voltage difference across the inner mitochondrial membrane, with the matrix of the mitochondrion negative and the intermembrane space positive. The corresponding free energy is given by the equation

$$\Delta G = -F\Delta V$$

where F is the Faraday constant and ΔV is the membrane potential. The additional free energy corresponding to the difference in proton concentration across the membrane is given by the equation

$$\Delta G = RT \ln \frac{[H^+]_i}{[H^+]_o}$$

where $[H^+]_i$ and $[H^+]_o$ refer, respectively, to the proton concentrations inside and outside the mitochondria.

In metabolically active cells, protons are typically pumped out of the matrix such that the proton gradient across the inner membrane corresponds to about one pH unit, or a tenfold lower concentration of protons within mitochondria (Figure 10.10). The pH of the mitochondrial matrix is therefore about 8, compared to the neutral pH (approximately 7) of the cytosol and intermembrane space. This gradient also generates an electric potential of approximately 0.14 V across the membrane, with the matrix negative. Both the pH gradient and the electric potential drive protons back into the matrix from the cytosol, so they combine to form an **electrochemical gradient** across the inner mitochondrial membrane, corresponding to a ΔG of approximately –5 kcal/mol per proton.

Because the phospholipid bilayer is impermeable to ions, protons are able to cross the membrane only through a protein channel. This restriction allows the energy in the electrochemical gradient to be harnessed and converted to ATP as a result of the action of the fifth complex involved in oxidative phosphorylation, complex V, or **ATP synthase** (see Figure 10.8). ATP synthase is organized into two structurally distinct components, F_0 and F_1, which are linked by a slender stalk (Figure 10.11). The F_0 portion spans the inner membrane and provides a channel through which protons are able to flow back from the intermembrane space to the matrix. The

Cytosol

Intermembrane space

pH 7

Concentration gradient

Electric potential

pH 8

Matrix

Figure 10.10
The electrochemical nature of the proton gradient   Since protons are positively charged, the proton gradient established across the inner mitochondrial membrane has both chemical and electric components. The chemical component is the proton concentration, or pH, gradient, which corresponds to about a tenfold higher concentration of protons on the cytosolic side of the inner mitochondrial membrane (a difference of one pH unit). In addition, there is an electric potential across the membrane, resulting from the net increase in positive charge on the cytosolic side.

energetically favorable return of protons to the matrix is coupled to ATP synthesis by the F_1 subunit, which catalyzes the synthesis of ATP from ADP and phosphate ions (P_i). Detailed structural studies have established the mechanism of ATP synthase action, which involves mechanical coupling between the F_0 and F_1 subunits. In particular, the flow of protons through F_0 drives the rotation of F_1, which acts as a rotary motor to drive ATP synthesis.

It appears that the flow of four protons back across the membrane through F_0 is required to drive the synthesis of one molecule of ATP by F_1, consistent with the proton transfers at complexes I, III, and IV each contributing sufficient free energy to the proton gradient to drive the synthesis of one ATP molecule. The oxidation of one molecule of NADH thus leads to the synthesis of three molecules of ATP, whereas the oxidation of $FADH_2$, which enters the electron transport chain at complex II, yields only two ATP molecules.

Transport of Metabolites across the Inner Membrane

In addition to driving the synthesis of ATP, the potential energy stored in the electrochemical gradient drives the transport of small molecules into and out of mitochondria. For example, the ATP synthesized within mitochondria has to be exported to the cytosol, while ADP and P_i need to be imported from the cytosol for ATP synthesis to continue. The electrochemical gradient generated by proton pumping provides energy required for the transport of these molecules and other metabolites that need to be concentrated within mitochondria (Figure 10.12).

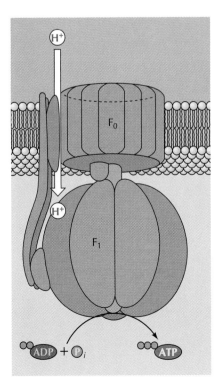

Figure 10.11
Structure of ATP synthase   The mitochondrial ATP synthase (complex V) consists of two multisubunit components, F_0 and F_1, which are linked by a slender stalk. F_0 spans the lipid bilayer, forming a channel through which protons can cross the membrane. F1 harvests the free energy derived from proton movement down the electrochemical gradient by catalyzing the synthesis of ATP.

Figure 10.12
Transport of metabolites across the mitochondrial inner membrane The transport of small molecules across the inner membrane is mediated by membrane-spanning transport proteins and driven by the electrochemical gradient. For example, ATP is exported from mitochondria to the cytosol by a transporter that exchanges it for ADP. The voltage component of the electrochemical gradient drives this exchange: ATP carries a greater negative charge (-4) than ADP (-3), so ATP is exported from the mitochondrial matrix to the cytosol while ADP is imported to mitochondria. In contrast, the transport of phosphate (P_i) and pyruvate is coupled to an exchange for hydroxyl ions (OH^-); in this case, the pH component of the electrochemical gradient drives the export of hydroxyl ions, coupled to the transport of P_i and pyruvate into mitochondria.

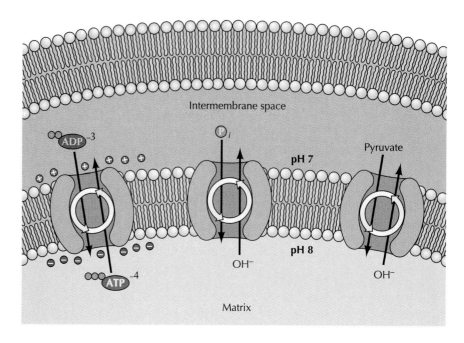

The transport of ATP and ADP across the inner membrane is mediated by an integral membrane protein, the adenine nucleotide translocator, which transports one molecule of ADP into the mitochondrion in exchange for one molecule of ATP transferred from the mitochondrion to the cytosol. Because ATP carries more negative charge than ADP (-4 compared to -3), this exchange is driven by the voltage component of the electrochemical gradient. Since the proton gradient establishes a positive charge on the cytosolic side of the membrane, the export of ATP in exchange for ADP is energetically favorable.

The synthesis of ATP within the mitochondrion requires phosphate ions (P_i) as well as ADP, so P_i must also be imported from the cytosol. This is mediated by another membrane transport protein, which imports phosphate ($H_2PO_4^-$) and exports hydroxyl ions (OH^-). This exchange is electrically neutral because both phosphate and hydroxyl ions have a charge of -1. However, the exchange is driven by the proton concentration gradient; the higher pH within mitochondria corresponds to a higher concentration of hydroxyl ions, favoring their translocation to the cytosolic side of the membrane.

Energy from the electrochemical gradient is similarly used to drive the transport of other metabolites into mitochondria. For example, the import of pyruvate from the cytosol (where it is produced by glycolysis) is mediated by a transporter that exchanges pyruvate for hydroxyl ions. Other intermediates of the citric acid cycle are able to shuttle between mitochondria and the cytosol by similar exchange mechanisms.

Chloroplasts and Other Plastids

Chloroplasts, the organelles responsible for photosynthesis, are in many respects similar to mitochondria. Both chloroplasts and mitochondria function to generate metabolic energy, evolved by endosymbiosis, contain their own genetic systems, and replicate by division. However, chloroplasts are larger and more complex than mitochondria, and they perform several critical tasks in addition to the generation of ATP. Most importantly, chloroplasts are responsible for the photosynthetic conversion of CO_2 to carbohydrates. In addition, chloroplasts synthesize amino acids, fatty acids, and the lipid components of their own membranes. The reduction of nitrite (NO_2^-) to ammonia (NH_3), an essential step in the incorporation of nitrogen into organic compounds, also occurs in chloroplasts. Moreover, chloroplasts are only one of several types of related organelles (plastids) that play a variety of roles in plant cells.

The Structure and Function of Chloroplasts

Plant chloroplasts are large organelles (5 to 10 μm long) that, like mitochondria, are bounded by a double membrane called the chloroplast envelope (Figure 10.13). In addition to the inner and outer membranes of the envelope, chloroplasts have a third internal membrane system, called the thylakoid membrane. The **thylakoid membrane** forms a network of flattened discs called thylakoids, which are frequently arranged in stacks called grana. Because of this three-membrane structure, the internal organization of chloroplasts is more complex than that of mitochondria. In particular, their three membranes divide chloroplasts into three distinct internal compartments: (1) the intermembrane space between the two membranes of the chloroplast envelope; (2) the **stroma**, which lies inside the envelope but outside the thylakoid membrane; and (3) the thylakoid lumen.

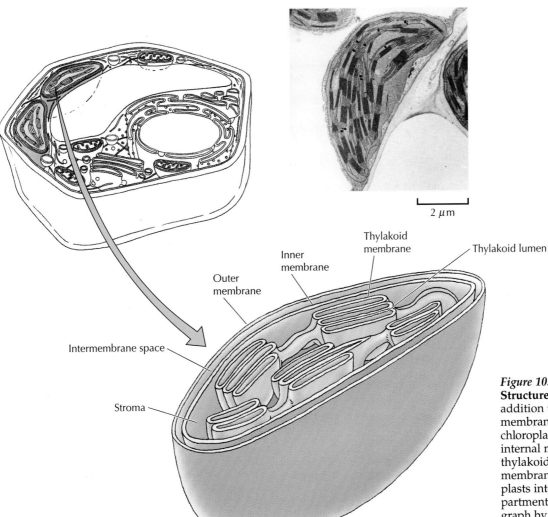

2 μm

Outer membrane

Inner membrane

Thylakoid membrane

Thylakoid lumen

Intermembrane space

Stroma

Figure 10.13
Structure of a chloroplast   In addition to the inner and outer membranes of the envelope, chloroplasts contain a third internal membrane system: the thylakoid membrane. These membranes divide chloroplasts into three internal compartments. (Electron micrograph by E. H. Newcombe/ Biological Photo Service.)

Despite this greater complexity, the membranes of chloroplasts have clear functional similarities with those of mitochondria—as expected, given the role of both organelles in the chemiosmotic generation of ATP. The outer membrane of the chloroplast envelope, like that of mitochondria, contains porins and is therefore freely permeable to small molecules. In contrast, the inner membrane is impermeable to ions and metabolites, which are therefore able to enter chloroplasts only via specific membrane transporters. These properties of the inner and outer membranes of the chloroplast envelope are similar to the inner and outer membranes of mitochondria: In both cases the inner membrane restricts the passage of molecules between the cytosol and the interior of the organelle. The chloroplast stroma is also equivalent in function to the mitochondrial matrix: It contains the chloroplast genetic system and a variety of metabolic enzymes, including those responsible for the critical conversion of CO_2 to carbohydrates during photosynthesis.

The major difference between chloroplasts and mitochondria, in terms of both structure and function, is the thylakoid membrane. This membrane is of central importance in chloroplasts, where it fills the role of the inner mitochondrial membrane in electron transport and the chemiosmotic generation of ATP (Figure 10.14). The inner membrane of the chloroplast envelope (which is not folded into cristae) does not function in photosynthesis. Instead, the chloroplast electron transport system is located in the thylakoid membrane, and protons are pumped across this membrane from the stroma to the thylakoid lumen. The resulting electrochemical gradient then drives ATP synthesis as protons cross back into the stroma. In terms of its role in generation of metabolic energy, the thylakoid membrane of chloroplasts is thus equivalent to the inner membrane of mitochondria.

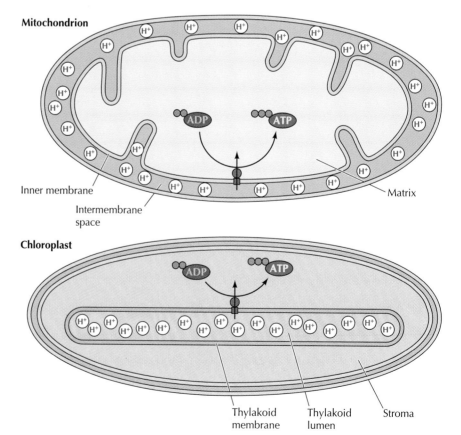

Figure 10.14
Chemiosmotic generation of ATP in chloroplasts and mitochondria In mitochondria, electron transport generates a proton gradient across the inner membrane, which is then used to drive ATP synthesis in the matrix. In chloroplasts, the proton gradient is generated across the thylakoid membrane and used to drive ATP synthesis in the stroma.

The Chloroplast Genome

Like mitochondria, chloroplasts contain their own genetic system, reflecting their evolutionary origins from photosynthetic bacteria. The genomes of chloroplasts are similar to those of mitochondria in that they consist of circular DNA molecules present in multiple copies per organelle. However, chloroplast genomes are larger and more complex than those of mitochondria, ranging from 120 to 160 kb and containing approximately 120 genes.

The chloroplast genomes of several plants have been completely sequenced, leading to the identification of many of the genes contained in the organelle DNAs. These chloroplast genes encode both RNAs and proteins involved in gene expression, as well as a variety of proteins that function in photosynthesis (Table 10.2). Both the ribosomal and transfer RNAs used for translation of chloroplast mRNAs are encoded by the organelle genome. These include four rRNAs (23S, 16S, 5S, and 4.5S) and 30 tRNA species. In contrast to the smaller number of tRNAs encoded by the mitochondrial genome, the chloroplast tRNAs are sufficient to translate all the mRNA codons according to the universal genetic code. In addition to these RNA components of the translation system, the chloroplast genome encodes about 20 ribosomal proteins, which represent approximately a third of the proteins of chloroplast ribosomes. Some subunits of RNA polymerase are also encoded by chloroplasts, although additional RNA polymerase subunits and other factors needed for chloroplast gene expression are encoded in the nucleus.

The chloroplast genome also encodes approximately 30 proteins that are involved in photosynthesis, including components of photosystems I and II, of the cytochrome *bf* complex, and of ATP synthase. In addition, one of the subunits of ribulose bisphosphate carboxylase (rubisco) is encoded by chloroplast DNA. Rubisco is the critical enzyme that catalyzes the addition of CO_2 to ribulose-1,5-bisphosphate during the Calvin cycle (see Figure 2.39). Not only is it the major protein component of the chloroplast stroma, but it is also thought to be the single most abundant protein on Earth, so it is noteworthy that one of its subunits is encoded by the chloroplast genome.

TABLE 10.2 Genes Encoded by Chloroplast DNA

Function	Number of genes
Genes for the genetic apparatus	
rRNAs (23S, 16S, 5S, 4.5S)	4
tRNAs	30
Ribosomal proteins	21
RNA polymerase subunits	4
Genes for photosynthesis	
Photosystem I	5
Photosystem II	12
Cytochrome *bf* complex	4
ATP synthase	6
Ribulose bisphosphate carboxylase	1

Sequence analysis indicates that chloroplast genomes contain about 30 genes in addition to those listed here. Some of these encode proteins involved in respiration, but most remain to be identified.

Import and Sorting of Chloroplast Proteins

Although chloroplasts encode more of their own proteins than mitochondria, about 90% of chloroplast proteins are still encoded by nuclear genes. As with mitochondria, these proteins are synthesized on cytosolic ribosomes and then imported into chloroplasts as completed polypeptide chains. They must then be sorted to their appropriate location within chloroplasts—an even more complicated task than protein sorting in mitochondria, since chloroplasts contain three separate membranes that divide them into three distinct internal compartments.

Protein import into chloroplasts generally resembles mitochondrial protein import (Figure 10.15). Proteins are targeted for import into chloroplasts by N-terminal sequences of 30 to 100 amino acids, called **transit peptides**, which direct protein translocation across the two membranes of the chloroplast envelope and are then removed by proteolytic cleavage. The transit peptides are recognized by the translocation complex of the chloroplast outer member (the Toc complex), and proteins are transported through this complex across the membrane. They are then transferred to

Figure 10.15
Protein import into the chloroplast stroma Proteins are targeted for import into chloroplasts by a transit peptide at their amino terminus. The transit peptide directs polypeptide translocation through the Toc complex in the chloroplast outer membrane and the Tic complex in the chloroplast inner membrane. This peptide is then removed by proteolytic cleavage within the stroma. Both cytosolic and chloroplast chaperones (Hsp60 and Hsp70) are required for protein import.

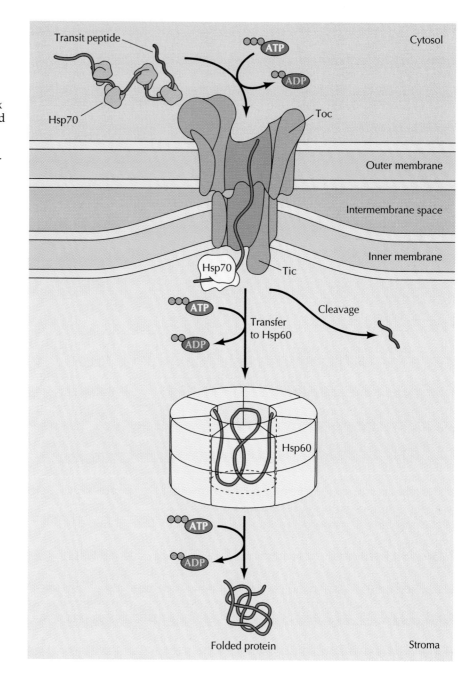

the translocation complex of the inner membrane (the Tic complex) and transported across the inner membrane to the stroma. As in mitochondria, molecular chaperones on both the cytosolic and stromal sides of the envelope are required for protein import, which requires energy in the form of ATP. In contrast to the presequences of mitochondrial import, however, transit peptides are not positively charged and the translocation of polypeptide chains into chloroplasts does not require an electric potential across the membrane.

Proteins incorporated into the thylakoid lumen are transported to their destination in two steps (Figure 10.16). They are first imported into the stroma, as already described, and are then targeted for translocation across the thylakoid membrane by a second hydrophobic signal sequence, which is exposed following cleavage of the transit peptide. The hydrophobic sig-

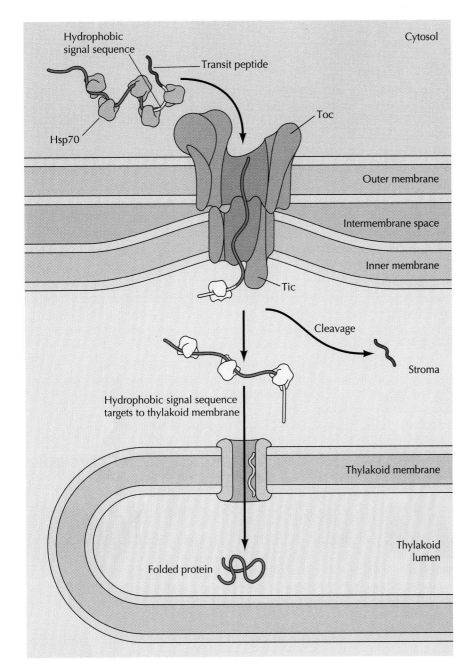

Cytosol

Hydrophobic signal sequence

Transit peptide

Hsp70

Toc

Outer membrane

Intermembrane space

Inner membrane

Tic

Cleavage

Stroma

Hydrophobic signal sequence targets to thylakoid membrane

Thylakoid membrane

Thylakoid lumen

Folded protein

Figure 10.16
Import of proteins into the thylakoid lumen Proteins are imported into the thylakoid lumen in two steps. The first step is import into the chloroplast stroma, as illustrated in Figure 10.15. Cleavage of the transit peptide then exposes a second hydrophobic signal sequence, which directs protein translocation across the thylakoid membrane.

nal sequence directs translocation of the polypeptide across the thylakoid membrane and is finally removed by a second proteolytic cleavage within the lumen.

The pathways of protein sorting to the other four compartments of chloroplasts—the inner and outer membranes, thylakoid membrane, and intermembrane space—are less well established. As with mitochondria, proteins appear to be inserted directly into the outer membrane of the chloroplast envelope. In contrast, proteins destined for either the thylakoid membrane or the inner membrane of the chloroplast envelope are initially targeted for import into the stroma by N-terminal transit peptides. Following cleavage of the transit peptides, these proteins are then targeted for insertion into the appropriate membrane by other sequences, which are not yet well characterized. Finally, neither the sequences that target proteins to the intermembrane space nor the pathways by which they travel to that destination have been identified.

Other Plastids

Chloroplasts are only one, albeit the most prominent, member of a larger family of plant organelles called **plastids**. All plastids contain the same genome as chloroplasts, but they differ in both structure and function. Chloroplasts are specialized for photosynthesis and are unique in that they contain the internal thylakoid membrane system. Other plastids, which are involved in different aspects of plant cell metabolism, are bounded by the two membranes of the plastid envelope but lack both the thylakoid membranes and other components of the photosynthetic apparatus.

The different types of plastids are frequently classified according to the kinds of pigments they contain. Chloroplasts are so named because they contain chlorophyll. **Chromoplasts** (Figure 10.17A) lack chlorophyll but contain carotenoids; they are responsible for the yellow, orange, and red colors of some flowers and fruits, although their precise function in cell metabolism is not clear. **Leucoplasts** are nonpigmented plastids, which store a variety of energy sources in nonphotosynthetic tissues. **Amyloplasts** (Figure 10.17B) and **elaioplasts** are examples of leucoplasts that store starch and lipids, respectively.

All plastids, including chloroplasts, develop from **proplastids**, small (0.5 to 1 μm in diameter) undifferentiated organelles present in the rapidly dividing cells of plant roots and shoots. Proplastids then develop into the various types of mature plastids according to the needs of differentiated cells. In addition, mature plastids are able to change from one type to another. Chromoplasts develop from chloroplasts, for example, during the ripening of fruit (e.g., tomatoes). During this process, chlorophyll and the thylakoid membranes break down, while new types of carotenoids are synthesized.

An interesting feature of plastids is that their development is controlled both by environmental signals and by intrinsic programs of cell differentiation. In the photosynthetic cells of leaves, for example, proplastids develop into chloroplasts (Figure 10.18). During this process, the thylakoid membrane is formed by vesicles budding from the inner membrane of the plastid envelope and the various components of the photosynthetic apparatus are synthesized and assembled. However, chloroplasts develop only in the presence of light. If plants are kept in the dark, the development of pro-

Figure 10.17
Electron micrographs of chromoplasts and amyloplasts (A) Chromoplasts contain lipid droplets in which carotenoids are stored. (B) Amyloplasts contain large starch granules. (A, Biophoto Associates/Photo Researchers, Inc.; B, Dr. Jeremy Burgess/Photo Researchers, Inc.)

(A)

1 μm

(B)

1 μm

Proplastid

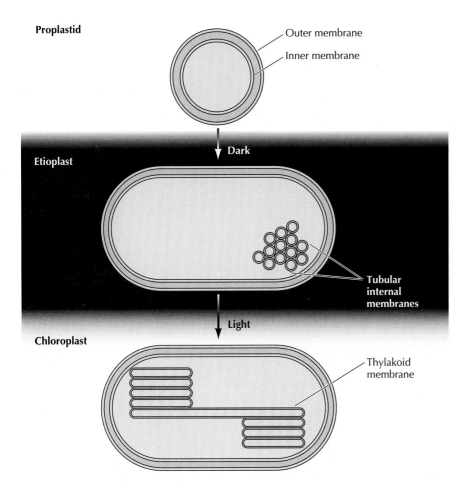

Outer membrane

Inner membrane

Dark

Etioplast

Tubular internal membranes

Light

Chloroplast

Thylakoid membrane

Figure 10.18
Development of chloroplasts
Chloroplasts develop from proplastids in the photosynthetic cells of leaves. Proplastids contain only the inner and outer envelope membranes; the thylakoid membrane is formed by vesicle budding from the inner membrane during chloroplast development. If the plant is kept in the dark, chloroplast development is arrested at an intermediate stage (etioplasts). Etioplasts lack chlorophyll and contain semicrystalline arrays of membrane tubules. In the presence of light, they continue their development to chloroplasts.

plastids in leaves is arrested at an intermediate stage (called **etioplasts**), in which a semicrystalline array of tubular internal membranes has formed but chlorophyll has not been synthesized (Figure 10.19). If dark-grown plants are then exposed to light, the etioplasts continue their development to chloroplasts. It is noteworthy that this dual control of plastid development involves the coordinated expression of genes within both the plastid and nuclear genomes. The mechanisms responsible for such coordinated gene expression are largely unknown, and their elucidation represents a challenging problem in plant molecular biology.

Photosynthesis

During photosynthesis, energy from sunlight is harvested and used to drive the synthesis of glucose from CO_2 and H_2O. By converting the energy of sunlight to a usable form of potential chemical energy, photosynthesis is the ultimate source of metabolic energy for all biological systems. Photosynthesis takes place in two distinct stages. In the light reactions, energy from sunlight drives the synthesis of ATP and NADPH, coupled to the formation of O_2 from H_2O. In the dark reactions, so named because they do not require sunlight, the ATP and NADPH produced by the light reactions drive glucose synthesis. In eukaryotic cells, both the light and dark reactions of photosynthesis occur within chloroplasts—the light reactions in the thylakoid membrane and the dark reactions within the

Figure 10.19
Electron micrograph of an etioplast
(John N. A. Lott/Biological Photo Service.)

1 μm

stroma. This section discusses the light reactions of photosynthesis, which are related to oxidative phosphorylation in mitochondria. The dark reactions were discussed in detail in Chapter 2.

Electron Flow through Photosystems I and II

Sunlight is absorbed by photosynthetic pigments, the most abundant of which in plants are the **chlorophylls**. Absorption of light excites an electron to a higher energy state, thus converting the energy of sunlight to potential chemical energy. The photosynthetic pigments are organized into **photocenters** in the thylakoid membrane, each of which contains hundreds of pigment molecules (Figure 10.20). The many pigment molecules in each photocenter act as antennae to absorb light and transfer the energy of their excited electrons to a chlorophyll molecule that serves as a reaction center. The reaction center chlorophyll then transfers its high-energy electron to an acceptor molecule in an electron transport chain. High-energy electrons are then transferred through a series of membrane carriers, coupled to the synthesis of ATP and NADPH.

The best characterized photosynthetic reaction center is that of the bacterium *Rhodopseudomonas viridis*, the structure of which was determined by Johann Deisenhofer, Hartmut Michel, Robert Huber, and their colleagues in 1985 (Figure 10.21). The reaction center consists of three transmembrane polypeptides, bound to a *c*-type cytochrome on the exterior side of the membrane. Energy from sunlight is captured by a pair of chlorophyll molecules known as the special pair. Electrons are then transferred from the special pair to another pair of chlorophylls and from there to other prosthetic groups (pheophytins and quinones). From there the electrons are transferred to a cytochrome *bc* complex in which electron transport is coupled to the generation of a proton gradient. The electrons are then transferred to the reaction center cytochrome and finally returned to the chlorophyll special pair. The reaction center thus converts the energy of sunlight to high-energy electrons, the potential energy of which is converted to a proton gradient by the cytochrome *bc* complex.

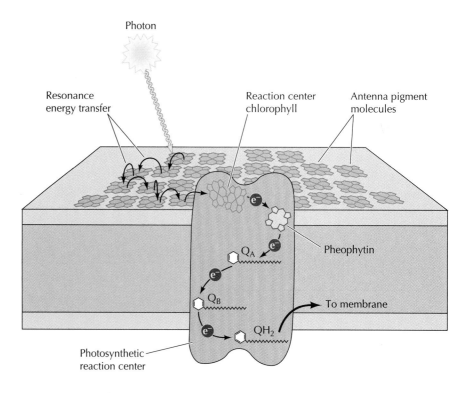

Figure 10.20
Organization of a photocenter Each photocenter consists of hundreds of antenna pigment molecules, which absorb photons and transfer energy to a reaction center chlorophyll. The reaction center chlorophyll then transfers its excited electron to an acceptor in the electron transport chain. The reaction center illustrated is that of photosystem II, in which electrons are transferred from the reaction center chlorophyll to pheophytin and then to quinones (Q_A, Q_B, and QH_2).

Figure 10.21
Structure of a photosynthetic reaction center The reaction center of *R. viridis* consists of three transmembrane proteins (purple, blue, and beige) and a *c*-type cytochrome (green). Chlorophylls and other prosthetic groups are colored yellow. (Courtesy of Johann Deisenhofer, University of Texas Medical Center and The Nobel Foundation, 1989.)

The proteins involved in the light reactions of photosynthesis in plants are organized into five complexes in the thylakoid membrane (Figure 10.22). Two of these complexes are photosystems (**photosystems I** and **II**), in which light is absorbed and transferred to reaction center chlorophylls. High-energy electrons are then transferred through a series of carriers in both photosystems and in a third protein complex, the **cytochrome *bf* complex**. As in mitochondria, these electron transfers are coupled to the transfer of protons into the thylakoid lumen, thereby establishing a proton gradient across the thylakoid membrane. The energy stored in this proton gradient is then harvested by a fourth protein complex in the thylakoid membrane, ATP synthase, which (like the mitochondrial enzyme) couples proton flow back across the membrane to the synthesis of ATP.

Figure 10.22
Electron transport and ATP synthesis during photosynthesis Five protein complexes in the thylakoid membrane function in electron transport and the synthesis of ATP and NADPH. Photons are absorbed by complexes of pigment molecules associated with photosystems I and II (PS I and PS II). At photosystem II, energy derived from photon absorption is used to split a water molecule within the thylakoid lumen. Electrons are then carried by plastoquinone (PQ) to the cytochrome *bf* complex, where they are transferred to a lower energy state and protons are pumped into the thylakoid lumen. Electrons are then transferred to photosystem I by plastocyanin (PC). At photosystem I, energy derived from light absorption again generates high-energy electrons, which are transferred to ferrodoxin (Fd) and used to reduce NADP$^+$ to NADPH in the stroma. ATP synthase then uses the energy stored in the proton gradient to convert ADP to ATP.

One important difference between electron transport in chloroplasts and that in mitochondria is that the energy derived from sunlight during photosynthesis not only is converted to ATP but also is used to generate the NADPH required for subsequent conversion of CO_2 to carbohydrates. This is accomplished by the use of two different photosystems in the light reactions of photosynthesis, one to generate ATP and the other to generate NADPH. Electrons are transferred sequentially between the two photosystems, with photosystem I acting to generate NADPH and photosystem II acting to generate ATP.

The pathway of electron flow starts at photosystem II, which is homologous to the photosynthetic reaction center of *R. viridis* already described. However, at photosystem II the energy derived from absorption of photons is used to split water molecules to molecular oxygen and protons (see Figure 10.22). This reaction takes place within the thylakoid lumen, so the release of protons from H_2O establishes a proton gradient across the thylakoid membrane. The high-energy electrons derived from this process are transferred through a series of carriers to plastoquinone, a lipid-soluble carrier similar to coenzyme Q (ubiquinone) of mitochondria. Plastoquinone carries electrons from photosystem II to the cytochrome *bf* complex, within which electrons are transferred to plastocyanin and additional protons are pumped into the thylakoid lumen. Electron transport through photosystem II is thus coupled to establishment of a proton gradient, which drives the chemiosmotic synthesis of ATP.

From photosystem II, electrons are carried by plastocyanin (a peripheral membrane protein) to photosystem I, where the absorption of additional photons again generates high-energy electrons. Photosystem I, however, does not act as a proton pump; instead, it uses these high-energy electrons to reduce $NADP^+$ to NADPH. The reaction center chlorophyll of photosystem I transfers its excited electrons through a series of carriers to ferrodoxin, a small protein on the stromal side of the thylakoid membrane. The enzyme **NADP reductase** then transfers electrons from ferrodoxin to $NADP^+$, generating NADPH. The passage of electrons through photosystems I and II thus generates both ATP and NADPH, which are used by the Calvin cycle enzymes in the chloroplast stroma to convert CO_2 to carbohydrates (see Figure 2.39).

Cyclic Electron Flow

A second electron transport pathway, called **cyclic electron flow**, produces ATP without the synthesis of NADPH, thereby supplying additional ATP for other metabolic processes. In cyclic electron flow, light energy harvested at photosystem I is used for ATP synthesis rather than NADPH synthesis (Figure 10.23). Instead of being transferred to $NADP^+$, high-energy electrons from photosystem I are transferred to the cytochrome *bf* complex. Electron transfer through the cytochrome *bf* complex is then coupled, as in photosystem II, to the establishment of a proton gradient across the thylakoid membrane. Plastocyanin then returns these electrons to photosystem I in a lower energy state, completing a cycle of electron transport in which light energy harvested at photosystem I is used to pump protons at the cytochrome *bf* complex. Electron transfer from photosystem I can thus generate either ATP or NADPH, depending on the metabolic needs of the cell.

ATP Synthesis

The ATP synthase of the thylakoid membrane is similar to the mitochondrial enzyme. However, the energy stored in the proton gradient across the

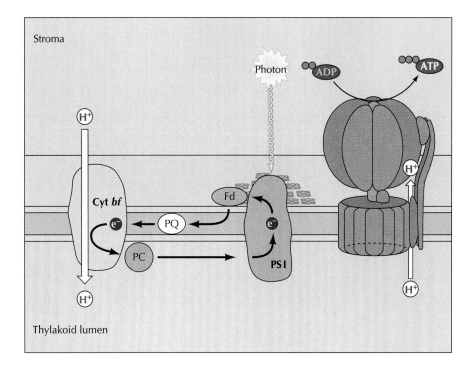

Figure 10.23
The pathway of cyclic electron flow Light energy absorbed at photosystem I (PS I) is used for ATP synthesis rather than NADPH synthesis. High-energy electrons generated by photon absorption are transferred to the cytochrome *bf* complex rather than to NADP$^+$. At the cytochrome *bf* complex, electrons are transferred to a lower energy state and protons are pumped into the thylakoid lumen. The electrons are then returned to photosystem I by plastocyanin (PC).

thylakoid membrane, in contrast to the inner mitochondrial membrane, is almost entirely chemical in nature. This is because the thylakoid membrane, although impermeable to protons, differs from the inner mitochondrial membrane in being permeable to other ions, particularly Mg^{2+} and Cl^-. The free passage of these ions neutralizes the voltage component of the proton gradient, so the energy derived from photosynthesis is conserved mainly as the difference in proton concentration (pH) across the thylakoid membrane. However, because the thylakoid lumen is a closed compartment, this difference in proton concentration can be quite large, corresponding to a differential of more than three pH units between the stroma and the thylakoid lumen. Because of the magnitude of this pH differential, the total free energy stored across the thylakoid membrane is similar to that stored across the inner mitochondrial membrane.

For each pair of electrons transported, two protons are transferred across the thylakoid membrane at photosystem II and two to four protons at the cytochrome *bf* complex. Since four protons are needed to drive the synthesis of one molecule of ATP, passage of each pair of electrons through photosystems I and II by noncyclic electron flow yields between 1 and 1.5 ATP molecules. Cyclic electron flow has a lower yield, corresponding to between 0.5 and 1 ATP molecules per pair of electrons.

Peroxisomes

Peroxisomes are small, membrane-enclosed organelles (Figure 10.24) that contain enzymes involved in a variety of metabolic reactions, including several aspects of energy metabolism. Although peroxisomes are morphologically similar to lysosomes, they are assembled, like mitochondria and chloroplasts, from proteins that are synthesized on free ribosomes and then imported into peroxisomes as completed polypeptide chains. Although peroxisomes do not contain their own genomes, they are similar to mitochondria and chloroplasts in that they replicate by division.

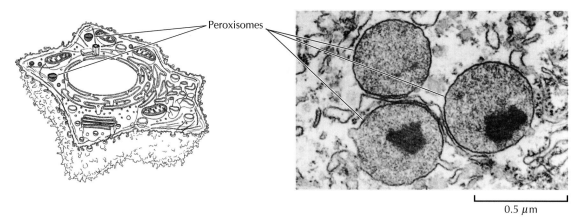

Peroxisomes

Figure 10.24
Electron micrograph of peroxisomes Three peroxisomes from rat liver are shown. Two contain dense regions, which are paracrystalline arrays of the enzyme urate oxidase. (Don Fawcett/Photo Researchers, Inc.)

Functions of Peroxisomes

Peroxisomes contain at least 50 different enzymes, which are involved in a variety of biochemical pathways in different types of cells. Peroxisomes originally were defined as organelles that carry out oxidation reactions leading to the production of hydrogen peroxide. Because hydrogen peroxide is harmful to the cell, peroxisomes also contain the enzyme **catalase**, which decomposes hydrogen peroxide either by converting it to water or by using it to oxidize another organic compound. A variety of substrates are broken down by such oxidative reactions in peroxisomes, including uric acid, amino acids, and fatty acids. The oxidation of fatty acids (Figure 10.25) is a particularly important example, since it provides a major source of metabolic energy. In animal cells, fatty acids are oxidized in both peroxisomes and mitochondria, but in yeasts and plants fatty acid oxidation is restricted to peroxisomes.

In addition to providing a compartment for oxidation reactions, peroxisomes are involved in lipid biosynthesis. In animal cells, cholesterol and dolichol are synthesized in peroxisomes as well as in the ER. In the liver, peroxisomes are also involved in the synthesis of bile acids, which are derived from cholesterol. In addition, peroxisomes contain enzymes required for the synthesis of **plasmalogens**—a family of phospholipids in which one of the hydrocarbon chains is joined to glycerol by an ether bond rather than an ester bond (Figure 10.26). Plasmalogens are important membrane components in some tissues, particularly heart and brain, although they are absent in others.

Figure 10.25
Fatty acid oxidation in peroxisomes The oxidation of a fatty acid is accompanied by the production of hydrogen peroxide (H_2O_2) from oxygen. The hydrogen peroxide is decomposed by catalase, either by conversion to water or by oxidation of another organic compound (designated AH_2).

$$R-CH_2-CH_2-\overset{O}{\overset{\|}{C}}-S-\textbf{CoA} + O_2 \longrightarrow R-CH=CH-\overset{O}{\overset{\|}{C}}-S-\textbf{CoA} + H_2O_2$$

$$2\,H_2O_2 \xrightarrow{\text{Catalase}} 2\,H_2O + O_2$$

or

$$H_2O_2 + AH_2 \xrightarrow{\text{Catalase}} 2\,H_2O + A$$

Figure 10.26
Structure of a plasmalogen   The plasmalogen shown is analogous to phosphatidylcholine. However, one of the fatty acid chains is joined to glycerol by an ether, rather than an ester, bond.

Peroxisomes play two particularly important roles in plants. First, peroxisomes in seeds are responsible for the conversion of stored fatty acids to carbohydrates, which is critical to providing energy and raw materials for growth of the germinating plant. This occurs via a series of reactions termed the **glyoxylate cycle**, which is a variant of the citric acid cycle (Figure 10.27). The peroxisomes in which this takes place are sometimes called **glyoxysomes**.

Second, peroxisomes in leaves are involved in **photorespiration**, which serves to metabolize a side product formed during photosynthesis (Figure 10.28). CO_2 is converted to carbohydrates during photosynthesis via a series of reactions called the Calvin cycle (see Figure 2.39). The first step is the addition of CO_2 to the five-carbon sugar ribulose-1,5-bisphosphate, yielding two molecules of 3-phosphoglycerate (three carbons each). However, the enzyme involved (ribulose bisphosphate carboxylase or rubisco) sometimes catalyzes the addition of O_2 instead of CO_2, producing one molecule of 3-phosphoglycerate and one molecule of phosphoglycolate (two carbons). This is a side reaction, and phosphoglycolate is not a useful metabolite. It is first converted to glycolate and then transferred to peroxisomes, where it is oxidized and converted to glycine. Glycine is then transferred to mitochondria, where two molecules of glycine are converted to

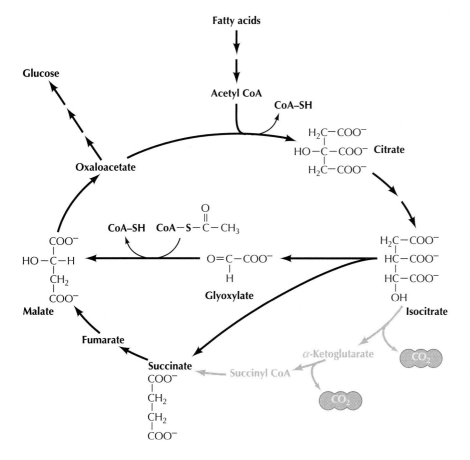

Figure 10.27
The glyoxylate cycle   Plants are capable of synthesizing carbohydrates from fatty acids via the glyoxylate cycle, which is a variant of the citric acid cycle (see Figure 2.34). As in the citric acid cycle, acetyl CoA combines with oxaloacetate to form citrate, which is converted to isocitrate. However, instead of being degraded to CO_2 and α-ketoglutarate, isocitrate is converted to succinate and glyoxylate. Glyoxylate then reacts with another molecule of acetyl CoA to yield malate, which is converted to oxaloacetate and used for glucose synthesis.

Figure 10.28
Role of peroxisomes in photores-
piration   During photosynthesis, CO_2
is converted to carbohydrates by the
Calvin cycle, which initiates with the
addition of CO_2 to the five-carbon
sugar ribulose-1,5-bisphosphate.
However, the enzyme involved sometimes catalyzes the addition of O_2
instead, resulting in production of the
two-carbon compound phosphoglycolate. Phosphoglycolate is converted to
glycolate, which is then transferred to
peroxisomes, where it is oxidized and
converted to glycine. Glycine is then
transferred to mitochondria and converted to serine. The serine is returned
to peroxisomes and converted to
glycerate, which is transferred back to
chloroplasts.

one molecule of serine, with the loss of CO_2 and NH_3. The serine is then
returned to peroxisomes, where it is converted to glycerate. Finally, the
glycerate is transferred back to chloroplasts, where it reenters the Calvin
cycle. Photorespiration does not appear to be beneficial for the plant, since
it is essentially the reverse of photosynthesis—O_2 is consumed and CO_2 is
released without any gain of ATP. However, the occasional utilization of O_2
in place of CO_2 appears to be an inherent property of rubisco, so photorespiration is a general accompaniment of photosynthesis. Peroxisomes thus
play an important role by allowing most of the carbon in glycolate to be
recovered and utilized.

Peroxisome Assembly

As already noted, the assembly of peroxisomes is fundamentally similar to
that of mitochondria and chloroplasts, rather than to that of the endoplasmic reticulum, Golgi apparatus, and lysosomes. Proteins destined for peroxisomes are translated on free cytosolic ribosomes and then transported
into peroxisomes as completed polypeptide chains (Figure 10.29). Phospholipids are also imported to peroxisomes, via phospholipid transfer proteins, from their major site of synthesis in the ER. The import of proteins
and phospholipids results in peroxisome growth, and new peroxisomes are
then formed by division of old ones.

Proteins are targeted to the interior of peroxisomes by at least two pathways, which are conserved from yeasts to humans. Most proteins are targeted to peroxisomes by the simple amino acid sequence Ser-Lys-Leu at
their carboxy terminus (peroxisome targeting signal 1, or PTS1). Other proteins are targeted by a sequence of nine amino acids (PTS2) at their amino

Figure 10.29
Assembly of peroxisomes   Proteins destined for peroxisomes are synthesized on free ribosomes and imported into preexisting peroxisomes as completed polypeptide chains. Protein import results in peroxisome growth and the formation of new peroxisomes by division of old ones.

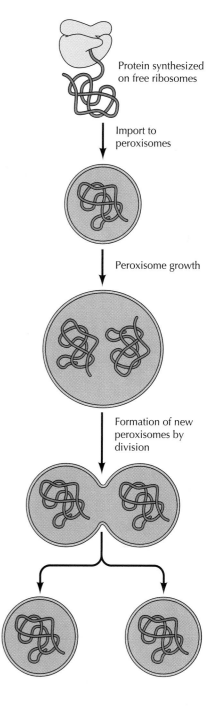

Protein synthesized on free ribosomes

Import to peroxisomes

Peroxisome growth

Formation of new peroxisomes by division

terminus, and some proteins may be targeted by alternative signals that have not yet been well defined.

PTS1 and PTS2 are recognized by distinct receptors and then transferred to a translocation complex that mediates their transport across the peroxisome membrane. However, the mechanism of protein import into peroxisomes is less well characterized than the mechanisms of protein translocation across the membranes of other subcellular organelles. In contrast to the translocation of polypeptide chains across the membranes of the endoplasmic reticulum, mitochondria, and chloroplasts, targeting signals are usually not cleaved during the import of proteins into peroxisomes. Cytosolic Hsp70 has been implicated in protein import to peroxisomes, but the possible role of molecular chaperones within peroxisomes is unclear. Moreover, it appears that proteins can be transported into peroxisomes in at least partially folded conformations, rather than as extended polypeptide chains.

Some peroxisome membrane proteins are similarly synthesized on cytosolic ribosomes and targeted to the peroxisome membrane by distinct internal signals. However, other experiments suggest that some peroxisomal membrane proteins may be synthesized on membrane-bound polysomes of the endoplasmic reticulum and then transported to peroxisomes, suggesting a role for the endoplasmic reticulum in peroxisome maintenance. The import of proteins into peroxisomes thus appears to have several novel features, making it an active area of investigation.

Interestingly, some components of peroxisome import pathways have been identified not only as mutants of yeasts but also as mutations associated with serious human diseases involving disorders of peroxisomes. In some such diseases, only a single peroxisomal enzyme is deficient. However, in other diseases resulting from defects in peroxisome function, multiple peroxisomal enzymes fail to be imported to peroxisomes, instead being localized in the cytosol. The latter group of diseases results from deficiencies in the PTS1 or PTS2 pathways responsible for peroxisomal protein import. The prototypical example is Zellweger syndrome, which is lethal within the first ten years of life. Zellweger syndrome can result from mutations in at least ten different genes affecting peroxisomal protein import, one of which has been identified as the gene encoding the receptor for the peroxisome targeting signal PTS1.

SUMMARY

MITOCHONDRIA

Organization and Function of Mitochondria: Mitochondria, which play a critical role in the generation of metabolic energy, are surrounded by a double-membrane system. The matrix contains the enzymes of the citric acid cycle; the inner membrane contains protein complexes involved in electron transport and oxidative phosphorylation. In contrast to the inner membrane, the outer membrane is freely permeable to small molecules.

KEY TERMS

mitochondria, crista, matrix, porin

endosymbiosis

The Genetic System of Mitochondria: Mitochondria contain their own genomes, which encode rRNAs, tRNAs, and some of the proteins that are involved in oxidative phosphorylation.

presequence, phospholipid transfer protein, cardiolipin

Protein Import and Mitochondrial Assembly: Most mitochondrial proteins are encoded by the nuclear genome. These proteins are translated on free ribosomes and imported into mitochondria as completed polypeptide chains. Positively charged presequences target proteins for import to the mitochondrial matrix. Phospholipids are carried to mitochondria from the endoplasmic reticulum by phospholipid transfer proteins.

THE MECHANISM OF OXIDATIVE PHOSPHORYLATION

oxidative phosphorylation, electron transport chain, coenzyme Q, ubiquinone, cytochrome *c*, cytochrome oxidase

The Electron Transport Chain: Most of the energy derived from oxidative metabolism comes from the transfer of electrons from NADH and $FADH_2$ to O_2. In order to harvest this energy in usable form, electrons are transferred through a series of carriers organized into four protein complexes in the inner mitochondrial membrane.

chemiosmotic coupling, electrochemical gradient, ATP synthase

Chemiosmotic Coupling: The energy-yielding reactions of electron transport are coupled to the generation of a proton gradient across the inner mitochondrial membrane. The potential energy stored in this gradient is harvested by a fifth protein complex, ATP synthase, which couples ATP synthesis to the energetically favorable return of protons to the mitochondrial matrix.

Transport of Metabolites across the Inner Membrane: In addition to driving ATP synthesis, potential energy stored in the proton gradient drives the transport of ATP, ADP, and other metabolites into and out of mitochondria.

CHLOROPLASTS AND OTHER PLASTIDS

chloroplast, thylakoid membrane, stroma

The Structure and Function of Chloroplasts: Chloroplasts are large organelles that function in photosynthesis and a variety of other metabolic activities. Like mitochondria, chloroplasts are bounded by a double-membrane envelope. In addition, chloroplasts have an internal thylakoid membrane, which is the site of electron transport and the chemiosmotic generation of ATP.

The Chloroplast Genome: Chloroplast genomes contain more than 100 genes, including genes encoding rRNAs, tRNAs, some ribosomal proteins, and some proteins involved in photosynthesis.

transit peptide

Import and Sorting of Chloroplast Proteins: Most chloroplast proteins are synthesized on free ribosomes in the cytosol and targeted for import to chloroplasts by amino-terminal transit peptides. Proteins incorporated into the thylakoid lumen are first imported into the chloroplast stroma and then targeted for transport across the thylakoid membrane by a second hydrophobic signal sequence.

plastid, chromoplast, leucoplast, amyloplast, elaioplast, proplastid, etioplast

Other Plastids: The chloroplast is only one member of a family of related organelles, all of which contain the same genome. Other plastids serve to store energy sources, such as starch and lipids, and function in other aspects of plant metabolism.

PHOTOSYNTHESIS

Electron Flow through Photosystems I and II: During photosynthesis, energy from sunlight is harvested and converted to usable forms of potential chemical energy. Absorption of light by chlorophylls excites electrons to a higher energy state. These high-energy electrons are then transferred through a series of carriers organized into two photosystems and the cytochrome *bf* complex in the thylakoid membrane. The sequential flow of electrons through both photosystems is coupled to the synthesis of ATP at photosystem II and the reduction of NADP$^+$ to NADPH at photosystem I. Both ATP and NADPH are then used in the synthesis of carbohydrates from CO_2, which takes place in the chloroplast stroma.

chlorophyll, photocenter, photosystem I, photosystem II, cytochrome *bf* complex, NADP reductase

Cyclic Electron Flow: The alternative pathway of cyclic electron flow allows light energy harvested at photosystem I to be converted to ATP, rather than NADPH.

cyclic electron flow

ATP Synthesis: The chemiosmotic synthesis of ATP is driven by a proton gradient across the thylakoid membrane.

PEROXISOMES

Functions of Peroxisomes: Peroxisomes are small organelles, bounded by a single membrane, that contain enzymes involved in a variety of metabolic reactions, including fatty acid oxidation, the glyoxylate cycle, and photorespiration.

peroxisome, catalase, plasmalogen, glyoxylate cycle, glyoxysome, photorespiration

Peroxisome Assembly: Peroxisomal proteins are synthesized on free ribosomes in the cytosol and imported to peroxisomes as complete polypeptide chains. At least two types of signals target proteins to the interior of peroxisomes, but the mechanism of protein import is not well understood.

Questions

1. How does the electrochemical gradient across the inner mitochondrial membrane contribute to protein import?

2. Assume that the electric potential across the inner mitochondrial membrane is dissipated, so the electrochemical gradient is composed solely of a proton concentration gradient corresponding to one pH unit. Calculate the free energy stored in this gradient. For your calculation, use $R = 1.98 \times 10^{-3}$ kcal/mol/deg, $T = 298K$ (25°C), and note that $\ln(x) = 2.3 \log_{10}(x)$.

3. What is the topological relationship between the thylakoid lumen and the intermembrane space of chloroplasts?

4. Why are the transit peptides of chloroplast proteins, in contrast to the presequences of mitochondrial proteins, not positively charged?

5. How many high-energy electrons are required to drive the synthesis of one molecule of glucose during photosynthesis, coupled to the formation of six molecules of O_2? How many molecules of ATP and NADPH are generated by passage of these electrons through photosystems I and II?

6. What fraction of the carbon atoms converted to glycolate during photorespiration are salvaged by peroxisomes?

References and Further Reading

Mitochondria

Andersson, S. G. E., A. Zomorodipour, J. O. Andersson, T. Sicheritz-Ponten, U. C. M. Alsmark, R. M. Podowski, A. K. Naslund, A.-S. Eriksson, H. H. Winkler and C. G. Kurland. 1998. The genome sequence of *Rickettsia prowazekii* and the origin of mitochondria. *Nature* 396: 133–140. [P]

Clayton, D. A. 1991. Replication and transcription of vertebrate mitochondrial DNA. *Ann. Rev. Cell Biol.* 7: 453–478. [R]

Dietrich, A., J. H. Weil and L. Marechal-Drouard. 1992. Nuclear-encoded transfer RNAs in plant mitochondria. *Ann. Rev. Cell Biol.* 8: 115–131. [R]

Gray, M. W., G. Burger and B. F. Lang. 1999. Mitochondrial evolution. *Science* 283: 1476–1481. [R]

Koehler, C. M., S. Merchant and G. Schatz. 1999. How membrane proteins travel across the mitochondrial intermembrane space. *Trends Biochem. Sci.* 24: 428–432. [R]

Lang, B. F., G. Burger, C. J. O'Kelly, R. Cedergren, G. B. Golding, C. Lemieux, D. Sankoff, M. Turmel and M. W. Gray. 1997. An ancestral mitochondrial DNA resembling a eubacterial genome in miniature. *Nature* 387: 493–497. [P]

Luft, R. 1994. The development of mitochondrial medicine. *Proc. Natl. Acad. Sci. USA* 91: 8731–8738. [R]

Michel, H. 1998. The mechanism of proton pumping by cytochrome *c* oxidase. *Proc. Natl. Acad. Sci. USA* 95: 12819–12824. [P]

Michikawa, Y., F. Mazzucchelli, N. Bresolin, G. Scarlato and G. Attardi. 1999. Aging-dependent large accumulation of point mutations in the human mtDNA control region for replication. *Science* 286: 774–779. [P]

Neupert, W. 1997. Protein import into mitochondria. *Ann. Rev. Biochem.* 66: 863–917. [R]

Pfanner, N., E. A. Craig and A. Honlinger. 1997. Mitochondrial preprotein translocase. *Ann. Rev. Cell Dev. Biol.* 13:25–51. [R]

Schatz, G. 1996. The protein import system of mitochondria. *J. Biol. Chem.* 271: 31763–31766. [R]

Schatz, G. and B. Dobberstein. 1996. Common principles of protein translocation across membranes. *Science* 271: 1519–1526. [R]

Shadel, G. S. and D. A. Clayton. 1997. Mitochondrial DNA maintenance in vertebrates. *Ann. Rev. Biochem.* 66: 409–435. [R]

Wallace, D. C. 1999. Mitochondrial diseases in man and mouse. *Science* 283: 1482–1488. [R]

The Mechanism of Oxidative Phosphorylation

Boyer, P. D. 1997. The ATP synthase—a splendid molecular machine. *Ann. Rev. Biochem.* 66: 717–749. [R]

Hatefi, Y. 1985. The mitochondrial electron transport and oxidative phosphorylation system. *Ann. Rev. Biochem.* 54: 1015–1069. [R]

Kinosita, K., Jr., R. Yasuda, H. Noji, S. Ishiwata and M. Yoshida. 1998. F_1-ATPase: A rotary motor made of a single molecule. *Cell* 93: 21–24. [R]

Mitchell, P. 1979. Keilin's respiratory chain concept and its chemiosmotic consequences. *Science* 206: 1148–1159. [R]

Nicholls, D. G. and S. J. Ferguson. 1992. *Bioenergetics 2*. London: Academic Press.

Racker, E. 1980. From Pasteur to Mitchell: A hundred years of bioenergetics. *Fed. Proc.* 39: 210–215. [R]

Rastogi, V. and M. E. Girvin. 1999. Structural changes linked to proton translocation by subunit *c* of the ATP synthase. *Nature* 402: 263–268. [P]

Saraste, M. 1999. Oxidative phosphorylation at the *fin de siecle*. *Science* 283: 1488–1493. [R]

Stock, D., A. G. W. Leslie and J. E. Walker. 1999. Molecular architecture of the rotary motor in ATP synthase. *Science* 286: 1700–1704. [P]

Chloroplasts and Other Plastids

Chen, X. and D. J. Schnell. 1999. Protein import into chloroplasts. *Trends Cell Biol.* 9: 222–227. [R]

Cline, K. and R. Henry. 1996. Import and routing of nucleus-encoded chloroplast proteins. *Ann. Rev. Cell Dev. Biol.* 12: 1–26. [R]

Douce, R. and J. Joyard. 1990. Biochemistry and function of the plastid envelope. *Ann. Rev. Cell Biol.* 6: 173–216. [R]

Ellis, R. J. 1990. Molecular chaperones: The plant connection. *Science* 250: 954–959. [R]

Gray, J. C. and P. E. Row. 1995. Protein translocation across chloroplast envelope membranes. *Trends Cell Biol.* 5: 243–247. [R]

Gruissem, W. 1989. Chloroplast gene expression: How plants turn their plastids on. *Cell* 56: 161–170. [R]

Rochaix, J.-D. 1992. Post-transcriptional steps in the expression of chloroplast genes. *Ann. Rev. Cell Biol.* 8: 1–28. [R]

Subramanian, A. R. 1993. Molecular genetics of chloroplast ribosomal proteins. *Trends Biochem. Sci.* 18: 177–180. [R]

Sugiura, M. 1992. The chloroplast genome. *Plant Mol. Biol.* 19: 149–168. [R]

Photosynthesis

Arnon, D. I. 1984. The discovery of photosynthetic phosphorylation. *Trends Biochem. Sci.* 9: 258–262. [R]

Barber, J. and B. Andersson. 1994. Revealing the blueprint of photosynthesis. *Nature* 370: 31–34. [R]

Bennett, J. 1979. The protein that harvests sunlight. *Trends Biochem. Sci.* 4: 268–271. [R]

Deisenhofer, J., O. Epp, K. Miki, R. Huber and H. Michel. 1985. Structure of the protein subunits in the photosynthetic reaction centre of *Rhodopseudomonas viridis*. *Nature* 318: 618–624. [P]

Deisenhofer, J. and H. Michel. 1991. Structures of bacterial photosynthetic reaction centers. *Ann. Rev. Cell Biol.* 7: 1–23. [R]

Kuhlbrandt, W., D. N. Wang and Y. Fujiyoshi. 1994. Atomic model of plant light-harvesting complex by electron crystallography. *Nature* 367: 614–621. [P]

Nicholls, D. G. and S. J. Ferguson. 1992. *Bioenergetics 2*. San Diego, CA: Academic Press.

Rhee, K.-H., E. P. Morris, J. Barber and W. Kuhlbrandt. 1998. Three-dimensional structure of the plant photosystem II reaction centre at 8 Å resolution. *Nature* 396: 283–286. [P]

Rogner, M., E. J. Boekema and J. Barber. 1996. How does photosystem 2 split water? The structural basis of efficient energy conversion. *Trends Biochem. Sci.* 21: 44–49. [R]

Youvan, D. C. and B. L. Marrs. 1987. Molecular mechanisms of photosynthesis. *Sci. Am.* 256(6): 42–48. [R]

Peroxisomes

McNew, J. A. and J. M. Goodman. 1996. The targeting and assembly of peroxisomal proteins: Some old rules do not apply. *Trends Biochem. Sci.* 21: 54–58. [R]

Subramani, S. 1996. Protein translocation into peroxisomes. *J. Biol. Chem.* 271: 32483–32486. [R]

Tabak, H. F., I. Braakman and B. Distel. 1999. Peroxisomes: Simple in function but complex in maintenance. *Trends Cell Biol.* 9: 447–453. [R]

Titorenko, V. I. and R. A. Rachubinski. 1998. The endoplasmic reticulum plays an essential role in peroxisome biogenesis. *Trends Biochem. Sci.* 23: 231–233. [R]

Tolbert, N. E. 1981. Metabolic pathways in peroxisomes and glyoxysomes. *Ann. Rev. Biochem.* 50: 133–157. [R]

Van den Bosch, H., R. B. H. Schutgens, R. J. A. Wanders and J. M. Tager. 1992. Biochemistry of peroxisomes. *Ann. Rev. Biochem.* 61: 157–197. [R]

11

The Cytoskeleton and Cell Movement

T HE MEMBRANE-ENCLOSED ORGANELLES discussed in the preceding chapters constitute one level of the organizational substructure of eukaryotic cells. A further level of organization is provided by the cytoskeleton, which consists of a network of protein filaments extending throughout the cytoplasm of all eukaryotic cells. The cytoskeleton provides a structural framework for the cell, serving as a scaffold that determines cell shape and the general organization of the cytoplasm. In addition to playing this structural role, the cytoskeleton is responsible for cell movements. These include not only the movements of entire cells, but also the internal transport of organelles and other structures (such as mitotic chromosomes) through the cytoplasm. Importantly, the cytoskeleton is much less rigid and permanent than its name implies. Rather, it is a dynamic structure that is continually reorganized as cells move and change shape, for example, during cell division.

The cytoskeleton is composed of three principal types of protein filaments: actin filaments, intermediate filaments, and microtubules, which are held together and linked to subcellular organelles and the plasma membrane by a variety of accessory proteins. This chapter discusses the structure and organization of each of these three major components of the cytoskeleton, as well as their roles in cell motility, organelle transport, cell division, and other types of cell movements.

Structure and Organization of Actin Filaments

The major cytoskeletal protein of most cells is **actin**, which polymerizes to form actin filaments—thin, flexible fibers approximately 7 nm in diameter and up to several micrometers in length (Figure 11.1). Within the cell, actin filaments (also called **microfilaments**) are organized into higher-order structures, forming bundles or three-dimensional networks with the properties of semisolid gels. The assembly and disassembly of actin filaments, their crosslinking into bundles and networks, and their association with other cell structures (such as the plasma membrane) are regulated by a variety of actin-binding proteins, which are critical components of the actin

Figure 11.1
Actin filaments Electron micrograph of actin filaments. (Courtesy of Roger Craig, University of Massachusetts Medical Center.)

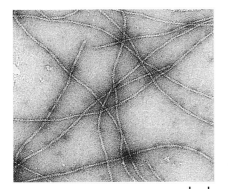

50 nm

cytoskeleton. Actin filaments are particularly abundant beneath the plasma membrane, where they form a network that provides mechanical support, determines cell shape, and allows movement of the cell surface, thereby enabling cells to migrate, engulf particles, and divide.

Assembly and Disassembly of Actin Filaments

Actin was first isolated from muscle cells, in which it constitutes approximately 20% of total cell protein, in 1942. Although actin was initially thought to be uniquely involved in muscle contraction, it is now known to be an extremely abundant protein (typically 5 to 10% of total protein) in all types of eukaryotic cells. Yeasts have only a single actin gene, but higher eukaryotes have several distinct types of actin, which are encoded by different members of the actin gene family. Mammals, for example, have at least six distinct actin genes: Four are expressed in different types of muscle and two are expressed in nonmuscle cells. All of the actins, however, are very similar in amino acid sequence and have been highly conserved throughout the evolution of eukaryotes. Yeast actin, for example, is 90% identical in amino acid sequence to the actins of mammalian cells.

The three-dimensional structures of both individual actin molecules and actin filaments were determined in 1990 by Kenneth Holmes, Wolfgang Kabsch, and their colleagues. Individual actin molecules are globular proteins of 375 amino acids (43 kd). Each actin monomer (**globular [G] actin**) has tight binding sites that mediate head-to-tail interactions with two other actin monomers, so actin monomers polymerize to form filaments (**filamentous [F] actin**) (Figure 11.2). Each monomer is rotated by 166° in the filaments, which

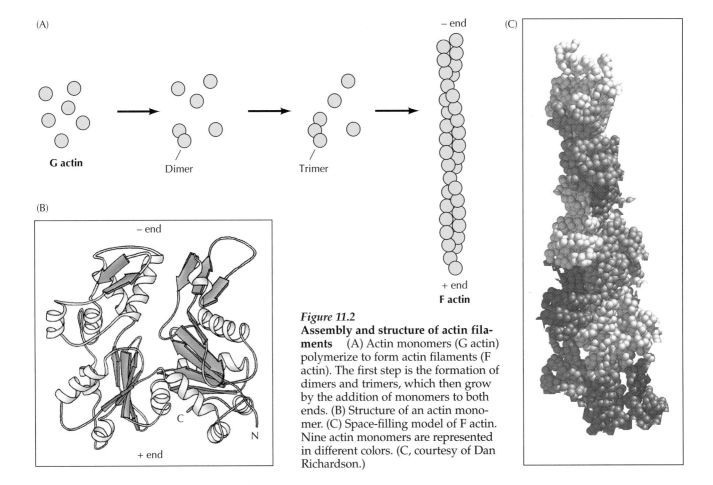

Figure 11.2
Assembly and structure of actin filaments (A) Actin monomers (G actin) polymerize to form actin filaments (F actin). The first step is the formation of dimers and trimers, which then grow by the addition of monomers to both ends. (B) Structure of an actin monomer. (C) Space-filling model of F actin. Nine actin monomers are represented in different colors. (C, courtesy of Dan Richardson.)

therefore have the appearance of a double-stranded helix. Because all the actin monomers are oriented in the same direction, actin filaments have a distinct polarity and their ends (called the plus and minus ends) are distinguishable from one another. This polarity of actin filaments is important both in their assembly and in establishing a unique direction of myosin movement relative to actin, as discussed later in the chapter.

The assembly of actin filaments can be studied *in vitro* by regulation of the ionic strength of actin solutions. In solutions of low ionic strength, actin filaments depolymerize to monomers. Actin then polymerizes spontaneously if the ionic strength is increased to physiological levels. The first step in actin polymerization (called nucleation) is the formation of a small aggregate consisting of three actin monomers. Actin filaments are then able to grow by the reversible addition of monomers to both ends, but one end (the plus end) elongates five to ten times faster than the minus end. The actin monomers also bind ATP, which is hydrolyzed to ADP following filament assembly. Although ATP is not required for polymerization, actin monomers to which ATP is bound polymerize more readily than those to which ADP is bound. As discussed below, ATP binding and hydrolysis play a key role in regulating the assembly and dynamic behavior of actin filaments.

Because actin polymerization is reversible, filaments can depolymerize by the dissociation of actin subunits, allowing actin filaments to be broken down when necessary (Figure 11.3). Thus, an apparent equilibrium exists between actin monomers and filaments, which is dependent on the concentration of free monomers. The rate at which actin monomers are incorporated into filaments is proportional to their concentration, so there is a critical concentration of actin monomers at which the rate of their polymerization into filaments equals the rate of dissociation. At this critical concentration, monomers and filaments are in apparent equilibrium.

As noted earlier, the two ends of an actin filament grow at different rates, with monomers being added to the fast-growing end (the plus end) five to ten times faster than to the slow-growing (minus) end. Because ATP-actin dissociates less readily than ADP-actin, this results in a difference in the critical concentration of monomers needed for polymerization at the two ends. This difference can result in the phenomenon known as **treadmilling**, which illustrates the dynamic behavior of actin filaments (Figure 11.4). For the sys-

Figure 11.3
Reversible polymerization of actin monomers Actin polymerization is a reversible process, in which monomers both associate with and dissociate from the ends of actin filaments. The rate of subunit dissociation (k_{off}) is independent of monomer concentration, while the rate of subunit association is proportional to the concentration of free monomers and given by $C \times k_{on}$ (C = concentration of free monomers). An apparent equilibrium is reached at the critical concentration of monomers (C_c), where $k_{off} = C_c \times k_{on}$.

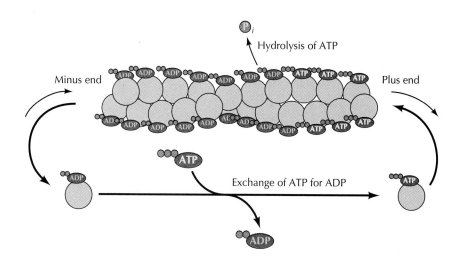

Figure 11.4
Treadmilling The minus ends grow less rapidly than the plus ends of actin filaments. This difference in growth rate is reflected in a difference in the critical concentration for addition of monomers to the two ends of the filament. Actin bound to ATP associates with the rapidly growing plus ends, and the ATP bound to actin is then hydrolyzed to ADP. Because ADP-actin dissociates from filaments more readily than ATP-actin, the critical concentration of actin monomers is higher for addition to the minus end than to the plus end of actin filaments. Treadmilling takes place at monomer concentrations intermediate between the critical concentrations for the plus and minus ends. Under these conditions, there is a net dissociation of monomers (bound to ADP) from the minus end, balanced by the addition of monomers (bound to ATP) to the plus end.

tem to be at an overall steady state, the concentration of free actin monomers must be intermediate between the critical concentrations required for polymerization at the plus and minus ends of the actin filaments. Under these conditions, there is a net loss of monomers from the minus end, which is balanced by a net addition to the plus end. Treadmilling requires ATP, with ATP-actin polymerizing at the plus end of filaments while ADP-actin dissociates from the minus end. Although the role of treadmilling in the cell is unclear, it may reflect the dynamic assembly and disassembly of actin filaments required for cells to move and change shape.

It is noteworthy that several drugs useful in cell biology act by binding to actin and affecting its polymerization. For example, the **cytochalasins** bind to the plus ends of actin filaments and block their elongation. This results in changes in cell shape as well as inhibition of some types of cell movements (e.g., cell division following mitosis), indicating that actin polymerization is required for these processes. Another drug, **phalloidin**, binds tightly to actin filaments and prevents their dissociation into individual actin molecules. Phalloidin labeled with a fluorescent dye is frequently used to visualize actin filaments by fluorescence microscopy.

Within the cell, both the assembly and disassembly of actin filaments are regulated by **actin-binding proteins** (Figure 11.5). The turnover of actin filaments is about 100 times faster within the cell than it is *in vitro*, and this rapid turnover of actin plays a critical role in a variety of cell movements. The key protein responsible for actin filament disassembly within the cell is **cofilin**, which binds to actin filaments and enhances the rate of dissociation of actin monomers from the minus end. In addition, cofilin can sever actin filaments, generating more ends and further enhancing filament disassembly.

Cofilin preferentially binds to ADP-actin, so it remains bound to actin monomers following filament disassembly and sequesters them in the ADP-bound form, preventing their reincorporation into filaments. However, another actin-binding protein, **profilin**, can reverse this effect of cofilin and stimulate the incorporation of actin monomers into filaments. Profilin acts by stimulating the exchange of bound ADP for ATP, resulting in the formation of ATP-actin monomers, which dissociate from cofilin and are then available for assembly into filaments. Other proteins (**Arp2/3 proteins**) can serve as nucleation sites to initiate the assembly of new filaments, so cofilin, profilin, and the Arp2/3 proteins (as well as other actin-binding proteins) can act together to promote the rapid turnover of actin filaments and remodeling of the actin cytoskeleton which is required for a variety of cell movements and changes in cell shape. As might be expected, the activities of cofilin, profilin, and Arp2/3 proteins are controlled by a variety of cell

Figure 11.5
Effects of actin-binding proteins on filament turnover Cofilin binds to actin filaments and increases the rate of dissociation of actin monomers (bound to ADP) from the minus end. Cofilin remains bound to the ADP-actin monomers, preventing their reassembly into filaments. However, profilin can stimulate the exchange of bound ADP for ATP, resulting in the formation of ATP-actin monomers that can be repolymerized into filaments, including new filaments nucleated by the Arp2/3 proteins.

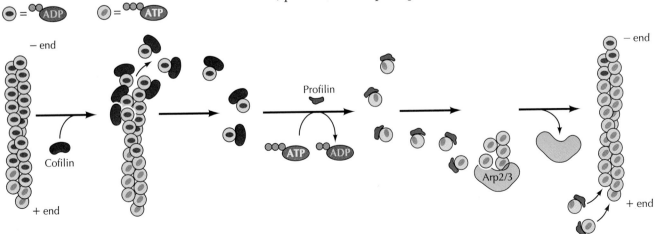

signaling mechanisms (discussed in Chapter 13), allowing actin polymerization to be appropriately regulated in response to environmental stimuli.

Organization of Actin Filaments

Individual actin filaments are assembled into two general types of structures, called **actin bundles** and **actin networks**, which play different roles in the cell (Figure 11.6). In bundles, the actin filaments are crosslinked into closely packed parallel arrays. In networks, the actin filaments are loosely crosslinked in orthogonal arrays that form three-dimensional meshworks with the properties of semisolid gels. The formation of these structures is governed by a variety of actin-binding proteins that crosslink actin filaments in distinct patterns.

All of the actin-binding proteins involved in crosslinking contain at least two domains that bind actin, allowing them to bind and crosslink two different actin filaments. The nature of the association between these filaments is then determined by the size and shape of the crosslinking proteins (see Figure 11.6). The proteins that crosslink actin filaments into bundles (called **actin-bundling proteins**) usually are small rigid proteins that force the filaments to align closely with one another. In contrast, the proteins that organize actin filaments into networks tend to be large flexible proteins that can crosslink perpendicular filaments. These actin-crosslinking proteins appear to be modular proteins consisting of related structural units. In particular, the actin-binding domains of many of these proteins are similar in structure. They are separated by spacer sequences that vary in length and flexibility, and it is these differences in the spacer sequences that are responsible for the distinct crosslinking properties of different actin-binding proteins.

(A)

0.1 μm

Figure 11.6
Actin bundles and networks   (A) Electron micrograph of actin bundles (arrowheads) projecting from the actin network (arrows) underlying the plasma membrane of a macrophage. The bundles support cell surface projections called microspikes or filopodia (see Figure 11.17). (B) Schematic organization of bundles and networks. Actin filaments in bundles are crosslinked into parallel arrays by small proteins that align the filaments closely with one another. In contrast, networks are formed by large flexible proteins that crosslink orthogonal filaments. (A, courtesy of John H. Hartwig, Brigham & Women's Hospital.)

(B)

Parallel bundle

Actin filaments

Ca²⁺-binding domain

ABD   ABD

Fimbrin

Contractile bundle

Actin filament

α-helical spacer domain

ABD

α-actinin   ABD

Figure 11.7
Actin-bundling proteins Actin filaments are associated into two types of bundles by different actin-bundling proteins. Fimbrin has two adjacent actin-binding domains (ABD) and crosslinks actin filaments into closely packed parallel bundles in which the filaments are approximately 14 nm apart. In contrast, the two separated actin-binding domains of α-actinin dimers crosslink filaments into more loosely spaced contractile bundles in which the filaments are separated by 40 nm. Both fimbrin and α-actinin contain two related Ca²⁺-binding domains, and α-actinin contains four repeated α-helical spacer domains.

There are two structurally and functionally distinct types of actin bundles, involving different actin-bundling proteins (Figure 11.7). The first type of bundle, containing closely spaced actin filaments aligned in parallel, supports projections of the plasma membrane, such as microvilli (see Figures 11.15 and 11.16). In these bundles, all the filaments have the same polarity, with their plus ends adjacent to the plasma membrane. An example of a bundling protein involved in the formation of these structures is **fimbrin**, which was first isolated from intestinal microvilli and later found in surface projections of a wide variety of cell types. Fimbrin is a 68-kd protein, containing two adjacent actin-binding domains. It binds to actin filaments as a monomer, holding two parallel filaments close together.

The second type of actin bundle is composed of filaments that are more loosely spaced and are capable of contraction, such as the actin bundles of the contractile ring that divides cells in two following mitosis. The looser structure of these bundles (which are called **contractile bundles**) reflects the properties of the crosslinking protein α-**actinin**. In contrast to fimbrin, α-actinin binds to actin as a dimer, each subunit of which is a 102-kd protein containing a single actin-binding site. Filaments crosslinked by α-actinin are consequently separated by a greater distance than those crosslinked by fimbrin (40 nm apart instead of 14 nm). The increased spacing between filaments allows the motor protein myosin to interact with the actin filaments in these bundles, which (as discussed later) enables them to contract.

The actin filaments in networks are held together by large actin-binding proteins, such as **filamin** (Figure 11.8). Filamin (also called actin-binding protein or ABP-280) binds actin as a dimer of two 280-kd subunits. The actin-binding domains and dimerization domains are at opposite ends of each subunit, so the filamin dimer is a flexible V-shaped molecule with actin-binding domains at the ends of each arm. As a result, filamin forms cross-links between orthogonal actin filaments, creating a loose three-dimensional meshwork. As discussed in the next section, such networks of actin filaments underlie the plasma membrane and support the surface of the cell.

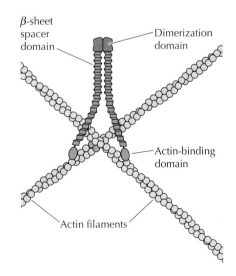

β-sheet spacer domain

Dimerization domain

Actin-binding domain

Actin filaments

Figure 11.8
Actin networks and filamin Filamin is a dimer of two large (280-kd) subunits, forming a flexible V-shaped molecule that crosslinks actin filaments into orthogonal networks. The carboxy-terminal dimerization domain is separated from the amino-terminal actin-binding domain by repeated β-sheet spacer domains.

Figure 11.9
Morphology of red blood cells
Scanning electron micrograph of red blood cells illustrating their biconcave shape. (Omikron/Photo Researchers, Inc.)

5 μm

Association of Actin Filaments with the Plasma Membrane

Actin filaments are highly concentrated at the periphery of the cell, where they form a three-dimensional network beneath the plasma membrane (see Figure 11.6). This network of actin filaments and associated actin-binding proteins (called the **cell cortex**) determines cell shape and is involved in a variety of cell surface activities, including movement. The association of the actin cytoskeleton with the plasma membrane is thus central to cell structure and function.

Red blood cells (erythrocytes) have proven particularly useful for studies of both the plasma membrane (discussed in the next chapter) and the cortical cytoskeleton. The principal advantage of red blood cells for these studies is that they contain no nucleus or internal organelles, so their plasma membrane and associated proteins can be easily isolated without contamination by the various internal membranes that are abundant in other cell types. In addition, human erythrocytes lack other cytoskeletal components (microtubules and intermediate filaments), so the cortical cytoskeleton is the principal determinant of their distinctive shape as biconcave discs (Figure 11.9).

The major protein that provides the structural basis for the cortical cytoskeleton in erythrocytes is the actin-binding protein **spectrin**, which is related to filamin (Figure 11.10). Erythrocyte spectrin is a tetramer consisting of two distinct polypeptide chains, called α and β, with molecular weights of 240 and 220 kd, respectively. The β chain has a single actin-binding domain at its amino terminus. The α and β chains associate laterally to form dimers, which then join head to head to form tetramers with two actin-binding domains separated by approximately 200 nm. The ends of

Figure 11.10
Structure of spectrin   Spectrin is a tetramer consisting of two α and two β chains. Each β chain has a single actin-binding domain (ABD) at its amino terminus. Both α and β chains contain multiple repeats of α-helical spacer domains, which separate the two actin-binding domains of the tetramer. The α chain has two Ca^{2+} binding domains at its carboxy terminus.

Spectrin tetramer

ABD
β chain
α-helical domain
β-sheet domain
α chain
α chain
β chain
ABD
Ca^{2+}-binding domain

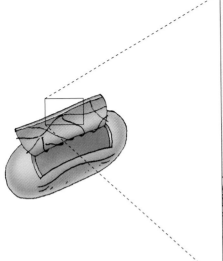

Figure 11.11
Association of the erythrocyte cortical cytoskeleton with the plasma membrane The plasma membrane is associated with a network of spectrin tetramers crosslinked by short actin filaments in association with protein 4.1. The spectrin-actin network is linked to the membrane by ankyrin, which binds to both spectrin and the abundant transmembrane protein band 3. An additional link is provided by the binding of protein 4.1 to glycophorin.

the spectrin tetramers then associate with short actin filaments, resulting in the spectrin-actin network that forms the cortical cytoskeleton of red blood cells (Figure 11.11). The major link between the spectrin-actin network and the plasma membrane is provided by a protein called **ankyrin**, which binds both to spectrin and to the cytoplasmic domain of an abundant transmembrane protein called band 3. An additional link between the spectrin-actin network and the plasma membrane is provided by protein 4.1, which binds to spectrin-actin junctions as well as recognizing the cytoplasmic domain of glycophorin (another abundant transmembrane protein).

Other types of cells contain linkages between the cortical cytoskeleton and the plasma membrane that are similar to those observed in red blood cells. Proteins related to spectrin (nonerythroid spectrin is also called **fodrin**), ankyrin, and protein 4.1 are expressed in a wide range of cell types, where they fulfill functions analogous to those described for erythrocytes. For example, a family of proteins related to protein 4.1 (the **ERM proteins**) link actin filaments to the plasma membranes of many different kinds of cells and the spectrin-related protein filamin (see Figure 11.8) constitutes a major link between actin filaments and the plasma membrane of blood platelets. Another member of this group of spectrin-related proteins is **dystrophin**, which is of particular interest because it is the product of the gene responsible for two types of muscular dystrophy (Duchenne's and Becker's). These X-linked inherited diseases result in progressive degeneration of skeletal muscle, and patients with the more severe form of the disease (Duchenne's muscular dystrophy) usually die in their teens or early twenties. Molecular cloning of the gene responsible for this disorder revealed that it encodes a large protein (427 kd) that is either absent or abnormal in patients with Duchenne's or Becker's muscular dystrophy, respectively. The sequence of dystrophin further indicated that it is related to spectrin, with a single actin-binding domain at its amino terminus and a membrane-binding domain at its carboxy terminus. Like spectrin, dystrophin forms dimers that link actin filaments to transmembrane proteins of the muscle cell plasma membrane. These transmembrane proteins in turn link the cytoskeleton to the extracellular matrix, which plays an important role in maintaining cell stability during muscle contraction.

Molecular Medicine

Muscular Dystrophy and the Cytoskeleton

The Disease

The muscular dystrophies are a group of hereditary diseases characterized by the progressive loss of muscle cells. Duchenne's muscular dystrophy (DMD) is the most common and severe form, affecting approximately one out of every 3500 male children. Children with DMD are usually unaffected until three to five years of age, when muscular weakness is first observed. Progressive loss of muscle then relentlessly continues, usually leaving affected children confined to wheelchairs by age 12 and frequently leading to death as a result of respiratory failure by the late teens or early twenties. Becker's muscular dystrophy (BMD) is a less common form of the disease, with an incidence of one in 30,000 male births. BMD is also less severe than DMD, and many BMD patients are able to live normal lives with minimal restrictions.

Molecular and Cellular Basis

The much higher incidence of DMD and BMD in boys than in girls initially suggested that both diseases result from recessive sex-linked genes. This hypothesis was confirmed by genetic studies, which localized the DMD/BMD gene to a specific region of the X chromosome. On the basis of its chromosomal position, the gene responsible for DMD and BMD was cloned by the research groups of Lou Kunkel and Ron Worton in 1986. Sequence analysis established that it

encodes a 427-kd protein, called dystrophin, which is related to spectrin. Dystrophin is linked to the plasma membrane of muscle cells by a complex of transmembrane proteins. These transmembrane proteins in turn bind to components of the extracellular matrix, so dystrophin plays a key role in anchoring the cytoskeleton of muscle cells to the extracellular matrix. This anchorage is thought to stabilize the plasma membrane and enable the cell to withstand the stress of muscle contraction. The mutations responsible for DMD or BMD result either in the absence of dystrophin or in the expression of an abnormal protein, respectively, consistent with the severity of disease in DMD and BMD patients.

Prevention and Treatment

Identification of the DMD/BMD gene has allowed the development of sensitive diagnostic tests, such as the detection of mutations by the polymerase chain reaction. Women carrying mutant alleles of the DMD/BMD gene can thus be identified, and their pregnancies monitored to determine if the mutant gene has been transmitted to the fetus. Because of the sensitivity of PCR analysis, mutations can also be detected in early embryos derived by *in vitro* fertilization procedures prior to their return to the uterus. The combination of genetic counseling and prenatal diagnosis thus has the potential of preventing the transmission of inherited DMD/BMD cases.

Unfortunately, the effectiveness of this approach is limited by the fact that approximately one-third of DMD cases are not inherited, but result instead from new mutations. These cases cannot be predicted by genetic analysis, so development of an effective therapy remains a compelling goal. Present efforts are focused on the use of gene therapy to restore dystrophin expression in muscle or to increase the expression of other dystrophin-related proteins that might compensate for its absence. Although treatment of muscular dystrophy remains a goal of future research, understanding its molecular basis has revolutionized diagnosis and permitted rational approaches to the development of new therapies.

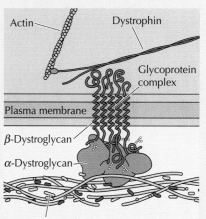

Dystrophin binds to actin and to a complex of transmembrane proteins, which link the cytoskeleton of muscle cells to the extracellular matrix.

In contrast to the uniform surface of red blood cells, most cells have specialized regions of the plasma membrane that form contacts with adjacent cells, tissue components, or other substrates (such as the surface of a culture dish). These regions also serve as attachment sites for bundles of actin filaments that anchor the cytoskeleton to areas of cell contact. These attachments of actin filaments are particularly evident in fibroblasts maintained in tissue culture (Figure 11.12). Such cultured fibroblasts secrete extracellu-

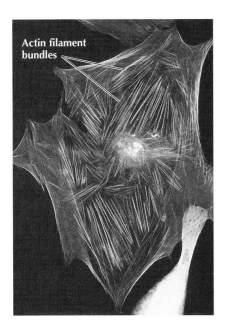

Actin filament bundles

Figure 11.12
Stress fibers and focal adhesions   Fluorescence microscopy of a human fibroblast in which actin filaments have been been stained with a fluorescent dye. Stress fibers are revealed as bundles of actin filaments anchored at sites of cell attachment to the culture dish surface (focal adhesions). (Don Fawcett/Photo Researchers, Inc.)

lar matrix proteins (discussed in Chapter 12) that stick to the plastic surface of the culture dish. The fibroblasts then attach to the culture dish via the binding of transmembrane proteins (called **integrins**) to the extracellular matrix. The sites of attachment are discrete regions (called **focal adhesions**) that also serve as attachment sites for large bundles of actin filaments called **stress fibers**.

Stress fibers are contractile bundles of actin filaments, crosslinked by α-actinin, that anchor the cell and exert tension against the substratum. They are attached to the plasma membrane at focal adhesions via interactions with integrin. These associations, which are complex and not well understood, may be mediated by several other proteins, including **talin** and **vinculin** (Figure 11.13). For example, both talin and α-actinin bind to the cytoplasmic domains of integrins. Talin also binds to vinculin, which in turn interacts with actin. Other proteins found at focal adhesions may also participate in the attachment of actin filaments, and a combination of these interactions may be responsible for the linkage of actin filaments to the plasma membrane.

The actin cytoskeleton is similarly anchored to regions of cell-cell contact called **adherens junctions** (Figure 11.14). In sheets of epithelial cells, these junctions form a continuous beltlike structure (called an **adhesion belt**) around each cell in which an underlying contractile bundle of actin filaments is linked to the plasma membrane. Contact between cells at adherens junctions is mediated by transmembrane proteins called **cadherins**, which are discussed further in Chapter 12. The cadherins form a complex with cytoplasmic proteins called **catenins**, which associate with actin filaments.

Protrusions of the Cell Surface

The surfaces of most cells have a variety of protrusions or extensions that are involved in cell movement, phagocytosis, or specialized functions such

Figure 11.13
Attachment of stress fibers to the plasma membrane at focal adhesions Focal adhesions are mediated by the binding of integrins to proteins of the extracellular matrix. Stress fibers (bundles of actin filaments crosslinked by α-actinin) are then bound to the cytoplasmic domain of integrins by complex associations involving a number of proteins. Two possible associations are illustrated: 1) talin binds to both integrin and vinculin, which in turn binds to actin, and 2) integrin binds to α-actinin. A number of other proteins (not shown) are also present at focal adhesions and may be involved in anchoring stress fibers to the plasma membrane.

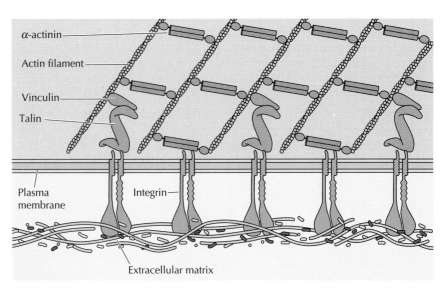

α-actinin

Actin filament

Vinculin

Talin

Plasma membrane

Integrin

Extracellular matrix

Figure 11.14
Attachment of actin filaments to adherens junctions
Cell-cell contacts at adherens junctions are mediated by cadherins, which serve as sites of attachment of actin bundles. In sheets of epithelial cells, these junctions form a continuous belt of actin filaments around each cell. The cadherins are transmembrane proteins that bind β-catenin to their cytoplasmic domains. β-catenin interacts with α-catenin, which serves as a link to actin filaments.

as absorption of nutrients. Most of these cell surface extensions are based on actin filaments, which are organized into either relatively permanent or rapidly rearranging bundles or networks.

The best-characterized of these actin-based cell surface protrusions are **microvilli,** fingerlike extensions of the plasma membrane that are particularly abundant on the surfaces of cells involved in absorption, such as the epithelial cells lining the intestine (Figure 11.15). The microvilli of these cells form a layer on the apical surface (called a **brush border**) that consists of approximately a thousand microvilli per cell and increases the exposed surface area available for absorption by 10- to 20-fold. In addition to their role in absorption, specialized forms of microvilli, the **stereocilia** of auditory hair cells, are responsible for hearing by detecting sound vibrations.

Their abundance and ease of isolation have facilitated detailed structural analysis of intestinal microvilli, which contain closely packed parallel bundles of 20 to 30 actin filaments (Figure 11.16). The filaments in these bundles are crosslinked in part by fimbrin, an actin-bundling protein (discussed earlier) that is present in surface projections of a variety of cell types. However, the major actin-bundling protein in intestinal microvilli is **villin,** a 95-kd protein present in microvilli of only a few specialized types of cells, such as those lining the intestine and kidney tubules. Along their length, the actin bundles of microvilli are attached to the plasma membrane by lateral arms consisting of the calcium-binding protein calmodulin in association with myosin I, which may be involved in movement of the plasma membrane along the actin bundle of the microvillus. At their base, the actin bundles are anchored in a spectrin-rich region of the actin cortex called the terminal web, which crosslinks and stabilizes the microvilli.

In contrast to microvilli, many surface protrusions are transient structures that form in response to environmental stimuli. Several types of these structures extend from the leading edge of a moving cell and are involved in cell locomotion (Figure 11.17). **Pseudopodia** are extensions of moderate width, based on actin filaments crosslinked into a three-dimensional net-

Figure 11.15
Electron micrograph of microvilli
The microvilli (arrows) of intestinal epithelial cells are fingerlike projections of the plasma membrane. They are supported by actin bundles anchored in a dense region of the cortex called the terminal web. (Fred E. Hossler/Visuals Unlimited.)

1 μm

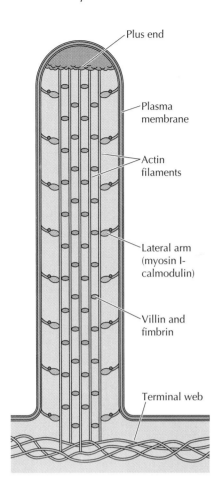

Plus end

Plasma membrane

Actin filaments

Lateral arm (myosin I-calmodulin)

Villin and fimbrin

Terminal web

Figure 11.16
Organization of microvilli   The core actin filaments of microvilli are crosslinked into closely packed bundles by fimbrin and villin. They are attached to the plasma membrane along their length by lateral arms, consisting of myosin I and calmodulin. The plus ends of the actin filaments are embedded in a cap of unidentified proteins at the tip of the microvillus.

work, that are responsible for phagocytosis and for the movement of amoebas across a surface. **Lamellipodia** are broad, sheetlike extensions at the leading edge of fibroblasts, which similarly contain a network of actin filaments. Many cells also extend **microspikes** or **filopodia**, thin projections of the plasma membrane supported by actin bundles. The formation and retraction of these structures is based on the regulated assembly and disassembly of actin filaments, as discussed in the following section.

Actin, Myosin, and Cell Movement

Actin filaments, usually in association with **myosin**, are responsible for many types of cell movements. Myosin is the prototype of a **molecular motor**—a protein that converts chemical energy in the form of ATP to mechanical energy, thus generating force and movement. The most striking variety of such movement is muscle contraction, which has provided the model for understanding actin-myosin interactions and the motor activity of myosin molecules. However, interactions of actin and myosin are responsible not only for muscle contraction but also for a variety of movements of nonmuscle cells, including cell division, so these interactions play a central

Figure 11.17
Examples of cell surface projections involved in phagocytosis and movement
(A) Scanning electron micrograph showing pseudopodia of a macrophage engulfing a tumor cell during phagocytosis. (B) An amoeba with several extended pseudopodia. (C) A tissue culture cell illustrating lamellipodia (L) and filopodia (arrow). (A, K. Wassermann/Visuals Unlimited; B, Stanley Flegler/Visuals Unlimited; C, Don Fawcett/Photo Researchers, Inc.)

(A)

(B)

(C)

role in cell biology. Moreover, the actin cytoskeleton is responsible for the crawling movements of cells across a surface, which appear to be driven directly by actin polymerization as well as actin-myosin interactions.

Muscle Contraction

Muscle cells are highly specialized for a single task, contraction, and it is this specialization in structure and function that has made muscle the prototype for studying movement at the cellular and molecular levels. There are three distinct types of muscle cells in vertebrates: skeletal muscle, which is responsible for all voluntary movements; cardiac muscle, which pumps blood from the heart; and smooth muscle, which is responsible for involuntary movements of organs such as the stomach, intestine, uterus, and blood vessels. In both skeletal and cardiac muscle, the contractile elements of the cytoskeleton are present in highly organized arrays that give rise to characteristic patterns of cross-striations. It is the characterization of these structures in skeletal muscle that has led to our current understanding of muscle contraction, and other actin-based cell movements, at the molecular level.

Skeletal muscles are bundles of **muscle fibers**, which are single large cells (approximately 50 μm in diameter and up to several centimeters in length) formed by the fusion of many individual cells during development (Figure 11.18). Most of the cytoplasm consists of **myofibrils**, which are cylindrical bundles of two types of filaments: thick filaments of myosin (about 15 nm in diameter) and thin filaments of actin (about 7 nm in diameter). Each myofibril is organized as a chain of contractile units called **sarcomeres**, which are responsible for the striated appearance of skeletal and cardiac muscle.

The sarcomeres (which are approximately 2.3 μm long) consist of several distinct regions, discernible by electron microscopy, which provided critical

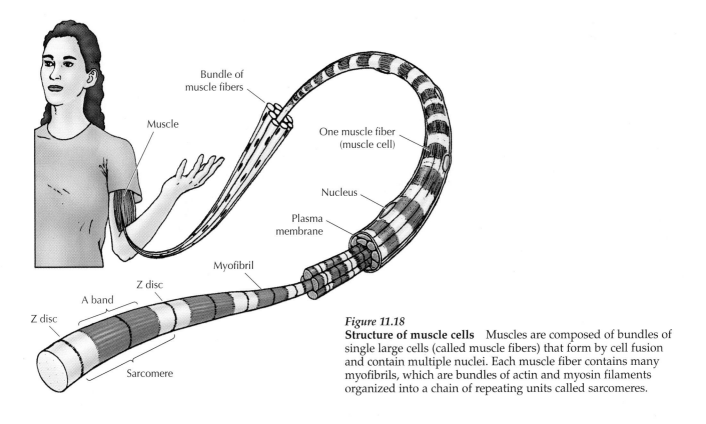

Figure 11.18
Structure of muscle cells   Muscles are composed of bundles of single large cells (called muscle fibers) that form by cell fusion and contain multiple nuclei. Each muscle fiber contains many myofibrils, which are bundles of actin and myosin filaments organized into a chain of repeating units called sarcomeres.

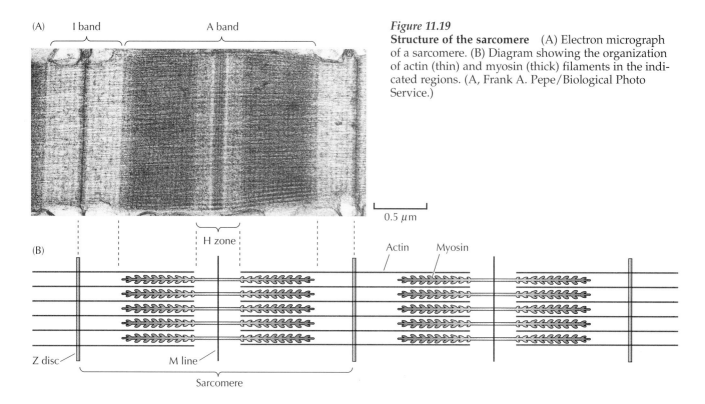

(A) I band   A band

0.5 µm

(B)

H zone

Actin   Myosin

Z disc   M line

Sarcomere

Figure 11.19
Structure of the sarcomere   (A) Electron micrograph of a sarcomere. (B) Diagram showing the organization of actin (thin) and myosin (thick) filaments in the indicated regions. (A, Frank A. Pepe/Biological Photo Service.)

insights into the mechanism of muscle contraction (Figure 11.19). The ends of each sarcomere are defined by the Z disc. Within each sarcomere, dark bands (called A bands because they are *a*nisotropic when viewed with polarized light) alternate with light bands (called I bands for *i*sotropic). These bands correspond to the presence or absence of myosin filaments. The I bands contain only thin (actin) filaments, whereas the A bands contain thick (myosin) filaments. The myosin and actin filaments overlap in peripheral regions of the A band, whereas a middle region (called the H zone) contains only myosin. The actin filaments are attached at their plus ends to the Z disc, which includes the crosslinking protein α-actinin. The myosin filaments are anchored at the M line in the middle of the sarcomere.

Two additional proteins (**titin** and **nebulin**) also contribute to sarcomere structure and stability (Figure 11.20). Titin is an extremely large protein (3000 kd), and single titin molecules extend from the M line to the Z disc. These long molecules of titin are thought to act like springs that keep the myosin filaments centered in the sarcomere and maintain the resting tension that allows a muscle to snap back if overextended. Nebulin filaments are associated with actin and are thought to regulate the assembly of actin filaments by acting as rulers that determine their length.

Figure 11.20
Titin and nebulin   Molecules of titin extend from the Z disc to the M line and act as springs to keep myosin filaments centered in the sarcomere. Molecules of nebulin extend from the Z disc and are thought to determine the length of associated actin filaments.

Z disc   Actin   Nebulin   Myosin   Titin

M line

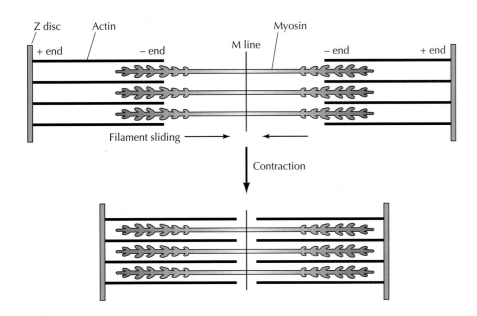

Figure 11.21
Sliding-filament model of muscle contraction   The actin filaments slide past the myosin filaments toward the middle of the sarcomere. The result is shortening of the sarcomere without any change in filament length.

The basis for understanding muscle contraction is the **sliding filament model**, first proposed in 1954 both by Andrew Huxley and Ralph Niedergerke and by Hugh Huxley and Jean Hanson (Figure 11.21). During muscle contraction, each sarcomere shortens, bringing the Z discs closer together. There is no change in the width of the A band, but both the I bands and the H zone almost completely disappear. These changes are explained by the actin and myosin filaments sliding past one another, so that the actin filaments move into the A band and H zone. Muscle contraction thus results from an interaction between the actin and myosin filaments that generates their movement relative to one another. The molecular basis for this interaction is the binding of myosin to actin filaments, allowing myosin to function as a motor that drives filament sliding.

The type of myosin present in muscle (**myosin II**) is a very large protein (about 500 kd) consisting of two identical heavy chains (about 200 kd each) and two pairs of light chains (about 20 kd each) (Figure 11.22). Each heavy chain consists of a globular head region and a long α-helical tail. The α-helical tails of two heavy chains twist around each other in a coiled-coil structure to form a dimer, and two light chains associate with the neck of each head region to form the complete myosin II molecule.

The thick filaments of muscle consist of several hundred myosin molecules, associated in a parallel staggered array by interactions between their tails (Figure 11.23). The globular heads of myosin bind actin, forming crossbridges between the thick and thin filaments. It is important to note that the orientation of myosin molecules in the thick filaments reverses at the

Figure 11.22
Myosin II   The myosin II molecule consists of two heavy chains and two pairs of light chains (called the essential and regulatory light chains). The heavy chains have globular head regions and long α-helical tails, which coil around each other to form dimers.

Figure 11.23
Organization of myosin thick filaments Thick filaments are formed by the association of several hundred myosin II molecules in a staggered array. The globular heads of myosin bind actin, forming cross-bridges between the myosin and actin filaments. The orientation of both actin and myosin filaments reverses at the M line, so their relative polarity is the same on both sides of the sarcomere.

M line of the sarcomere. The polarity of actin filaments (which are attached to Z discs at their plus ends) similarly reverses at the M line, so the relative orientation of myosin and actin filaments is the same on both halves of the sarcomere. As discussed later, the motor activity of myosin moves its head groups along the actin filament in the direction of the plus end. This movement slides the actin filaments from both sides of the sarcomere toward the M line, shortening the sarcomere and resulting in muscle contraction.

In addition to binding actin, the myosin heads bind and hydrolyze ATP, which provides the energy to drive filament sliding. This translation of chemical energy to movement is mediated by changes in the shape of myosin resulting from ATP binding. The generally accepted model (the swinging-cross-bridge model) is that ATP hydrolysis drives repeated cycles of interaction between myosin heads and actin. During each cycle, conformational changes in myosin result in the movement of myosin heads along actin filaments.

Although the molecular mechanisms are still not fully understood, a plausible working model for myosin function has been derived both from *in vitro* studies of myosin movement along actin filaments (a system developed by James Spudich and Michael Sheetz) and from determination of the three-dimensional structure of myosin by Ivan Rayment and his colleagues (Figure 11.24). The cycle starts with myosin (in the absence of ATP) tightly bound to actin. ATP binding dissociates the myosin-actin complex and the hydrolysis of ATP then induces a conformational change in myosin. This change affects the neck region of myosin that binds the light chains (see Figure 11.22), which acts as a lever arm to displace the myosin head by about 5 nm. The products of hydrolysis (ADP and P_i) remain bound to the myosin head, which is said to be in the "cocked" position. The myosin head then rebinds at a new position on the actin filament, resulting in the release of ADP and P_i and triggering the "power stroke," in which the myosin head returns to its initial conformation, thereby sliding the actin filament toward the M line of the sarcomere.

Figure 11.24
Model for myosin action The binding of ATP dissociates myosin from actin. ATP hydrolysis then induces a conformational change that displaces the myosin head group. This is followed by binding of the myosin head to a new position on the actin filament and release of ADP and P_i. The return of the myosin head to its original conformation drives actin filament sliding.

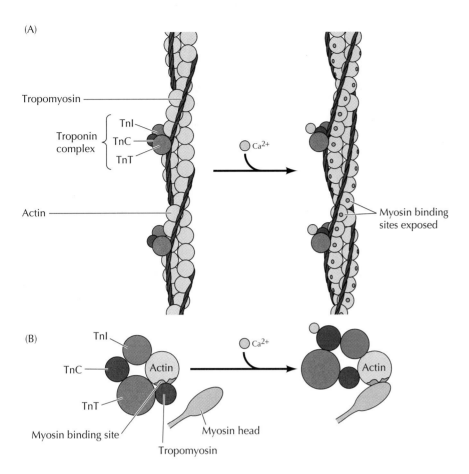

(A)

Tropomyosin

Troponin complex { TnI, TnC, TnT

Actin

Ca²⁺

Myosin binding sites exposed

(B)

TnI

TnC

Actin

TnT

Ca²⁺

Actin

Myosin binding site

Myosin head

Tropomyosin

Figure 11.25
Association of tropomyosin and troponins with actin filaments
(A) Tropomyosin binds lengthwise along actin filaments and, in striated muscle, is associated with a complex of three troponins: troponin I (TnI), troponin C (TnC), and troponin T (TnT). In the absence of Ca²⁺, the tropomyosin-troponin complex blocks the binding of myosin to actin. Binding of Ca²⁺ to TnC shifts the complex, relieving this inhibition and allowing contraction to proceed. (B) Cross-sectional view.

The contraction of skeletal muscle is triggered by nerve impulses, which stimulate the release of Ca^{2+} from the **sarcoplasmic reticulum**—a specialized network of internal membranes, similar to the endoplasmic reticulum, that stores high concentrations of Ca^{2+} ions. The release of Ca^{2+} from the sarcoplasmic reticulum increases the concentration of Ca^{2+} in the cytosol from approximately 10^{-7} to 10^{-5} M. The increased Ca^{2+} concentration signals muscle contraction via the action of two accessory proteins bound to the actin filaments: **tropomyosin** and **troponin** (Figure 11.25). Tropomyosin is a fibrous protein that binds lengthwise along the groove of actin filaments. In striated muscle, each tropomyosin molecule is bound to troponin, which is a complex of three polypeptides: troponin C (Ca^{2+}-binding), troponin I (inhibitory), and troponin T (tropomyosin-binding). When the concentration of Ca^{2+} is low, the complex of the troponins with tropomyosin blocks the interaction of actin and myosin, so the muscle does not contract. At high concentrations, Ca^{2+} binding to troponin C shifts the position of the complex, relieving this inhibition and allowing contraction to proceed.

Contractile Assemblies of Actin and Myosin in Nonmuscle Cells

Contractile assemblies of actin and myosin, resembling small-scale versions of muscle fibers, are present also in nonmuscle cells. As in muscle, the actin filaments in these contractile assemblies are interdigitated with bipolar filaments of myosin II, consisting of 15 to 20 myosin II molecules, which produce contraction by sliding the actin filaments relative to one another (Figure 11.26). The actin filaments in contractile bundles in non-

Figure 11.26
Contractile assemblies in nonmuscle cells   Bipolar filaments of myosin II produce contraction by sliding actin filaments in opposite directions.

Figure 11.27
Cytokinesis   Following completion of mitosis (nuclear division), a contractile ring consisting of actin filaments and myosin II divides the cell in two.

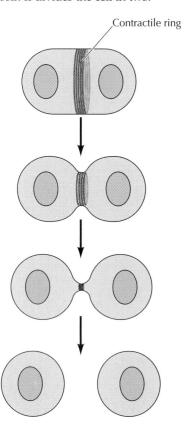

muscle cells are also associated with tropomyosin, which facilitates their interaction with myosin II, probably by competing with filamin for binding sites on actin.

Two examples of contractile assemblies in nonmuscle cells, stress fibers and adhesion belts, were discussed earlier with respect to attachment of the actin cytoskeleton to regions of cell-substrate and cell-cell contacts (see Figures 11.13 and 11.14). The contraction of stress fibers produces tension across the cell, allowing the cell to pull on a substrate (e.g., the extracellular matrix) to which it is anchored. The contraction of adhesion belts alters the shape of epithelial cell sheets: a process that is particularly important during embryonic development, when sheets of epithelial cells fold into structures such as tubes.

The most dramatic example of actin-myosin contraction in nonmuscle cells, however, is provided by **cytokinesis**—the division of a cell into two following mitosis (Figure 11.27). Toward the end of mitosis in animal cells, a **contractile ring** consisting of actin filaments and myosin II assembles just underneath the plasma membrane. Its contraction pulls the plasma membrane progressively inward, constricting the center of the cell and pinching it in two. Interestingly, the thickness of the contractile ring remains constant as it contracts, implying that actin filaments disassemble as contraction proceeds. The ring then disperses completely following cell division.

The regulation of actin-myosin contraction in striated muscle, discussed earlier, is mediated by the binding of Ca^{2+} to troponin. In nonmuscle cells and in smooth muscle, however, contraction is regulated primarily by phosphorylation of one of the myosin light chains, called the regulatory light chain (Figure 11.28). Phosphorylation of the regulatory light chain in these cells has at least two effects: It promotes the assembly of myosin into filaments, and it increases myosin catalytic activity, enabling contraction to proceed. The enzyme that catalyzes this phosphorylation, called **myosin light-chain kinase**, is itself regulated by association with the Ca^{2+}-binding protein **calmodulin**. Increases in cytosolic Ca^{2+} promote the binding of calmodulin to the kinase, resulting in phosphorylation of the myosin regulatory light chain. Increases in cytosolic Ca^{2+} are thus responsible, albeit indirectly, for activating myosin in smooth muscle and nonmuscle cells, as well as in striated muscle.

Unconventional Myosins

In addition to myosin II ("conventional" two-headed myosin), several other types of myosin are found in nonmuscle cells. In contrast to myosin II, these "unconventional" myosins do not form filaments and therefore are not involved in contraction. They may, however, be involved in a variety of

Figure 11.28
Regulation of myosin by phosphorylation   Ca²⁺ binds to calmodulin, which in turn binds to myosin light-chain kinase (MLCK). The active calmodulin-MLCK complex then phosphorylates the myosin II regulatory light chain, converting myosin from an inactive to an active state.

other kinds of cell movements, such as the transport of membrane vesicles and organelles along actin filaments, phagocytosis, and extension of pseudopods in amoebae (see Figure 11.17).

The best-studied of these unconventional myosins are members of the **myosin I** family (Figure 11.29). The myosin I proteins contain a globular head group that acts as a molecular motor, like that of myosin II. However, members of the myosin I family are much smaller molecules (about 110 kd in mammalian cells) that lack the long tail of myosin II and do not form dimers. Their tails can instead bind to other structures, such as membrane vesicles or organelles. The movement of myosin I along an actin filament can then transport its attached cargo. One function of myosin I, discussed earlier, is to form the lateral arms that link actin bundles to the plasma membrane of intestinal microvilli (see Figure 11.16). In these structures, the motor activity of myosin I may move the plasma membrane along the actin bundles, toward the tip of the microvillus. Additional functions of myosin I may be in the transport of vesicles and organelles along actin filaments and in movement of the plasma membrane during phagocytosis and pseudopod extension.

In addition to myosins I and II, at least 12 other classes of unconventional myosins (III through XIV) have been identified. Some of these unconventional myosins are two-headed like myosin II, whereas others are one-headed like myosin I. The functions of most of these unconventional myosins remain to be determined, but some have been clearly shown to play important roles in organelle movement (myosins V and VI) and in sensory functions such as vision (myosin III) and hearing (myosins VI and VII).

Cell Crawling

The crawling movements of cells across a surface represent a basic form of cell locomotion, employed by a wide variety of different kinds of cells. Examples include the movements of amoebas, the migration of embryonic cells during development, the invasion of tissues by white blood cells to fight infection, the migration of cells involved in wound healing, and the

Figure 11.29
Myosin I   Myosin I contains a head group similar to myosin II, but it has a comparatively short tail and does not form dimers or filaments. Although it cannot induce contraction, myosin I can move along actin filaments (toward the plus end), carrying a variety of cargoes (such as membrane vesicles) attached to its tail.

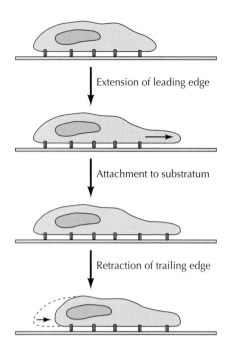

Figure 11.30
Cell crawling The crawling movements of cells across a surface can be viewed as three stages of coordinated movements: (1) extension of the leading edge, (2) attachment of the leading edge to the substratum, and (3) retraction of the rear of the cell into the cell body.

spread of cancer cells during the metastasis of malignant tumors. Similar types of movement are also responsible for phagocytosis and for the extension of nerve cell processes during development of the nervous system. All of these movements are based on the dynamic properties of the actin cytoskeleton, although the detailed mechanisms involved remain to be fully understood.

Cell crawling involves a coordinated cycle of movements, which can be viewed in three stages. First, protrusions such as pseudopodia, lamellipodia, or microspikes (see Figure 11.17) must be extended from the leading edge of the cell (Figure 11.30). Second, these extensions must attach to the substratum across which the cell is migrating. Finally, the trailing edge of the cell must dissociate from the substratum and retract into the cell body.

A variety of experiments indicate that extension of the leading edge involves the polymerization and crosslinking of actin filaments. For example, inhibition of actin polymerization (e.g., by treatment with cytochalasin) blocks the formation of cell surface protrusions. The regulated turnover of actin filaments, as illustrated in Figure 11.5, leads to the extension of processes such as filopodia and lamellipodia at the leading edge of the cell, and both cofilin and Arp2/3 proteins appear to be involved in this process. Unconventional myosins may also participate in the extension of processes at the leading edge: Myosin I is required for pseudopod extension in the amoeba *Dictyostelium* and Myosin V for extension of filopodia in neurons.

Following their extension, protrusions from the leading edge must attach to the substratum in order to function in cell locomotion. For slow-moving cells, such as fibroblasts, attachment involves the formation of focal adhesions (see Figure 11.13). Cells moving more rapidly, such as amoebas or white blood cells, form more diffuse contacts with the substratum, the molecular composition of which is not known.

The third stage of cell crawling, retraction of the trailing edge, is the least understood. The attachments of the trailing edge to the substratum are broken, and the rear of the cell recoils into the cell body. The process appears to require the development of tension between the front and rear of the cell, generating contractile force that eventually pulls the rear of the cell forward. This aspect of cell locomotion is impaired in mutants of *Dictyostelium* lacking myosin II, consistent with a role for myosin II in contracting the actin cortex and generating the force required for retraction of the trailing edge.

Intermediate Filaments

Intermediate filaments have a diameter of about 10 nm, which is intermediate between the diameters of the two other principal elements of the cytoskeleton, actin filaments (about 7 nm) and microtubules (about 25 nm). In contrast to actin filaments and microtubules, the intermediate filaments are not directly involved in cell movements. Instead, they appear to play basically a structural role by providing mechanical strength to cells and tissues.

Intermediate Filament Proteins

Whereas actin filaments and microtubules are polymers of single types of proteins (actin and tubulin, respectively), intermediate filaments are composed of a variety of proteins that are expressed in different types of cells. More than 50 different intermediate filament proteins have been identified and classified into six groups based on similarities between their amino acid sequences (Table 11.1). Types I and II consist of two groups of **keratins**, each consisting of about 15 different proteins, which are expressed in epithelial

TABLE 11.1 Intermediate Filament Proteins

Type	Protein	Size (kd)	Site of expression
I	Acidic keratins (~15 proteins)	40–60	Epithelial cells
II	Neutral or basic keratins (~15 proteins)	50–70	Epithelial cells
III	Vimentin	54	Fibroblasts, white blood cells, and other cell types
	Desmin	53	Muscle cells
	Glial fibrillary acidic protein	51	Glial cells
	Peripherin	57	Peripheral neurons
IV	Neurofilament proteins		
	NF-L	67	Neurons
	NF-M	150	Neurons
	NF-H	200	Neurons
	α-Internexin	66	Neurons
V	Nuclear lamins	60–75	Nuclear lamina of all cell types
VI	Nestin	200	Stem cells of central nervous system

cells. Each type of epithelial cell synthesizes at least one type I (acidic) and one type II (neutral/basic) keratin, which copolymerize to form filaments. Some type I and II keratins (called **hard keratins**) are used for production of structures such as hair, nails, and horns. The other type I and II keratins (**soft keratins**) are abundant in the cytoplasm of epithelial cells, with different keratins being expressed in various differentiated cell types.

The type III intermediate filament proteins include **vimentin**, which is found in a variety of different kinds of cells, including fibroblasts, smooth muscle cells, and white blood cells. Another type III protein, **desmin**, is specifically expressed in muscle cells, where it connects the Z discs of individual contractile elements. A third type III intermediate filament protein is specifically expressed in glial cells, and a fourth is expressed in neurons of the peripheral nervous system.

The type IV intermediate filament proteins include the three **neurofilament (NF) proteins** (designated NF-L, NF-M, and NF-H for *l*ight, *m*edium, and *h*eavy, respectively). These proteins form the major intermediate filaments of many types of mature neurons. They are particularly abundant in the axons of motor neurons and are thought to play a critical role in supporting these long, thin processes, which can extend more than a meter in length. Another type IV protein (α-internexin) is expressed at an earlier stage of neuron development, prior to expression of the neurofilament proteins. The single type VI intermediate filament protein (nestin) is expressed even earlier during the development of neurons, in stem cells of the central nervous system.

The type V intermediate filament proteins are the nuclear lamins, which are found in most eukaryotic cells. Rather than being part of the cytoskeleton, the nuclear lamins are components of the nuclear envelope (see Figure 8.3). They also differ from the other intermediate filament proteins in that they assemble to form an orthogonal meshwork underlying the nuclear membrane.

Despite considerable diversity in size and amino acid sequence, the various intermediate filament proteins share a common structural organiza-

Figure 11.31
Structure of intermediate filament proteins Intermediate filament proteins contain a central α-helical rod domain of approximately 310 amino acids (350 amino acids in the nuclear lamins). The N-terminal head and C-terminal tail domains vary in size and shape.

tion (Figure 11.31). All of the intermediate filament proteins have a central α-helical rod domain of approximately 310 amino acids (350 amino acids in the nuclear lamins). This central rod domain is flanked by amino- and carboxy-terminal domains, which vary among the different intermediate filament proteins in size, sequence, and secondary structure. As discussed next, the α-helical rod domain plays a central role in filament assembly, while the variable head and tail domains presumably determine the specific functions of the different intermediate filament proteins.

Assembly of Intermediate Filaments

The first stage of filament assembly is the formation of dimers in which the central rod domains of two polypeptide chains are wound around each other in a coiled-coil structure, similar to that formed by myosin II heavy chains (Figure 11.32). The dimers then associate in a staggered antiparallel fashion to form tetramers, which can assemble end to end to form protofilaments. The final intermediate filament contains approximately eight protofilaments wound around each other in a ropelike structure. Because they are assembled from antiparallel tetramers, both ends of intermediate filaments are equivalent. Consequently, in contrast to actin filaments and microtubules, intermediate filaments are apolar; they do not have distinct plus and minus ends.

Filament assembly requires interactions between specific types of intermediate filament proteins. For example, keratin filaments are always assembled from heterodimers containing one type I and one type II polypeptide. In contrast, the type III proteins can assemble into filaments containing only a single polypeptide (e.g., vimentin) or consisting of two dif-

Figure 11.32
Assembly of intermediate filaments The central rod domains of two polypeptides wind around each other in a coiled-coil structure to form dimers. Dimers then associate in a staggered antiparallel fashion to form tetramers. Tetramers associate end to end to form protofilaments and laterally to form filaments. Each filament contains approximately eight protofilaments wound around each other in a ropelike structure.

ferent type III proteins (e.g., vimentin plus desmin). The type III proteins do not, however, form copolymers with the keratins. Among the type IV proteins, α-internexin can assemble into filaments by itself, whereas the three neurofilament proteins copolymerize to form heteropolymers. Intermediate filaments are generally more stable than actin filaments or microtubules and do not exhibit the dynamic behavior associated with these other elements of the cytoskeleton (e.g., the treadmilling of actin filaments illustrated in Figure 11.4). However, intermediate filament proteins are frequently modified by phosphorylation, which can regulate their assembly and disassembly within the cell. The clearest example is phosphorylation of the nuclear lamins (see Figure 8.31), which results in disassembly of the nuclear lamina and breakdown of the nuclear envelope during mitosis. Cytoplasmic intermediate filaments, such as vimentin, are also phosphorylated at mitosis, which can lead to their disassembly and reorganization in dividing cells.

Intracellular Organization of Intermediate Filaments

Intermediate filaments form an elaborate network in the cytoplasm of most cells, extending from a ring surrounding the nucleus to the plasma membrane (Figure 11.33). Both keratin and vimentin filaments attach to the nuclear envelope, apparently serving to position and anchor the nucleus within the cell. In addition, intermediate filaments can associate not only with the plasma membrane but also with the other elements of the cytoskeleton, actin filaments and microtubules. Intermediate filaments thus provide a scaffold that integrates the components of the cytoskeleton and organizes the internal structure of the cell.

The keratin filaments of epithelial cells are tightly anchored to the plasma membrane at two areas of specialized cell contacts, **desmosomes** and **hemidesmosomes** (Figure 11.34). Desmosomes are junctions between adjacent cells, at which cell-cell contacts are mediated by transmembrane proteins related to the cadherins. On their cytoplasmic side, desmosomes are associated with a characteristic dense plaque of intracellular proteins, to which keratin filaments are attached. These attachments are mediated by desmoplakin, a member of a family of proteins called **plakins** that bind intermediate filaments and link them to other cellular structures. Hemidesmosomes are morphologically similar junctions between epithelial cells and underlying connective tissue, at which keratin filaments are linked by different members of the plakin family (e.g., plectin) to integrins. Desmosomes and hemidesmosomes thus anchor intermediate filaments to regions of cell-cell and cell-substratum contact, respectively, similar to the attachment of the actin cytoskeleton to the plasma membrane at adherens junctions and focal adhesions. It is important to note that the keratin filaments anchored to both sides of desmosomes serve as a mechanical link between

10 µm

Figure 11.33
Intracellular organization of keratin filaments   Micrograph of epithelial cells stained with fluorescent antibodies to keratin (green). Nuclei are stained blue. The keratin filaments extend from a ring surrounding the nucleus to the plasma membrane. (Nancy Kedersha/Immunogen/Photo Researchers, Inc.)

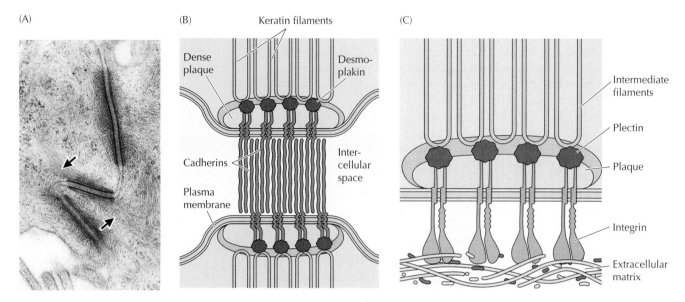

(A)

(B)

Keratin filaments

Dense plaque

Desmoplakin

Cadherins

Intercellular space

Plasma membrane

(C)

Intermediate filaments

Plectin

Plaque

Integrin

Extracellular matrix

Figure 11.34
Attachment of intermediate filaments to desmosomes and hemidesmosomes
(A) Electron micrograph illustrating keratin filaments (arrows) attached to the dense plaques of intracellular protein on both sides of a desmosome. (B) Schematic of a desmosome. Intermediate filaments are anchored to sites of cell-cell adhesion by desmoplaskin. (C) Schematic of a hemidesmosome. Intermediate filaments are anchored to an integrin by plectin. (A, Don Fawcett/Photo Researchers, Inc.)

adjacent cells in an epithelial layer, thereby providing mechanical stability to the entire tissue.

In addition to linking intermediate filaments to cell junctions, some plakins link intermediate filaments to other elements of the cytoskeleton. Plectin, for example, binds actin filaments and microtubules in addition to intermediate filaments, so it can provide bridges between these cytoskeletal components (Figure 11.35). These bridges to intermediate filaments are thought to brace and stabilize actin filaments and microtubules, thereby increasing the mechanical stability of the cell.

Two types of intermediate filaments, desmin and the neurofilaments, play specialized roles in muscle and nerve cells, respectively. Desmin connects the individual actin-myosin assemblies of muscle cells both to one another and to the plasma membrane, thereby linking the actions of individual contractile elements. Neurofilaments are the major intermediate filaments in most mature neurons. They are particularly abundant in the long axons of motor neurons, where they appear to be anchored to actin filaments and microtubules by neuronal members of the plakin family. Neu-

Figure 11.35
Electron micrograph of plectin bridges between intermediate filaments and microtubules Micrograph of a fibroblast stained with antibody against plectin. The micrograph has been artificially colored to show plectin (green), antibodies against plectin (yellow), intermediate filaments (blue) and microtubules (red). (Courtesy of Tatyana Svitkina and Gary Borisy, University of Wisconsin/ Madison.)

rofilaments are thought to play an important role in providing mechanical support and stabilizing other elements of the cytoskeleton in these long, thin extensions of nerve cells.

Functions of Keratins and Neurofilaments: Diseases of the Skin and Nervous System

Although intermediate filaments have long been thought to provide structural support to the cell, direct evidence for their function has only recently been obtained. Some cells in culture make no intermediate filament proteins, indicating that these proteins are not required for the growth of cells *in vitro*. Similarly, injection of cultured cells with antibody against vimentin disrupts intermediate filament networks without affecting cell growth or movement. Therefore, it has been thought that intermediate filaments are most needed to strengthen the cytoskeleton of cells in the tissues of multicellular organisms, where they are subjected to a variety of mechanical stresses that do not affect cells in the isolated environment of a culture dish.

Experimental evidence for such an *in vivo* role of intermediate filaments was first provided in 1991 by studies in the laboratory of Elaine Fuchs. These investigators used transgenic mice to investigate the *in vivo* effects of expressing a keratin deletion mutant encoding a truncated polypeptide that disrupted the formation of normal keratin filaments (Figure 11.36). This mutant keratin gene was introduced into transgenic mice, where it was expressed in basal cells of the epidermis and disrupted formation of a normal keratin cytoskeleton. This resulted in the development of severe skin abnormalities, including blisters due to epidermal cell lysis following mild mechanical trauma, such as rubbing of the skin. The skin abnormalities of these transgenic mice thus provided direct support for the presumed role of keratins in providing mechanical strength to epithelial cells in tissues.

These experiments also pointed to the molecular basis of a human genetic disease, epidermolysis bullosa simplex (EBS). Like the transgenic mice expressing mutant keratin genes, patients with this disease develop skin blisters resulting from cell lysis after minor trauma. This similarity prompted studies of the keratin genes in EBS patients, leading to the demonstration that EBS is caused by keratin gene mutations that interfere

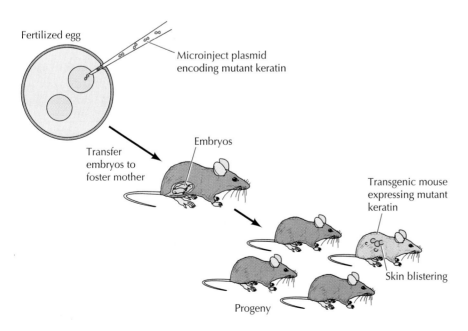

Figure 11.36
Experimental demonstration of keratin function   A plasmid encoding a mutant keratin that interferes with the normal assembly of keratin filaments was microinjected into one pronucleus of a fertilized egg. Microinjected embryos were then transferred to a foster mother, and some of the offspring were found to have incorporated the mutant keratin gene into their genome. Expression of the mutant gene in these transgenic mice disrupted the keratin cytoskeleton of cells of the epidermis, resulting in severe skin blistering due to cell lysis following mild mechanical stress.

with the normal assembly of keratin filaments. Thus, both experimental studies in transgenic mice and molecular analysis of a human genetic disease have demonstrated the role of keratins in allowing skin cells to withstand mechanical stress. Continuing studies have shown that mutations in other keratins are responsible for several other inherited skin diseases, which are similarly characterized by abnormal fragility of epidermal cells.

Other studies in transgenic mice have implicated abnormalities of neurofilaments in diseases of motor neurons, particularly amyotrophic lateral sclerosis (ALS). ALS, known as Lou Gehrig's disease and the disease afflicting the renowned physicist Stephen Hawking, results from progressive loss of motor neurons, which in turn leads to muscle atrophy, paralysis, and eventual death. ALS and other types of motor neuron disease are characterized by the accumulation and abnormal assembly of neurofilaments, suggesting that neurofilament abnormalities might contribute to these pathologies. Consistent with this possibility, overexpression of NF-L or NF-H in transgenic mice has been found to result in the development of a condition similar to ALS. Although the mechanism involved remains to be understood, these experiments clearly suggest the involvement of neurofilaments in the pathogenesis of motor neuron disease.

Microtubules

Microtubules, the third principal component of the cytoskeleton, are rigid hollow rods approximately 25 nm in diameter. Like actin filaments, microtubules are dynamic structures that undergo continual assembly and disassembly within the cell. They function both to determine cell shape and in a variety of cell movements, including some forms of cell locomotion, the intracellular transport of organelles, and the separation of chromosomes during mitosis.

Structure, Assembly, and Dynamic Instability of Microtubules

In contrast to intermediate filaments, which are composed of a variety of different fibrous proteins, microtubules are composed of a single type of globular protein, called **tubulin**. Tubulin is a dimer consisting of two closely related 55-kd polypeptides, α-tubulin and β-tubulin. Like actin, both α- and β-tubulin are encoded by small families of related genes. In addition, a third type of tubulin (γ-tubulin) is specifically localized to the centrosome, where it plays a critical role in initiating microtubule assembly (discussed shortly).

Tubulin dimers polymerize to form microtubules, which generally consist of 13 linear protofilaments assembled around a hollow core (Figure 11.37). The protofilaments, which are composed of head-to-tail arrays of tubulin dimers, are arranged in parallel. Consequently, microtubules (like actin filaments) are polar structures with two distinct ends: a fast-growing plus end and a slow-growing minus end. This polarity is an important consideration in determining the direction of movement along microtubules, just as the polarity of actin filaments defines the direction of myosin movement.

Tubulin dimers can depolymerize as well as polymerize, and microtubules can undergo rapid cycles of assembly and disassembly. Both α- and β-tubulin bind GTP, which functions analogously to the ATP bound to actin to regulate polymerization. In particular, the GTP bound to β-tubulin (though not that bound to α-tubulin) is hydrolyzed to GDP during or shortly after polymerization. This GTP hydrolysis weakens the binding affinity of tubulin for adjacent molecules, thereby favoring

Figure 11.37
Structure of microtubules Dimers of α- and β-tubulin polymerize to form microtubules, which are composed of 13 protofilaments assembled around a hollow core.

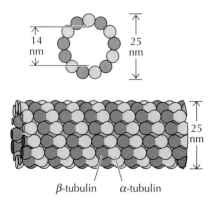

β-tubulin α-tubulin

depolymerization and resulting in the dynamic behavior of microtubules. Like actin filaments (see Figure 11.4), microtubules undergo treadmilling, a dynamic behavior in which tubulin molecules bound to GDP are continually lost from the minus end and replaced by the addition of tubulin molecules bound to GTP to the plus end of the same microtubule. In microtubules, GTP hydrolysis also results in the behavior known as **dynamic instability**, in which individual microtubules alternate between cycles of growth and shrinkage (Figure 11.38). Whether a microtubule grows or shrinks is determined by the rate of tubulin addition relative to the rate of GTP hydrolysis. As long as new GTP-bound tubulin molecules are added more rapidly than GTP is hydrolyzed, the microtubule retains a GTP cap at its plus end and microtubule growth continues. However, if the rate of polymerization slows, the GTP bound to tubulin at the plus end of the microtubule will be hydrolyzed to GDP. If this occurs, the GDP-bound tubulin will dissociate, resulting in rapid depolymerization and shrinkage of the microtubule.

Dynamic instability, described by Tim Mitchison and Marc Kirschner in 1984, results in the continual and rapid turnover of most microtubules, which have half-lives of only several minutes within the cell. As discussed later, this rapid turnover of microtubules is particularly critical for the remodeling of the cytoskeleton that occurs during mitosis. Because of the central role of microtubules in mitosis, drugs that affect microtubule assembly are useful not only as experimental tools in cell biology but also in the treatment of cancer. **Colchicine** and **colcemid** are examples of com-

Figure 11.38
Dynamic instability of microtubules Dynamic instability results from the hydrolysis of GTP bound to β-tubulin during or shortly after polymerization, which reduces its binding affinity for adjacent molecules. Growth of microtubules continues as long as there is a high concentration of tubulin bound to GTP. New GTP-bound tubulin molecules are then added more rapidly than GTP is hydrolyzed, so a GTP cap is retained at the growing end. However, if GTP is hydrolyzed more rapidly than new subunits are then added, the presence of GDP-bound tubulin at the end of the microtubule leads to disassembly and shrinkage. Only the plus ends of microtubules are illustrated.

High concentration of tubulin bound to GTP

Low concentration of tubulin bound to GTP

Interphase cell

Centrosome

Mitotic cell

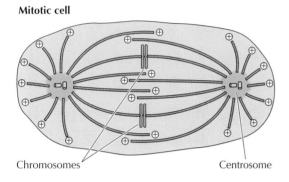

Chromosomes   Centrosome

Figure 11.39
Intracellular organization of microtubules   The minus ends of microtubules are anchored in the centrosome. In interphase cells, the centrosome is located near the nucleus and microtubules extend outward to the cell periphery. During mitosis, duplicated centrosomes separate and microtubules reorganize to form the mitotic spindle.

monly used experimental drugs that bind tubulin and inhibit microtubule polymerization, which in turn blocks mitosis. Two related drugs (**vincristine** and **vinblastine**) are used in cancer chemotherapy because they selectively inhibit rapidly dividing cells. Another useful drug, **taxol**, stabilizes microtubules rather than inhibiting their assembly. Such stabilization also blocks cell division, and taxol is used as an anticancer agent as well as an experimental tool.

The Centrosome and Microtubule Organization

The microtubules in most cells extend outward from a **microtubule-organizing center**, in which the minus ends of microtubules are anchored. In animal cells, the major microtubule-organizing center is the **centrosome**, which is located adjacent to the nucleus near the center of interphase (nondividing) cells (Figure 11.39). During mitosis, microtubules similarly extend outward from duplicated centrosomes to form the mitotic spindle, which is responsible for the separation and distribution of chromosomes to daughter cells. The centrosome thus plays a key role in determining the intracellular organization of microtubules, although most details of its function remain a mystery.

The centrosome serves as the initiation site for the assembly of microtubules, which grow outward from the centrosome toward the periphery of the cell. This can be clearly visualized in cells that have been treated with colcemid to disassemble their microtubules (Figure 11.40). When the drug is removed, the cells recover and new microtubules can be seen growing outward from the centrosome. Importantly, the initiation of microtubule growth at the centrosome establishes the polarity of microtubules within the

(A)   (B)

10 µm

Figure 11.40
Growth of microtubules from the centrosome   Microtubules in mouse fibroblasts are visualized by immunofluorescence microscopy using an antibody against tubulin. (A) The distribution of microtubules in a normal interphase cell. (B) This cell was treated with colcemid for one hour to disassemble microtubules. The drug was then removed and the cell allowed to recover for 30 minutes, allowing the visualization of new microtubules growing out of the centrosome. (From M. Osborn and K. Weber, 1976. *Proc. Natl. Acad. Sci. USA* 73: 867.)

(A)

(B)

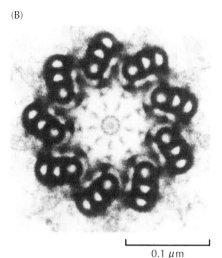

0.1 µm

Figure 11.41
Structure of centrosomes   (A) Electron micrograph of a centrosome showing microtubules radiating from the pericentriolar material that surrounds a pair of centrioles. (B) Transverse section of a centriole illustrating its nine triplets of microtubules. (A, © Cytographics; B, Don Fawcett, Photo Researchers, Inc.)

cell. In particular, microtubules grow by the addition of tubulin to their plus ends, which extend outward from the centrosome toward the cell periphery.

The centrosomes of most animal cells contain a pair of **centrioles**, oriented perpendicular to each other, surrounded by amorphous **pericentriolar material** (Figure 11.41). The centrioles are cylindrical structures consisting of nine triplets of microtubules, similar to the basal bodies of cilia and flagella (discussed later in the chapter). Although centrioles are probably the precursors of basal bodies, they appear to be dispensable for the function of the centrosome. Centrioles do not appear to be required for the assembly or organization of microtubules, and they are not found in plant cells, many unicellular eukaryotes, and some animal cells (such as mouse eggs). The microtubules that emanate from the centrosome terminate in the pericentriolar material, not the centrioles, and it is the pericentriolar material that initiates microtubule assembly.

The key protein in the centrosome that nucleates assembly of microtubules is γ-tubulin, a minor species of tubulin first identified in fungi. Complexes of γ-tubulin form ring structures that contain 10 to 13 γ-tubulin molecules and have diameters similar to those of microtubules. These γ-tubulin rings serve as nucleation sites for the assembly of microtubules and may remain bound to their minus ends.

Reorganization of Microtubules during Mitosis

As noted earlier, microtubules completely reorganize during mitosis, providing a dramatic example of the importance of their dynamic instability. The microtubule array present in interphase cells disassembles and the free tubulin subunits are reassembled to form the **mitotic spindle**, which is responsible for the separation of daughter chromosomes (Figure 11.42). This restructuring of the microtubule cytoskeleton is directed by duplica-

2 µm

Figure 11.42
Electron micrograph of the mitotic spindle   The spindle microtubules are attached to condensed chromosomes at metaphase. (From C. L. Rieder and S. S. Bowser, 1985. *J. Histochem. Cytochem.* 33: 165/Biological Photo Service.)

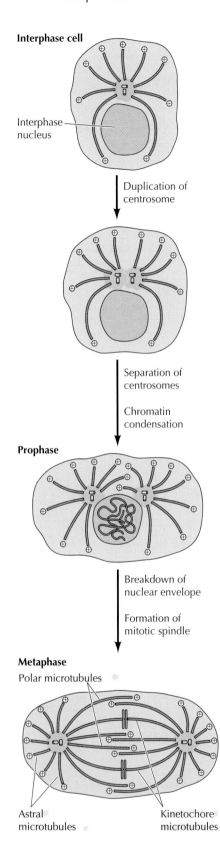

Interphase cell

Interphase nucleus

Duplication of centrosome

Separation of centrosomes

Chromatin condensation

Prophase

Breakdown of nuclear envelope

Formation of mitotic spindle

Metaphase

Polar microtubules

Astral microtubules

Kinetochore microtubules

Figure 11.43
Formation of the mitotic spindle   The centrioles and centrosomes duplicate during interphase. During prophase of mitosis, the duplicated centrosomes separate and move to opposite sides of the nucleus. The nuclear envelope then disassembles, and microtubules reorganize to form the mitotic spindle. Kinetochore microtubules are attached to the condensed chromosomes, polar microtubules overlap with each other in the center of the cell, and astral microtubules extend outward to the cell periphery. At metaphase, the condensed chromosomes are aligned at the center of the spindle.

tion of the centrosome to form two separate microtubule-organizing centers at opposite poles of the mitotic spindle.

The centrioles and other components of the centrosome are duplicated in interphase cells, but they remain together on one side of the nucleus until the beginning of mitosis (Figure 11.43). The two centrosomes then separate and move to opposite sides of the nucleus, forming the two poles of the mitotic spindle. As the cell enters mitosis, the dynamics of microtubule assembly and disassembly also change dramatically. First, the rate of microtubule disassembly increases about tenfold, resulting in overall depolymerization and shrinkage of microtubules. At the same time, the number of microtubules emanating from the centrosome increases by five- to tenfold. In combination, these changes result in disassembly of the interphase microtubules and the outgrowth of large numbers of short microtubules from the centrosomes.

As first proposed by Marc Kirschner and Tim Mitchison in 1986, formation of the mitotic spindle involves the selective stabilization of some of the microtubules radiating from the centrosomes. These microtubules are of three types, two of which make up the mitotic spindle. **Kinetochore microtubules** attach to the condensed chromosomes of mitotic cells at their centromeres, which are associated with specific proteins to form the kinetochore (see Figure 4.16). Attachment to the kinetochore stabilizes these microtubules, which, as discussed below, play a critical role in separation of the mitotic chromosomes. The second type of microtubules found in the mitotic spindle (**polar microtubules**) are not attached to chromosomes. Instead, the polar microtubules emanating from the two centrosomes are stabilized by overlapping with each other in the center of the cell. **Astral microtubules** extend outward from the centrosomes to the cell periphery and have freely exposed plus ends. As discussed later, both the polar and astral microtubules also contribute to chromosome movement by pushing the spindle poles apart.

As mitosis proceeds, the condensed chromosomes first align on the metaphase plate and then separate, with the two chromatids of each chromosome being pulled to opposite poles of the spindle. Chromosome movement is mediated by motor proteins associated with the spindle microtubules, as will be discussed shortly. In the final stage of mitosis, nuclear envelopes re-form, the chromosomes decondense, and cytokinesis takes place. Each daughter cell then contains one centrosome, which nucleates the formation of a new network of interphase microtubules.

Stabilization of Microtubules and Cell Polarity

Because of their inherent dynamic instability, most microtubules are frequently disassembled within the cell. This dynamic behavior can, however, be modified by the interactions of microtubules with other proteins. Some cellular proteins act to disassemble microtubules, either by severing microtubules or by increasing the rate of tubulin depolymerization from micro-tubule ends. Other proteins (called **microtubule-associated proteins** or **MAPs**) bind to microtubules and increase their stability. Such interactions allow the cell to stabilize microtubules in particular locations and provide an important mechanism for determining cell shape and polarity.

A large number of MAPs have been identified, and they vary depending on the type of cell. The best-characterized are MAP-1, MAP-2, and tau, iso-lated from neuronal cells, and MAP-4, which is present in all non-neuronal vertebrate cell types. The tau protein has been extensively studied because it is the main component of the characteristic lesions found in the brains of Alzheimer patients. The activity of MAPs is regulated by phosphorylation, allowing the cell to control microtubule stability.

A good example of the role of stable microtubules in determining cell polarity is provided by nerve cells, which consist of two distinct types of processes (axons and dendrites) extending from a cell body (Figure 11.44). Both axons and dendrites are supported by stable microtubules, together with the neurofilaments discussed in the preceding section of this chapter. However, the microtubules in axons and dendrites are organized differently and associated with distinct MAPs. In axons, the microtubules are all ori-ented with their plus ends away from the cell body, similar to the general ori-entation of microtubules in other cell types. The minus ends of most of the microtubules in axons, however, are not anchored in the centrosome; instead, both the plus and minus ends of these microtubules terminate in the cyto-plasm of the axon. In dendrites, the microtubules are oriented in both direc-tions; some plus ends point toward the cell body and some point toward the cell periphery. These distinct microtubule arrangements are paralleled by differences in MAPs: Axons contain tau proteins, but no MAP-2, whereas dendrites contain MAP-2, but no tau proteins, and it appears that these dif-ferences in MAP-2 and tau distribution are responsible for the distinct orga-nization of stable microtubules in axons and dendrites.

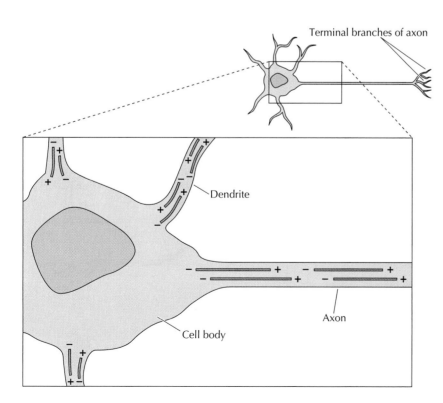

Figure 11.44
Organization of microtubules in nerve cells Two distinct types of processes extend from the cell body of nerve cells (neurons). Dendrites are short processes that receive stimuli from other nerve cells. The single long axon then carries impulses from the cell body to other cells, which may be either other neurons or an effector cell, such as a muscle. Stable microtubules in both axons and dendrites terminate in the cytoplasm rather than being anchored in the centrosome. In dendrites, microtu-bules are oriented in both directions, with their plus ends pointing both toward and away from the cell body. In contrast, all of the axon microtubules are oriented with their plus ends pointing toward the tip of the axon.

Microtubule Motors and Movements

Microtubules are responsible for a variety of cell movements, including the intracellular transport and positioning of membrane vesicles and organelles, the separation of chromosomes at mitosis, and the beating of cilia and flagella. As discussed for actin filaments earlier in this chapter, movement along microtubules is based on the action of motor proteins that utilize energy derived from ATP hydrolysis to produce force and movement. Members of two large families of motor proteins—the **kinesins** and the **dyneins**—are responsible for powering the variety of movements in which microtubules participate.

Identification of Microtubule Motor Proteins

Kinesin and dynein, the prototypes of microtubule motor proteins, move along microtubules in opposite directions—kinesin toward the plus end and dynein toward the minus end (Figure 11.45). The first of these microtubule motor proteins to be identified was dynein, which was isolated by Ian Gibbons in 1965. The purification of this form of dynein (called **axonemal dynein**) was facilitated because it is a highly abundant protein in cilia, just as the abundance of myosin facilitated its isolation from muscle cells. The identification of other microtubule-based motors, however, was more problematic because the proteins responsible for processes such as chromosome movement and organelle transport are present at comparatively low concentrations in the cytoplasm. Isolation of these proteins therefore depended on the development of new experimental methods to detect the activity of molecular motors in cell-free systems.

The development of *in vitro* assays for cytoplasmic motor proteins was based on the use of **video-enhanced microscopy**, developed by Robert Allen and Shinya Inoué in the early 1980s, to study the movement of membrane vesicles and organelles along microtubules in squid axons. In this method, a video camera is used to increase the contrast of images obtained with the light microscope, substantially improving the detection of small objects and allowing the movement of organelles to be followed in living cells. Using this approach, Allen, Scott Brady, and Ray Lasek demonstrated that organelle movements also took place in a cell-free system in which the plasma membrane had been removed and a cytoplasmic extract had been

Figure 11.45
Microtubule motor proteins Kinesin and dynein move in opposite directions along microtubules, toward the plus and minus ends, respectively. Kinesin consists of two heavy chains, wound around each other in a coiled-coil structure, and two light chains. The globular head domains of the heavy chains bind microtubules and are the motor domains of the molecule. Dynein consists of two or three heavy chains (two are shown here) in association with multiple light and intermediate chains. The globular head domains of the heavy chains are the motor domains.

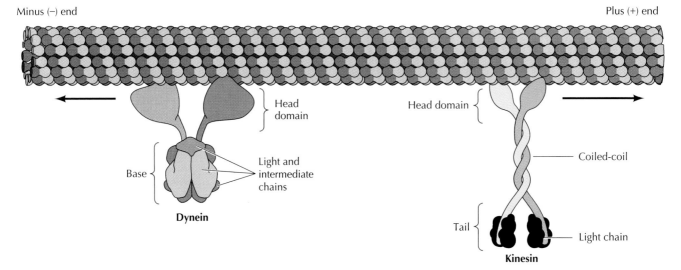

Minus (−) end

Plus (+) end

Head domain

Base

Light and intermediate chains

Dynein

Head domain

Coiled-coil

Tail

Light chain

Kinesin

spread on a glass slide. These observations led to the development of an *in vitro* reconstructed system, which provided an assay capable of detecting cellular proteins responsible for organelle movement. In 1985 Brady, as well as Ronald Vale, Thomas Reese, and Michael Sheetz, capitalized on these developments to identify kinesin as a novel microtubule motor protein, present in both squid axons and bovine brain.

Further studies demonstrated that kinesin translocates along microtubules in only a single direction—toward the plus end. Because the plus ends of microtubules in axons are all oriented away from the cell body (see Figure 11.44), the movement of kinesin in this direction transports vesicles and organelles away from the cell body, toward the tip of the axon. Within intact axons, however, vesicles and organelles also had been observed to move back toward the cell body, implying that a different motor protein might be responsible for movement along microtubules in the opposite direction—toward the minus end. Consistent with this prediction, further experiments showed that a protein previously identified as the microtubule-associated protein MAP-1C was in fact a motor protein that moved along microtubules in the minus end direction. Subsequent analysis demonstrated that MAP-1C is related to the dynein isolated from cilia (axonemal dynein), so MAP-1C is now referred to as **cytoplasmic dynein**.

Kinesin is a molecule of approximately 380 kd, consisting of two heavy chains (120 kd each) and two light chains (64 kd each) (see Figure 11.45). The heavy chains have long α-helical regions that wind around each other in a coiled-coil structure. The amino-terminal globular head domains of the heavy chains are the motor domains of the molecule: They bind to both microtubules and ATP, the hydrolysis of which provides the energy required for movement. Although the motor domain of kinesin (approximately 340 amino acids) is much smaller than that of myosin (about 850 amino acids), X-ray crystallography indicates that the kinesin and myosin motor domains are structurally similar, suggesting that kinesin and myosin evolved from a common ancestor. The tail portion of the kinesin molecule consists of the light chains in association with the carboxy-terminal domains of the heavy chains. This portion of kinesin is responsible for binding to other cell components (such as membrane vesicles and organelles) that are transported along microtubules by the action of kinesin motors.

Dynein is an extremely large molecule (up to 2000 kd), which consists of two or three heavy chains (each about 500 kd) complexed with a variable number of light and intermediate polypeptides, which range from 14 to 120 kd (see Figure 11.45). As in kinesin, the heavy chains form globular ATP-binding motor domains that are responsible for movement along microtubules. The basal portion of the molecule, including the light and intermediate chains, is thought to bind to other subcellular structures, such as organelles and vesicles.

Like the myosins, both kinesin and dynein define families of related motor proteins. Following the initial isolation of kinesin in 1985, a variety of kinesin-related proteins have been identified. Eighteen different kinesins are encoded in the genome of *C. elegans*, and it is thought that there may be as many as 100 different members of the kinesin family in humans. Some members of the kinesin family, like kinesin itself, move along microtubules in the plus end direction (see Figure 11.45). Other members of the kinesin family, however, move in the opposite direction, toward the minus end. Different members of the kinesin family vary in the sequences of their carboxy-terminal tails and are responsible for the movements of different types of "cargo," including vesicles, organelles, and chromosomes, along

Key Experiment

The Isolation of Kinesin

Identification of a Novel Force-Generating Protein, Kinesin, Involved in Microtubule-Based Motility

Ronald D. Vale, Thomas S. Reese, and Michael P. Sheetz
National Institute of Neurological and Communicative Disorders and Stroke, Marine Biological Laboratory, Woods Hole, MA; University of Connecticut Health Center, Farmington, CT; Stanford University School of Medicine, Stanford, CA
Cell, Volume 42, 1985, pages 39–50

reported at the same time by Scott Brady (*Nature* 317: 73–75, 1985).

The Experiments

Two experimental strategies were key to the isolation of kinesin. The first, based on the work of Allen and his colleagues, was the use of an *in vitro* system in which motor protein activity

The Context

The transport and positioning of cytoplasmic organelles is key to the organization of eukaryotic cells, so understanding the mechanisms responsible for vesicle and organelle transport is a fundamental question in cell biology. In 1982, Robert Allen, Scott Brady, Ray Lasek, and their colleagues used video-enhanced microscopy to visualize the movement of organelles along cytoplasmic filaments in squid giant axons, both *in vivo* and in a cell-free system. These filaments were then identified as microtubules by electron microscopy, but the motor proteins responsible for organelle movement were unknown: The only microtubule motor identified at the time was axonemal dynein, which was present only in cilia and flagella. In 1985,

Ronald Vale, Thomas Reese, and Michael Sheetz described the isolation of a novel motor protein, kinesin, which was responsible for the movement of organelles along microtubules. Similar experiments were

Binding of a motor protein to microtubules in the presence of AMP-PNP. Protein samples were analyzed by electrophoresis through a polyacrylamide gel, stained, and photographed. Molecular weights of marker proteins are indicated in kilodaltons at the left. Lane a represents purified microtubules, in which only tubulin (55 kd) is detected. Lane b is a soluble extract of squid axon cytoplasm, containing many different polypeptides. This soluble extract was incubated with microtubules either without (lane c) or with (lane d) AMP-PNP. Microtubules were then recovered and incubated with ATP to release bound proteins, which were subjected to electrophoresis. Note that a 110-kd polypeptide was specifically bound to microtubules in the presence of AMP-PNP and then released by ATP. The proteins bound in the presence of AMP-PNP and released by ATP were also found to induce microtubule movement (indicated by "+" above the gel).

microtubules. There are also several types of axonemal dynein, as well as multiple cytoplasmic dyneins. All members of the dynein family move toward the minus ends of microtubules, but different cytoplasmic dyneins may transport different cargoes.

Organelle Transport and Intracellular Organization

One of the major roles of microtubules is to transport membrane vesicles and organelles through the cytoplasm of eukaryotic cells. As already discussed, such cytoplasmic organelle transport is particularly evident in nerve cell axons, which may extend more than a meter in length. Ribosomes are present only in the cell body and dendrites, so proteins, membrane vesicles, and organelles (e.g., mitochondria) must be transported from the cell body to the axon. Via video-enhanced microscopy, the transport of membrane vesicles and organelles in both directions can be visual-

The Isolation of Kinesin (continued)

could be detected. Vale, Reese, and Sheetz used a system in which proteins in the cytoplasm of axons were found to power the movement of microtubules across the surface of glass coverslips. This cell-free system provided a sensitive and rapid functional assay for the activity of kinesin as a molecular motor.

The second important approach used in the isolation of kinesin was to take advantage of its binding to microtubules. Lasek and Brady (*Nature* 316: 645–647, 1985) had made the key observation that *in vitro* movements of microtubules require ATP and are inhibited by an ATP analog (adenylyl imidodiphosphate, AMP-PNP) that cannot be hydrolyzed and therefore does not provide a usable source of energy. In the presence of AMP-PNP, organelles remained attached to microtubules, suggesting that the motor protein responsible for organelle movement might also remain bound to microtubules under these conditions.

On the basis of these findings, Vale and colleagues incubated microtubules with cytoplasmic proteins from squid axon in the presence of AMP-PNP. The microtubules were then recovered and incubated with ATP to release proteins that were specifically bound in the presence of the nonhydrolyzable ATP analog. This experiment identified a 110-kd polypeptide that was bound to microtubules in the presence of AMP-PNP

and released by subsequent incubation with ATP (see figure). Furthermore, the proteins bound to microtubules and then released by ATP were also shown to support the movement of microtubules *in vitro*. Binding to microtubules in the presence of AMP-PNP thus provided an efficient approach to isolation of the motor protein, which was shown by further biochemical studies to contain the 110-kd polypeptide complexed to polypeptides of 60 to 70 kd. By a similar approach, a related protein was also purified from bovine brain. The authors concluded that these proteins "represent a novel class of motility proteins that are structurally as well as enzymatically distinct from dynein, and we propose to call these translocators kinesin (from the Greek *kinein*, to move)."

The Impact

The movement of vesicles and organelles along microtubules is fundamental to the organization of eukaryotic cells, so the motor proteins responsible for these movements play critical roles in cell biology. Subsequent experiments using *in vitro* assays similar to those described here established that kinesin moves along microtubules in the plus end direction, whereas cytoplasmic dynein is responsible for the transport of vesicles and organelles toward microtubule minus ends. Moreover, a large

R. D. Vale

T. S. Reese

M. P. Sheetz

family of kinesin-related proteins has since been identified. In addition to their roles in vesicle and organelle transport, members of these families of motor proteins are responsible for the separation and distribution of chromosomes during mitosis. The identification of kinesin thus opened the door to understanding a variety of microtubule-based movements that are critical to the structure and function of eukaryotic cells.

ized along axon microtubules, where kinesin and dynein carry their cargoes to and from the tips of the axons, respectively. For example, secretory vesicles containing neurotransmitters are carried from the Golgi apparatus to the terminal branches of the axon by kinesin. In the reverse direction, cytoplasmic dynein transports endocytic vesicles from the axon back to the cell body.

Microtubules similarly transport membrane vesicles and organelles in other types of cells. Because microtubules are usually oriented with their minus end anchored in the centrosome and their plus end extending toward the cell periphery, different members of the kinesin and dynein families are thought to transport vesicles and organelles in opposite directions through the cytoplasm (Figure 11.46). Conventional kinesin and other plus end–directed members of the kinesin family carry their cargo toward the cell periphery, whereas cytoplasmic dyneins and minus end–directed

Figure 11.46
Transport of vesicles along microtubules Kinesin and other plus end–directed members of the kinesin family transport vesicles and organelles in the direction of microtubule plus ends, which extend toward the cell periphery. In contrast, dynein and minus end-directed members of the kinesin family carry their cargo in the direction of microtubule minus ends, which are anchored in the center of the cell.

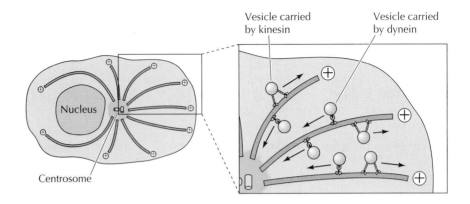

members of the kinesin family transport materials toward the center of the cell. In addition to transporting membrane vesicles in the endocytic and secretory pathways, microtubules and associated motor proteins position membrane-enclosed organelles (such as the endoplasmic reticulum, Golgi apparatus, lysosomes, and mitochondria) within the cell. For example, the endoplasmic reticulum extends to the periphery of the cell in association with microtubules (Figure 11.47). Drugs that depolymerize microtubules cause the endoplasmic reticulum to retract toward the cell center, indicating that association with microtubules is required to maintain the endoplasmic reticulum in its extended state. This positioning of the endoplasmic reticulum appears to involve the action of kinesin (or possibly multiple members of the kinesin family), which pulls the endoplasmic reticulum

(A)

(B)

Figure 11.47
Association of the endoplasmic reticulum with microtubules Fluorescence microscopy of the endoplasmic reticulum (A) and microtubules (B) in an epithelial cell. The endoplasmic reticulum is stained with a fluorescent dye and microtubules with an antibody against tubulin. Note the close correlation between the endoplasmic reticulum and microtubules at the periphery of the cell. (From M. Terasaki, L. B. Chen and K. Fujiwara, 1986. *J. Cell Biol.* 103: 1557.)

10 μm

along microtubules in the plus end direction, toward the cell periphery. Similarly, kinesin appears to play a key role in the positioning of lysosomes away from the center of the cell, and three different members of the kinesin family have been implicated in the movements of mitochondria.

Conversely, cytoplasmic dynein is thought to play a role in positioning the Golgi apparatus. The Golgi apparatus is located in the center of the cell, near the centrosome. If microtubules are disrupted, either by a drug or when the cell enters mitosis, the Golgi breaks up into small vesicles that disperse throughout the cytoplasm. When the microtubules re-form, the Golgi apparatus also reassembles, with the Golgi vesicles apparently being transported to the center of the cell (toward the minus end of microtubules) by cytoplasmic dynein. Movement along microtubules is thus responsible not only for vesicle transport, but also for establishing the positions of membrane-enclosed organelles within the cytoplasm of eukaryotic cells.

Separation of Mitotic Chromosomes

As discussed earlier in this chapter, microtubules reorganize at the beginning of mitosis to form the mitotic spindle, which plays a central role in cell division by distributing the duplicated chromosomes to daughter nuclei. This critical distribution of the genetic material takes place during anaphase of mitosis, when sister chromatids separate and move to opposite poles of the spindle. Chromosome movement proceeds by two distinct mechanisms, referred to as **anaphase A** and **anaphase B,** which involve different types of spindle microtubules and associated motor proteins.

Anaphase A consists of the movement of chromosomes toward the spindle poles along the kinetochore microtubules, which shorten as chromosome movement proceeds (Figure 11.48). This type of chromosome movement appears to be driven principally by kinetochore-associated motor proteins that translocate chromosomes along the spindle microtubules in the minus end direction, toward the centrosomes. Cytoplasmic dynein is associated with kinetochores and may play a role in poleward chromosome movement, as may minus end–directed members of the kinesin family. The action of these kinetochore motor proteins is coupled to disassembly and shortening of the kinetochore microtubules, which may be mediated by some members of the kinesin family that act as microtubule-destabilizing enzymes.

Anaphase B refers to the separation of the spindle poles themselves (Figure 11.49). Spindle-pole separation is accompanied by elongation of the polar microtubules and is similar to the initial separation of duplicated centrosomes to form the spindle poles at the beginning of mitosis (see Figure 11.43). During anaphase B the overlapping polar microtubules slide against one another, pushing the spindle poles apart. This type of movement has been found to result from the action of several plus end–directed members of the kinesin family, which crosslink polar microtubules and move them toward the plus end of their overlapping microtubule—away from the opposite spindle pole. In addition, the spindle poles may be pulled apart by the astral microtubules. The mechanism responsible for this type of movement has not been established, but it could result from the action of

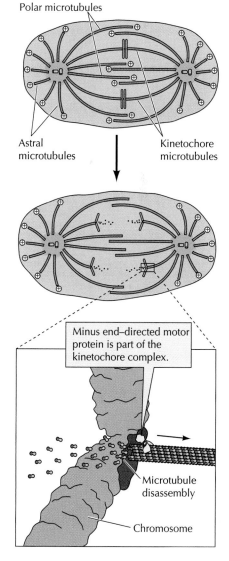

Polar microtubules

Astral microtubules

Kinetochore microtubules

Minus end–directed motor protein is part of the kinetochore complex.

Microtubule disassembly

Chromosome

Figure 11.48
Anaphase A chromosome movement   Chromosomes move toward the spindle poles along the kinetochore microtubules. Chromosome movement is thought to be driven by minus end–directed motor proteins associated with the kinetochore. The action of these motor proteins is coupled to disassembly and shortening of the kinetochore microtubules.

Figure 11.49
Spindle pole separation in anaphase B
The separation of spindle poles results from two types of movement. First, overlapping polar microtubules slide past each other to push the spindle poles apart, probably as a result of the action of plus end–directed motor proteins. Second, the spindle poles are pulled apart by the astral microtubules. The driving force could be either a minus end–directed motor anchored to a cytoplasmic structure, such as the cell cortex, or a plus end–directed motor associated with the spindle pole.

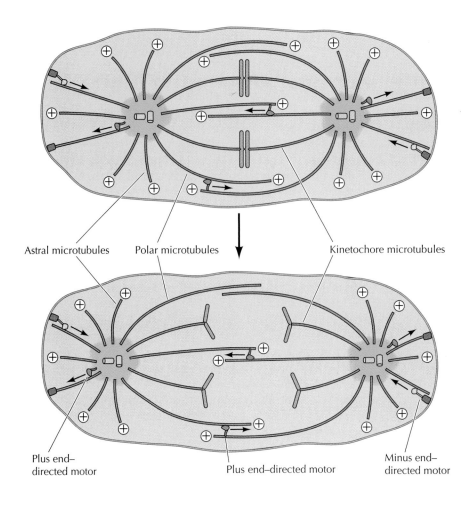

Astral microtubules   Polar microtubules   Kinetochore microtubules

Plus end–directed motor   Plus end–directed motor   Minus end–directed motor

cytoplasmic dynein anchored to the cell cortex or another structure in the cytoplasm. The translocation of such an anchored dynein motor along astral microtubules in the minus end direction would have the effect of pulling the spindle poles apart, toward the periphery of the cell. Alternatively, a motor protein associated with the spindle poles could move along astral microtubules in the plus end direction, which would also pull the spindle poles toward the cell periphery.

Cilia and Flagella

Cilia and **flagella** are microtubule-based projections of the plasma membrane that are responsible for movement of a variety of eukaryotic cells. Many bacteria also have flagella, but these prokaryotic flagella are quite different from those of eukaryotes. Bacterial flagella (which are not discussed further here) are protein filaments projecting from the cell surface, rather than projections of the plasma membrane supported by microtubules.

Eukaryotic cilia and flagella are very similar structures, each with a diameter of approximately 0.25 μm (Figure 11.50). Many cells are covered by numerous cilia, which are about 10 μm in length. Cilia beat in a coordinated back-and-forth motion, which either moves the cell through fluid or moves fluid over the surface of the cell. For example, the cilia of some protozoans (such as *Paramecium*) are responsible both for cell motility and for sweeping food organisms over the cell surface and into the oral cavity. In animals, an important function of cilia is to move fluid or mucus over the surface of epithelial cell sheets. A good example is provided by the ciliated cells lining the respiratory tract, which clear mucus and dust from the res-

(A)

(C)

20 μm

(B)

5 μm

10 μm

Figure 11.50
Examples of cilia and flagella
(A) Scanning electron micrograph showing numerous cilia covering the surface of *Paramecium*. (B) Scanning electron micrograph of ciliated epithelial cells lining the surface of a trachea. (C) Multiple-flash photograph (500 flashes per second) showing the wavelike movement of a sea urchin sperm flagellum. (A, Karl Aufderheide/ Visuals Unlimited; B, Fred E. Hossler/ Visuals Unlimited; C, C. J. Brokaw, California Institute of Technology.)

Figure 11.51
Structure of the axoneme of cilia and flagella (A) Computer-enhanced electron micrograph of a cross section of the axoneme of a rat sperm flagellum. (B) Schematic cross section of an axoneme. The nine outer doublets consist of one complete (A) and one incomplete (B) microtubule, containing only 10 or 11 protofilaments. The outer doublets are joined to each other by nexin links and to the central pair of microtubules by radial spokes. Each outer microtubule doublet is associated with inner and outer dynein arms. (A, K. G. Murti/Visuals Unlimited.)

piratory passages. Flagella differ from cilia in their length (they can be as long as 200 μm) and in their wavelike pattern of beating. Cells usually have only one or two flagella, which are responsible for the locomotion of a variety of protozoans and of sperm.

The fundamental structure of both cilia and flagella is the **axoneme**, which is composed of microtubules and their associated proteins (Figure 11.51). The microtubules are arranged in a characteristic "9 + 2" pattern in which a central pair of microtubules is surrounded by nine outer microtubule doublets. The two fused microtubules of each outer doublet are distinct: One (called the A tubule) is a complete microtubule consisting of 13 protofilaments; the other (the B tubule) is incomplete, containing only 10 or

(A)

(B)

0.1 μm

(A)

(B)

0.1 μm

0.1 μm

Figure 11.52
Electron micrographs of basal bodies
(A) A longitudinal view of cilia anchored in basal bodies. (B) A cross section of basal bodies. Each basal body consists of nine triplets of microtubules. (A, Conly L. Reider/Biological Photo Service; B, W. L. Dentler, Biological Photo Service.)

11 protofilaments fused to the A tubule. The outer microtubule doublets are connected to the central pair by radial spokes and to each other by links of a protein called **nexin**. In addition, two arms of dynein are attached to each A tubule, and it is the motor activity of these axonemal dyneins that drives the beating of cilia and flagella.

The minus ends of the microtubules of cilia and flagella are anchored in a **basal body**, which is similar in structure to a centriole and contains nine triplets of microtubules (Figure 11.52). Centrioles were discussed earlier as components of the centrosome, in which their function is uncertain. Basal bodies, however, play a clear role in organization of the axoneme microtubules. Namely, each of the outer microtubule doublets of the axoneme is formed by extension of two of the microtubules present in the triplets of the basal body. Basal bodies thus serve to initiate the growth of axonemal microtubules, as well as anchoring cilia and flagella to the surface of the cell.

The movements of cilia and flagella result from the sliding of outer microtubule doublets relative to one another, powered by the motor activity of axonemal dynein (Figure 11.53). The dynein bases bind to the A tubules while the dynein head groups bind to the B tubules of adjacent doublets. Movement of the dynein head group in the minus end direction

Figure 11.53
Movement of microtubules in cilia and flagella   The bases of dynein arms are attached to A tubules, and the motor head groups interact with the B tubules of adjacent doublets. Movement of the dynein head groups in the minus end direction (toward the base of the cilium) then causes the A tubule of one doublet to slide toward the base of the adjacent B tubule. Because both microtubule doublets are connected by nexin links, this sliding movement forces them to bend.

then causes the A tubule of one doublet to slide toward the basal end of the adjacent B tubule. Because the microtubule doublets in an axoneme are connected by nexin links, the sliding of one doublet along another causes them to bend, forming the basis of the beating movements of cilia and flagella. It is apparent, however, that the activities of dynein molecules in different regions of the axoneme must be carefully regulated to produce the coordinated beating of cilia and the wavelike oscillations of flagella—a process about which little is currently understood.

SUMMARY

STRUCTURE AND ORGANIZATION OF ACTIN FILAMENTS

Assembly and Disassembly of Actin Filaments: Actin filaments are formed by the head-to-tail polymerization of actin monomers into a helix. A variety of actin-binding proteins regulate the assembly and disassembly of actin filaments within the cell.

Organization of Actin Filaments: Actin filaments are crosslinked by actin-binding proteins to form bundles or three-dimensional networks.

Association of Actin Filaments with the Plasma Membrane: A network of actin filaments and other cytoskeletal proteins underlies the plasma membrane and determines cell shape. Actin bundles also attach to the plasma membrane and anchor the cell at regions of cell-cell and cell-substratum contact.

Protrusions of the Cell Surface: Actin filaments support permanent protrusions of the cell surface, such as microvilli, as well as transient extensions that are responsible for phagocytosis and cell locomotion.

ACTIN, MYOSIN, AND CELL MOVEMENT

Muscle Contraction: Studies of muscle established the role of myosin as a motor protein that uses the energy derived from ATP hydrolysis to generate force and movement. Muscle contraction results from the sliding of actin and myosin filaments past each other. ATP hydrolysis drives repeated cycles of interaction between myosin and actin, during which conformational changes result in movement at the myosin head group along actin filaments.

Contractile Assemblies of Actin and Myosin in Nonmuscle Cells: Assemblies of actin and myosin II are responsible for a variety of movements of nonmuscle cells, including cytokinesis.

Unconventional Myosins: Other types of myosin that do not function in contraction serve to transport membrane vesicles and organelles along actin filaments.

Cell Crawling: Cell crawling is a complex process in which extensions of the plasma membrane are formed by polymerization of actin filaments at the leading edge of the cell. These extensions then attach to the substratum, and the trailing edge retracts into the cell body.

KEY TERMS

actin, microfilament, globular (G) actin, filamentous (F) actin, treadmilling, cytochalasin, phalloidin, actin-binding protein, cofilin, profilin, Arp2/3 proteins

actin bundle, actin network, actin-bundling protein, fimbrin, contractile bundle, α-actinin, filamin

cell cortex, spectrin, ankyrin, fodrin, ERM proteins, dystrophin, integrin, focal adhesion, stress fiber, talin, vinculin, adherens junction, adhesion belt, cadherin, catenins

microvillus, brush border, stereocilium, villin, pseudopodium, lamellipodium, microspike, filopodium

myosin, molecular motor, muscle fiber, myofibril, sarcomere, titin, nebulin, sliding filament model, myosin II, sarcoplasmic reticulum, tropomyosin, troponin

cytokinesis, contractile ring, myosin light-chain kinase, calmodulin

myosin I

INTERMEDIATE FILAMENTS

intermediate filament, keratin,
hard keratin, soft keratin,
vimentin, desmin,
neurofilament (NF) protein

Intermediate Filament Proteins: Intermediate filaments are polymers of more than 50 different proteins that are expressed in various types of cells. They are not involved in cell movement, but provide mechanical support to cells and tissues.

Assembly of Intermediate Filaments: Intermediate filaments are formed from dimers of two polypeptide chains wound around each other in a coiled-coil structure. The dimers then associate to form tetramers, which assemble into protofilaments. Intermediate filaments are formed from protofilaments wound around one another in a ropelike structure.

desmosome, hemidesmosome,
plakin

Intracellular Organization of Intermediate Filaments: Intermediate filaments form a network extending from a ring surrounding the nucleus to the plasma membrane of most cell types. In epithelial cells, intermediate filaments are anchored to the plasma membrane at regions of specialized cell contacts (desmosomes and hemidesmosomes). Intermediate filaments also play specialized roles in muscle and nerve cells.

Functions of Keratins and Neurofilaments: Diseases of the Skin and Nervous System: The importance of intermediate filaments in providing mechanical strength to cells in tissues has been demonstrated by the introduction of mutant keratin genes into transgenic mice. Similar keratin gene mutations are responsible for some human skin diseases, and abnormalities of neurofilaments have been implicated in the development of motor neuron disease.

MICROTUBULES

microtubule, tubulin, dynamic
instability, colchicine, colcemid,
vincristine, vinblastine, taxol

Structure, Assembly, and Dynamic Instability of Microtubules: Microtubules are formed by the reversible polymerization of tubulin. They display dynamic instability and undergo continual cycles of assembly and disassembly as a result of GTP hydrolysis following tubulin polymerization.

microtubule-organizing center,
centrosome, centriole,
pericentriolar material

The Centrosome and Microtubule Organization: The microtubules in most cells extend outward from a microtubule-organizing center, or centrosome, located near the center of the cell. In animal cells, the centrosome usually contains a pair of centrioles surrounded by pericentriolar material. The growth of microtubules is initiated in the pericentriolar material, which then serves to anchor their minus ends.

mitotic spindle, kinetochore
microtubule, polar microtubule,
astral microtubule

Reorganization of Microtubules during Mitosis: Microtubules reorganize at the beginning of mitosis to form the mitotic spindle, which is responsible for chromosome separation.

microtubule-associated protein
(MAP)

Stabilization of Microtubules and Cell Polarity: Selective stabilization of microtubules by microtubule-associated proteins can determine cell shape and polarity, such as the extension of nerve cell axons and dendrites.

MICROTUBULE MOTORS AND MOVEMENTS

kinesin, dynein, axonemal dynein,
video-enhanced microscopy,
cytoplasmic dynein

Identification of Microtubule Motor Proteins: Two families of motor proteins, the kinesins and the dyneins, are responsible for movement along microtubules. Kinesin and most kinesin-related proteins move in the plus end direction, whereas the dyneins and some members of the kinesin family move toward microtubule minus ends.

Organelle Transport and Intracellular Organization: Movement along microtubules transports membrane vesicles and organelles through the cytoplasm, as well as positioning cytoplasmic organelles within the cell.

Separation of Mitotic Chromosomes: During anaphase of mitosis, daughter chromosomes separate and move to opposite poles of the mitotic spindle. Chromosome separation results from several types of movements in which different classes of spindle microtubules and motor proteins participate.

anaphase A, anaphase B

Cilia and Flagella: Cilia and flagella are microtubule-based extensions of the plasma membrane. Their movements result from the sliding of microtubules, driven by the action of dynein motors.

cilium, flagellum, axoneme, nexin, basal body

Questions

1. Would cytochalasin affect the movement of cells that extend pseudopodia or that utilize flagella for locomotion?

2. Why is the polarity of actin filaments important to muscle contraction?

3. Would you expect mutations of keratin genes to affect fibroblasts?

4. Which aspect of cell division would be affected by colchicine: chromosome segregation or cytokinesis?

5. How would elimination of nexin affect the beating of cilia?

References and Further Reading

General References

Amos, L. A. and W. B. Amos. 1991. *Molecules of the Cytoskeleton.* New York: Guilford Press.

Bray, D. 1992. *Cell Movements.* New York: Garland Publishing.

Structure and Organization of Actin Filaments

Bamburg, J. R. 1999. Proteins of the ADF/cofilin family: Essential regulators of actin dynamics. *Ann. Rev. Cell Dev. Biol.* 15: 185–230. [R]

Bamburg, J. R., A. McGough and S. Ono. 1999. Putting a new twist on actin: ADF/cofilins modulate actin dynamics. *Trends Cell Biol.* 9: 364–370. [R]

Bennett, V. and D. M. Gilligan. 1993. The spectrin-based membrane skeleton and micron-scale organization of the plasma membrane. *Ann. Rev. Cell Biol.* 9: 27–66. [R]

Bretscher, A. 1991. Microfilament structure and function in the cortical cytoskeleton. *Ann. Rev. Cell Biol.* 7: 337–374. [R]

Burridge, K. and M. Chrzanowska-Wodnicka. 1996. Focal adhesions, contractility, and signaling. *Ann. Rev. Cell Dev. Biol.* 12: 463–519. [R]

Campbell, K. P. 1995. Three muscular dystrophies: Loss of cytoskeleton-extracellular matrix linkage. *Cell* 80: 675–679. [R]

Holmes, K. C., D. Popp, W. Gebhard and W. Kabsch. 1990. Atomic model of the actin filament. *Nature* 347: 44–49. [P]

Jokusch, B. M., P. Bubeck, K. Giehl, M. Kroemker, J. Moschner, M. Roghkegel, M. Rudiger, K. Schluter, G. Stanke and J. Winkler. 1995. The molecular architecture of focal adhesions. *Ann. Rev. Cell Dev. Biol.* 11: 379–416. [R]

Kabsch, W., H. G. Mannherz, D. Suck, E. F. Pai and K. C. Holmes. 1990. Atomic structure of the actin: DNase I complex. *Nature* 347: 37–44. [P]

Luna, E. J. and A. L. Hitt. 1992. Cytoskeleton-plasma membrane interactions. *Science* 258: 955–964. [R]

Matsudaira, P. 1991. Modular organization of actin crosslinking proteins. *Trends Biochem. Sci.* 16: 87–92. [R]

Pollard, T. D., S. Almo, S. Quirk, V. Vinson and E. E. Lattman. 1994. Structure of actin binding proteins: Insights about function at atomic resolution. *Ann. Rev. Cell Biol.* 10: 207–249. [R]

Pumplin, D. W. and R. J. Bloch. 1993. The membrane skeleton. *Trends Cell Biol.* 3: 113–117. [R]

Schmidt, A. and M. N. Hall. 1998. Signaling to the actin cytoskeleton. *Ann. Rev. Cell Dev. Biol.* 14: 305–338. [R]

Theriot, J. A. and T. J. Mitchison. 1993. The three faces of profilin. *Cell* 75: 835–838. [R]

Tuskita, S., S. Yonemura and S. Tsukita. 1997. ERM proteins: Head-to-tail regulation of actin-plasma membrane interaction. *Trends Biochem. Sci.* 22: 53–58. [R]

Welch, M. D. 1999. The world according to Arp: Regulation of actin nucleation by the Arp2/3 complex. *Trends Cell Biol.* 9: 423–427. [R]

Yap, A. S., W. M. Brieher and B. M. Gumbiner. 1997. Molecular and functional analysis of cadherin-based adherens junctions. *Ann. Rev. Cell Dev. Biol.* 13: 119–146. [R]

Actin, Myosin, and Cell Movement

Erickson, H. P. 1997. Stretching single protein molecules: Titin is a weird spring. *Science* 276: 1090–1092. [R]

Finer, J. T., R. M. Simmons and J. A. Spudich. 1994. Single myosin molecule mechanics: Piconewton forces and nanometre steps. *Nature* 368: 113–119. [P]

Goldman, Y. E. 1998. Wag the tail: Structural dynamics of actomyosin. *Cell* 93: 1–4. [R]

Geeves, M. A. and K. C. Holmes. 1999. Structural mechanism of muscle contraction. *Ann. Rev. Biochem.* 68: 687–728. [R]

Horwitz, A. F. and J. T. Parsons. 1999. Cell migration—movin' on. *Science* 286: 1102–1103. [R]

Huxley, A. F. and R. Niedergerke. 1954. Interference microscopy of living muscle fibres. *Nature* 173: 971–973. [P]

Huxley, H. E. 1969. The mechanism of muscle contraction. *Science* 164: 1356–1366. [R]

Huxley, H.E. and J. Hanson. 1954. Changes in the cross-striations of muscle contraction and their structural interpretation. *Nature* 173: 973–976. [P]

Lauffenburger, D. A. and A. F. Horwitz. 1996. Cell migration: A physically integrated molecular process. *Cell* 84: 359–369. [R]

Littlefield, R. and V. M. Fowler. 1998. Defining actin filament length in striated muscle: Rulers and caps or dynamic stability? *Ann. Rev. Cell Dev. Biol.* 14: 487–525. [R]

Mermall, V., P. L. Post and M. S. Moosekar. 1998. Unconventional myosins in cell movement, membrane traffic, and signal transduction. *Science* 279: 527–533. [R]

Mitchison, T. J. and L. P. Cramer. 1996. Actin-based cell motility and cell locomotion. *Cell* 84: 371–379. [R]

Rayment, I. and H. M. Holden. 1994. The three-dimensional structure of a molecular motor. *Trends Biochem. Sci.* 19: 129–134. [R]

Rayment, I., H. M. Holden, M. Whittaker, C. B. Yohn, M. Lorenz, K. C. Kolmes and R. A. Milligan. 1993. Structure of the actin-myosin complex and its implications for muscle contraction. *Science* 261: 58–65. [P]

Rayment, I., W. R. Rypniewski, K. Schmidt-Base, R. Smith, D. R. Tomchick, M. M. Benning, D. A. Winkelmann, G. Wesenberg and H. M. Holden. 1993. Three-dimensional structure of myosin subfragment-1: A molecular motor. *Science* 261: 50–58. [P]

Ruppel, K. M. and J. A. Spudich. 1996. Structure-function analysis of the motor domain of myosin. *Ann. Rev. Cell Dev. Biol.* 12: 543–573. [R]

Spudich, J. A. 1994. How molecular motors work. *Nature* 372: 515– 518. [R]

Stossel, T. P. 1993. On the crawling of animal cells. *Science* 260: 1086–1094. [R]

Tan, J. L., S. Ravid and J. A. Spudich. 1992. Control of nonmuscle myosins by phosphorylation. *Ann. Rev. Biochem.* 61: 721–759. [R]

Trinick, J. and L. Tskhovrebova. 1997. Titin: A molecular control freak. *Trends Cell Biol.* 9: 377–380. [R]

Zot, A. S. and J. D. Potter. 1987. Structural aspects of troponin-tropomyosin regulation of skeletal muscle contraction. *Ann. Rev. Biophys. Biophys. Chem.* 16: 535–559. [R]

Intermediate Filaments

Bonifas, J. M., A. L. Rothman and E. H. Epstein, Jr. 1991. Epidermolysis bullosa simplex: Evidence in two families for keratin gene abnormalities. *Science* 254: 1202–1205. [P]

Brown, R. H., Jr. 1995. Amyotrophic lateral sclerosis: Recent insights from genetics and transgenic mice. *Cell* 80: 687–692. [R]

Coulombe, P. A., M. E. Hutton, A. Letai, A. Hebert, A. S. Paller and E. Fuchs. 1991. Point mutations in human keratin 14 genes of epidermolysis bullosa simplex patients: Genetic and functional analyses. *Cell* 66: 1301–1311. [P]

Fuchs, E. and D. W. Cleveland. 1998. A structural scaffolding of intermediate filaments in health and disease. *Science* 279: 514–519. [R]

Fuchs, E. and Y. Yang. 1999. Crossroads on cytoskeletal highways. *Cell* 98: 547–550. [R]

Schwarz, M. A., K. Owaribe, J. Kartenbeck and W. W. Franke. 1990. Desmosomes and hemidesmosomes: Constitutive molecular components. *Ann. Rev. Cell Biol.* 6: 461–491. [R]

Skalli, O., Y.-H. Chou and R. D. Goldman. 1992. Intermediate filaments: Not so tough after all. *Trends Cell Biol.* 2: 308–312. [R]

Steinert, P. M. and D. R. Roop. 1988. Molecular and cellular biology of intermediate filaments. *Ann. Rev. Biochem.* 57: 593–625. [R]

Vassar, R., P. A. Coulombe, L. Degenstein, K. Albers and E. Fuchs. 1991. Mutant keratin expression in transgenic mice causes marked abnormalities resembling a human genetic skin disease. *Cell* 64: 365–380. [P]

Microtubules

Brinkley, R. R. 1985. Microtubule organizing centers. *Ann. Rev. Cell Biol.* 1: 145–172. [R]

Desai, A. and T. J. Mitchison. 1997. Microtubule polymerization dynamics. *Ann. Rev. Cell Dev. Biol.* 13: 83–117. [R]

Drewes, G., A. Ebneth and E.-M. Mandelkow. 1998. MAPs, MARKs and microtubule dynamics. *Trends Biochem. Sci.* 23: 307–311. [R]

Hyman, A. A. and E. Karsenti. 1996. Morphogenetic properties of microtubules and mitotic spindle assembly. *Cell* 84: 401–410 [R].

Kellogg, D. R., M. Moritz and B. M. Alberts. 1994. The centrosome and cellular organization. *Ann. Rev. Biochem.* 63: 639–674. [R]

Kirschner, M. and T. Mitchison. 1986. Beyond self-assembly: From microtubules to morphogenesis. *Cell* 45: 329–342. [R]

Lange, B. M. H. and K. Gull. 1996. Structure and function of the centriole in animal cells: Progress and questions. *Trends Cell Biol.* 6: 348–352. [R]

Mitchison, T. and M. Kirschner. 1984. Dynamic instability of microtubule growth. *Nature* 312: 237–242. [P]

Mitchison, T. and M. Kirschner. 1984. Microtubule assembly nucleated by isolated centrosomes. *Nature* 312: 232–237. [P]

Nogales, E., M. Whittaker, R.A. Milligan and K. H. Downing. 1999. High-resolution model of the microtubule. *Cell* 96: 79–88. [P]

Oakley, B. R., C. E. Oakley, Y. Yoon and M. K. Jung. 1990. γ-Tubulin is a component of the spindle pole body that is essential for microtubule function in *Aspergillus nidulans*. *Cell* 61: 1289–1301. [P]

Osborn, M. and K. Weber. 1976. Cytoplasmic microtubules in tissue culture cells appear to grow from an organizing structure towards the plasma membrane. *Proc. Natl. Acad. Sci. USA* 73: 867–871. [P]

Zheng, Y., M. L. Wong, B. Alberts and T. Mitchison. 1995. Nucleation of microtubule assembly by a γ-tubulin-containing ring complex. *Nature* 378: 578–583. [P]

Microtubule Motors and Movements

Asai, D. J. and C. J. Brokaw. 1993. Dynein heavy chain isoforms and axonemal motility. *Trends Cell Biol.* 3: 398–403. [R]

Brady, S. T. 1985. A novel brain ATPase with properties expected for the fast axonal motor. *Nature* 317: 73–75. [P]

Brady, S. T., R. J. Lasek and R. D. Allen. 1982. Fast axonal transport in extruded axoplasm from squid giant axon. *Science* 218: 1129–1131. [P]

Desai, A., S. Verma, T. J. Mitchison and C. E. Walczak. 1999. Kin I kinesins are microtubule-destabilizing enzymes. *Cell* 96: 69–78. [P]

Gibbons, I. R. 1981. Cilia and flagella of eukaryotes. *J. Cell Biol.* 91: 107s–124s. [R]

Gibbons, I. R., and A. Rowe. 1965. Dynein: A protein with adenosine triphosphatase activity from cilia. *Science* 149: 424–426. [P]

Goldstein, L. S. B. and A. V. Philp. 1999. The road less traveled: Emerging principles of kinesin motor utilization. *Ann. Rev. Cell Dev. Biol.* 15: 141–183. [R]

Haimo, L. T. 1995. Regulation of kinesin-directed movements. *Trends Cell Biol.* 5: 165–168. [R]

Hirokawa, N. 1998. Kinesin and dynein superfamily proteins and the mechanism of organelle transport. *Science* 279: 519–526. [R]

Holzbaur, E. L. F. and R. B. Vallee. 1994. Dyneins: Molecular structure and cellular function. *Ann. Rev. Cell Biol.* 10: 339–372. [R]

Lasek, R. J. and S. T. Brady. 1985. Attachment of transported vesicles to microtubules in axoplasm is facilitated by AMP-PNP. *Nature* 316: 645–647. [P]

Mandelkow, E. and K. A. Johnson. 1998. The structural and mechanochemical cycle of kinesin. *Trends Biochem. Sci.* 23: 429–433. [R]

Porter, M. E. and K. A. Johnson. 1991. Dynein structure and function. *Ann. Rev. Cell Biol.* 5: 119–151. [R]

Rieder, C. L. and E. D. Salmon. 1998. The vertebrate cell kinetochore and its roles during mitosis. *Trends Cell Biol.* 8: 310–318. [R]

Salmon, E. D. 1995. VE-DIC light microscopy and the discovery of kinesin. *Trends Cell Biol.* 5: 154–157. [R]

Saunders, W. S. 1993. Mitotic spindle pole separation. *Trends Cell Biol.* 3: 432–437. [R]

Vale, R. D. and R. J. Fletterick. 1997. The design plan of kinesin motors. *Ann. Rev. Cell Dev. Biol.* 13: 745–777. [R]

Vale, R. D., T. S. Reese and M. P. Sheetz. 1985. Identification of a novel force-generating protein, kinesin, involved in microtubule-based motility. *Cell* 42: 39–50. [P]

Vallee, R. B. and M. P. Sheetz. 1996. Targeting of motor proteins. *Science* 271: 1539–1544. [R]

The Cell Surface

ALL CELLS—BOTH PROKARYOTIC AND EUKARYOTIC—are surrounded by a plasma membrane, which defines the boundary of the cell and separates its internal contents from the environment. By serving as a selective barrier to the passage of molecules, the plasma membrane determines the composition of the cytoplasm. This ultimately defines the very identity of the cell, so the plasma membrane is one of the most fundamental structures of cellular evolution. Indeed, as discussed in Chapter 1, the first cell is thought to have arisen by the enclosure of self-replicating RNA in a membrane of phospholipids.

The plasma membranes of present-day cells are composed of both lipids and proteins. The basic structure of the plasma membrane is the phospholipid bilayer, which is impermeable to most water-soluble molecules. The passage of ions and most biological molecules across the plasma membrane is therefore mediated by proteins, which are responsible for the selective traffic of molecules into and out of the cell. Other proteins of the plasma membrane control the interactions between cells of multicellular organisms and serve as sensors through which the cell receives signals from its environment. The plasma membrane thus plays a dual role: It both isolates the cytoplasm and mediates interactions between the cell and its environment.

Structure of the Plasma Membrane

Like all other cellular membranes, the plasma membrane consists of both lipids and proteins. The fundamental structure of the membrane is the phospholipid bilayer, which forms a stable barrier between two aqueous compartments. In the case of the plasma membrane, these compartments are the inside and the outside of the cell. Proteins embedded within the phospholipid bilayer carry out the specific functions of the plasma membrane, including selective transport of molecules and cell-cell recognition.

The Phospholipid Bilayer
The plasma membrane is the most thoroughly studied of all cell membranes, and it is largely through investigations of the plasma membrane

that our current concepts of membrane structure have evolved. The plasma membranes of mammalian red blood cells (erythrocytes) have been particularly useful as a model for studies of membrane structure. Mammalian red blood cells do not contain nuclei or internal membranes, so they represent a source from which pure plasma membranes can be easily isolated for biochemical analysis. Indeed, studies of the red blood cell plasma membrane provided the first evidence that biological membranes consist of lipid bilayers. In 1925, two Dutch scientists (E. Gorter and R. Grendel) extracted the membrane lipids from a known number of red blood cells, corresponding to a known surface area of plasma membrane. They then determined the surface area occupied by a monolayer of the extracted lipid spread out at an air-water interface. The surface area of the lipid monolayer turned out to be twice that occupied by the erythrocyte plasma membranes, leading to the conclusion that the membranes consisted of lipid bilayers rather than monolayers.

The bilayer structure of the erythrocyte plasma membrane is clearly evident in high-magnification electron micrographs (Figure 12.1). The plasma membrane appears as two dense lines separated by an intervening space—a morphology frequently referred to as a "railroad track" appearance. This image results from the binding of the electron-dense heavy metals used as stains in transmission electron microscopy (see Chapter 1) to the polar head groups of the phospholipids, which therefore appear as dark lines. These dense lines are separated by the lightly stained interior portion of the membrane, which contains the hydrophobic fatty acid chains.

As discussed in Chapter 2, the plasma membranes of animal cells contain four major phospholipids (**phosphatidylcholine, phosphatidylethanolamine, phosphatidylserine**, and **sphingomyelin**), which together account for more than half of the lipid in most membranes. These phospholipids are asymmetrically distributed between the two halves of the membrane bilayer (Figure 12.2). The outer leaflet of the plasma membrane consists mainly of phosphatidylcholine and sphingomyelin, whereas phosphatidylethanolamine and phosphatidylserine are the predominant phospholipids of the inner leaflet. A fifth phospholipid, **phosphatidylinositol**, is also localized to the inner half of the plasma membrane. Although phosphatidylinositol is a quantitatively minor membrane component, it plays an important role in cell signaling, as discussed in the next chapter. The head groups of both phosphatidylserine and phosphatidylinositol are negatively charged, so their predominance in the inner leaflet results in a net negative charge on the cytosolic face of the plasma membrane.

Membrane

20 nm

Figure 12.1
Bilayer structure of the plasma membrane   Electron micrograph of a human red blood cell. Note the railroad track appearance of the plasma membrane. (Courtesy of J. David Robertson, Duke University Medical Center.)

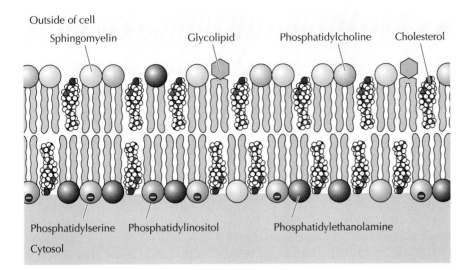

Outside of cell

Sphingomyelin   Glycolipid   Phosphatidylcholine   Cholesterol

Phosphatidylserine   Phosphatidylinositol   Phosphatidylethanolamine

Cytosol

Figure 12.2
Lipid components of the plasma membrane   The outer leaflet consists predominantly of phosphatidyl-choline, sphingomyelin, and glycolip-ids, whereas the inner leaflet contains phosphatidylethanolamine, phospha-tidylserine, and phosphatidylinositol. Cholesterol is distributed in both leaflets. The net negative charge of the head groups of phosphatidylserine and phosphatidylinositol is indicated. The structures of phospholipids, gly-colipids, and cholesterol are shown in Figures 2.7, 2.8, and 2.9, respectively.

In addition to the phospholipids, the plasma membranes of animal cells contain **glycolipids** and **cholesterol**. The glycolipids are found exclusively in the outer leaflet of the plasma membrane, with their carbohydrate por-tions exposed on the cell surface. They are relatively minor membrane components, constituting only about 2% of the lipids of most plasma membranes. Cholesterol, on the other hand, is a major membrane con-stituent of animal cells, being present in about the same molar amounts as the phospholipids.

Two general features of phospholipid bilayers are critical to membrane function. First, the structure of phospholipids is responsible for the basic function of membranes as barriers between two aqueous compartments. Because the interior of the phospholipid bilayer is occupied by hydropho-bic fatty acid chains, the membrane is impermeable to water-soluble mole-cules, including ions and most biological molecules. Second, bilayers of the naturally occurring phospholipids are viscous fluids, not solids. The fatty acids of most natural phospholipids have one or more double bonds, which introduce kinks into the hydrocarbon chains and make them difficult to pack together. The long hydrocarbon chains of the fatty acids therefore move freely in the interior of the membrane, so the membrane itself is soft and flexible. In addition, both phospholipids and proteins are free to dif-fuse laterally within the membrane—a property that is critical for many membrane functions.

Because of its rigid ring structure, cholesterol plays a distinct role in membrane structure. Cholesterol will not form a membrane by itself, but inserts into a bilayer of phospholipids with its polar hydroxyl group close to the phospholipid head groups (see Figure 12.2). Depending on the tem-perature, cholesterol has distinct effects on membrane fluidity. At high temperatures, cholesterol interferes with the movement of the phospho-lipid fatty acid chains, making the outer part of the membrane less fluid and reducing its permeability to small molecules. At low temperatures, however, cholesterol has the opposite effect: By interfering with interac-tions between fatty acid chains, cholesterol prevents membranes from freezing and maintains membrane fluidity. Although cholesterol is not present in bacteria, it is an essential component of animal cell plasma membranes. Plant cells also lack cholesterol, but they contain related com-pounds (sterols) that fulfill a similar function.

Recent studies suggest that not all lipids diffuse freely in the plasma membrane. Instead, discrete membrane domains appear to be enriched in cholesterol and the sphingolipids (sphingomyelin and glycolipids). These clusters of sphingolipids and cholesterol are thought to form "rafts" that move laterally within the plasma membrane and may associate with specific membrane proteins. Although the functions of lipid rafts remain to be understood, they may play important roles in processes such as cell signaling and the uptake of extracellular molecules by endocytosis.

Membrane Proteins

While lipids are the fundamental structural elements of membranes, proteins are responsible for carrying out specific membrane functions. Most plasma membranes consist of approximately 50% lipid and 50% protein by weight, with the carbohydrate portions of glycolipids and glycoproteins constituting 5 to 10% of the membrane mass. Since proteins are much larger than lipids, this percentage corresponds to about one protein molecule per every 50 to 100 molecules of lipid. In 1972, Jonathan Singer and Garth Nicolson proposed the **fluid mosaic model** of membrane structure, which is now generally accepted as the basic paradigm for the organization of all biological membranes. In this model, membranes are viewed as two-dimensional fluids in which proteins are inserted into lipid bilayers (Figure 12.3).

Singer and Nicolson distinguished two classes of membrane-associated proteins, which they called **peripheral** and **integral membrane proteins**. Peripheral membrane proteins were operationally defined as proteins that dissociate from the membrane following treatments with polar reagents, such as solutions of extreme pH or high salt concentration, that do not disrupt the phospholipid bilayer. Once dissociated from the membrane, peripheral membrane proteins are soluble in aqueous buffers. These proteins are not inserted into the hydrophobic interior of the lipid bilayer. Instead, they are indirectly associated with membranes through protein-

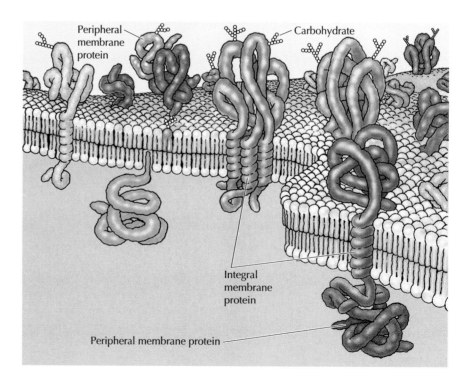

Figure 12.3
Fluid mosaic model of the plasma membrane Integral membrane proteins are inserted into the lipid bilayer, whereas peripheral proteins are bound to the membrane indirectly by protein-protein interactions. Most integral membrane proteins are transmembrane proteins, with portions exposed on both sides of the lipid bilayer. The extracellular portions of these proteins are usually glycosylated, as are the peripheral membrane proteins bound to the external face of the membrane.

protein interactions. These interactions frequently involve ionic bonds, which are disrupted by extreme pH or high salt.

In contrast to the peripheral membrane proteins, integral membrane proteins can be released only by treatments that disrupt the phospholipid bilayer. Portions of these integral membrane proteins are inserted into the lipid bilayer, so they can be dissociated only by reagents that disrupt hydrophobic interactions. The most commonly used reagents for solubilization of integral membrane proteins are detergents, which are small amphipathic molecules containing both hydrophobic and hydrophilic groups (Figure 12.4). The hydrophobic portions of detergents displace the membrane lipids and bind to the hydrophobic portions of integral membrane proteins. Because the other end of the detergent molecule is hydrophilic, the detergent-protein complexes are soluble in aqueous solutions.

Many integral proteins are **transmembrane proteins**, which span the lipid bilayer with portions exposed on both sides of the membrane. These proteins can be visualized in electron micrographs of plasma membranes prepared by the freeze-fracture technique (see Figure 1.35). In these specimens, the membrane is split and separates into its two leaflets. Transmembrane proteins are then apparent as particles on the internal faces of the membrane (Figure 12.5).

Figure 12.4
Solubilization of integral membrane proteins by detergents Detergents (e.g., octyl glucoside) are amphipathic molecules containing hydrophilic head groups and hydrophobic tails. The hydrophobic tails bind to the hydrophobic regions of integral membrane proteins, forming detergent-protein complexes that are soluble in aqueous solution.

0.2 μm

Figure 12.5
Freeze-fracture electron micrograph of human red blood cell membranes The particles in the membrane are transmembrane proteins. (Harold H. Edwards/Visuals Unlimited.)

The membrane-spanning portions of transmembrane proteins are usually α helices of 20 to 25 hydrophobic amino acids that are inserted into the membrane of the endoplasmic reticulum during synthesis of the polypeptide chain (see Figures 9.11, 9.12, and 9.13). These proteins are then transported in membrane vesicles from the endoplasmic reticulum to the Golgi apparatus, and from there to the plasma membrane. Carbohydrate groups are added to the polypeptide chains in both the endoplasmic reticulum and Golgi apparatus, so most transmembrane proteins of the plasma membrane are glycoproteins with their oligosaccharides exposed on the surface of the cell.

Studies of red blood cells have provided good examples of both peripheral and integral proteins associated with the plasma membrane. The membranes of human erythrocytes contain about a dozen major proteins, which were originally identified by gel electrophoresis of membrane preparations. Most of these are peripheral membrane proteins that have been identified as components of the cortical cytoskeleton, which underlies the plasma membrane and determines cell shape (see Chapter 11). For example, the most abundant peripheral membrane protein of red blood cells is spectrin, which is the major cytoskeletal protein of erythrocytes. Other peripheral membrane proteins of red blood cells include actin, ankyrin, and band 4.1. Ankyrin serves as the principal link between the plasma membrane and the cytoskeleton by binding to both spectrin and the integral membrane protein band 3 (see Figure 11.11). An additional link between the membrane and the cytoskeleton is provided by band 4.1, which binds to the junctions of spectrin and actin, as well as to glycophorin (the other major integral membrane protein of erythrocytes).

The two major integral membrane proteins of red blood cells, glycophorin and band 3, provide well-studied examples of transmembrane protein structure (Figure 12.6). Glycophorin is a small glycoprotein of 131 amino acids, with a molecular weight of about 30,000, half of which is protein and half carbohydrate. Glycophorin crosses the membrane with a single membrane-spanning α helix of 23 amino acids, with its glycosylated amino-terminal portion exposed on the cell surface. Although glycophorin was one of the first transmembrane proteins to be characterized, its precise function remains unknown. In contrast, the function of the other major transmembrane protein of red blood cells is well understood. This protein, originally known as band 3, is the anion transporter responsible for the passage of bicarbonate (HCO_3^-) and chloride (Cl^-) ions across the red blood

Figure 12.6
Integral membrane proteins of red blood cells   Glycophorin (131 amino acids) contains a single transmembrane α helix. It is heavily glyocosylated, with oligosaccharides attached to 16 sites on the extracellular portion of the polypeptide chain. Band 3 (929 amino acids) has multiple transmembrane α helices and is thought to cross the membrane 14 times.

cell membrane. The band 3 polypeptide chain is 929 amino acids and is thought to have 14 membrane-spanning α-helical regions. Within the membrane, dimers of band 3 form globular structures containing internal channels through which ions are able to travel across the lipid bilayer.

Because of their amphipathic character, transmembrane proteins have proved difficult to crystallize, as required for three-dimensional structural analysis by X-ray diffraction. The first transmembrane protein to be analyzed by X-ray crystallography was the photosynthetic reaction center of the bacterium *Rhodopseudomonas viridis*, whose structure was reported in 1985 (Figure 12.7). The reaction center contains three transmembrane proteins, designated L, M, and H (light, medium, and heavy) according to their apparent sizes indicated by gel electrophoresis. The L and M subunits each have five membrane-spanning α helices. The H subunit has only a single transmembrane α helix, with the bulk of the polypeptide chain on the cytosolic side of the membrane. The fourth subunit of the reaction center is a cytochrome, which is a peripheral membrane protein bound to the complex by protein-protein interactions.

Although most transmembrane proteins span the membrane by α-helical regions, this is not always the case. A well-characterized exception is provided by the **porins**—a class of proteins that form channels in the outer membranes of some bacteria. Many bacteria, including *E. coli*, have a dual membrane system in which the plasma membrane (or inner membrane) is surrounded by the cell wall and a distinct outer membrane (Figure 12.8). In contrast to the plasma membrane, the outer membrane is highly permeable to ions and small polar molecules (in the case of *E. coli*, with molecular weights up to 600). This permeability results from the porins, which form open aqueous channels through the lipid bilayer. As discussed in Chapter 10, proteins related to the bacterial porins are also found in the outer membranes of mitochondria and chloroplasts.

Structural analysis has indicated that the porins do not contain hydrophobic α-helical regions. Instead, they cross the membrane as β barrels, in which 16 β sheets fold up into a barrel-like structure enclosing an aqueous pore (Figure 12.9). The side chains of polar amino acids line the pore, whereas side chains of hydrophobic amino acids interact with the interior of the membrane. The porin monomers associate to form stable trimers,

Figure 12.7
A bacterial photosynthetic reaction center   The reaction center consists of three transmembrane proteins, designated L (red), M (yellow), and H (green). The L and M subunits each have five transmembrane α helices, whereas the H subunit has only one. The fourth subunit of the reaction center is a cytochrome (white), which is a peripheral membrane protein.

Figure 12.9
Structure of a porin monomer   Each monomer is a β barrel consisting of 16 antiparallel β strands (arrows). The top end of the molecule faces the external medium. (From H. Nikaido, 1994. *J. Biol. Chem.* 269: 3905.)

Figure 12.8
Bacterial outer membranes   The plasma membrane of some bacteria is surrounded by a cell wall and a distinct outer membrane. The outer membrane contains porins, which form open aqueous channels allowing the free passage of ions and small molecules.

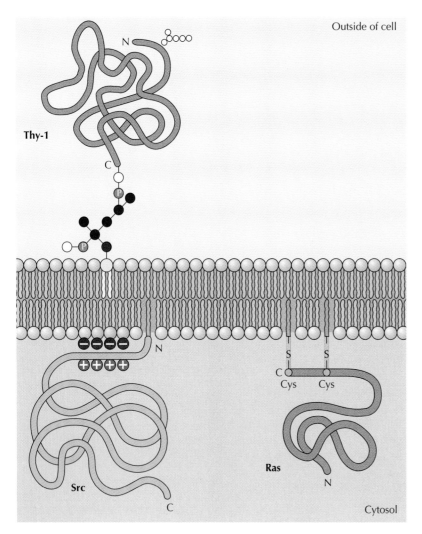

Figure 12.10
Examples of proteins anchored in the plasma membrane by lipids and glycolipids   Some proteins (e.g., the lymphocyte protein Thy-1) are anchored in the outer leaflet of the plasma membrane by GPI anchors added to their C terminus in the endoplasmic reticulum. These proteins are glycosylated and exposed on the cell surface. Other proteins are anchored in the inner leaflet of the plasma membrane following their translation on free cytosolic ribosomes. The Ras protein illustrated is anchored by a prenyl group attached to the side chain of a C-terminal cysteine and by a palmitoyl group attached to a cysteine located five amino acids upstream. The Src protein is anchored by a myristoyl group attached to its N terminus. A positively charged region of Src also plays a role in membrane association, perhaps by interacting with the negatively charged head groups of phosphatidylserine. The structures of these lipid and glycolipid groups are illustrated in Figures 7.29 through 7.32.

each of which contains three open channels through which polar molecules can diffuse across the membrane.

In contrast to transmembrane proteins, a variety of proteins (many of which behave as integral membrane proteins) are anchored in the plasma membrane by covalently attached lipids or glycolipids (Figure 12.10). Members of one class of these proteins are inserted into the outer leaflet of the plasma membrane by **glycosylphosphatidylinositol (GPI) anchors**. GPI anchors are added to certain proteins that have been transferred into the endoplasmic reticulum and are anchored in the membrane by a C-terminal transmembrane region (see Figure 9.16). The transmembrane region is cleaved as the GPI anchor is added, so these proteins remain attached to the membrane only by the glycolipid. Since the polypeptide chains of GPI-anchored proteins are transferred into the endoplasmic reticulum, they are glycosylated and exposed on the surface of the cell following transport to the plasma membrane.

Other proteins are anchored in the inner leaflet of the plasma membrane by covalently attached lipids. Rather than being processed through the secretory pathway, these proteins are synthesized on free cytosolic ribosomes and then modified by the addition of lipids. These modifications include the addition of myristic acid (a 14-carbon fatty acid) to the amino terminus of the polypeptide chain, the addition of palmitic acid (16 carbons) to the side chains of cysteine residues, and the addition of prenyl groups (15 or 20 carbons) to the side chains of carboxy-terminal cysteine residues (see Figures 7.29, 7.30, and 7.31). In some cases, these proteins (many of which behave as peripheral membrane proteins) are targeted to the plasma membrane by positively charged regions of the polypeptide chain as well as by the attached lipids. These positively charged protein domains may interact with the negatively charged head groups of phosphatidylserine on the cytosolic face of the plasma membrane. It is noteworthy that many of the proteins anchored in the inner leaflet of the plasma membrane (including the Src and Ras proteins illustrated in Figure 12.10) play important roles in the transmission of signals from cell surface receptors to intracellular targets, as discussed in the next chapter.

Mobility of Membrane Proteins

Membrane proteins and phospholipids are unable to move back and forth between the inner and outer leaflets of the membrane at an appreciable rate. However, because they are inserted into a fluid lipid bilayer, both proteins and lipids are able to diffuse laterally through the membrane. This lateral movement was first shown directly in an experiment reported by Larry

Frye and Michael Edidin in 1970, which provided support for the fluid mosaic model. Frye and Edidin fused human and mouse cells in culture to produce human-mouse cell hybrids (Figure 12.11). They then analyzed the distribution of proteins in the membranes of these hybrid cells using antibodies that specifically recognize proteins of human and mouse origin. These antibodies were labeled with different fluorescent dyes, so the human and mouse proteins could be distinguished by fluorescence microscopy. Immediately after fusion, human and mouse proteins were localized to different halves of the hybrid cells. However, after a brief period of incubation at 37°C, the human and mouse proteins were completely intermixed over the cell surface, indicating that they moved freely through the plasma membrane.

However, not all proteins are able to diffuse freely through the membrane. In some cases, the mobility of membrane proteins is restricted by their association with the cytoskeleton. For example, a fraction of band 3 in the red blood cell membrane is immobilized as a result of its association with ankyrin and spectrin. In other cases, the mobility of membrane proteins may be restricted by their associations with other membrane proteins, with proteins on the surface of adjacent cells, or with the extracellular matrix.

In contrast to blood cells, epithelial cells are polarized when they are organized into tissues, with different parts of the cell responsible for performing distinct functions. Consequently, the plasma membranes of many epithelial cells are divided into distinct **apical** and **basolateral domains** that differ in function and protein composition (Figure 12.12). For example, epithelial cells of the small intestine function to absorb nutrients from the digestive tract. The apical surface of these cells, which faces the intestinal lumen, is therefore covered by microvilli and specialized for nutrient absorption. The basolateral surface, which faces underlying connective tissue and the blood supply, is specialized to mediate the transfer of absorbed nutrients into the circulation. In order to maintain these distinct functions,

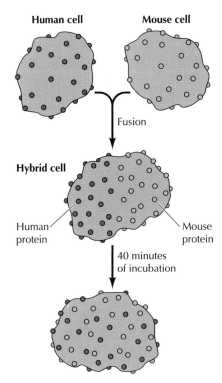

Figure 12.11
Mobility of membrane proteins
Human and mouse cells were fused to produce hybrid cells. The distribution of cell surface proteins was then analyzed using anti-human and anti-mouse antibodies labeled with different fluorescent dyes (red and green, respectively). The human and mouse proteins were detected in different halves of the hybrid cells immediately after fusion but had intermingled over the cell surface following 40 minutes of incubation.

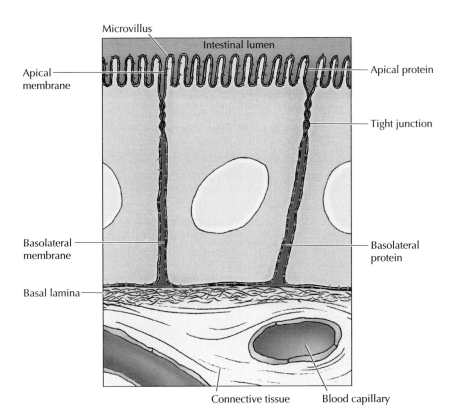

Figure 12.12
A polarized intestinal epithelial cell
The apical surface of the cell contains microvilli and is specialized for absorption of nutrients from the intestinal lumen. The basolateral surface is specialized for the transfer of absorbed nutrients to the underlying connective tissue, which contains blood capillaries. Tight junctions separate the apical and basolateral domains of the plasma membrane. Membrane proteins are free to diffuse within each domain but are not able to cross from one domain to the other.

Figure 12.13
The glycocalyx An electron micrograph of intestinal epithelium illustrating the glycocalyx (arrows). (Don Fawcett/Visuals Unlimited.)

1 μm

the mobility of plasma membrane proteins must be restricted to the appropriate domains of the cell surface. At least part of the mechanism by which this occurs involves the formation of tight junctions (which are discussed later in this chapter) between adjacent cells of the epithelium. These junctions not only seal the space between cells but also serve as barriers to the movement of membrane lipids and proteins. As a result, proteins are able to diffuse within either the apical or basolateral domains of the plasma membrane but are not able to cross from one domain to the other.

The Glycocalyx

As already discussed, the extracellular portions of plasma membrane proteins are generally glycosylated. Likewise, the carbohydrate portions of glycolipids are exposed on the outer face of the plasma membrane. Consequently, the surface of the cell is covered by a carbohydrate coat, known as the **glycocalyx**, formed by the oligosaccharides of glycolipids and transmembrane glycoproteins (Figure 12.13).

Part of the role of the glycocalyx is to protect the cell surface. In addition, the oligosaccharides of the glycocalyx serve as markers for a variety of cell-cell interactions. A well-studied example of these interactions is the adhesion of white blood cells (leukocytes) to the endothelial cells that line blood vessels—a process that allows the leukocytes to leave the circulatory system and mediate the inflammatory response in injured tissues. The initial step in adhesion between leukocytes and endothelial cells is mediated by a family of transmembrane proteins called **selectins**, which recognize specific carbohydrates on the cell surface (Figure 12.14). Two members of the selectin family (E-selectin and P-selectin), expressed by endothelial cells and platelets, bind to specific oligosaccharides expressed on the surface of leukocytes. A different selectin (L-selectin) is expressed by leukocytes and recognizes an oligosaccharide on the surface of endothelial cells. The oligosaccharides exposed on the cell surface thus provide a set of markers that help identify the distinct cell types of multicellular organisms.

Transport of Small Molecules

The internal composition of the cell is maintained because the plasma membrane is selectively permeable to small molecules. Most biological molecules are unable to diffuse through the phospholipid bilayer, so the plasma membrane forms a barrier that blocks the free exchange of molecules between the cytoplasm and the external environment of the cell. Specific transport proteins (carrier proteins and channel proteins) then mediate

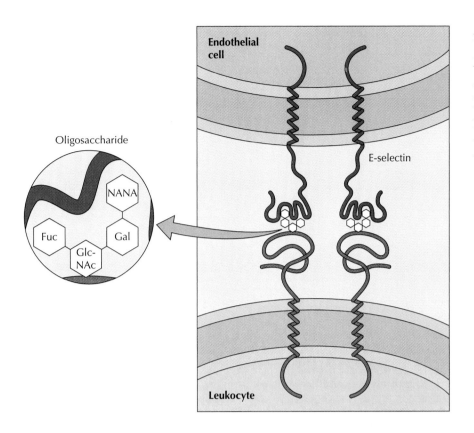

Figure 12.14
Binding of selectins to oligosaccharides   E-selectin is a transmembrane protein expressed by endothelial cells that binds to an oligosaccharide expressed on the surface of leukocytes. The oligosaccharide recognized by E-selectin contains *N*-acetylglucosamine (GlcNAc), fucose (Fuc), galactose (Gal), and sialic acid (*N*-acetylneuraminic acid, NANA).

the selective passage of small molecules across the membrane, allowing the cell to control the composition of its cytoplasm.

Passive Diffusion

The simplest mechanism by which molecules can cross the plasma membrane is **passive diffusion**. During passive diffusion, a molecule simply dissolves in the phospholipid bilayer, diffuses across it, and then dissolves in the aqueous solution at the other side of the membrane. No membrane proteins are involved and the direction of transport is determined simply by the relative concentrations of the molecule inside and outside of the cell. The net flow of molecules is always down their concentration gradient—from a compartment with a high concentration to one with a lower concentration of the molecule.

Passive diffusion is thus a nonselective process by which any molecule able to dissolve in the phospholipid bilayer is able to cross the plasma membrane and equilibrate between the inside and outside of the cell. Importantly, only small, relatively hydrophobic molecules are able to diffuse across a phospholipid bilayer at significant rates (Figure 12.15). Thus, gases (such as O_2 and CO_2), hydrophobic molecules (such as benzene), and small polar but uncharged molecules (such as H_2O and ethanol) are able to diffuse across the plasma membrane. Other biological molecules, however, are unable to dissolve in the hydrophobic interior of the phospholipid bilayer. Consequently, larger uncharged polar molecules such as glucose are unable to cross the plasma membrane by passive diffusion, as are charged molecules of any size (including small ions such as H^+, Na^+, K^+, and Cl^-). The passage of these molecules across the membrane instead requires the activity of specific transport and channel proteins, which therefore control the traffic of most biological molecules into and out of the cell.

Figure 12.15
Permeability of phospholipid bilayers   Gases, hydrophobic molecules, and small polar uncharged molecules can diffuse through phospholipid bilayers. Larger polar molecules and charged molecules cannot.

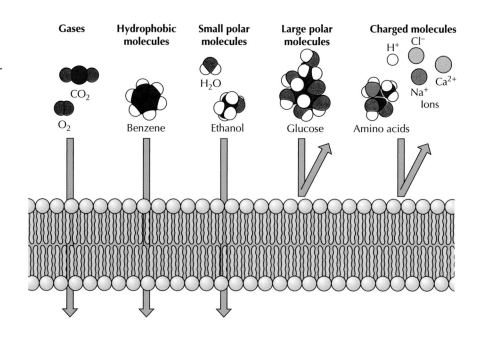

Facilitated Diffusion and Carrier Proteins

Facilitated diffusion, like passive diffusion, involves the movement of molecules in the direction determined by their relative concentrations inside and outside of the cell. No external source of energy is provided, so molecules travel across the membrane in the direction determined by their concentration gradients and, in the case of charged molecules, by the electric potential across the membrane. However, facilitated diffusion differs from passive diffusion in that the transported molecules do not dissolve in the phospholipid bilayer. Instead, their passage is mediated by proteins that enable the transported molecules to cross the membrane without directly interacting with its hydrophobic interior. Facilitated diffusion therefore allows polar and charged molecules, such as carbohydrates, amino acids, nucleosides, and ions, to cross the plasma membrane.

Two classes of proteins that mediate facilitated diffusion are generally distinguished: carrier proteins and channel proteins. **Carrier proteins** bind specific molecules to be transported on one side of the membrane. They then undergo conformational changes that allow the molecule to pass through the membrane and be released on the other side. In contrast, **channel proteins** (see the next section) form open pores through the membrane, allowing the free diffusion of any molecule of the appropriate size and charge.

Carrier proteins are responsible for the facilitated diffusion of sugars, amino acids, and nucleosides across the plasma membranes of most cells. The uptake of glucose, which serves as a primary source of metabolic energy, is one of the most important transport functions of the plasma membrane, and the glucose transporter provides a well-studied example of a carrier protein. The glucose transporter was initially identified as a 55-kd protein in human red blood cells, in which it represents approximately 5% of total membrane protein. Subsequent isolation and sequence analysis of a cDNA clone revealed that the glucose transporter has 12 α-helical transmembrane segments—a structure typical of many carrier proteins (Figure 12.16). These transmembrane α helices contain predominantly hydrophobic amino acids, but several also contain polar amino acid residues that are thought to form the glucose-binding site in the interior of the protein.

As with many membrane proteins, the three-dimensional structure of the glucose transporter is not known, so the molecular mechanism of glu-

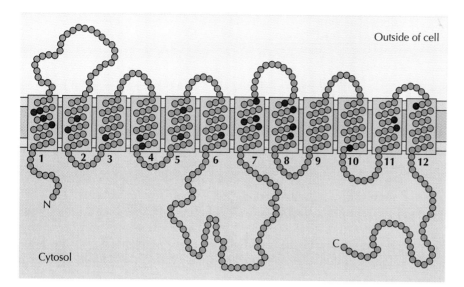

Figure 12.16
Structure of the glucose transporter
The glucose transporter has 12 transmembrane α helices. Polar amino acid residues located within the phospholipid bilayer are indicated as dark purple circles. (Adapted from G. I. Bell, C. F. Burant, J. Takeda and G. W. Gould, 1993. *J. Biol. Chem.* 268: 19161.)

cose transport remains an open question. However, kinetic studies indicate that the glucose transporter functions by alternating between two conformational states (Figure 12.17). In the first conformation, a glucose-binding site faces the outside of the cell. The binding of glucose to this exterior site induces a conformational change in the transporter, such that the glucose-binding site now faces the interior of the cell. Glucose can then be released into the cytosol, followed by the return of the transporter to its original conformation.

Most cells, including erythrocytes, are exposed to extracellular glucose concentrations that are higher than those inside the cell, so facilitated diffusion results in the net inward transport of glucose. Once glucose is taken up by these cells it is rapidly metabolized, so intracellular glucose concentrations remain low and glucose continues to be transported into the cell from the extracellular fluids. Because the conformational changes of the glucose transporter are reversible, however, glucose can be transported in the opposite direction simply by reversing the steps in Figure 12.17. Such

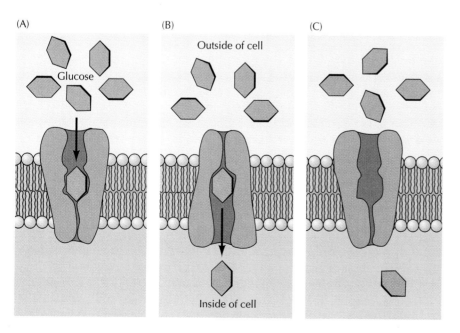

Figure 12.17
Model for the facilitated diffusion of glucose The glucose transporter alternates between two conformations in which a glucose-binding site is alternately exposed on the outside and the inside of the cell. In the first conformation shown (A), glucose binds to a site exposed on the outside of the plasma membrane. The transporter then undergoes a conformational change such that the glucose-binding site faces the inside of the cell and glucose is released into the cytosol (B). The transporter then returns to its original conformation (C).

Figure 12.18
Model of an ion channel In the closed conformation, the flow of ions is blocked by a gate. Opening of the gate allows ions to flow rapidly through the channel. The channel contains a narrow pore that restricts passage to ions of the appropriate size and charge.

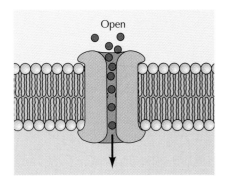

TABLE 12.1	Extracellular and Intracellular Ion Concentrations	
	Concentration (mM)	
Ion	**Intracellular**	**Extracellular**
Squid axon		
K⁺	400	20
Na⁺	50	440
Cl⁻	40–150	560
Ca²⁺	0.0001	10
Mammalian cell		
K⁺	140	5
Na⁺	5–15	145
Cl⁻	4	110
Ca²⁺	0.0001	2.5–5

reverse flow occurs, for example, in liver cells, in which glucose is synthesized and released into the circulation.

Ion Channels

In contrast to carrier proteins, channel proteins simply form open pores in the membrane, allowing small molecules of the appropriate size and charge to pass freely through the lipid bilayer. One group of channel proteins, discussed earlier, is the porins, which permit the free passage of ions and small polar molecules through the outer membranes of bacteria (see Figure 12.8). Channel proteins also permit the passage of molecules between cells connected at gap junctions, which are discussed later in the chapter. The plasma membranes of many cells also contain water channel proteins (aquaporins), through which water molecules are able to cross the membrane much more rapidly than they can diffuse through the phospholipid bilayer. The best-characterized channel proteins, however, are the **ion channels**, which mediate the passage of ions across plasma membranes. Although ion channels are present in the membranes of all cells, they have been especially well studied in nerve and muscle, where their regulated opening and closing is responsible for the transmission of electric signals.

Three properties of ion channels are central to their function (Figure 12.18). First, transport through channels is extremely rapid. More than a million ions per second flow through open channels—a flow rate approximately a thousand times greater than the rate of transport by carrier proteins. Second, ion channels are highly selective because narrow pores in the channel restrict passage to ions of the appropriate size and charge. Thus, specific channel proteins allow the passage of Na⁺, K⁺, Ca²⁺, and Cl⁻ across the membrane. Third, most ion channels are not permanently open. Instead, the opening of ion channels is regulated by "gates" that transiently open in response to specific stimuli. Some channels (called **ligand-gated channels**) open in response to the binding of neurotransmitters or other signaling molecules; others (**voltage-gated channels**) open in response to changes in electric potential across the plasma membrane.

The fundamental role of ion channels in the transmission of electric impulses was elucidated through a series of elegant experiments reported by Alan Hodgkin and Andrew Huxley in 1952. These investigators used the giant nerve cells of the squid as a model. The axons of these giant neurons have a diameter of about 1 mm, making it possible to insert electrodes and measure the changes in membrane potential that take place during the transmission of nerve impulses. Using this approach, Hodgkin and Huxley demonstrated that these changes in membrane potential result from the regulated opening and closing of Na⁺ and K⁺ channels in the plasma membrane. It subsequently became possible to study the activity of individual ion channels, using the **patch clamp technique** developed by Erwin Neher and Bert Sakmann in 1976 (Figure 12.19). In this method, a micropipette with a tip diameter of about 1 μm is used to isolate a small patch of membrane, allowing the flow of ions through a single channel to be analyzed and greatly increasing the precision with which the activities of ion channels can be studied.

The flow of ions through membrane channels is dependent on the establishment of ion gradients across the plasma membrane. All cells, including nerve and muscle, contain ion pumps (discussed in the next section) that use energy derived from ATP hydrolysis to actively transport ions across the plasma membrane. As a result, the ionic composition of the cytoplasm is substantially different from that of extracellular fluids (Table 12.1). For example, Na⁺ is actively pumped out of cells while K⁺ is pumped in. In the squid axon, therefore, the concentration of Na$^+$ is about 10 times higher in extracellular fluids than inside the cell, whereas the concentration of K$^+$ is approximately 20 times higher in the cytosol than in the surrounding medium.

Because ions are electrically charged, their transport results in the establishment of an electric gradient across the plasma membrane. With resting squid axons there is an electric potential of about 60 mV across the plasma membrane, with the inside of the cell negative with respect to the outside (Figure 12.20). This electric potential arises both from ion pumps and from the flow of ions through channels that are open in the resting cell plasma membrane. The plasma membrane of resting squid axons contains open K$^+$ channels, so it is more permeable to K$^+$ than to Na$^+$ or other ions. Consequently, the flow of K$^+$ makes the largest contribution to the resting membrane potential.

As discussed in Chapter 10, the flow of ions across a membrane is driven by both the concentration and voltage components of an electrochemical gradient. For example, the 20-fold higher concentration of K$^+$ inside the squid axon as compared to the extracellular fluid drives the flow of K$^+$ out of the cell. However, because K$^+$ is positively charged, this efflux of K$^+$ from the cell generates an electric potential across the membrane, with the inside of the cell becoming negatively charged. This membrane potential opposes the continuing flow of K$^+$ out of the cell, and the system approaches the equilibrium state, in which the membrane potential balances the K$^+$ concentration gradient.

Quantitatively, the relationship between ion concentration and membrane potential is given by the **Nernst equation**:

$$V = \frac{RT}{zF} \ln \frac{C_o}{C_i}$$

where V is the equilibrium potential in volts, R is the gas constant, T is the absolute temperature, z is the charge of the ion, F is Faraday's constant, and C_o and C_i are the concentrations of the ion outside and inside of the cell, respectively. An equilibrium potential exists separately for each ion, and the membrane potential is determined by the flow of all the ions that cross the plasma membrane. However, because resting squid axons are more permeable to K$^+$ than to Na$^+$ or other ions (including Cl$^-$), the resting mem-

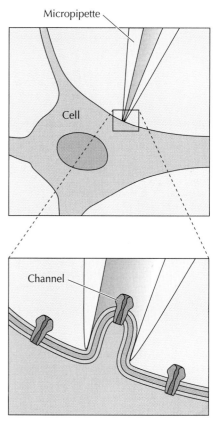

Figure 12.19
The patch clamp technique A small patch of membrane is isolated in the tip of a micropipette. Stimuli can then be applied from within the pipette, allowing the behavior of the trapped channel to be measured. (Adapted from E. Neher and B. Sakmann, 1992. *Sci. Am.* 266(3): 44.)

Figure 12.20
Ion gradients and resting membrane potential of the giant squid axon Only the concentrations of Na$^+$ and K$^+$ are shown, because these are the ions that function in the transmission of nerve impulses. Na$^+$ is pumped out of the cell while K$^+$ is pumped in, so the concentration of Na$^+$ is higher outside than inside of the axon, whereas the concentration of K$^+$ is higher inside than out. The resting membrane is more permeable to K$^+$ than to Na$^+$ or other ions because it contains open K$^+$ channels. The flow of K$^+$ through these channels makes the major contribution to the resting membrane potential of −60 mV, which is therefore close to the K$^+$ equilibrium potential.

(A)

Figure 12.21
Membrane potential and ion channels during an action potential
(A) Changes in membrane potential at one point on a squid giant axon following a stimulus. E_{Na} and E_K are the equilibrium potentials for Na^+ and K^+, respectively. (B) The membrane potential first increases as voltage-gated Na^+ channels open. The membrane potential then falls below its resting value as the Na^+ channels are inactivated and voltage-gated K^+ channels open. The voltage-gated K^+ channels are then inactivated, and the membrane potential returns to its resting value.

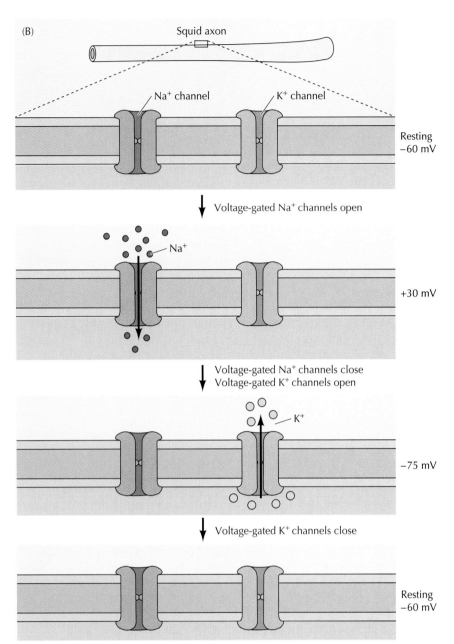

brane potential (–60 mV) is close to the equilibrium potential determined by the intracellular and extracellular K^+ concentrations (–75 mV).

As nerve impulses (**action potentials**) travel along axons, the membrane depolarizes (Figure 12.21). The membrane potential changes from –60 mV to approximately +30 mV in less than a millisecond, after which it becomes negative again and returns to its resting value. These changes result from the rapid sequential opening and closing of voltage-gated Na^+ and K^+ channels. Relatively small initial changes in membrane potential (from –60 to about –40 mV) lead to the rapid opening of Na^+ channels. This allows Na^+ to flow into the cell, driven by both its concentration gradient and the membrane potential. The sudden entry of Na^+ leads to a large change in membrane potential, which increases to nearly +30 mV, approaching the Na^+ equilibrium potential of approximately +50 mV. At this time, the Na^+ channels are inactivated and voltage-gated K^+ channels open, substantially

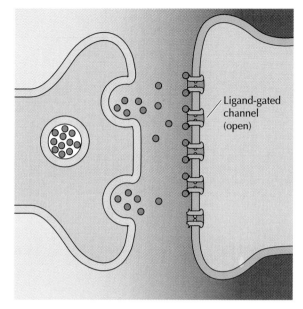

Figure 12.22
Signaling by neurotransmitter release at a synapse The arrival of a nerve impulse at the terminus of the neuron signals the fusion of synaptic vesicles with the plasma membrane, resulting in the release of neurotransmitter from the presynaptic cell into the synaptic cleft. The neurotransmitter binds to receptors and opens ligand-gated ion channels in the target cell plasma membrane.

increasing the permeability of the membrane to K^+. K^+ then flows rapidly out of the cell, driven by both the membrane potential and the K^+ concentration gradient, leading to a rapid decrease in membrane potential to about –75 mV (the K^+ equilibrium potential). The voltage-gated K^+ channels are then inactivated and the membrane potential returns to its resting level of –60 mV, determined by the flow of K^+ and other ions through the channels that remain open in unstimulated cells.

Depolarization of adjacent regions of the plasma membrane allows action potentials to travel down the length of nerve cell axons as electric signals, resulting in the rapid transmission of nerve impulses over long distances. For example, the axons of human motor neurons can be more than a meter long. The arrival of action potentials at the terminus of most neurons then signals the release of neurotransmitters, such as acetylcholine, which carry signals between cells at a synapse (Figure 12.22). Neurotransmitters released from presynaptic cells bind to receptors on the membranes of postsynaptic cells, where they act to open ligand-gated ion channels. One of the best-characterized of these channels is the acetylcholine receptor of muscle cells. Binding of acetylcholine opens a channel that is permeable to both Na^+ and K^+. This permits the rapid influx of Na^+, which depolarizes the muscle cell membrane and triggers an action potential. The action potential then results in the opening of voltage-gated Ca^{2+} channels, leading to the increase in intracellular Ca^{2+} that signals contraction (see Figure 11.25).

The acetylcholine receptor, initially isolated from the electric organ of *Torpedo* rays in the 1970s, is the prototype of ligand-gated channels. The receptor consists of five subunits arranged as a cylinder in the membrane (Figure 12.23). In its closed state, the channel pore is thought to be blocked by the side chains of hydrophobic amino acids. The binding of acetylcholine induces a conformational change in the receptor such that these hydrophobic side chains shift out of the channel, opening a pore that

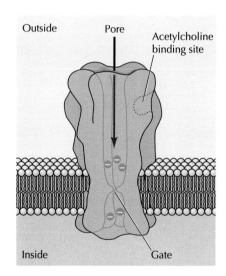

Figure 12.23
Model of the acetylcholine receptor The receptor consists of five subunits arranged around a central pore. The binding of acetylcholine to a site in the extracellular region of the receptor induces allosteric changes that open the channel gate. The channel is lined by negatively charged amino acids that prevent the flow of negatively charged ions. (Adapted from N. Unwin, 1993. *Cell* 72/*Neuron* 10 (Suppl.): 31.)

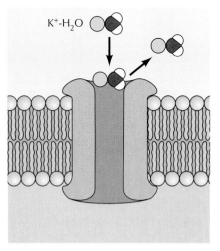

Figure 12.24
Ion selectivity of Na+ channels
A narrow pore permits the passage of Na+ bound to a single water molecule but interferes with the passage of K+ or larger ions.

allows the passage of positively charged ions, including Na+ and K+. However, the channel remains impermeable to negatively charged ions, such as Cl−, because it is lined by negatively charged amino acids.

A greater degree of ion selectivity is displayed by the voltage-gated Na+ and K+ channels. Na+ channels are more than ten times more permeable to Na+ than to K+, whereas K+ channels are more than a thousand times more permeable to K+ than to Na+. The selectivity of the Na+ channel can be explained, at least in part, on the basis of a narrow pore that acts as a size filter. The ionic radius of Na+ (0.95 Å) is smaller than that of K+ (1.33 Å), and it is thought that the Na+ channel pore is narrow enough to interfere with the passage of K+ or larger ions (Figure 12.24).

K+ channels also have narrow pores, which prevent the passage of larger ions. However, since Na+ has a smaller ionic radius, this does not account for the selective permeability of these channels to K+. Selectivity of the K+ channel is based on a different mechanism, which was elucidated with the determination of the three-dimensional structure of a K+ channel by X-ray crystallography in 1998 (Figure 12.25). The channel pore contains a narrow selectivity filter that is lined with carbonyl oxygen (C=O) atoms from the polypeptide backbone. When a K+ ion enters the selectivity filter, interactions with these carbonyl oxygens displace the water molecules to which K+ is bound, allowing dehydrated K+ to pass through the pore. In contrast, a dehydrated Na+ is too small to interact with these carbonyl oxygens in the selectivity filter, which is held rigidly open. Consequently, Na+ remains

Figure 12.25
Selectivity of K+ channels The K+ channel contains a narrow selectivity filter lined with carbonyl oxygen (C=O) atoms. The pore is just wide enough to allow the passage of dehydrated K+ from which all associated water molecules have been displaced as a result of interactions between K+ and these carbonyl oxygens. Na+ is too small to interact with the carbonyl oxygens of the selectivity filter, so it remains bound to water in a complex that is too large to pass through the channel pore.

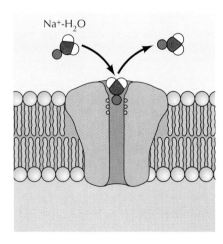

bound to water molecules in a hydrated complex that is too large to pass through the channel.

Voltage-gated Na^+, K^+, and Ca^{2+} channels all belong to a large family of related proteins (Figure 12.26). For example, the genome sequence of *C. elegans* has revealed nearly 200 genes encoding ion channels, which presumably are needed to play diverse roles in cell signaling. K^+ channels consist of four identical subunits, each containing either two or six transmembrane α helices. Na^+ and Ca^{2+} channels consist of a single polypeptide chain, but each polypeptide contains four repeated domains that correspond to the K^+ channel subunits. Voltage gating is mediated by one of the transmembrane α helices, which contains multiple positively charged amino acids. Membrane depolarization induces the movement of these positive charges toward the outside of the cell, shifting the position of this transmembrane segment and opening the channel. Rapid inactivation of Na^+ and K^+ channels during the propagation of action potentials is then mediated by cytoplasmic portions of the polypeptide chain, which bind to the cytoplasmic mouth of the channel pore and prevent further ion flow (Figure 12.27).

Figure 12.26
Structures of voltage-gated cation channels The K^+, Na^+, and Ca^{2+} channels belong to a family of related proteins. The K^+ channel is formed from the association of four identical subunits, one of which is shown. The Na^+ channel consists of a single polypeptide chain containing four repeated domains, each of which is similar to one K^+ channel subunit. The Ca^{2+} channel is similar to the Na^+ channel. Each subunit or domain contains six membrane-spanning α helices. The α helix designated 4 contains multiple positively charged amino acids and acts as the voltage sensor that mediates channel opening in response to changes in membrane potential.

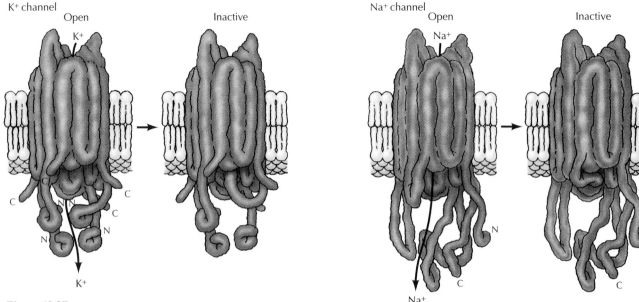

K⁺ channel

Open   Inactive

Na⁺ channel

Open   Inactive

Figure 12.27
Inactivation of K⁺ and Na⁺ channels
Following voltage-gated opening, the K⁺ and Na⁺ channels are rapidly inactivated by the binding of cytoplasmic portions of the polypeptide chains to the pore. For the K⁺ channel, inactivation is mediated by a ball-and-chain mechanism with the ball corresponding to the amino terminus of the polypeptide chain. For the Na⁺ channel, inactivation is mediated by the intracellular loop connecting domains III and IV.

A wide variety of ion channels (including Ca^{2+} and Cl^- channels) respond to different neurotransmitters or open and close with different kinetics following membrane depolarization. The concerted actions of these multiple channels are responsible for the complexities of signaling in the nervous system. Moreover, as discussed in the next chapter, the roles of ion channels are not restricted to the electrically excitable cells of nerve and muscle; they also play critical roles in signaling in other cell types. The regulated opening and closing of ion channels thus provides cells with a sensitive and versatile mechanism for responding to a variety of environmental stimuli.

Active Transport Driven by ATP Hydrolysis

The net flow of molecules by facilitated diffusion, through either carrier proteins or channel proteins, is always energetically downhill in the direction determined by electrochemical gradients across the membrane. In many cases, however, the cell must transport molecules against their concentration gradients. In **active transport**, energy provided by another coupled reaction (such as the hydrolysis of ATP) is used to drive the uphill transport of molecules in the energetically unfavorable direction.

The **ion pumps** responsible for maintaining gradients of ions across the plasma membrane provide important examples of active transport driven directly by ATP hydrolysis. As discussed earlier (see Table 12.1), the concentration of Na⁺ is approximately ten times higher outside than inside of cells, whereas the concentration of K⁺ is higher inside than out. These ion gradients are maintained by the **Na⁺-K⁺ pump** (also called the **Na⁺-K⁺ ATPase**), which uses energy derived from ATP hydrolysis to transport Na⁺ and K⁺ against their electrochemical gradients. This process is a result of ATP-driven conformational changes in the pump (Figure 12.28). First, Na⁺ ions bind to high-affinity sites inside the cell. This binding stimulates the hydrolysis of ATP and phosphorylation of the pump, inducing a conformational change that exposes the Na⁺-binding sites to the outside of the cell and reduces their affinity for Na⁺. Consequently, the bound Na⁺ is released into the extracellular fluids. At the same time, high-affinity K⁺-binding sites are exposed on the cell surface. The binding of extracellular K⁺ to these sites then stimulates hydrolysis of the phosphate group bound to the pump, which induces a second conformational change,

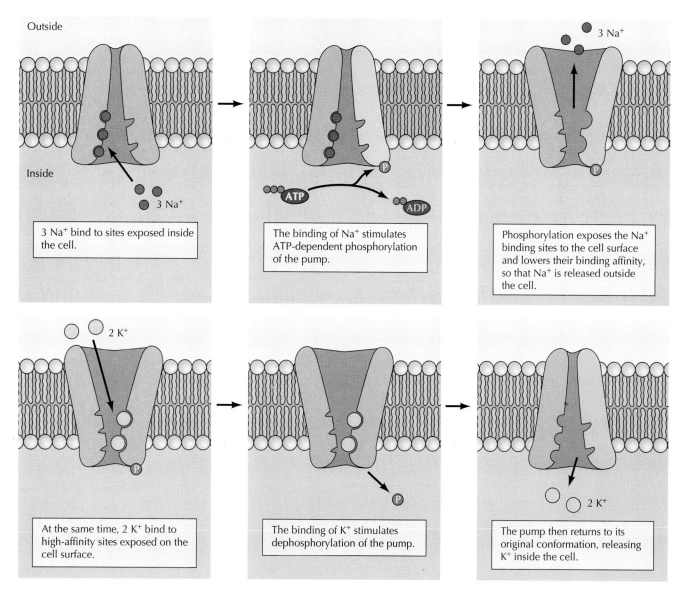

Outside

Inside

3 Na⁺ bind to sites exposed inside the cell.

The binding of Na⁺ stimulates ATP-dependent phosphorylation of the pump.

Phosphorylation exposes the Na⁺ binding sites to the cell surface and lowers their binding affinity, so that Na⁺ is released outside the cell.

At the same time, 2 K⁺ bind to high-affinity sites exposed on the cell surface.

The binding of K⁺ stimulates dephosphorylation of the pump.

The pump then returns to its original conformation, releasing K⁺ inside the cell.

Figure 12.28
Model for operation of the Na⁺-K⁺ pump

exposing the K^+-binding sites to the cytosol and lowering their binding affinity so that K^+ is released inside the cell. The pump has three binding sites for Na^+ and two for K^+, so each cycle transports three Na^+ and two K^+ across the plasma membrane at the expense of one molecule of ATP.

The importance of the Na^+-K^+ pump is indicated by the fact that it is estimated to consume nearly 25% of the ATP utilized by many animal cells. One critical role of the Na^+ and K^+ gradients established by the pump is the propagation of electric signals in nerve and muscle. As will be discussed shortly, the Na^+ gradient established by the pump is also utilized to drive the active transport of a variety of other molecules. Yet another important role of the Na^+-K^+ pump in most animal cells is to maintain osmotic balance and cell volume. The cytoplasm contains a high concentration of organic molecules, including macromolecules, amino acids, sugars, and nucleotides. In the absence of a counterbalance, this would drive the inward flow of water by osmosis, which if unchecked would result in swelling and eventual bursting of the cell. The required counterbalance is provided by the ion gradients established by the Na^+-K^+ pump (Figure 12.29). In particular, the pump establishes a higher concentration of Na^+ outside than inside the cell. As already discussed, the flow of K^+ through

Figure 12.29
Ion gradients across the plasma membrane of a typical mammalian cell
The concentrations of Na^+ and Cl^- are higher outside than inside the cell, whereas the concentration of K^+ is higher inside than out. The low concentrations of Na^+ and Cl^- balance the high intracellular concentration of organic compounds, equalizing the osmotic pressure and preventing the net influx of water.

open channels further establishes an electric potential across the plasma membrane. This membrane potential in turn drives Cl^- out of the cell, so the concentration of Cl^- (like that of Na^+) is about ten times higher in extracellular fluids than in the cytoplasm. These differences in ion concentrations balance the high concentrations of organic molecules inside cells, equalizing the osmotic pressure and preventing the net influx of water.

The active transport of Ca^{2+} across the plasma membrane is driven by a Ca^{2+} pump that is structurally related to the Na^+-K^+ pump and is similarly powered by ATP hydrolysis. The Ca^{2+} pump transports Ca^{2+} out of the cell, so intracellular Ca^{2+} concentrations are extremely low: approximately 0.1 μM, in comparison to extracellular concentrations of about 1 mM. This low intracellular concentration of Ca^{2+} makes the cell sensitive to small increases in intracellular Ca^{2+} levels. Such transient increases in intracellular Ca^{2+} play important roles in cell signaling, as noted already with respect to muscle contraction (see Figure 11.25) and discussed further in the next chapter.

Similar ion pumps in the plasma membranes of bacteria, yeasts, and plant cells are responsible for the active transport of H^+ out of the cell. In addition, H^+ is actively pumped out of cells lining the stomach, resulting in the acidity of gastric fluids. Structurally distinct pumps are responsible for the active transport of H^+ into lysosomes and endosomes (see Figure 9.35). Yet a third type of H^+ pump is exemplified by the ATP synthases of mitochondria and chloroplasts: In these cases the pumps can be viewed as operating in reverse, with the movement of ions down the electrochemical gradient being used to drive ATP synthesis.

The largest family of membrane transporters consists of the **ABC transporters**, so called because they are characterized by highly conserved ATP-binding domains or *ATP-binding cassettes* (Figure 12.30). More than a hundred family members have been identified in both prokaryotic and eukaryotic cells. In bacteria, ABC transporters utilize energy derived from ATP hydrolysis to transport a wide range of molecules, including ions, sugars, and amino acids. In eukaryotic cells, the first ABC transporter was discovered as the product of a gene (called the multidrug resistance, or *mdr*, gene) that makes cancer cells resistant to a variety of drugs used in chemotherapy. Two MDR transporters have now been identified. They are normally expressed in a variety of cells, where they function to remove potentially toxic foreign compounds. For example, expression of an MDR

Figure 12.30
Structure of an ABC transporter
The basic structural unit of ABC transporters consists of six transmembrane domains followed by an ATP-binding cassette. In plasma membrane transporters (such as the MDR transporter illustrated), two of these units are fused so that the transporter contains 12 transmembrane domains and two ATP-binding sites.

transporter in capillary endothelial cells of the brain appears to play an important role in protecting the brain from toxic chemicals. Unfortunately, the MDR transporters are frequently expressed at high levels in cancer cells where they recognize a variety of drugs and pump them out of cells, thereby making the cancer cells resistant to a broad spectrum of chemotherapeutic agents and posing a major obstacle to successful cancer treatment.

Another medically important member of the ABC transporter family is the gene responsible for cystic fibrosis. Although it is a member of the ABC family, the product of this gene (called the cystic fibrosis transmembrane conductance regulator, or CFTR) functions as a Cl^- channel in epithelial cells, and defective Cl^- transport is characteristic of the disease. The CFTR Cl^- channel is also unusual in that it appears to require both ATP hydrolysis and cAMP-dependent phosphorylation in order to open. The structural basis for the function of CFTR as a regulated ion channel remains to be elucidated by future research.

Active Transport Driven by Ion Gradients

The ion pumps and ABC transporters discussed in the previous section utilize energy derived directly from ATP hydrolysis to transport molecules against their electrochemical gradients. Other molecules are transported against their concentration gradients using energy derived not from ATP hydrolysis but from the coupled transport of a second molecule in the energetically favorable direction. The Na^+ gradient established by the Na^+-K^+ pump provides a source of energy that is frequently used to power the active transport of sugars, amino acids, and ions in mammalian cells. The H^+ gradients established by the H^+ pumps of bacteria, yeast, and plant cells play similar roles.

The epithelial cells lining the intestine provide a good example of active transport driven by the Na^+ gradient. These cells use active-transport systems in the apical domains of their plasma membranes to take up dietary sugars and amino acids from the lumen of the intestine. The uptake of glucose, for example, is carried out by a transporter that coordinately transports two Na^+ and one glucose into the cell (Figure 12.31). The flow of Na^+ down

Intestinal lumen

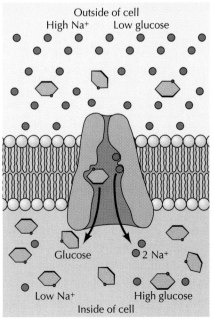

Figure 12.31
Active transport of glucose Active transport driven by the Na^+ gradient is responsible for the uptake of glucose from the intestinal lumen. The transporter coordinately binds and transports one glucose and two Na^+ into the cell. The transport of Na^+ in the energetically favorable direction drives the uptake of glucose against its concentration gradient.

Molecular Medicine

Cystic Fibrosis

The Disease

Cystic fibrosis is a recessive genetic disease affecting children and young adults. It is the most common lethal inherited disease of Caucasians, with approximately one in 2500 newborns affected, although it is rare in other races. The characteristic dysfunction in cystic fibrosis is the production of abnormally thick sticky mucus by several types of epithelial cells, including the cells lining the respiratory and gastrointestinal tracts. The primary clinical manifestation is respiratory disease resulting from obstruction of the pulmonary airways by thick plugs of mucus, followed by the development of recurrent bacterial infections. In most patients, the pancreas is also involved because the pancreatic ducts are blocked by mucus. Sweat glands also function abnormally, and the presence of excessive salt in sweat is diagnostic of cystic fibrosis.

Current management of the disease includes physical therapy to promote bronchial drainage, antibiotic administration, and pancreatic enzyme replacement. Although such treatment has extended the survival of affected individuals to about 30 years of age, cystic fibrosis is ultimately fatal, with lung disease being responsible for 95% of mortality.

Molecular and Cellular Basis

The hallmark of cystic fibrosis is defective Cl⁻ transport in affected epithelia, including sweat ducts and the cells lining the respiratory tract. In 1984, it was demonstrated that Cl⁻ channels fail to function normally in epithelial cells from cystic fibrosis patients. The molecular basis of the disease was then elucidated in 1989, with the isolation of the cystic fibrosis gene as a molecular clone. The sequence of the gene revealed that it encodes a protein (called CFTR for *cystic fibrosis transmembrane conductance regulator*) belonging to the ABC transporter family. A variety of subsequent studies then demonstrated that CFTR functions as a Cl⁻ channel and that the inherited mutations responsible for cystic fibrosis result directly in defective Cl⁻ transport.

Prevention and Treatment

As with other inherited diseases, isolation of the cystic fibrosis gene opens the possibility of genetic screening to identify individuals carrying mutant alleles. In some populations, the frequency of heterozygote carriers of mutant genes is as high as one in 25 individuals, suggesting the possibility of general population screening to identify couples at risk and provide genetic counseling. In addition, understanding the function of CFTR as a Cl⁻ channel has suggested new approaches to treatment. One possibility is the use of drugs that stimulate the opening of other Cl⁻ channels in affected epithelia. Alternatively, gene therapy provides the potential of replacing normal CFTR genes in the respiratory epithelium of cystic fibrosis patients.

The potential of gene therapy has been supported by experiments demonstrating that introduction of a normal CFTR gene into cultured cells of cystic fibrosis patients is sufficient to restore Cl⁻ channel function. The possible application of gene therapy to cystic fibrosis is also enhanced by the accessibility of the epithelial cells lining the airway to aerosol delivery systems. Studies with experimental animals have demonstrated that viral vectors can transmit CFTR cDNA to the respiratory epithelium, and the first human trial of gene therapy for cystic fibrosis was initiated in 1993. Trials to date have demonstrated that the CFTR can be safely delivered to and expressed in the bronchial epithelial cells of cystic fibrosis patients. So far, however, the efficiency of gene transfer has been low and expression of the transferred CFTR cDNA has persisted for less than a month. Thus, the principle of successful gene transfer has been established, but significant obstacles in developing an effective gene therapy protocol remain to be overcome.

References

Collins, F. S. 1992. Cystic fibrosis: Molecular biology and therapeutic implications. *Science* 256: 774–779.

Welsh, M. J. 1999. Gene transfer for cystic fibrosis. *J. Clin. Invest.* 104: 1165–1166.

ATP binding sites

Model of the cystic fibrosis transmembrane conductance regulator (CFTR).

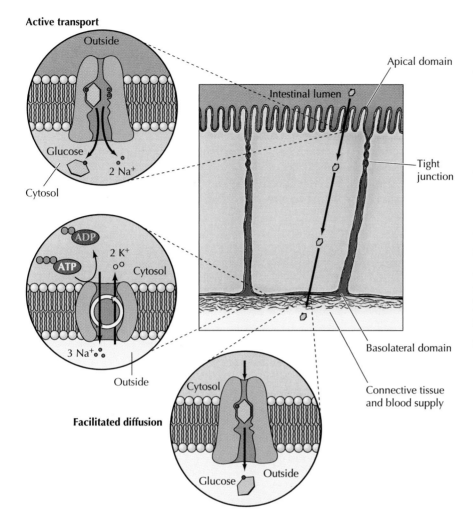

Active transport

Outside

Glucose

Cytosol

2 Na⁺

ADP

ATP

2 K⁺

Cytosol

3 Na⁺

Outside

Facilitated diffusion

Cytosol

Glucose

Outside

Intestinal lumen

Apical domain

Tight junction

Basolateral domain

Connective tissue and blood supply

Figure 12.32
Glucose transport by intestinal epithelial cells   A transporter in the apical domain of the plasma membrane is responsible for the active uptake of glucose (by cotransport with Na⁺) from the intestinal lumen. As a result, dietary glucose is absorbed and concentrated within intestinal epithelial cells. Glucose is then transferred from these cells to the underlying connective tissue and blood supply by facilitated diffusion, mediated by a transporter in the basolateral domain of the plasma membrane. The system is driven by the Na⁺-K⁺ pump, which is also found in the basolateral domain. Note that the uptake of glucose from the digestive tract and its transfer to the circulation is dependent on the restricted localization of glucose transporters mediating active transport and facilitated diffusion to the apical and basolateral domains of the plasma membrane, respectively.

its electrochemical gradient provides the energy required to take up dietary glucose and to accumulate high intracellular glucose concentrations. Glucose is then released into the underlying connective tissue (which contains blood capillaries) at the basolateral surface of the intestinal epithelium, where it is transported down its concentration gradient by facilitated diffusion (Figure 12.32). The uptake of glucose from the intestinal lumen and its release into the circulation thus provides a good example of the polarized function of epithelial cells, which results from the specific localization of active transport and facilitated diffusion carriers to the apical and basolateral domains of the plasma membrane, respectively.

The coordinate uptake of glucose and Na⁺ is an example of **symport**, the transport of two molecules in the same direction. In contrast, the facilitated diffusion of glucose is an example of **uniport**, the transport of only a single molecule. Active transport can also take place by **antiport**, in which two molecules are transported in opposite directions (Figure 12.33). For example, Ca²⁺ is exported from cells not only by the Ca²⁺ pump but also by an Na⁺-Ca²⁺ antiporter that transports Na⁺ into the cell and Ca²⁺ out. Another example is provided by the Na⁺-H⁺ exchange protein, which functions in the regulation of intracellular pH. The Na⁺-H⁺ antiporter couples the transport of Na⁺ into the cell with the export of H⁺, thereby removing excess H⁺ produced by metabolic reactions and preventing acidification of the cytoplasm.

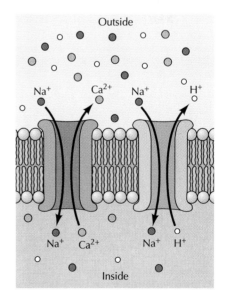

Outside

Na⁺   Ca²⁺   Na⁺   H⁺

Na⁺   Ca²⁺   Na⁺   H⁺

Inside

Figure 12.33
Examples of antiport   Ca²⁺ and H⁺ are exported from cells by antiporters, which couple their export to the energetically favorable import of Na⁺.

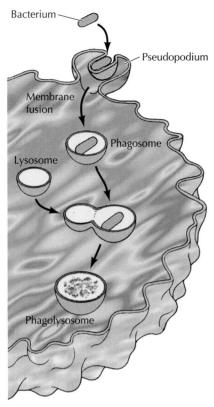

Figure 12.34
Phagocytosis Binding of a bacterium to the cell surface stimulates the extension of a pseudopodium, which eventually engulfs the bacterium. Fusion of the pseudopodium membranes then results in formation of a large intracellular vesicle (a phagosome). The phagosome fuses with lysosomes to form a phagolysosome, within which the ingested bacterium is digested.

Endocytosis

The carrier and channel proteins discussed in the preceding section transport small molecules through the phospholipid bilayer. Eukaryotic cells are also able to take up macromolecules and particles from the surrounding medium by a distinct process called **endocytosis**. In endocytosis, the material to be internalized is surrounded by an area of plasma membrane, which then buds off inside the cell to form a vesicle containing the ingested material. The term "endocytosis" was coined by Christian deDuve in 1963 to include both the ingestion of large particles (such as bacteria) and the uptake of fluids or macromolecules in small vesicles. The former of these activities is known as **phagocytosis** (cell eating) and the latter as **pinocytosis** (cell drinking).

Phagocytosis

During phagocytosis, cells engulf large particles such as bacteria, cell debris, or even intact cells (Figure 12.34). Binding of the particle to receptors on the surface of the phagocytic cell triggers the extension of pseudopodia—an actin-based movement of the cell surface, discussed in Chapter 11. The pseudopodia eventually surround the particle and their membranes fuse to form a large intracellular vesicle ($>0.25\ \mu$m in diameter) called a **phagosome**. The phagosomes then fuse with lysosomes, producing **phagolysosomes** in which the ingested material is digested by the action of lysosomal acid hydrolases (see Chapter 9). During maturation of the phagolysosome, some of the internalized membrane proteins are recycled to the plasma membrane, as discussed in the next section for receptor-mediated endocytosis.

The ingestion of large particles by phagocytosis plays distinct roles in different kinds of cells (Figure 12.35). Many amoebas use phagocytosis to capture food particles, such as bacteria or other protozoans. In multicellular animals, the major roles of phagocytosis are to provide a defense against invading microorganisms and to eliminate aged or damaged cells from the body. In mammals, phagocytosis is the function of primarily two types of white blood cells, macrophages and neutrophils, which are fre-

Figure 12.35
Examples of phagocytic cells (A) An amoeba engulfing another protist. (B) Macrophages ingesting red blood cells. False color has been added to the micrograph. (A, R. N. Band and H. S. Pankratz/Biological Photo Service; B, courtesy of Joel Swanson.)

(A)

(B)

(A)

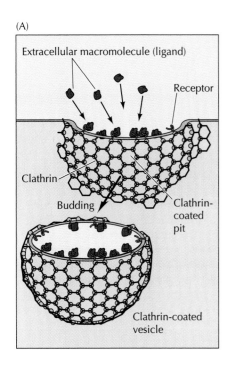

Extracellular macromolecule (ligand)

Receptor

Clathrin

Budding

Clathrin-coated pit

Clathrin-coated vesicle

(B)

Stage 1

Stage 2

Stage 3

Stage 4

0.2 μm

Figure 12.36
Clathrin-coated vesicle formation
(A) Extracellular macromolecules (ligands) bind to cell surface receptors that are concentrated in clathrin-coated pits. These pits bud from the plasma membrane to form intracellular clathrin-coated vesicles. (B) Electron micrographs showing four stages in the formation of a clathrin-coated vesicle from a clathrin-coated pit. (B, M. M. Perry, 1979. *J. Cell Science* 34: 266.)

quently referred to as "professional phagocytes." Both macrophages and neutrophils play critical roles in the body's defense systems by eliminating microorganisms from infected tissues. In addition, macrophages eliminate aged or dead cells from tissues throughout the body. A striking example of the scope of this activity is provided by the macrophages of the human spleen and liver, which are responsible for the disposal of more than 10^{11} aged blood cells on a daily basis.

Receptor-Mediated Endocytosis

In contrast to phagocytosis, which plays only specialized roles, pinocytosis is common among eukaryotic cells. The best-characterized form of this process is **receptor-mediated endocytosis**, which provides a mechanism for the selective uptake of specific macromolecules (Figure 12.36). The macromolecules to be internalized first bind to specific cell surface receptors. These receptors are concentrated in specialized regions of the plasma membrane, called **clathrin-coated pits**. These pits bud from the membrane to form small **clathrin-coated vesicles** containing the receptors and their bound macromolecules (**ligands**). The clathrin-coated vesicles then fuse with early endosomes, in which their contents are sorted for transport to lysosomes or recycling to the plasma membrane.

The uptake of cholesterol by mammalian cells has provided a key model for understanding receptor-mediated endocytosis at the molecular level. Cholesterol is transported through the bloodstream in the form of lipoprotein particles, the most common of which is called **low-density lipoprotein**, or **LDL** (Figure 12.37). Studies in the laboratories of Michael Brown and Joseph

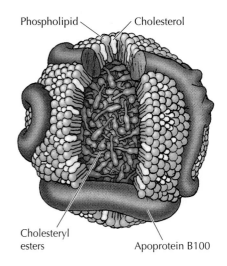

Phospholipid

Cholesterol

Cholesteryl esters

Apoprotein B100

Figure 12.37
Structure of LDL　Each particle of LDL contains approximately 1500 molecules of cholesteryl esters in an oily core. The core is surrounded by a coat containing 500 molecules of cholesterol, 800 molecules of phospholipid, and one molecule of apoprotein B100.

Key Experiment

The LDL Receptor

Familial Hypercholesterolemia: Defective Binding of Lipoproteins to Cultured Fibroblasts Associated with Impaired Regulation of 3-Hydroxy-3-Methylglutaryl Coenzyme A Reductase Activity

Michael S. Brown and Joseph L. Goldstein
University of Texas Southwestern Medical School, Dallas
Proceedings of the National Academy of Science USA, 1974, Volume 71, pages 788–792

The Context

Familial hypercholesterolemia (FH) is a genetic disease in which patients have greatly elevated levels of serum cholesterol and suffer from heart attacks early in life. Michael Brown and Joseph Goldstein began their efforts to understand this disease in 1972 with the idea that cholesterol overproduction results from a defect in the control mechanisms that normally regulate cholesterol biosynthesis. Consistent with this hypothesis, they found that addition of LDL to the culture medium of normal human fibroblasts inhibits the activity of 3-hydroxy-3-methylglutaryl coenzyme A reductase (HMG-CoA reductase), the rate-limiting enzyme in the cholesterol biosynthetic pathway. In contrast, HMG-CoA reductase activity is unaffected by addition of LDL to the cells of FH patients, resulting in overproduction of cholesterol by FH cells.

Perhaps surprisingly, subsequent experiments indicated that this abnormality in HMG-CoA reductase regulation is not the result of a mutation in the HMG-CoA reductase gene. Instead, the abnormal regulation of HMG-CoA reductase appeared to be due to the inability of FH cells to extract cholesterol from LDL. In 1974, Brown and Goldstein demonstrated that the lesion in FH cells is a defect in LDL binding to a receptor on the cell surface. This identification of the LDL receptor led to a series of groundbreaking experiments in which Brown, Goldstein, and their colleagues delineated the pathway of receptor-mediated endocytosis.

The Experiments

In their 1974 paper, Brown and Goldstein reported the results of experiments in which they investigated the binding of radiolabeled LDL to fibroblasts from either normal individuals or FH patients. Small amounts of radioactive LDL were added to the culture media, and the amount of radioactivity bound to the cells was determined after varying times of incubation (see figure). Increasing amounts of radioactive LDL bound to normal cells as a func-

tion of incubation time. Importantly, addition of excess unlabeled LDL reduced the binding of radioactive LDL, indicating that binding was due to a specific interaction of LDL with a limited number of sites on the cell surface. The specificity of the interaction was further demonstrated by the failure of excess amounts of other lipoproteins to interfere with LDL binding.

In contrast to these results with normal fibroblasts, the cells of FH patients failed to bind specifically radioactive LDL. It therefore appeared that normal fibroblasts possessed a specific LDL receptor that was absent or defective in FH cells. Brown and Goldstein concluded that the defect in LDL binding observed in FH cells "may represent the primary genetic lesion in this disorder," accounting for the inability of LDL to inhibit HMG-CoA reductase and the resultant overproduction of cholesterol. Additional experiments showed that the LDL bound to nor-

Time course of the binding of radioactive LDL to normal and FH cells. Cells were incubated with radioactive [^{125}I]LDL in the presence (open symbols) or absence (closed symbols) of excess unlabeled LDL. Cells were then harvested, and the amount of radioactive LDL bound was determined. Data are presented as nanograms of LDL bound per milligram of cell protein.

The LDL Receptor *(continued)*

mal fibroblasts is associated with the membrane fraction of the cell, suggesting that the LDL receptor is a cell surface protein.

The Impact

Following this identification of the LDL receptor, Brown and Goldstein demonstrated that LDL bound to the cell surface is rapidly internalized and degraded to free cholesterol in lysosomes. In collaboration with Richard Anderson, they further established that the LDL receptor is internalized by endocytosis from coated pits. In addition, their early studies

demonstrated that the LDL receptor is recycled to the plasma membrane after dissociation of its ligand inside the cell. Experiments that had been initiated with the goal of understanding the regulation of cholesterol biosynthesis thus led to the elucidation of a major pathway by which eukaryotic cells internalize specific macromolecules—a striking example of the way in which science and scientists can proceed in unanticipated but exciting new directions.

Michael Brown

Joseph Goldstein

Goldstein demonstrated that the uptake of LDL by mammalian cells requires the binding of LDL to a specific cell surface receptor that is concentrated in clathrin-coated pits and internalized by endocytosis. As discussed in the next section, the receptor is then recycled to the plasma membrane while LDL is transported to lysosomes, where cholesterol is released for use by the cell.

The key insights into this process came from studies of patients with the inherited disease known as familial hypercholesterolemia. Patients with this disease have very high levels of serum cholesterol and suffer heart attacks early in life. Brown and Goldstein found that cells of these patients are unable to internalize LDL from extracellular fluids, resulting in the accumulation of high levels of cholesterol in the circulation. Further experiments demonstrated that cells of normal individuals possess a receptor for LDL, which is concentrated in coated pits, and that familial hypercholesterolemia results from inherited mutations in the LDL receptor. These mutations are of two types. Cells from most patients with familial hypercholesterolemia simply failed to bind LDL, demonstrating that a specific cell surface receptor was required for LDL uptake. In addition, a few patients were identified whose cells bound LDL but were unable to internalize it. The LDL receptors of these patients failed to concentrate in coated pits, providing direct evidence for the central role of coated pits in receptor-mediated endocytosis.

The mutations that prevent the LDL receptor from concentrating in coated pits lie within the cytoplasmic tail of the receptor and can be as subtle as the change of a single tyrosine to cysteine (Figure 12.38). Further studies have defined the internalization signal of the LDL receptor as a sequence of six amino acids, including the essential tyrosine. Similar internalization signals, frequently including tyrosine residues, are found in the cytoplasmic tails of other receptors taken up via clathrin-coated pits. These internalization signals bind to adaptor proteins, which in turn bind clathrin on the cytosolic side of the membrane (Figure 12.39), similar to the way in which clathrin-coated vesicles form during the transport of lysosomal hydrolases from the *trans* Golgi network (see Figure 9.31). Clathrin assembles into a basketlike structure that distorts the membrane, forming invaginated pits. A GTP-binding protein, called dynamin, assembles into rings around the necks of these

LDL-binding domain

N-linked oligosaccharide

O-linked oligosaccharide

Plasma membrane

Cytoplasmic tail

Internalization signal

Mutation

Figure 12.38
The LDL receptor   The LDL receptor includes 700 extracellular amino acids, a transmembrane α helix of 22 amino acids, and a cytoplasmic tail of 50 amino acids. The N-terminal 292 amino acids constitute the LDL-binding domain. Six amino acids within the cytoplasmic tail define the internalization signal, first recognized because the mutation of Tyr to Cys in a case of familial hypercholesterolemia prevented concentration of the receptor in coated pits.

Figure 12.39
Formation of clathrin-coated pits
(A) Adaptor proteins bind both to clathrin and to the internalization signals present in the cytoplasmic tails of receptors. (B) Electron micrograph of a clathrin-coated pit showing the basket-like structure of the clathrin network. (B, courtesy of John E. Heuser, Washington University School of Medicine.)

invaginated pits, eventually leading to the release of coated vesicles inside the cell.

Receptor-mediated endocytosis is a major activity of the plasma membranes of eukaryotic cells. More than 20 different receptors have been shown to be selectively internalized by this pathway. Extracellular fluids are also incorporated into coated vesicles as they bud from the plasma

(A)

(B)

Ligand

Receptor

Adaptor protein

Clathrin

0.1 μm

membrane, so receptor-mediated endocytosis results in the nonselective uptake of extracellular fluids and their contents (**fluid phase endocytosis**), in addition to the internalization of specific macromolecules. Coated pits typically occupy 1 to 2% of the surface area of the plasma membrane and are estimated to have a lifetime of 1 to 2 minutes. From these figures, one can calculate that receptor-mediated endocytosis results in the internalization of an area of cell surface equivalent to the entire plasma membrane approximately every 2 hours.

A variety of studies indicate that cells also possess clathrin-independent endocytosis pathways. For example, fluids and some membrane-bound molecules continue to be endocytosed under experimental conditions that inhibit endocytosis from clathrin-coated pits. One pathway of clathrin-independent endocytosis involves the uptake of molecules in small invaginations of the plasma membrane (50 to 80 nm in diameter) called **caveolae** (Figure 12.40). Caveolae are enriched in lipid rafts of cholesterol and sphingolipids and possess a distinct coat formed by a protein called caveolin. They have been implicated in cell signaling and a variety of transport processes, including endocytosis. In addition, large vesicles (0.15 to 5.0 μm in diameter) can mediate the uptake of fluids in a process known as **macropinocytosis**. Thus, while clathrin-dependent endocytosis clearly provides a major pathway for the uptake of both fluids and specific macromolecules, cells also use several possible clathrin-independent mechanisms.

Protein Trafficking in Endocytosis

Following their internalization, clathrin-coated vesicles rapidly shed their coats and fuse with early **endosomes,** which are vesicles with tubular extensions located at the periphery of the cell. The specificity of fusion of endocytic vesicles with endosomes is determined by interactions between complementary pairs of transmembrane proteins of the vesicle and target membranes (v-SNAREs and t-SNAREs) and by Rab GTP-binding proteins, as discussed in Chapter 9. The early endosomes serve as a sorting compartment, from which molecules taken up by endocytosis are either recycled to the plasma membrane or transported to lysosomes for degradation. In addition, the early endosomes of polarized cells can transfer endocytosed proteins between different domains of the plasma membrane—for example, between the apical and basolateral domains of epithelial cells.

An important feature of early endosomes is that they maintain an acidic internal pH (about 6.0 to 6.2) as the result of the action of a membrane H$^+$

Figure 12.40
Caveolae Electron micrographs of caveolae. (A, courtesy of R. G. W. Anderson/University of Texas Southwestern Medical School-Dallas; B, courtesy of John E. Heuser, Washington University School of Medicine.)

(A)

(B)

0.2 μm

0.1 μm

Figure 12.41
Sorting in early endosomes LDL bound to its receptor is internalized in clathrin-coated vesicles, which shed their coats and fuse with early endosomes. At the acidic pH of early endosomes, LDL dissociates from its receptor and the endocytosed materials are sorted for degradation in lysosomes or recycling to the plasma membrane. LDL is transported from early to late endosomes in large carrier vesicles that move along microtubules. Transport vesicles carrying lysosomal hydrolases from the Golgi apparatus then fuse with late endosomes, which mature to lysosomes where LDL is degraded and cholesterol is released. In contrast, the LDL receptor is recycled from early endosomes to the plasma membrane.

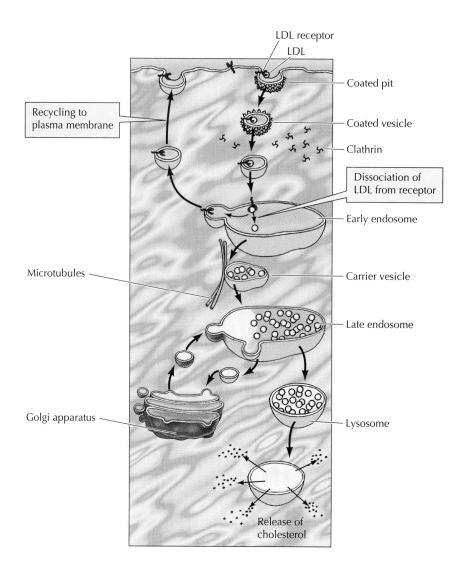

pump. This acidic pH leads to the dissociation of many ligands from their receptors within the early endosome compartment. Following this uncoupling, the receptors and their ligands can be transported to different intracellular destinations. A classic example is provided by LDL, which dissociates from its receptor within early endosomes (Figure 12.41). The receptor is then returned to the plasma membrane via transport vesicles that bud from the tubular extensions of endosomes. In contrast, LDL is transported (along with other soluble contents of the endosome) to lysosomes, where its degradation releases cholesterol.

Recycling to the plasma membrane is the major fate of membrane proteins taken up by receptor-mediated endocytosis, with many receptors (like the LDL receptor) being returned to the plasma membrane following dissociation of their bound ligands in early endosomes. The recycling of these receptors results in the continuous internalization of their ligands. Each LDL receptor, for example, makes a round-trip from the plasma membrane to endosomes and back approximately every 10 minutes. The importance of the recycling pathway is further emphasized by the magnitude of membrane traffic resulting from endocytosis. As already noted, approximately 50% of the plasma membrane is internalized by receptor-mediated endocytosis every hour and must therefore be replaced at an equivalent rate.

Most of this replacement is the result of receptor recycling; only about 5% of the cell surface is newly synthesized per hour.

Ligands and membrane proteins destined for degradation in lysosomes are transported from early endosomes to late endosomes, which are located near the nucleus (see Figure 12.41). Transport from early to late endosomes is mediated by the movement of large endocytic carrier vesicles along microtubules. The late endosomes are more acidic than early endosomes (pH about 5.5 to 6.0) and, as discussed in Chapter 9, are able to fuse with transport vesicles carrying lysosomal hydrolases from the Golgi apparatus. Late endosomes then mature into lysosomes as they acquire a full complement of lysosomal enzymes and become still more acidic (pH about 5). Within lysosomes, the endocytosed materials are degraded by the action of acid hydrolases.

Although many receptors (like the LDL receptor) are recycled to the plasma membrane, others follow different fates. Some are transported to lysosomes and degraded along with their ligands. For example, the cell surface receptors for several growth factors (discussed in the next chapter) are internalized following growth factor binding and eventually degraded in lysosomes. The effect of this process is to remove the receptor-ligand complexes from the plasma membrane, thereby terminating the response of the cell to growth factor stimulation—a phenomenon known as **receptor down-regulation**.

A specialized kind of recycling from endosomes plays an important role in the transmission of nerve impulses across synapses (Figure 12.42). As discussed earlier in this chapter, the arrival of an action potential at the terminus of most neurons signals the fusion of synaptic vesicles with the plasma membrane, releasing the neurotransmitters that carry the signal to postsynaptic cells. The empty synaptic vesicles are then recovered from the plasma membrane in clathrin-coated vesicles, which fuse with early endosomes. The synaptic vesicles are then regenerated directly by budding from endosomes. They accumulate a new supply of neurotransmitter and recycle to the plasma membrane, ready for the next cycle of synaptic transmission.

In polarized cells (e.g., epithelial cells), internalized receptors can also be transferred across the cell to the opposite domain of the plasma mem-

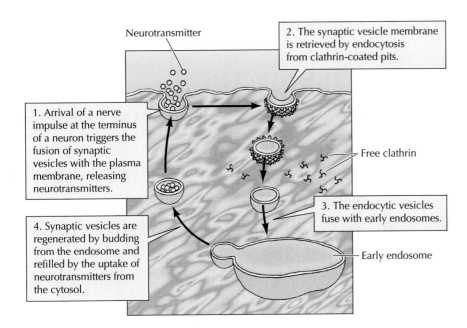

Neurotransmitter

2. The synaptic vesicle membrane is retrieved by endocytosis from clathrin-coated pits.

1. Arrival of a nerve impulse at the terminus of a neuron triggers the fusion of synaptic vesicles with the plasma membrane, releasing neurotransmitters.

Free clathrin

3. The endocytic vesicles fuse with early endosomes.

4. Synaptic vesicles are regenerated by budding from the endosome and refilled by the uptake of neurotransmitters from the cytosol.

Early endosome

Figure 12.42
Recycling of synaptic vesicles

Figure 12.43
Protein sorting by transcytosis
A protein destined for the apical domain of the plasma membrane is first transported from the Golgi apparatus to the basolateral domain. It is then endocytosed and selectively transported to the apical domain from early endosomes.

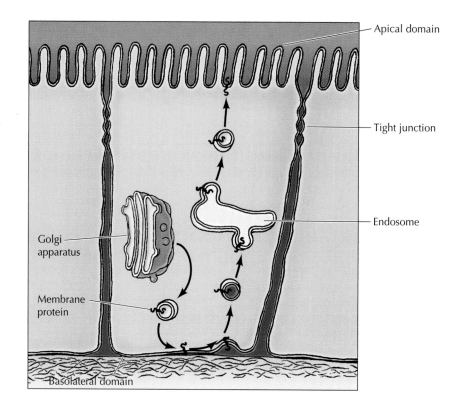

Apical domain

Tight junction

Endosome

Golgi apparatus

Membrane protein

Basolateral domain

brane—a process called **transcytosis**. For example, a receptor endocytosed from the basolateral domain of the plasma membrane can be sorted in early endosomes for transport to the apical membrane. In some cells, this is an important mechanism for sorting membrane proteins (Figure 12.43). Rather than being sorted for delivery to the apical or basolateral domains in the *trans* Golgi network (see Figure 9.28), proteins are initially delivered to the basolateral membrane. Proteins targeted for the apical membrane are then transferred to that site by transcytosis. In addition, transcytosis provides a mechanism for the transfer of extracellular macromolecules across epithelial cell sheets. For example, many kinds of epithelial cells transport antibodies from the blood to a variety of secreted fluids, such as milk. The antibodies bind to receptors on the basolateral surface and are then transcytosed along with their receptors to the apical surface. The receptors are then cleaved, releasing the antibodies into extracellular secretions.

Cell Walls and the Extracellular Matrix

Although cell boundaries are defined by the plasma membrane, many cells are surrounded by an insoluble array of secreted macromolecules. Cells of bacteria, fungi, algae, and higher plants are surrounded by rigid cell walls, which are an integral part of the cell. Although not encased in cell walls, animal cells in tissues are closely associated with an extracellular matrix composed of proteins and polysaccharides. The extracellular matrix not only provides structural support to cells and tissues, but also plays important roles in regulating the behavior of cells in multicellular organisms.

Bacterial Cell Walls

The rigid cell walls of bacteria determine cell shape and prevent the cell from bursting as a result of osmotic pressure. The structure of their cell walls divides bacteria into two broad classes that can be distinguished by a staining procedure known as the Gram stain, developed by Christian Gram

Figure 12.44
Bacterial cell walls   The plasma membrane of Gram-negative bacteria is surrounded by a thin cell wall beneath the outer membrane. Gram-positive bacteria lack outer membranes and have thick cell walls.

Gram-negative

Outer membrane

Cell wall

Plasma membrane

Cytosol

Gram-positive

Cell wall

Plasma membrane

Cytosol

in 1884 (Figure 12.44). As described earlier in this chapter, Gram-negative bacteria (such as *E. coli*) have a dual membrane system, in which the plasma membrane is surrounded by a permeable outer membrane. These bacteria have thin cell walls located between their inner and outer membranes. In contrast, Gram-positive bacteria (such as the common human pathogen *Staphylococcus aureus*) have only a single plasma membrane, which is surrounded by a much thicker cell wall.

Despite these structural differences, the principal component of the cell walls of both Gram-positive and Gram-negative bacteria is a **peptidoglycan** (Figure 12.45) consisting of linear polysaccharide chains crosslinked by short peptides. Because of this crosslinked structure, the peptidoglycan forms a strong covalent shell around the entire bacterial cell. Interestingly,

Figure 12.45
The peptidoglycan of *E. coli* Polysaccharide chains consist of alternating *N*-acetylglucosamine (NAG) and *N*-acetylmuramic acid (NAM) residues joined by $\beta(1\rightarrow4)$ glycosidic bonds. Parallel chains are crosslinked by tetrapeptides attached to the NAM residues. The amino acids forming the tetrapeptides vary in different species of bacteria.

(A)   **Chitin**

$\beta(1\rightarrow 4)$ linkage

Cellulose

(B)

Microfibril

Cellulose chains

Figure 12.46
Polysaccharides of fungal and plant cell walls   (A) Chitin (the principal polysaccharide of fungal cell walls) is a linear polymer of *N*-acetylglucosamine residues, whereas cellulose is a linear polymer of glucose. The carbohydrate monomers are joined by $\beta(1\rightarrow 4)$ linkages, allowing the polysaccharides to form long, straight chains. (B) Parallel chains of cellulose associate to form microfibrils.

the unique structure of their cell walls also makes bacteria vulnerable to some antibiotics. Penicillin, for example, inhibits the enzyme responsible for forming cross-links between different strands of the peptidoglycan, thereby interfering with cell wall synthesis and blocking bacterial growth.

Plant Cell Walls

In contrast to bacteria, the cell walls of eukaryotes (including fungi, algae, and higher plants) are composed principally of polysaccharides (Figure 12.46). The basic structural polysaccharide of fungal cell walls is **chitin** (a polymer of *N*-acetylglucosamine residues), which also forms the exoskeleton of arthropods (e.g., the shells of crabs). The cell walls of most algae and higher plants are composed principally of **cellulose**, which is the single most abundant polymer on Earth. Cellulose is a linear polymer of glucose residues, often containing more than 10,000 glucose monomers. The glucose residues are joined by $\beta(1\rightarrow 4)$ linkages, which allow the polysaccharide to form long straight chains. Several dozen such chains then associate in parallel with one another to form **cellulose microfibrils**, which can extend for many micrometers in length.

Within the cell wall, cellulose microfibrils are embedded in a matrix consisting of proteins and two other types of polysaccharides: **hemicelluloses** and **pectins** (Figure 12.47). Hemicelluloses are highly branched polysac-

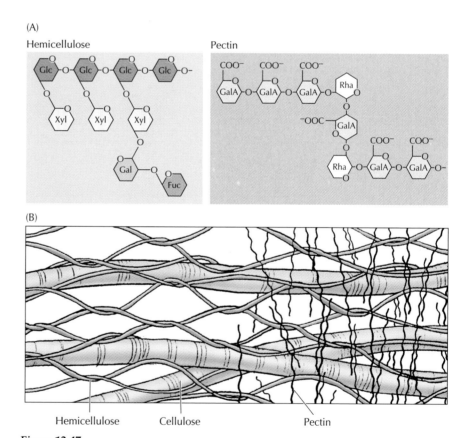

Figure 12.47
Model of a plant cell wall   (A) Structures of a representative hemicellulose (xyloglucan) and pectin (rhamnogalacturonan). Xyloglucan consists of a backbone of glucose (Glc) residues with side chains of xylose (Xyl), galactose (Gal), and fucose (Fuc). The backbone of rhamnogalacturonan contains galacturonic acid (GalA) and rhamnose (Rha) residues, to which numerous side chains are also attached. (B) Hemicelluloses bind to the surface of cellulose microfibrils, forming a fibrous network that is embedded in a gel-like matrix of pectins.

Figure 12.48
Primary and secondary cell walls Secondary cell walls are laid down between the primary cell wall and the plasma membrane. Secondary walls frequently consist of three layers, which differ in the orientation of their cellulose microfibrils. Electron micrographs show cellulose microfibrils in primary and secondary cell walls. (Primary wall, courtesy of F. C. Steward; secondary wall, Biophoto Associates/Photo Researchers, Inc.)

Primary wall

charides that are hydrogen-bonded to the surface of cellulose microfibrils. This crosslinks the cellulose microfibrils into a network of tough, fibrous molecules, which is responsible for the mechanical strength of plant cell walls. Pectins are branched polysaccharides containing a large number of negatively charged galacturonic acid residues. Because of these multiple negative charges, pectins bind positively charged ions (such as Ca^{2+}) and trap water molecules to form gels. An illustration of their gel-forming properties is provided by the fact that jams and jellies are produced by the addition of pectins to fruit juice. In the cell wall, the pectins form a gel-like network that is interlocked with the crosslinked cellulose microfibrils. In addition, cell walls contain a variety of glycoproteins that are incorporated into the matrix and are thought to provide further structural support.

Both the structure and function of cell walls change as plant cells develop. The walls of growing plant cells (called **primary cell walls**) are relatively thin and flexible, allowing the cell to expand in size. Once cells have ceased growth, they frequently lay down **secondary cell walls** between the plasma membrane and the primary cell wall (Figure 12.48). Such secondary cell walls, which are both thicker and more rigid than primary walls, are particularly important in cell types responsible for conducting water and providing mechanical strength to the plant.

Primary and secondary cell walls differ in composition as well as in thickness. Primary cell walls contain approximately equal amounts of cellulose, hemicelluloses, and pectins. In contrast, the more rigid secondary walls generally lack pectin and contain 50 to 80% cellulose. Many secondary walls are further strengthened by **lignin**, a complex polymer of phenolic residues that is responsible for much of the strength and density of wood. The orientation of cellulose microfibrils also differs in primary and secondary cell walls. The cellulose fibers of primary walls appear to be randomly arranged, whereas those of secondary walls are highly ordered (see Figure 12.48). Secondary walls are frequently laid down in layers in which the cellulose fibers differ in orientation, forming a laminated structure that greatly increases cell wall strength.

One of the critical functions of plant cell walls is to prevent cell swelling as a result of osmotic pressure. In contrast to animal cells, plant cells do not maintain an osmotic balance between their cytosol and extracellular fluids. Consequently, osmotic pressure continually drives the flow of water into the cell. This water influx is tolerated by plant cells because their rigid cell walls prevent swelling and bursting. Instead, an internal hydrostatic pressure (called **turgor pressure**) builds up within the cell, eventually equalizing the osmotic pressure and preventing the further influx of water.

Turgor pressure is responsible for much of the rigidity of plant tissues, as is readily apparent from examination of a dehydrated, wilted plant. In addition, turgor pressure provides the basis for a form of cell growth that is unique to plants. In particular, plant cells frequently expand by taking up water without synthesizing new cytoplasmic components (Figure 12.49). Cell expansion by this mechanism is signaled by plant hormones (**auxins**) that weaken a region of the cell wall, allowing turgor pressure to drive the expansion of the cell in that direction. As this occurs, the water that flows

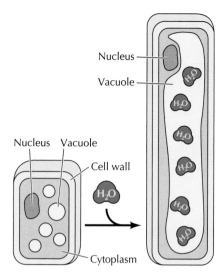

Figure 12.49
Expansion of plant cells Turgor pressure drives the expansion of plant cells by the uptake of water, which is accumulated in a large central vacuole.

into the cell accumulates within a large central vacuole, so the cell expands without increasing the volume of its cytosol. Such expansion can result in a 10- to 100-fold increase in the size of plant cells during development.

As cells expand, new components of the cell wall are deposited outside the plasma membrane. Matrix components, including hemicelluloses and pectins, are synthesized in the Golgi apparatus and secreted. Cellulose, however, is synthesized by a plasma membrane enzyme complex (**cellulose synthase**). In expanding cells, the newly synthesized cellulose microfibrils are deposited at right angles to the direction of cell elongation—an orientation that is thought to play an important role in determining the direction of further cell expansion (Figure 12.50). Interestingly, the cellulose microfibrils in elongating cell walls are laid down in parallel to cortical microtubules underlying the plasma membrane. These microtubules appear to define the orientation of newly synthesized cellulose microfibrils, possibly by determining the direction of movement of the cellulose synthase complexes in the membrane. The cortical microtubules thus define the direction of cell wall growth, which in turn determines the direction of cell expansion and ultimately the shape of the entire plant.

The Extracellular Matrix

Although animal cells are not surrounded by cell walls, many of the cells in tissues of multicellular organisms are embedded in an **extracellular matrix** consisting of secreted proteins and polysaccharides. The extracellular matrix fills the spaces between cells and binds cells and tissues together. One type of extracellular matrix is exemplified by the thin, sheetlike **basal laminae,** or **basement membranes,** upon which layers of epithelial cells rest (Figure 12.51). In addition to supporting sheets of epithelial cells, basal laminae surround muscle cells, adipose cells, and peripheral nerves. Extracellular matrix, however, is most abundant in connective tissues. For exam-

Figure 12.50
Cellulose synthesis during cell elongation New cellulose microfibrils, synthesized by a plasma membrane enzyme complex (cellulose synthase), are laid down at right angles to the direction of cell elongation. The direction of cellulose synthesis is parallel to microtubules beneath the plasma membrane.

ple, the loose connective tissue beneath epithelial cell layers consists predominantly of an extracellular matrix in which fibroblasts are distributed. Other types of connective tissue, such as bone, tendon, and cartilage, similarly consist largely of extracellular matrix, which is principally responsible for their structure and function.

Extracellular matrices are composed of tough fibrous proteins embedded in a gel-like polysaccharide ground substance—a design basically similar to that of plant cell walls. In addition to fibrous structural proteins and polysaccharides, the extracellular matrix contains adhesion proteins that link components of the matrix both to one another and to attached cells. The differences between the various types of extracellular matrix result from variations on this general theme. For example, tendons contain a high proportion of fibrous proteins, whereas cartilage contains a high concentration of polysaccharides that form a firm compression-resistant gel. In bone, the extracellular matrix is hardened by deposition of calcium phosphate crystals. The sheetlike structure of basal laminae also results from the utilization of matrix components that differ from those found in connective tissues.

The major structural protein of the extracellular matrix is **collagen**, which is the single most abundant protein in animal tissues. The collagens are a large family of proteins, containing at least 19 different members. They are characterized by the formation of triple helices in which three polypeptide chains are wound tightly around one another in a ropelike structure (Figure 12.52). The triple helix domains of the collagens consist of repeats of the amino acid sequence Gly-X-Y. A glycine (the smallest amino acid, with a side chain consisting only of a hydrogen) is required in every third position in order for the polypeptide chains to pack together close enough to form the collagen triple helix. Proline is frequently found in the X position and hydroxyproline in the Y position; because of their ring structure, these amino acids stabilize the helical conformations of the polypeptide chains.

The unusual amino acid hydroxyproline is formed within the endoplasmic reticulum by modification of proline residues that have already been incorporated into collagen polypeptide chains (Figure 12.53). Lysine residues in collagen are also frequently converted to hydroxylysines. The hydroxyl groups of these modified

Figure 12.51
Examples of extracellular matrix
Sheets of epithelial cells rest on a thin layer of extracellular matrix called a basal lamina. Beneath the basal lamina is loose connective tissue, which consists largely of extracellular matrix secreted by fibroblasts. The extracellular matrix contains fibrous structural proteins embedded in a gel-like polysaccharide ground substance.

(A)

(B) **Amino acid sequence**

Figure 12.52
Structure of collagen   (A) Three polypeptide chains coil around one another in a characteristic triple helix structure. (B) The amino acid sequence of a collagen triple helix domain consists of Gly-X-Y repeats, in which X is frequently proline and Y is frequently hydroxyproline (Hyp).

Proline

Prolyl hydroxylase

Hydroxyproline

Figure 12.53
Formation of hydroxyproline Prolyl hydroxylase converts proline residues in collagen to hydroxyproline.

TABLE 12.2 Representative Members of the Collagen Family

Collagen class	Types	Tissue distribution
Fibril-forming	I	Most connective tissues
	II	Cartilage and vitreous humor
	III	Extensible connective tissues (e.g., skin and lung)
	V	Tissues containing collagen I
	XI	Tissues containing collagen II
Fibril-associated	IX	Tissues containing collagen II
	XII	Tissues containing collagen I
	XIV	Tissues containing collagen I
	XVI	Many tissues
Network-forming	IV	Basal laminae
Anchoring filaments	VII	Attachments of basal laminae to underlying connective tissue

amino acids are thought to stabilize the collagen triple helix by forming hydrogen bonds between polypeptide chains. These amino acids are rarely found in other proteins, although hydroxyproline is also common in some of the glycoproteins of plant cell walls.

The most abundant type of collagen (type I collagen) is one of the fibril-forming collagens that are the basic structural components of connective tissues (Table 12.2). The polypeptide chains of these collagens consist of approximately a thousand amino acids or 330 Gly-X-Y repeats. After being secreted from the cell, these collagens assemble into **collagen fibrils** in which the triple helical molecules are associated in regular staggered arrays (Figure 12.54). These fibrils do not form within the cell, because the fibril-forming collagens are synthesized as soluble precursors (**procollagens**) that contain nonhelical segments at both ends of the polypeptide chain. Procollagen is cleaved to collagen after its secretion, so the assembly of collagen into fibrils takes place only outside the cell. The association of collagen molecules in fibrils is further strengthened by the formation of covalent cross-links between the side chains of lysine and hydroxylysine residues. Frequently, the fibrils further associate with one another to form collagen fibers, which can be several micrometers in diameter.

Several other types of collagen do not form fibrils but play distinct roles in various kinds of extracellular matrices. In addition to the fibril-forming collagens, connective tissues contain fibril-associated collagens, which bind to the surface of collagen fibrils and link them both to one another and to other matrix components. Basal laminae form from a different type of collagen (type IV collagen), which is a network-forming collagen (Figure 12.55). The Gly-X-Y repeats of these collagens are frequently interrupted by short nonhelical sequences. Because of these interruptions, the network-forming collagens are more flexible than the fibril-forming collagens. Con-

(A)

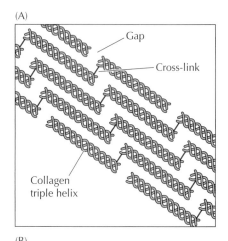

Gap

Cross-link

Collagen triple helix

(B)

1 μm

Figure 12.54
Collagen fibrils (A) Collagen molecules assemble in a regular staggered array to form fibrils. The molecules overlap by one-fourth of their length, and there is a short gap between the N terminus of one molecule and the C terminus of the next. The assembly is strengthened by covalent cross-links between side chains of lysine or hydroxylysine residues, primarily at the ends of the molecules. (B) Electron micrograph of collagen fibrils. The staggered arrangement of collagen molecules and the gaps between them are responsible for the characteristic cross-striations in the fibrils. (B, J. Gross, F. O. Sahmitt, and D. Fawcett/Visuals Unlimited.)

Figure 12.55
Type IV collagen (A) The Gly-X-Y repeat structure of type IV collagen (yellow) is interrupted by multiple nonhelical sequences (bars). (B) Electron micrograph of a type IV collagen network. (B, P. D. Yurchenco and J. C. Schittny, 1990. *FASEB J.* 4: 1577.)

(A)

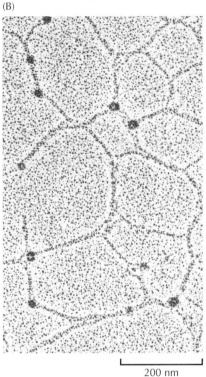

(B)

200 nm

sequently, they assemble into two-dimensional crosslinked networks instead of fibrils. Yet another type of collagen forms anchoring fibrils, which link some basal laminae to underlying connective tissues.

Connective tissues also contain **elastic fibers**, which are particularly abundant in organs that regularly stretch and then return to their original shape. The lungs, for example, stretch each time a breath is inhaled and return to their original shape with each exhalation. Elastic fibers are composed principally of a protein called **elastin**, which is crosslinked into a network by covalent bonds formed between the side chains of lysine residues (similar to those found in collagen). This network of crosslinked elastin chains behaves like a rubber band, stretching under tension and then snapping back when the tension is released.

The fibrous structural proteins of the extracellular matrix are embedded in gels formed from polysaccharides called **glycosaminoglycans**, or **GAGs**, which consist of repeating units of disaccharides (Figure 12.56). One sugar of the disaccharide is either *N*-acetylglucosamine or *N*-acetylgalactosamine and the second is usually acidic (either glucuronic acid or iduronic acid). With the exception of hyaluronan, these sugars are modified by the addition of sulfate groups. Consequently, GAGs are highly negatively charged. Like the pectins of plant cell walls, they bind positively charged ions and trap water molecules to form hydrated gels, thereby providing mechanical support to the extracellular matrix.

Hyaluronan is the only GAG that occurs as a single long polysaccharide chain. All of the other GAGs are linked to proteins to form **proteoglycans**, which can consist of up to 95% polysaccharide by weight. Proteoglycans can contain as few as one or as many as more than a hundred GAG chains attached to serine residues of a core protein. A variety of core proteins (ranging from 10 to >500 kd) have been identified, so the proteoglycans are

Figure 12.56
Major types of glycosaminoglycans Glycosaminoglycans consist of repeating disaccharide units. With the exception of hyaluronan, the sugars frequently contain sulfate. Heparan sulfate is similar to heparin except that it contains fewer sulfate groups.

Figure 12.57
Complexes of aggrecan and hyaluronan   Aggrecan is a large proteoglycan consisting of more than 100 chondroitin sulfate chains joined to a core protein. Multiple aggrecan molecules bind to long chains of hyaluronan, forming large complexes in the extracellular matrix of cartilage. This association is stabilized by link proteins.

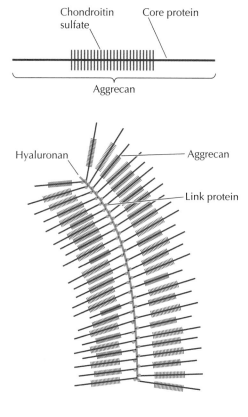

a diverse group of macromolecules. In addition to being components of the extracellular matrix, some proteoglycans are cell surface proteins that function in cell adhesion.

A number of proteoglycans interact with hyaluronan to form large complexes in the extracellular matrix. A well-characterized example is aggrecan, the major proteoglycan of cartilage (Figure 12.57). More than a hundred chains of chondroitin sulfate are attached to a core protein of about 250 kd, forming a proteoglycan of about 3000 kd. Multiple aggrecan molecules then associate with chains of hyaluronan, forming large aggregates (>100,000 kd) that become trapped in the collagen network. Proteoglycans also interact with both collagen and other matrix proteins to form gel-like networks in which the fibrous structural proteins of the extracellular matrix are embedded. For example, perlecan (the major heparan sulfate proteoglycan of basal laminae) binds to both type IV collagen and to the adhesion protein laminin, which is discussed shortly.

Adhesion proteins, the third class of extracellular matrix constituents, are responsible for linking the components of the matrix both to one another and to the surfaces of cells. The prototype of these molecules is **fibronectin**, which is the principal adhesion protein of connective tissues. Fibronectin is a dimeric glycoprotein consisting of two polypeptide chains, each containing nearly 2500 amino acids (Figure 12.58). In the extracellular matrix, fibronectin is further crosslinked into fibrils by disulfide bonds. Fibronectin has binding sites for both collagen and GAGs, so it crosslinks these matrix components. A distinct site on the fibronectin molecule is recognized by cell surface receptors and is thus responsible for the attachment of cells to the extracellular matrix.

Figure 12.58
Structure of fibronectin   Fibronectin is a dimer of similar polypeptide chains joined by disulfide bonds near the C terminus. Sites for binding to proteoglycans, cells, and collagen are indicated. The molecule also contains additional binding sites that are not shown.

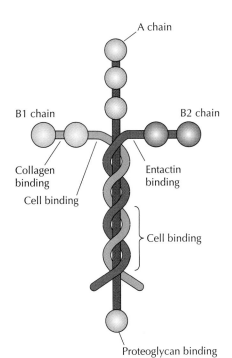

Figure 12.59
Structure of laminin Laminin consists of three polypeptide chains designated A, B1, and B2. Some of the binding sites for entactin, type IV collagen, proteoglycans, and cell surface receptors are indicated.

A chain

B1 chain   B2 chain

Collagen
binding
Cell binding

Entactin
binding

Cell binding

Proteoglycan binding

Basal laminae contain a distinct adhesion protein called **laminin** (Figure 12.59). Like type IV collagen, laminins can self-assemble into meshlike polymers. Such laminin networks are the major structural components of the basal laminae synthesized in very early embryos, which do not contain collagen. The laminins also have binding sites for cell surface receptors, type IV collagen, and perlecan. In addition, laminins are tightly associated with another adhesion protein, called **entactin** or **nidogen**, which also binds to type IV collagen. As a result of these multiple interactions, laminin, entactin, type IV collagen, and perlecan form crosslinked networks in the basal lamina.

The major cell surface receptors responsible for the attachment of cells to the extracellular matrix are the **integrins**. The integrins are a family of transmembrane proteins consisting of two subunits, designated α and β (Figure 12.60). More than 20 different integrins, formed from combinations of 18 known α subunits and 8 known β subunits, have been identified. The integrins bind to short amino acid sequences present in multiple components of the extracellular matrix, including collagen, fibronectin, and laminin. The first such integrin-binding site to be characterized was the sequence Arg-Gly-Asp, which is recognized by several members of the integrin family. Other integrins, however, bind to distinct peptide sequences, including recognition sequences in collagens and laminin. Transmembrane proteoglycans on the surface of a variety of cells also bind to components of the extracellular matrix and modulate cell-matrix interactions.

In addition to attaching cells to the extracellular matrix, the integrins serve as anchors for the cytoskeleton (Figure 12.61). The resulting linkage of the cytoskeleton to the extracellular matrix is responsible for the stability of cell-matrix junctions. Distinct interactions between integrins and the cytoskeleton are found at two types of cell-matrix junctions, **focal adhesions** and **hemidesmosomes**, which were discussed in Chapter 11. Focal adhesions attach a variety of cells, including fibroblasts, to the extracellular matrix. The cytoplasmic domains of the β subunits of integrins at these cell-matrix junctions anchor the actin cytoskeleton by associating with bundles

Figure 12.60
Structure of integrins The integrins are heterodimers of two transmembrane subunits, designated α and β. The α subunit binds divalent cations (M^{2+}). The matrix-binding region is composed of portions of both subunits.

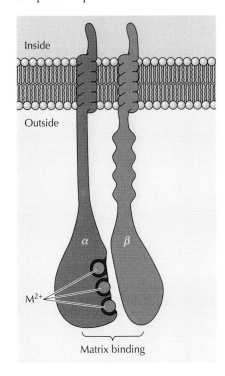

Inside

Outside

α β

M^{2+}

Matrix binding

Figure 12.61
Junctions between cells and the extracellular matrix   Integrins mediate two types of stable junctions in which the cytoskeleton is linked to the extracellular matrix. In focal adhesions, bundles of actin filaments are anchored to the β subunits of most integrins via associations with a number of other proteins, including α-actinin, talin, and vinculin (see Figure 11.13). In hemidesmosomes, $\alpha_6\beta_4$ integrin links the basal lamina to intermediate filaments via plectin.

of actin filaments. Hemidesmosomes are specialized sites of epithelial cell attachment at which a specific integrin (designated $\alpha_6\beta_4$) interacts with intermediate filaments instead of with actin. The $\alpha_6\beta_4$ integrin binds to laminin, so hemidesmosomes anchor epithelial cells to the basal lamina.

Cell-Cell Interactions

Direct interactions between cells, as well as between cells and the extracellular matrix, are critical to the development and function of multicellular organisms. Some cell-cell interactions are transient, such as the interactions between cells of the immune system and the interactions that direct white blood cells to sites of tissue inflammation. In other cases, stable cell-cell junctions play a key role in the organization of cells in tissues. For example, several different types of stable cell-cell junctions are critical to the maintenance and function of epithelial cell sheets. Plant cells also associate with their neighbors not only by interactions between their cell walls, but also by specialized junctions between their plasma membranes.

Cell Adhesion Proteins

Cell-cell adhesion is a selective process, such that cells adhere only to other cells of specific types. This selectivity was first demonstrated in classical studies of embryo development, which showed that cells from one tissue (e.g., liver) specifically adhere to cells of the same tissue rather than to cells of a different tissue (e.g., brain). Such selective cell-cell adhesion is mediated by transmembrane proteins called **cell adhesion molecules**, which can be divided into four major groups: the **selectins**, the **integrins**, the **im-**

TABLE 12.3 Cell Adhesion Molecules

Family	Ligands recognized	Stable cell junctions
Selectins	Carbohydrates	No
Integrins	Extracellular matrix	Focal adhesions and hemidesmosomes
	Members of Ig superfamily	No
Ig superfamily	Integrins	No
	Homophilic interactions	No
Cadherins	Homophilic interactions	Adherens junctions and desmosomes

munoglobulin (Ig) superfamily (so named because they contain structural domains similar to immunoglobulins), and the **cadherins** (Table 12.3). Cell adhesion mediated by the selectins, integrins, and cadherins requires Ca^{2+} or Mg^{2+}, so many adhesive interactions between cells are Ca^{2+}- or Mg^{2+}-dependent.

The selectins mediate transient interactions between leukocytes and endothelial cells or blood platelets. There are three members of the selectin family: L-selectin, which is expressed on leukocytes; E-selectin, which is expressed on endothelial cells; and P-selectin, which is expressed on platelets. As discussed earlier in this chapter, the selectins recognize cell surface carbohydrates (see Figure 12.14). One of their critical roles is to initiate the interactions between leukocytes and endothelial cells during the migration of leukocytes from the circulation to sites of tissue inflammation (Figure 12.62). The selectins mediate the initial adhesion of leukocytes to endothelial cells. This is followed by the formation of more stable adhesions, in which integrins on the surface of leukocytes bind to intercellular adhesion molecules (ICAMs), which are members of the Ig superfamily expressed on the surface of endothelial cells. The firmly attached leukocytes are then able to penetrate the walls of capillaries and enter the underlying tissue by migrating between endothelial cells.

The binding of ICAMs to integrins is an example of a **heterophilic interaction**, in which an adhesion molecule on the surface of one cell (e.g., an ICAM) recognizes a different molecule on the surface of another cell (e.g., an integrin). Other members of the Ig superfamily mediate **homophilic interactions**, in which an adhesion molecule on the surface of one cell binds to the

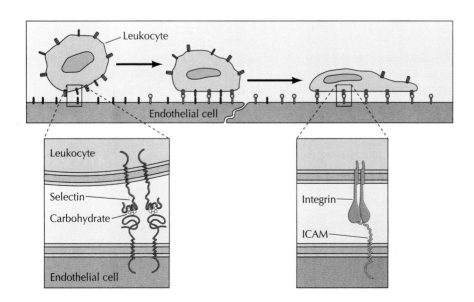

Figure 12.62
Adhesion between leukocytes and endothelial cells   Leukocytes leave the circulation at sites of tissue inflammation by interacting with the endothelial cells of capillary walls. The first step in this interaction is the binding of leukocyte selectins to carbohydrate ligands on the endothelial cell surface. This step is followed by more stable interactions between leukocyte integrins and members of the Ig superfamily (ICAMs) on endothelial cells.

same molecule on the surface of another cell. Such homophilic binding leads to selective adhesion between cells of the same type. For example, nerve cell adhesion molecules (N-CAMs) are members of the Ig superfamily expressed on nerve cells, and homophilic binding between N-CAMs contributes to the formation of selective associations between nerve cells during development. There are more than 100 members of the Ig superfamily, which mediate a variety of cell-cell interactions.

The fourth group of cell adhesion molecules, the cadherins, also display homophilic binding specificities. They are not only involved in selective adhesion between embryonic cells but are also primarily responsible for the formation of stable junctions between cells in tissues. For example, E-cadherin is expressed on epithelial cells, so homophilic interactions between E-cadherins lead to the selective adhesion of epithelial cells to one another. It is noteworthy that loss of E-cadherin can lead to the development of cancers arising from epithelial cells, illustrating the importance of cell-cell interactions in controlling cell behavior. Different members of the cadherin family, such as N-cadherin (neural cadherin) and P-cadherin (placental cadherin), mediate selective adhesion of other cell types.

About twenty different classic cadherins, such as E-cadherin, have been identified. In addition, a distinct subfamily of cadherins, called protocadherins, are expressed in the central nervous system where they appear to play a role in adhesion between neurons at synapses. Intriguingly, different neurons appear to express different protocadherins, suggesting that the protocadherins may play a role in the establishment of specific connections between neurons. About 50 human protocadherin genes have been identified and shown to be organized into three gene clusters. Each cluster contains multiple exons encoding the N-terminal extracellular and transmembrane protocadherin domains, but only a single set of three exons encoding the C-terminal cytoplasmic domain (Figure 12.63). The protocadherin gene clusters thus appear to consist of a variable region, encoding multiple extracellular and transmembrane domains, linked to a constant region encoding a single cytoplasmic domain. This organization of protocadherin genes strikingly resembles that of immunoglobulin and T-cell receptor genes (see Figures 5.42 and 5.43), in which multiple variable region exons are joined to a single constant region exon. In the immunoglobulin and T-cell receptor genes, this occurs as a result of DNA rearrangements that generate diversity in the immune system. It remains to be determined whether the variable and constant regions of protocadherins are joined at the DNA or the RNA level (for example, by alternative splicing) and to what extent rearrangements of protocadherin genes might contribute to the establishment of specific synaptic connections in the brain.

In contrast to the stable cell-matrix junctions discussed in the preceding section, the cell-cell interactions mediated by the selectins, integrins, and members of the Ig superfamily are transient adhesions in which the cyto-

Figure 12.63
Organization of protocadherin gene clusters The human protocadherin genes are organized into three clusters. In the cluster illustrated, 15 different variable regions encoding extracellular and transmembrane domains are linked to a single constant region, consisting of three exons encoding the cytoplasmic domain. Protocadherin mRNAs contain one of the variable region exons linked to the constant region.

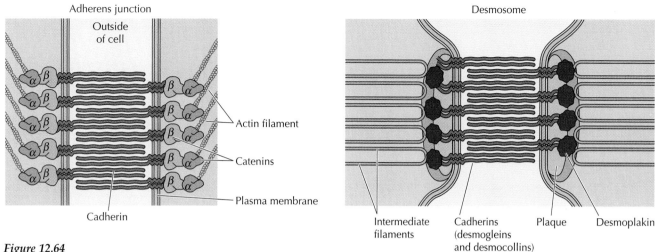

Figure 12.64
Stable cell-cell junctions mediated by the cadherins   Homophilic interactions between cadherins mediate two types of stable cell-cell adhesions. In adherens junctions, the cadherins are linked to bundles of actin filaments via the catenins (see Figure 11.14). In desmosomes, desmoplakin links members of the cadherin superfamily (desmogleins and desmocollins) to intermediate filaments (see Figure 11.34).

skeletons of adjacent cells are not linked to one another. Stable adhesion junctions involving the cytoskeletons of adjacent cells are instead mediated by the cadherins. As discussed in Chapter 11, these cell-cell junctions are of two types: **adherens junctions** and **desmosomes**, in which cadherins or related proteins (desmogleins and desmocollins) are linked to actin bundles and intermediate filaments, respectively (Figure 12.64). The role of the cadherins in linking the cytoskeletons of adjacent cells is thus analogous to that of the integrins in forming stable junctions between cells and the extracellular matrix.

Tight Junctions

In addition to the adhesion junctions mediated by the cadherins, two other types of specialized cell-cell junctions play key roles in animal tissues. **Tight junctions**, which are usually associated with adherens junctions and desmosomes in a **junctional complex** (Figure 12.65), are critically important to the function of epithelial cell sheets as barriers between fluid compartments. For example, the intestinal epithelium separates the lumen of the intestine from the underlying connective tissue, which contains blood capillaries. Tight junctions play two roles in allowing epithelia to fulfill such barrier functions. First, tight junctions form seals that prevent the free passage of molecules (including ions) between the cells of epithelial sheets. Second, tight junctions separate the apical and basolateral domains of the plasma membrane by preventing the free diffusion of lipids and membrane proteins between them. Consequently, specialized transport systems in the apical and basolateral domains are able to control the traffic of molecules between distinct extracellular compartments, such as the transport of glucose between the intestinal lumen and the blood supply (see Figure 12.32).

Tight junctions are the closest known contacts between adjacent cells. They were originally described as sites of apparent fusion between the outer leaflets of the plasma membranes, although it is now clear that the membranes do not fuse. Instead, tight junctions appear to be formed by a network of protein strands that continues around the entire circumference

(A)

(B)

Figure 12.65
Tight junctions (A) Electron micrograph of epithelial cells joined by a junctional complex, including a tight junction, an adherens junction, and a desmosome. (B) Tight junctions are formed by interactions between strands of transmembrane proteins (occludin and claudins) on adjacent cells. (A, Don Fawcett/Photo Researchers, Inc.)

of the cell (see Figure 12.65). Each strand in these networks is thought to be composed of transmembrane proteins (claudins and occludin) that bind to similar proteins on adjacent cells, thereby sealing the space between their plasma membranes.

Gap Junctions

Gap junctions, which are found in most animal tissues, serve as direct connections between the cytoplasms of adjacent cells. They provide open channels through the plasma membrane, allowing ions and small molecules (less than approximately a thousand daltons) to diffuse freely between neighboring cells, but preventing the passage of proteins and nucleic acids. Consequently, gap junctions couple both the metabolic activities and the electric responses of the cells they connect. Most cells in animal tissues—including epithelial cells, endothelial cells, and the cells of cardiac and smooth muscle—communicate by gap junctions. In electrically excitable cells, such as heart muscle cells, the direct passage of ions through gap junctions couples and synchronizes the contractions of neighboring cells. Gap junctions also allow the passage of some intracellular signaling molecules, such as cAMP and Ca^{2+}, between adjacent cells, potentially coordinating the responses of cells in tissues.

Gap junctions are constructed of transmembrane proteins called **connexins** (Figure 12.66). Six connexins assemble to form a cylinder with an open aqueous pore in its center. Such an assembly of connexins in the plasma membrane of one cell then aligns with the connexins of an adjacent cell, forming an open channel between the two cytoplasms. The plasma membranes of the two cells are separated by a gap corresponding to the space occupied by the connexin extracellular domains—hence the term "gap junction," which was coined by electron microscopists.

(A)

(B)

Connexin

Figure 12.66
Gap junctions (A) Electron micrograph of a gap junction (arrows) between two liver cells. (B) Gap junctions consist of assemblies of six connexins, which form open channels through the plasma membranes of adjacent cells. (A, Don Fawcett and R. Wood/Photo Researchers, Inc.)

Plant Cell Adhesion and Plasmodesmata

Adhesion between plant cells is mediated by their cell walls rather than by transmembrane proteins. In particular, a specialized pectin-rich region of the cell wall called the **middle lamella** acts as a glue to hold adjacent cells together. Because of the rigidity of plant cell walls, stable associations between plant cells do not require the formation of cytoskeletal links, such as those provided by the desmosomes and adherens junctions of animal cells. However, adjacent plant cells communicate with each other through cytoplasmic connections called **plasmodesmata** (singular, plasmodesma), which function analogously to animal cell gap junctions.

Despite their similarities in function, plasmodesmata are structurally unrelated to gap junctions. At each plasmodesma the plasma membrane of one cell is continuous with that of its neighbor, creating an open channel between the two cytosols (Figure 12.67). An extension of the smooth endoplasmic reticulum passes through the pore, leaving a ring of surrounding cytoplasm through which ions and small molecules are able to pass freely between the cells. In addition, plasmodesmata can expand in response to appropriate

(B)

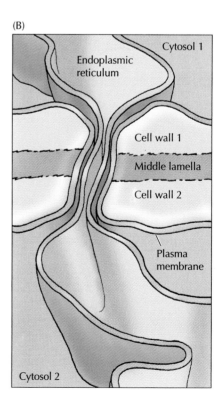

Cytosol 1

Endoplasmic reticulum

Cell wall 1

Middle lamella

Cell wall 2

Plasma membrane

Cytosol 2

(A)

0.1 μm

Figure 12.67
Plasmodesmata (A) Electron micrograph of plasmodesmata (arrows). (B) At plasmodesmata, the plasma membranes of neighboring cells are continuous, forming cytoplasmic channels through the adjacent cell walls. An extension of the endoplasmic reticulum usually passes through the channel. (A, E. H. Newcomb, University of Wisconsin/Biological Photo Service.)

stimuli, permitting the regulated passage of macromolecules between adjacent cells. Plasmodesmata may thus play a key role in plant development by controlling the trafficking of regulatory molecules, such as transcription factors or RNAs, between cells.

STRUCTURE OF THE PLASMA MEMBRANE

The Phospholipid Bilayer: The fundamental structure of the plasma membrane is a phospholipid bilayer, which also contains glycolipids and cholesterol.

phosphatidylcholine, phosphatidylethanolamine, phosphatidylserine, sphingomyelin, phosphatidylinositol, glycolipids, cholesterol

Membrane Proteins: Associated proteins are responsible for carrying out specific membrane functions. Membranes are viewed as fluid mosaics in which proteins are inserted into phospholipid bilayers.

fluid mosaic model, peripheral membrane proteins, integral membrane proteins, transmembrane proteins, porins, glycosylphosphatidylinositol (GPI) anchor

Mobility of Membrane Proteins: Proteins are free to diffuse laterally through the phospholipid bilayer. However, the mobility of some proteins is restricted by their associations with other molecules. In addition, tight junctions prevent proteins from moving between distinct plasma membrane domains of epithelial cells.

apical domain, basolateral domain

The Glycocalyx: The cell surface is covered by a carbohydrate coat called the glycocalyx. Cell surface carbohydrates serve as markers for cell-cell recognition.

glycocalyx, selectin

TRANSPORT OF SMALL MOLECULES

Passive Diffusion: Small hydrophobic molecules are able to cross the plasma membrane by diffusing through the phospholipid bilayer.

passive diffusion

Facilitated Diffusion and Carrier Proteins: The passage of most biological molecules is mediated by carrier or channel proteins that allow polar and charged molecules to cross the plasma membrane without interacting with its hydrophobic interior.

facilitated diffusion, carrier protein, channel protein

Ion Channels: Ion channels mediate the rapid passage of selected ions across the plasma membrane. They are particularly well characterized in nerve and muscle cells, where they are responsible for the transmission of electric signals.

ion channel, ligand-gated channel, voltage-gated channel, patch clamp technique, Nernst equation, action potential

Active Transport Driven by ATP Hydrolysis: Energy derived from ATP hydrolysis can drive the transport of molecules against their electrochemical gradients.

active transport, ion pump, Na⁺-K⁺ pump, Na⁺-K⁺ ATPase, ABC transporter

Active Transport Driven by Ion Gradients: Ion gradients are frequently used as a source of energy to drive the active transport of other molecules.

symport, uniport, antiport

ENDOCYTOSIS

Phagocytosis: Cells ingest large particles, such as bacteria and cell debris, by phagocytosis.

endocytosis, phagocytosis, pinocytosis, phagosome, phagolysosome

Receptor-Mediated Endocytosis: The best-characterized form of endocytosis is receptor-mediated endocytosis, which provides a mechanism for the selective uptake of specific macromolecules.

receptor-mediated endocytosis, clathrin-coated pit, clathrin-coated vesicle, ligand, low-density lipo-protein (LDL), fluid phase endocy-tosis, caveolae, macropinocytosis

Protein Trafficking in Endocytosis: Molecules taken up by endocytosis are transported to endosomes, where they are sorted for recycling to the plasma membrane or degradation in lysosomes.

endosome, receptor down-regulation, transcytosis

CELL WALLS AND THE EXTRACELLULAR MATRIX

Bacterial Cell Walls: The principal component of bacterial cell walls is a peptidoglycan consisting of polysaccharide chains crosslinked by short peptides.

peptidoglycan

Plant Cell Walls: The cell walls of fungi, algae, and higher plants are composed of fibrous polysaccharides (e.g., cellulose) embedded in a gel-like matrix of polysaccharides and proteins. Their rigid cell walls allow plant cells to expand rapidly by the uptake of water.

chitin, cellulose, cellulose microfibril, hemicellulose, pectin, primary cell wall, secondary cell wall, lignin, turgor pressure, auxin, cellulose synthase

The Extracellular Matrix: Animal cells in tissues are surrounded by an extracellular matrix consisting of secreted proteins and polysaccharides. Cell surface receptors bind to the extracellular matrix and anchor the cytoskeleton at cell-matrix junctions.

extracellular matrix, basal lamina, basement membrane, collagen, collagen fibril, procollagen, elastic fiber, elastin, glycosaminoglycan (GAG), proteoglycan, fibronectin, laminin, entactin, nidogen, integrin, focal adhesion, hemidesmosome

CELL-CELL INTERACTIONS

Cell Adhesion Proteins: Selective cell-cell interactions are mediated by four major groups of cell adhesion proteins: selectins, integrins, immunoglobulin superfamily members, and cadherins. The cadherins link the cytoskeletons of adjacent cells at stable cell-cell junctions.

cell adhesion molecule, selectin, integrin, immunoglobulin (Ig) superfamily, cadherin, heterophilic interaction, homophilic interaction, adherens junction, desmosome

Tight Junctions: Tight junctions prevent the free passage of molecules between epithelial cells and separate the apical and basolateral domains of the plasma membrane.

tight junction, junctional complex

Gap Junctions: Gap junctions are open channels connecting the cyto-plasms of adjacent cells.

gap junction, connexin

Plant Cell Adhesion and Plasmodesmata: Adjacent plant cells are linked by cytoplasmic connections called plasmodesmata.

middle lamella, plasmodesma

Questions

1. What would be the consequences of incorporating porins into the plasma membrane instead of into the outer membrane of bacteria?

2. The concentration of K⁺ is about 20 times higher inside squid axons than in extracellular fluids, generating an equilibrium membrane potential of –75 mV. What would be the expected equilibrium membrane potential if the K⁺ concentration were only 10 times higher inside than outside the cell? Why does the actual resting membrane potential (–60 mV) differ from the K⁺ equilibrium potential of –75 mV?

3. An important function of the Na⁺-K⁺ pump in animal cells is the maintenance of osmotic equilibrium. Why is this unnecessary for plant cells?

4. An anticipated problem in gene therapy for cystic fibrosis is that normal CFTR genes would probably be successfully transferred to only a fraction of the cells in the patient's respiratory epithelium. How is this potential pitfall affected by the presence of gap junctions?

5. What is the importance of selectively targeting different glucose transporters to the apical and basolateral domains of the plasma membrane of intestinal epithelial cells? What is the role of tight junctions in this process?

References and Further Reading

Structure of the Plasma Membrane

Bennett, V. and D. M. Gilligan. 1993. The spectrin-based membrane skeleton and micron-scale organization of the plasma membrane. *Ann. Rev. Cell Biol.* 9: 27–66. [R]

Branton, D., C. M. Cohen and J. Tyler. 1981. Interaction of cytoskeletal proteins on the human erythrocyte membrane. *Cell* 24: 24–32. [P]

Bretscher, M. 1985. The molecules of the cell membrane. *Sci. Am.* 253(4): 100–108. [R]

Bretscher, M. S. and S. Munro. 1993. Cholesterol and the Golgi apparatus. *Science* 261: 1280–1281. [R]

Brown, D. A. and E. London. 1998. Functions of lipid rafts in biological membranes. *Ann. Rev. Cell Dev. Biol.* 14: 111–136. [R]

Diesenhofer, J., O. Epp, K. Miki, R. Huber and H. Michel. 1985. The structure of the protein subunits in the photosynthetic reaction centre of *Rhodopseudomonas viridis* at 3-Å resolution. *Nature* 318: 618–624. [P]

Dowham, W. 1997. Molecular basis for membrane phospholipid diversity: Why are there so many lipids? *Ann. Rev. Biochem.* 66: 199–212. [R]

Englund, P. T. 1993. The structure and biosynthesis of glycosyl phosphatidylinositol anchors. *Ann. Rev. Biochem.* 62: 121–138. [R]

Frye, L. D. and M. Edidin. 1970. The rapid intermixing of cell surface antigens after formation of mouse-human heterokaryons. *J. Cell Sci.* 7: 319–335. [P]

Gahmberg, C. G. and M. Tolvanen. 1996. Why mammalian cell surface proteins are glycoproteins. *Trends Biochem. Sci.* 21: 308–311. [R]

Jacobson, K., E. D. Sheets and R. Simson. 1995. Revisiting the fluid mosaic model of membranes. *Science* 268: 1441–1442. [R]

Jay, D. G. 1996. Role of band 3 in homeostasis and cell shape. *Cell* 86: 853–854. [R]

Lasky, L. A. 1995. Selectin-carbohydrate interactions and the initiation of the inflammatory response. *Ann. Rev. Biochem.* 64: 113–139. [R]

Magee, T. and C. Marshall. 1999. New insights into the interaction of Ras with the plasma membrane. *Cell* 98: 9–12. [R]

Nikaido, H. 1994. Porins and specific diffusion channels in bacterial outer membranes. *J. Biol. Chem.* 269: 3905–3908. [R]

Rees, D. C., H. Komiya, T. O. Yeates, J. P. Allen and G. Feher. 1989. The bacterial photosynthetic reaction center as a model for membrane proteins. *Ann. Rev. Biochem.* 58: 607–633. [R]

Sharon, N. and H. Lis. 1993. Carbohydrates in cell recognition. *Sci. Am.* 268(1): 82–89. [R]

Simons, K. and E. Ikonen. 1997. Functional rafts in cell membranes. *Nature* 387: 569–572. [R]

Singer, S. J. 1990. The structure and insertion of integral proteins in membranes. *Ann. Rev. Cell Biol.* 6: 247–296. [R]

Singer, S. J. and G. L. Nicolson. 1972. The fluid mosaic model of the structure of cell membranes. *Science* 175: 720–731. [P]

Zhang, F. L. and P. J. Casey. 1996. Protein prenylation: Molecular mechanisms and functional consequences. *Ann. Rev. Biochem.* 65: 241–269. [R]

Transport of Small Molecules

Bell, G. I., C. F. Burant, J. Takeda and G. W. Gould. 1993. Structure and function of mammalian facilitative sugar transporters. *J. Biol. Chem.* 268: 19161–19164. [R]

Borgnia, M., S. Nielsen, A. Engel and P. Agre. 1999. Cellular and molecular biology of the aquaporin water channels. *Ann. Rev. Biochem.* 68: 425–458. [R]

Borst, P. and A. H. Schinkel. 1997. Genetic dissection of the function of mammalian P-glycoproteins. *Trends Genet.* 13: 217–222. [R]

Carafoli, E. 1992. The Ca^{2+} pump of the plasma membrane. *J. Biol. Chem.* 267: 2115–2118. [R]

Catterall, W. A. 1995. Structure and function of voltage-gated ion channels. *Ann. Rev. Biochem.* 64: 493–531. [R]

Clapham, D. E. 1999. Unlocking family secrets: K^+ channel transmembrane domains. *Cell* 97: 547–550. [R]

Cole, S. P. and R. G. Deeley. 1998. Multidrug resistance mediated by the ATP-binding cassette transporter protein MRP. *Bioessays* 20: 931–940. [R]

Cooper, E. C. and L. Y. Jan. 1999. Ion channel genes and human neurological disease: Recent progress, prospects, and challenges. *Proc. Natl. Acad. Sci. USA* 96: 4759–4766. [R]

Doyle, D. A., J. M. Cabral, R. A. Pfuetzner, A. Kuo, J. M. Gulbis, S. L. Cohen, B. T. Chait and R. MacKinnon. 1998. The structure of the potassium channel: Molecular basis of K^+ conductivity and selectivity. *Science* 280: 69–77. [P]

Gottesman, M. M. and I. M. Pastan. 1993. Biochemistry of multidrug resistance mediated by the multidrug transporter. *Ann. Rev. Biochem.* 62: 385–427. [R]

Higgins, C. F. 1992. ABC transporters: From microorganisms to man. *Ann. Rev. Cell Biol.* 8: 67–113. [R]

Hille, B. 1992. *Ionic Channels of Excitable Membranes.* 2nd ed. Sunderland, MA: Sinauer Associates.

Hodgkin, A. L. and A. F. Huxley. 1952. A quantitative description of membrane current and its application to conduction and excitation in nerve. *J. Physiol.* 117: 500–544. [P]

Lanyi, J. K. 1995. Bacteriorhodopsin as a model for proton pumps. *Nature* 375: 461–463. [R]

Lingrel, J. B. and T. Kuntzweiler. 1994. Na^+, K^+-ATPase. *J. Biol. Chem.* 269: 19659–19662. [R]

MacLennan, D. H., W. J. Rice and N. M. Green. 1997. The mechanism of Ca^{2+} transport by sarco(endo)plasmic reticulum Ca^{2+}-ATPases. *J. Biol. Chem.* 272: 28815–28818. [R]

Neher, E. and B. Sakmann. 1992. The patch clamp technique. *Sci. Am.* 266(3): 44–51. [R]

Perozo, E., D. M. Cortes and L. G. Cuello. 1999. Structural rearrangements underlying K^+-channel activation gating. *Science* 285: 73–78. [P]

Sakmann, B. 1992. Elementary steps in synaptic transmission revealed by currents through single ion channels. *Science* 256: 503–512. [R]

Unwin, N. 1993. Neurotransmitter action: Opening of ligand-gated ion channels. *Cell* 72/*Neuron* 10 (Suppl.): 31–41. [R]

Walmsley, A. R., M. P. Barrett, F. Bringaud and G. W. Gould. 1998. Sugar transporters from bacteria, parasites and mammals: Structure-activity relationships. *Trends Biochem. Sci.* 23: 476–481. [R]

Welsh, M. J. and A. E. Smith. 1993. Molecular mechanisms of CFTR chloride channel dysfunction in cystic fibrosis. *Cell* 73: 1251–1254. [R]

Endocytosis

Anderson, R. G. W. 1998. The caveolae membrane system. *Ann. Rev. Biochem.* 67: 199–225. [R]

Beron, W., C. Alvarez-Dominguez, L. Mayorga and P. D. Stahl. 1995. Membrane trafficking along the phagocytic pathway. *Trends Cell Biol.* 5: 100–104. [R]

Brown, M. S. and J. L. Goldstein. 1986. A receptor-mediated pathway for cholesterol homeostasis. *Science* 232: 34–47. [R]

Gruenberg, J. and K. E. Howell. 1989. Membrane traffic in endocytosis: Insights from cell-free assays. *Ann. Rev. Cell Biol.* 5: 453–481. [R]

Kirchhausen, T. 1999. Adaptors for clathrin-mediated traffic. *Ann. Rev. Cell Dev. Biol.* 15: 705–732. [R]

Marsh, M. and H. T. McMahon. 1999. The structural era of endocytosis. *Science* 285: 215–220. [R]

McNiven, M. A. 1998. Dynamin: A molecular motor with pinchase action. *Cell* 94: 151–154. [R]

Mostov, K. E. and M. H. Cardone. 1995. Regulation of protein traffic in polarized epithelial cells. *Bioessays* 17: 129–138. [R]

Nelson, W. J. 1992. Regulation of cell surface polarity from bacteria to mammals. *Science* 258: 948–955. [R]

Rabinovitch, M. 1995. Professional and nonprofessional phagocytes: An introduction. *Trends Cell Biol.* 5: 85–87. [R]

Schmid, S. L. 1997. Clathrin-coated vesicle formation and protein sorting: An integrated process. *Ann. Rev. Biochem.* 66: 511–548. [R]

Smith, C. J. and B. M. Pearse. 1999. Clathrin: Anatomy of a coat protein. *Trends Cell Biol.* 9: 335–338. [R]

Sudhof, T. C. 1995. The synaptic vesicle cycle: A cascade of protein-protein interactions. *Nature* 375: 645–653. [R]

Swanson, J. A. and C. Watts. 1995. Macropinocytosis. *Trends Cell Biol.* 5: 424–428. [R]

Trowbridge, I. S., J. F. Collawn and C. R. Hopkins. 1993. Signal-dependent membrane protein trafficking in the endocytic pathway. *Ann. Rev. Cell Biol.* 9: 129–161. [R]

Van Ijzendoorn, S. C. D. and D. Hoekstra. 1999. The subapical compartment: A novel sorting centre? *Trends Cell Biol.* 9: 144–149. [R]

Cell Walls and the Extracellular Matrix

Bernfield, M., M. Gotte, P. W. Park, O. Reizes, M. L. Fitzgerald, J. Lincecum and M. Zako. 1999. Functions of cell surface heparan sulfate proteoglycans. *Ann. Rev. Biochem.* 68: 729–777. [R]

Carpita, N. C. and D. M. Gibeaut. 1993. Structural models of primary cell walls in flowering plants: Consistency of molecular structure with the physical properties of the walls during growth. *Plant J.* 3: 1–30. [R]

Chothia, C. and E. Y. Jones. 1997. The molecular structure of cell adhesion molecules. *Ann. Rev. Biochem.* 66: 823–862. [R]

Cosgrove, D. J. 1997. Assembly and enlargement of the primary cell wall in plants. *Ann. Rev. Cell Dev. Biol.* 13: 171–201. [R]

Cyr, R. J. 1994. Microtubules in plant morphogenesis: Role of the cortical array. *Ann. Rev. Cell Biol.* 10: 153–180. [R]

Engel, J. 1992. Laminins and other strange proteins. *Biochemistry* 31: 10643–10651. [R]

Hynes, R. O. 1992. Integrins: Versatility, modulation, and signaling in cell adhesion. *Cell* 69: 11–25. [R]

Iozzo, R. V. 1998. Matrix proteoglycans: From molecular design to cellular function. *Ann. Rev. Biochem.* 67: 609–652. [R]

Kjellen, L. and U. Lindahl. 1991. Proteoglycans: Structures and interactions. *Ann. Rev. Biochem.* 60: 443–475. [R]

Loftus, J. C., J. W. Smith and M. H. Ginsberg. 1994. Integrin-mediated cell adhesion: The extracellular face. *J. Biol. Chem.* 269: 25235–25238. [R]

Lukashev, M. E. and Z. Werb. 1998. ECM signalling: Orchestrating cell behaviour and misbehaviour. *Trends Cell Biol.* 8: 437–441. [R]

McDonald, J. A. 1988. Extracellular matrix assembly. *Ann. Rev. Cell Biol.* 4: 183–207. [R]

Mecham, R. P. 1991. Laminin receptors. *Ann. Rev. Cell Biol.* 7: 71–91. [R]

Prockop, D. J. and K. I. Kivirikko. 1995. Collagens: Molecular biology, diseases, and potentials for therapy. *Ann. Rev. Biochem.* 64: 403–434. [R]

Ruoslahti, E. 1988. Structure and biology of proteoglycans. *Ann. Rev. Cell Biol.* 4: 229–255. [R]

Ruoslahti, E. 1996. RGD and other recognition sequences for integrins. *Ann. Rev. Cell Dev. Biol.* 12: 697–715. [R]

Varner, J. E. and L.-S. Lin. 1989. Plant cell wall architecture. *Cell* 56: 231–239. [R]

Vertel, B. M. 1995. The ins and outs of aggrecan. *Trends Cell Biol.* 5: 458–464. [R]

Vuorio, E. and B. de Crombrugghe. 1990. The family of collagen genes. *Ann. Rev. Biochem.* 59: 837–872. [R]

Woods, A. and J. R. Couchman. 1998. Syndecans: Synergistic activators of cell adhesion. *Trends Cell Biol.* 8: 189–192. [R]

Yurchenko, P. D. and J. C. Schnitty. 1990. Molecular architecture of basement membranes. *FASEB J.* 4: 1577–1590. [R]

Cell-Cell Interactions

Chothia, C. and E. Y. Jones. 1997. The molecular structure of cell adhesion molecules. *Ann. Rev. Biochem.* 66: 823–862. [R]

Christofori, G. and H. Semb. 1999. The role of the cell-adhesion molecule E-cadherin as a tumour-suppressor gene. *Trends Biochem. Sci.* 24: 73–76. [R]

Gumbiner, B. M. 1996. Cell adhesion: The molecular basis of tissue architecture and morphogenesis. *Cell* 84: 345–357. [R]

Humphries, M. J. and P. Newham. 1998. The structure of cell-adhesion molecules. *Trends Cell Biol.* 8: 78–83. [R]

Hynes, R. O. 1999. Cell adhesion: Old and new questions. *Trends Cell Biol.* 9: M33–M42. [R]

Kragler, F., W. J. Lucas and J. Monzer. 1998. Plasmodesmata: Dynamics, domains and patterning. *Ann. Botany* 81: 1–10. [R]

Pickard, B. G. and R. N. Beachy. 1999. Intercellular connections are developmentally controlled to help move molecules through the plant. *Cell* 98: 5–8. [R]

Simon, A. M. and D. A. Goodenough. 1998. Diverse functions of vertebrate gap junctions. *Trends Cell Biol.* 8: 477–483. [R]

Springer, T. A. 1994. Traffic signals for lymphocyte recirculation and leukocyte emigration: The multistep paradigm. *Cell* 76: 301–314. [R]

Stevenson, B. R. and B. H. Keon. 1998. The tight junction: Morphology to molecules. *Ann. Rev. Cell Biol.* 14: 89–109. [R]

Takeichi, M. 1991. Cadherin cell adhesion receptors as a morphogenetic regulator. *Science* 251: 1451–1455. [R]

Tsukita, S. and M. Furuse. 1999. Occludin and claudins in tight-junction strands: Leading or supporting players? *Trends Cell Biol.* 9: 268–273. [R]

Uemura, T. 1998. The cadherin superfamily at the synapse: More members, more missions. *Cell* 93: 1095–1098. [R]

Wu, Q. and T. Maniatis. 1999. A striking organization of a large family of human neural cadherin-like cell adhesion genes. *Cell* 97: 779–790. [P]

Yap, A. S., W. M. Brieher and B. M. Gumbiner. 1997. Molecular and functional analysis of cadherin-based adherens junctions. *Ann. Rev. Cell Dev. Biol.* 13: 119–146. [R]

PART IV

Cell Regulation

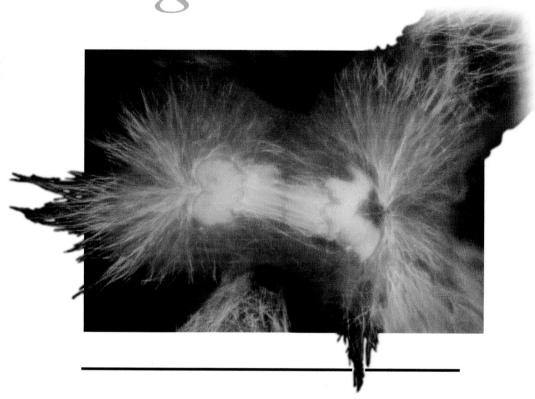

Cell Signaling

ALL CELLS RECEIVE AND RESPOND TO SIGNALS from their surroundings. Even the simplest bacteria sense and swim toward high concentrations of nutrients, such as glucose or amino acids. Many unicellular eukaryotes also respond to signaling molecules secreted by other cells, allowing cell-cell communication. Mating between yeast cells, for example, is signaled by peptides that are secreted by one cell and bind to receptors on the surface of another. It is in multicellular organisms, however, that cell-cell communication reaches its highest level of sophistication. Whereas the cells of prokaryotes and unicellular eukaryotes are largely autonomous, the behavior of each individual cell in multicellular plants and animals must be carefully regulated to meet the needs of the organism as a whole. This is accomplished by a variety of signaling molecules that are secreted or expressed on the surface of one cell and bind to receptors expressed by other cells, thereby integrating and coordinating the functions of the many individual cells that make up organisms as complex as human beings.

The binding of most signaling molecules to their receptors initiates a series of intracellular reactions that regulate virtually all aspects of cell behavior, including metabolism, movement, proliferation, survival, and differentiation. Understanding the molecular mechanisms responsible for these pathways of cell signaling has thus become a major area of active research. Interest in this area is further heightened by the fact that many cancers arise as a result of a breakdown in the signaling pathways that control normal cell proliferation and survival. Conversely, many of our current insights into cell signaling mechanisms have come from the study of cancer cells—a striking example of the fruitful interplay between medicine and basic research in cell and molecular biology.

Signaling Molecules and Their Receptors

Many different kinds of molecules transmit information between the cells of multicellular organisms. Although all these molecules act as ligands that bind to receptors expressed by their target cells, there is considerable variation in the structure and function of the different types of molecules that

Direct Cell-Cell Signaling

Signaling by Secreted Molecules

(A) Endocrine signaling

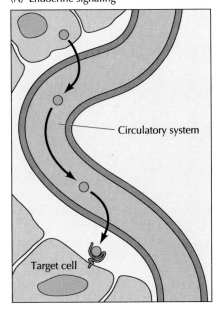

Circulatory system

Target cell

(B) Paracrine signaling

(C) Autocrine signaling

serve as signal transmitters. Structurally, the signaling molecules used by plants and animals range in complexity from simple gases to proteins. Some of these molecules carry signals over long distances, whereas others act locally to convey information between neighboring cells. In addition, signaling molecules differ in their mode of action on their target cells. Some signaling molecules are able to cross the plasma membrane and bind to intracellular receptors in the cytoplasm or nucleus, whereas most bind to receptors expressed on the target cell surface. The sections that follow discuss the major types of signaling molecules and the receptors with which they interact. Subsequent discussion in this chapter focuses on the mechanisms by which cell surface receptors then function to regulate cell behavior.

Modes of Cell-Cell Signaling

Cell signaling can result either from the direct interaction of a cell with its neighbor or from the action of secreted signaling molecules (Figure 13.1). Signaling by direct cell-cell (or cell-matrix) interactions plays a critical role in regulating the behavior of cells in animal tissues. For example, the integrins and cadherins (which were discussed in the previous chapter) function not only as cell adhesion molecules but also as signaling molecules that regulate cell proliferation and survival in response to cell-cell and cell-matrix contacts. In addition, cells express a variety of cell surface receptors that interact with signaling molecules on the surface of neighboring cells. Signaling via such direct cell-cell interactions plays a critical role in regulating the many interactions between different types of cells that take place during embryonic development, as well as in the maintenance of adult tissues.

The multiple varieties of signaling by secreted molecules are frequently divided into three general categories based on the distance over which signals are transmitted. In **endocrine signaling**, the signaling molecules (**hormones**) are secreted by specialized endocrine cells and carried through the circulation to act on target cells at distant body sites. A classic example is provided by the steroid hormone estrogen, which is produced by the ovary and stimulates development and maintenance of the female reproductive system and secondary sex characteristics. In animals, more than 50 different hormones are produced by endocrine glands, including the pituitary, thyroid, parathyroid, pancreas, adrenal glands, and gonads.

In contrast to hormones, some signaling molecules act locally to affect the behavior of nearby cells. In **paracrine signaling**, a molecule released by one cell acts on neighboring target cells. An example is provided by the action of neurotransmitters in carrying signals between nerve cells at a synapse. Finally, some cells respond to signaling molecules that they themselves produce. One important example of such **autocrine signaling** is the response of cells of the vertebrate immune system to foreign antigens. Certain types of T lymphocytes respond to antigenic stimulation by synthesizing a growth factor that drives their own proliferation, thereby increasing the number of responsive T lymphocytes and amplifying the immune response. It is also noteworthy that abnormal autocrine signaling frequently contributes to the uncontrolled growth of cancer cells (see Chapter

Figure 13.1
Modes of cell-cell signaling Cell signaling can take place either through direct cell-cell contacts or through the action of secreted signaling molecules. (A) In endocrine signaling, hormones are carried through the circulatory system to act on distant target cells. (B) In paracrine signaling, a molecule released from one cell acts locally to affect nearby target cells. (C) In autocrine signaling, a cell produces a signaling molecule to which it also responds.

15). In this situation, a cancer cell produces a growth factor to which it also responds, thereby continuously driving its own unregulated proliferation.

Steroid Hormones and the Steroid Receptor Superfamily

As already noted, all signaling molecules act by binding to receptors expressed by their target cells. In many cases, these receptors are expressed on the target cell surface, but some receptors are intracellular proteins located in the cytosol or the nucleus. These intracellular receptors respond to small hydrophobic signaling molecules that are able to diffuse across the plasma membrane. The **steroid hormones** are the classic examples of this group of signaling molecules, which also includes thyroid hormone, vitamin D_3, and retinoic acid (Figure 13.2).

The steroid hormones (including testosterone, estrogen, progesterone, the corticosteroids, and ecdysone) are all synthesized from cholesterol. **Testosterone, estrogen,** and **progesterone** are the sex steroids, which are produced by the gonads. The **corticosteroids** are produced by the adrenal gland. They include the **glucocorticoids**, which act on a variety of cells to stimulate production of glucose, and the **mineralocorticoids**, which act on the kidney to regulate salt and water balance. **Ecdysone** is an insect hormone that plays a key role in development by triggering the metamorphosis of larvae to adults.

Although thyroid hormone, vitamin D_3, and retinoic acid are both structurally and functionally distinct from the steroids, they share a common mechanism of action in their target cells. **Thyroid hormone** is synthesized from tyrosine in the thyroid gland; it plays important roles in development

Figure 13.2
Structure of steroid hormones, thyroid hormone, vitamin D_3, and retinoic acid The steroids include the sex hormones (testosterone, estrogen, and progesterone), glucocorticoids, and mineralocorticoids.

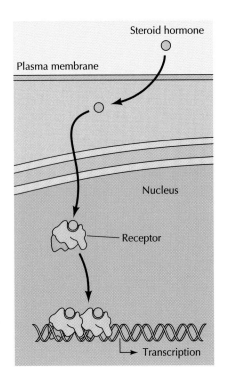

Figure 13.3
Action of steroid hormones   The steroid hormones diffuse across the plasma membrane and bind to nuclear receptors, which directly stimulate transcription of their target genes. The steroid hormone receptors bind DNA as dimers.

and regulation of metabolism. **Vitamin D$_3$** regulates Ca^{2+} metabolism and bone growth. **Retinoic acid** and related compounds (**retinoids**) synthesized from vitamin A play important roles in vertebrate development.

Because of their hydrophobic character, the steroid hormones, thyroid hormone, vitamin D$_3$, and retinoic acid are able to enter cells by diffusing across the plasma membrane (Figure 13.3). Once inside the cell, they bind to intracellular receptors that are expressed by the hormonally responsive target cells. These receptors, which are members of a family of proteins known as the **steroid receptor superfamily**, are transcription factors that contain related domains for ligand binding, DNA binding, and transcriptional activation. Ligand binding regulates their function as activators or repressors of their target genes, so the steroid hormones and related molecules directly regulate gene expression.

Ligand binding has distinct effects on different receptors. Some members of the steroid receptor superfamily, such as the estrogen and glucocorticoid receptors, are unable to bind to DNA in the absence of hormone. The binding of hormone induces a conformational change in the receptor, allowing it to bind to regulatory DNA sequences and activate transcription of target genes. In other cases, the receptor binds DNA in either the presence or absence of hormone, but hormone binding alters the activity of the receptor as a transcriptional regulatory molecule. For example, thyroid hormone receptor acts as a repressor in the absence of hormone, but hormone binding converts it to an activator that stimulates transcription of thyroid hormone-inducible genes (Figure 13.4).

Nitric Oxide and Carbon Monoxide

The simple gas nitric oxide (NO) is a major paracrine signaling molecule in the nervous, immune, and circulatory systems. Like the steroid hormones, NO is able to diffuse directly across the plasma membrane of its target cells. The molecular basis of NO action, however, is distinct from that of steroid action; rather than binding to a receptor that regulates transcription, NO alters the activity of intracellular target enzymes.

Nitric oxide is synthesized from the amino acid arginine by the enzyme nitric oxide synthase (Figure 13.5). Once synthesized, NO diffuses out of the cell and can act locally to affect nearby cells. Its action is restricted to such local effects because NO is extremely unstable, with a half-life of only a few seconds. One well-characterized example of NO action is signaling the dilation of blood vessels. The first step in this process is the release of neurotransmitters, such as acetylcholine, from the terminus of nerve cells in the blood vessel wall. These neurotransmitters act on endothelial cells to stimulate NO synthesis. NO then diffuses to neighboring smooth muscle cells where it reacts with iron bound to the active site of the enzyme

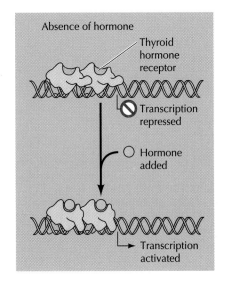

Figure 13.4
Gene regulation by the thyroid hormone receptor   Thyroid hormone receptor binds DNA in either the presence or absence of hormone. However, hormone binding changes the function of the receptor from a repressor to an activator of target gene transcription.

Figure 13.5
Synthesis of nitric oxide The enzyme nitric oxide synthase (NOS) catalyzes the formation of nitric oxide from arginine.

guanylyl cyclase. This increases enzymatic activity, resulting in synthesis of the second messenger cyclic GMP (discussed later in this chapter), which induces muscle cell relaxation and blood vessel dilation. For example, NO is responsible for signaling the dilation of blood vessels that leads to penile erection. It is also interesting to note that the medical use of nitroglycerin in treatment of heart disease is based on its conversion to NO, which dilates coronary blood vessels and increases blood flow to the heart.

Another simple gas, carbon monoxide (CO), also functions as a signaling molecule in the nervous system. CO is closely related to NO and appears to act similarly as a neurotransmitter and mediator of blood vessel dilation. The synthesis of CO in brain cells, like that of NO, is stimulated by neurotransmitters. In addition, CO can stimulate guanylate cyclase, which may also represent the major physiological target of CO signaling.

Neurotransmitters

The **neurotransmitters** carry signals between neurons or from neurons to other types of target cells (such as muscle cells). They are a diverse group of small hydrophilic molecules including acetylcholine, dopamine, epinephrine (adrenaline), serotonin, histamine, glutamate, glycine, and γ-aminobutyric acid (GABA) (Figure 13.6). The release of neurotransmitters is signaled by the arrival of an action potential at the terminus of a neuron (see Figure 12.22). The neurotransmitters then diffuse across the synaptic cleft and bind to receptors on the target cell surface. Note that some neurotransmitters can also act as hormones. For example, epinephrine functions both as a neurotransmitter and as a hormone produced by the adrenal gland to signal glycogen breakdown in muscle cells.

Because the neurotransmitters are hydrophilic molecules, they are unable to cross the plasma membrane of their target cells. Therefore, in contrast to steroid hormones and NO or CO, the neurotransmitters act by binding to cell surface receptors. Many neurotransmitter receptors are ligand-gated ion channels, such as the acetylcholine receptor discussed in the preceding chapter (see Figure 12.23). Neurotransmitter binding to these receptors induces a conformational change that opens ion channels,

Figure 13.6
Structure of representative neurotransmitters The neurotransmitters are hydrophilic molecules that bind to cell surface receptors.

Acetylcholine

$$CH_3-\overset{\displaystyle O}{\overset{\displaystyle \|}{C}}-O-CH_2-CH_2-N^+-(CH_3)_3$$

Glycine

$$H_3\overset{+}{N}-CH_2-\overset{\displaystyle O}{\overset{\displaystyle \|}{C}}-O^-$$

Glutamate

$$H_3\overset{+}{N}-\underset{\underset{\displaystyle O^-}{\overset{\displaystyle |}{\underset{\displaystyle C=O}{|}}}}{CH}-CH_2-CH_2-\overset{\displaystyle O}{\overset{\displaystyle \|}{C}}-O^-$$

Dopamine

$$-CH_2-CH_2-NH_3^+$$

Norepinephrine

$$-\underset{\underset{\displaystyle OH}{|}}{CH}-CH_2-NH_3^+$$

Epinephrine

$$-\underset{\underset{\displaystyle OH}{|}}{CH}-CH_2-\overset{+}{NH_2}-CH_3$$

Serotonin

$$-CH_2-CH_2-NH_3^+$$

Histamine

$$HC=C-CH_2-CH_2-NH_3^+$$

γ-Aminobutyric acid (GABA)

$$H_3\overset{+}{N}-CH_2-CH_2-CH_2-\overset{\displaystyle O}{\overset{\displaystyle \|}{C}}-O^-$$

directly resulting in changes in ion flux in the target cell. Other neurotransmitter receptors are coupled to G proteins—a major group of signaling molecules (discussed later in this chapter) that link cell surface receptors to a variety of intracellular responses. In the case of neurotransmitter receptors, the associated G proteins frequently act to indirectly regulate ion channel activity.

Peptide Hormones and Growth Factors

The widest variety of signaling molecules in animals are peptides, ranging in size from only a few to more than a hundred amino acids. This group of signaling molecules includes peptide hormones, neuropeptides, and a diverse array of polypeptide growth factors (Table 13.1). Well-known examples of **peptide hormones** include insulin, glucagon, and the hormones produced by the pituitary gland (growth hormone, follicle-stimulating hormone, prolactin, and others).

Neuropeptides are secreted by some neurons instead of the small-molecule neurotransmitters discussed in the previous section. Some of these peptides, such as the **enkephalins** and **endorphins**, function not only as neurotransmitters at synapses but also as **neurohormones** that act on dis-

TABLE 13.1 Representative Peptide Hormones, Neuropeptides, and Growth Factors

Signaling molecule	Size[a]	Activities[b]
Peptide hormones		
Insulin	A = 21, B = 30	Regulation of glucose uptake; stimulation of cell proliferation
Glucagon	29	Stimulation of glucose synthesis
Growth hormone	191	General stimulation of growth
Follicle-stimulating hormone (FSH)	$\alpha = 92, \beta = 118$	Stimulation of the growth of oocytes and ovarian follicles
Prolactin	198	Stimulation of milk production
Neuropeptides and neurohormones		
Substance P	11	Sensory synaptic transmission
Oxytocin	9	Stimulation of smooth muscle contraction
Vasopressin	9	Stimulation of water reabsorption in the kidney
Enkephalins	5	Analgesics
β-Endorphin	31	Analgesic
Growth factors		
Nerve growth factor (NGF)	118	Differentiation and survival of neurons
Epidermal growth factor (EGF)	53	Proliferation of many types of cells
Platelet-derived growth factor (PDGF)	A = 125, B = 109	Proliferation of fibroblasts and other cell types
Interleukin-2	133	Proliferation of T lymphocytes
Erythropoietin	166	Development of red blood cells

[a] Size is indicated in number of amino acids. Some hormones and growth factors consist of two different polypeptide chains, which are designated either A and B or α and β.

[b] Most of these hormones and growth factors possess other activities in addition to those indicated.

tant cells. The enkephalins and endorphins have been widely studied because of their activity as natural analgesics that decrease pain responses in the central nervous system. Discovered during studies of drug addiction, they are naturally occurring compounds that bind to the same receptors on the surface of brain cells as morphine does.

The polypeptide **growth factors** include a wide variety of signaling molecules that control animal cell growth and differentiation. The first of these factors (**nerve growth factor,** or **NGF**) was discovered by Rita Levi-Montalcini in the 1950s. NGF is a member of a family of polypeptides (called **neurotrophins**) that regulate the development and survival of neurons. During the course of experiments on NGF, Stanley Cohen serendipitously discovered an unrelated factor (called **epidermal growth factor,** or **EGF**) that stimulates cell proliferation. EGF, a 53-amino-acid polypeptide (Figure 13.7), has served as the prototype of a large array of growth factors that play critical roles in controlling animal cell proliferation, both during embryonic development and in adult organisms.

A good example of growth factor action is provided by the activity of **platelet-derived growth factor** (**PDGF**) in wound healing. PDGF is stored in blood platelets and released during blood clotting at the site of a wound. It then stimulates the proliferation of fibroblasts in the vicinity of the clot, thereby contributing to regrowth of the damaged tissue. Members of another large group of polypeptide growth factors (called **cytokines**) regulate the development and differentiation of blood cells and control the activities of lymphocytes during the immune response. Other polypeptide growth factors (**membrane-anchored growth factors**) remain associated with the plasma membrane rather than being secreted into extracellular fluids, therefore functioning specifically as signaling molecules during direct cell-cell interactions.

Peptide hormones, neuropeptides, and growth factors are unable to cross the plasma membrane of their target cells, so they act by binding to cell surface receptors, as discussed later in this chapter. As might be expected from the critical roles of polypeptide growth factors in controlling cell proliferation, abnormalities in growth factor signaling are the basis for a variety of diseases, including many kinds of cancer. For example, abnormal expression of a close relative of the EGF receptor is an important factor in the development of many human breast and ovarian cancers.

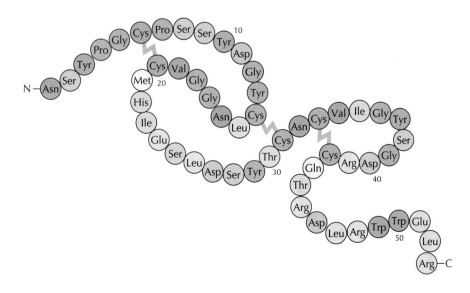

Figure 13.7
Structure of epidermal growth factor (EGF) EGF is a single polypeptide chain of 53 amino acids. Disulfide bonds between cysteine residues are indicated. (After G. Carpenter and S. Cohen, 1979. *Ann. Rev. Biochem.* 48: 193.)

Eicosanoids

Several types of lipids serve as signaling molecules that, in contrast to the steroid hormones, act by binding to cell surface receptors. The most important of these molecules are members of a class of lipids called the **eicosanoids**, which includes **prostaglandins**, **prostacyclin**, **thromboxanes**, and **leukotrienes** (Figure 13.8). The eicosanoids are rapidly broken down and therefore act locally in autocrine or paracrine signaling pathways. They stimulate a variety of responses in their target cells, including blood platelet aggregation, inflammation, and smooth-muscle contraction.

All eicosanoids are synthesized from arachidonic acid, which is formed from phospholipids. The first step in the pathway leading to synthesis of either prostaglandins or thromboxanes is the conversion of arachidonic acid to prostaglandin H_2. Interestingly, the enzyme that catalyzes this reaction (cyclooxygenase) is the target of aspirin and other nonsteroidal anti-inflammatory drugs. By inhibiting synthesis of the prostaglandins, aspirin reduces inflammation and pain. By inhibiting synthesis of thromboxane, aspirin also reduces platelet aggregation and blood clotting. Because of this activity, small daily doses of aspirin are frequently prescribed for prevention of strokes. In addition, aspirin and nonsteroidal anti-inflammatory drugs have been found to reduce the frequency of colon cancer in both animal models and humans, apparently by inhibiting the synthesis of prostaglandins that act to stimulate cell proliferation and promote cancer development.

Figure 13.8
Synthesis and structure of eicosanoids The eicosanoids include the prostaglandins, prostacyclin, thromboxanes, and leukotrienes. They are synthesized from arachidonic acid, which is formed by the hydrolysis of phospholipids catalyzed by phospholipase A_2 (PLA$_2$). Arachidonic acid can then be metabolized via two alternative pathways; one pathway leads to synthesis of prostaglandins, prostacyclin, and thromboxanes, while the other pathway leads to synthesis of leukotrienes.

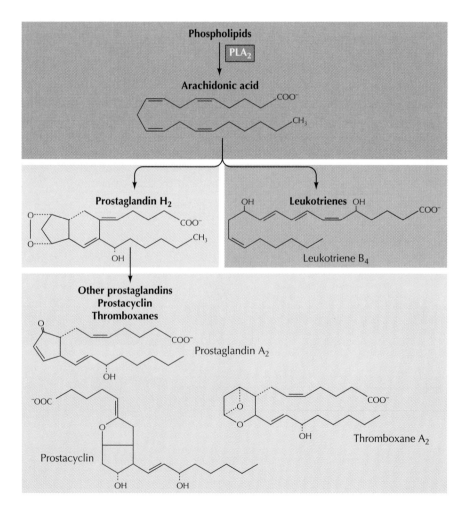

Plant Hormones

Plant growth and development are regulated by a group of small molecules called **plant hormones**. The levels of these molecules within the plant are typically modified by environmental factors, such as light or infection, so they coordinate the responses of tissues in different parts of the plant to environmental signals.

The plant hormones are generally divided into five major classes: **auxins**, **gibberellins**, **cytokinins**, **abscisic acid**, and **ethylene** (Figure 13.9), although several additional plant hormones have recently been discovered. The first plant hormone to be identified was auxin, with the early experiments leading to its discovery having been performed by Charles Darwin in the 1880s. One of the effects of auxins is to induce plant cell elongation by weakening the cell wall (see Figure 12.49). In addition, auxins regulate many other aspects of plant development, including cell division and differentiation. The other plant hormones likewise have multiple effects in their target tissues, including stem elongation (gibberellins), fruit ripening (ethylene), cell division (cytokinins), and the onset of dormancy (abscisic acid).

Our understanding of the molecular mechanisms of plant hormone action is less advanced than comparable studies of animal cells, and the receptors for plant hormones are just beginning to be identified and characterized. One area of noteworthy progress has been in understanding the mechanism by which plant cells respond to ethylene. Using the small weed *Arabidopsis* as a model, several of the genes required for ethylene responsiveness have been identified. These include genes encoding the ethylene receptor, which is similar to a family of receptors commonly found in bacteria and yeast. Additional genes that have been identified in the ethylene signaling pathway include a protein related to the Raf protein kinase, which plays a key role in animal cell signaling pathways (discussed later in this chapter), and transcription factors, which regulate the expression of ethylene-responsive genes.

Gibberellic acid
(a gibberellin)

Indole-3-acetic acid
(an auxin)

Ethylene

Zeatin
(a cytokinin)

Abscisic acid

Figure 13.9
Structure of plant hormones

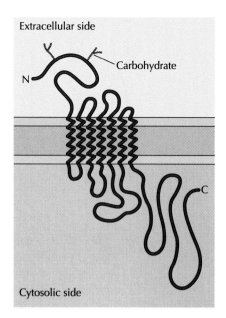

Extracellular side

Carbohydrate

N

C

Cytosolic side

Figure 13.10
Structure of a G protein-coupled receptor The G protein-coupled receptors are characterized by seven transmembrane α helices.

Functions of Cell Surface Receptors

As already reviewed, most ligands responsible for cell-cell signaling (including neurotransmitters, peptide hormones, and growth factors) bind to receptors on the surface of their target cells. Consequently, a major challenge in understanding cell-cell signaling is unraveling the mechanisms by which cell surface receptors transmit the signals initiated by ligand binding. As discussed in Chapter 12, some neurotransmitter receptors are ligand-gated ion channels that directly control ion flux across the plasma membrane. Other cell surface receptors, including the receptors for peptide hormones and growth factors, act instead by regulating the activity of intracellular proteins. These proteins then transmit signals from the receptor to a series of additional intracellular targets, frequently including transcription factors. Ligand binding to a receptor on the surface of the cell thus initiates a chain of intracellular reactions, ultimately reaching the target cell nucleus and resulting in programmed changes in gene expression. The functions of the major classes of cell surface receptors are discussed here, with the pathways of intracellular signaling downstream of these receptors being considered in the next section of this chapter.

G Protein-Coupled Receptors

The largest family of cell surface receptors transmit signals to intracellular targets via the intermediary action of guanine nucleotide-binding proteins called **G proteins**. More than a thousand such **G protein-coupled receptors** have been identified, including the receptors for many neurotransmitters, neuropeptides, and peptide hormones. In addition, the G protein-coupled receptor family includes a large number of receptors that are responsible for smell, sight, and taste.

The G protein-coupled receptors are structurally and functionally related proteins characterized by seven membrane-spanning α helices (Figure 13.10). The binding of ligands to the extracellular domain of these receptors induces a conformational change that allows the cytosolic domain of the receptor to bind to a G protein associated with the inner face of the plasma membrane. This interaction activates the G protein, which then dissociates from the receptor and carries the signal to an intracellular target, which may be either an enzyme or an ion channel.

The discovery of G proteins came from studies of hormones (such as epinephrine) that regulate the synthesis of cyclic AMP (cAMP) in their target cells. As discussed later in the chapter, cAMP is an important second messenger that mediates cellular responses to a variety of hormones. In the 1970s, Martin Rodbell and his colleagues made the key observation that GTP is required for hormonal stimulation of adenylyl cyclase (the enzyme responsible for cAMP formation). This finding led to the discovery that a guanine nucleotide-binding protein (called a G protein) is an intermediary in adenylyl cyclase activation (Figure 13.11). Since then, an array of G proteins have been found to act as physiological switches that regulate the activities of a variety of intracellular targets in response to extracellular signals.

G proteins consist of three subunits, designated α, β, and γ (Figure 13.12). They are frequently called **heterotrimeric G proteins** to distinguish them from other guanine nucleotide-binding proteins, such as the Ras proteins discussed later in the chapter. The α subunit binds guanine nucleotides, which regulate G protein activity. In the resting state, α is bound to GDP in a complex with β and γ. Hormone binding induces a conformational change in the receptor, such that the cytosolic domain of the receptor interacts with the G protein and stimulates the release of bound GDP and

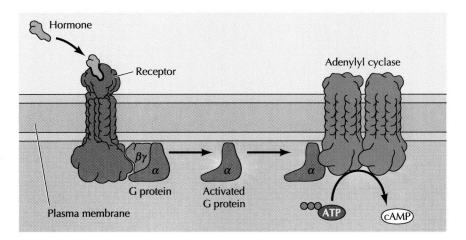

Figure 13.11
Hormonal activation of adenylyl cyclase   Binding of hormone promotes the interaction of the receptor with a G protein. The activated G protein α subunit then dissociates from the receptor and stimulates adenylyl cyclase, which catalyzes the conversion of ATP to cAMP.

its exchange for GTP. The activated GTP-bound α subunit then dissociates from β and γ, which remain together and function as a $\beta\gamma$ complex. Both the active GTP-bound α subunit and the $\beta\gamma$ complex then interact with their targets to elicit an intracellular response. The activity of the α subunit is terminated by hydrolysis of the bound GTP, and the inactive α subunit (now with GDP bound) then reassociates with the $\beta\gamma$ complex, ready for the cycle to start anew.

Mammalian genomes encode at least 20 different α subunits, 6 β subunits, and 12 γ subunits. Different G proteins associate with different receptors, so this array of G proteins couples receptors to distinct intracellular targets. For example, the G protein associated with the epinephrine recep-

Figure 13.12
Regulation of G proteins

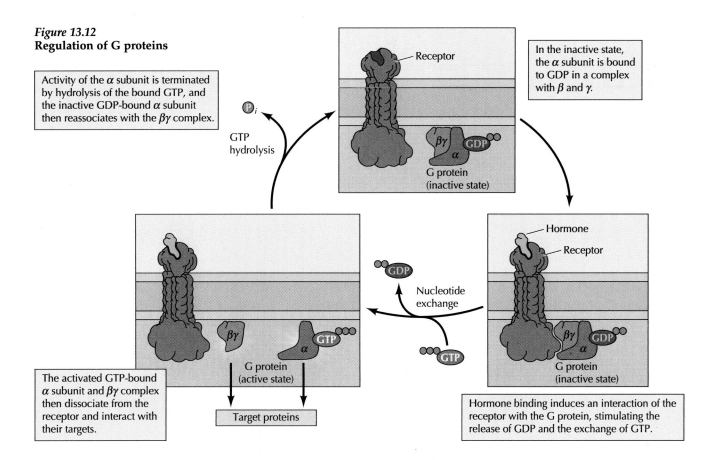

Activity of the α subunit is terminated by hydrolysis of the bound GTP, and the inactive GDP-bound α subunit then reassociates with the $\beta\gamma$ complex.

GTP hydrolysis

In the inactive state, the α subunit is bound to GDP in a complex with β and γ.

The activated GTP-bound α subunit and $\beta\gamma$ complex then dissociate from the receptor and interact with their targets.

Nucleotide exchange

Hormone binding induces an interaction of the receptor with the G protein, stimulating the release of GDP and the exchange of GTP.

Key Experiment

The Src Protein-Tyrosine Kinase

Transforming Gene Product of Rous Sarcoma Virus Phosphorylates Tyrosine

Tony Hunter and Bartholomew M. Sefton
The Salk Institute, San Diego, CA
Proceedings of the National Academy of Science, USA, 1980, Volume 77, pages 1311–1315

The Context

Following its isolation in 1911, Rous sarcoma virus (RSV) became the first virus that was generally accepted to cause tumors in animals (see the Molecular Medicine box in Chapter 1). Several features of RSV then made it an attractive model for studying the development of cancer. In particular, the small size of the RSV genome offered the hope of identifying specific viral genes responsible for inducing the abnormal proliferation that is characteristic of cancer cells. This goal was reached in the 1970s, when it was established that a single RSV gene (called *src* for *sarcoma*) is required for tumor induction. Importantly, a closely related *src* gene was also found to be part of the normal genetic complement of a variety of vertebrates, including humans. Since the viral Src protein is responsible for driving the uncontrolled proliferation of cancer cells, it appeared that understanding Src function would yield crucial insights into the molecular bases of both cancer induction and the regulation of normal cell proliferation.

In 1977, Ray Erikson and his colleagues identified the Src protein by immunoprecipitation (see Figure 3.32) with antisera from animals bearing RSV-induced tumors. Shortly thereafter, it was found that incubation of Src immunoprecipitates with radioactive ATP resulted in phosphorylation of the immunoglobulin molecules. Src therefore appeared to be a protein kinase, clearly implicating protein phosphorylation in the

control of cell proliferation.

All previously studied protein kinases phosphorylated serine or threonine residues, which were also the only phosphoamino acids to have been detected in animal cells. However, Walter Eckhardt and Tony Hunter had observed in 1979 that the oncogenic protein of another animal tumor virus (polyomavirus) was phosphorylated on a tyrosine residue. Hunter and Sefton therefore tested the possibility that Src might phosphorylate tyrosine, rather than serine/threonine, residues in its substrate proteins. Their experiments demonstrated that Src does indeed function as a protein-tyrosine kinase—an activity now recognized as playing a central role in cell signaling pathways.

The Experiments

Hunter and Sefton identified the amino acid phosphorylated by Src by incubating Src immunoprecipitates with ^{32}P-labeled ATP. The amino acid that was phosphorylated by Src in the substrate protein (in this case, immunoglobulin) therefore became radioactively labeled. The immunoglobulin was then isolated and hydrolyzed to yield individual amino acids, which were analyzed by electrophoresis and chromatography methods that separated phosphotyrosine, phosphoserine, and phosphothreonine (see figure). The radioactive amino acid detected in these experiments was phosphotyrosine, indicating that Src specifically phosphorylates tyrosine residues.

Further experiments showed that the normal cell Src protein, as well as viral Src, functions as a protein-tyrosine kinase in immunoprecipitation assays. In addition, Hunter and Sefton extended these *in vitro* experiments by demonstrating the presence of phosphotyrosine in proteins extracted from whole cells. In normal cells, phosphotyrosine accounted for only about 0.03% of total phospho-amino acids (the rest being phosphoserine and phosphothreonine), explaining why it had previously escaped detection. However, phosphotyrosine was about ten times more abundant in cells that were infected with RSV, suggesting that increased protein-tyrosine kinase activity of the viral Src protein was responsible for its ability to induce abnormal cell proliferation.

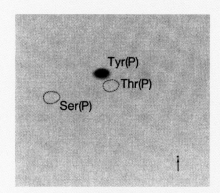

Identification of phosphotyrosine in immunoglobulin phosphorylated by Src. An immunoprecipitate containing RSV Src was incubated with [^{32}P]-ATP. The immunoglobulin was then isolated and hydrolyzed. Amino acids in the hydrolysate were separated by electrophoresis and chromatography on a cellulose thin-layer plate. The positions of ^{32}P-labeled amino acids were determined by exposing the plate to X-ray film. Broken lines indicate the positions of unlabeled phosphoamino acids that were included as markers. Note that the principal ^{32}P-labeled amino acid is phosphotyrosine.

The Src Protein-Tyrosine Kinase (continued)

The Impact

The discovery that Src was a protein-tyrosine kinase both identified a new protein kinase activity and established this activity as being related to the control of cell proliferation. The results of Hunter and Sefton were followed by demonstrations that many other tumor virus proteins also function as protein-tyrosine kinases, generalizing the link between protein-tyrosine phosphorylation and the abnormal proliferation of cancer cells. Stanley Cohen and his colleagues further found that the EGF receptor is a protein-tyrosine kinase, directly implicating protein-tyrosine phosphorylation in the control of normal cell proliferation. Continuing studies have identified numerous additional receptor and nonreceptor protein-tyrosine kinases that function in a variety of cell signaling pathways. Studies of the mechanism by which a virus causes cancer in chickens thus revealed a previously unknown enzymatic activity that plays a central role in the signaling pathways that regulate animal cell growth and differentiation.

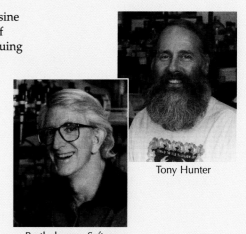

Tony Hunter

Bartholomew Sefton

tor is called G_s because its α subunit *s*timulates adenylyl cyclase (see Figure 13.11). Other G protein α and $\beta\gamma$ subunits act instead to inhibit adenylyl cyclase or to regulate the activities of other target enzymes.

In addition to regulating target enzymes, both the α and $\beta\gamma$ subunits of some G proteins directly regulate ion channels. A good example is provided by the action of the neurotransmitter acetylcholine on heart muscle, which is distinct from its effects on nerve and skeletal muscle. The acetylcholine receptor on nerve and skeletal muscle cells is a ligand-gated ion channel (see Figure 12.23). Heart muscle cells have a different acetylcholine receptor, which is G protein-coupled. This G protein is designated G_i because its α subunit *i*nhibits adenylyl cyclase. In addition, the G_i $\beta\gamma$ subunits act directly to open K^+ channels in the plasma membrane, which has the effect of slowing heart muscle contraction.

Receptor Protein-Tyrosine Kinases

In contrast to the G protein-coupled receptors, other cell surface receptors are directly linked to intracellular enzymes. The largest family of such enzyme-linked receptors are the **receptor protein-tyrosine kinases**, which phosphorylate their substrate proteins on tyrosine residues. This family includes the receptors for most polypeptide growth factors, so protein-tyrosine phosphorylation has been particularly well studied as a signaling mechanism involved in the control of animal cell growth and differentiation. Indeed, the first protein-tyrosine kinase was discovered in 1980 during studies of the oncogenic proteins of animal tumor viruses, in particular Rous sarcoma virus, by Tony Hunter and Bartholomew Sefton. The EGF receptor was then found to function as a protein-tyrosine kinase by Stanley Cohen and his colleagues, clearly establishing protein-tyrosine phosphorylation as a key signaling mechanism in the response of cells to growth factor stimulation.

By now more than 50 receptor protein-tyrosine kinases have been identified, including the receptors for EGF, NGF, PDGF, insulin, and many other growth factors. All these receptors share a common structural organization: an N-terminal extracellular ligand-binding domain, a single transmembrane α helix, and a cytosolic C-terminal domain with protein-tyro-

Figure 13.13
Organization of receptor protein-tyrosine kinases Each receptor consists of an N-terminal extracellular ligand-binding domain, a single transmembrane α helix, and a cytosolic C-terminal domain with protein-tyrosine kinase activity. The structures of three distinct subfamilies of receptor protein-tyrosine kinases are shown. The EGF receptor and insulin receptor both have cysteine-rich extracellular domains, whereas the PDGF receptor has immunoglobulin (Ig)-like domains. The PDGF receptor is also noteworthy in that its kinase domain is interrupted by an insert of approximately a hundred amino acids unrelated to those found in most other protein-tyrosine kinase catalytic domains. The insulin receptor is unusual in being a dimer of two pairs of polypeptide chains (designated α and β).

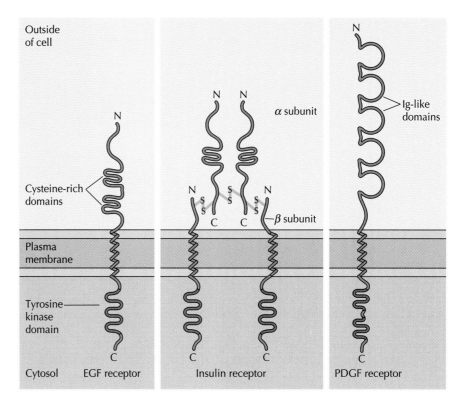

Figure 13.14
Dimerization and autophosphorylation of receptor protein-tyrosine kinases Growth factor binding induces receptor dimerization, which results in receptor autophosphorylation as the two polypeptide chains phosphorylate one another.

sine kinase activity (Figure 13.13). Most of the receptor protein-tyrosine kinases consist of single polypeptides, although the insulin receptor and some related receptors are dimers consisting of two pairs of polypeptide chains. The binding of ligands (e.g., growth factors) to the extracellular domains of these receptors activates their cytosolic kinase domains, resulting in phosphorylation of both the receptors themselves and intracellular target proteins that propagate the signal initiated by growth factor binding.

The first step in signaling from most receptor protein-tyrosine kinases is ligand-induced receptor dimerization (Figure 13.14). Some growth factors, such as PDGF and NGF, are themselves dimers consisting of two identical polypeptide chains; these growth factors directly induce dimerization by simultaneously binding to two different receptor molecules. Other growth

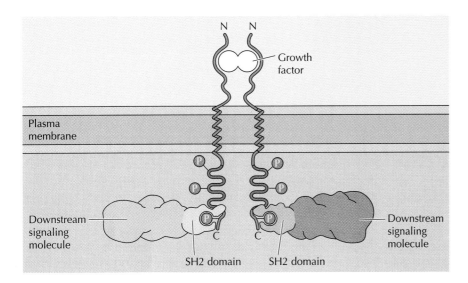

Figure 13.15
Association of downstream signaling molecules with receptor protein-tyrosine kinases SH2 domains bind to specific phosphotyrosine-containing peptides of the activated receptors.

factors (such as EGF) are monomers but have two distinct receptor binding sites that serve to crosslink receptors.

Ligand-induced dimerization then leads to **autophosphorylation** of the receptor as the dimerized polypeptide chains cross-phosphorylate one another (see Figure 13.14). Such autophosphorylation plays two key roles in signaling from these receptors. First, phosphorylation of tyrosine residues within the catalytic domain may play a regulatory role by increasing receptor protein kinase activity. Second, phosphorylation of tyrosine residues outside of the catalytic domain creates specific binding sites for additional proteins that transmit intracellular signals downstream of the activated receptors.

The association of these downstream signaling molecules with receptor protein-tyrosine kinases is mediated by protein domains that bind to specific phosphotyrosine-containing peptides (Figure 13.15). The best-characterized of these domains are called **SH2 domains** (for *Src h*omology 2) because they were first recognized in protein-tyrosine kinases related to Src, the oncogenic protein of Rous sarcoma virus. SH2 domains consist of approximately a hundred amino acids and bind to specific short peptide

Figure 13.16
Complex between an SH2 domain and a phosphotyrosine peptide The polypeptide chain of the Src SH2 domain is shown in red with its surface indicated by green dots. Purple spheres indicate a groove on the surface. The three amino acid residues that interact with the phosphotyrosine are shown in blue. The phosphotyrosine-containing peptide is shown as a space-filling model. Yellow and white spheres indicate the backbone and side-chain atoms, respectively, and the phosphate group is shown in red. (From G. Waksman and 13 others, 1992. *Nature* 358: 646.)

sequences containing phosphotyrosine residues (Figure 13.16). The resulting association of SH2-containing proteins with activated receptor protein-tyrosine kinases can have several effects: It localizes the SH2-containing proteins to the plasma membrane, leads to their association with other proteins, promotes their phosphorylation, and stimulates their enzymatic activities. The association of these proteins with autophosphorylated receptors thus represents the first step in the intracellular transmission of signals initiated by the binding of growth factors to the cell surface.

Cytokine Receptors and Nonreceptor Protein-Tyrosine Kinases

Rather than possessing intrinsic enzymatic activity, many receptors act by stimulating intracellular protein-tyrosine kinases with which they are noncovalently associated. This family of receptors (called the **cytokine receptor superfamily**) includes the receptors for most cytokines (e.g., interleukin-2 and erythropoietin) and for some polypeptide hormones (e.g., growth hormone). Like receptor protein-tyrosine kinases, the cytokine receptors contain N-terminal extracellular ligand-binding domains, single transmembrane α helices, and C-terminal cytosolic domains. However, the cytosolic domains of the cytokine receptors are devoid of any known catalytic activity. Instead, the cytokine receptors function in association with **nonreceptor protein-tyrosine kinases**, which are activated as a result of ligand binding.

The first step in signaling from cytokine receptors is thought to be ligand-induced receptor dimerization and cross-phosphorylation of the associated nonreceptor protein-tyrosine kinases (Figure 13.17). These activated kinases then phosphorylate the receptor, providing phosphotyrosine-binding sites for the recruitment of downstream signaling molecules that contain SH2 domains. Combinations of cytokine receptors plus associated nonreceptor protein-tyrosine kinases thus function analogously to the receptor protein-tyrosine kinases discussed in the previous section.

The nonreceptor protein-tyrosine kinases associated with the cytokine receptors fall into two major families. Many of these kinases are members of the **Src** family, which consists of Src and eight closely related proteins. As already noted, Src was initially identified as the oncogenic protein of Rous sarcoma virus and was the first protein shown to possess protein-tyrosine kinase activity, so it has played a pivotal role in experiments leading to our current understanding of cell signaling. In addition to Src family members, the cytokine receptors are associated with nonreceptor protein-tyrosine kinases belonging to the **Janus kinase**, or **JAK**, family. Members of the JAK family appear to be universally required for signaling from cytokine receptors, indicating that JAK family kinases play a critical role in coupling these receptors to the tyrosine phosphorylation of intracellular targets. In contrast, members of the Src family play key roles in signaling from antigen receptors on B and T lymphocytes but do not appear to be required for signaling from most cytokine receptors.

Receptors Linked to Other Enzymatic Activities

Although the vast majority of enzyme-linked receptors stimulate protein-tyrosine phosphorylation, some receptors are associated with other enzymatic activities. These receptors include protein-tyrosine phosphatases, protein-ser-

Figure 13.17
Signaling from cytokine receptors Ligand binding induces receptor dimerization and leads to the activation of associated nonreceptor protein-tyrosine kinases as a result of cross-phosphorylation. The activated kinases then phosphorylate tyrosine residues of the receptor, creating phosphotyrosine-binding sites for downstream signaling molecules.

ine/threonine kinases, and guanylyl cyclases. The functions of most of these receptors are less well understood than those of either the G protein-coupled receptors or the receptors associated with protein-tyrosine kinase activity.

Protein-tyrosine phosphatases remove phosphate groups from phosphotyrosine residues, thus acting to counterbalance the effects of protein-tyrosine kinases. In many cases, protein-tyrosine phosphatases play negative regulatory roles in cell signaling pathways by terminating the signals initiated by protein-tyrosine phosphorylation. However, some protein-tyrosine phosphatases are cell surface receptors whose enzymatic activities play a positive role in cell signaling. A good example is provided by a receptor called CD45, which is expressed on the surface of T and B lymphocytes. Following antigen stimulation, CD45 is thought to dephosphorylate a specific phosphotyrosine that inhibits the enzymatic activity of Src family members. Thus, the CD45 protein-tyrosine phosphatase acts (somewhat paradoxically) to stimulate nonreceptor protein-tyrosine kinases.

The receptors for **transforming growth factor β (TGF-β)** and related polypeptides are protein kinases that phosphorylate serine or threonine, rather than tyrosine, residues on their substrate proteins. TGF-β is the prototype of a family of polypeptide growth factors that control proliferation and differentiation of a variety of cell types, generally inhibiting proliferation of their target cells. The cloning of the first receptor for a member of the TGF-β family in 1991 revealed that it is the prototype of a unique receptor family with a cytosolic **protein-serine/threonine kinase** domain. Since then, receptors for additional TGF-β family members have similarly been found to be protein-serine/threonine kinases. The binding of ligand to these receptors results in the association of two distinct polypeptide chains, which are encoded by different members of the TGF-β receptor family, to form heterodimers in which the receptor kinases cross-phosphorylate one another. The activated TGF-β receptors then phosphorylate members of a family of transcription factors called SMADs, which translocate to the nucleus and stimulate expression of target genes.

Some peptide ligands bind to receptors whose cytosolic domains are guanylyl cyclases, which catalyze formation of cyclic GMP. As discussed earlier, nitric oxide also acts by stimulating guanylyl cyclase, but the target of nitric oxide is an intracellular enzyme rather than a transmembrane receptor. The receptor **guanylyl cyclases** have an extracellular ligand-binding domain, a single transmembrane α helix, and a cytosolic domain with catalytic activity. Ligand binding stimulates cyclase activity, leading to the formation of cyclic GMP—a second messenger whose intracellular effects are discussed in the next section of this chapter.

Other receptors bind to cytoplasmic proteins with additional biochemical activities. For example, the cytokine tumor necrosis factor (TNF) induces cell death, perhaps (as discussed later in this chapter) as a way of eliminating damaged or unwanted cells from tissues. The receptors for TNF and related death-signaling molecules are associated with specific proteases, which are activated in response to ligand binding. Activation of these receptor-associated proteases triggers the activation of additional downstream proteases, ultimately leading to degradation of a variety of intracellular proteins and death of the cell.

Pathways of Intracellular Signal Transduction

Most cell surface receptors stimulate intracellular target enzymes, which may be either directly linked or indirectly coupled to receptors by G proteins. These intracellular enzymes serve as downstream signaling elements that propagate and amplify the signal initiated by ligand binding. In most

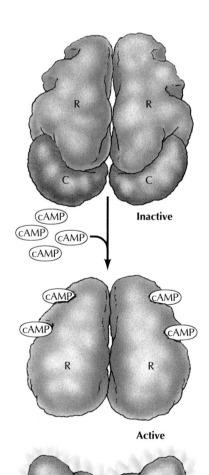

Figure 13.18
Synthesis and degradation of cAMP
Cyclic AMP is synthesized from ATP by adenylyl cyclase and degraded to AMP by cAMP phosphodiesterase.

cases, a chain of reactions transmits signals from the cell surface to a variety of intracellular targets—a process called **intracellular signal transduction**. The targets of such signaling pathways frequently include transcription factors that function to regulate gene expression. Intracellular signaling pathways thus connect the cell surface to the nucleus, leading to changes in gene expression in response to extracellular stimuli.

The cAMP Pathway: Second Messengers and Protein Phosphorylation

Intracellular signaling was first elucidated by studies of the action of hormones such as epinephrine, which signals the breakdown of glycogen to glucose in anticipation of muscular activity. In 1958, Earl Sutherland discovered that the action of epinephrine was mediated by an increase in the intracellular concentration of **cyclic AMP (cAMP)**, leading to the concept that cAMP is a **second messenger** in hormonal signaling (the first messenger being the hormone itself). Cyclic AMP is formed from ATP by the action of **adenylyl cyclase** and degraded to AMP by **cAMP phosphodiesterase** (Figure 13.18). As discussed earlier, the epinephrine receptor is coupled to adenylyl cyclase via a G protein that stimulates enzymatic activity, thereby increasing the intracellular concentration of cAMP (see Figure 13.11).

How does cAMP then signal the breakdown of glycogen? This and most other effects of cAMP in animal cells are mediated by the action of **cAMP-dependent protein kinase**, or **protein kinase A**, an enzyme discovered by Donal Walsh and Ed Krebs in 1968. The inactive form of protein kinase A is a tetramer consisting of two catalytic and two regulatory subunits (Figure 13.19). Cyclic AMP binds to the regulatory subunits, leading to their dissociation from the catalytic subunits. The free catalytic subunits are then enzymatically active and able to phosphorylate serine residues on their target proteins.

In the regulation of glycogen metabolism, protein kinase A phosphorylates two key target enzymes (Figure 13.20). The first is another protein kinase, phosphorylase kinase, which is phosphorylated and activated by protein kinase A. Phosphorylase kinase in turn phosphorylates and activates glycogen phosphorylase, which catalyzes the breakdown of glycogen to glucose-1-phosphate. In addition, protein kinase A phosphorylates the enzyme glycogen synthase, which catalyzes glycogen synthesis. In this case, however, phosphorylation inhibits enzymatic activity. Elevation of cAMP and activation of protein kinase A thus blocks further glycogen synthesis at the same time as it stimulates glycogen breakdown.

The chain of reactions leading from the epinephrine receptor to glycogen phosphorylase provides a good illustration of signal amplification during intracellular signal transduction. Each molecule of epinephrine activates

Figure 13.19
Regulation of protein kinase A   The inactive form of protein kinase A consists of two regulatory (R) and two catalytic (C) subunits. Binding of cAMP to the regulatory subunits induces a conformational change that leads to dissociation of the catalytic subunits, which are then enzymatically active.

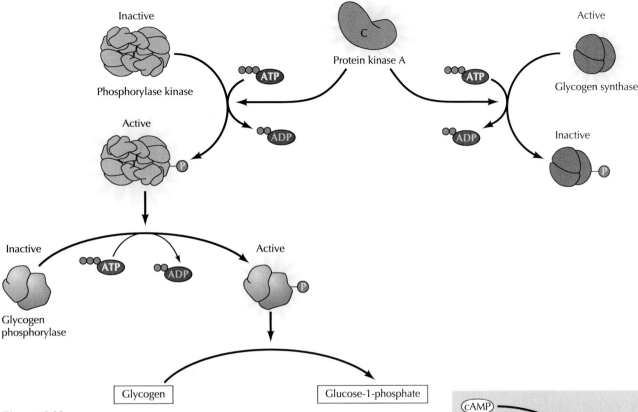

Figure 13.20
Regulation of glycogen metabolism by protein kinase A   Protein kinase A phosphorylates both glycogen synthase and phosphorylase kinase. Glycogen synthase (which catalyzes glycogen synthesis) is inhibited by this phosphorylation, whereas phosphorylase kinase is activated. Phosphorylase kinase then phosphorylates and activates glycogen phosphorylase, which catalyzes the breakdown of glycogen to glucose-1-phosphate.

only a single receptor. However, each receptor may activate up to a hundred molecules of G_s. Each molecule of G_s then stimulates the enzymatic activity of adenylyl cyclase, which can catalyze the synthesis of many molecules of cAMP. Signal amplification continues as each molecule of protein kinase A phosphorylates many molecules of phosphorylase kinase, which in turn phosphorylate many molecules of glycogen phosphorylase. Hormone binding to a small number of receptors thus leads to activation of a much larger number of intracellular target enzymes.

In many animal cells, increases in cAMP activate the transcription of specific target genes that contain a regulatory sequence called the **cAMP response element**, or **CRE** (Figure 13.21). In this case, the signal is carried from the cytoplasm to the nucleus by the catalytic subunit of protein kinase A, which is able to enter the nucleus following its release from the regulatory subunit. Within the nucleus, protein kinase A phosphorylates a transcription factor called **CREB** (for CRE-*b*inding protein), leading to the activation of cAMP-inducible genes. Such regulation of gene expression by

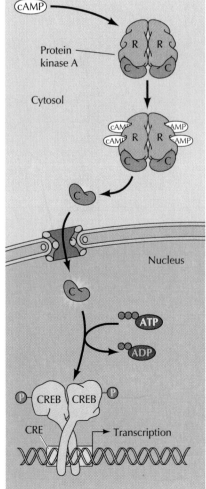

Figure 13.21
Cyclic AMP-inducible gene expression   The free catalytic subunit of protein kinase A translocates to the nucleus and phosphorylates the transcription factor CREB (CRE-binding protein), leading to expression of cAMP-inducible genes.

cAMP plays important roles in controlling the proliferation, survival, and differentiation of a wide variety of animal cells.

It is important to recognize that protein kinases, such as protein kinase A, do not function in isolation within the cell. To the contrary, protein phosphorylation is rapidly reversed by the action of protein phosphatases. Some protein phosphatases are transmembrane receptors, as discussed in the preceding section. A number of others are cytosolic enzymes that remove phosphate groups from either phosphorylated tyrosine or serine/threonine residues in their substrate proteins. These protein phosphatases serve to terminate the responses initiated by receptor activation of protein kinases. For example, the serine residues of proteins that are phosphorylated by protein kinase A are usually dephosphorylated by the action of a phosphatase called protein phosphatase 1 (Figure 13.22). The levels of phosphorylation of protein kinase A substrates (such as phosphorylase kinase and CREB) are thus determined by a balance between the intracellular activities of protein kinase A and protein phosphatases.

Although most effects of cAMP are mediated by protein kinase A, cAMP can also directly regulate ion channels, independent of protein phosphorylation. Cyclic AMP functions in this way as a second messenger involved in sensing smells. Many of the odorant receptors in sensory neurons in the nose are G protein-coupled receptors that stimulate adenylyl cyclase, leading to an increase in intracellular cAMP. Rather than stimulating protein kinase A, cAMP in this system directly opens Na^+ channels in the plasma membrane, leading to membrane depolarization and initiation of a nerve impulse.

Cyclic GMP

Cyclic GMP (cGMP) is also an important second messenger in animal cells, although its roles are not as clearly understood as those of cAMP. Cyclic GMP is formed from GTP by guanylyl cyclases and degraded to GMP by a phosphodiesterase. As discussed earlier in this chapter, different types of guanylyl cyclases are activated by both nitric oxide and peptide ligands. Stimulation of these guanylyl cyclases leads to elevated levels of cGMP, which then mediate biological responses, such as blood vessel dilation. The action of cGMP is frequently mediated by activation of a cGMP-dependent protein kinase, although cGMP can also act to regulate other targets, including ion channels.

The best-characterized role of cGMP is in the vertebrate eye, where it serves as the second messenger responsible for converting the visual signals received as light to nerve impulses. The photoreceptor in rod cells of the retina is a G protein-coupled receptor called **rhodopsin** (Figure 13.23). Rhodopsin is activated as a result of the absorption of light by the associated small molecule 11-*cis*-retinal, which then isomerizes to all-*trans*-retinal,

Figure 13.22
Regulation of protein phosphorylation by protein kinase A and protein phosphatase 1 The phosphorylation of target proteins by protein kinase A is reversed by the action of protein phosphatase 1.

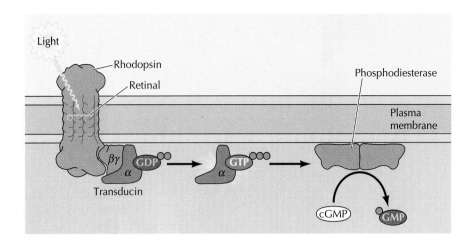

Figure 13.23
Role of cGMP in photoreception
Absorption of light by retinal activates
the G protein-coupled receptor
rhodopsin. The α subunit of transducin then stimulates cGMP phosphodiesterase, leading to a decrease in
intracellular levels of cGMP.

inducing a conformational change in the rhodopsin protein. Rhodopsin then
activates the G protein **transducin**, and the α subunit of transducin stimulates the activity of **cGMP phosphodiesterase**, leading to a decrease in the
intracellular level of cGMP. This change in cGMP level in retinal rod cells is
translated to a nerve impulse by a direct effect of cGMP on ion channels in
the plasma membrane, similar to the action of cAMP in sensing smells.

Phospholipids and Ca²⁺

One of the most widespread pathways of intracellular signaling is based
on the use of second messengers derived from the membrane phospholipid **phosphatidylinositol 4,5-bisphosphate (PIP₂)**. PIP_2 is a minor component of the plasma membrane, localized to the inner leaflet of the phospholipid bilayer (see Figure 12.2). A variety of hormones and growth
factors stimulate the hydrolysis of PIP_2 by **phospholipase C**—a reaction
that produces two distinct second messengers, **diacylglycerol** and **inositol
1,4,5-trisphosphate (IP₃)** (Figure 13.24). Diacylglycerol and IP_3 stimulate
distinct downstream signaling pathways (protein kinase C and Ca^{2+} mobilization, respectively), so PIP_2 hydrolysis triggers a two-armed cascade of
intracellular signaling.

Figure 13.24
Hydrolysis of PIP₂   Phospholipase C
(PLC) catalyzes the hydrolysis of
phosphatidylinositol 4,5-bisphosphate
(PIP_2) to yield diacylglycerol (DAG)
and inositol trisphosphate (IP_3).
Diacylglycerol activates members of
the protein kinase C family, and IP_3
signals the release of Ca^{2+} from intracellular stores.

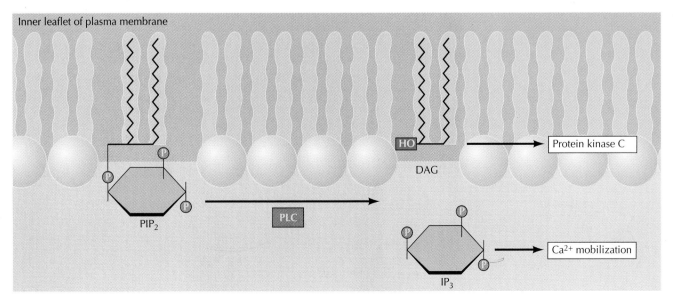

Figure 13.25
Activation of phospholipase C by protein-tyrosine kinases Phospholipase C-γ (PLC-γ) binds to activated receptor protein-tyrosine kinases via its SH2 domains. Tyrosine phosphorylation increases PLC-γ activity, stimulating the hydrolysis of PIP_2.

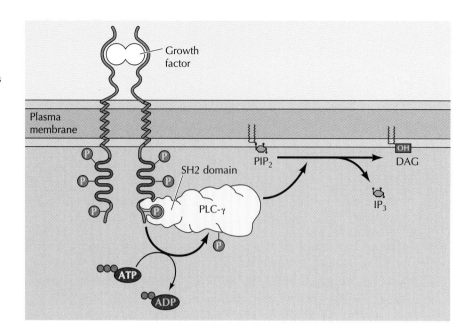

It is noteworthy that the hydrolysis of PIP_2 is activated downstream of both G protein-coupled receptors and protein-tyrosine kinases. This occurs because one form of phospholipase C (PLC-β) is stimulated by G proteins, whereas a second (PLC-γ) contains SH2 domains that mediate its association with activated receptor protein-tyrosine kinases (Figure 13.25). This interaction localizes PLC-γ to the plasma membrane as well as leading to its tyrosine phosphorylation, which increases its catalytic activity.

The diacylglycerol produced by hydrolysis of PIP_2 activates protein-serine/threonine kinases belonging to the **protein kinase C** family, many of which play important roles in the control of cell growth and differentiation. A good illustration of this role of protein kinase C is provided by the action of **phorbol esters** (Figure 13.26), which have been studied extensively because they promote the growth of tumors in animals. This tumor-promoting activity of the phorbol esters is based on their ability to stimulate protein kinase C by acting as analogs of diacylglycerol. Protein kinase C then activates other intracellular targets, including a cascade of protein kinases known as the MAP kinase pathway (discussed in detail in the next section), leading to transcription factor phosphorylation, changes in gene expression, and stimulation of cell proliferation.

Whereas diacylglycerol remains associated with the plasma membrane, the other second messenger produced by PIP_2 cleavage, IP_3, is a small polar molecule that is released into the cytosol, where it acts to signal the release of Ca^{2+} from intracellular stores (Figure 13.27). As noted in Chapter 12, the cytosolic concentration of Ca^{2+} is maintained at an extremely low level (about 0.1 μM) as a result of Ca^{2+} pumps that actively export Ca^{2+} from the cell. Ca^{2+} is pumped not only across the plasma membrane, but also into the endoplasmic reticulum, which therefore serves as an intracellular Ca^{2+} store. IP_3 acts to release Ca^{2+} from the endoplasmic reticulum by binding to

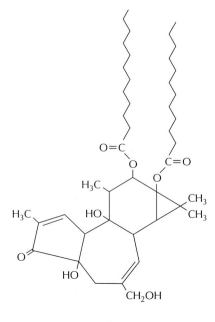

Figure 13.26
Structure of a phorbol ester Phorbol esters stimulate protein kinase C by acting as analogs of diacylglycerol.

Figure 13.27
Ca²⁺ mobilization by IP₃ Ca^{2+} is pumped from the cytosol into the endoplasmic reticulum, which therefore serves as an intracellular Ca^{2+} store. IP₃ binds to receptors that are ligand-gated Ca^{2+} channels in the endoplasmic reticulum membrane, thereby allowing the efflux of Ca^{2+} to the cytosol.

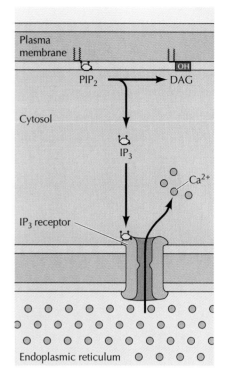

receptors that are ligand-gated Ca^{2+} channels. As a result, cytosolic Ca^{2+} levels increase to about 1 μM, which affects the activities of a variety of target proteins, including protein kinases and phosphatases. For example, some members of the protein kinase C family require Ca^{2+} as well as diacylglycerol for their activation, so these protein kinases are regulated jointly by both arms of the PIP₂ signaling pathway.

Many of the effects of Ca^{2+} are mediated by the Ca^{2+}-binding protein **calmodulin**, which is activated by Ca^{2+} binding when the concentration of cytosolic Ca^{2+} increases to about 0.5 μM (Figure 13.28). Ca^{2+}/calmodulin then binds to a variety of target proteins, including protein kinases. One example of such a Ca^{2+}/calmodulin-dependent protein kinase is myosin light-chain kinase, which signals actin-myosin contraction by phosphorylating one of the myosin light chains (see Figure 11.28). Other protein kinases that are activated by Ca^{2+}/calmodulin include members of the **CaM kinase** family, which phosphorylate a number of different proteins, including metabolic enzymes, ion channels, and transcription factors. One form of CaM kinase is particularly abundant in the nervous system, where it regulates the synthesis and release of neurotransmitters. In addition, CaM kinases can regulate gene expression by phosphorylating transcription factors. Interestingly, one of the transcription factors phosphorylated by CaM kinase is CREB, which (as discussed earlier) is phosphorylated at the same site by protein kinase A. This phosphorylation of CREB illustrates one of many intersections between the Ca^{2+} and cAMP signaling pathways. Other examples include the regulation of adenylyl cyclases and phosphodiesterases by Ca^{2+}/calmodulin, the regulation of Ca^{2+} channels by cAMP, and the phosphorylation of a number of target proteins by both protein kinase A and Ca^{2+}/calmodulin-dependent protein kinases. The cAMP and Ca^{2+} signaling pathways thus function coordinately to regulate many cellular responses.

Ca^{2+} is an extremely common second messenger, and it is important to note that IP₃-mediated release of Ca^{2+} from the endoplasmic reticulum is not the only mechanism by which the intracellular concentration of Ca^{2+} can be increased. One alternative pathway involves the entry of extracellular Ca^{2+} through Ca^{2+} channels in the plasma membrane. In many cells, the transient increase in intracellular Ca^{2+} resulting from production of IP₃ is followed by a more sustained increase resulting from extracellular Ca^{2+} entry. The entry of extracellular Ca^{2+} is particularly important in the electrically excitable cells of nerve and muscle, in which voltage-gated Ca^{2+} channels in the plasma membrane are opened by membrane depolarization (Figure 13.29). The resulting increases in intracellular Ca^{2+} then trigger the further release of Ca^{2+} from intracellular stores by activating distinct Ca^{2+} channels known as **ryanodine receptors**. One effect of increases in intracel-

Figure 13.28
Function of calmodulin   Calmodulin is a dumbbell-shaped protein with four Ca^{2+}-binding sites. The active Ca^{2+}/calmodulin complex binds to a variety of target proteins, including Ca^{2+}/calmodulin-dependent protein kinases.

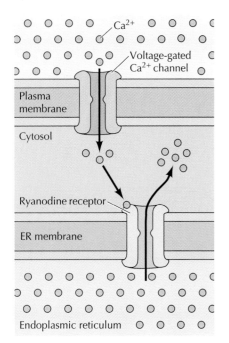

Figure 13.29
Regulation of intracellular Ca²⁺ in electrically excitable cells   Membrane depolarization leads to the opening of voltage-gated Ca^{2+} channels in the plasma membrane, causing the influx of Ca^{2+} from extracellular fluids. The resulting increase in intracellular Ca^{2+} then signals the further release of Ca^{2+} from intracellular stores by opening distinct Ca^{2+} channels (ryanodine receptors) in the endoplasmic reticulum membrane. In muscle cells, ryanodine receptors in the sarcoplasmic reticulum may also be opened directly in response to membrane depolarization.

lular Ca^{2+} in neurons is to trigger the release of neurotransmitters, so Ca^{2+} plays a critical role in converting electric to chemical signals in the nervous system. In muscle cells, Ca^{2+} is stored in the sarcoplasmic reticulum, from which it is released by the opening of ryanodine receptors in response to changes in membrane potential. This release of stored Ca^{2+} leads to large increases in cytosolic Ca^{2+}, which trigger muscle contraction (see Chapter 11). Cells thus utilize a variety of mechanisms to regulate intracellular Ca^{2+} levels, making Ca^{2+} a versatile second messenger that controls a wide range of cellular processes.

PIP_2 not only serves as the source of diacylglycerol and IP_3, but is also the starting point of a distinct second messenger pathway that plays a key role in regulating cell survival. In this pathway, PIP_2 is phosphorylated on the 3 position of inositol by the enzyme **phosphatidylinositide (PI) 3-kinase** (Figure 13.30). Like phospholipase C, one form of PI 3-kinase is activated by G proteins, while a second has SH2 domains and is activated by association with receptor protein-tyrosine kinases. Phosphorylation of PIP_2 yields **phosphatidylinositol 3,4,5-trisphosphate (PIP_3)**, which functions as a distinct second messenger. A key target of PIP_3, which is critical for signaling cell survival, is a protein-serine/threonine kinase called **Akt**. PIP_3 binds to a domain of Akt known as the pleckstrin homology domain (Figure 13.31). This interaction recruits Akt to the inner face of the plasma membrane, where it is phosphorylated and activated by other protein kinases (called PDKs) that also contain pleckstrin homology domains and bind PIP_3. The formation of PIP_3 thus results in the association of both Akt and PDKs with the plasma membrane, leading to phosphorylation and activation of Akt. Once activated, Akt phosphorylates a number of target proteins, including proteins that are direct regulators of cell survival, transcription factors, and other protein kinases that regulate cell metabolism and protein synthesis.

Second messengers can also be derived from other phospholipids. The hydrolysis of phosphatidylcholine is stimulated by a variety of growth factors, providing a second source of diacylglycerol, in addition to that derived from PIP_2. While PIP_2 hydrolysis is a transient response to growth factor stimulation, the hydrolysis of phosphatidylcholine typically persists for several hours, providing a sustained source of diacylglycerol that may be important in signaling long-term responses, such as cell proliferation. Sphingomyelin is also cleaved in response to a variety of extracellular stimuli, resulting in the formation of ceramide. Although its targets remain to be fully elucidated, ceramide regulates a number of protein kinases and phosphatases that can affect cell proliferation and survival.

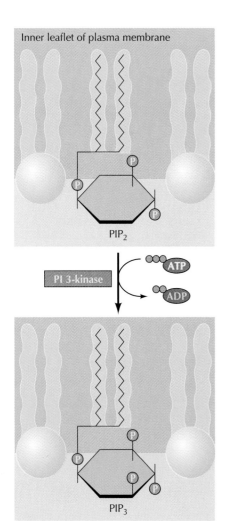

Inner leaflet of plasma membrane

PIP₂

PI 3-kinase

ATP

ADP

PIP₃

Figure 13.30
Activity of PI 3-kinase   PI 3-kinase phosphorylates the 3 position of inositol, converting PIP_2 to PIP_3.

Figure 13.31
Activation of the Akt protein kinase
Akt is recruited to the plasma membrane by binding to PIP$_3$ via its pleckstrin homology (PH) domain. It is then activated as a result of phosphorylation by another protein kinase (PDK) that also binds PIP$_3$.

PH domain

Akt

PH domain

PDK

Akt

Ras, Raf, and the MAP Kinase Pathway

The MAP kinase pathway refers to a cascade of protein kinases that are highly conserved in evolution and play central roles in signal transduction in all eukaryotic cells, ranging from yeasts to humans. The central elements in the pathway are a family of protein-serine/threonine kinases called the **MAP kinases** (for *m*itogen-*a*ctivated *p*rotein kinases) that are activated in response to a variety of growth factors and other signaling molecules. In yeasts, MAP kinase pathways control a variety of cellular responses, including mating, cell shape, and sporulation. In higher eukaryotes (including *C. elegans, Drosophila*, frogs, and mammals), MAP kinases are ubiquitous regulators of cell growth and differentiation.

The best-characterized forms of MAP kinase in mammalian cells belong to the **ERK** (*e*xtracellular signal-*r*egulated *k*inase) family. ERK activation plays a central role in signaling cell proliferation induced by growth factors that act through either protein-tyrosine kinase or G protein-coupled receptors. As already noted, protein kinase C can also activate the ERK pathway, which appears to be responsible for the stimulation of cell proliferation induced by phorbol ester tumor promoters. In addition, both the Ca^{2+} and

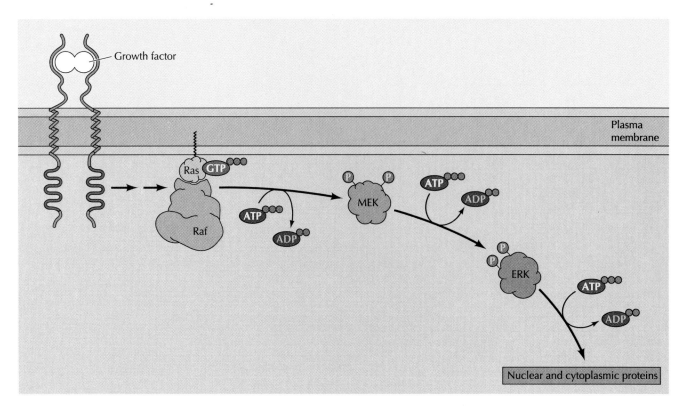

Figure 13.32
Activation of the ERK MAP kinases
Stimulation of growth factor receptors leads to activation of the small GTP-binding protein Ras, which interacts with the Raf protein kinase. Raf phosphorylates and activates MEK, a dual-specificity protein kinase that activates ERK by phosphorylation on both threonine and tyrosine residues (Thr-183 and Tyr-185). ERK then phosphorylates a variety of nuclear and cytoplasmic target proteins.

cAMP pathways intersect with ERK signaling, either stimulating or inhibiting the ERK pathway in different types of cells.

Activation of ERK is mediated by two upstream protein kinases, which are coupled to growth factor receptors by a GTP-binding protein called **Ras** (Figure 13.32). Activation of Ras leads to activation of the **Raf** protein-serine/threonine kinase, which phosphorylates and activates a second protein kinase called **MEK** (for *M*AP kinase/*E*RK *k*inase). MEK is a dual-specificity protein kinase that activates members of the ERK family by phosphorylation of both threonine and tyrosine residues separated by one amino acid (e.g., threonine-183 and tyrosine-185 of ERK2). Once activated, ERK phosphorylates a variety of targets, including other protein kinases and transcription factors.

The central role of the ERK pathway in mammalian cells emerged from studies of the Ras proteins, which were first identified as the oncogenic proteins of tumor viruses that cause sarcomas in rats (hence the name Ras, from *ra*t *s*arcoma virus). Interest in Ras intensified considerably in 1982, when mutations in *ras* genes were first implicated in the development of human cancers (discussed in Chapter 15). The importance of Ras in intracellular signaling was then indicated by experiments showing that microinjection of active Ras protein directly induces proliferation of normal mammalian cells. Conversely, interference with Ras function by either microinjection of anti-Ras antibody or expression of a dominant negative Ras mutant blocks growth factor-induced cell proliferation. Thus, Ras is not only capable of inducing the abnormal growth characteristic of cancer cells, but also appears to be required for the response of normal cells to growth factor stimulation.

The Ras proteins are guanine nucleotide-binding proteins that function analogously to the α subunits of G proteins, alternating between inactive GDP-bound and active GTP-bound forms (Figure 13.33). In contrast to the G protein α subunits, however, Ras functions as a monomer rather than in

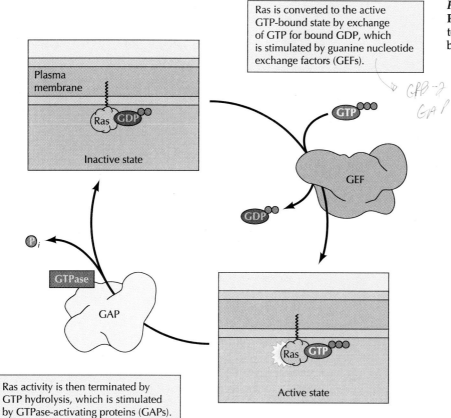

Ras is converted to the active GTP-bound state by exchange of GTP for bound GDP, which is stimulated by guanine nucleotide exchange factors (GEFs).

Ras activity is then terminated by GTP hydrolysis, which is stimulated by GTPase-activating proteins (GAPs).

Figure 13.33
Regulation of Ras proteins Ras proteins alternate between inactive GDP-bound and active GTP-bound states.

association with $\beta\gamma$ subunits. Ras activation is mediated by **guanine nucleotide exchange factors** that stimulate the release of bound GDP and its exchange for GTP. Activity of the Ras-GTP complex is then terminated by GTP hydrolysis, which is stimulated by the interaction of Ras-GTP with **GTPase-activating proteins**. It is interesting to note that the mutations of *ras* genes in human cancers have the effect of inhibiting GTP hydrolysis by the Ras proteins. These mutated Ras proteins therefore remain continuously in the active GTP-bound form, driving the unregulated proliferation of cancer cells even in the absence of growth factor stimulation.

The Ras proteins are prototypes of a large family of approximately 50 related proteins, frequently called **small GTP-binding proteins** because Ras and its relatives are about half the size of G protein α subunits. While the Ras proteins regulate cell growth and differentiation, other subfamilies of small GTP-binding proteins control distinct cellular activities. For example, the largest subfamily of small GTP-binding proteins (the Rab proteins) function to regulate vesicle trafficking, as discussed in Chapter 9. Other small GTP-binding proteins are involved in nuclear protein import (the Ran protein, discussed in Chapter 8) and organization of the cytoskeleton (the Rho subfamily, discussed later in this chapter).

The best understood mode of Ras activation is that mediated by receptor protein-tyrosine kinases (Figure 13.34). Autophosphorylation of these receptors results in their association with Ras guanine nucleotide exchange factors as a result of SH2-mediated protein interactions. One well-characterized example is provided by the guanine nucleotide exchange factor Sos, which is bound to the SH2-containing protein Grb2 in the cytosol of

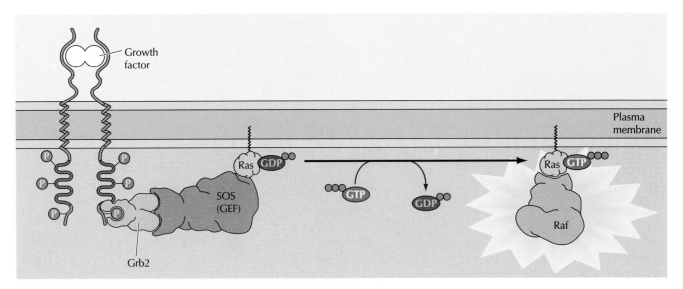

Figure 13.34
Ras activation downstream of receptor protein-tyrosine kinases A complex of Grb2 and the guanine nucleotide exchange factor Sos binds to a phosphotyrosine-containing sequence in the activated receptor via the Grb2 SH2 domain. This interaction recruits Sos to the plasma membrane, where it can stimulate Ras GDP/GTP exchange. The activated Ras-GTP complex then binds to the Raf protein kinase.

unstimulated cells. Tyrosine phosphorylation of receptors (or of other receptor-associated proteins) creates a binding site for the Grb2 SH2 domains. Association of Grb2 with activated receptors localizes Sos to the plasma membrane, where it is able to interact with Ras proteins, which are anchored to the inner leaflet of the plasma membrane by lipids attached to the Ras C terminus (see Figure 12.10). Sos then stimulates guanine nucleotide exchange, resulting in formation of the active Ras-GTP complex. In its active GTP-bound form, Ras interacts with a number of effector proteins, including the Raf protein-serine/threonine kinase. This interaction with Ras recruits Raf from the cytosol to the plasma membrane, where it is activated as a result of phosphorylation by both protein-tyrosine and protein-serine/threonine kinases.

As already noted, activation of Raf initiates a protein kinase cascade leading to ERK activation. ERK then phosphorylates a variety of target proteins, including other protein kinases. Importantly, a fraction of activated ERK translocates to the nucleus, where it regulates transcription factors by phosphorylation (Figure 13.35). In this regard, it is notable that a primary response to growth factor stimulation is the rapid transcriptional induction of a family of approximately 100 genes called **immediate-early genes**. The induction of a number of immediate-early genes is mediated by a regulatory sequence called the **serum response element (SRE)**, which is recognized by a complex of transcription factors including the **serum response factor (SRF)** and **Elk-1**. ERK phosphorylates and activates Elk-1, providing a direct link between the ERK family of MAP kinases and immediate-early gene induction. Many immediate-early genes themselves encode transcription factors, so their induction in response to growth factor stimulation leads to altered expression of a battery of other downstream genes, thereby establishing new programs of gene expression.

Figure 13.35
Induction of immediate-early genes by ERK Activated ERK translocates to the nucleus, where it phosphorylates the transcription factor Elk-1. Elk-1 binds to the serum response element (SRE) in a complex with serum response factor (SRF). Phosphorylation stimulates the activity of Elk-1 as a transcriptional activator, leading to immediate-early gene induction.

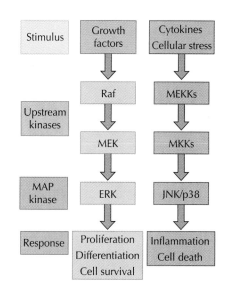

Figure 13.36
Pathways of MAP kinase activation in mammalian cells In addition to ERK, mammalian cells contain JNK and p38 MAP kinases. Activation of JNK and p38 is mediated by protein kinase cascades parallel to that responsible for ERK activation. The protein kinase cascades leading to JNK and p38 activation appear to be preferentially activated by cytokines or cellular stress and generally lead to inflammation and cell death.

Both yeasts and mammalian cells have multiple MAP kinase pathways that control distinct cellular responses. Each cascade consists of three protein kinases: a terminal MAP kinase and two upstream kinases (analogous to Raf and MEK) that regulate its activity. In the yeast *S. cerevisiae*, five different MAP kinase cascades regulate mating, sporulation, filamentation, cell wall remodeling, and response to high osmolarity. In mammalian cells, at least five MAP kinases have been identified. In addition to members of the ERK family, these include the JNK and p38 MAP kinases, which are preferentially activated in response to inflammatory cytokines and cellular stress (e.g., ultraviolet irradiation) (Figure 13.36). Whereas ERK signaling principally leads to cell proliferation, survival, and differentiation, the JNK and p38 MAP kinase pathways often lead to inflammation and cell death. Like ERK, the JNK and p38 MAP kinases can translocate to the nucleus and phosphorylate transcription factors that regulate gene expression. Multiple MAP kinase pathways thus function in all types of eukaryotic cells to control cellular responses to diverse environmental signals.

The JAK/STAT Pathway

The MAP kinase pathway provides an indirect connection between the cell surface and the nucleus, in which a cascade of protein kinases ultimately leads to transcription factor phosphorylation. An alternative pathway, known as the **JAK/STAT pathway**, provides a much more immediate connection between protein-tyrosine kinases and transcription factors. In this pathway protein-tyrosine phosphorylation directly affects transcription factor localization and function (Figure 13.37).

The key elements in this pathway are the **STAT proteins** (signal *trans*ducers and *a*ctivators of *t*ranscription), which were originally identified in studies of cytokine receptor signaling. The STAT proteins are a family of transcription factors that contain SH2 domains. They are inactive in unstimulated cells, where they are localized to the cytoplasm. Stimulation of cytokine receptors leads to recruitment of STAT proteins, which bind via their SH2 domains to phosphotyrosine-containing sequences in the cytoplasmic domains of receptor polypeptides. Following their association with activated receptors, the STAT proteins are phosphorylated by members of the JAK family of nonreceptor protein-tyrosine kinases, which are associated with cytokine receptors. Tyrosine phosphorylation promotes the dimerization of STAT proteins, which then translocate to the nucleus, where they stimulate transcription of their target genes.

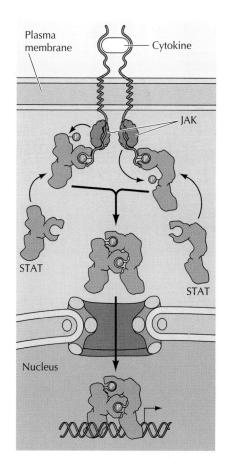

Figure 13.37
The JAK/STAT pathway The STAT proteins are transcription factors that contain SH2 domains that mediate their binding to phosphotyrosine-containing sequences. In unstimulated cells, STAT proteins are inactive in the cytosol. Stimulation of cytokine receptors leads to the binding of STAT proteins, where they are phosphorylated by the receptor-associated JAK protein-tyrosine kinases. The phosphorylated STAT proteins then dimerize and translocate to the nucleus, where they activate the transcription of target genes.

Molecular Medicine

Cancer, Signal Transduction, and the ras Oncogenes

The Disease

Cancer claims the lives of approximately one out of every four Americans, accounting for 550,000 deaths each year in the United States. There are more than a hundred different kinds of cancer, but some are more common than others. In this country, the most common lethal cancers are those of the lung and colon, which together account for about 40% of all cancer deaths. Other major contributors to cancer mortality include cancers of the breast, prostate, and pancreas, which are responsible for approximately 8%, 7%, and 5% of U.S. cancer deaths, respectively.

The common feature of all cancers is the unrestrained proliferation of cancer cells, which eventually spread throughout the body, invading normal tissues and organs and leading to death of the patient. Surgery and radiotherapy are effective treatments for localized cancers but are unable to reach cancer cells that have spread to distant body sites. Treatment of these cancers therefore requires the use of chemotherapeutic drugs. Unfortunately, the currently available chemotherapeutic agents are not specific for cancer cells. Most act by either damaging DNA or interfering with DNA synthesis, so they also kill rapidly dividing normal cells, such as the epithelial cells that line the digestive tract and the blood-forming cells of the bone marrow. The resulting toxicity of these drugs limits their effectiveness, and many cancers are not eliminated by doses of chemotherapy that can be tolerated by the patient. Consequently, although major progress has been made in cancer treatment, nearly half of all patients diagnosed with cancer ultimately die of their disease.

Molecular and Cellular Basis

The identification of viral genes that can convert normal cells to cancer cells, such as the *src* gene of RSV, provided the first demonstration that cancers can result from the action of specific genes (oncogenes). The subsequent discovery that viral oncogenes are related to genes of normal cells then engendered the hypothesis that non-virus-induced cancers (including most human cancers) might arise as a result of mutations in normal cell genes, giving rise to oncogenes of cellular rather than viral origin. Such cellular oncogenes were first identified in human cancers in 1981. Shortly thereafter, human oncogenes of bladder, lung, and colon cancers were found to be related to the *ras* genes previously identified in rat sarcoma viruses.

Although many different genes are now known to play critical roles in cancer development, mutations of the *ras* genes remain one of the most common genetic abnormalities in human tumors. Mutated *ras* oncogenes are found in about 20–30% of all human cancers, including approximately 25% of lung cancers, 50% of colon cancers, and more than 90% of pancreatic can-cers. Moreover, the action of *ras* oncogenes has clearly linked the development of human cancer to abnormalities in the signaling pathways that regulate cell proliferation. The mutations that convert normal *ras* genes to oncogenes substantially decrease GTP hydrolysis by the Ras proteins. Consequently, the mutated oncogenic Ras proteins remain locked in the active GTP-bound form, rather than alternating normally between inactive and active states in response to extracellular signals. The oncogenic Ras proteins thus continuously stimulate the MAP kinase pathway and drive cell proliferation, even in the absence of the growth factors that would be required to activate Ras and signal proliferation of normal cells.

A human colon cancer. The *ras* oncogenes contribute to the development of about half of these cancers. (John Cabisco/ Visuals Unlimited).

Further studies have shown that STAT proteins are also activated downstream of receptor protein-tyrosine kinases, where their phosphorylation may be catalyzed either by the receptors themselves or by associated nonreceptor kinases. The STAT transcription factors thus serve as direct links between both cytokine and growth factor receptors on the cell surface and regulation of gene expression in the nucleus.

Cancer, Signal Transduction, and the ras Oncogenes *(continued)*

Prevention and Treatment

The discovery of mutated oncogenes in human cancers raises the possibility of developing drugs specifically targeted against the oncogene proteins. In principle, such drugs might act selectively against cancer cells with less toxicity toward normal cells than that of conventional chemotherapeutic agents. Because *ras* is frequently mutated in human cancers, the Ras proteins have attracted considerable interest as potential drug targets.

An important aspect of Ras function is the targeting of Ras proteins to the plasma membrane by the post-translational addition of lipid (a farnesyl isoprenoid) to their C terminus (see Figure 7.30). Although farnesylation is not unique to Ras, it is a relatively uncommon modification of cellular proteins, leading several research groups to develop inhibitors of the enzyme farnesyl transferase as potential Ras-targeted drugs. Such drugs have now been found not only to inhibit Ras membrane localization and

function, but also to display substantial activity against human tumor cells. At least four farnesyl transferase inhibitors are currently being evaluated in clinical trials, so direct data on the potential of these drugs in the treatment of human cancers should be forthcoming in the near future.

Reference

Oliff, A. 1999. Farnesyltransferase inhibitors: targeting the molecular basis of cancer. *Biochim. Biophys. Acta* 1423: C19–C30.

Signal Transduction and the Cytoskeleton

The preceding sections focused on signaling pathways that regulate changes in metabolism or gene expression in response to hormones and growth factors. However, the functions of most cells are also directly affected by cell adhesion and the organization of the cytoskeleton. The receptors responsible for cell adhesion thus act to initiate intracellular signaling pathways that regulate other aspects of cell behavior, including gene expression. Conversely, growth factors frequently act to induce cytoskeletal alterations resulting in cell movement or changes in cell shape. Components of the cytoskeleton thus act as both receptors and targets in cell signaling pathways, integrating cell shape and movement with other cellular responses.

Integrins and Signal Transduction

As discussed in Chapters 11 and 12, the integrins are the major receptors responsible for the attachment of cells to the extracellular matrix. At two types of cell-matrix junctions (focal adhesions and hemidesmosomes), the integrins also interact with components of the cytoskeleton to provide a stable linkage between the extracellular matrix and adherent cells (see Figure 12.61). In addition to this structural role, the integrins serve as receptors that activate intracellular signaling pathways, thereby controlling gene expression and other aspects of cell behavior in response to adhesive interactions.

Like members of the cytokine receptor superfamily, the integrins have short cytoplasmic tails that lack any intrinsic enzymatic activity. However, protein-tyrosine phosphorylation is an early response to the interaction of integrins with extracellular matrix components, suggesting that the integrins are linked to nonreceptor protein-tyrosine kinases. In particular, a nonreceptor protein-tyrosine kinase called **FAK** (*focal adhesion kinase*) plays a key role in integrin signaling (Figure 13.38). As its name implies, FAK is localized to focal adhesions and rapidly becomes tyrosine-phosphorylated following the binding of integrin to extracellular matrix components, such as fibronectin. Like other protein-tyrosine kinases, the activation of FAK is thought to involve autophosphorylation induced by the clustering of integrins bound to the extracellular matrix, although details of the mechanism of FAK activation remain to be established.

Figure 13.38
Model for signaling from the FAK protein-tyrosine kinase   Binding of integrins to the extracellular matrix stimulates FAK activity, leading to its autophosphorylation. Src then binds to the FAK autophosphorylation site and phosphorylates FAK on additional tyrosine residues. These phosphotyrosines serve as binding sites for the Grb2-Sos complex, leading to activation of Ras and the MAP kinase cascade, as well as for additional downstream signaling molecules, including PI 3-kinase.

In addition to FAK, members of the Src family of nonreceptor protein-tyrosine kinases also associate with focal adhesions and are involved in integrin signaling. Interestingly, Src and FAK appear to function in association with each other as a result of the binding of the Src SH2 domain to an autophosphorylation site of FAK (see Figure 13.38). Src then phosphorylates additional sites on FAK, so these two nonreceptor protein-tyrosine kinases act jointly in signaling from integrin receptors.

As discussed earlier for growth factor receptors, tyrosine phosphorylation of FAK creates binding sites for the SH2 domains of other downstream signaling molecules, including PI 3-kinase and the Grb2-Sos complex. Recruitment of the Sos guanine nucleotide exchange factor leads to activation of Ras, which in turn couples integrins to activation of the ERK MAP kinase signaling pathway. Integrin activation of the FAK and Src nonreceptor protein-tyrosine kinases thus links cell adhesion to changes in gene expression and cell behavior that are analogous to those induced by the binding of growth factors to their cell surface receptors.

Regulation of the Actin Cytoskeleton

Cellular responses to extracellular signals, including growth factors, frequently include changes in cell movement and cell shape. For example, growth factor-induced alterations in cell motility (as well as in cell proliferation) play critical roles in processes such as wound healing and embryonic development. As discussed in Chapter 11, these aspects of cell behavior are governed principally by the actin cytoskeleton. In particular, many types of cell movement are based on the dynamic assembly and disassembly of actin filaments underlying the plasma membrane. Remodeling of the actin cytoskeleton therefore represents a key element of the response of many cells to growth factors and other extracellular stimuli.

Members of the Rho subfamily of small GTP-binding proteins (including **Rho**, **Rac**, and **Cdc42**) play central roles in regulating the organization of the actin cytoskeleton and thus control a variety of cell processes, including cell motility, cell adhesion, and cytokinesis. The role of Rho family members in regulating different aspects of actin remodeling was first elucidated by studies of the response of fibroblasts to growth factor stimulation (Figure 13.39). The cytoskeletal alterations resulting from growth fac-

Figure 13.39
Regulation of actin remodeling by Rho family proteins Different members of the Rho family regulate the polymerization of actin to produce filopodia (Cdc42), lamellipodia (Rac), and focal adhesions and stress fibers (Rho). Fluorescence micrographs illustrate the distribution of actin following microinjection of fibroblasts with Cdc42, Rac, and Rho. (From C. D. Nobes and A. Hall, 1995. *Cell* 81: 53.)

tor stimulation include the production of cell surface protrusions (filopodia, lamellipodia, and membrane ruffles) as well as the formation of focal adhesions and stress fibers. Microinjection of cells with specific mutants of different Rho family members has shown that Cdc42 induces the formation of microspikes or filopodia, Rac mediates the formation of lamellipodia and membrane ruffles, and Rho is responsible for the formation of focal adhesions and stress fibers.

Further studies have demonstrated that the activities of Rho family members are not restricted to fibroblasts: they play similar roles in regulating the actin cytoskeleton in all types of eukaryotic cells. For example, Rho is required for cytokinesis (cell division following mitosis), which is mediated by an actin-myosin contractile ring (see Figure 11.27). In neurons, Rac, Cdc42 and Rho regulate the extension and retraction of axons during development of the nervous system. In smooth muscle cells, Rho contributes to regulation of contraction. In epithelial cells, Rho family members regulate the formation of adherens junctions, which involve the linkage of cadherins to the actin cytoskeleton (see Figure 12.64). Members of the Rho family thus serve as universal regulators of the actin cytoskeleton, linking extracellular signals to changes in cell shape and movement. In addition, Rho family members can activate MAP kinase signaling pathways, so these small GTP-binding proteins function as dual regulators of cytoskeletal remodeling and gene expression.

A large number of proteins have been identified as potential targets of Rho, Rac, and Cdc42, and the cellular roles of many of these candidate target proteins remain to be elucidated. However, recent experiments have clarified some of the pathways by which Rho family members regulate cytoskeletal alterations (Figure 13.40). A key target of Rho appears to be a protein-serine/threonine kinase called Rho kinase. Activation of Rho kinase increases the phosphorylation of the light chain of myosin II by two mechanisms: Rho not only directly phosphorylates the myosin light chain but also phosphorylates and inhibits myosin light chain phosphatase. The resulting increase in myosin light chain phosphorylation activates myosin and leads to the assembly of actin-myosin filaments, resulting in cytoskeletal alterations such as the formation of stress fibers

Figure 13.40
Targets of Rho family members   Rho
activates Rho kinase, which phosphorylates the regulatory light chain of
myosin II and inhibits myosin light
chain phosphatase. The resulting
increase in phosphorylation of the
light chain activates myosin II, leading
to assembly of actin-myosin filaments
and the formation of stress fibers and
focal adhesions. Conversely, Rac and
Cdc42 stimulate the PAK protein
kinase, which inhibits myosin light
chain kinase, leading to a decrease in
myosin II phosphorylation. Both Rho
and Rac also stimulate LIM-kinase,
which regulates actin remodeling by
phosphorylating cofilin.

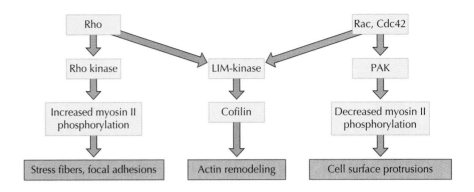

Figure 13.40
Targets of Rho family members   Rho
activates Rho kinase, which phosphorylates the regulatory light chain of
myosin II and inhibits myosin light
chain phosphatase. The resulting
increase in phosphorylation of the
light chain activates myosin II, leading
to assembly of actin-myosin filaments
and the formation of stress fibers and
focal adhesions. Conversely, Rac and
Cdc42 stimulate the PAK protein
kinase, which inhibits myosin light
chain kinase, leading to a decrease in
myosin II phosphorylation. Both Rho
and Rac also stimulate LIM-kinase,
which regulates actin remodeling by
phosphorylating cofilin.

and focal adhesions, adherens junctions, and cytokinesis. In addition to
affecting phosphorylation of myosin light chain, Rho kinase phosphorylates another protein kinase (LIM-kinase), which in turn phosphorylates
the actin-binding protein cofilin. Cofilin is a key regulator of actin filament disassembly (see Figure 11.5), so its phosphorylation by LIM-kinase
directly links Rho to regulation of actin filament dynamics.

A key target of both Rac and Cdc42 is the protein-serine/threonine
kinase called PAK. Like Rho kinase, PAK also affects phosphorylation of
the myosin light chain, but in the opposite direction. PAK phosphorylates
and inhibits myosin light chain kinase, leading to a decrease in myosin
light chain phosphorylation. This results in inactivation of myosin II and
a decrease in actin-myosin interaction, consistent with the ability of Rac
and Cdc42 to stimulate the formation of cell surface protrusions (filopodia and lamellipodia) rather than stress fibers and focal adhesions. Activation of Rac also leads to stimulation of LIM-kinase and phosphorylation of cofilin, linking Rac to remodeling of the actin cytoskeleton. Other
targets of Rac and Cdc42 also participate in the formation of filopodia
and lamellopodia, as well as in the regulation of cell-cell interactions at
adherens junctions.

Signaling in Development and Differentiation

Understanding the molecular mechanisms that govern animal development is one of the major challenges of contemporary cell and molecular
biology. Starting from only a single cell, the fertilized egg, all the diverse
cell types of the body are produced and organized into tissues and organs.
Both cell differentiation and the development of body structures must be
regulated by intricate pathways of cell-cell signaling that coordinate the
activities of individual cells and ultimately give rise to organisms as complex as human beings. Although a comprehensive discussion of developmental biology is beyond the scope of this book, it is noteworthy that considerable progress has been made in elucidating the signaling pathways
responsible for some aspects of development and differentiation. Three
examples of such signaling pathways are discussed here.

The Receptor Tyrosine Kinase/Ras/MAP Kinase Pathway in
Drosophila *and* C. elegans

Signaling by receptor tyrosine kinases that activate the Ras/Raf/MAP
kinase pathway regulates development and differentiation of many types
of cells. A well-studied example in vertebrates is provided by the differentiation of neurons, which is mediated by activation of the nerve growth factor receptor (a receptor tyrosine kinase) and subsequent stimulation of the

(A)

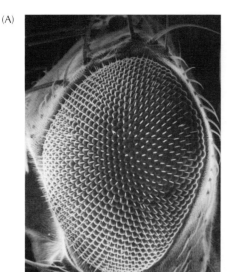

Figure 13.41
The *Drosophila* compound eye (A) Scanning electron micrograph showing the compound eye, composed of about 800 individual units. (B) Each unit contains eight photoreceptor neurons (designated R1 through R8) that develop in a fixed order and pattern. (A, courtesy of T. Venkatesh, University of Michigan.)

(B)

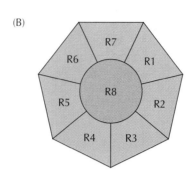

Ras/Raf/ERK pathway. However, the key role of this pathway in development has been demonstrated most clearly by genetic analysis of the model organisms *Drosophila* and *C. elegans*.

Signaling by the Ras/Raf/MAP kinase pathway plays a key role in development of the compound eye of *Drosophila*, which also illustrates the role of direct cell-cell signaling in differentiation. The *Drosophila* compound eye consists of about 800 individual units, each of which contains eight photoreceptor neurons (R1 through R8) and 12 lens cells (Figure 13.41). The photoreceptor neurons develop in a fixed order, beginning with the differentiation of R8. R8 then induces two neighboring cells to become the R2 and R5 photoreceptors. Next, R2 induces neighboring cells to become R1 and R3, and R5 induces neighboring cells to become R4 and R6. The final step is differentiation of R7, which is induced by interaction with R8. Lens cells then develop from those cells that do not differentiate into photoreceptors.

The signaling pathway leading to development of the R7 cell has been characterized in detail, based on the isolation of mutant flies in which R7 fails to develop (Figure 13.42). One of these mutants (*sevenless*) results from defects in a gene encoding a receptor protein-tyrosine kinase that is expressed by precursors of R7 cells. Another mutant (called *boss*, which is short for *bride-of-sevenless*) results from defects in a gene encoding a cell surface protein expressed by R8 cells. Boss is the ligand for Sevenless, so direct cell-cell interaction between R8 and a precursor cell activates the Sevenless protein-tyrosine kinase. Further studies have shown that cell differentiation induced by signaling from Sevenless also requires Ras and Raf, leading to activation of the ERK MAP kinase pathway and resulting phosphorylation of transcription factors that mediate R7 differentiation.

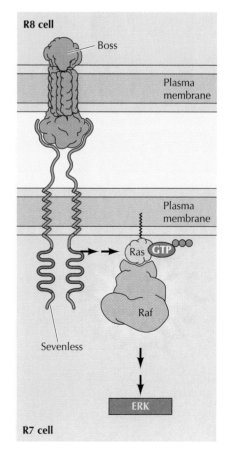

Figure 13.42
Induction of R7 differentiation Differentiation of the R7 photoreceptor neuron is induced by contact of a precursor cell with R8. The Boss protein on the R8 cell surface is the ligand for the Sevenless receptor protein-tyrosine kinase, which is expressed by R7 precursor cells. Stimulation of Sevenless activates the Ras/Raf/ERK MAP kinase pathway.

Figure 13.43
Induction of the vulva in *C. elegans*
The gonadal anchor cell secretes Lin-3, which is related to EGF. Lin-3 stimulates Let-23, a receptor protein-tyrosine kinase expressed by vulval precursor cells. Activated Let-23 stimulates the Ras/Raf/ERK pathway. ERK phosphorylates two transcription factors (Lin-1 and Lin-31), which induce three vulval precursor cells to proliferate and differentiate, forming the 22 cells of the vulva.

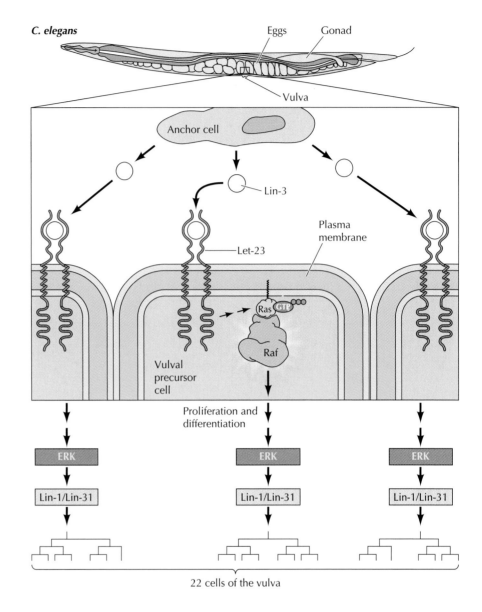

Development of the vulva in the nematode *C. elegans* is another example in which the role of the Ras/Raf/MAP kinase pathway has been elucidated by genetic analysis. In this system, a single cell first differentiates to become a gonadal anchor cell, which attaches the vulva to the uterus. The anchor cell then induces the differentiation of three precursor cells, which proliferate to form the 22 cells of the vulva.

Isolation of mutants in which the vulva fails to develop has allowed the characterization of several genes that are necessary for vulval induction, thereby delineating the pathway by which the anchor cell signals differentiation of the vulval precursor cells (Figure 13.43). One of these genes, *lin-3*, encodes a protein related to the mammalian growth factor EGF. The Lin-3 protein is secreted by anchor cells and binds to a receptor (Let-23) expressed on the surface of vulval precursor cells. Let-23 is a receptor protein-tyrosine kinase related to the mammalian EGF receptor. Other genes required for vulval induction include *let-60* and *lin-45*, which encode *C. elegans* Ras and Raf proteins, respectively. Vulval development in *C. elegans* thus involves growth factor stimulation of a receptor protein-tyrosine kinase and subsequent activation of the Ras/Raf signaling pathway. This leads to activation of the ERK MAP kinase, which phosphorylates two transcription factors (Lin-1 and Lin-31) responsible for vulval induction.

Hedgehog and Wingless

The **Hedgehog** and **Wingless** pathways are closely connected signaling systems that play key roles in determining cell fate during embryonic development. Both Hedgehog and Wingless pathways were first described in *Drosophila*, but members of the Hedgehog and Wingless (Wnt) families have been found to control a wide range of events that establish cell patterning during the development of both vertebrate and invertebrate embryos. Examples of the processes regulated by these signaling pathways include the determination of cell types and establishment of cell patterning during the development of limbs, the nervous system, the skeleton, lungs, hair, teeth, and gonads.

The *hedgehog* genes (one in *Drosophila* and at least three in vertebrates) encode secreted proteins that are modified by the addition of cholesterol. This modification appears to anchor Hedgehog polypeptides to the cell surface and restrict their range of signaling activity. The functional receptor for Hedgehog consists of two transmembrane proteins, Patched and Smoothened (Figure 13.44). Hedgehog binds to Patched, which acts a negative regulator of Smoothened. The binding of Hedgehog to Patched allows Smoothened to propagate an intracellular signal. Smoothened has seven transmembrane α helices, and is thus similar in structure to G protein-coupled receptors (see Figure 13.10). However, Smoothened is not linked to a G protein, and the mechanism of signaling from Smoothened remains to be understood. In *Drosophila*, the target for Smoothened signaling is a transcription factor called Cubitus Interruptus (Ci), which is in a complex with a protein kinase called Fused and a kinesin-related protein called Coastal. In the absence of Smoothened signaling, this complex is associated with microtubules, presumably by the interaction of Coastal with tubulin. Activation of Smoothened leads to dissociation of the complex from microtubules and translocation of Ci to the nucleus, where it activates transcription of its target genes. Interestingly, the targets of Ci include genes encoding members of the Wnt family, providing a direct link between the Hedgehog and Wnt signaling pathways.

The Wnt proteins are a family of secreted growth factors that bind to receptors of the Frizzled family (Figure 13.45). The Frizzled receptors are related to Smoothened and similarly have seven transmembrane α helices, although they are not coupled to G proteins. Signaling from Frizzled leads to phosphorylation of a cytoplasmic protein called Dishevelled and inhibition of the protein kinase glycogen synthase kinase-3 (GSK-3). GSK-3 phosphorylates and promotes the degradation of β-catenin, which links cadherins to actin at adherens junctions (see Figure 11.14), so Wnt signaling leads to an increase in the concentration of β-catenin in the cytoplasm. Importantly, linking cadherins to actin is only one role of β-catenin. In Wnt

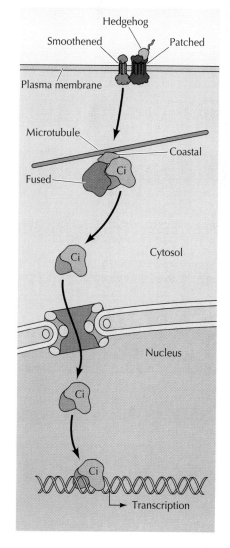

Figure 13.44
Hedgehog signaling The Hedgehog polypeptide, modified by addition of cholesterol, binds to Patched on the surface of a target cell. This relieves the inhibition of Smoothened by Patched, allowing Smoothened to propagate an intracellular signal. Signaling from Smoothened disrupts a complex in which the transcription factor Cubitis Interruptus (Ci) is anchored to microtubules by the kinesin-related protein Coastal, in association with the protein kinase Fused. Ci then translocates to the nucleus where it activates transcription of its target genes.

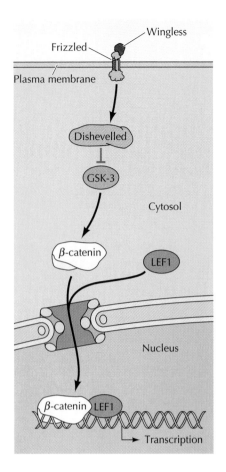

Figure 13.45
The Wingless pathway Wingless (Wnt) polypeptides bind to cell surface receptors of the Frizzled family. Signaling from Frizzled leads to inhibition of the protein kinase GSK-3, resulting in stabilization of β-catenin and an increase in the concentration of β-catenin in the cytosol. Free β-catenin then forms a complex with the transcription factor LEF1 and the β-catenin/LEF1 complex translocates to the nucleus and drives gene expression.

signaling, β-catenin acts as a direct regulator of gene expression by forming a complex with the transcription factor LEF1. The target genes activated by the β-catenin/LEF1 complex include genes encoding other cell signaling molecules and a variety of transcription factors that control cell fate. The Wnt pathway thus provides a novel link between the cytoskeleton and cell signaling, in which a component of cell-cell junctional complexes also acts as a transcriptional regulatory protein.

Notch Signaling

The **Notch** pathway is another highly conserved signaling pathway that controls cell fate during animal development. Like the signaling pathway leading to differentiation of the R7 photoreceptor neuron in *Drosophila*, Notch signaling is an example of direct cell-cell interactions during development. It functions at all stages of development to regulate cell proliferation, survival, and differentiation in organisms ranging from *Drosophila* and *C. elegans* to humans.

Notch is a large protein with a single transmembrane domain that serves as a receptor for signaling by transmembrane proteins (e.g., Delta) on the surface of adjacent cells (Figure 13.46). Stimulation of Notch initiates a novel and direct pathway of transcriptional activation. In particular, ligand binding leads to proteolytic cleavage of Notch, and the intracellular domain of Notch is then translocated into the nucleus. The Notch intracellular domain then interacts with another transcription factor (called Su(H) in *Drosophila* or CBF-1 in mammals) and induces expression of its target genes. As in the Wnt signaling pathway, the Notch target genes include genes encoding other transcriptional regulatory proteins, which act to determine cell fate.

Regulation of Programmed Cell Death

Programmed cell death is a normal physiological form of cell death that plays a key role both in the maintenance of adult tissues and in embryonic development. In adults, programmed cell death is responsible for balancing cell proliferation and maintaining constant cell numbers in tissues undergoing cell turnover. For example, about 5×10^{11} blood cells are eliminated by programmed cell death daily in humans, balancing their continual production in the bone marrow. In addition, programmed cell death provides a defense mechanism by which damaged and potentially dangerous cells can be eliminated for the good of the organism as a whole. Virus-infected cells frequently undergo programmed cell death, thereby preventing the production of new virus particles and limiting spread of

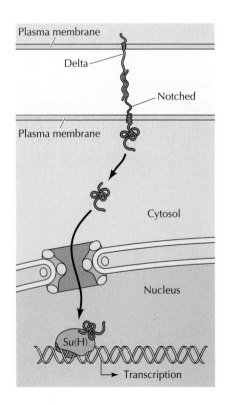

Figure 13.46
Notch signaling Notch serves as a receptor for direct cell-cell signaling by transmembrane proteins (e.g., Delta) on neighboring cells. The binding of Delta leads to proteolytic cleavage of Notch, releasing the Notch intracellular domain, which translocates to the nucleus and interacts with a transcription factor [Su(H) or CBF-1] to induce gene expression.

the virus through the host organism. Other types of insults, such as DNA damage, also induce programmed cell death. In the case of DNA damage, programmed cell death may eliminate cells carrying potentially harmful mutations, including cells with mutations that might lead to the development of cancer.

During development, programmed cell death plays a key role by eliminating unwanted cells from a variety of tissues. For example, programmed cell death is responsible for the elimination of larval tissues during amphibian and insect metamorphosis, as well as for the elimination of tissue between the digits during the formation of fingers and toes. Another well-characterized example of programmed cell death is provided by development of the mammalian nervous system. Neurons are produced in excess, and up to 50% of developing neurons are eliminated by programmed cell death. Those that survive are selected for having made the correct connections with their target cells, which secrete growth factors that signal cell survival by blocking the neuronal cell death program. The survival of many other types of cells in animals is similarly dependent on growth factors or contacts with neighboring cells or the extracellular matrix, so programmed cell death is thought to play an important role in regulating the associations between cells in tissues. Regulation of programmed cell death is mediated by the integrated activity of a variety of signaling pathways, some acting to induce cell death and others to promote cell survival.

Caspases and Apoptosis

In contrast to the accidental death of cells that results from an acute injury, programmed cell death is an active process characterized by a distinct morphological change known as **apoptosis** (Figure 13.47). During apoptosis, chromosomal DNA is usually fragmented as a result of cleavage between nucleosomes. The chromatin condenses and the nucleus then breaks up into small pieces. Finally, the cell itself shrinks and breaks up into membrane-enclosed fragments called apoptotic bodies. Such apoptotic cells and cell fragments are readily recognized and phagocytosed by both macrophages and neighboring cells, so cells that die by apoptosis are efficiently removed from tissues. In contrast, cells that die as a result of acute injury swell and lyse, releasing their contents into the extracellular space and causing inflammation.

Studies of programmed cell death during the development of *C. elegans* have identified three genes that play key roles in regulating and executing apoptosis. During normal nematode development, 131 somatic cells out of a total of 1090 are eliminated by programmed cell death. Two genes, *ced-3* and *ced-4*, are required for apoptosis to occur; if either of these genes is inactivated, the normal programmed cell deaths do not take place. A third gene, *ced-9*, functions as a negative regulator of apoptosis. If *ced-9* is inactivated by mutation, the cells that would normally survive fail to do so. Instead, they also undergo apoptosis, leading to death of the developing animal. Conversely, if *ced-9* is expressed at an abnormally high level, the normal programmed cell deaths fail to occur.

Figure 13.47
Apoptosis (A) Diagrammatic representation of the events of apoptosis. (B) Light micrographs of normal and apoptotic human leukemia cells, illustrating chromatin condensation and nuclear fragmentation during apoptosis. (B, courtesy of D. R. Green/La Jolla Institute for Allergy and Immunology.)

(A)

DNA fragmentation

Chromatin condensation

Fragmentation of nucleus

Fragmentation of the cell

(B)

Normal

Apoptotic

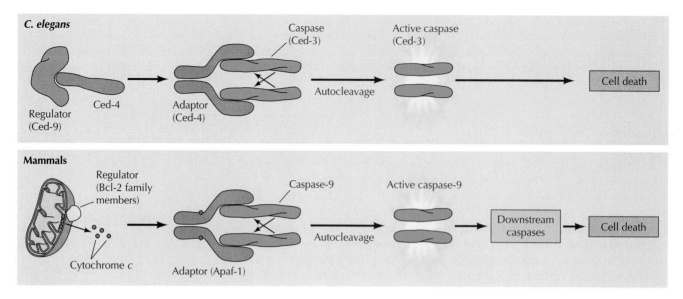

Figure 13.48
Regulators and effectors of apoptosis
Many cell death signals induce apoptosis via a conserved pathway of regulators, adaptors, and caspases. In *C. elegans*, the negative regulator Ced-9 inhibits apoptosis by binding to the adaptor Ced-4. In the absence of inhibition by Ced-9, Ced-4 binds two molecules of the caspase Ced-3, resulting in autocleavage and caspase activation. In mammals, regulators of the Bcl-2 family (Ced-9 homologs) act at the mitochondria to control release of cytochrome *c*, which is required for the binding of caspase-9 to the adaptor Apaf-1 (the Ced-4 homolog). Release of cytochrome *c* from mitochondria thus signals the activation of caspase-9, which then activates downstream caspases to induce apoptosis.

Genes related to *ced-3*, *ced-4*, and *ced-9* have been identified in mammals and found to encode proteins that represent conserved effectors and regulators of apoptosis induced by a variety of stimuli (Figure 13.48). Ced-3 is the prototype of a family of more than a dozen proteases, known as **caspases** because they have cysteine (C) residues at their active sites and cleave after aspartic acid (Asp) residues in their substrate proteins. The caspases are the ultimate effectors or executioners of programmed cell death, bringing about the events of apoptosis by cleaving more than 40 different cell target proteins. Key targets of the caspases include an inhibitor of a DNase which, when activated, is responsible for fragmentation of nuclear DNA. In addition, caspases cleave nuclear lamins, leading to fragmentation of the nucleus, and cytoskeletal proteins, leading to disruption of the cytoskeleton, membrane blebbing, and cell fragmentation.

The caspases are synthesized as inactive precursors that are converted to the active form by proteolytic cleavage, catalyzed by other caspases. The initial activation of a caspase therefore starts off a chain reaction leading to activation of additional caspases and death of the cell. Regulation of caspase activation is thus key to determining cell survival. Ced-4 and its mammalian homolog (Apaf-1) bind to caspases and promote their activation by forming a complex in which two caspases can cleave and activate one another. In contrast, Ced-9 inhibits caspase activation. Mammals encode a whole family of proteins (called the **Bcl-2** family) that are related to Ced-9. Some members of the Bcl-2 family, including Bcl-2 itself, function analogously to Ced-9 as inhibitors of caspase activation and programmed cell death. Other members of the Bcl-2 family, however, promote cell death.

In *C. elegans*, Ced-9 directly regulates caspase activation by binding to Ced-4 and preventing it from activating Ced-3. In mammalian cells, however, members of the Bcl-2 family regulate caspase activation by a less direct mechanism. In particular, Bcl-2 family members appear to act principally at mitochondria, which play a central role in controlling programmed cell death in mammalian cells. One of the key caspases in mammalian cells (caspase-9) is activated, like Ced-3 in *C. elegans*, by forming a complex with the Ced-4 homolog Apaf-1. However, in contrast to *C. elegans*, formation of a complex with Apaf-1 is not sufficient for activation of caspase-9. Instead, activation of caspase-9 requires formation of a complex consisting of cas-

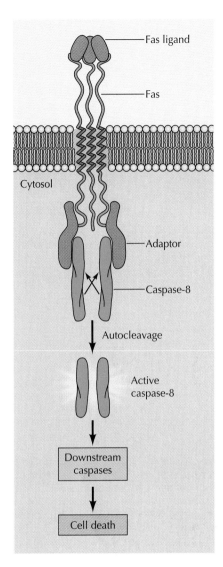

Figure 13.49
Cell death receptors   Binding of ligand to the Fas receptor induces apoptosis by direct activation of caspase-8. Fas ligand consists of three polypeptide chains, so its binding induces receptor trimerization. Caspase-8 bound to the receptor via adaptor molecules is then activated by autocleavage, leading to activation of downstream caspases and cell death.

pase-9, Apaf-1, and cytochrome *c*. Under normal conditions of cell survival, cytochrome *c* is localized to the mitochondrial intermembrane space (see Figure 10.8) while Apaf-1 and caspase-9 are found in the cytosol, so caspase-9 remains inactive. However, many stimuli that trigger cell death, including DNA damage and growth factor deprivation, lead to damage of the mitochondria and release of cytochrome *c* to the cytosol. In the cytosol, cytochrome *c* can form a complex with caspase-9 and Apaf-1, leading to caspase-9 activation and cell death. Members of the Bcl-2 family act in the mitochondrial membrane to regulate mitochondrial integrity and cytochrome *c* release. Bcl-2 family members that inhibit apoptosis (such as Bcl-2 itself) prevent cytochrome *c* release, whereas Bcl-2 family members that promote cell death act by inducing mitochondrial damage, cytochrome *c* release, and caspase activation. As discussed below, both caspases and members of the Bcl-2 family are critical targets of the signaling pathways that control survival of mammalian cells.

Cell Death Receptors and Caspase Activation

Some secreted polypeptides signal programmed cell death by activating receptors that directly induce apoptosis of the target cell. These cell death signals are polypeptides belonging to the **tumor necrosis factor (TNF)** family. They bind to members of the TNF receptor family, which can signal apoptosis in a variety of cell types. One of the best characterized members of this family is the cell surface receptor called Fas, which plays important roles in controlling cell death in the immune system. For example, apoptosis induced by activation of Fas is responsible for killing target cells of the immune system, such as cancer cells or virus-infected cells, as well as for eliminating excess lymphocytes at the end of an immune response.

The cell death receptors signal apoptosis by directly activating caspases (Figure 13.49). TNF and related family members consist of three identical polypeptide chains, and their binding induces receptor trimerization. The cytoplasmic portions of the receptors bind adaptor molecules that in turn bind an upstream caspase called caspase-8. This leads to activation of caspase-8 as a result of self-cleavage, and the activated molecules of caspase-8 can then activate other downstream caspases, thereby initiating a caspase cascade that results in death of the cell.

Caspase-8 not only cleaves other caspases, but it also cleaves a member of the Bcl-2 family called Bid. Bid is one of the Bcl-2 family members that induces rather than protects against apoptosis. Normally, it is retained in inactive form in the cytosol. However, cleavage by caspase-8 allows Bid to translocate to mitochondria where it disrupts the membrane and releases cytochrome *c* into the cytosol. This leads to activation of caspase-9, further amplifying the caspase cascade initiated by direct activation of caspase-8 at cell death receptors.

Signaling Cell Survival

Signaling by TNF and related polypeptides is an active process in which stimulation of cell death receptors induces apoptosis. Other signaling pathways act in the opposite direction to promote cell survival by inhibiting apoptosis. These signaling pathways control the fate of a wide variety of cells whose survival is dependent on extracellular growth factors or cell-cell interactions. Indeed, most cells in higher animals are programmed to undergo apoptosis unless cell death is actively suppressed by survival signals from other cells.

As already noted, a well-characterized example of programmed cell death in development is provided by the vertebrate nervous system. About 50% of neurons die by apoptosis, with the survivors having received sufficient amounts of survival signals from their target cells. These survival signals are polypeptide growth factors related to nerve growth factor (NGF), which induces both neuronal survival and differentiation by activating a receptor protein-tyrosine kinase. Other types of cells are similarly dependent upon growth factors or cell contacts that activate signaling molecules (including nonreceptor protein-tyrosine kinases) associated with integrins.

One of the major intracellular signaling pathways responsible for promoting cell survival is initiated by the enzyme PI 3-kinase, which is activated by either protein-tyrosine kinases or G protein-coupled receptors. PI 3-kinase phosphorylates the membrane phospholipid PIP_2 to form PIP_3, which activates the protein-serine/threonine kinase Akt (see Figure 13.31). Akt then phosphorylates a number of proteins that regulate apoptosis (Figure 13.50). One substrate for Akt is a member of the Bcl-2 family called Bad. Bad is one of the Bcl-2 family members (like Bid) that induces cell death by stimulating the release of cytochrome c from mitochondria. Phosphorylation of Bad by Akt creates a binding site for proteins that sequester Bad in the cytosol, thereby preventing the translocation of Bad to the mitochondrial membrane. Akt may also directly block caspase activation by phosphorylating caspase-9. In addition to these direct effects on components of the cell death machinery, Akt phosphorylates transcription factors that regulate cell survival and another protein kinase (GSK-3) that affects apoptosis as

Figure 13.50
The PI 3-kinase pathway and cell survival   Survival factors such as NGF activate receptor protein-tyrosine kinases, leading to activation of PI 3-kinase and formation of PIP_3. PIP_3 recruits the protein kinase Akt to the plasma membrane where it is activated as a result of phosphorylation by PDK. Akt then appears to phosphorylate a number of proteins that contribute to cell survival. The targets of Akt that have been implicated in suppression of apoptosis include the Bcl-2 family member Bad, caspase-9, several transcription factors, and the protein kinase GSK-3, which affects cell metabolism and protein synthesis.

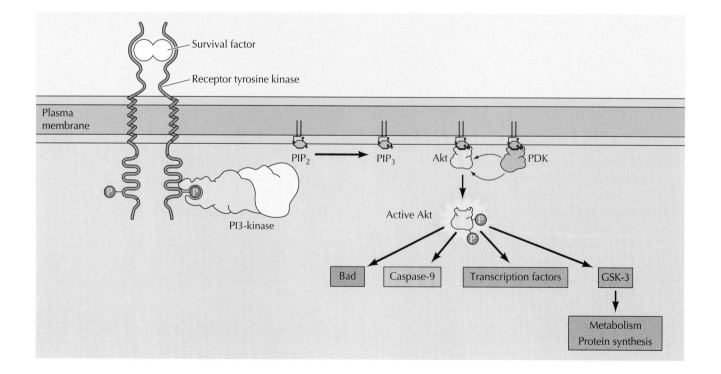

well as regulating cell metabolism and protein synthesis. The PI 3-kinase/ Akt pathway thus regulates cell survival through phosphorylation of a variety of downstream targets, more of which are likely to be identified by future research.

Cell survival is mediated not only by PI 3-kinase/Akt signaling, but also by other signaling pathways including the Ras/Raf/MAP kinase pathway. One mechanism through which this pathway inhibits apoptosis involves the phosphorylation and activation of a protein kinase called RSK by ERK. Like Akt, RSK phosphorylates the Bcl-2 family member Bad, so Bad serves as a site of convergence for the PI 3-kinase/Akt and MAP kinase pathways in signaling cell survival. In addition, ERK and RSK may act by phosphorylating transcription factors that affect the expression of genes that regulate apoptosis. In *Drosophila*, the MAP kinase pathway suppresses apoptosis by regulating the activity of proteins that directly inhibit caspases, but it is not known if this mechanism is also operative in mammalian cells. Understanding the signals and mechanisms that control cell survival thus remains an active and exciting area of ongoing research, with many questions still to be answered.

SUMMARY

SIGNALING MOLECULES AND THEIR RECEPTORS

Modes of Cell-Cell Signaling: Most signaling molecules are secreted by one cell and bind to receptors expressed by a target cell. Cell-cell signaling is divided into three general categories (endocrine, paracrine, and autocrine signaling) based on the distance over which signals are transmitted.

Steroid Hormones and the Steroid Receptor Superfamily: The steroid hormones, thyroid hormone, vitamin D_3, and retinoic acid are small hydrophobic molecules that diffuse across the plasma membrane of their target cells and bind to intracellular receptors. Members of the steroid receptor superfamily function as transcription factors to directly regulate gene expression in response to ligand binding.

Nitric Oxide and Carbon Monoxide: The simple gases nitric oxide and carbon monoxide are important paracrine signaling molecules in the nervous system and other cell types.

Neurotransmitters: Neurotransmitters are small hydrophilic molecules that carry signals between neurons or between neurons and other target cells at a synapse. Many neurotransmitters bind to ligand-gated ion channels.

Peptide Hormones and Growth Factors: The widest variety of signaling molecules in animals are peptides, ranging from only a few to more than a hundred amino acids. This group of molecules includes peptide hormones, neuropeptides, and growth factors.

KEY TERMS

endocrine signaling, hormone, paracrine signaling, autocrine signaling

steroid hormone, testosterone, estrogen, progesterone, corticosteroid, glucocorticoid, mineralocorticoid, ecdysone, thyroid hormone, vitamin D_3, retinoic acid, retinoid, steroid receptor superfamily

neurotransmitter

peptide hormone, neuropeptide, enkephalin, endorphin, neurohormone, growth factor, nerve growth factor (NGF), neurotrophin, epidermal growth factor (EGF), platelet-derived growth factor (PDGF), cytokine, membrane-anchored growth factor eicosanoid, prostaglandin, prosta-cyclin, thromboxane, leukotriene

Eicosanoids: The eicosanoids are a class of lipids that function in paracrine and autocrine signaling.

plant hormone, auxin, gibberellin, cytokinin, abscisic acid, ethylene

Plant Hormones: Small molecules known as plant hormones regulate plant growth and development.

FUNCTIONS OF CELL SURFACE RECEPTORS

G protein, G protein-coupled receptor, heterotrimeric G protein

G Protein-Coupled Receptors: The largest family of cell surface receptors, including the receptors for many hormones and neurotransmitters, transmit signals to intracellular targets via the intermediary action of G proteins.

receptor protein-tyrosine kinase, autophosphorylation, SH2 domain

Receptor Protein-Tyrosine Kinases: The receptors for most growth factors are protein-tyrosine kinases.

cytokine receptor superfamily, nonreceptor protein-tyrosine kinase, Src, Janus kinase (JAK)

Cytokine Receptors and Nonreceptor Protein-Tyrosine Kinases: The receptors for many cytokines act in association with nonreceptor protein-tyrosine kinases.

protein-tyrosine phosphatase, transforming growth factor β (TGF-β), protein-serine/threonine kinase, guanylyl cyclase

Receptors Linked to Other Enzymatic Activities: Other kinds of cell surface receptors include protein-tyrosine phosphatases, protein-serine/threonine kinases, and guanylyl cyclases.

PATHWAYS OF INTRACELLULAR SIGNAL TRANSDUCTION

intracellular signal transduction, cyclic AMP (cAMP), second messenger, adenylyl cyclase, cAMP phosphodiesterase, cAMP-dependent protein kinase (protein kinase A), cAMP response element (CRE), CREB

The cAMP Pathway: Second Messengers and Protein Phosphorylation: Cyclic AMP is an important second messenger in the response of animal cells to a variety of hormones and odorants. Most actions of cAMP are mediated by protein kinase A, which phosphorylates both metabolic enzymes and the transcription factor CREB.

cyclic GMP (cGMP), rhodopsin, transducin, cGMP phosphodiesterase

Cyclic GMP: Cyclic GMP is also an important second messenger in animal cells. Its best-characterized role is in visual reception in the vertebrate eye.

phosphatidylinositol 4,5-bisphosphate (PIP$_2$), phospholipase C, diacylglycerol, inositol 1,4,5-trisphosphate (IP$_3$), protein kinase C, phorbol ester, calmodulin, CaM kinase, ryanodine receptor, phosphatidylinositide (PI) 3-kinase, phosphatidylinositol 3,4,5-trisphosphate (PIP$_3$), Akt

Phospholipids and Ca^{2+}: Phospholipids and Ca^{2+} are common second messengers activated downstream of both G protein-coupled receptors and protein-tyrosine kinases. Hydrolysis of phosphatidylinositol 4,5-bisphosphate (PIP$_2$) yields diacylglycerol and inositol 1,4,5-trisphosphate (IP$_3$), which activate protein kinase C and mobilize Ca^{2+} from intracellular stores, respectively. Increased levels of cytosolic Ca^{2+} then activate a variety of target proteins, including Ca^{2+}/calmodulin-dependent protein kinases. In electrically excitable cells of nerve and muscle, levels of cytosolic Ca^{2+} are increased by the opening of voltage-gated Ca^{2+} channels in the plasma membrane and ryanodine receptors in the endoplasmic and sarcoplasmic reticulum. In addition to being cleaved into diacylglycerol and IP$_3$, PIP$_2$ can be phosphorylated to the distinct second messenger PIP$_3$.

MAP kinase, ERK, Ras, Raf, MEK, guanine nucleotide exchange factor, GTPase-activating protein, small GTP-binding protein, immediate-early gene, serum response element (SRE), serum response factor (SRF), Elk-1

Ras, Raf, and the MAP Kinase Pathway: The MAP kinase pathway is a conserved chain of protein kinases activated downstream of a variety of extracellular signals. In animal cells, the best-characterized forms of MAP kinase are coupled to growth factor receptors by the small GTP-binding protein Ras, which initiates a protein kinase cascade leading to MAP kinase (ERK) activation. ERK then phosphorylates a variety of cytosolic and nuclear proteins, including transcription factors that mediate immediate-early gene induction.

The JAK/STAT Pathway: STAT proteins are transcription factors that contain SH2 domains and are activated directly by protein-tyrosine kinases associated with cytokine and growth factor receptors.

JAK/STAT pathway, STAT protein

SIGNAL TRANSDUCTION AND THE CYTOSKELETON

Integrins and Signal Transduction: Binding of integrins to the extracellular matrix stimulates the FAK nonreceptor protein-tyrosine kinase, leading to activation of MAP kinase and other downstream signaling pathways.

FAK

Regulation of the Actin Cytoskeleton: Growth factors induce alterations in cell movement and cell shape by remodeling the actin cytoskeleton. These cytoskeletal alterations are mediated by members of the Rho subfamily of small GTP-binding proteins.

Rho, Rac, Cdc42

SIGNALING IN DEVELOPMENT AND DIFFERENTIATION

The Receptor Tyrosine Kinase/Ras/MAP Kinase Pathway in **Drosophila** *and* **C. elegans***:* The role of the Ras/Raf/MAP kinase pathway in development has been elucidated by studies of the differentiation of photoreceptor neurons in *Drosophila* and vulval induction in *C. elegans*.

Hedgehog and Wingless: The Hedgehog and Wingless signaling pathways play key roles in determination of cell fate and patterning during development of both invertebrate and vertebrate embryos.

Hedgehog, Wingless

Notch Signaling: The Notch pathway controls cell fate by direct cell-cell interactions during animal development.

Notch

REGULATION OF PROGRAMMED CELL DEATH

Caspases and Apoptosis: Programmed cell death plays a key role both in the maintenance of adult tissues and embryonic development. In contrast to the accidental death of cells from an acute injury, programmed cell death takes place by the active process of apoptosis. Genes responsible for the regulation and execution of apoptosis are conserved from *C. elegans* to humans. These components of the cell death machinery include a family of proteases (caspases) that are the effectors of apoptosis as well as proteins that regulate caspase activation.

programmed cell death, apoptosis, caspase, Bcl-2

Cell Death Receptors and Caspase Activation: Some secreted polypeptides induce programmed cell death by activating receptors that are directly linked to caspases.

tumor necrosis factor (TNF)

Signaling Cell Survival: Many cells are dependent on survival signals from secreted factors or cell-cell contacts that suppress apoptosis. The PI 3-kinase/Akt pathway is a major signaling pathway responsible for promoting cell survival.

Questions

1. The proliferation of thyroid cells is stimulated by hormones that activate a receptor coupled to G_s. How would inhibitors of cAMP phosphodiesterase affect the proliferation of these cells?

2. The epinephrine receptor is coupled to G_s, whereas the acetylcholine receptor (on heart muscle cells) is coupled to G_i. Suppose you were to construct a recombinant molecule containing the extracellular sequences of the epinephrine receptor joined to the cytosolic sequences of the acetylcholine receptor. What effect would epinephrine have on cAMP levels in cells expressing such a recombinant receptor? What would be the effect of acetylcholine?

3. Platelet-derived growth factor (PDGF) is a dimer of two polypeptide chains. What would be the predicted effect of PDGF monomers on signaling from the PDGF receptor?

4. How would overexpression of protein phosphatase 1 affect the induction of cAMP-inducible genes in response to hormone stimulation of appropriate target cells? Would protein phosphatase 1 affect the function of cAMP-gated ion channels involved in odorant reception?

5. Protein kinase C-α (PKC-α) and protein kinase C-ε (PKC-ε) are two different members of the protein kinase C family, which differ in their regulation. PKC-α requires both Ca^{2+} and diacylglycerol for activation, whereas PKC-ε requires only diacylglycerol. How would hydrolysis of the phospholipids PIP_2 and phosphatidylcholine by phospholipase C affect the activities of these different PKC family members?

6. Dominant negative mutants of both Ras and Raf block growth factor-stimulated cell proliferation. The inhibitory effects of dominant negative Ras are overcome by expression of activated Raf. Would you expect activated Ras similarly to overcome the inhibitory effects of dominant negative Raf? How about activated MEK?

References and Further Reading

Signaling Molecules and Their Receptors

Arai, K., F. Lee, A. Miyajima, S. Miyatake, N. Arai and T. Yokota. 1990. Cytokines: Coordinators of immune and inflammatory responses. *Ann. Rev. Biochem.* 59: 783–836. [R]

Beato, M., Herrlich, P. and G. Schutz. 1995. Steroid hormone receptors: Many actors in search of a plot. *Cell* 83: 851–857. [R]

Burgess, W. H. and T. Maciag. 1989. The heparin-binding (fibroblast) growth factor family of proteins. *Ann. Rev. Biochem.* 58: 575–606. [R]

Carpenter, G. and S. Cohen. 1990. Epidermal growth factor. *J. Biol. Chem.* 265: 7709–7712. [R]

Denninger, J. W. and M. A. Marletta. 1999. Guanylate cyclase and the NO/cGMP signaling pathway. *Biochim. Biophys. Acta* 1411: 334–350. [R]

Ecker, J. R. 1997. BRI-ghtening the pathway to steroid hormone signaling events in plants. *Cell* 90: 825–827. [R]

Jaffrey, S. R. and S. H. Snyder. 1995. Nitric oxide: A neural messenger. *Ann. Rev. Cell Dev. Biol.* 11: 417–440. [R]

Jessell, T. M. and E. R. Kandel. 1993. Synaptic transmission: A bidirectional and self-modifiable form of cell-cell communication. *Cell* 72/*Neuron* 10 (Suppl.): 1–30. [R]

Johnson, P. R. and J. R. Ecker. 1998. The ethylene gas signal transduction pathway: A molecular perspective. *Ann. Rev. Genet.* 32: 227–254. [R]

Kastner, P., M. Mark and P. Chambon. 1995. Nonsteroid nuclear receptors: What are genetic studies telling us about their role in real life? *Cell* 83: 859–869. [R]

Levi-Montalcini, R. 1987. The nerve growth factor 35 years later. *Science* 237: 1154–1162. [R]

Mangelsdorf, D. J., C. Thummel, M. Beato, P. Herrlich, G. Schütz, K. Umesono, B. Blumberg, P. Kastner, M. Mark, P. Chambon and R. M. Evans. 1995. The nuclear receptor superfamily: The second decade. *Cell* 83: 835–839. [R]

Massagué, J. and A. Pandiella. 1993. Membrane-anchored growth factors. *Ann. Rev. Biochem.* 62: 515–541. [R]

Mayer, B. and B. Hemmens. 1997. Biosynthesis and action of nitric oxide in mammalian cells. *Trends Biochem. Sci.* 22: 477–481. [R]

Ross, R., E. W. Raines and D. F. Bowen-Pope. 1986. The biology of platelet-derived growth factor. *Cell* 46: 155–169. [R]

Smalley, W. and R. N. DuBois. 1997. Colorectal cancer and nonsteroidal anti-inflammatory drugs. *Adv. Pharmacol.* 39: 1–20. [R]

Snyder, S. H., S. R. Jaffrey and R. Zakhary. 1998. Nitric oxide and carbon monoxide: Parallel roles as neural messengers. *Brain Res. Rev.* 26: 167–175. [R]

Vane, J. and R. Botting. 1987. Inflammation and the mechanism of action of anti-inflammatory drugs. *FASEB J.* 1: 89–96. [R]

Functions of Cell Surface Receptors

Burke, D., D. Wilkes, T. L. Blundell and S. Malcolm. 1998. Fibroblast growth factor receptors: Lessons from the genes. *Trends Biochem. Sci.* 23: 59–62. [R]

Derynck, R., Y. Zhang and X.-H. Feng. 1998. Smads: Transcriptional activators of TGF-β responses. *Cell* 95: 737–740. [R]

Garbers, D. L. and D. G. Lowe. 1994. Guanylyl cyclase receptors. *J. Biol. Chem.* 269: 30741–30744. [R]

Gether, U. and B. K. Kobilka. 1998. G protein-coupled receptors. II. Mechanism of agonist activation. *J. Biol. Chem.* 273: 17979–17982. [R]

Hepler, J. R. and A. G. Gilman. 1992. G proteins. *Trends Biochem. Sci.* 17: 383–387. [R]

Hubbard, S. R., M. Mohammadi and J. Schlessinger. 1998. Autoregulatory mechanisms in protein-tyrosine kinases. *J. Biol. Chem.* 273: 11987–11990. [R]

Hunter, T. and B. M. Sefton. 1980. Transforming gene product of Rous sarcoma virus phosphorylates tyrosine. *Proc. Natl. Acad. Sci. USA* 77: 1311–1315. [P]

Ji, T. H., M. Grossmann and I. Ji. 1998. G protein-coupled receptors. I. Diversity of receptor-ligand interactions. *J. Biol. Chem.* 273: 17299–17302. [R]

Massagué, J. 1998. TGF-β signal transduction. *Ann. Rev. Biochem.* 67: 753–791. [R]

Mombaerts, P. 1999. Seven-transmembrane proteins as odorant and chemosensory receptors. *Science* 286: 707–711. [R]

Neer, E. J. 1995. Heterotrimeric G proteins: Organizers of transmembrane signals. *Cell* 80: 249–257. [R]

Pawson, T. 1995. Protein modules and signalling networks. *Nature* 373: 573–580. [R]

Rodbell, M. 1980. The role of hormone receptors and GTP-regulatory proteins in membrane transduction. *Nature* 284: 17–22. [R]

Strader, C. D., T. M. Fong, M. R. Tota, D. Underwood and R. A. F. Dixon. 1994. Structure and function of G protein-coupled receptors. *Ann. Rev. Biochem.* 63: 101–132. [R]

Taniguchi, T. 1995. Cytokine signaling through nonreceptor protein tyrosine kinases. *Science* 268: 251–255. [R]

Thomas, S. M. and J. S. Brugge. 1997. Cellular functions regulated by Src family kinases. *Ann. Rev. Cell Dev. Biol.* 13: 513-609. [R]

Van der Geer, P., T. Hunter and R. A. Lindberg. 1994. Receptor protein-tyrosine kinases and their signal transduction pathways. *Ann. Rev. Cell Biol.* 10: 251–337. [R]

Watowich, S. S., H. Wu, M. Socolovsky, U. Klingmuller, S. N. Constantinescu and H. F. Lodish. 1998. Cytokine receptor signal transduction and the control of hematopoietic cell development. *Ann. Rev. Cell Dev. Biol.* 12: 91-128. [R]

Weiss, A. and D. R. Littman. 1994. Signal transduction by lymphocyte antigen receptors. *Cell* 76: 263–274. [R]

Weiss, A. and J. Schlessinger. 1998. Switching signals on or off by receptor dimerization. *Cell* 94: 27–280. [R]

Williams, J. C., R. K. Wierenga and M. Saraste. 1998. Insights into Src kinase functions: Structural comparisons. *Trends Biochem. Sci.* 23: 179–184. [R]

Pathways of Intracellular Signal Transduction

Clapham, D. E. 1995. Calcium signaling. *Cell* 80: 259–268. [R]

Cohen, P. T. W. 1997. Novel protein serine/threonine phosphatases: Variety is the spice of life. *Trends Biochem. Sci.* 22: 245–251. [R]

Darnell, J. E. Jr. 1997. STATs and gene regulation. *Science* 277: 1630–1635. [R]

Datta, S. R., A. Brunet and M. E. Greenberg. 1999. Cellular survival: A play in three Akts. *Genes Dev.* 13: 290–2927. [R]

Divecha, N. and R. F. Irvine. 1995. Phospholipid signaling. *Cell* 80: 269–278. [R]

Ghosh, A. and M. E. Greenberg. 1995. Calcium signaling in neurons: Molecular mechanisms and cellular consequences. *Science* 268: 239–247. [R]

Hannun, Y. A. 1996. Functions of ceramide in coordinating cellular responses to stress. *Science* 274: 1855–1859. [R]

Hill, C. S. and R. Treisman. 1995. Transcriptional regulation by extracellular signals: Mechanisms and specificity. *Cell* 80: 199–211. [R]

Houslay, M. D. and G. Milligan. 1997. Tailoring cAMP-signalling responses through isoform multiplicity. *Trends Biochem. Sci.* 22: 217–224. [R]

Hunter, T. 1997. Oncoprotein networks. *Cell* 88: 333–346. [R]

Hunter, T. 1995. Protein kinases and phosphatases: The yin and yang of protein phosphorylation and signaling. *Cell* 80: 225–236. [R]

Hurley, J. H. 1999. Structure, mechanism, and regulation of mammalian adenylyl cyclase. *J. Biol. Chem.* 274: 7599–7602. [R]

Ihle, J. N. 1995. Cytokine receptor signalling. *Nature* 377: 591–594. [R]

Lohmann, S. M., A. B. Vaandrager, A. Smolenski, U. Walter and H. R. De Jonge. 1997. Distinct and specific functions of cGMP-dependent protein kinases. *Trends Biochem. Sci.* 22: 307–312. [R]

Lowy, D. R. and B. M. Willumsen. 1993. Function and regulation of Ras. *Ann. Rev. Biochem.* 62: 851–891. [R]

Montminy, M. 1997. Transcriptional regulation by cyclic AMP. *Ann. Rev. Biochem.* 66: 807–823. [R]

Nishizuka, Y. 1992. Intracellular signaling by hydrolysis of phospholipids and activation of protein kinase C. *Science* 258: 607–614. [R]

Pawson, T. and J. D. Scott. 1997. Signaling through scaffold, anchoring, and adaptor proteins. *Science* 278: 2075–2080. [R]

Rameh, L. E. and L. C. Cantley. 1999. The role of phosphoinositide 3-kinase lipid products in cell function. *J. Biol. Chem.* 274: 8347–8350. [R]

Ree, S. G. and Y. S. Bae. 1997. Regulation of phosphoinositide-specific phospholipase C isozymes. *J. Biol. Chem.* 272: 5045–15048. [R]

Schaeffer, H. J. and M. J. Weber. 1999. Mitogen-activated protein kinases: Specific messages from ubiquitous messengers. *Mol. Cell. Biol.* 19: 2435–2444. [R]

Schlessinger, J. 1993. How receptor tyrosine kinases activate Ras. *Trends Biochem. Sci.* 18: 273–275. [R]

Shaywitz, A. J. and M. E. Greenberg. 1999. CREB: A stimulus-induced transcription factor activated by a diverse array of extracellular signals. *Ann. Rev. Biochem.* 68: 821–861. [R]

Soderling, T. R. 1999. The Ca^{2+}-calmodulin-dependent protein kinase cascade. *Trends Biochem. Sci.* 24: 232–236. [R]

Stryer, L. 1991. Visual excitation and recovery. *J. Biol. Chem.* 266: 10711–10714. [R]

Vojtek, A. B. and C. J. Der. 1998. Increasing complexity of the Ras signaling pathway. *J. Biol. Chem.* 273: 19925–19928. [R]

Whitmarsh, A. J. and R. J. Davis. 1998. Structural organization of MAP-kinase signaling modules by scaffold proteins in yeast and mammals. *Trends Biochem. Sci.* 23: 481–485. [R]

Yarfitz, S. and J. B. Hurley. 1994. Transduction mechanisms of vertebrate and invertebrate photoreceptors. *J. Biol. Chem.* 269: 14329–14332. [R]

Zufall, F., S. Firestein and G. M. Shepherd. 1994. Cyclic nucleotide-gated ion channels and sensory transduction in olfactory receptor neurons. *Ann. Rev. Biophys. Biomol. Struct.* 23: 577–607. [R]

Signal Transduction and the Cytoskeleton

Arber, S., F. A. Barbayannis, H. Hanser, C. Schneider, C. A. Stanyon, O. Bernard and P. Caroni. 1998. Regulation of actin dynamics through phosphorylation of cofilin by LIM-kinase. *Nature* 393: 805–809. [P]

Giancotti, F G. and E. Ruoslahti. 1999. Integrin signaling. *Science* 285: 1028–1032. [R]

Hall, A. 1998. Rho GTPases and the actin cytoskeleton. *Science* 279: 509–514. [R]

Kaibuchi, K., S. Kuroda and M. Amano. 1999. Regulation of the cytoskeleton and cell adhesion by the Rho family GTPases in mammalian cells. *Ann. Rev. Biochem.* 68: 459–486. [R]

Maekawa, M., T. Ishizaki, S. Boku, N. Watanabe, A. Fujita, A. Iwamatsu, T. Obinata, K. Ohashi, K. Mizuno and S. Narumiya. 1999. Signaling from Rho to the actin cytoskeleton through protein kinases ROCK and LIM-kinase. *Science* 285: 895–898. [P]

Nobles, C. D. and A. Hall. 1995. Rho, Rac, and Cdc42 GTPases regulate the assembly of multimolecular focal complexes associated with actin stress fibers, lamellipodia, and filopodia. *Cell* 81: 53–62. [P]

Sanders, L. C., F. Matsumura, G. M. Bokoch and P. de Lanerolle. 1999. Inhibition of myosin light chain kinase by p21-activated kinase. *Science* 283: 2083–2085. [P]

Schlaepfer, D. D. and T. Hunter. 1998. Integrin signalling and tyrosine phosphorylation: Just the FAKs? *Trends Cell Biol.* 8: 151–157. [R]

Yang, N., O. Higuchi, K. Ohashi, K. Nagata, A. Wada, K. Kangawa, E. Nishida and K. Mizuno. 1998. Cofilin phosphorylation by LIM-kinase 1 and its role in Rac-mediated actin reorganization. *Nature* 393: 809–812. [P]

Signaling in Development and Differentiation

Artavanis-Tsakonas, S., M. D. Rand and R. J. Lake. 1999. Notch signaling: Cell fate control and signal integration in development. *Science* 284: 770–776. [R]

Aza-Blanc, P. and T. B. Kornberg. 1999. Ci: A complex transducer of the Hedgehog signal. *Trends Genet.* 15: 458–462. [R]

Dickson, B. 1995. Nuclear factors in Sevenless signalling. *Trends Genet.* 11: 106–111. [R]

Ingham, P. W. 1998. Transducing Hedgehog: The story so far. *EMBO J.* 17: 3505–3511. [R]

Kimble, J. and P. Simpson. 1997. The LIN-12/Notch signaling pathway and its regulation. *Ann. Rev. Cell Dev. Biol.* 13: 333–361. [R]

Ruiz I Altaba, A. 1999. Gli proteins and Hedgehog signaling: Development and cancer. *Trends Genet.* 15: 418–425. [R]

Sternberg, P. W. and M. Han. 1998. Genetics of RAS signaling in *C. elegans*. *Trends Genet.* 14: 466–472. [R]

Tabin, C. J. and A. P. McMahon. 1997. Recent advances in Hedgehog signaling. *Trends Cell Biol.* 7: 442–446. [R]

Thomas, B. J. and D. A. Wassarman. 1999. A fly's eye view of biology. *Trends Genet.* 15: 184–190. [R]

Wicking, C., I. Smyth and A. Bale. 1999. The Hedgehog signalling pathway in tumorigenesis and development. *Oncogene* 18: 7844–7851. [R]

Wodarz, A. and R. Nusse. 1998. Mechanisms of Wnt signaling in development. *Ann. Rev. Cell Dev. Biol.* 14: 59–88. [R]

Zipursky, S. L. and G. M. Rubin. 1994. Determination of neuronal cell fate: Lessons from the R7 neuron of *Drosophila*. *Ann. Rev. Neurosci.* 17: 373–397. [R]

Regulation of Programmed Cell Death

Adams, J. M. and S. Cory. 1998. The Bcl-2 protein family: Arbiters of cell survival. *Science* 281: 1322–1326. [R]

Arch, R. H. and C. B. Thompson. 1999. Lymphocyte survival—the struggle against death. *Ann. Rev. Cell Dev. Biol.* 15: 113–140. [R]

Ashkenazi, A. and V. M. Dixit. 1998. Death receptors: Signaling and modulation. *Science* 281: 1305–1308. [R]

Bonni, A., A. Brunet, A. E. West, S. R. Datta, M. A. Takasu and M. E. Greenberg. 1999. Cell survival promoted by the Ras-MAPK signaling pathway by transcription-dependent and –independent mechanisms. *Science* 286: 1358–1362. [P]

Budihardjo, I., H. Oliver, M. Lutter, X. Luo and X. Wang. 1999. Biochemical pathways of caspase activation during apoptosis. *Ann. Rev. Cell Dev. Biol.* 15: 269–290. [R]

Datta, S. R., A. Brunet and M. E. Greenberg. 1999. Cellular survival: A play in three Akts. *Genes Dev.* 13: 290–2927. [R]

Earnshaw, W. C., L. M. Martins and S. H. Kaufmann. 1999. Mammalian caspases: Structure, activation, substrates, and functions during apoptosis. *Ann. Rev. Biochem.* 68: 383–424. [R]

Green, D. R. 1998. Apoptotic pathways: The roads to ruin. *Cell* 94: 695–698. [R]

Green, D. R. and J. C. Reed. 1998. Mitochondria and apoptosis. *Science* 281: 1309–1312. [R]

Jacobson, M. D., M. Weil and M. C. Raff. 1997. Programmed cell death in animal development. *Cell* 88: 347–354. [R]

Nagata, S. 1997. Apoptosis by death factor. *Cell* 88: 355–365. [R]

Raff, M. 1998. Cell suicide for beginners. *Nature* 396: 119–122. [R]

Song, Z. and H. Steller. 1999. Death by design: Mechanism and control of apoptosis. *Trends Cell Biol.* 9: M49–M52. [R]

Thornberry, N. A. and Y. Lazebnik. 1998. Caspases: Enemies within. *Science* 281: 1312–1316. [R]

Vaux, D. L. and S. J. Korsmeyer. 1999. Cell death in development. *Cell* 96: 245–254. [R]

14

The Cell Cycle

SELF-REPRODUCTION IS PERHAPS THE MOST FUNDAMENTAL characteristic of cells—as may be said for all living organisms. All cells reproduce by dividing in two, with each parental cell giving rise to two daughter cells on completion of each cycle of cell division. These newly formed daughter cells can themselves grow and divide, giving rise to a new cell population formed by the growth and division of a single parental cell and its progeny. In the simplest case, such cycles of growth and division allow a single bacterium to form a colony consisting of millions of progeny cells during overnight incubation on a plate of nutrient agar medium. In a more complex case, repeated cycles of cell growth and division result in the development of a single fertilized egg into the more than 10^{13} cells that make up the human body.

The division of all cells must be carefully regulated and coordinated with both cell growth and DNA replication in order to ensure the formation of progeny cells containing intact genomes. In eukaryotic cells, progression through the cell cycle is controlled by a series of protein kinases that have been conserved from yeasts to mammals. In higher eukaryotes, this cell cycle machinery is itself regulated by the growth factors that control cell proliferation, allowing the division of individual cells to be coordinated with the needs of the organism as a whole. Not surprisingly, defects in cell cycle regulation are a common cause of the abnormal proliferation of cancer cells, so studies of the cell cycle and cancer have become closely interconnected, similar to the relationship between studies of cancer and the cell signaling pathways discussed in Chapter 13.

The Eukaryotic Cell Cycle

The division cycle of most cells consists of four coordinated processes: cell growth, DNA replication, distribution of the duplicated chromosomes to daughter cells, and cell division. In bacteria, cell growth and DNA replication take place throughout most of the cell cycle, and duplicated chromosomes are distributed to daughter cells in association with the plasma membrane. In eukaryotes, however, the cell cycle is more complex and consists of four discrete phases. Although cell growth is usually a continuous process, DNA is synthesized during only one phase of the cell cycle, and the replicated chromosomes are then distributed to daughter nuclei by a complex series of events preceding cell division. Progression between these stages of the cell cycle is controlled by a conserved regulatory apparatus, which not only coordinates the different events of the cell cycle but also links the cell cycle with extracellular signals that control cell proliferation.

Phases of the Cell Cycle

A typical eukaryotic cell cycle is illustrated by human cells in culture, which divide approximately every 24 hours. As viewed in the microscope, the cell cycle is divided into two basic parts: **mitosis** and **interphase**. Mitosis (nuclear division) is the most dramatic stage of the cell cycle, corresponding to the separation of daughter chromosomes and usually ending with cell division (**cytokinesis**). However, mitosis and cytokinesis last only about an hour, so approximately 95% of the cell cycle is spent in interphase—the period between mitoses. During interphase, the chromosomes are decondensed and distributed throughout the nucleus, so the nucleus appears morphologically uniform. At the molecular level, however, interphase is the time during which both cell growth and DNA replication occur in an orderly manner in preparation for cell division.

The cell grows at a steady rate throughout interphase, with most dividing cells doubling in size between one mitosis and the next. In contrast, DNA is synthesized during only a portion of interphase. The timing of DNA synthesis thus divides the cycle of eukaryotic cells into four discrete phases (Figure 14.1). The **M phase** of the cycle corresponds to mitosis, which is usually followed by cytokinesis. This phase is followed by the **G_1 phase** (gap 1), which corresponds to the interval (gap) between mitosis and initiation of DNA replication. During G_1, the cell is metabolically active and continuously grows but does not replicate its DNA. G_1 is followed by **S phase** (synthesis), during which DNA replication takes place. The completion of DNA synthesis is followed by the **G_2 phase** (gap 2), during which cell growth continues and proteins are synthesized in preparation for mitosis.

The duration of these cell cycle phases varies considerably in different kinds of cells. For a typical rapidly proliferating human cell with a total cycle time of 24 hours, the G_1 phase might last about 11 hours, S phase about 8 hours, G_2 about 4 hours, and M about 1 hour. Other types of cells, however, can divide much more rapidly. Budding yeasts, for example, can progress through all four stages of the cell cycle in only about 90 minutes. Even shorter cell cycles (30 minutes or less) occur in early embryo cells shortly after fertilization of the egg (Figure 14.2). In this case, however, cell growth does not take place. Instead, these early embryonic cell cycles rapidly divide the egg cytoplasm into smaller cells. There is no G_1 or G_2 phase, and DNA replication occurs very rapidly in these early embryonic cell cycles, which therefore consist of very short S phases alternating with M phases.

In contrast to the rapid proliferation of embryonic cells, some cells in adult animals cease division altogether (e.g., nerve cells) and many other cells divide

Figure 14.1
Phases of the cell cycle The division cycle of most eukaryotic cells is divided into four discrete phases: M, G_1, S, and G_2. M phase (mitosis) is usually followed by cytokinesis. S phase is the period during which DNA replication occurs. The cell grows throughout interphase, which includes G_1, S, and G_2. The relative lengths of the cell cycle phases shown here are typical of rapidly replicating mammalian cells.

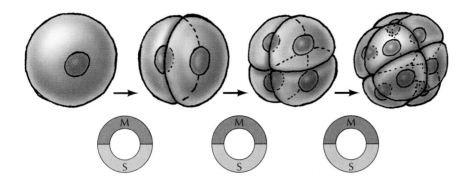

Figure 14.2
Embryonic cell cycles Early embryonic cell cycles rapidly divide the cytoplasm of the egg into smaller cells. The cells do not grow during these cycles, which lack G_1 and G_2 and consist simply of short S phases alternating with M phases.

only occasionally, as needed to replace cells that have been lost because of injury or cell death. Cells of the latter type include skin fibroblasts, as well as the cells of many internal organs, such as the liver, kidney, and lung. As discussed further in the next section, these cells exit G_1 to enter a quiescent stage of the cycle called G_0, where they remain metabolically active but no longer proliferate unless called on to do so by appropriate extracellular signals.

Analysis of the cell cycle requires identification of cells at the different stages discussed above. Although mitotic cells can be distinguished microscopically, cells in other phases of the cycle (G_1, S, and G_2) must be identified by biochemical criteria. Cells in S phase can be readily identified because they incorporate radioactive thymidine, which is used exclusively for DNA synthesis (Figure 14.3). For example, if a population of rapidly proliferating human cells in culture is exposed to radioactive thymidine for a short period of time (e.g., 15 minutes) and then analyzed by autoradiography, about a third of the cells will be found to be radioactively labeled, corresponding to the fraction of cells in S phase.

Variations of such cell labeling experiments can also be used to determine the length of different stages of the cell cycle. For example, consider an experiment in which cells are exposed to radioactive thymidine for 15 minutes, after which the radioactive thymidine is removed and the cells are

Figure 14.3
Identification of S phase cells by incorporation of radioactive thymidine The cells were exposed to radioactive thymidine and analyzed by autoradiography. Labeled cells are indicated by arrows. (From D. W. Stacey et al., 1991. *Mol. Cell Biol.* 11: 4053.)

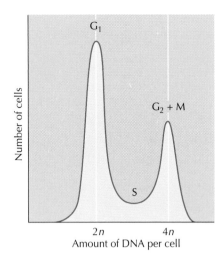

Figure 14.4
Determination of cellular DNA content   A population of cells is labeled with a fluorescent dye that binds DNA. The cells are then passed through a flow cytometer, which measures the fluorescence intensity of individual cells. The data are plotted as cell number versus fluorescence intensity, which is proportional to DNA content. The distribution shows two peaks, corresponding to cells with DNA contents of $2n$ and $4n$; these cells are in the G_1 and G_2/M phases of the cycle, respectively. Cells in S phase have DNA contents between $2n$ and $4n$ and are distributed between these two peaks.

cultured for varying lengths of time prior to autoradiography. Radioactively labeled interphase cells that were in S phase during the time of exposure to radioactive thymidine will be observed for several hours as they progress through the remainder of S and G_2. In contrast, radioactively labeled mitotic cells will not be observed until 4 hours after labeling. This 4-hour lag time corresponds to the length of G_2—the minimum time required for a cell that incorporated radioactive thymidine at the end of S phase to enter mitosis.

Cells at different stages of the cell cycle can also be distinguished by their DNA content (Figure 14.4). For example, animal cells in G_1 are diploid (containing two copies of each chromosome), so their DNA content is referred to as $2n$ (n designates the haploid DNA content of the genome). During S phase, replication increases the DNA content of the cell from $2n$ to $4n$, so cells in S have DNA contents ranging from $2n$ to $4n$. DNA content then remains at $4n$ for cells in G_2 and M, decreasing to $2n$ after cytokinesis. Experimentally, cellular DNA content can be determined by incubation of cells with a fluorescent dye that binds to DNA, followed by analysis of the fluorescence intensity of individual cells in a **flow cytometer** or **fluorescence-activated cell sorter**, thereby distinguishing cells in the G_1, S, and G_2/M phases of the cell cycle.

Regulation of the Cell Cycle by Cell Growth and Extracellular Signals

The progression of cells through the division cycle is regulated by extracellular signals from the environment, as well as by internal signals that monitor and coordinate the various processes that take place during different cell cycle phases. An example of cell cycle regulation by extracellular signals is provided by the effect of growth factors on animal cell proliferation. In addition, different cellular processes, such as cell growth, DNA replication, and mitosis, all must be coordinated during cell cycle progression. This is accomplished by a series of control points that regulate progression through various phases of the cell cycle.

A major cell cycle regulatory point in many types of cells occurs late in G_1 and controls progression from G_1 to S. This regulatory point was first defined by studies of budding yeast (*Saccharomyces cerevisiae*), where it is known as **START** (Figure 14.5). Once cells have passed START, they are committed to entering S phase and undergoing one cell division cycle. However, passage through START is a highly regulated event in the yeast cell cycle, where it is controlled by external signals, such as the availability of nutrients, as well as by cell size. For example, if yeasts are faced with a shortage of nutrients, they arrest their cell cycle at START and enter a resting state rather than proceeding to S phase. Thus, START represents a decision point at which the cell determines whether sufficient nutrients are available to support progression through the rest of the division cycle. Polypeptide factors that signal yeast mating also arrest the cell cycle at START, allowing haploid yeast cells to fuse with one another instead of progressing to S phase.

In addition to serving as a decision point for monitoring extracellular signals, START is the point at which cell growth is coordinated with DNA replication and cell division. The importance of this regulation is particularly evident in budding yeasts, in which cell division produces progeny cells of very different sizes: a large mother cell and a small daughter cell. In order for yeast cells to maintain a constant size, the small daughter cell must grow more than the large mother cell does before they divide again. Thus, cell size must be monitored in order to coordinate cell growth with other cell cycle events. This regulation is accomplished by a control mech-

(A)

(B)

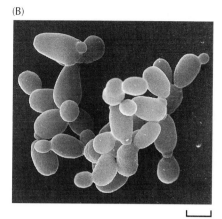

Figure 14.5
Regulation of the cell cycle of budding yeast (A) The cell cycle of *Saccharomyces cerevisiae* is regulated primarily at a point in late G_1 called START. Passage through START is controlled by the availability of nutrients, mating factors, and cell size. Note that these yeasts divide by budding. Buds form just after START and continue growing until they separate from the mother cell after mitosis. The daughter cell formed from the bud is smaller than the mother cell and therefore requires more time to grow during the G_1 phase of the next cell cycle. Although G_1 and S phases occur normally, the mitotic spindle begins to form during S phase, so the cell cycle of budding yeast lacks a distinct G_2 phase. (B) Scanning electron micrograph of *S. cerevisiae*. The size of the bud reflects the position of the cell in the cycle. (B, David M. Phillips/ Visuals Unlimited.)

anism that requires each cell to reach a minimum size before it can pass START. Consequently, the small daughter cell spends a longer time in G_1 and grows more than the mother cell.

The proliferation of most animal cells is similarly regulated in the G_1 phase of the cell cycle. In particular, a decision point in late G_1, called the **restriction point** in animal cells, functions analogously to START in yeasts (Figure 14.6). In contrast to yeasts, however, the passage of animal cells through the cell cycle is regulated primarily by the extracellular growth factors that signal cell proliferation, rather than by the availability of nutrients. In the presence of the appropriate growth factors, cells pass the restriction point and enter S phase. Once it has passed through the restriction point, the cell is committed to proceed through S phase and the rest of the cell cycle, even in the absence of further growth factor stimulation. On the other hand, if appropriate growth factors are not available in G_1, progression through the cell cycle stops at the restriction point. Such arrested cells then enter a quiescent stage of the cell cycle called G_0, in which they can remain for long periods of time without proliferating. G_0 cells are metabolically active, although they cease growth and have reduced rates of protein synthesis. As already noted, many cells in animals remain in G_0 unless called on to proliferate by appropriate growth factors or other extracellular signals. For example, skin fibroblasts are arrested in G_0 until they are stimulated to divide as required to repair damage resulting from a wound. The proliferation of these cells is triggered by platelet-derived growth factor, which is released from blood platelets during clotting and signals the proliferation of fibroblasts in the vicinity of the injured tissue.

Although the proliferation of most cells is regulated primarily in G_1, some cell cycles are instead controlled principally in G_2. One example is the cell cycle of the fission yeast *Schizosaccharomyces pombe* (Figure 14.7). In contrast to *Saccharomyces cerevisiae*, the cell cycle of *S. pombe* is regulated pri-

Figure 14.6
Regulation of animal cell cycles by growth factors The availability of growth factors controls the animal cell cycle at a point in late G_1 called the restriction point. If growth factors are not available during G_1, the cells enter a quiescent stage of the cycle called G_0.

(A)

(B)

1 2 3 4 5 6 7 8

5 μm

Figure 14.7
Cell cycle of fission yeast (A) Fission yeasts grow by elongating at both ends and divide by forming a wall through the middle of the cell. In contrast to the cycle of budding yeasts, the cell cycle of fission yeasts has normal G_1, S, G_2, and M phases. Note that cytokinesis occurs in G_1. The length of the cell indicates its position in the cycle. (B) Light micrographs showing successive stages of mitosis and cytokinesis of *Schizosaccharomyces pombe*. (B, courtesy of C. F. Robinow, University of Western Ontario.)

Figure 14.8
Cell cycle checkpoints Several checkpoints function to ensure that complete genomes are transmitted to daughter cells. One major checkpoint arrests cells in G_2 in response to damaged or unreplicated DNA. The presence of damaged DNA also leads to cell cycle arrest at a checkpoint in G_1. Another checkpoint, in M phase, arrests mitosis if the daughter chromosomes are not properly aligned on the mitotic spindle.

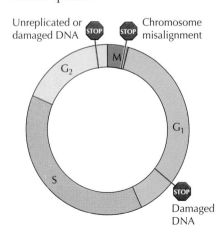

Unreplicated or damaged DNA · **STOP**   Chromosome misalignment · **STOP**

M

G_2

G_1

S

STOP

Damaged DNA

marily by control of the transition from G_2 to M, which is the principal point at which cell size and nutrient availability are monitored. In animals, the primary example of cell cycle control in G_2 is provided by oocytes. Vertebrate oocytes can remain arrested in G_2 for long periods of time (several decades in humans) until their progression to M phase is triggered by hormonal stimulation. Extracellular signals can thus control cell proliferation by regulating progression from the G_2 to M as well as the G_1 to S phases of the cell cycle.

Cell Cycle Checkpoints

The controls discussed in the previous section regulate cell cycle progression in response to cell size and extracellular signals, such as nutrients and growth factors. In addition, the events that take place during different stages of the cell cycle must be coordinated with one another so that they occur in the appropriate order. For example, it is critically important that the cell not begin mitosis until replication of the genome has been completed. The alternative would be a catastrophic cell division, in which the daughter cells failed to inherit complete copies of the genetic material. In most cells, this coordination between different phases of the cell cycle is dependent on a system of checkpoints and feedback controls that prevent entry into the next phase of the cell cycle until the events of the preceding phase have been completed.

Several cell cycle checkpoints function to ensure that incomplete or damaged chromosomes are not replicated and passed on to daughter cells (Figure 14.8). One of the most clearly defined of these checkpoints occurs in G_2 and prevents the initiation of mitosis until DNA replication is completed. This G_2 checkpoint senses unreplicated DNA, which generates a signal that leads to cell cycle arrest. Operation of the G_2 checkpoint therefore prevents the initiation of M phase before completion of S phase, so cells remain in G_2 until the genome has been completely replicated. Only then is the inhibition of G_2 progression relieved, allowing the cell to initiate mitosis and distribute the completely replicated chromosomes to daughter cells.

Progression through the cell cycle is also arrested at the G_2 checkpoint in response to DNA damage, such as that resulting from irradiation. This arrest allows time for the damage to be repaired, rather than being passed on to daughter cells. Studies of yeast mutants have shown that the same cell cycle checkpoint is responsible for G_2 arrest induced by either unreplicated or damaged DNA, both of which signal cell cycle arrest through related pathways.

DNA damage not only arrests the cell cycle in G_2, but also slows the progression of cells through S phase and arrests cell cycle progression at a checkpoint in G_1. This G_1 arrest may allow repair of the damage to take place before the cell enters S phase, where the damaged DNA would be replicated. In mammalian cells, arrest at the G_1 checkpoint is mediated by the action of a protein known as **p53**, which is rapidly induced in response to damaged DNA (Figure 14.9). Interestingly, the gene encoding p53 is frequently mutated in human cancers. Loss of p53 function as a result of these mutations prevents G_1 arrest in response to DNA damage, so the damaged DNA is replicated and passed on to daughter cells instead of being repaired. This inheritance of damaged DNA results in an increased frequency of mutations and general instability of the cellular genome, which contributes to cancer development. Mutations in the *p53* gene are the most common genetic alterations in human cancers (see Chapter 15), illustrating the critical importance of cell cycle regulation in the life of multicellular organisms.

Another important cell cycle checkpoint that maintains the integrity of the genome occurs toward the end of mitosis (see Figure 14.8). This checkpoint monitors the alignment of chromosomes on the mitotic spindle, thus ensuring that a complete set of chromosomes is distributed accurately to the daughter cells. For example, the failure of one or more chromosomes to align properly on the spindle causes mitosis to arrest at metaphase, prior to the segregation of the newly replicated chromosomes to daughter nuclei. As a result of this checkpoint, the chromosomes do not separate until a complete complement of chromosomes has been organized for distribution to each daughter cell.

Coupling of S Phase to M Phase

The G_2 checkpoint prevents the initiation of mitosis prior to the completion of S phase, thereby ensuring that incompletely replicated DNA is not distributed to daughter cells. It is equally important to ensure that the genome is replicated only once per cell cycle. Thus, once DNA has been replicated, control mechanisms must exist to prevent initiation of a new S phase prior to mitosis. These controls prevent cells in G_2 from reentering S phase and block the initiation of another round of DNA replication until after mitosis, at which point the cell has entered the G_1 phase of the next cell cycle.

Initial insights into this dependence of S phase on M phase came from cell fusion experiments of Potu Rao and Robert Johnson in 1970 (Figure 14.10). These investigators isolated cells in different phases of the cycle and then fused these cells to each other to form cell hybrids. When G_1 cells were fused with S phase cells, the G_1 nucleus immediately began to synthesize DNA. Thus, the cytoplasm of S phase cells contained factors that initiated DNA

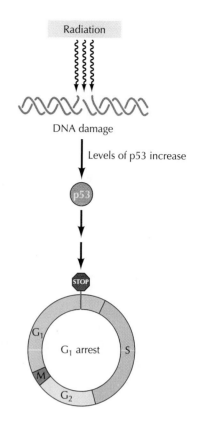

Figure 14.9
Role of p53 in G_1 arrest induced by DNA damage DNA damage, such as that resulting from irradiation, leads to rapid increases in p53 levels. The protein p53 then signals cell cycle arrest at the G_1 checkpoint.

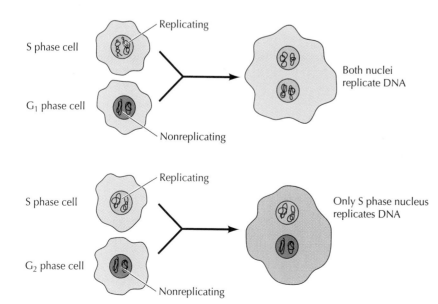

Figure 14.10
Cell fusion experiments demonstrating the dependence of S phase on M phase Cells in S phase were fused either to cells in G_1 or to cells in G_2. When G_1 cells were fused with S phase cells, the G_1 nucleus immediately began to replicate DNA. In contrast, when G_2 cells were fused with S phase cells, only the S phase nucleus continued DNA replication. It therefore appeared that the G_2 nucleus had to pass through M and enter G_1 before another round of DNA replication could be initiated.

G$_1$

S

G$_2$

M

ORC

MCM

MCM ORC

MCM

Figure 14.11
Restriction of DNA replication DNA replication is restricted to once per cell cycle by MCM proteins that bind to origins of replication together with ORC (origin replication complex) proteins and are required for the initiation of DNA replication. MCM proteins are only able to bind to DNA in G$_1$, allowing DNA replication to initiate in S phase. Once initiation has occurred, the MCM proteins are displaced, so that replication cannot initiate again until after mitosis.

synthesis in the G$_1$ nucleus. Fusing G$_2$ cells with S phase cells, however, yielded a quite different result: The G$_2$ nucleus was unable to initiate DNA synthesis even in the presence of an S phase cytoplasm. It thus appeared that DNA synthesis in the G$_2$ nucleus was prevented by a mechanism that blocked rereplication of the genome until after mitosis had taken place.

The molecular mechanism that restricts DNA replication to once per cell cycle involves the action of a family of proteins (called MCM proteins) that bind to replication origins together with the origin replication complex (ORC) proteins (see Figure 5.17). The MCM proteins act as "licensing factors" that allow replication to initiate (Figure 14.11). Their binding to DNA is regulated during the cell cycle such that the MCM proteins are only able to bind to replication origins during G$_1$, allowing DNA replication to initiate when the cell enters S phase. Once initiation has occurred, however, the MCM proteins are displaced from the origin, so replication cannot initiate again until the cell passes through mitosis and enters G$_1$ phase of the next cell cycle.

Regulators of Cell Cycle Progression

One of the most exciting developments of the last decade has been the elucidation of the molecular mechanisms that control the progression of eukaryotic cells through the division cycle. Our current understanding of cell cycle regulation has emerged from a convergence of results obtained through experiments on organisms as diverse as yeasts, sea urchins, frogs, and mammals. These studies have revealed that the cell cycle of all eukaryotes is controlled by a conserved set of protein kinases, which are responsible for triggering the major cell cycle transitions.

MPF: A Dimer of Cdc2 and Cyclin

Three initially distinct experimental approaches contributed to identification of the key molecules responsible for cell cycle regulation. The first of these avenues of investigation originated with studies of frog oocytes (Figure 14.12). These oocytes are arrested in the G$_2$ phase of the cell cycle until hormonal stimulation triggers their entry into the M phase of meiosis (discussed later in this chapter). In 1971, two independent teams of researchers (Yoshio Masui and Clement Markert, as well as Dennis Smith and Robert Ecker) found that oocytes arrested in G$_2$ could be induced to enter M phase by microinjection of cytoplasm from oocytes that had been hormonally stimulated. It thus appeared that a cytoplasmic factor present in hormone-treated oocytes was sufficient to trigger the transition from G$_2$ to M in oocytes that had not been exposed to hormone. Because the entry of oocytes into meiosis is frequently referred to as oocyte maturation, this cytoplasmic factor was called **maturation promoting factor (MPF)**. Further studies showed, however, that the activity of MPF is not restricted to the entry of oocytes into meiosis. To the contrary, MPF is also present in somatic cells, where it induces entry into M phase of the mitotic cycle. Rather than being specific to oocytes, MPF thus appeared to act as a general regulator of the transition from G$_2$ to M.

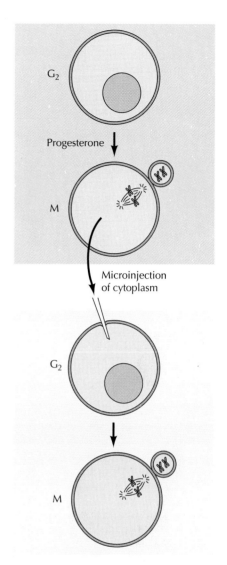

Figure 14.12
Identification of MPF Frog oocytes are arrested in the G₂ phase of the cell cycle, and entry into M phase of meiosis is triggered by the hormone progesterone. In the experiment diagrammed here, G₂-arrested oocytes were microinjected with cytoplasm extracted from oocytes that had undergone the transition from G₂ to M. Such cytoplasmic transfers induced the G₂ to M transition in the absence of hormonal stimulation, demonstrating that a cytoplasmic factor (MPF) is sufficient to induce entry into the M phase of meiosis.

The second approach to understanding cell cycle regulation was the genetic analysis of yeasts, pioneered by Lee Hartwell and his colleagues in the early 1970s. Studying the budding yeast *Saccharomyces cerevisiae*, these investigators identified temperature-sensitive mutants that were defective in cell cycle progression. The key characteristic of these mutants (called *cdc* for *cell division cycle* mutants) was that they underwent growth arrest at specific points in the cell cycle. For example, a particularly important mutant designated *cdc28* caused the cell cycle to arrest at START, indicating that the Cdc28 protein is required for passage through this critical regulatory point in G₁ (Figure 14.13). A similar collection of cell cycle mutants was isolated in the fission yeast *Schizosaccharomyces pombe* by Paul Nurse and his collaborators. These mutants included *cdc2*, which arrests the *S. pombe* cell cycle both in G₁ and at the G₂ to M transition (the major regulatory point in fission yeast). Further studies showed that *S. cerevisiae cdc28* and *S. pombe cdc2* are functionally homologous genes, which are required for passage through START as well as for entry into mitosis in both species of yeasts. To avoid confusion resulting from the difference in genetic nomenclature between *S. cerevisiae* and *S. pombe*, the protein encoded by both genes will be called **Cdc2** in this text. Further studies of *cdc2* yielded two important insights. First, molecular cloning and nucleotide sequencing revealed that *cdc2* encodes a protein kinase—the first indication of the prominent role of protein phosphorylation in regulating the cell cycle.

Figure 14.13
Properties of *S. cerevisiae cdc28* mutants The temperature-sensitive *cdc28* mutant replicates normally at the permissive temperature. At the nonpermissive temperature, however, progression through the cell cycle is blocked at START.

Key Experiment

The Discovery of MPF

Cytoplasmic Control of Nuclear Behavior during Meiotic Maturation of Frog Oocytes

Yoshio Masui and Clement L. Markert

Yale University, New Haven, CT

Journal of Experimental Zoology, 1971, Volume 177, Pages 129–146

The Context

Nuclear transplantation and cell fusion experiments performed in the 1960s indicated that nuclei transferred into cells at different stages of the mitotic cell cycle generally adopt the behavior of the host cell. Thus, it appeared that mitotic activity of the nucleus was regulated by the cytoplasm. However, the existence of postulated cytoplasmic factors that control the mitotic activity of nuclei remained to be demonstrated by a direct experimental approach. This demonstration was provided by the studies of Masui and Markert, who investigated the role of cytoplasmic factors in regulating nuclear behavior during meiosis of frog oocytes.

Several features of frog oocyte meiosis suggested that it is controlled by cytoplasmic factors. In particular, meiosis of frog oocytes is arrested at the end of the prophase of meiosis I. Treatment with the hormone progesterone triggers the resumption of meiosis, which is equivalent to the G_2 to M transition of somatic cells. The oocytes then undergo a second arrest at the metaphase of meiosis II, where they remain until fertilization. Masui and Markert hypothesized that the effects of both hormone treatment and fertilization on meiosis were due to changes in the cytoplasm that secondarily controlled behavior of the nucleus. They tested this hypothesis directly by transferring cytoplasm from hormone-stimulated oocytes into unstimulated oocytes. These experiments demonstrated that a cytoplasmic factor, which Masui and

Markert named maturation promoting factor (MPF), is responsible for the induction of meiosis following hormone treatment.

The Experiments

Because of both their large size and their ability to survive injection with glass micropipettes, frog oocytes appeared to provide a particularly suitable experimental system for testing the activity of cytoplasmic factors. The basic design of Masui and Markert's experiments was to remove cytoplasm from donor oocytes that had been treated with progesterone to induce the resumption of meiosis. Varying amounts of this cytoplasm were then injected into untreated recipient oocytes. The key result was that cytoplasm removed six or more hours after hormone treatment of donor oocytes induced the resumption of meiosis in injected recipients (see figure). In contrast, injection of cytoplasm from control oocytes that had not been exposed to progesterone had no effect on the recipients. It thus appeared that hormone-treated oocytes contained a cytoplasmic factor that could induce the resumption of meiosis in recipients that had never been exposed to progesterone.

Control experiments ruled out the possibility that progesterone itself is the meiosis-inducing factor in donor cytoplasm. In particular, it was demonstrated that the injection of progesterone into recipient oocytes fails to induce meiosis. Only external application of the hormone is effective, indicating that progesterone acts

Recipient oocytes were injected with the indicated amounts of cytoplasm from oocytes that had been treated with progesterone. Donor cytoplasm was either withdrawn from the central region of oocytes with a micropipette (solid line) or prepared by homogenization of whole oocytes (dashed line). Results are presented as the percentage of injected oocytes that were induced to resume meiosis.

on a cell surface receptor to activate a distinct cytoplasmic factor. Similar experiments performed independently by Dennis Smith and Robert Ecker (The interaction of steroids with *Rana pipiens* oocytes in the induction of maturation. *Dev. Biol.* 25: 232–247, 1971) led to the same conclusion. Interestingly, the action of progesterone in this system is distinct from its action in most cells, where it diffuses across the plasma membrane and binds to an intracellular receptor (see Chapter 13). In oocytes, however, progesterone clearly acts on the cell surface to activate a distinct factor in the oocyte cytoplasm. Since the resumption of oocyte meiosis is commonly referred to as oocyte maturation, Masui and Markert coined the

The Discovery of MPF *(continued)*

term "maturation promoting factor" for their newly discovered regulator of meiosis.

The Impact

Following its discovery in frog oocytes, MPF was also found to be present in somatic cells, where it induces the G_2 to M transition of mitosis. Thus, MPF appeared to be a general regulator of entry into the M phase of both mitotic and meiotic cell cycles. The eventual purification of MPF from frog oocytes in 1988 then converged with yeast genetics and studies of sea urchin embryos to reveal the identity of this critical cell cycle regulator. Namely, MPF was found to be a dimer of cyclin B and the Cdc2 protein kinase. Further studies have established that both cyclin B and Cdc2 are members of large families of proteins, with different cyclins and Cdc2-related protein kinases functioning analogously to MPF in the regulation of other cell cycle transitions. The discovery of MPF in frog oocytes thus paved the way to understanding a cell cycle regulatory apparatus that is conserved throughout all eukaryotes.

Yoshio Masui

Clement Markert

Second, a human gene related to *cdc2* was identified and shown to function in yeasts, providing a dramatic demonstration of the conserved activity of this cell cycle regulator.

The third line of investigation that eventually converged with the identification of MPF and yeast genetics emanated from studies of protein synthesis in early sea urchin embryos. Following fertilization, these embryos go through a series of rapid cell divisions. Intriguingly, studies with protein synthesis inhibitors had revealed that entry into M phase of these embryonic cell cycles requires new protein synthesis. In 1983, Tim Hunt and his colleagues identified two proteins that display a periodic pattern of accumulation and degradation in sea urchin and clam embryos. These proteins accumulate throughout interphase and are then rapidly degraded toward the end of each mitosis (Figure 14.14). Hunt called these proteins **cyclins** (the two proteins were designated cyclin A and cyclin B) and suggested that they might function to induce mitosis, with their periodic accumulation and destruction controlling entry and exit from M phase. Direct support for such a role of cyclins was provided in 1986, when Joan Ruderman and her colleagues showed that microinjection of cyclin A into frog oocytes is sufficient to trigger the G_2 to M transition.

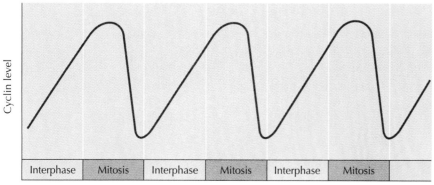

Figure 14.14
Accumulation and degradation of cyclins in sea urchin embryos The cyclins were identified as proteins that accumulate throughout interphase and are rapidly degraded toward the end of mitosis.

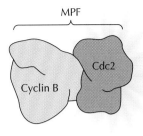

Figure 14.15
Structure of MPF   MPF is a dimer consisting of cyclin B and the Cdc2 protein kinase.

These initially independent approaches converged dramatically in 1988, when MPF was purified from frog eggs in the laboratory of James Maller. Molecular characterization of MPF in several laboratories then showed that this conserved regulator of the cell cycle is composed of two key subunits: Cdc2 and cyclin B (Figure 14.15). Cyclin B is a regulatory subunit required for catalytic activity of the Cdc2 protein kinase, consistent with the notion that MPF activity is controlled by the periodic accumulation and degradation of cyclin B during cell cycle progression.

A variety of further studies have confirmed this role of cyclin B, as well as demonstrating the regulation of MPF by phosphorylation and dephosphorylation of Cdc2 (Figure 14.16). In mammalian cells, cyclin B synthesis begins in S phase. Cyclin B then accumulates and forms complexes with Cdc2 throughout S and G_2. As these complexes form, Cdc2 is phosphorylated at two critical regulatory positions. One of these phosphorylations occurs on threonine-161 and is required for Cdc2 kinase activity. The second is a phosphorylation of tyrosine-15, and of the adjacent threonine-14 in vertebrates. Phosphorylation of tyrosine-15, catalyzed by a protein kinase called Wee1, inhibits Cdc2 activity and leads to the accumulation of inactive Cdc2/cyclin B complexes throughout S and G_2. The transition from G_2 to M is then brought about by activation of the Cdc2/cyclin B complex as a result of dephosphorylation of threonine-14 and tyrosine-15 by a protein phosphatase called Cdc25.

Once activated, the Cdc2 protein kinase phosphorylates a variety of target proteins that initiate the events of M phase, which are discussed later in this chapter. In addition, Cdc2 activity triggers the degradation of cyclin B, which occurs as a result of ubiquitin-mediated proteolysis. This proteolytic destruction of cyclin B then inactivates Cdc2, leading the cell to exit mitosis, undergo cytokinesis, and return to interphase.

Figure 14.16
MPF regulation   Cdc2 forms complexes with cyclin B during S and G_2. Cdc2 is then phosphorylated on threonine-161, which is required for Cdc2 activity, as well as on tyrosine-15 (and threonine-14 in vertebrate cells), which inhibits Cdc2 activity. Dephosphorylation of Thr14 and Tyr15 activates MPF at the G_2 to M transition. MPF activity is then terminated toward the end of mitosis by proteolytic degradation of cyclin B.

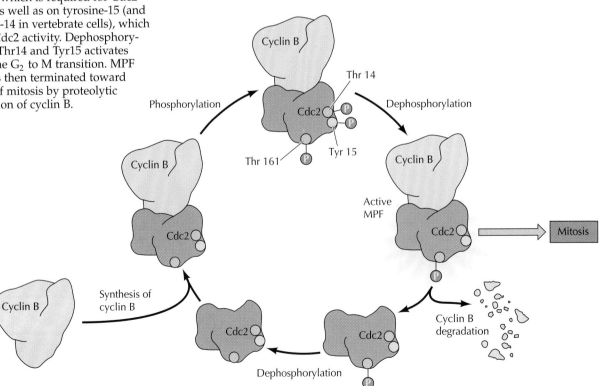

Families of Cyclins and Cyclin-Dependent Kinases

The structure and function of MPF (Cdc2/cyclin B) provide not only a molecular basis for understanding entry and exit from M phase, but also the foundation for elucidating the regulation of other cell cycle transitions. The insights provided by characterization of the Cdc2/cyclin B complex have thus had a sweeping impact on understanding cell cycle regulation. In particular, further research has established that both Cdc2 and cyclin B are members of large families of related proteins, with different members of these families controlling progression through distinct phases of the cell cycle.

As discussed earlier, Cdc2 controls passage through START as well as entry into mitosis in yeasts. It does so, however, in association with distinct cyclins (Figure 14.17). In particular, the G_2 to M transition is driven by Cdc2 in association with the mitotic B-type cyclins (Clb1, Clb2, Clb3, and Clb4). Passage through START, however, is controlled by Cdc2 in association with a distinct class of cyclins called **G_1 cyclins** or **Cln's**. Cdc2 then associates with different B-type cyclins (Clb5 and Clb6), which are required for progression through S phase. These associations of Cdc2 with distinct B-type and G_1 cyclins direct Cdc2 to phosphorylate different substrate proteins, as required for progression through specific phases of the cell cycle.

The cell cycles of higher eukaryotes are controlled not only by multiple cyclins, but also by multiple Cdc2-related protein kinases. These Cdc2-related kinases are known as **Cdk's** (for *cyclin-dependent kinases*). As the original member of this family, Cdc2 is also known as Cdk1, with other currently identified family members being designated Cdk2 through Cdk8.

These multiple members of the Cdk family associate with specific cyclins to drive progression through the different stages of the cell cycle (see Figure 14.17). For example, progression from G_1 to S is regulated principally by Cdk2 and Cdk4 (and in some cells Cdk6) in association with cyclins D and E. Complexes of Cdk4 and Cdk6 with the D-type cyclins (cyclin D1, D2, and D3) play a critical role in progression through the restriction point in G_1. Cyclin E is expressed later in G_1, and Cdk2/cyclin E complexes are required for the G_1 to S transition and initiation of DNA synthesis. Complexes of Cdk2 with cyclin A function in the progression of cells through S phase. As already discussed, the transition from G_2 to M is then driven by complexes of Cdc2 with cyclin B.

The activity of Cdk's during cell cycle progression is regulated by at least four molecular mechanisms (Figure 14.18). As already discussed for Cdc2, the first level of regulation involves the association of Cdk's with their cyclin partners. Thus, the formation of specific Cdk/cyclin complexes is controlled by cyclin synthesis and degradation. Second, activation of Cdk/ cyclin complexes requires phosphorylation of a conserved Cdk threonine residue around position 160. This activating phosphorylation of the Cdk's is catalyzed by an enzyme

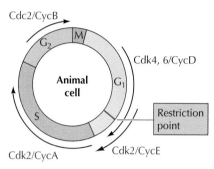

Figure 14.17
Complexes of cyclins and cyclin-dependent kinases In yeast, passage through START is controlled by Cdc2 in association with G_1 cyclins (Cln1, Cln2, and Cln3). Complexes of Cdc2 with distinct B-type cyclins (Clb's) then regulate progression through S phase and entry into mitosis. In animal cells, progression through the G_1 restriction point is controlled by complexes of Cdk4 and Cdk6 with D-type cyclins. Cdk2/cyclin E complexes function later in G_1 and are required for the G_1 to S transition. Cdk2/cyclin A complexes are then required for progression through S phase, and Cdc2/cyclin B complexes drive the G_2 to M transition.

Figure 14.18
Mechanisms of Cdk regulation
The activities of Cdk's are regulated by four molecular mechanisms.

1. Association with cyclins.

Cyclin

4. Association with Cdk inhibitors (CKI's).

CKI

2. Activating phosphorylation of threonine around position 160.

Cdk

3. Inhibitory phosphorylation of threonine 14 and tyrosine 15.

Thr

Tyr

Thr

TABLE 14.1 Cdk Inhibitors

Inhibitor	Cdk/cyclin complex	Cell cycle phase affected
Animal cells		
Cip/Kip family (p21, p27, p57)	Cdk4/cyclin D	G_1
	Cdk6/cyclin D	G_1
	Cdk2/cyclin E	G_1/S
	Cdk2/cyclin A	S
Ink4 family (p15, p16, p18, p19)	Cdk4/cyclin D	G_1
	Cdk6/cyclin D	G_1
Yeast (*S. cerevisiae*)		
Far1	Cdc2/Cln	G_1
Sic1	Cdc2/Clb5	S
	Cdc2/Clb6	S

called CAK (for *C*dk-*a*ctivating *k*inase), which may itself be composed of a Cdk (Cdk7) complexed with cyclin H. Complexes of Cdk7 and cyclin H are also associated with the transcription factor TFIIH, which is required for initiation of transcription by RNA polymerase II (see Chapter 5). It thus appears that this member of the Cdk family may participate in transcription as well as cell cycle regulation.

In contrast to the activating phosphorylation by CAK, the third mechanism of Cdk regulation involves inhibitory phosphorylation of tyrosine residues near the Cdk amino terminus, catalyzed by the Wee1 protein kinase. In particular, both Cdc2 and Cdk2 are inhibited by phosphorylation of tyrosine-15, and the adjacent threonine-14 in vertebrates. These Cdk's are then activated by dephosphorylation of these residues by members of the Cdc25 family of protein phosphatases.

In addition to regulation of the Cdk's by phosphorylation, their activities can also be controlled by the binding of inhibitory proteins (called **Cdk inhibitors** or **CKIs**) to Cdk/cyclin complexes. In mammalian cells, two families of Cdk inhibitors are responsible for regulating different Cdk/cyclin complexes (Table 14.1). Members of the Cip/Kip family regulate all stages of progression through G_1 and S phase by inhibiting complexes of Cdk2, 4, and 6 with cyclins A, D, and E. In contrast, members of the Ink4 family are specific for complexes of Cdk4 and 6 with cyclin D, so the Ink4 CKIs only regulate progression through the restriction point in G_1. In yeast, different CKIs similarly regulate distinct stages of cell cycle progression by inhibiting specific Cdk/cyclin complexes. Control of Cdk inhibitors thus provides an additional mechanism for regulating Cdk activity. The combined effects of these multiple modes of Cdk regulation are responsible for controlling cell cycle progression in response both to checkpoint controls and to the variety of extracellular stimuli that regulate cell proliferation.

Growth Factors and the D-Type Cyclins

As discussed earlier, the proliferation of animal cells is regulated largely by a variety of extracellular growth factors that control the progression of cells through the restriction point in late G_1. In the absence of growth factors, cells are unable to pass the restriction point and become quiescent, frequently entering the resting state known as G_0, from which they can reenter the cell cycle in response to growth factor stimulation. This control of cell cycle progression by extracellular growth factors implies that the intracellular signaling pathways stimulated downstream of growth factor receptors (discussed in the preceding chapter) ultimately act to regulate components of the cell cycle machinery.

One critical link between growth factor signaling and cell cycle progression is provided by the D-type cyclins (Figure 14.19). Cyclin D synthesis is induced in response to growth factor stimulation as a result of signaling through the Ras/Raf/ERK pathway, and the D-type cyclins continue to be synthesized as long as growth factors are present. However, the D-type cyclins are also rapidly degraded, so their intracellular concentrations rapidly fall if growth factors are removed. Thus, as long as growth factors are present through G_1, complexes of Cdk4, 6/cyclin D drive cells through the restriction point. On the other hand, if growth factors are removed prior to this key regulatory point in the cell cycle, the levels of cyclin D rapidly fall and cells are unable to progress through G_1 to S, instead becoming quiescent and entering G_0. The inducibility and rapid turnover of D-type cyclins thus integrates growth factor signaling with the cell cycle machinery, allowing the availability of extracellular growth factors to control the progression of cells through G_1.

Since cyclin D is a critical target of growth factor signaling, it might be expected that defects in cyclin D regulation could contribute to the loss of growth regulation characteristic of cancer cells. Consistent with this expectation, many human cancers have been found to arise as a result of defects in cell cycle regulation, just as many others result from abnormalities in the intracellular signaling pathways activated by growth factor receptors (see Chapter 13). For example, mutations resulting in continual unregulated expression of cyclin D1 contribute to the development of a variety of human cancers, including lymphomas and breast cancers. Similarly, mutations that inactivate the Ink4 Cdk inhibitors (e.g., p16) that bind to Cdk4, 6/cyclin D complexes are commonly found in human cancer cells.

The connection between cyclin D, growth control, and cancer is further fortified by the fact that a key substrate protein of Cdk4, 6/cyclin D complexes is itself frequently mutated in a wide array of human tumors. This protein, designated **Rb**, was first identified as the product of a gene responsible for retinoblastoma, a rare inherited childhood eye tumor (see Chapter 15). Further studies then showed that mutations resulting in the absence of functional Rb protein are not restricted to retinoblastoma but also contribute to a variety of common human cancers. Rb is the prototype of a **tumor suppressor gene**—a gene whose inactivation leads to tumor development. Whereas oncogene proteins such as Ras (see Chapter 13) and cyclin D drive cell proliferation, the proteins encoded by tumor suppressor genes act as brakes that slow down cell cycle progression. Additional examples of cell cycle regulators encoded by tumor suppressor genes include the Ink4 Cdk inhibitors that bind Cdk4, 6/cyclin D complexes and the important growth regulator p53, which was discussed earlier in this chapter.

Further studies of Rb have revealed that it plays a key role in coupling the cell cycle machinery to the expression of genes required for cell cycle progression and DNA synthesis (Figure 14.20). The activity of Rb is regulated by changes in its phosphorylation as cells progress through the cycle. In particular, Rb becomes phosphorylated by Cdk4, 6/cyclin D complexes as cells pass through the restriction point in G_1. In its underphosphorylated form (present in G_0 or early G_1), Rb binds to members of the **E2F** family of transcription factors, which regulate expression of several genes involved

Figure 14.19
Induction of D-type cyclins Growth factors regulate cell cycle progression through the G_1 restriction point by inducing synthesis of D-type cyclins via the Ras/Raf/ERK signaling pathway.

Figure 14.20
Cell cycle regulation of Rb and E2F
In its underphosphorylated form, Rb binds to members of the E2F family, repressing transcription of E2F-regulated genes. Phosphorylation of Rb by Cdk4, 6/cyclin D complexes results in its dissociation from E2F in late G_1. E2F then stimulates expression of its target genes, which encode proteins required for cell cycle progression.

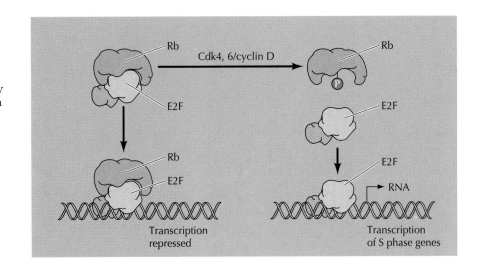

in cell cycle progression, including the gene encoding cyclin E. E2F binds to its target sequences in either the presence or absence of Rb. However, Rb acts as a repressor, so the Rb/E2F complex suppresses transcription of E2F-regulated genes. Phosphorylation of Rb by Cdk4, 6/cyclin D complexes results in its dissociation from E2F, which then activates transcription of its target genes. Rb thus acts as a molecular switch that converts E2F from a repressor to an activator of genes required for cell cycle progression. The control of Rb via Cdk4, 6/cyclin D phosphorylation in turn couples this critical regulation of gene expression to the availability of growth factors in G_1.

Inhibitors of Cell Cycle Progression

Cell proliferation is regulated not only by growth factors but also by a variety of signals that act to inhibit cell cycle progression. For example, agents that damage DNA result in cell cycle arrest, presumably to allow time for the cell to repair the damage. In addition, cell contacts and a variety of extracellular factors act to inhibit rather than stimulate proliferation of their target cells. The effects of such inhibitory signals are also mediated by regulators of the cell cycle machinery, frequently via the induction of Cdk inhibitors.

A good example of the action of Cdk inhibitors is provided by cell cycle arrest in response to DNA damage, which is mediated by the protein p53 (discussed earlier in this chapter). The p53 protein is a transcriptional regulator that functions, at least in part, to stimulate expression of the Cdk inhibitor p21 (Figure 14.21). The p21 protein inhibits several Cdk/cyclin complexes, and its induction by p53 appears to represent at least one mechanism of p53-dependent cell cycle arrest following DNA damage. In addition to inhibiting cell cycle progression via its interaction with Cdk's, p21 may directly inhibit DNA replication. In particular, p21 binds to proliferating cell nuclear antigen (PCNA), which, as discussed in Chapter 5, is a subunit of DNA polymerase δ. Thus, p21 may play a dual role in cell cycle arrest

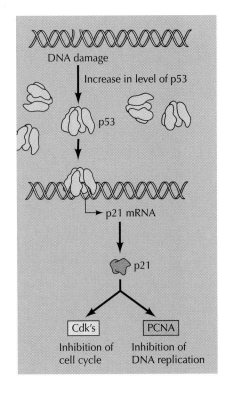

Figure 14.21
Induction of p21 by DNA damage DNA damage results in the elevation of intracellular levels of p53, which activates transcription of the gene encoding the Cdk inhibitor p21. In addition to inhibiting cell cycle progression by binding to Cdk/cyclin complexes, p21 may directly inhibit DNA synthesis by interacting with PCNA (a subunit of DNA polymerase δ).

Replication completed

Unreplicated or damaged DNA

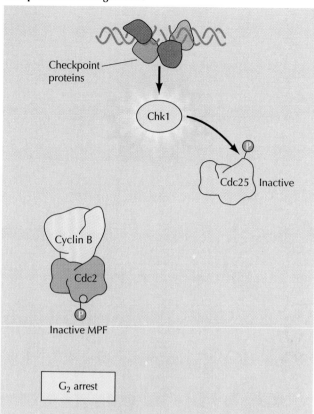

Figure 14.22
Control of the G$_2$ checkpoint
A complex of checkpoint proteins recognizes unreplicated or damaged DNA and activates the protein kinase Chk1, which phosphorylates and inhibits the Cdc25 protein phosphatase. Inhibition of Cdc25 prevents dephosphorylation and activation of Cdc2.

induced by DNA damage, not only blocking cell cycle progression by inhibiting Cdk's but also directly inhibiting DNA replication in S phase cells.

The best-characterized extracellular inhibitor of animal cell proliferation is TGF-β—a polypeptide factor that inhibits the proliferation of a variety of types of epithelial cells by arresting cell cycle progression in G$_1$. This action of TGF-β appears to be mediated by induction of the Cdk inhibitor p15, which binds to Cdk4, 6/cyclin D complexes. In the resulting absence of Cdk4 activity, Rb phosphorylation is blocked and the cell cycle is arrested in G$_1$.

A different molecular mechanism is used to control cell cycle progression through the G$_2$ checkpoint, which prevents entry into mitosis in the presence of unreplicated or damaged DNA (Figure 14.22). Arrest of the cell cycle at this checkpoint is mediated by a protein kinase called Chk1, which is activated in response to DNA damage or incomplete replication. Chk1 phosphorylates the protein phosphatase Cdc25, thereby preventing Cdc25 from dephosphorylating and activating Cdc2. In the absence of Cdc2 activation, progression to mitosis is blocked and the cell remains arrested in G$_2$.

The Events of M Phase

M phase is the most dramatic period of the cell cycle, involving a major reorganization of virtually all cell components. During mitosis (nuclear division), the chromosomes condense, the nuclear envelope of most cells breaks down, the cytoskeleton reorganizes to form the mitotic spindle, and the chromosomes move to opposite poles. Chromosome segregation is then

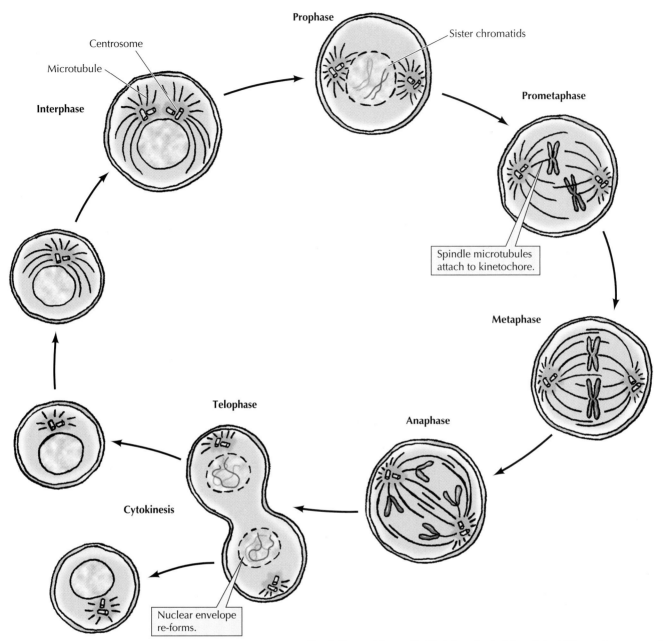

Figure 14.23
Stages of mitosis in an animal cell During prophase, the chromosomes condense and centrosomes move to opposite sides of the nucleus, initiating formation of the mitotic spindle. Breakdown of the nuclear envelope then allows spindle microtubules to attach to the kinetochores of chromosomes. During prometaphase, the chromosomes shuffle back and forth between the centrosomes and the center of the cell, eventually aligning in the center of the spindle (metaphase). At anaphase, the sister chromatids separate and move to opposite poles of the spindle. Mitosis then ends with re-formation of nuclear envelopes and chromosome decondensation during telophase, and cytokinesis yields two interphase daughter cells. Note that each daughter cell receives one centrosome, which duplicates prior to the next mitosis.

usually followed by cell division (cytokinesis). Although many of these events have been discussed in previous chapters with respect to the structure and function of the nucleus and cytoskeleton, they are reviewed here in the context of a coordinated view of M phase and the action of MPF.

Stages of Mitosis

Although many of the details of mitosis vary among different organisms, the fundamental processes that ensure the faithful segregation of sister chromatids are conserved in all eukaryotes. These basic events of mitosis include chromosome condensation, formation of the mitotic spindle, and attachment of chromosomes to the spindle microtubules. Sister chromatids then separate from each other and move to opposite poles of the spindle, followed by the formation of daughter nuclei.

Mitosis is conventionally divided into four stages—**prophase**, **metaphase**, **anaphase**, and **telophase**—which are illustrated for an animal cell in Figures 14.23 and 14.24. The beginning of prophase is marked by the appearance of condensed chromosomes, each of which consists of two sister chromatids (the daughter DNA molecules produced in S phase). These newly replicated DNA molecules remain intertwined throughout S and G_2, becoming untangled during the process of chromatin condensation. The condensed sister chromatids are then held together at the **centromere**, which (as discussed in Chapter 4) is a DNA sequence to which proteins bind to form the **kinetochore**—the site of eventual attachment of the spindle microtubules. In addition to chromosome condensation, cytoplasmic changes leading to the development of the mitotic spindle initiate during prophase. The **centrosomes** (which had duplicated during interphase) separate and move to opposite sides of the nucleus. There they serve as the two poles of the **mitotic spindle**, which begins to form during late prophase.

In higher eukaryotes the end of prophase corresponds to the breakdown of the nuclear envelope. As discussed in Chapter 8, however, nuclear envelope breakdown is not a universal feature of mitosis. In particular, yeasts and many other unicellular eukaryotes undergo "closed mitosis," in which the nuclear envelope remains intact (see Figure 8.30). In these cells the spin-

Mitosis

Interphase

Early prophase

Late prophase

Prometaphase

Metaphase

Early anaphase

Late anaphase

Telophase

Figure 14.24
Fluorescence micrographs of chromatin, keratin, and microtubules during mitosis of newt lung cells
Chromatin is stained blue, keratin is stained red, and microtubules are stained green. (Conly L. Rieder/Biological Photo Service.)

dle pole bodies are embedded within the nuclear envelope, and the nucleus divides in two following migration of daughter chromosomes to opposite poles of the spindle.

Following completion of prophase, the cell enters **prometaphase**—a transition period between prophase and metaphase. During prometaphase the microtubules of the mitotic spindle attach to the kinetochores of condensed chromosomes. The kinetochores of sister chromatids are oriented on opposite sides of the chromosome, so they attach to microtubules emanating from opposite poles of the spindle. The chromosomes shuffle back and forth until they eventually align on the metaphase plate in the center of the spindle. At this stage, the cell has reached metaphase.

Most cells remain only briefly at metaphase before proceeding to anaphase. The transition from metaphase to anaphase is triggered by breakage of the link between sister chromatids, which then separate and move to opposite poles of the spindle. Mitosis ends with telophase, during which nuclei re-form and the chromosomes decondense. Cytokinesis usually begins during late anaphase and is almost complete by the end of telophase, resulting in the formation of two interphase daughter cells.

MPF and Progression to Metaphase

Mitosis involves dramatic changes in multiple cellular components, leading to a major reorganization of the entire structure of the cell. As discussed earlier in this chapter, these events are initiated by activation of the MPF protein kinase (Cdc2/cyclin B). It appears that MPF not only acts as a master regulator of the M phase transition, phosphorylating and activating other downstream protein kinases, but also acts directly by phosphorylating some of the structural proteins involved in this cellular reorganization (Figure 14.25).

The condensation of interphase chromatin to form the compact chromosomes of mitotic cells is a key event in mitosis, critical in enabling the chromosomes to move along the mitotic spindle without becoming broken or tangled with one another. As discussed in Chapter 4, the chromatin in interphase nuclei condenses nearly a thousand fold during the formation of metaphase chromosomes. Such highly condensed chromatin cannot be transcribed, so transcription ceases as chromatin condensation takes place.

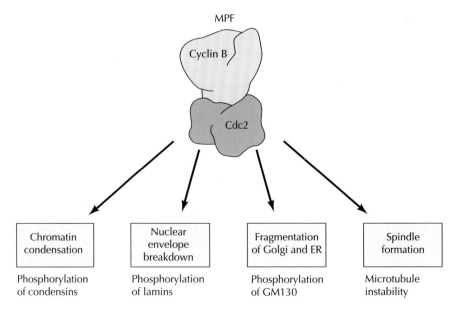

Figure 14.25
Targets of MPF MPF induces multiple nuclear and cytoplasmic changes at the onset of M phase, both by activating other protein kinases and by phosphorylating proteins such as condensins and the nuclear lamins.

Despite the fundamental importance of this event, we do not fully understand either the structure of metaphase chromosomes or the molecular mechanism of chromatin condensation. However, protein complexes called condensins have recently been found to drive chromosome condensation by wrapping DNA around itself, compacting chromosomes into the condensed mitotic structure. The condensins are phosphorylated directly by the Cdc2 protein kinase, which drives chromatin condensation by activating condensins as cells enter mitosis. One molecular alteration that generally accompanies chromosome condensation is phosphorylation of histone H1, so it is noteworthy that histone H1 is also a substrate for Cdc2. However, histone H1 phosphorylation is not required for mitotic chromosome condensation, so the significance of H1 phosphorylation by Cdc2 is unclear. In contrast, chromosome condensation has been shown to require phosphorylation of histone H3. Perhaps surprisingly, however, histone H3 is not phosphorylated by Cdc2 and the kinase responsible for H3 phosphorylation in mitotic cells remains to be identified.

Breakdown of the nuclear envelope, which is one of the most dramatic events of mitosis, represents the most clearly defined target for MPF action. As discussed in Chapter 8, Cdc2 phosphorylates the lamins, leading directly to depolymerization of the nuclear lamina (see Figure 8.31). This is followed by fragmentation of the nuclear membrane into small vesicles, which eventually fuse to form new daughter nuclei at telophase. The endoplasmic reticulum and Golgi apparatus similarly fragment into small vesicles, which can then be distributed to daughter cells at cytokinesis. The breakdown of these membranes is also induced by MPF, and may in part be mediated by Cdc2 phosphorylation of the Golgi matrix protein GM130, which is required for the docking of COPI-coated vesicles to the Golgi membrane. Phosphorylation and inactivation of GM130 by Cdc2 inhibits vesicle docking and fusion, leading to fragmentation of the Golgi apparatus. However, additional targets of Cdc2 may also be involved, and the mechanisms by which MPF leads to membrane fragmentation remain to be fully elucidated.

The reorganization of the cytoskeleton that culminates in formation of the mitotic spindle results from the dynamic instability of microtubules (see Chapter 11). At the beginning of prophase, the centrosomes move to opposite sides of the nucleus. The rise in MPF activity then induces a dramatic change in the dynamic behavior of microtubules. First, the rate of microtubule disassembly increases, resulting in depolymerization and shrinkage of the interphase microtubules. This disassembly is thought to result from phosphorylation of microtubule-associated proteins, either by MPF itself or by other MPF-activated protein kinases. In addition, the number of microtubules emanating from the centrosomes increases, so the interphase microtubules are replaced by large numbers of short microtubules radiating from the centrosomes.

The breakdown of the nuclear envelope then allows some of the spindle microtubules to attach to chromosomes at their kinetochores (Figure 14.26), initiating the process of chromosome movement that characterizes prometaphase. The proteins assembled at the kinetochore include microtubule motors that direct the movement of chromosomes toward the minus ends of the spindle microtubules, which are anchored in the centrosome. The action of these proteins, which draw chromosomes toward the centrosome, is opposed by the growth of the spindle microtubules, which pushes the chromosomes away from the spindle poles. Consequently, the chromosomes in prometaphase shuffle back and forth between the centrosomes and the center of the spindle.

Figure 14.26
Electron micrograph of microtubules attached to the kinetochore of a chromosome. (Conly L. Rieder/ Biological Photo Service.)

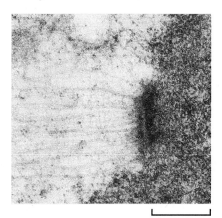

0.5 μm

Figure 14.27
The metaphase spindle (A) The
spindle consists of three kinds of
microtubules. Kinetochore microtubules are attached to chromosomes,
polar microtubules overlap in the center of the cell, and astral microtubules
radiate from the centrosome to the cell
periphery. (B) A whitefish cell at metaphase. (B, Michael Abbey/Photo
Researchers, Inc.)

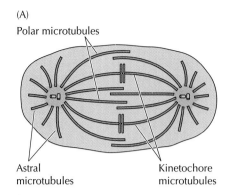

(A)

Polar microtubules

Astral
microtubules

Kinetochore
microtubules

(B)

10 μm

Figure 14.27
The metaphase spindle (A) The
spindle consists of three kinds of
microtubules. Kinetochore microtubules are attached to chromosomes,
polar microtubules overlap in the center of the cell, and astral microtubules
radiate from the centrosome to the cell
periphery. (B) A whitefish cell at metaphase. (B, Michael Abbey/Photo
Researchers, Inc.)

Microtubules from opposite poles of the spindle eventually attach to the
two kinetochores of sister chromatids (which are located on opposite sides
of the chromosome), and the balance of forces acting on the chromosomes
leads to their alignment on the metaphase plate in the center of the spindle
(Figure 14.27). As discussed in Chapter 11, the spindle consists of both
kinetochore microtubules, which are attached to the chromosomes, and
polar microtubules, which overlap with one another in the center of the
cell. In addition, short **astral microtubules** radiate outward from the centrosomes toward the cell periphery.

Proteolysis and the Inactivation of MPF: Anaphase and Telophase

As discussed earlier in this chapter, an important cell cycle checkpoint
monitors the alignment of chromosomes on the metaphase spindle. Once
this has been accomplished, the cell proceeds to initiate anaphase and
complete mitosis. The progression from metaphase to anaphase results
from ubiquitin-mediated proteolysis of key regulatory proteins, triggered by activation of a ubiquitin ligase (see Figure 7.39) called the **anaphase-
promoting complex**. Activation of the anaphase-promoting complex is
induced by MPF at the beginning of mitosis, so MPF ultimately triggers its
own destruction. The anaphase-promoting complex remains inhibited,
however, until the cell passes the metaphase checkpoint, after which activation of the ubiquitin degradation system brings about the transition from
metaphase to anaphase and progression through the rest of mitosis.

Activation of the anaphase-promoting complex leads to the degradation
of at least two key regulatory proteins (Figure 14.28). The onset of anaphase
results from proteolytic degradation of a protein called Scc1, a component of
a complex of proteins called **cohesins** that maintain the connection between
sister chromatids while they are aligned on the metaphase plate. Degradation of Scc1 is not catalyzed directly by the anaphase-promoting complex,
which instead degrades a regulatory protein called Pds1. Degradation of
Pds1 in turn activates another protein, called Esp1, which leads to proteolysis of the cohesin Scc1. Cleavage of Scc1 breaks the linkage between sister
chromatids, allowing them to segregate by moving to opposite poles of the

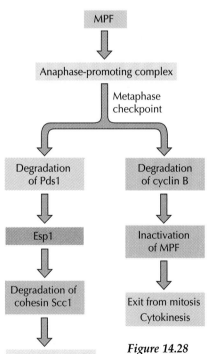

Figure 14.28
Targets of the cyclin B proteolysis system The anaphase-promoting complex is a ubiquitin ligase
that is activated following passage through the metaphase checkpoint. Its activation brings about the
transition from metaphase to anaphase by leading to degradation of the cohesin Scc1, which breaks
the link between sister chromatids. The anaphase-promoting complex also targets cyclin B for degradation, leading to inactivation of MPF, exit from mitosis, and cytokinesis.

spindle (Figure 14.29). The separation of chromosomes during anaphase then proceeds as a result of the action of several types of motor proteins associated with the spindle microtubules (see Figures 11.48 and 11.49).

The other key regulatory protein targeted for ubiquitination and degradation by the anaphase-promoting complex is cyclin B. Degradation of cyclin B leads to inactivation of MPF, which is required for the cell to exit mitosis and return to interphase. Many of the cellular changes involved in these transitions are simply the reversal of the events induced by MPF during entry into mitosis. For example, reassembly of the nuclear envelope, chromatin decondensation, and the return of microtubules to an interphase state probably result directly from loss of MPF activity and dephosphorylation of proteins that had been phosphorylated by MPF at the beginning of mitosis. As discussed next, inactivation of MPF also triggers cytokinesis.

Cytokinesis

The completion of mitosis is usually accompanied by cytokinesis, giving rise to two daughter cells. Cytokinesis usually initiates in late anaphase and is triggered by the inactivation of MPF, thereby coordinating nuclear and cytoplasmic division of the cell. As discussed in Chapter 11, cytokinesis of animal cells is mediated by a **contractile ring** of actin and myosin II filaments that forms beneath the plasma membrane (Figure 14.30). The location of this ring is determined by the position of the mitotic spindle, so the cell is eventually cleaved in a plane that passes through the metaphase plate perpendicular to the spindle. Cleavage proceeds as contraction of the actin-myosin filaments pulls the plasma membrane inward, eventually pinching the cell in half.

The mechanism of cytokinesis is different for higher plant cells, which are surrounded by rigid cell walls. Rather than being pinched in half by a contractile ring, these cells divide by forming new cell walls and plasma membranes inside the cell (Figure 14.31). In early telophase, vesicles carrying cell wall precursors from the Golgi apparatus associate with spindle microtubules and accumulate at the former site of the metaphase plate. These vesicles then fuse to form a large, membrane-enclosed, disclike structure, and their polysaccharide contents assemble to form the matrix of a new cell wall (called a **cell plate**). The cell plate expands outward, perpendicular to the spindle, until it reaches the plasma membrane. The membrane surrounding the cell plate then fuses with the parental plasma membrane, dividing the cell in two.

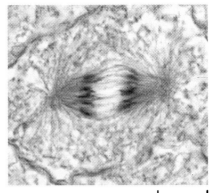

Figure 14.29   10 μm
A whitefish cell at anaphase
(Michael Abbey/Photo Researchers, Inc.)

(A)

Contractile ring

(B)

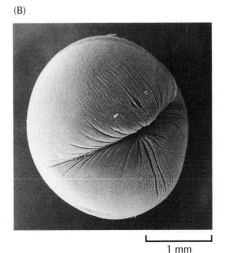

1 mm

Figure 14.30
Cytokinesis of animal cells
(A) Cytokinesis results from contraction of a ring of actin and myosin filaments, which pinches the cell in two.
(B) Scanning electron micrograph of a frog egg undergoing cytokinesis. (B, David M. Phillips/Visuals Unlimited).

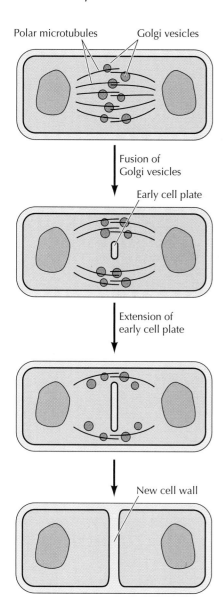

Polar microtubules   Golgi vesicles

Fusion of
Golgi vesicles

Early cell plate

Extension of
early cell plate

New cell wall

Figure 14.31
Cytokinesis in higher plants   Golgi vesicles carrying cell wall precursors associate with polar microtubules at the former site of the metaphase plate. Fusion of these vesicles yields a membrane-enclosed, disclike structure (the early cell plate) that expands outward and fuses with the parental plasma membrane.

Meiosis and Fertilization

The somatic cell cycles discussed so far in this chapter result in diploid daughter cells with identical genetic complements. Meiosis, in contrast, is a specialized kind of cell cycle that reduces the chromosome number by half, resulting in the production of haploid daughter cells. Unicellular eukaryotes, such as yeasts, can undergo meiosis as well as reproducing by mitosis. Diploid *Saccharomyces cerevisiae*, for example, undergo meiosis and produce spores when faced with unfavorable environmental conditions. In multicellular plants and animals, however, meiosis is restricted to the germ cells, where it is key to sexual reproduction. Whereas somatic cells undergo mitosis to proliferate, the germ cells undergo meiosis to produce haploid gametes (the sperm and the egg). The development of a new progeny organism is then initiated by the fusion of these gametes at fertilization.

The Process of Meiosis

In contrast to mitosis, **meiosis** results in the division of a diploid parental cell into haploid progeny, each containing only one member of the pair of homologous chromosomes that were present in the diploid parent (Figure 14.32). This reduction in chromosome number is accomplished by two sequential rounds of nuclear and cell division (called meiosis I and meiosis II), which follow a single round of DNA replication. Like mitosis, meiosis I initiates after S phase has been completed and the parental chromosomes have replicated to produce identical sister chromatids. The pattern of chromosome segregation in meiosis I, however, is dramatically different from that of mitosis. During meiosis I, homologous chromosomes first pair with one another and then segregate to different daughter cells. Sister chromatids remain together, so completion of meiosis I results in the formation of daughter cells containing a single member of each chromosome pair (consisting of two sister chromatids). Meiosis I is followed by meiosis II, which resembles mitosis in that the sister chromatids separate and segregate to different daughter cells. Completion of meiosis II thus results in the production of four haploid daughter cells, each of which contains only one copy of each chromosome.

The pairing of homologous chromosomes after DNA replication is not only a key event underlying meiotic chromosome segregation, but also allows recombination between chromosomes of paternal and maternal origin. This critical pairing of homologous chromosomes takes place during an extended prophase of meiosis I, which is divided into five stages (**leptotene**, **zygotene**, **pachytene**, **diplotene**, and **diakinesis**) on the basis of chromosome morphology (Figure 14.33). The initial association of homologous chromosomes is thought to be mediated by base pairing between complementary DNA strands during the leptotene stage, before the chromatin becomes highly condensed. The close association of homologous chromosomes (**synapsis**) begins during the zygotene stage. During this stage, a zipperlike protein structure, called the **synaptomenal complex**, forms along the length of the paired chromosomes (Figure 14.34). This complex keeps the homologous chromosomes closely associated and aligned with one another

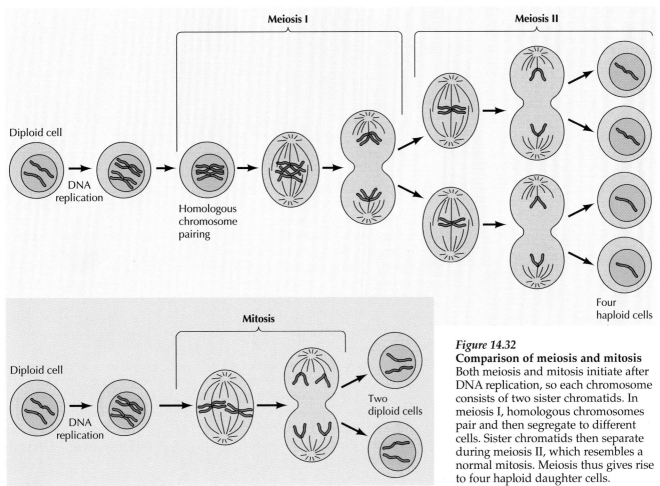

Meiosis I

Meiosis II

Diploid cell

DNA replication

Homologous chromosome pairing

Four haploid cells

Mitosis

Diploid cell

DNA replication

Two diploid cells

Figure 14.32
Comparison of meiosis and mitosis
Both meiosis and mitosis initiate after DNA replication, so each chromosome consists of two sister chromatids. In meiosis I, homologous chromosomes pair and then segregate to different cells. Sister chromatids then separate during meiosis II, which resembles a normal mitosis. Meiosis thus gives rise to four haploid daughter cells.

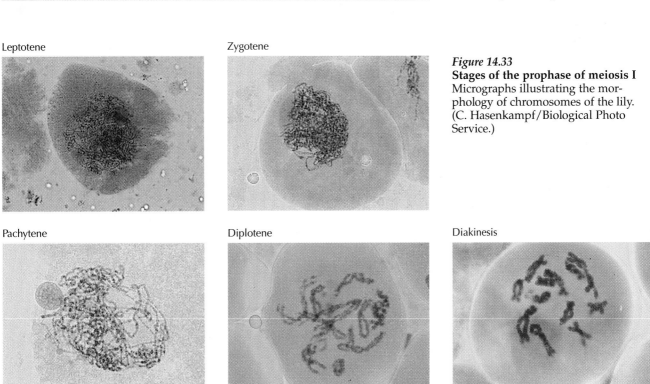

Leptotene

Zygotene

Pachytene

Diplotene

Diakinesis

Figure 14.33
Stages of the prophase of meiosis I
Micrographs illustrating the morphology of chromosomes of the lily. (C. Hasenkampf/Biological Photo Service.)

Figure 14.34
The synaptonemal complex
Chromatin loops are attached to the lateral elements, which are joined to each other by a zipperlike central element.

Chromatin

Central element

Lateral element

through the pachytene stage, which can persist for several days. Recombination between homologous chromosomes is completed during their association at the pachytene stage, leaving the chromosomes linked at the sites of crossing over (**chiasmata**; singular, chiasma). The synaptonemal complex disappears at the diplotene stage and the homologous chromosomes separate along their length. Importantly, however, they remain associated at the chiasmata, which is critical for their correct alignment at metaphase. At this stage, each chromosome pair (called a bivalent) consists of four chromatids with clearly evident chiasmata (Figure 14.35). Diakinesis, the final stage of prophase I, represents the transition to metaphase, during which the chromosomes become fully condensed.

At metaphase I, the bivalent chromosomes align on the spindle. In contrast to mitosis (see Figure 14.27), the kinetochores of sister chromatids are adjacent to each other and oriented in the same direction, while the kinetochores of homologous chromosomes are pointed toward opposite spindle poles (Figure 14.36). Consequently, microtubules from the same pole of the spindle attach to sister chromatids, while microtubules from opposite poles attach to homologous chromosomes. Anaphase I is initiated by disruption of the chiasmata at which homologous chromosomes are joined. The homologous chromosomes then separate, while sister chromatids remain associated at their centromeres. At completion of meiosis I, each daughter cell has therefore acquired one member of each homologous pair, consisting of two sister chromatids.

Meiosis II initiates immediately after cytokinesis, usually before the chromosomes have fully decondensed. In contrast to meiosis I, meiosis II resembles a normal mitosis. At metaphase II, the chromosomes align on the spindle with microtubules from opposite poles of the spindle attached to the kinetochores of sister chromatids. The link between the centromeres of sister chromatids is broken at anaphase II, and sister chromatids segregate to opposite poles. Cytokinesis then follows, giving rise to haploid daughter cells.

Regulation of Oocyte Meiosis

Vertebrate oocytes (developing eggs) have been particularly useful models for research on the cell cycle, in part because of their large size and ease of manipulation in the laboratory. A notable example, discussed earlier in this chapter, is provided by the discovery and subsequent purification of MPF from frog oocytes. Meiosis of these oocytes, like those of other species, is

Figure 14.35
A bivalent chromosome at the diplotene stage   The bivalent chromosome consists of paired homologous chromosomes. Sister chromatids of each chromosome are joined at the centromere. Chromatids of homologous chromosomes are joined at chiasmata, which are the sites at which genetic recombination has occurred. (John Cabisco/Visuals Unlimited.)

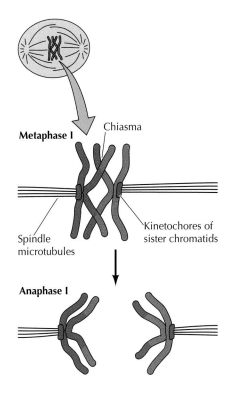

Figure 14.36
Chromosome segregation in meiosis I   At metaphase I, the kinetochores of sister chromatids are either fused or adjacent to one another. Microtubules from the same pole of the spindle therefore attach to the kinetochores of sister chromatids, while microtubules from opposite poles attach to the kinetochores of homologous chromosomes. Chiasmata are disrupted at anaphase I, and homologous chromosomes move to opposite poles of the spindle.

regulated at two unique points in the cell cycle, and studies of oocyte meiosis have illuminated novel mechanisms of cell cycle control.

The first regulatory point in oocyte meiosis is in the diplotene stage of the first meiotic division (Figure 14.37). Oocytes can remain arrested at this stage for long periods of time—up to 40 to 50 years in humans. During this diplotene arrest, the oocyte chromosomes decondense and are actively transcribed. This transcriptional activity is reflected in the tremendous growth of oocytes during this period. Human oocytes, for example, are about 100 μm in diameter (more than a hundred times the volume of a typical somatic cell). Frog oocytes are even larger, with diameters of approximately 1 mm. During this period of cell growth, the oocytes accumulate stockpiles of materials, including RNAs and proteins, that are needed to support early development of the embryo. As noted earlier in this chapter, early embryonic cell cycles then occur in the absence of cell growth, rapidly dividing the fertilized egg into smaller cells (see Figure 14.2).

Oocytes of different species vary as to when meiosis resumes and fertilization takes place. In some animals, oocytes remain arrested at the diplotene stage until they are fertilized, only then proceeding to complete meiosis. However, the oocytes of most vertebrates (including frogs, mice, and humans) resume meiosis in response to hormonal stimulation and proceed through meiosis I prior to fertilization. Cell division following meiosis I is asymmetric, resulting in the production of a small **polar body** and an oocyte that retains its large size. The oocyte then proceeds to enter meiosis II without having re-formed a nucleus or decondensed its chromosomes. Most vertebrate oocytes are then arrested again at metaphase II, where they remain until fertilization.

Like the M phase of somatic cells, the meiosis of oocytes is controlled by MPF. The regulation of MPF during oocyte meiosis, however, displays unique features that are responsible for metaphase II arrest (Figure 14.38). Hormonal stimulation of diplotene-arrested oocytes initially triggers the resumption of meiosis by activating MPF, as at the G_2 to M transition of somatic cells. As in mitosis, MPF then induces chromosome condensation, nuclear envelope breakdown, and formation of the spindle. Activation of the anaphase-promoting complex B then leads to the metaphase to anaphase transition of meiosis I, accompanied by a decrease in the activity of MPF. Following cytokinesis, however, MPF activity again rises and remains high while the egg is arrested at metaphase II. A regulatory mechanism unique to oocytes thus acts to maintain MPF activity during metaphase II arrest, preventing the metaphase to anaphase transition of meiosis II and the inactivation of MPF that would result from cyclin B proteolysis during a normal M phase.

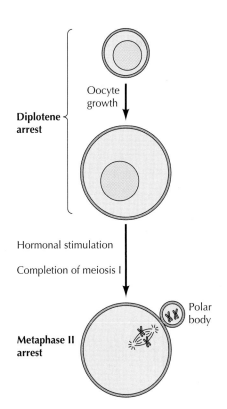

Figure 14.37
Meiosis of vertebrate oocytes   Meiosis is arrested at the diplotene stage, during which oocytes grow to a large size. Oocytes then resume meiosis in response to hormonal stimulation and complete the first meiotic division, with asymmetric cytokinesis giving rise to a small polar body. Most vertebrate oocytes are then arrested again at metaphase II.

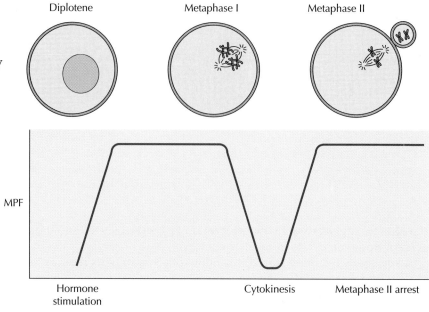

Figure 14.38
Activity of MPF during oocyte meiosis Hormonal stimulation of diplotene oocytes activates MPF, resulting in progression to metaphase I. MPF activity then falls at the transition from metaphase I to anaphase I. Following completion of meiosis I, MPF activity again rises and remains high during metaphase II arrest.

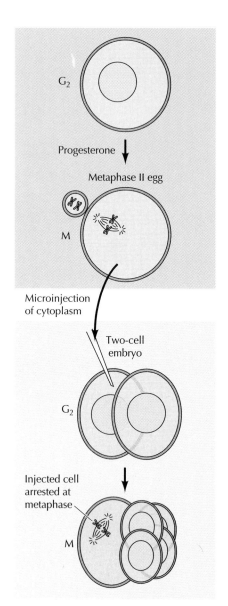

The factor responsible for metaphase II arrest was first identified by Yoshio Masui and Clement Markert in 1971, in the same series of experiments that led to the discovery of MPF. In this case, however, cytoplasm from an egg arrested at metaphase II was injected into an early embryo cell that was undergoing mitotic cell cycles (Figure 14.39). This injection of egg cytoplasm caused the embryonic cell to arrest at metaphase, indicating that metaphase arrest was induced by a cytoplasmic factor present in the egg. Because this factor acted to arrest mitosis, it was called **cytostatic factor (CSF)**.

More recent experiments have identified a protein-serine/threonine kinase known as **Mos** as an essential component of CSF. Mos is specifically synthesized in oocytes around the time of completion of meiosis I and is then required both for the increase in MPF activity during meiosis II and for the maintenance of MPF activity during metaphase II arrest. The action of Mos results from activation of the ERK MAP kinase, which plays a central role in the cell signaling pathways discussed in the previous chapter. In oocytes, however, ERK plays a different role; it activates another protein kinase called Rsk, which inhibits action of the anaphase-promoting complex and arrests meiosis at metaphase II (Figure 14.40). Oocytes can remain arrested at this point in the meiotic cell cycle for several days, awaiting fertilization.

Fertilization

At **fertilization**, the sperm binds to a receptor on the surface of the egg and fuses with the egg plasma membrane, initiating the development of a new diploid organism containing genetic information derived from both parents (Figure 14.41). Not only does fertilization lead to the mixing of paternal and maternal chromosomes, but it also induces a number of changes in the egg cytoplasm that are critical for further development. These alterations

Figure 14.39
Identification of cytostatic factor Cytoplasm from a metaphase II egg is microinjected into one cell of a two-cell embryo. The injected embryo cell arrests at metaphase, while the uninjected cell continues to divide. A factor in metaphase II egg cytoplasm (cytostatic factor) therefore has induced metaphase arrest of the injected embryo cell.

Figure 14.40
Maintenance of metaphase II arrest by the Mos protein kinase
The Mos protein kinase maintains metaphase II arrest by inhibiting the anaphase-promoting complex. The action of Mos is mediated by MEK, ERK, and Rsk protein kinases.

activate the egg, leading to the completion of oocyte meiosis and initiation of the mitotic cell cycles of the early embryo.

A key signal resulting from the binding of a sperm to its receptor on the plasma membrane of the egg is an increase in the level of Ca^{2+} in the egg cytoplasm, probably as a consequence of stimulation of the hydrolysis of phosphatidylinositol 4,5-bisphosphate (PIP_2) (see Figure 13.27). One effect of this elevation in intracellular Ca^{2+} is the induction of surface alterations that prevent additional sperm from entering the egg. Because eggs are usually exposed to large numbers of sperm at one time, this is a critical event in ensuring the formation of a normal diploid embryo. These surface alterations are thought to result, at least in part, from the Ca^{2+}-induced exocytosis of secretory vesicles that are present in large numbers beneath the egg plasma membrane. Release of the contents of these vesicles alters the extracellular coat of the egg so as to block the entry of additional sperm.

The increase in cytosolic Ca^{2+} following fertilization also signals the completion of meiosis (Figure 14.42). In eggs arrested at metaphase II, the metaphase to anaphase transition is triggered by a Ca^{2+}-dependent activation of the anaphase-promoting complex. The resultant inactivation of MPF leads to completion of the second meiotic division, with asymmetric cytokinesis (as in meiosis I) giving rise to a second small polar body.

Following completion of oocyte meiosis, the fertilized egg (now called a **zygote**) contains two haploid nuclei (called **pronuclei**), one derived from each parent. In mammals, the two pronuclei then enter S phase and replicate their DNA as they migrate toward each other. As they meet, the zygote enters M phase of its first mitotic division. The two nuclear envelopes break down, and the condensed chromosomes of both paternal and maternal origin align on a common spindle. Completion of mitosis then gives rise to two embryonic cells, each containing a new diploid genome. These cells then commence the series of embryonic cell divisions that eventually lead to the development of a new organism.

Cell Proliferation in Development and Differentiation

Early development is characterized by the rapid proliferation of embryonic cells, which then differentiate to produce the many specialized types of cells that make up the tissues and organs of multicellular animals. As cells differentiate, their rate of proliferation usually decreases, and most cells in adult animals are arrested in the G_0 stage of the cell cycle. A few types of differentiated cells never divide again, but most cells are able to resume proliferation as required to replace cells that have been lost as a result of injury or cell death. In addition, some cells divide continuously throughout life to replace cells that have a high rate of turnover in adult animals. Cell

Figure 14.41
Fertilization Scanning electron micrograph of a human sperm fertilizing an egg. (David M. Philips/Visuals Unlimited.)

10 μm

(A)

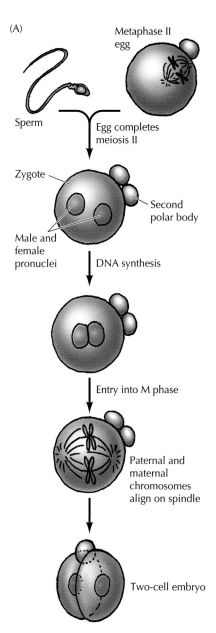

Sperm

Metaphase II
egg

Egg completes
meiosis II

Zygote

Male and
female
pronuclei

Second
polar body

DNA synthesis

Entry into M phase

Paternal and
maternal
chromosomes
align on spindle

Two-cell embryo

(B)

Metaphase II egg   Zygote   Two-cell embryo

20 μm

Figure 14.42
Fertilization and completion of meiosis   (A) Fertilization induces the transition from metaphase II to anaphase II, leading to completion of oocyte meiosis and emission of a second polar body (which usually degenerates). The sperm nucleus decondenses, so the fertilized egg (zygote) contains two haploid nuclei (male and female pronuclei). In mammals, the pronuclei replicate DNA as they migrate toward each other. They then initiate mitosis, with male and female chromosomes aligning on a common spindle. Completion of mitosis and cytokinesis thus gives rise to a two-cell embryo, with each cell containing a diploid genome. (B) Micrographs of a mouse metaphase II egg, zygote, and two-cell embryo. (B, courtesy of Ann A. Kiessling, Harvard Medical School.)

proliferation is thus carefully balanced with cell death to maintain a constant number of cells in adult tissues and organs.

Proliferation of Differentiated Cells

The cells of adult animals can be grouped into three general categories with respect to cell proliferation. A few types of differentiated cells, such as cardiac muscle cells in humans, are no longer capable of cell division. These cells are produced during embryonic development, differentiate, and are then retained throughout the life of the organism. If they are lost because of injury (e.g., the death of cardiac muscle cells during a heart attack), they can never be replaced.

In contrast, most cells in adult animals enter the G_0 stage of the cell cycle but resume proliferation as needed to replace cells that have been injured or have died. Cells of this type include skin fibroblasts, smooth muscle cells, the endothelial cells that line blood vessels, and the epithelial cells of most internal organs, such as the liver, pancreas, kidney, lung, prostate, and breast. One example of the controlled proliferation of these cells, discussed earlier in this chapter, is the rapid proliferation of skin fibroblasts to repair damage resulting from a cut or wound. Another striking example is provided by liver cells, which normally divide only rarely. However, if large numbers of liver cells are lost (e.g., by surgical removal of part of the liver), the remaining cells are stimulated to proliferate to replace the missing tissue. For example, surgical removal of two-thirds of the liver of a rat is followed by rapid cell proliferation, leading to regeneration of the entire liver within a few days.

Stem Cells

Other types of differentiated cells, including blood cells, epithelial cells of the skin, and the epithelial cells lining the digestive tract, have short life spans and must be replaced by continual cell proliferation in adult animals. In these cases, the fully differentiated cells do not themselves proliferate. Instead, they are replaced via the proliferation of cells that are less differen-

Figure 14.43
Stem cell proliferation Stem cells divide to form one daughter cell that remains a stem cell and a second that differentiates (e.g., to an intestinal epithelial cell).

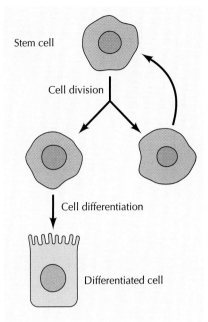

tiated, called **stem cells** (Figure 14.43). Stem cells divide to produce daughter cells that can either differentiate or remain as stem cells, thereby serving as a source for the production of differentiated cells throughout life.

A good example of the continual proliferation of stem cells is provided by blood cell differentiation. There are several distinct types of blood cells with specialized functions: Erythrocytes (red blood cells) transport O_2 and CO_2; granulocytes and macrophages are phagocytic cells; platelets (which are fragments of megakaryocytes) function in blood coagulation; and lymphocytes are responsible for the immune response. All these cells have limited life spans, ranging from less than a day to a few months, and are continually produced by the division of a common stem cell (the pluripotent stem cell) in the bone marrow (Figure 14.44). Descendants of the pluripotent stem cell then become committed to specific differentiation pathways.

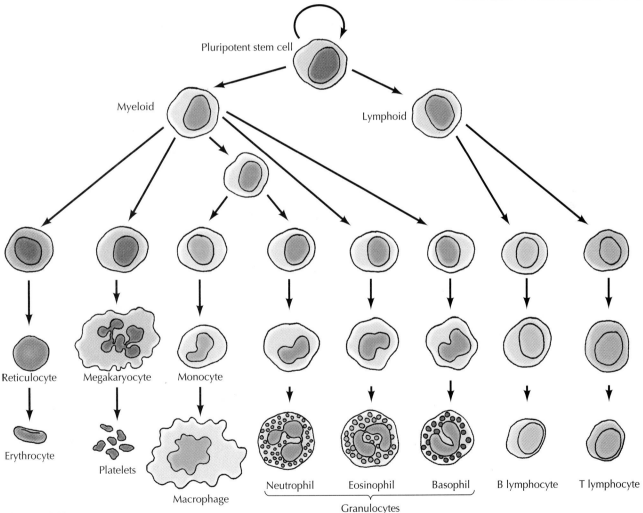

Figure 14.44
Formation of blood cells All of the different types of blood cells develop from a pluripotent stem cell in the bone marrow. The precursors of differentiated cells undergo several rounds of cell division as they mature, but cell proliferation ceases at the terminal stages of differentiation.

Molecular Medicine

In Vitro *Fertilization*

The Disease

Reproduction is a basic drive of all species of animals, and humans are no exception. Indeed, producing children and raising a family is a fundamental aspect of the lives of most people. The inability to reproduce (infertility) is not only a devastating problem for many affected couples, but it is also common. For example, one frequently used clinical definition of subnormal fertility is the failure to achieve a pregnancy after a two-year period of trying. According to this criterion, it is estimated that approximately 20% of couples in the United States are subfertile.

Infertility can result from a variety of causes, which can affect either the female or the male partner. Many cases are due to anatomical defects in the Fallopian tubes, through which oocytes must travel from the ovary to the uterus. Some women are infertile (or subfertile) because of hormonal disorders that result in the failure to produce normal metaphase II eggs. Other cases of infertility or subfertility result from a failure of the male partner to produce or ejaculate adequate numbers of normal sperm. *In vitro* fertilization (IVF), which was originally developed as a treatment for tubal defects, is now widely used in the treatment of both male and female reproductive disorders.

Molecular and Cellular Bases

The basic procedure of *in vitro* fertilization is to recover metaphase II eggs from the ovary and culture them together with sperm. Fertilization is assessed 12 to 18 hours later by microscopic observation to detect the formation of two pronuclei. Fertilized eggs can then be returned to the Fallopian tube or uterus. Alternatively, they can be maintained in culture for a few additional days, during which the early embryonic cell divisions take place.

In vitro fertilization was initially developed to bypass defects in the Fallopian tube, which present an obvious anatomical block to a natural pregnancy. As already noted, however, IVF is also effective for other reproductive disorders. For example, hormonal defects resulting in the failure to produce metaphase II eggs can be treated by administration of appropriate hormones to the female partner. This treatment results in the production of multiple eggs, which can then be fertilized *in vitro*, followed by the return of successfully fertilized eggs to the uterus.

IVF is also effective in treating cases of infertility resulting from failure of the male partner to ejaculate sufficient functional sperm to achieve fertilization *in vivo*. Moreover, a new procedure in which a single sperm is directly injected into an egg (intracytoplasmic sperm injection, or ICSI) can be applied to cases in which fertilization does not result from the standard *in vitro* coculture of sperm and eggs. Interestingly, successful fertilization by the ICSI procedure

also requires stimulation of the surface of the egg in order to achieve egg activation.

The rate of pregnancy resulting from fertilization *in vitro* is similar to that observed *in vivo*. In particular, it is estimated that about 25% of ovulated human eggs naturally give rise to offspring in a normal fertile couple. Similar pregnancy rates are obtained following *in vitro* fertilization of eggs of young women with an anatomical block to pregnancy. Although lower pregnancy rates are obtained with eggs of older women or of women with hormonal defects, these reduced rates can be compensated for by the return of multiple fertilized eggs to the uterus. ICSI also yields a pregnancy rate of about 25%, suggesting that there are intrinsic limitations on the developmental capacity of human eggs, both *in vitro* and *in vivo*.

A human egg containing two pronuclei after fertilization *in vitro*. (Courtesy of Ann A. Kiessling, Harvard Medical School.) 50 μm

These cells continue to proliferate and undergo several rounds of division as they differentiate. Once they become fully differentiated, however, they cease proliferation, so the maintenance of differentiated blood cell populations is dependent on continual proliferation of the pluripotent stem cell.

Because stem cells can replicate as well as differentiating to give rise to a variety of cell types, they are of considerable interest with respect to potential medical applications. For example, it may be possible to use stem cells

In Vitro *Fertilization* (continued)

Prevention and Treatment

The first attempt at *in vitro* fertilization of human eggs was reported in 1942, but not until 1978 was the first IVF baby born. Since then, a variety of technical improvements have made IVF a standard treatment for infertility, and IVF has allowed tens of thousands of otherwise infertile couples to bear children. Although not successful for many couples, IVF has clearly had a major impact on a widespread medical problem.

In addition to its direct impact on the treatment of infertility, IVF offers new opportunities for the diagnosis and prevention of genetic disease. It is possible to remove one or two cells of an early embryo developing *in vitro* (usually at the eight-cell stage) without harming further embryonic development. Consequently, single embryo cells can be isolated and tested either for chromosomal abnormalities or for mutant alleles of any inherited disease gene that can be identified by PCR. It is then possible to select only the embryos carrying normal gene copies for transfer to the uterus. In combination with these methods of preimplantation embryo diagnosis, IVF thus allows couples carrying genes for inherited diseases to prevent the transmission of these diseases to their offspring.

Reference

Handyside, A. H. and J. D. A. Delhanty. 1997. Preimplantation genetic diagnosis: Strategies and surprises. *Trends Genet.* 13: 270–275.

to treat human diseases and repair damaged tissues. The stem cells with the broadest differentiative capacity are the embryonal stem cells (ES cells) that are present in early embryos and can give rise to all of the differentiated cell types of adult organisms. As discussed in Chapter 3, these cells can be cultured from mouse embryos and used to introduce altered genes into mice (see Figure 3.38). In 1998, two groups of researchers reported the isolation of ES cells from human embryos, raising the possibility that these human stem cells could be used for medical applications. Notably, this advance followed the first demonstration, in 1997, that the nucleus of an adult cell could give rise to a viable animal (in this case, a lamb) following transplantation into an oocyte. It is thus theoretically possible that the nucleus of an adult human cell could be used to give rise to ES cells, which could then provide a source of tissue for treatment of that individual. In addition, stem cells have been isolated from adult tissues that give rise not only to blood cells but also to many other cell types, including neurons and cells of connective tissues such as bone, cartilage, fat, and muscle. Therapeutic applications of these adult stem cells may avoid the ethical issues associated with the use of embryos as a source of ES cells. Continuing research on stem cells may thus open new approaches to the treatment of a broad array of human diseases.

SUMMARY

KEY TERMS

THE EUKARYOTIC CELL CYCLE

Phases of the Cell Cycle: Eukaryotic cell cycles are divided into four discrete phases: M, G_1, S, and G_2. M phase consists of mitosis, which is usually followed by cytokinesis. S phase is the period of DNA replication.

Regulation of the Cell Cycle by Cell Growth and Extracellular Signals: Extracellular signals and cell size regulate progression through specific control points in the cell cycle.

Cell Cycle Checkpoints: Checkpoints and feedback controls coordinate the events that take place during different phases of the cell cycle and arrest cell cycle progression if DNA is damaged.

mitosis, interphase, cytokinesis, M phase, G_1 phase, S phase, G_2 phase, G_0, flow cytometer, fluorescence-activated cell sorter

START, restriction point

p53

maturation promoting factor (MPF), Cdc2, cyclin

Coupling of S Phase to M Phase: Once DNA replication has taken place, initiation of a new S phase is prevented until the cell has passed through mitosis.

REGULATORS OF CELL CYCLE PROGRESSION

MPF: A Dimer of Cdc2 and Cyclin: MPF is the key molecule responsible for regulating the G_2 to M transition in all eukaryotes. MPF is a dimer of cyclin B and the Cdc2 protein kinase.

G_1 cyclin, Cln, Cdk, Cdk inhibitor (CKI)

Families of Cyclins and Cyclin-Dependent Kinases: Distinct pairs of cyclins and Cdc2-related protein kinases regulate progression through different stages of the cell cycle. The activity of Cdk's is regulated by association with cyclins, activating and inhibitory phosphorylations, and the binding of Cdk inhibitors.

Rb, tumor suppressor gene, E2F

Growth Factors and the D-Type Cyclins: Growth factors stimulate animal cell proliferation by inducing synthesis of the D-type cyclins. Cdk4, 6/cyclin D complexes then act to drive cells through the restriction point in G_1. A key substrate of Cdk4, 6/cyclin D complexes is the tumor suppressor protein Rb, which regulates transcription of genes required for cell cycle progression.

Inhibitors of Cell Cycle Progression: DNA damage and a variety of extracellular signals inhibit cell cycle progression. These inhibitory factors frequently act by inducing synthesis of Cdk inhibitors.

THE EVENTS OF M PHASE

prophase, metaphase, anaphase, telophase, centromere, kinetochore, centrosome, mitotic spindle, prometaphase

Stages of Mitosis: Mitosis is conventionally divided into four stages: prophase, metaphase, anaphase, and telophase. The basic events of mitosis include chromosome condensation, formation of the mitotic spindle, nuclear envelope breakdown, and attachment of spindle microtubules to chromosomes at the kinetochore. Sister chromatids then separate and move to opposite poles of the spindle. Finally, nuclei re-form, the chromosomes decondense, and cytokinesis divides the cell in half.

kinetochore microtubule, polar microtubule, astral microtubule

MPF and Progression to Metaphase: M phase is initiated by activation of MPF, which phosphorylates other protein kinases, as well as the nuclear lamins and condensins. Activation of MPF is responsible for chromatin condensation, nuclear envelope breakdown, fragmentation of the endoplasmic reticulum and Golgi apparatus, and reorganization of microtubules to form the mitotic spindle. The attachment of spindle microtubules to the kinetochores of sister chromatids then leads to their alignment on the metaphase plate.

anaphase-promoting complex, cohesin

Proteolysis and the Inactivation of MPF: Anaphase and Telophase: Activation of a ubiquitin ligase called the anaphase-promoting complex leads to degradation of key regulatory proteins at the metaphase to anaphase transition. Ubiquitin-mediated proteolysis initiated by the anaphase-promoting complex leads to the degradation of a cohesin, breaking the link between sister chromatids at the onset of anaphase. Activation of the anaphase-promoting complex also leads to the degradation of cyclin B and inactivation of MPF, resulting in re-formation of nuclear envelopes, chromatin decondensation, and the return of microtubules to the interphase state.

Cytokinesis: Inactivation of MPF also triggers cytokinesis. In animal cells, cytokinesis results from contraction of a ring of actin and myosin filaments. In higher plant cells, cytokinesis results from the formation of a new cell wall and plasma membrane inside the cell.

contractile ring, cell plate

MEIOSIS AND FERTILIZATION

The Process of Meiosis: Meiosis is a specialized cell cycle that gives rise to haploid daughter cells. A single round of DNA synthesis is followed by two sequential cell divisions. During meiosis I, homologous chromosomes first pair and then segregate to different daughter cells. Meiosis II then resembles a normal mitosis, in which sister chromatids separate.

meiosis, leptotene, zygotene, pachytene, diplotene, diakinesis, synapsis, synaptonemal complex, chiasma

Regulation of Oocyte Meiosis: Meiosis of vertebrate oocytes is regulated at two unique points in the cell cycle: the diplotene stage of meiosis I and metaphase of meiosis II. Metaphase II arrest results from inhibition of the anaphase-promoting complex by a protein kinase expressed in oocytes.

polar body, cytostatic factor (CSF), Mos

Fertilization: Fertilization triggers the resumption of oocyte meiosis by Ca^{2+}-dependent activation of the anaphase-promoting complex. The fertilized egg then contains two haploid nuclei, which form a new diploid genome and initiate embryonic cell divisions.

fertilization, zygote, pronucleus

CELL PROLIFERATION IN DEVELOPMENT AND DIFFERENTIATION

Proliferation of Differentiated Cells: Most cells in adult animals are arrested in the G_0 stage of the cell cycle. A few types of differentiated cells are no longer capable of proliferation, but most are able to proliferate as required to replace cells that have been lost because of injury or cell death.

Stem Cells: Some differentiated cells do not themselves proliferate but can be replaced via the proliferation of stem cells. Stem cells divide to produce one daughter cell that differentiates and another that remains a stem cell.

stem cell

Questions

1. Consider a mammalian cell that divides every 30 hours. Microscopic observation indicates that 3.3% of the cells are in mitosis at any given time. Analysis in the flow cytometer establishes that 53.3% of the cells have DNA contents of $2n$, 16.7% have DNA contents of $4n$, and 30% have DNA contents ranging between $2n$ and $4n$. What are the lengths of the G_1, S, G_2, and M phases of the cycle of these cells?

2. The metaphase checkpoint delays the onset of anaphase until all chromosomes are properly aligned on the spindle. What would be the result if a failure of this checkpoint permitted anaphase to initiate while one chromosome was still attached to microtubules from only a single centrosome?

3. The Cdk inhibitor p16 binds specifically to Cdk4, 6/cyclin D complexes. What would be the predicted effect of overexpression of p16 on cell cycle progression? Would overexpression of p16 affect a tumor cell lacking functional Rb protein?

4. *In vitro* mutagenesis of cloned lamin cDNAs has been used to generate mutants that cannot be phosphorylated by MPF. How would expression of these mutant lamins affect nuclear envelope breakdown at the end of prophase?

5. Mutants of cyclin B that are resistant to degradation by the cyclin B protease have been generated. How would expression of these mutant cyclin B's affect the events at the metaphase to anaphase transition?

6. Homologous recombination has been used to inactivate the *mos* gene in mice. What effect would you expect this to have on oocyte meiosis?

References and Further Reading

General Reference

Murray, A. and T. Hunt. 1993. *The Cell Cycle: An Introduction.* New York: W. H. Freeman.

The Eukaryotic Cell Cycle

Chong, J. P. J., P. Thommes and J. J. Blow. 1996. The role of MCM/P1 proteins in the licensing of DNA replication. *Trends Biochem. Sci.* 21: 102–107. [R]

Elledge, S. J. 1996. Cell cycle checkpoints: Preventing an identity crisis. *Science* 274: 1664–1672. [R]

Forsburg, S. L. and P. Nurse. 1991. Cell cycle regulation in the yeasts *Saccharomyces cerevisiae* and *Schizosaccharomyces pombe*. *Ann. Rev. Cell Biol.* 7: 227–256. [R]

Hartwell, L. H. and T. A. Weinert. 1989. Checkpoints: Controls that ensure the order of cell cycle events. *Science* 246: 629–634. [R]

Levine, A. J. 1997. p53, the cellular gatekeeper for growth and division. *Cell* 88: 323–331. [R]

Norbury, C. and P. Nurse. 1992. Animal cell cycles and their control. *Ann. Rev. Biochem.* 61: 441–470. [R]

Pardee, A. B. 1989. G_1 events and the regulation of cell proliferation. *Science* 246: 603–608. [R]

Paulovich, A. G., D. P. Tocyski and L. H. Hartwell. 1997. When checkpoints fail. *Cell* 88: 315–321. [R]

Rao, P. N. and R. T. Johnson. 1970. Mammalian cell fusion studies on the regulation of DNA synthesis and mitosis. *Nature* 225: 159–164. [P]

Russell, P. 1998. Checkpoints on the road to mitosis. *Trends Biochem. Sci.* 23: 399–402. [R]

Stillman, B. 1996. Cell cycle control of DNA replication. *Science* 274: 1659–1664. [R]

Tye, B. K. 1999. MCM proteins in DNA replication. *Ann. Rev. Biochem.* 68: 649–686. [R]

Regulators of Cell Cycle Progression

Beach, D., B. Durkacz and P. Nurse. 1982. Functionally homologous cell cycle control genes in budding and fission yeast. *Nature* 300: 706–709. [P]

Dynlacht, B. D. 1997. Regulation of transcription by proteins that control the cell cycle. *Nature* 389: 149–152. [R]

Elledge, S. J., J. Winston and J. W. Harper. 1996. A question of balance: The role of cyclin-kinase inhibitors in development and tumorigenesis. *Trends Cell Biol.* 6: 388–392. [R]

Evans, T., E. T. Rosenthal, J. Youngbloom, D. Distel and T. Hunt. 1983. Cyclin: A protein specified by maternal mRNA in sea urchin eggs that is destroyed at each cleavage division. *Cell* 33: 389–396. [P]

Hartwell, L. H., R. K. Mortimer, J. Culotti and M. Culotti. 1973. Genetic control of the cell division cycle in yeast: V. Genetic analysis of *cdc* mutants. *Genetics* 74: 267–287. [P]

King, R. W., R. J. Deshaies, J.-M. Peters and M. W. Kirschner. 1996. How proteolysis drives the cell cycle. *Science* 274: 1652–1659. [R]

Koepp, D. M., J. W. Harper and S. J. Elledge. 1999. How the cyclin became a cyclin: Regulated proteolysis in the cell cycle. *Cell* 97: 431–434. [R]

Levine, A. J. 1997. p53, the cellular gatekeeper for growth and division. *Cell* 88: 323–331. [R]

Lohka, M. J., M. K. Hayes and J. L. Maller. 1988. Purification of maturation-promoting factor, an intracellular regulator of early mitotic events. *Proc. Natl. Acad. Sci. USA* 85: 3009–3013. [P]

Masui, Y. and C. L. Markert. 1971. Cytoplasmic control of nuclear behavior during meiotic maturation of frog oocytes. *J. Exp. Zool.* 177: 129–146. [P]

Morgan, D. O. 1997. Cyclin-dependent kinases: Engines, clocks, and microprocessors. *Ann. Rev. Cell Dev. Biol.* 13: 261–291. [R]

Murray, A. 1995. Cyclin ubiquitination: The destructive end of mitosis. *Cell* 81: 149–152. [R]

Roberts, J. M. 1999. Evolving ideas about cyclins. *Cell* 98: 129–132. [R]

Russell, P. 1998. Checkpoints on the road to mitosis. *Trends Biochem. Sci.* 23: 399–402. [R]

Sherr, C. J. 1996. Cancer cell cycles. *Science* 274: 1672–1677. [R]

Smith, L. D. and R. E. Ecker. 1971. The interaction of steroids with *Rana pipiens* oocytes in the induction of maturation. *Dev. Biol.* 25: 232–247. [P]

Swenson, K. I., K. M. Farrell and J. V. Ruderman. 1986. The clam embryo protein cyclin A induces entry into M phase and the resumption of meiosis in *Xenopus* oocytes. *Cell* 47: 861–870. [P]

Weinberg, R. A. 1995. The retinoblastoma protein and cell cycle control. *Cell* 81: 323–330. [R]

The Events of M Phase

Earnshaw, W. C. and A. F. Pluta. 1994. Mitosis. *BioEssays* 16: 639–643. [R]

King, R. W., P. K. Jackson and M. W. Kirschner. 1994. Mitosis in transition. *Cell* 79: 563–571. [R]

Kimura, K., M. Hirano, R. Kobayashi and T. Hirano. 1998. Phosphorylation and activation of 13S condensin by Cdc2 *in vitro*. *Science* 282: 487–490. [P]

Lowe, M., C. Rabouille, N. Nakamura, R. Watson, M. Jackman, E. Jamsa, D. Rahman, D. J. C. Pappin and G. Warren. 1998. Cdc2 kinase directly phosphorylates the *cis*-Golgi matrix protein GM130 and is required for Golgi fragmentation in mitosis. *Cell* 94: 783–793. [P]

McIntosh, J. R. and G. E. Hering. 1991. Spindle fiber action and chromosome movement. *Ann. Rev. Cell Biol.* 7: 403–426. [R]

McIntosh, J. R. and M. P. Koonce. 1989. Mitosis. *Science* 246: 622–628. [R]

Murray, A. 1995. Cyclin ubiquitination: The destructive end of mitosis. *Cell* 81: 149–152. [R]

Murray, A. W. 1998. How to compact DNA. *Science* 282: 425–427. [R]

Nasmyth, K. 1999. Separating sister chromatids. *Trends Biochem. Sci.* 24: 98–104. [R]

Page, A. M. and P. Hieter. 1999. The anaphase-promoting complex: New subunits and regulators. *Ann. Rev. Biochem.* 68: 583–609. [R]

Rieder, C. L. and E. D. Salmon. 1998. The vertebrate cell kinetochore and its roles during mitosis. *Trends Cell Biol.* 8: 310–318. [R]

Roth, M. G. 1999. Inheriting the Golgi. *Cell* 99:559–562. [R]

Staehelin, L. A. and P. K. Hepler. 1996. Cytokinesis in higher plants. *Cell* 84: 821–824. [R]

Warren, G. 1993. Membrane partitioning during cell division. *Ann. Rev. Biochem.* 323–348. [R]

Wei, Y., L. Yu, J. Bowen, M. A. Gorovsky and C. D. Allis. 1999. Phosphorylation of histone H3 is required for proper chromosome condensation and segregation. *Cell* 97: 99–109. [P]

Meiosis and Fertilization

Bhatt, R. R. and J. E. Ferrell Jr. 1999. The protein kinase p90 Rsk as an essential mediator of cytostatic factor activity. *Science* 286: 1362–1365. [P]

Gilbert, S. F. 2000. *Developmental Biology*, 6th ed. Sunderland, MA: Sinauer Associates.

Gross, S. D., M. S. Schwab, A. L. Lewellyn and J. L. Maller. 1999. Induction of metaphase arrest in cleaving *Xenopus* embryos by the protein kinase p90^{Rsk}. *Science* 286: 1365–1367. [P]

Kleckner, N. 1996. Meiosis: How could it work? *Proc. Natl. Acad. Sci. USA* 93: 8167–8174. [R]

McKim, K. S. and R. S. Hawley. 1995. Chromosomal control of meiotic cell division. *Science* 270: 1595–1601. [R]

Roeder, G. S. 1997. Meiotic chromosomes: It takes two to tango. *Genes Devel.* 11: 2600–2621. [R]

Sagata, N. 1996. Meiotic metaphase arrest in animal oocytes: Its mechanisms and biological significance. *Trends Cell Biol.* 6: 22–28. [R]

Wassarman, P. M. 1999. Mammalian fertilization: Molecular aspects of gamete adhesion, exocytosis, and fusion. *Cell* 96: 175–183. [R]

Cell Proliferation in Development and Differentiation

Metcalf, D. 1992. Hemopoietic regulators. *Trends Biochem. Sci.* 17: 286–289. [R]

Michalopoulos, G. K. and M. C. DeFrances. 1997. Liver regeneration. *Science* 276: 60–66. [R]

Morrison, S. J., N. M. Shah and D. J. Anderson. 1997. Regulatory mechanisms in stem cell biology. *Cell* 88: 287–298. [R]

Morrison, S. J., P. M. White, C. Zock and D. J. Anderson. 1999. Prospective identification, isolation by flow cytometry, and *in vivo* self-renewal of multipotent mammalian neural crest stem cells. *Cell* 96: 737–749. [P]

Pittenger, M. F., A. M. Mackay, S. C. Beck, R. K. Jaiswal, R. Douglas, J. D. Mosca, M. A. Moorman, D. W. Simonetti, S. Craig and D. R. Marshak. 1999. Multilineage potential of adult human mesenchymal stem cells. *Science* 284: 143–147. [P]

Shamblott, M. J., J. Axelman, S. Wang, E. M. Bugg, J. W. Littlefield, P. J. Donovan, P. D. Blumenthal, G. R. Huggins and J. D. Gearhart. 1998. Derivation of pluripotent stem cells from cultured human primordial germ cells. *Proc. Natl. Acad. Sci. USA* 95: 13726–13731. [P]

Thomson, J. A. J. Itskovitz-Eldor, S. S. Shapiro, M. A. Waknotz, J. J. Swiergiel, V. S. Marshall and J. M. Jones. 1998. Embryonic stem cell lines derived from human blastocysts. *Science* 282: 1145–1147. [P]

Wilmut, I., A. E. Schnieke, J. McWhir, A. J. Kind and K. H. S. Campbell. 1997. Viable offspring derived from fetal and adult mammalian cells. *Nature* 385: 810–813. [P]

15 Cancer

CANCER IS A PARTICULARLY APPROPRIATE TOPIC for the concluding chapter of this book because it results from a breakdown of the regulatory mechanisms that govern normal cell behavior. As discussed in preceding chapters, the proliferation, differentiation, and survival of individual cells in multicellular organisms are carefully regulated to meet the needs of the organism as a whole. This regulation is lost in cancer cells, which grow and divide in an uncontrolled manner, ultimately spreading throughout the body and interfering with the function of normal tissues and organs.

Because cancer results from defects in fundamental cell regulatory mechanisms, it is a disease that ultimately has to be understood at the molecular and cellular levels. Indeed, understanding cancer has been an objective of molecular and cellular biologists for many years. Importantly, studies of cancer cells have also illuminated the mechanisms that regulate normal cell behavior. In fact, many of the proteins that play key roles in cell signaling, regulation of the cell cycle, and control of programmed cell death were first identified because abnormalities in their activities led to the uncontrolled proliferation of cancer cells. The study of cancer has thus contributed significantly to our understanding of normal cell regulation, as well as vice versa.

The Development and Causes of Cancer

The fundamental abnormality resulting in the development of cancer is the continual unregulated proliferation of cancer cells. Rather than responding appropriately to the signals that control normal cell behavior, cancer cells grow and divide in an uncontrolled manner, invading normal tissues and organs and eventually spreading throughout the body. The generalized loss of growth control exhibited by cancer cells is the net result of accumulated abnormalities in multiple cell regulatory systems and is reflected in several aspects of cell behavior that distinguish cancer cells from their normal counterparts.

609

Figure 15.1
A malignant tumor of the uterus
Micrographs of normal uterus (A) and a section of a uterine sarcoma (B). Note that the cancer cells (darkly stained) have invaded the surrounding normal tissue. (Cecil Fox/Molecular Histology, Inc.)

(A)

(B)

Types of Cancer

Cancer can result from abnormal proliferation of any of the different kinds of cells in the body, so there are more than a hundred distinct types of cancer, which can vary substantially in their behavior and response to treatment. The most important issue in cancer pathology is the distinction between benign and malignant tumors (Figure 15.1). A **tumor** is any abnormal proliferation of cells, which may be either benign or malignant. A **benign tumor**, such as a common skin wart, remains confined to its original location, neither invading surrounding normal tissue nor spreading to distant body sites. A **malignant tumor**, however, is capable of both invading surrounding normal tissue and spreading throughout the body via the circulatory or lymphatic systems (**metastasis**). Only malignant tumors are properly referred to as cancers, and it is their ability to invade and metastasize that makes cancer so dangerous. Whereas benign tumors can usually be removed surgically, the spread of malignant tumors to distant body sites frequently makes them resistant to such localized treatment.

Both benign and malignant tumors are classified according to the type of cell from which they arise. Most cancers fall into one of three main groups: carcinomas, sarcomas, and leukemias or lymphomas. **Carcinomas,**

TABLE 15.1 Ten Most Frequent Cancers in the United States

Cancer site	Cases per year	Deaths per year
Breast	184,200 (15.1%)	41,200 (7.5%)
Prostate	180,400 (14.8%)	31,900 (5.8%)
Lung	164,100 (13.4%)	156,900 (28.4%)
Colon/rectum	130,200 (10.7%)	56,300 (10.2%)
Lymphomas	62,300 (5.1%)	27,500 (5.0%)
Bladder	53,200 (4.4%)	12,200 (2.2%)
Uterus	48,900 (4.0%)	11,100 (2.0%)
Skin (melanoma)	47,700 (3.9%)	7,700 (1.4%)
Kidney	31,200 (2.6%)	11,900 (2.2%)
Leukemias	30,800 (2.5%)	12,100 (2.2%)
Subtotal	933,000 (76.5%)	368,800 (66.8%)
All sites	1,220,100 (100%)	552,200 (100%)

Source: American Cancer Society, *Cancer Facts and Figures—2000.*

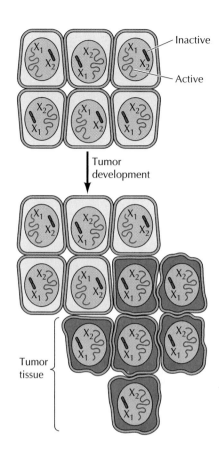

Figure 15.2
Tumor clonality   Normal tissue is a mosaic of cells in which different X chromosomes (X_1 and X_2) have been inactivated. Tumors develop from a single initially altered cell, so each tumor cell displays the same pattern of X inactivation (X_1 inactive, X_2 active).

which include approximately 90% of human cancers, are malignancies of epithelial cells. **Sarcomas**, which are rare in humans, are solid tumors of connective tissues, such as muscle, bone, cartilage, and fibrous tissue. **Leukemias** and **lymphomas**, which account for approximately 8% of human malignancies, arise from the blood-forming cells and from cells of the immune system, respectively. Tumors are further classified according to tissue of origin (e.g., lung or breast carcinomas) and the type of cell involved. For example, fibrosarcomas arise from fibroblasts, and erythroid leukemias from precursors of erythrocytes (red blood cells).

Although there are many kinds of cancer, only a few occur frequently (Table 15.1). More than a million cases of cancer are diagnosed annually in the United States, and more than 500,000 Americans die of cancer each year. Cancers of 10 different body sites account for more than 75% of this total cancer incidence. The four most common cancers, accounting for more than half of all cancer cases, are those of the breast, prostate, lung, and colon/rectum. Lung cancer, by far the most lethal, is responsible for nearly 30% of all cancer deaths.

The Development of Cancer

One of the fundamental features of cancer is tumor clonality, the development of tumors from single cells that begin to proliferate abnormally. The single-cell origin of many tumors has been demonstrated by analysis of X chromosome inactivation (Figure 15.2). As discussed in Chapter 8, one member of the X chromosome pair is inactivated by being converted to heterochromatin in female cells. X inactivation occurs randomly during embryonic development, so one X chromosome is inactivated in some cells, while the other X chromosome is inactivated in other cells. Thus, if a female is heterozygous for an X chromosome gene, different alleles will be expressed in different cells. Normal tissues are composed of mixtures of cells with different inactive X chromosomes, so expression of both alleles is detected in normal tissues of heterozygous females. In contrast, tumor tissues generally express only one allele of a heterozygous X chromosome gene. The implication is that all of the cells constituting such a tumor were derived from a single cell of origin, in which the pattern of X inactivation was fixed before the tumor began to develop.

The clonal origin of tumors does not, however, imply that the original progenitor cell that gives rise to a tumor has initially acquired all of the characteristics of a cancer cell. On the contrary, the development of cancer is a multistep process in which cells gradually become malignant through a progressive series of alterations. One indication of the multistep development of cancer is that most cancers develop late in life. The incidence of colon cancer, for example, increases more than tenfold between the ages of 30 and 50, and another tenfold between 50 and 70 (Figure 15.3). Such a dramatic increase of cancer incidence with age suggests that most cancers develop as a consequence of multiple abnormalities, which accumulate over periods of many years.

At the cellular level, the development of cancer is viewed as a multistep process involving mutation and selection for cells with progressively

Figure 15.3
Increased rate of colon cancer with age   Annual death rates from colon cancer in the United States. (Data from J. Cairns, 1978. *Cancer: Science and Society*, New York: W. H. Freeman.)

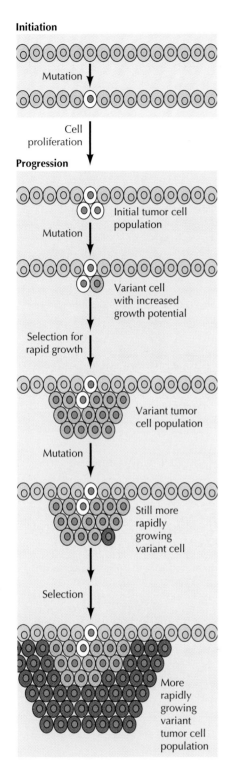

Initiation

Mutation

Cell proliferation

Progression

Initial tumor cell population

Mutation

Variant cell with increased growth potential

Selection for rapid growth

Variant tumor cell population

Mutation

Still more rapidly growing variant cell

Selection

More rapidly growing variant tumor cell population

Figure 15.4
Stages of tumor development The development of cancer initiates when a single mutated cell begins to proliferate abnormally. Additional mutations followed by selection for more rapidly growing cells within the population then result in progression of the tumor to increasingly rapid growth and malignancy.

increasing capacity for proliferation, survival, invasion, and metastasis (Figure 15.4). The first step in the process, **tumor initiation**, is thought to be the result of a genetic alteration leading to abnormal proliferation of a single cell. Cell proliferation then leads to the outgrowth of a population of clonally derived tumor cells. **Tumor progression** continues as additional mutations occur within cells of the tumor population. Some of these mutations confer a selective advantage to the cell, such as more rapid growth, and the descendants of a cell bearing such a mutation will consequently become dominant within the tumor population. The process is called clonal selection, since a new clone of tumor cells has evolved on the basis of its increased growth rate or other properties (such as survival, invasion, or metastasis) that confer a selective advantage. Clonal selection continues throughout tumor development, so tumors continuously become more rapid-growing and increasingly malignant.

Studies of colon carcinomas have provided a clear example of tumor progression during the development of a common human malignancy (Figure 15.5). The earliest stage in tumor development is increased proliferation of colon epithelial cells. One of the cells within this proliferative cell population is then thought to give rise to a small benign neoplasm (an **adenoma** or **polyp**). Further rounds of clonal selection lead to the growth of adenomas of increasing size and proliferative potential. Malignant carcinomas then arise from the benign adenomas, indicated by invasion of the tumor cells through the basal lamina into underlying connective tissue. The cancer cells then continue to proliferate and spread through the connective tissues of the colon wall. Eventually the cancer cells penetrate the wall of the colon and invade other abdominal organs, such as the bladder or small intestine. In addition, the cancer cells invade blood and lymphatic vessels, allowing them to metastasize throughout the body.

Causes of Cancer

Substances that cause cancer, called **carcinogens**, have been identified both by studies in experimental animals and by epidemiological analysis of cancer frequencies in human populations (e.g., the high incidence of lung cancer among cigarette smokers). Since the development of malignancy is a complex multistep process, many factors may affect the likelihood that cancer will develop, and it is overly simplistic to speak of single causes of most cancers. Nonetheless, many agents, including radiation, chemicals, and viruses, have been found to induce cancer in both experimental animals and humans.

Radiation and many chemical carcinogens (Figure 15.6) act by damaging DNA and inducing mutations. These carcinogens are generally referred to as initiating agents, since the induction of mutations in key target genes is thought to be the initial event leading to cancer development. Some of the initiating agents that contribute to human cancers include solar ultraviolet radiation (the major cause of skin cancer), carcinogenic chemicals in tobacco smoke, and aflatoxin (a potent liver carcinogen produced by some molds that contaminate improperly stored supplies of peanuts and other grains). The carcinogens in tobacco smoke (including benzo(*a*)pyrene,

Figure 15.5
Development of colon carcinomas   A single initially altered cell gives rise to a proliferative cell population, which progresses first to benign adenomas of increasing size and then to malignant carcinoma. The cancer cells invade the underlying connective tissue and penetrate blood and lymphatic vessels, thereby spreading throughout the body.

dimethylnitrosamine, and nickel compounds) are the major identified causes of human cancer. Smoking is the undisputed cause of 80 to 90% of lung cancers, as well as being implicated in cancers of the oral cavity, pharynx, larynx, esophagus, and other sites. In total, it is estimated that smoking is responsible for nearly one-third of all cancer deaths—an impressive toll for a single carcinogenic agent.

Other carcinogens contribute to cancer development by stimulating cell proliferation, rather than by inducing mutations. Such compounds are referred to as **tumor promoters**, since the increased cell division they induce is required for the outgrowth of a proliferative cell population during early stages of tumor development. The phorbol esters that stimulate cell proliferation by activating protein kinase C (see Figure 13.26) are classic examples. Their activity was defined by studies of chemical induction of skin tumors in mice (Figure 15.7). Tumorigenesis in this system can be initiated by a single treatment with a mutagenic carcinogen. Tumors do not develop, however, unless the mice are subsequently treated with a tumor promoter (usually a phorbol ester) to stimulate proliferation of the mutated cells.

Hormones, particularly estrogens, are important as tumor promoters in the development of some human cancers. The proliferation of cells of the uterine endometrium, for example, is stimulated by estrogen, and exposure to excess estrogen significantly increases the likelihood that a woman will develop endometrial cancer. The risk of endometrial cancer is therefore substantially increased by long-term postmenopausal estrogen replacement therapy with high doses of estrogen alone. Fortunately, this risk is minimized by administration of progesterone to counteract the stimulatory effect of estrogen on endometrial cell proliferation. However, long-term

Figure 15.6
Structure of representative chemical carcinogens

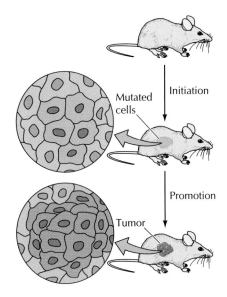

Figure 15.7
Induction of tumors in mouse skin
Tumors are initiated by mutations induced by a carcinogen. Development of a tumor then requires treatment with a tumor promoter to stimulate proliferation of the mutated cells.

therapy with combinations of estrogen and progesterone may lead to an increased risk of breast cancer.

In addition to chemicals and radiation, some viruses induce cancer both in experimental animals and in humans. The common human cancers caused by viruses include liver cancer and cervical carcinoma, which together account for 10 to 20% of worldwide cancer incidence. These viruses are important not only as causes of human cancer; as discussed later in this chapter, studies of tumor viruses have played a key role in elucidating the molecular events responsible for the development of cancers induced by both viral and nonviral carcinogens.

Properties of Cancer Cells

The uncontrolled growth of cancer cells results from accumulated abnormalities affecting many of the cell regulatory mechanisms that have been discussed in preceding chapters. This relationship is reflected in several aspects of cell behavior that distinguish cancer cells from their normal counterparts. Cancer cells typically display abnormalities in the mechanisms that regulate normal cell proliferation, differentiation, and survival. Taken together, these characteristic properties of cancer cells provide a description of malignancy at the cellular level.

The uncontrolled proliferation of cancer cells *in vivo* is mimicked by their behavior in cell culture. A primary distinction between cancer cells and normal cells in culture is that normal cells display **density-dependent inhibition** of cell proliferation (Figure 15.8). Normal cells proliferate until they reach a finite cell density, which is determined in part by the availability of growth factors added to the culture medium (usually in the form of serum). They then cease proliferating and become quiescent, arrested in the G_0 stage of the cell cycle (see Figure 14.6). The proliferation of most cancer cells, however, is not sensitive to density-dependent inhibition. Rather than responding to the signals that cause normal cells to cease proliferation and enter G_0, tumor cells generally continue growing to high cell densities in culture, mimicking their uncontrolled proliferation *in vivo*.

A related difference between normal cells and cancer cells is that many cancer cells have reduced requirements for extracellular growth factors. As discussed in Chapter 13, the proliferation of most cells is controlled, at least in part, by polypeptide growth factors. For some cell types, particularly fibroblasts, the availability of serum growth factors is the principal deter-

Figure 15.8
Density-dependent inhibition
Normal cells proliferate in culture until they reach a finite cell density, at which point they become quiescent. Tumor cells, however, continue to proliferate independent of cell density.

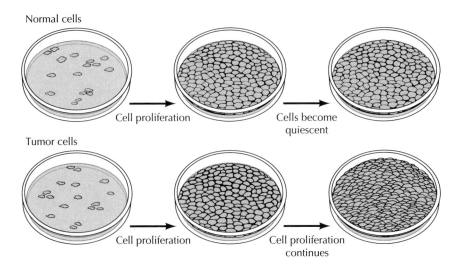

Normal cells

Cell proliferation   Cells become quiescent

Tumor cells

Cell proliferation   Cell proliferation continues

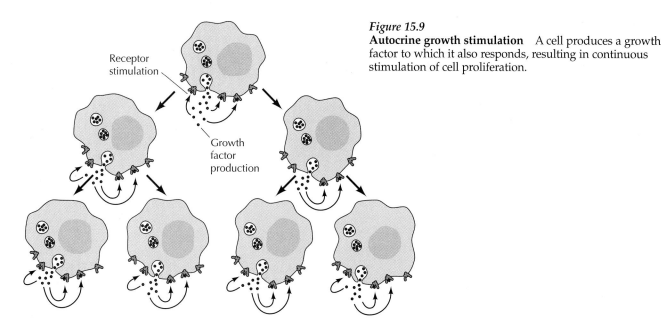

Figure 15.9
Autocrine growth stimulation   A cell produces a growth factor to which it also responds, resulting in continuous stimulation of cell proliferation.

Receptor stimulation

Growth factor production

minant of their proliferative capacity in culture. The growth factor requirements of these cells are closely related to the phenomenon of density-dependent inhibition, since the density at which normal fibroblasts become quiescent is proportional to the concentration of serum growth factors in the culture medium.

The growth factor requirements of many tumor cells are reduced compared to their normal counterparts, contributing to the unregulated proliferation of tumor cells both *in vitro* and *in vivo*. In some cases, cancer cells produce growth factors that stimulate their own proliferation (Figure 15.9). Such abnormal production of a growth factor by a responsive cell leads to continuous autostimulation of cell division (**autocrine growth stimulation**), and the cancer cells are therefore less dependent on growth factors from other, physiologically normal sources. In other cases, the reduced growth factor dependence of cancer cells results from abnormalities in intracellular signaling systems, such as unregulated activity of growth factor receptors or other proteins (e.g., Ras proteins or protein kinases) that were discussed in Chapter 13 as elements of signal transduction pathways leading to cell proliferation.

Cancer cells are also less stringently regulated than normal cells by cell-cell and cell-matrix interactions. Most cancer cells are less adhesive than normal cells, often as a result of reduced expression of cell surface adhesion molecules. For example, loss of E-cadherin, the principal adhesion molecule of epithelial cells, is important in the development of carcinomas (epithelial cancers). As a result of reduced expression of cell adhesion molecules, cancer cells are comparatively unrestrained by interactions with other cells and tissue components, contributing to the ability of malignant cells to invade and metastasize. The reduced adhesiveness of cancer cells also results in morphological and cytoskeletal alterations: Many tumor cells are rounder than normal, in part because they are less firmly attached to either the extracellular matrix or neighboring cells.

A striking difference in the cell-cell interactions displayed by normal cells and those of cancer cells is illustrated by the phenomenon of **contact inhibition** (Figure 15.10). Normal fibroblasts migrate across the surface of a culture dish until they make contact with a neighboring cell. Further cell migration is then inhibited, and normal cells adhere to each other, forming

Figure 15.10
Contact inhibition   Light micrographs (left) and scanning electron micrographs (right) of normal fibroblasts and tumor cells. The migration of normal fibroblasts is inhibited by cell contact, so they form an orderly side-by-side array on the surface of a culture dish. Tumor cells, however, are not inhibited by cell contact, so they migrate over one another and grow in a disordered, multilayered pattern. (Courtesy of Lan Bo Chen, Dana-Farber Cancer Institute.)

Normal cells

Tumor cells

an orderly array of cells on the culture dish surface. Tumor cells, in contrast, continue moving after contact with their neighbors, migrating over adjacent cells, and growing in disordered, multilayered patterns. Not only the movement but also the proliferation of many normal cells is inhibited by cell-cell contact, and cancer cells are characteristically insensitive to such contact inhibition of growth.

Two additional properties of cancer cells affect their interactions with other tissue components, thereby playing important roles in invasion and metastasis. First, malignant cells generally secrete proteases that digest extracellular matrix components, allowing the cancer cells to invade adjacent normal tissues. Secretion of collagenase, for example, appears to be an important determinant of the ability of carcinomas to digest and penetrate through basal laminae to invade underlying connective tissue (see Figure 15.5). Second, cancer cells secrete growth factors that promote the formation of new blood vessels (**angiogenesis**). Angiogenesis is needed to support the growth of a tumor beyond the size of about a million cells, at which point new blood vessels are required to supply oxygen and nutrients to the proliferating tumor cells. Such blood vessels are formed in response to growth factors, secreted by the tumor cells, that stimulate proliferation of endothelial cells in the walls of capillaries in surrounding tissue, resulting in the outgrowth of new capillaries into the tumor. The formation of such new blood vessels is important not only in supporting tumor growth, but also in metastasis. The actively growing new capillaries formed in response to angiogenic stimulation are easily penetrated by the tumor cells, providing a ready opportunity for cancer cells to enter the circulatory system and begin the metastatic process.

Another general characteristic of most cancer cells is that they fail to differentiate normally. Such defective differentiation is closely coupled to abnormal proliferation, since, as discussed in Chapter 14, most fully differentiated cells either cease cell division or divide only rarely. Rather than carrying out their normal differentiation program, cancer cells are usually blocked at an early stage of differentiation, consistent with their continued active proliferation.

The leukemias provide a particularly good example of the relationship between defective differentiation and malignancy. All of the different types of blood cells are derived from a common stem cell in the bone marrow (see Figure 14.44). Descendants of these cells then become committed to specific differentiation pathways. Some cells, for example, differentiate to form erythrocytes whereas others differentiate to form lymphocytes, granulocytes, or macrophages. Cells of each of these types undergo several rounds of division as they differentiate, but once they become fully differentiated, cell division ceases. Leukemic cells, in contrast, fail to undergo terminal differentiation (Figure 15.11). Instead, they become arrested at early stages of maturation at which they retain their capacity for proliferation and continue to reproduce.

As discussed in Chapter 13, **programmed cell death**, or **apoptosis**, is an integral part of the differentiation program of many cell types, including blood cells. Many cancer cells fail to undergo apoptosis, and therefore exhibit increased life spans compared to their normal counterparts. This failure of cancer cells to undergo programmed cell death contributes substantially to tumor development. For example, the survival of many normal cells is dependent on signals from growth factors or from the extracellular matrix that prevent apoptosis. In contrast, tumor cells are often able to survive in the absence of growth factors required by their normal counterparts. Such a failure of tumor cells to undergo apoptosis when deprived of normal environmental signals may be important not only in primary tumor development but also in the survival and growth of metastatic cells in abnormal tissue sites. Normal cells also undergo apoptosis following DNA damage, while many cancer cells fail to do so. In this case, the failure to undergo apoptosis contributes to the resistance of cancer

Leukemic cells fail to differentiate and continue to divide

Figure 15.11
Defective differentiation and leukemia Different types of blood cells develop from a multipotential (pluripotent) stem cell in the bone marrow. The precursors of differentiated cells undergo several rounds of cell division as they mature, but cell division ceases at the terminal stages of differentiation. The differentiation of leukemic cells is blocked at early stages of maturation, consistent with their continued proliferation.

Figure 15.12
The focus assay A focus of chicken embryo fibroblasts induced by Rous sarcoma virus. (From H. M. Temin and H. Rubin, 1958. *Virology* 6: 669.)

cells to irradiation and many chemotherapeutic drugs, which act by damaging DNA. Abnormal cell survival, as well as cell proliferation, thus plays a major role in the unrelenting growth of cancer cells in an animal.

Transformation of Cells in Culture

The study of tumor induction by radiation, chemicals, or viruses requires experimental systems in which the effects of a carcinogenic agent can be reproducibly observed and quantitated. Although the activity of carcinogens can be assayed in intact animals, such experiments are difficult to quantitate and control. The development of *in vitro* assays to detect the conversion of normal cells to tumor cells in culture, a process called **cell transformation**, therefore represented a major advance in cancer research. Such assays are designed to detect transformed cells, which display the *in vitro* growth properties of tumor cells, following exposure of a culture of normal cells to a carcinogenic agent. Their application has allowed experimental analysis of cell transformation to reach a level of sophistication that could not have been attained by studies in whole animals alone.

The first and most widely used assay of cell transformation is the focus assay, which was developed by Howard Temin and Harry Rubin in 1958. The focus assay is based on the ability to recognize a group of transformed cells as a morphologically distinct "focus" against a background of normal cells on the surface of a culture dish (Figure 15.12). The focus assay takes advantage of three properties of transformed cells: altered morphology, loss of contact inhibition, and loss of density-dependent inhibition of growth. The result is the formation of a colony of morphologically altered transformed cells that overgrow the background of normal cells in the culture. Such foci of transformed cells can usually be detected and quantified within a week or two after exposure to a carcinogenic agent. In general, cells transformed *in vitro* are able to form tumors following inoculation into susceptible animals, supporting *in vitro* transformation as a valid indicator of the formation of cancer cells.

Tumor Viruses

Members of six distinct families of animal viruses, called **tumor viruses**, are capable of directly causing cancer in either experimental animals or humans (Table 15.2). Viruses belonging to five of these families have DNA genomes and are referred to as DNA tumor viruses. Members of the sixth family of tumor viruses, the retroviruses, have RNA genomes in virus par-

TABLE 15.2 Tumor Viruses		
Virus family	**Human tumors**	**Genome size (kb)**
DNA tumor viruses		
Hepatitis B viruses	Liver cancer	3
SV40 and polyomavirus	None	5
Papillomaviruses	Cervical carcinoma	8
Adenoviruses	None	35
Herpesviruses	Burkitt's lymphoma, nasopharyngeal carcinoma, Kaposi's sarcoma	100–200
RNA tumor viruses		
Retroviruses	Adult T-cell leukemia	9

ticles but replicate via synthesis of a DNA provirus in infected cells. The viruses that cause human cancer include hepatitis B virus (liver cancer), papillomaviruses (cervical and other anogenital cancers), Epstein-Barr virus (Burkitt's lymphoma and nasopharyngeal carcinoma), Kaposi's sarcoma-associated herpesvirus (Kaposi's sarcoma), and human T-cell lymphotropic virus (adult T-cell leukemia). In addition, HIV is indirectly responsible for the cancers that develop in AIDS patients as a result of immunodeficiency, and hepatitis C virus (an RNA virus) is an indirect cause of liver cancers resulting from chronic tissue damage.

As already noted, tumor viruses not only are important as causes of human disease but have also played a critical role in cancer research by serving as models for cellular and molecular studies of cell transformation. The small size of their genomes has made tumor viruses readily amenable to molecular analysis, leading to the identification of viral genes responsible for cancer induction and paving the way to our current understanding of cancer at the molecular level.

Hepatitis B Viruses

The **hepatitis B viruses**, which have the smallest genomes (approximately 3 kb) of all animal DNA viruses, specifically infect liver cells of several species, including ducks, woodchucks, squirrels, and humans. Infection with hepatitis B virus usually results in acute liver damage. In 5 to 10% of cases, however, the acute infection is not resolved and a chronic infection of the liver develops. Such chronic infection is associated with more than a hundredfold increased risk of liver cancer. Hepatitis B virus infection is particularly common in parts of Asia and Africa, where it is associated with up to a million cases of liver cancer annually (approximately 10% of worldwide cancer incidence).

Cell transformation by hepatitis B virus is mediated by a viral gene (called the *X* gene) that affects expression of a variety of cellular genes that drive abnormal cell proliferation and survival. In addition, the development of cancers induced by hepatitis B virus is driven by the continual proliferation of liver cells that results from chronic tissue damage.

SV40 and Polyomavirus

The best studied DNA tumor viruses, from the standpoint of molecular biology, are probably **simian virus 40 (SV40)** and **polyomavirus**. Although neither of these viruses is associated with human cancer, they have been critically important as models for understanding the molecular basis of cell transformation. The utility of these viruses in cancer research has stemmed from the availability of good cell culture assays for both virus replication and transformation, as well as from the small size of their genomes (approximately 5 kb).

SV40 and polyomavirus do not induce tumors or transform cells of their natural host species—monkeys and mice, respectively. In cells of their natural hosts (permissive cells), infection leads to virus replication, cell lysis, and release of progeny virus particles (Figure 15.13). Since a permissive cell is killed as a consequence of virus replication, it cannot become transformed. The transforming potential of these viruses is revealed, however, by infection of nonpermissive cells, in which virus replication is blocked. In this case, the viral genome sometimes integrates into cellular DNA, and expression of specific viral genes results in transformation of the infected cell.

The SV40 and polyomavirus genes that lead to cell transformation have been identified by detailed molecular analyses. The viral genomes and mRNAs have been completely sequenced, viral mutants that are unable to

Figure 15.13
SV40 replication and transformation Infection of a permissive cell results in virus replication, cell lysis, and release of progeny virus particles. In a nonpermissive cell, virus replication is blocked, allowing some cells to become permanently transformed.

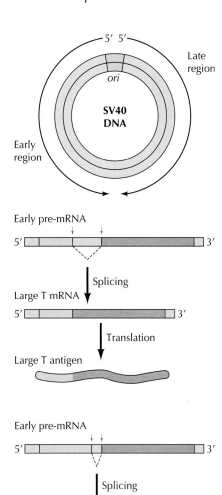

Figure 15.14
The SV40 genome   The genome is divided into early and late regions. Large and small T antigens are produced by alternative splicing of early-region pre-mRNA.

induce transformation have been isolated, and the transforming potentials of individual viral genes have been determined by gene transfer assays. Transformation by these viruses has thus been found to result from expression of the same viral genes that function in early stages of lytic infection. The genomes of SV40 and polyomavirus are divided into early and late regions. The early region is expressed immediately after infection and is required for synthesis of viral DNA. The late region is not expressed until after viral DNA replication has begun, and includes genes encoding structural components of the virus particle. The early region of SV40 encodes two proteins, called small and large T antigens, of about 17 kd and 94 kd, respectively (Figure 15.14). Their mRNAs are generated by alternative splicing of a single early-region primary transcript. Polyomavirus likewise encodes small and large T antigens, as well as a third early-region protein of about 55 kd, designated middle T. Transfection of cells with cDNAs for individual early-region proteins has shown that SV40 large T is sufficient to induce transformation, whereas middle T is primarily responsible for transformation by polyomavirus.

During lytic infection, these early-region proteins fulfill multiple functions required for virus replication. SV40 T antigen, for example, binds to the SV40 origin and initiates viral DNA replication (see Chapter 5). In addition, the early-region proteins of SV40 and polyomavirus stimulate host cell gene expression and DNA synthesis. Since virus replication is dependent on host cell enzymes (e.g., DNA polymerase), such stimulation of the host cell is a critical event in the viral life cycle. Most cells in an animal are nonproliferating, and therefore must be stimulated to divide in order to induce the enzymes needed for viral DNA replication. This stimulation of cell proliferation by the early gene products can lead to transformation if the viral DNA becomes stably integrated and expressed in a nonpermissive cell.

As discussed later in this chapter, both SV40 and polyomavirus early-region proteins induce transformation by interacting with host proteins that regulate cell proliferation. For example, SV40 T antigen binds to and inactivates the host cell tumor suppressor proteins Rb and p53, which are key regulators of cell proliferation and cell cycle progression (see Figures 14.20 and 14.21).

Papillomaviruses

The **papillomaviruses** are small DNA viruses (genomes of approximately 8 kb) that induce both benign and malignant tumors in humans and a variety of other animal species. Approximately 60 different types of human papillomaviruses, which infect epithelial cells of several tissues, have been identified. Some of these viruses cause only benign tumors (such as warts), whereas others are causative agents of malignant carcinomas, particularly cervical and other anogenital cancers. The mortality from cervical cancer is relatively low in the United States, in large part as a result of early detection and curative treatment made possible by the Pap smear. In other parts of the world, however, cervical cancer remains common; it is responsible for 5 to 10% of worldwide cancer incidence.

Cell transformation by human papillomaviruses results from expression of two early-region genes, *E6* and *E7* (Figure 15.15). The E6 and E7 proteins act analogously to SV40 T antigen by interfering with the function of the

Figure 15.15
**The genome of a human papilloma-
virus** Gene products are designated
E (early) or L (late). Transformation
results from the action of E6 and E7.

cellular Rb and p53 proteins. In particular, E7 binds to Rb, and E6 stimulates the degradation of p53 by ubiquitin-mediated proteolysis.

Adenoviruses

The **adenoviruses** are a large family of DNA viruses with genomes of
about 35 kb. In contrast to the papillomaviruses, the adenoviruses are not
associated with naturally occurring cancers in either humans or other animals. However, they are widely studied and important models in experimental cancer biology.

Like SV40 and polyomaviruses, the adenoviruses are lytic in cells of their
natural host species, but can induce transformation in nonpermissive hosts.
Transformation by the adenoviruses results from expression of two early
genes, *E1A* and *E1B*, which are required for virus replication in permissive
cells (Figure 15.16). These transforming proteins inactivate the Rb and p53
tumor suppressor proteins, with E1A binding to Rb and E1B binding to p53.
It thus appears that SV40, papillomaviruses, and adenoviruses all induce
transformation by a common pathway, in which altering regulation of the
cell cycle by interfering with the activities of Rb and p53 plays a central role.

Herpesviruses

The **herpesviruses** are among the most complex animal viruses, with
genomes of 100 to 200 kb. Several herpesviruses induce tumors in animal
species, including frogs, chickens, and monkeys. In addition, two members
of the herpesvirus family, **Kaposi's sarcoma-associated herpesvirus** and
Epstein-Barr virus, are associated with human cancers. Kaposi's sarcoma-
associated herpesvirus plays a critical role in the development of Kaposi's
sarcomas, and Epstein-Barr virus has been implicated in several human
malignancies, including Burkitt's lymphoma in some regions of Africa,
B-cell lymphomas in AIDS patients and other immunosuppressed individuals, and nasopharyngeal carcinoma in China.

In addition to its association with these human malignancies, Epstein-
Barr virus is able to transform human B lymphocytes in culture. Partly
because of the complexity of the genome, however, the molecular biology of
Epstein-Barr virus replication and transformation remains to be fully understood. Several viral genes required to induce transformation of lymphocytes
have been identified, but their functions have not been established.

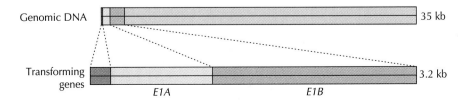

Figure 15.16
The adenovirus genome Two early-
region genes, *E1A* and *E1B*, are responsible for induction of transformation.

Figure 15.17
A typical retrovirus genome   The DNA provirus, integrated into cellular DNA, is transcribed to yield genome-length RNA. This primary transcript serves as the genomic RNA for progeny virus particles, and as mRNA for the *gag* and *pol* genes. In addition, the full-length RNA is spliced to yield mRNA for *env*. The *gag* gene encodes the viral protease and structural proteins of the virus particle, *pol* encodes reverse transcriptase and integrase, and *env* encodes envelope glycoproteins.

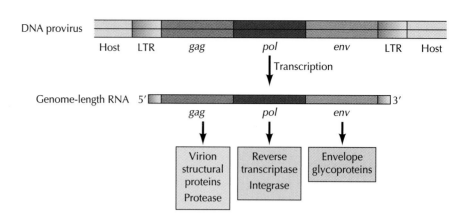

Retroviruses

Members of one family of RNA viruses, the **retroviruses**, cause cancer in a variety of animal species, including humans. One human retrovirus, human T-cell lymphotropic virus type I (HTLV-I), is the causative agent of adult T-cell leukemia, which is common in parts of Japan, the Caribbean, and Africa. Transformation of T lymphocytes by HTLV-I results from expression of the viral gene *tax*, which encodes a regulatory protein affecting expression of several cellular growth control genes. AIDS is caused by another retrovirus, HIV. In contrast to HTLV-I, HIV does not cause cancer by directly converting a normal cell into a tumor cell. However, AIDS patients suffer a high incidence of some malignancies, particularly lymphomas and Kaposi's sarcoma. These cancers, which are also common among other immunosuppressed individuals, apparently develop as a secondary consequence of immunosuppression in AIDS patients.

Different retroviruses differ substantially in their oncogenic potential. Most retroviruses contain only three genes (*gag*, *pol*, and *env*) that are required for virus replication but play no role in cell transformation (Figure 15.17). Retroviruses of this type induce tumors only rarely, if at all, as a consequence of mutations resulting from the integration of proviral DNA within or adjacent to cellular genes.

Other retroviruses, however, contain specific genes responsible for induction of cell transformation and are potent carcinogens. The prototype of these highly oncogenic retroviruses is **Rous sarcoma virus (RSV)**, first isolated from a chicken sarcoma by Peyton Rous in 1911. More than 50 years later, studies of RSV led to identification of the first viral oncogene, which has provided a model for understanding many aspects of tumor development at the molecular level.

Oncogenes

Cancer results from alterations in critical regulatory genes that control cell proliferation, differentiation, and survival. Studies of tumor viruses revealed that specific genes (called **oncogenes**) are capable of inducing cell transformation, thereby providing the first insights into the molecular basis of cancer. However, the majority (approximately 80%) of human cancers are not induced by viruses and apparently arise from other causes, such as radiation and chemical carcinogens. Therefore, in terms of our overall understanding of cancer, it has been critically important that studies of viral oncogenes also led to the identification of cellular oncogenes, which are involved in the development of non-virus-induced cancers. The key link between viral and cellular oncogenes was provided by studies of the highly oncogenic retroviruses.

Retroviral Oncogenes

Viral oncogenes were first defined in RSV, which transforms chicken embryo fibroblasts in culture and induces large sarcomas within 1 to 2 weeks after inoculation into chickens (Figure 15.18). In contrast, the closely related avian leukosis virus (ALV) replicates in the same cells as RSV without inducing transformation. This difference in transforming potential suggested the possibility that RSV contains specific genetic information responsible for transformation of infected cells. A direct comparison of the genomes of RSV and ALV was consistent with this hypothesis: The genomic RNA of RSV is about 10 kb, whereas that of ALV is smaller, about 8.5 kb.

In the early 1970s, Peter Vogt and Steven Martin isolated both deletion mutants and temperature-sensitive mutants of RSV that were unable to induce transformation. Importantly, these mutants still replicated normally in infected cells, indicating that RSV contains genetic information that is required for transformation but not for virus replication. Further analysis demonstrated that both the deletion and temperature-sensitive RSV mutants define a single gene responsible for the ability of RSV to induce tumors in birds and transform fibroblasts in culture. Because RSV causes sarcomas, its oncogene was called *src*. The *src* gene is an addition to the genome of RSV; it is not present in ALV (Figure 15.19). It encodes a 60-kd protein that was the first protein-tyrosine kinase to be identified (see the Key Experiment in Chapter 13).

More than 40 different highly oncogenic retroviruses have been isolated from a variety of animals, including chickens, turkeys, mice, rats, cats, and monkeys. All of these viruses, like RSV, contain at least one oncogene (in some cases two) that is not required for virus replication but is responsible for cell transformation. In some cases, different viruses contain the same oncogenes, but more than two dozen distinct oncogenes have been identified among this group of viruses (Table 15.3). Like *src*, many of these genes (such as *ras* and *raf*) encode proteins that are now recognized as key components of signaling pathways that stimulate cell proliferation (see Figure 13.32).

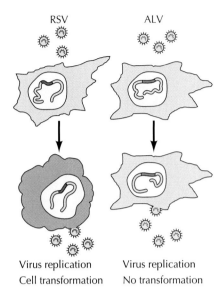

Figure 15.18
Cell transformation by RSV and ALV Both RSV and ALV infect and replicate in chicken embryo fibroblasts, but only RSV induces cell transformation.

Figure 15.19
The RSV genome RSV contains an additional gene, *src*, that is not present in ALV and encodes the Src protein-tyrosine kinase.

TABLE 15.3 Retroviral Oncogenes

Oncogene	Virus	Species
abl	Abelson leukemia	Mouse
akt	AKT8 virus	Mouse
cbl	Cas NS-1	Mouse
crk	CT10 sarcoma	Chicken
erbA	Avian erythroblastosis-ES4	Chicken
erbB	Avian erythroblastosis-ES4	Chicken
ets	Avian erythroblastosis-E26	Chicken
fes	Gardner-Arnstein feline sarcoma	Cat
fgr	Gardner-Rasheed feline sarcoma	Cat
fms	McDonough feline sarcoma	Cat
fos	FBJ murine osteogenic sarcoma	Mouse
fps	Fujinami sarcoma	Chicken
jun	Avian sarcoma-17	Chicken
kit	Hardy-Zuckerman feline sarcoma	Cat
maf	Avian sarcoma AS42	Chicken
mos	Moloney sarcoma	Mouse
mpl	Myeloproliferative leukemia	Mouse
myb	Avian myeloblastosis	Chicken
myc	Avian myelocytomatosis	Chicken
p3k	Avian sarcoma-16	Chicken
qin	Avian sarcoma-31	Chicken
raf	3611 murine sarcoma	Mouse
rasH	Harvey sarcoma	Rat
rasK	Kirsten sarcoma	Rat
rel	Reticuloendotheliosis	Turkey
ros	UR2 sarcoma	Chicken
sea	Avian erythroblastosis-S13	Chicken
sis	Simian sarcoma	Monkey
ski	Avian SK	Chicken
src	Rous sarcoma	Chicken
yes	Y73 sarcoma	Chicken

Proto-Oncogenes

An unexpected feature of retroviral oncogenes is their lack of involvement in virus replication. Since most viruses are streamlined to replicate as efficiently as possible, the existence of viral oncogenes that are not an integral part of the virus life cycle seems paradoxical. Scientists were thus led to question where the retroviral oncogenes had originated and how they had become incorporated into viral genomes—a line of investigation that ultimately led to the identification of cellular oncogenes in human cancers.

The first clue to the origin of oncogenes came from the way in which the highly oncogenic retroviruses were isolated. The isolation of Abelson leukemia virus is a typical example (Figure 15.20). More than 150 mice were inoculated with a nontransforming virus containing only the *gag*, *pol*, and *env* genes required for virus replication. One of these mice developed a lymphoma from which a new, highly oncogenic virus (Abelson leukemia virus), which now contained an oncogene (*abl*), was isolated. The scenario suggested the hypothesis that the retroviral oncogenes are derived from genes of the host cell, and that occasionally such a host cell gene becomes incorporated into a viral genome, yielding a new, highly oncogenic virus as the product of a virus-host recombination event.

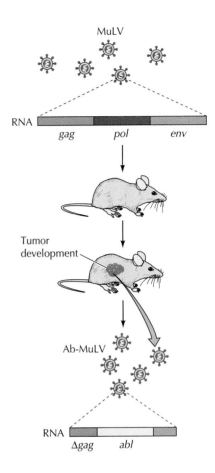

Figure 15.20
Isolation of Abelson leukemia virus The highly oncogenic virus Ab-MuLV was isolated from a rare tumor that developed in a mouse that had been inoculated with a nontransforming virus (Moloney murine leukemia virus, or MuLV). MuLV contains only the *gag*, *pol*, and *env* genes required for virus replication. In contrast, Ab-MuLV has acquired a new oncogene (*abl*), which is responsible for its transforming activity. The *abl* oncogene replaced some of the viral replicative genes and is fused with a partially deleted *gag* gene (designated Δ*gag*) in the Ab-MuLV genome.

The critical prediction of this hypothesis was that normal cells contain genes that are closely related to the retroviral oncogenes. This was definitively demonstrated in 1976 by Harold Varmus, J. Michael Bishop, and their colleagues, who showed that a cDNA probe for the *src* oncogene of RSV hybridized to closely related sequences in the DNA of normal chicken cells. Moreover, *src*-related sequences were also found in normal DNAs of a wide range of other vertebrates (including humans) and thus appeared to be highly conserved in evolution. Similar experiments with probes for the oncogenes of other highly oncogenic retroviruses have yielded comparable results, and it is now firmly established that the retroviral oncogenes were derived from closely related genes of normal cells.

The normal-cell genes from which the retroviral oncogenes originated are called **proto-oncogenes**. They are important cell regulatory genes, in many cases encoding proteins that function in the signal transduction pathways controlling normal cell proliferation (e.g., *src*, *ras*, and *raf*). The oncogenes are abnormally expressed or mutated forms of the corresponding proto-oncogenes. As a consequence of such alterations, the oncogenes induce abnormal cell proliferation and tumor development.

An oncogene incorporated into a retroviral genome differs in several respects from the corresponding proto-oncogene. First, the viral oncogene is transcribed under the control of viral promoter and enhancer sequences, rather than being controlled by the normal transcriptional regulatory sequences of the proto-oncogene. Consequently, oncogenes are usually expressed at much higher levels than the proto-oncogenes and are sometimes transcribed in inappropriate cell types. In some cases, such abnormalities of gene expression are sufficient to convert a normally functioning proto-oncogene into an oncogene that drives cell transformation.

In addition to such alterations in gene expression, oncogenes frequently encode proteins that differ in structure and function from those encoded by their normal homologs. Many oncogenes, such as *raf*, are expressed as fusion proteins with viral sequences at the amino terminus (Figure 15.21). Recombination events leading to the generation of such fusion proteins often occur during the capture of proto-oncogenes by retroviruses, and sequences from both the amino and carboxy termini of proto-oncogenes are frequently deleted during the process. Such deletions may result in the loss of regulatory domains that control the activity of the proto-oncogene proteins, thereby generating oncogene proteins that function in an unregulated manner. For example, the viral *raf* oncogene encodes a fusion protein in which amino-terminal sequences of the normal Raf protein have been deleted. These amino-terminal sequences are critical to the normal regulation of Raf protein kinase activity, and their deletion results in unregulated constitutive activity of the oncogene-encoded Raf protein. This unregulated Raf activity drives cell proliferation, resulting in transformation.

Many other oncogenes differ from the corresponding proto-oncogenes by point mutations, resulting in single amino acid substitutions in the oncogene products. In some cases, such amino acid substitutions (like the deletions already discussed) lead to unregulated activity of the oncogene

Figure 15.21
The Raf oncogene protein The Raf proto-oncogene protein consists of an amino-terminal regulatory domain and a carboxy-terminal protein kinase domain. In the viral Raf oncogene protein, the regulatory domain has been deleted and replaced by viral Gag sequences. As a result, the Raf kinase domain is constitutively active, causing cell transformation.

Raf proto-oncogene protein

| Regulatory domain | Kinase domain |

Raf oncogene protein

| Gag | Kinase domain |

Key Experiment

The Discovery of Proto-Oncogenes

DNA Related to the Transforming Gene(s) of Avian Sarcoma Viruses Is Present in Normal Avian DNA

Dominique Stehelin, Harold E. Varmus, J. Michael Bishop and Peter K. Vogt

Department of Microbiology, University of California, San Francisco (DS, HEV, and JMB) and Department of Microbiology, University of California, Los Angeles (PKV)

Nature, Volume 260, 1976, pages 170–173

The Context

Genetic analysis of RSV defined the first viral oncogene (*src*) as a gene that was specifically responsible for cell transformation but was not required for virus replication. The origin of highly oncogenic retroviruses from tumors of infected animals then suggested the hypothesis that retroviral oncogenes are derived from related genes of host cells. Consistent with this suggestion, normal cells of several species were found to contain retrovirus-related DNA sequences that could be detected by nucleic acid hybridization. However, it was unclear whether these sequences were related to the retroviral oncogenes or to the genes required for virus replication.

Harold Varmus, J. Michael Bishop, and their colleagues resolved this critical issue by exploiting the genetic characterization of the *src* oncogene. In particular, Peter Vogt had previously isolated transformation-defective mutants of RSV that had sustained

deletions of approximately 1.5 kb, corresponding to most or all of the *src* gene. Stehelin and collaborators used these mutants to prepare a cDNA probe that specifically represented *src* sequences. The use of this defined probe in nucleic acid hybridization experiments allowed them to definitively demonstrate that normal cells contain *src*-related DNA sequences.

The Experiments

First reverse transcriptase was used to synthesize a radioactive cDNA probe composed of short single-stranded DNA fragments complementary to the entire genomic RNA of RSV. This probe was then hybridized to an excess of RNA isolated from a transformation-defective deletion mutant. Fragments of cDNA that were complementary to the viral replication genes hybridized to the transformation-defective RSV RNA, forming RNA-DNA duplexes. In contrast, cDNA fragments that were complementary to *src* were unable to hybridize and

remained single-stranded. This single-stranded DNA was then isolated to provide a specific probe for *src* oncogene sequences. As predicted from the size of deletions in the transformation-defective RSV mutants, the *src*-specific probe was homologous to about 1.5 kb of RSV RNA.

The radioactive *src* cDNA was then used as a hybridization probe to attempt to detect related DNA sequences in normal avian cells. Strikingly, the *src* cDNA hybridized extensively to normal chicken DNA, as well as to DNA of other avian species (see figure). These experi-

Hybridization of *src*-specific cDNA to normal chicken, quail, and duck DNA.

proteins. An important example of such point mutations is provided by the *ras* oncogenes, which are discussed in the next section in terms of their role in human cancers.

Oncogenes in Human Cancer

Understanding the origin of retroviral oncogenes raised the question as to whether non-virus-induced tumors contain cellular oncogenes that were generated from proto-oncogenes by mutations or DNA rearrangements during tumor development. Direct evidence for the involvement of cellular oncogenes in human tumors was first obtained by gene transfer experi-

The Discovery of Proto-Oncogenes (continued)

ments thus demonstrated that normal cells contain DNA sequences that are closely related to the *src* oncogene, supporting the hypothesis that retroviral oncogenes originated from cellular genes that became incorporated into viral genomes.

The Impact

Stehelin and colleagues concluded their 1976 paper by raising the possibility that the cellular *src* sequences are "involved in the normal regulation of cell growth and development or the transformation of cell behavior by physical, chemical or viral agents." This prediction has been strikingly substantiated, and the discovery of cellular *src* sequences has opened new doors to understanding both the regu-

lation of normal cell proliferation and the molecular basis of human cancer. Studies of oncogene and proto-oncogene proteins, including Src itself, have proven critical in unraveling the signaling pathways that control the proliferation and differentiation of normal cells. The discovery of the *src* proto-oncogene further suggested that non-virus-induced tumors could arise as a result of mutations in related cellular genes, leading directly to the discovery of oncogenes in human tumors. By unifying studies of tumor viruses, normal cells, and non-virus-induced tumors, the results of Varmus, Bishop, and their colleagues have had an impact on virtually all aspects of cell regulation and cancer research.

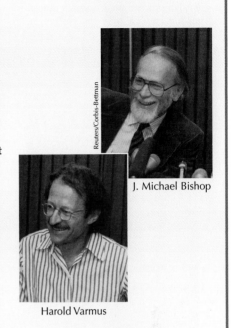

J. Michael Bishop

Harold Varmus

ments in the laboratories of Robert Weinberg and of the author in 1981. DNA of a human bladder carcinoma was found to efficiently induce transformation of recipient mouse cells in culture, indicating that the human tumor contained a biologically active cellular oncogene (Figure 15.22). Both gene transfer assays and alternative experimental approaches have since led to the detection of active cellular oncogenes in human tumors of many different types (Table 15.4).

Some of the oncogenes identified in human tumors are cellular homologs of oncogenes that were previously characterized in retroviruses, whereas others are new oncogenes first discovered in human cancers. The first human oncogene identified in gene transfer assays was subsequently identified as the human homolog of the *ras*H oncogene of Harvey sarcoma virus (see Table 15.3). Three closely related members of the **ras** gene family (*ras*H, *ras*K, and *ras*N) are the oncogenes most frequently encountered in human tumors. These genes are involved in approximately 20% of all human malignancies, including about 50% of colon and 25% of lung carcinomas.

The *ras* oncogenes are not present in normal cells; rather, they are generated in tumor cells as a consequence of mutations that occur during tumor development. The *ras* oncogenes differ from their proto-oncogenes by point mutations resulting in single amino acid substitutions at critical positions. The first such mutation discovered was the substitution of valine for glycine at position 12 (Figure 15.23). Other amino acid substitutions at position 12, as well as at positions 13 and 61, are also frequently encountered in *ras* oncogenes in human tumors. In animal models, it has been shown that

Figure 15.22
Detection of a human tumor oncogene by gene transfer   DNA extracted from a human bladder carcinoma induced transformation of recipient mouse cells in culture. Transformation resulted from integration and expression of an oncogene derived from the human tumor.

Figure 15.23
Point mutations in *ras* oncogenes A single nucleotide change, which alters codon 12 from GGC (Gly) to GTC (Val), is responsible for the transforming activity of the *ras*H oncogene detected in bladder carcinoma DNA.

mutations that convert *ras* proto-oncogenes to oncogenes are caused by chemical carcinogens, providing a direct link between the mutagenic action of carcinogens and cell transformation.

As discussed in Chapter 13, the *ras* genes encode guanine-nucleotide binding proteins that function in transduction of mitogenic signals from a variety of growth factor receptors. The activity of the Ras proteins is controlled by GTP or GDP binding, such that they alternate between active (GTP-bound) and inactive (GDP-bound) states (see Figure 13.33). The mutations characteristic of *ras* oncogenes have the effect of maintaining the Ras proteins constitutively in the active GTP-bound conformation. In large part, this effect is a result of nullifying the response of oncogenic Ras proteins to GAP (GTPase-activating protein), which stimulates hydrolysis of

TABLE 15.4 Representative Oncogenes of Human Tumors

Oncogene	Type of cancer	Activation mechanism
abl	Chronic myelogenous leukemia, acute lymphocytic leukemia	Translocation
akt	Ovarian and pancreatic carcinomas	Amplification
bcl-2	Follicular B-cell lymphoma	Translocation
*D*1	Parathyroid adenoma, B-cell lymphoma	Translocation
*D*1	Squamous cell, bladder, breast, esophageal, liver, and lung carcinomas	Amplification
*E2A/pbx*1	Acute lymphocytic leukemia	Translocation
*erb*B-2	Breast and ovarian carcinomas	Amplification
gip	Adrenal cortical and ovarian carcinomas	Point mutation
gli	Glioblastoma	Amplification
gsp	Pituitary and thyroid tumors	Point mutation
hox-11	Acute T-cell leukemia	Translocation
lyl	Acute T-cell leukemia	Translocation
c-*myc*	Burkitt's lymphoma	Translocation
c-*myc*	Breast and lung carcinomas	Amplification
L-*myc*	Lung carcinoma	Amplification
N-*myc*	Neuroblastoma, lung carcinoma	Amplification
PDGFR	Chronic myelomonocytic leukemia	Translocation
PML/RARα	Acute promyelocytic leukemia	Translocation
*ras*H	Thyroid carcinoma	Point mutation
*ras*K	Colon, lung, pancreatic, and thyroid carcinomas	Point mutation
*ras*N	Acute myelogenous and lymphocytic leukemias, thyroid carcinoma	Point mutation
ret	Multiple endocrine neoplasia types 2A and 2B	Point mutation
ret	Thyroid carcinoma	DNA rearrangement
SMO	Basal cell carcinoma	Point mutation

bound GTP by normal Ras. Because of the resulting decrease in their intracellular GTPase activity, the oncogenic Ras proteins remain in the active
GTP-bound state and drive unregulated cell proliferation.

Point mutations are only one of the ways in which proto-oncogenes are
converted to oncogenes in human tumors. Many cancer cells display abnormalities in chromosome structure, including translocations, duplications,
and deletions. The gene rearrangements resulting from chromosome translocations frequently lead to the generation of oncogenes. In some cases,
analysis of these rearrangements has implicated already known oncogenes
in tumor development. In other cases, novel oncogenes have been discovered by molecular cloning and analysis of rearranged DNA sequences.

The first characterized example of oncogene activation by chromosome
translocation was the involvement of the **c-*myc*** oncogene in human Burkitt's
lymphomas and mouse plasmacytomas, which are malignancies of anti-
body-producing B lymphocytes (Figure 15.24). Both of these tumors are characterized by chromosome translocations involving the genes that encode
immunoglobulins. For example, virtually all Burkitt's lymphomas have
translocations of a fragment of chromosome 8 to one of the immunoglobulin
gene loci, which reside on chromosomes 2 (κ light chain), 14 (heavy chain),
and 22 (λ light chain). The fact that the immunoglobulin genes are actively
expressed in these tumors suggested that the translocations activate a proto-
oncogene from chromosome 8 by inserting it into the immunoglobulin loci.
This possibility was investigated by analysis of tumor DNAs with probes for
known oncogenes, leading to the finding that the c-*myc* proto-oncogene was
the chromosome 8 translocation break point in Burkitt's lymphomas. These
translocations inserted c-*myc* into an immunoglobulin locus, where it was
expressed in an unregulated manner. Such uncontrolled expression of the
c-*myc* gene, which encodes a transcription factor normally induced in
response to growth factor stimulation, is sufficient to drive cell proliferation
and contribute to tumor development.

Translocations of other proto-oncogenes frequently result in rearrangements of coding sequences, leading to the
formation of abnormal gene products.
The prototype is translocation of the ***abl***
proto-oncogene from chromosome 9 to
chromosome 22 in chronic myelogenous
leukemia (Figure 15.25). This translocation
leads to fusion of *abl* with its translocation
partner, a gene called *bcr*, on chromosome
22. The result is production of a Bcr/Abl
fusion protein in which the normal amino
terminus of the Abl proto-oncogene protein has been replaced by Bcr amino acid
sequences. The fusion of Bcr sequences
results in unregulated activity of the Abl
protein-tyrosine kinase, leading to cell
transformation.

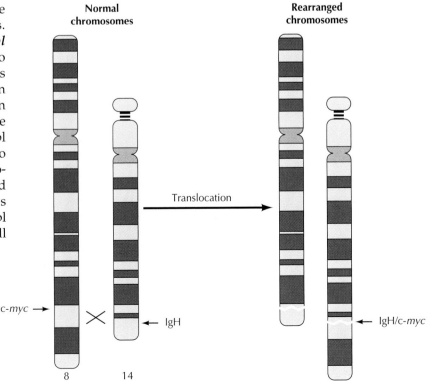

Normal chromosomes

Rearranged chromosomes

Translocation

c-*myc* →

← IgH

← IgH/c-*myc*

8   14

Figure 15.24
Translocation of c-*myc*   The c-*myc*
proto-oncogene is translocated from
chromosome 8 to the immunoglobulin
heavy-chain locus (IgH) on chromosome 14 in Burkitt's lymphomas, resulting in abnormal c-*myc* expression.

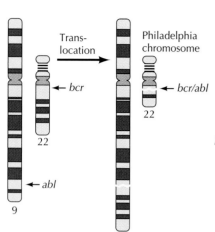

Figure 15.25
Translocation of *abl* The *abl* onco-gene is translocated from chromosome 9 to chromosome 22, forming the Philadelphia chromosome in chronic myelogenous leukemias. The *abl* proto-oncogene, which contains two alternative first exons (1A and 1B), is joined to the middle of the *bcr* gene on chromosome 22. Exon 1B is deleted as a result of the translocation. Transcription of the fused gene initiates at the *bcr* promoter and continues through *abl*. Splicing then generates a fused Bcr/Abl mRNA, in which *abl* exon 1A sequences are also deleted and *bcr* sequences are joined to *abl* exon 2. The Bcr/Abl mRNA is translated to yield a recombinant Bcr/Abl fusion protein.

A distinct mechanism by which oncogenes are activated in human tumors is gene amplification, which results in elevated gene expression. Gene amplification (see Figure 5.54) is common in tumor cells, occurring more than a thousand times more frequently than in normal cells, and amplification of oncogenes may play a role in the progression of many tumors to more rapid growth and increasing malignancy. Indeed, novel oncogenes have been identified by molecular cloning and characterization of DNA sequences that are amplified in tumors.

A prominent example of oncogene amplification is the involvement of the **N-*myc*** gene, which is related to c-*myc*, in neuroblastoma (a childhood tumor of embryonal neuronal cells). Amplified copies of N-*myc* are frequently present in rapidly growing aggressive tumors, indicating that N-*myc* amplification is associated with the progression of neuroblastomas to increasing malignancy. Amplification of another oncogene, ***erb*B-2**, which encodes a receptor protein-tyrosine kinase, is similarly related to progression of breast and ovarian carcinomas.

Functions of Oncogene Products

The viral and cellular oncogenes have defined a large group of genes (about 100 in total) that can contribute to the abnormal behavior of malignant cells. As already noted, many of the proteins encoded by proto-oncogenes regulate normal cell proliferation; in these cases, the elevated expression or activity of the corresponding oncogene proteins drives the uncontrolled proliferation of cancer cells. Other oncogene products contribute to other aspects of the behavior of cancer cells, such as defective differentiation and failure to undergo programmed cell death.

The majority of oncogene proteins function as elements of the signaling pathways that regulate cell proliferation and survival in response to growth factor stimulation. These oncogene proteins include polypeptide growth factors, growth factor receptors, elements of intracellular signaling pathways, and transcription factors (Figure 15.26).

The action of growth factors as oncogene proteins results from their abnormal expression, leading to a situation in which a tumor cell produces a growth factor to which it also responds. The result is autocrine stimulation of the growth factor–producing cell (see Figure 15.9), which drives

Figure 15.26
Oncogenes and signal transduction Oncogene proteins act as growth factors (e.g., EGF), growth factor receptors (e.g., ErbB), and intracellular signaling molecules (Ras and Raf). Ras and Raf activate the ERK MAP kinase pathway (see Figures 13.32 and 13.35), leading to the induction of additional genes (e.g., *fos*) that encode potentially oncogenic transcriptional regulatory proteins. Proteins with known oncogenic potential are highlighted with a yellow glow.

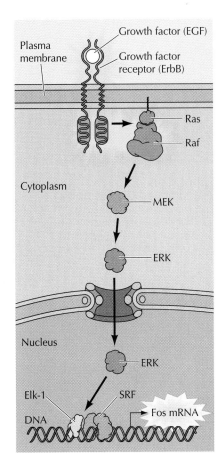

abnormal cell proliferation and contributes to the development of a wide variety of human tumors.

A large group of oncogenes encode growth factor receptors, most of which are protein-tyrosine kinases. These receptors are frequently converted to oncogene proteins by alterations of their amino-terminal domains, which would normally bind extracellular growth factors. For example, the receptor for platelet-derived growth factor (PDGF) is converted to an oncogene in some human leukemias by a chromosome translocation in which the normal amino terminus of the PDGF receptor is replaced by the amino terminal sequences of a transcription factor called Tel (Figure 15.27). The Tel sequences of the resulting Tel/PDGFR fusion protein dimerize in the absence of growth factor binding, resulting in constitutive activity of the intracellular kinase domain and unregulated production of a proliferative signal from the oncogene protein. Alternatively, genes that encode some receptor protein-tyrosine kinases, such as *erb*B-2, are activated by gene amplification. Other oncogenes (including *src* and *abl*) encode nonreceptor protein-tyrosine kinases that are constitutively activated by deletions or mutations of regulatory sequences.

Figure 15.27
Mechanism of *Tel/PDGFR* oncogene activation The normal PDGF receptor (PDGFR) is activated by dimerization induced by PDGF binding. The *Tel/PDGFR* oncogene encodes a fusion protein in which the normal extracellular domain of the PDGF receptor is replaced by the amino terminal sequences of the Tel transcription factor, which include its helix-loop-helix dimerization domain (see Figure 6.27). These sequences dimerize in the absence of PDGF, leading to constitutive activation of the oncogene protein kinase.

Figure 15.28
The AP-1 transcription factor Fos and Jun dimerize to form AP-1, which activates transcription of a variety of growth factor–inducible genes.

The Ras proteins play a key role in mitogenic signaling by coupling growth factor receptors to activation of the Raf protein-serine/threonine kinase, which initiates a protein kinase cascade leading to activation of the ERK MAP kinase (see Figure 13.32). As discussed earlier, the mutations that convert *ras* proto-oncogenes to oncogenes result in constitutive Ras activity, which leads to activation of the MAP kinase pathway. The *raf* gene can similarly be converted to an oncogene by deletions that result in loss of the amino-terminal regulatory domain of the Raf protein (see Figure 15.21). These deletions result in unregulated activity of the Raf protein kinase, which also leads to constitutive MAP kinase activation.

The MAP kinase pathway ultimately leads to the phosphorylation of transcription factors and alterations in gene expression. As might therefore be expected, many oncogenes encode transcriptional regulatory proteins that are normally induced in response to growth factor stimulation. For example, transcription of the *fos* proto-oncogene is induced as a result of phosphorylation of Elk-1 by the ERK MAP kinase (see Figure 15.26). **Fos** and the product of another proto-oncogene, **Jun**, are components of the AP-1 transcription factor, which activates transcription of a number of target genes in growth factor–stimulated cells (Figure 15.28). Constitutive activity of AP-1, resulting from unregulated expression of either the Fos or Jun oncogene proteins, is sufficient to drive abnormal cell proliferation, leading to cell transformation. The Myc proteins similarly function as transcription factors regulated by mitogenic stimuli, and abnormal expression of *myc* oncogenes contributes to the development of a variety of human tumors. Other transcription factors are frequently activated as oncogenes by chromosome translocations in human leukemias and lymphomas.

G protein-coupled receptors and G proteins also act as oncogenes in some human tumors (Figure 15.29). For example, mutations of the gene encoding the thyrotropin receptor convert it to an oncogene in thyroid tumors. Thyrotropin is a pituitary hormone that stimulates proliferation of thyroid cells through a G protein–coupled receptor that activates adenylyl cyclase. Mutations of the thyrotropin receptor in thyroid tumors result in constitutive activity of the receptor, which then drives cell proliferation via activation of the cAMP signaling pathway. Likewise, the genes encoding G proteins can act as oncogenes in some cell types. The *gsp* oncogene, which encodes the α subunit of G_s, is generated by point mutations similar to those found in *ras*. These mutations result in constitutive activity of the G_s α subunit, leading to unregulated stimulation of adenylyl cyclase. As might be expected, the *gsp* oncogene is involved in thyroid and pituitary tumors, where cAMP stimulates cell proliferation. Similar mutations con-

Figure 15.29
Oncogenic activity of G protein–coupled receptors and G proteins The thyrotropin receptor is coupled to adenylyl cyclase by G_s. The genes encoding both the receptor and the G_s α subunit ($G_s\alpha$) can act as oncogenes by stimulating thyroid cell proliferation.

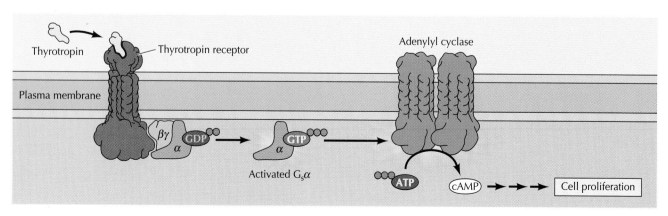

vert the genes encoding other G protein α subunits to oncogenes in other cell types, including adrenal and ovarian tumors.

The intracellular signaling pathways activated by growth factor stimulation ultimately regulate components of the cell cycle machinery that promote progression through the restriction point in G_1. The D-type cyclins are induced in response to growth factor stimulation and play a key role in coupling growth factor signaling to cell cycle progression (see Figure 14.19). Perhaps not surprisingly, the gene encoding cyclin D1 is a proto-oncogene, which can be activated as an oncogene (called **D1**) by chromosome translocation or gene amplification. These alterations lead to constitutive expression of cyclin D1, which then drives cell proliferation in the absence of normal growth factor stimulation.

Although most oncogenes stimulate cell proliferation, the oncogenic activity of some transcription factors instead results from inhibition of cell differentiation. As noted in Chapter 13, thyroid hormone and retinoic acid induce differentiation of a variety of cell types. These hormones diffuse through the plasma membrane and bind to intracellular receptors that act as transcriptional regulatory molecules. Mutated forms of both the thyroid hormone receptor (**ErbA**) and the retinoic acid receptor (**PML/RARα**) act as oncogene proteins in chicken erythroleukemia and human acute promyelocytic leukemia, respectively. In both cases, the mutated oncogene receptors appear to interfere with the action of their normal homologs, thereby blocking cell differentiation and maintaining the leukemic cells in an actively proliferating state (Figure 15.30). In the case of acute promyelocytic leukemia, high doses of retinoic acid can overcome the effect of the PML/RARα oncogene protein and induce differentiation of the leukemic

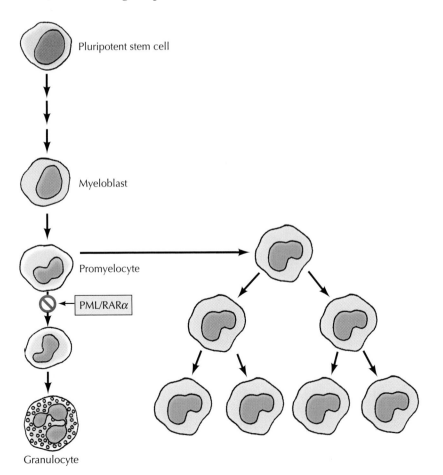

Figure 15.30
Action of the PML/RARα oncogene protein The PML/RARα fusion protein blocks the differentiation of promyelocytes to granulocytes.

cells. This biological observation has a direct clinical correlate: Patients with acute promyelocytic leukemia can be treated effectively by administration of retinoic acid, which induces differentiation and blocks continued cell proliferation.

As discussed earlier in this chapter, the failure of cancer cells to undergo programmed cell death, or apoptosis, is a critical factor in tumor development, and several oncogenes encode proteins that act to promote cell survival (Figure 15.31). The survival of most animal cells is dependent on growth factor stimulation, so those oncogenes that encode growth factors, growth factor receptors, and signaling proteins such as Ras act not only to stimulate cell proliferation, but also to prevent cell death. As discussed in Chapter 13, the PI 3-kinase/Akt signaling pathway plays a key role in preventing apoptosis of many growth factor–dependent cells, and the genes encoding **PI 3-kinase** and **Akt** act as oncogenes in retroviruses and some human tumors. The downstream targets of PI 3-kinase/Akt signaling include a member of the Bcl-2 family (Bad), and it is notable that **Bcl-2** was first discovered as an oncogene in human lymphomas. The *bcl*-2 oncogene is generated by a chromosome translocation that results in elevated expression of Bcl-2, which blocks apoptosis and maintains cell survival under conditions that normally induce cell death. The identification of *bcl*-2 as an

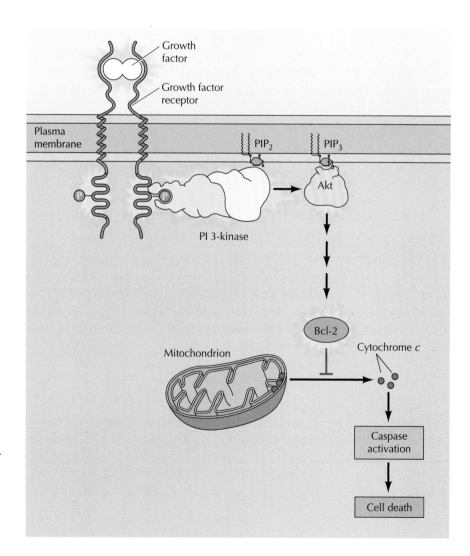

Figure 15.31
Oncogenes and cell survival The oncogene proteins that signal cell survival include growth factors, growth factor receptors, PI 3-kinase, and Akt. Signaling by the PI 3-kinase/Akt pathway regulates members of the Bcl-2 family, which promote cell survival by inhibiting the release of cytochrome *c* from mitochondria. Proteins with oncogenic potential are highlighted with a yellow glow.

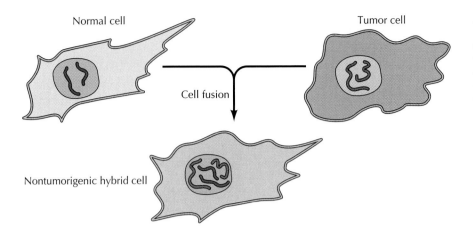

Normal cell

Tumor cell

Cell fusion

Nontumorigenic hybrid cell

Figure 15.32
Suppression of tumorigenicity by cell fusion Fusion of tumor cells with normal cells yields hybrids that contain chromosomes from both parents. Such hybrids are usually non-tumorigenic.

oncogene not only provided the first demonstration of the significance of programmed cell death in the development of cancer, but also led to the discovery of the role of Bcl-2 and related genes as central regulators of apoptosis in organisms ranging from *C. elegans* to humans.

Tumor Suppressor Genes

The activation of cellular oncogenes represents only one of two distinct types of genetic alterations involved in tumor development; the other is inactivation of tumor suppressor genes. Oncogenes drive abnormal cell proliferation as a consequence of genetic alterations that either increase gene expression or lead to uncontrolled activity of the oncogene-encoded proteins. **Tumor suppressor genes** represent the opposite side of cell growth control, normally acting to inhibit cell proliferation and tumor development. In many tumors, these genes are lost or inactivated, thereby removing negative regulators of cell proliferation and contributing to the abnormal proliferation of tumor cells.

Identification of Tumor Suppressor Genes

The first insight into the activity of tumor suppressor genes came from somatic cell hybridization experiments, initiated by Henry Harris and his colleagues in 1969. The fusion of normal cells with tumor cells yielded hybrid cells containing chromosomes from both parents (Figure 15.32). In most cases, such hybrid cells were not capable of forming tumors in animals. Therefore, it appeared that genes derived from the normal cell parent acted to inhibit (or suppress) tumor development. Definition of these genes at the molecular level came, however, from a different approach—the analysis of rare inherited forms of human cancer.

The first tumor suppressor gene was identified by studies of retinoblastoma, a rare childhood eye tumor. Provided that the disease is detected early, retinoblastoma can be successfully treated, and many patients survive to have families. Consequently, it was recognized that some cases of retinoblastoma are inherited. In these cases, approximately 50% of the children of an affected parent develop retinoblastoma, consistent with Mendelian transmission of a single dominant gene that confers susceptibility to tumor development (Figure 15.33).

Although susceptibility to retinoblastoma is transmitted as a dominant trait, inheritance of the susceptibility gene is not sufficient to transform a normal retinal cell into a tumor cell. All retinal cells in a patient inherit the

Figure 15.33
Inheritance of retinoblastoma Susceptibility to retinoblastoma is transmitted to approximately 50% of offspring. Affected and normal individuals are indicated by purple and green symbols, respectively.

Female   Male

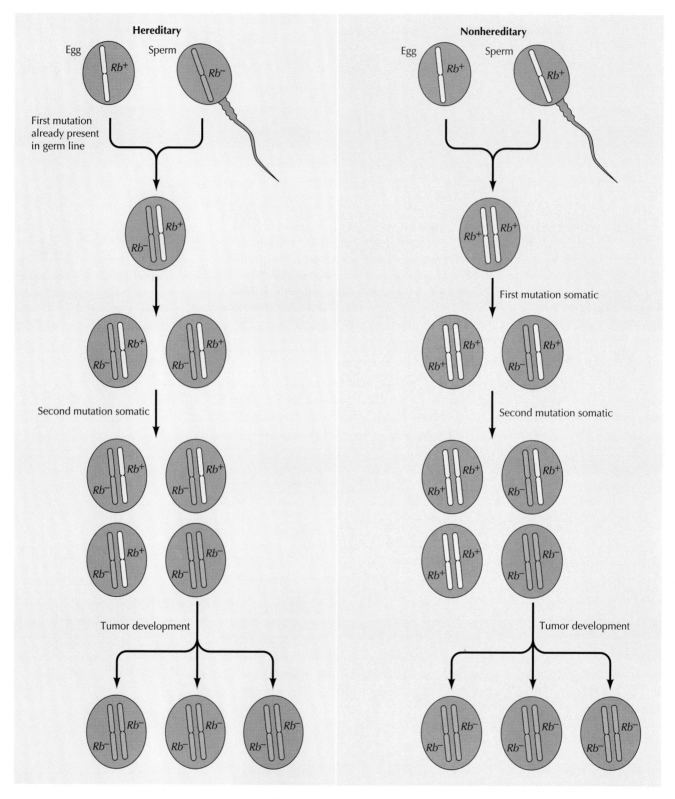

Figure 15.34
Mutations of *Rb* during retinoblastoma development In hereditary retinoblastoma, a defective copy of the *Rb* gene (*Rb⁻*) is inherited from the affected parent. A second somatic mutation, which inactivates the single normal *Rb⁺* copy in a retinal cell, then leads to the development of retinoblastoma. In nonhereditary cases, two normal *Rb⁺* genes are inherited, and retinoblastoma develops only if two somatic mutations inactivate both copies of *Rb* in the same cell.

susceptibility gene, but only a small fraction of these cells give rise to tumors. Thus, tumor development requires additional events beyond inheritance of tumor susceptibility. In 1971, Alfred Knudson proposed that the development of retinoblastoma requires two mutations, which are now known to correspond to the loss of both of the functional copies of the tumor susceptibility gene (the **Rb** tumor suppressor gene) that would be present on homologous chromosomes of a normal diploid cell (Figure 15.34). In inherited retinoblastoma, one defective copy of *Rb* is genetically transmitted. The loss of this single *Rb* copy is not by itself sufficient to trigger tumor development, but retinoblastoma almost always develops in these individuals as a result of a second somatic mutation leading to the loss of the remaining normal *Rb* allele. Noninherited retinoblastoma, in contrast, is rare, since its development requires two independent somatic mutations to inactivate both normal copies of *Rb* in the same cell.

The functional nature of the *Rb* gene as a negative regulator of tumorigenesis was initially indicated by observations of chromosome morphology. Visible deletions of chromosome 13q14 were found in some retinoblastomas, suggesting that loss (rather than activation) of the *Rb* gene led to tumor development (Figure 15.35). Gene-mapping studies further indicated that tumor development resulted from loss of normal *Rb* alleles in the tumor cells, consistent with the function of *Rb* as a tumor suppressor gene. Isolation of the *Rb* gene as a molecular clone in 1986 then firmly established that *Rb* is consistently lost or mutated in retinoblastomas. Gene transfer experiments also demonstrated that introduction of a normal *Rb* gene into retinoblastoma cells reverses their tumorigenicity, providing direct evidence for the activity of *Rb* as a tumor suppressor.

Although *Rb* was identified in a rare childhood cancer, it is also involved in some of the more common tumors of adults. In particular, studies of the cloned gene have established that *Rb* is lost or inactivated in many bladder, breast, and lung carcinomas. The significance of the *Rb* tumor suppressor gene thus extends beyond retinoblastoma, apparently contributing to development of a substantial fraction of more common human cancers. In addition, as noted earlier in this chapter, the Rb protein is a key target for the oncogene proteins of several DNA tumor viruses, including SV40, adenoviruses, and human papillomaviruses, which bind to Rb and inhibit its activity (Figure 15.36). Transformation by these viruses thus results, at least in part, from inactivation of Rb at the protein level rather than from mutational inactivation of the *Rb* gene.

Characterization of *Rb* as a tumor suppressor gene served as the prototype for the identification of additional tumor suppressor genes that contribute to the development of many different human cancers (Table 15.5). Some of these genes were identified as the causes of rare inherited cancers, playing a role similar to that of *Rb* in hereditary retinoblastoma. Other tumor suppressor genes have been identified as genes that are frequently deleted or mutated in common noninherited cancers of adults, such as colon carcinoma. In either case, it appears that most tumor suppressor genes are involved in the development of both inherited and noninherited forms of cancer. Indeed, mutations of some tumor suppressor genes appear to be the most common molecular alterations leading to human tumor development.

The second tumor suppressor gene to have been identified is *p53*, which is frequently inactivated in a wide variety of human cancers, including leukemias, lymphomas, sarcomas, brain tumors, and carcinomas of many tissues, including breast, colon, and lung. In total, mutations of *p53* may play a role in up to 50% of all cancers, making it the most common target of genetic alterations in human malignancies. It is also of interest that inherited

Figure 15.35
Rb deletions in retinoblastoma
Many retinoblastomas have deletions of the chromosomal locus (13q14) that contains the *Rb* gene.

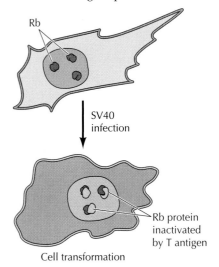

Figure 15.36
Interaction of Rb with oncogene proteins of DNA tumor viruses The oncogene proteins of several DNA tumor viruses (e.g., SV40 T antigen) induce transformation by binding to and inactivating Rb protein.

TABLE 15.5 Tumor Suppressor Genes

Gene	Type of cancer
APC	Colon/rectum carcinoma
BRCA1	Breast and ovarian carcinomas
BRCA2	Breast carcinoma
DPC4	Pancreatic carcinoma
INK4	Melanoma, lung carcinoma, brain tumors, leukemias, lymphomas
MADR2	Colon/rectum carcinoma
NF1	Neurofibrosarcoma
NF2	Meningioma
p53	Brain tumors; breast, colon/rectum, esophageal, liver, and lung carcinomas; sarcomas; leukemias and lymphomas
PTC	Basal cell carcinoma
PTEN	Brain tumors; melanoma; prostate, endometrial, kidney, and lung carcinomas
Rb	Retinoblastoma; sarcomas; bladder, breast, and lung carcinomas
VHL	Renal cell carcinoma
WT1	Wilms' tumor

mutations of *p53* are responsible for genetic transmission of a rare hereditary cancer syndrome, in which affected individuals develop any of several different types of cancer. In addition, the p53 protein (like Rb) is a target for the oncogene proteins of SV40, adenoviruses, and human papillomaviruses.

Like *p53*, the *INK4* and *PTEN* tumor suppressor genes are very frequently mutated in several common cancers, including lung cancer, prostate cancer, and melanoma. Two other tumor suppressor genes (*APC* and *MADR2*) are frequently deleted or mutated in colon cancers. In addition to being involved in noninherited cases of this common adult cancer, inherited mutations of the *APC* gene are responsible for a rare hereditary form of colon cancer, called familial adenomatous polyposis. Individuals with this condition develop hundreds of benign colon adenomas (polyps), some of which almost inevitably progress to malignancy. Inherited mutations of two other tumor suppressor genes, *BRCA1* and *BRCA2*, are responsible for hereditary cases of breast cancer, which account for 5 to 10% of the total breast cancer incidence.

Additional tumor suppressor genes have been implicated in the development of brain tumors, pancreatic cancers, and basal cell skin carcinomas, as well as in several rare inherited cancer syndromes, such as Wilms' tumor. The number of identified tumor suppressor genes is rapidly expanding, and the characterization of these genes remains an active area of cancer research.

Functions of Tumor Suppressor Gene Products

In contrast to proto-oncogene and oncogene proteins, the proteins encoded by most tumor suppressor genes inhibit cell proliferation or survival. Inactivation of tumor suppressor genes therefore leads to tumor development by eliminating negative regulatory proteins. In several cases, tumor suppressor proteins inhibit the same cell regulatory pathways that are stimulated by the products of oncogenes.

The protein encoded by the **PTEN** tumor suppressor gene is an interesting example of antagonism between oncogene and tumor suppressor gene products (Figure 15.37). The PTEN protein is a lipid phosphatase that dephosphorylates the 3 position of phosphatidylinositides, such as phosphatidylinositol 3,4,5-bisphosphate (PIP_3). By dephosphorylating PIP_3, PTEN antagonizes the activities of PI 3-kinase and Akt, which can act as oncogenes by promoting cell survival. Conversely, inactivation or loss of the PTEN tumor suppressor protein can contribute to tumor development as a result of increased levels of PIP_3, activation of Akt, and inhibition of programmed cell death.

Proteins encoded by both oncogenes and tumor suppressor genes also function in the Hedgehog signaling pathway (see Figure 13.44). The receptor Smoothened is an oncogene in basal cell carcinomas, whereas Patched (the negative regulator of Smoothened) is a tumor suppressor gene. In addition, the Gli proteins (the mammalian homologs of the *Drosophila* Ci transcription factor activated by Smoothened) were first identified as the products of an amplified oncogene.

Several tumor suppressor genes encode transcriptional regulatory proteins. A good example is provided by the product of *WT1*, which is frequently inactivated in Wilms' tumors (a childhood kidney tumor). The

Inner leaflet of plasma membrane

PIP$_2$

PI 3-Kinase

PTEN

PIP$_3$

Akt

Cell survival

Figure 15.37
Suppression of cell survival by PTEN The tumor suppressor protein PTEN is a lipid phosphatase that dephosphorylates PIP$_3$ at the 3 position of inositol, yielding PIP$_2$. PTEN thus counters the action of the oncogenes PI 3-kinase and Akt, which promote cell survival.

WT1 protein is a repressor that appears to suppress transcription of a number of growth factor–inducible genes. One of the targets of WT1 is thought to be the gene that encodes insulin-like growth factor II, which is overexpressed in Wilms' tumors and may contribute to tumor development by acting as an autocrine growth factor. Inactivation of WT1 may thus lead to abnormal growth factor expression, which in turn drives tumor cell proliferation. Two other tumor suppressor genes, *DPC4* and *MADR2*, encode SMAD family transcription factors that are activated by TGF-β signaling and lead to inhibition of cell proliferation.

The products of the *Rb* and *INK4* tumor suppressor genes regulate cell cycle progression at the same point as that affected by cyclin D1 (Figure 15.38). Rb inhibits passage through the restriction point in G$_1$ by repressing transcription of a number of genes involved in cell cycle progression and DNA synthesis (see Figure 14.20). In normal cells, passage through the restriction point is regulated by Cdk4/cyclin D

Figure 15.38
Inhibition of cell cycle progression by Rb and p16 Rb inhibits progression past the restriction point in G$_1$. Cdk4/cyclin D complexes promote passage through the restriction point by phosphorylating and inactivating Rb. The activity of Cdk4/cyclin D is inhibited by p16.

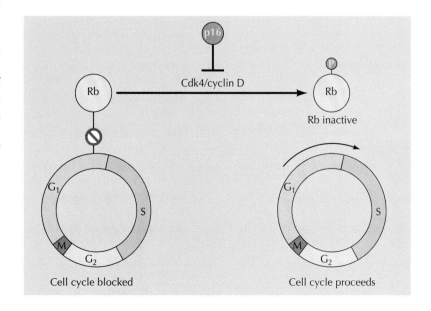

p16

Rb

Cdk4/cyclin D

Rb

Rb inactive

G$_1$

S

G$_2$

M

Cell cycle blocked

G$_1$

S

G$_2$

M

Cell cycle proceeds

Molecular Medicine

Herceptin: Cancer Treatment Targeted against the erbB-2 Oncogene

The Disease

Breast cancer is the most common cancer among women, accounting for nearly one-third of all cancers diagnosed in women in the United States. In the year 2000, it is estimated that more than 180,000 United States women will be diagnosed with breast cancer and more than 40,000 will die of this disease. Although men can also develop breast cancer, it is 100-fold less frequent in men than in women.

The incidence of breast cancer increases with age, and more than 75% of breast cancers are diagnosed in women over 50. Other factors that increase the risk of breast cancer include a family history of the disease and exposure to hormones that stimulate the proliferation of breast epithelial cells. Early menarche (before age 12), late menopause (after age 55), and a late age for first childbirth (after 30) are all associated with an increased risk of breast cancer as the result of effects on reproductive hormones. The risk of breast cancer is also increased by recent use of oral contraceptives

and long-term postmenopausal estrogen replacement therapy.

Early detection is critically important in reducing breast cancer mortality, and screening for breast cancer by self-examination and mammography is recommended by the American Cancer Society and the National Cancer Institute. It is estimated that annual screening by mammography for women over age 40 reduces breast cancer mortality by about 30%. The benefits of early detection are clearly apparent in survival rates for women diagnosed with breast cancer. The cure rate for localized breast cancer is approximately 90%, but this drops to about 70% for cancers that have spread to regional lymph nodes and to less than 20% for cancers that have metastasized to distant body sites.

Molecular and Cellular Basis

Breast cancers develop as a result of mutations in a variety of oncogenes and tumor suppressor genes. Approximately 5 to 10% of breast cancers are inherited as a result of mutations in the *BRCA1* and *BRCA2* tumor sup-

pressor genes. Mutations in *BRCA1* and *BRCA2* are not generally found in noninherited breast cancers, which may instead involve mutations in several other oncogenes and tumor suppressor genes, including *erb*B-2, *c-myc*, *D*1, *p53*, and *Rb*.

The *erb*B-2 gene, which encodes a receptor protein-tyrosine kinase, is amplified and expressed at high levels in about 30% of breast cancers. Elevated expression of *erb*B-2 is characteristic of rapidly progressing

Light micrograph of human breast cancer cells. (Jonathan Ashton/Science Photo Library/Photo Researchers, Inc.)

Figure 15.39
Action of p53 Wild-type p53 is required for both cell cycle arrest and apoptosis induced by DNA damage.

complexes, which phosphorylate and inactivate Rb. Mutational inactivation of *Rb* in tumors thus removes a key negative regulator of cell cycle progression. The *INK4* tumor suppressor gene, which encodes the Cdk inhibitor p16, also regulates passage through the restriction point. As discussed in Chapter 14, p16 inhibits Cdk4/cyclin D activity. Inactivation of *INK4* therefore leads to elevated activity of Cdk4/cyclin D complexes, resulting in uncontrolled phosphorylation of Rb.

The *p53* gene product regulates both cell cycle progression and apoptosis (Figure 15.39). DNA damage leads to rapid induction of p53, which activates transcription of the Cdk inhibitor p21 (see Figure 14.21). The inhibitor p21 blocks cell cycle progression, both by acting as a general inhibitor of Cdk/cyclin complexes and by inhibiting DNA replication by binding to PCNA (proliferating cell nuclear antigen). The resulting cell cycle arrest presumably allows time for damaged DNA to be repaired before it is replicated. Loss of p53 prevents this damage-induced cell cycle arrest, leading to increased mutation frequencies and a general instability of the cell genome. Such genetic instability is a common property of cancer cells, and

Herceptin: Cancer Treatment Targeted against the **erbB-2** *Oncogene* (continued)

metastatic cancers with a poor prognosis. Perhaps ironically, then, the ErbB-2 protein has served as the first oncogene target against which a new anticancer drug has been approved for clinical use by the FDA.

A mammogram showing a breast tumor. (Courtesy of Cooley-Dickenson Hospital, Mammography Department.)

Prevention and Treatment

Like other cancers, breast cancer is generally treated by surgery, radiation, and chemotherapy with cytotoxic drugs that damage DNA or act as general inhibitors of cell division. In addition, some breast cancers respond to hormone therapy, which blocks the growth-promoting effects of estrogens on the cancer cells. Herceptin, a monoclonal antibody against the ErbB-2 oncogene protein, has now been added to the arsenal of drugs used to treat breast cancer, specifically for those women whose cancers overexpress ErbB-2.

The development of Herceptin was initiated because it was found that the growth of tumor cells expressing high levels of ErbB-2 could be inhibited by an antibody that recognized its extracellular domain. The antibody was first tested in a small clinical trial in 46 women with advanced metastatic breast cancer, with encouraging results. Herceptin was then evaluated in two large clinical trials, involving 222 and 469 patients. The results of these trials clearly demonstrated that Herceptin was of benefit to patients whose tumors overexpressed ErbB-2. Either when used alone or in combination with conventional chemotherapeutic drugs, Herceptin significantly reduced tumor growth and prolonged the survival of some patients.

Based on these results, the FDA approved Herceptin for the treatment of metastatic breast cancer in 1998. It is used specifically for those patients whose tumors express elevated levels of the ErbB-2 oncogene protein, and may be used either alone or in combination with other anticancer drugs. Although far from curative, Herceptin is the first drug targeted specifically against an oncogene protein to have achieved FDA approval for clinical use. The successful development of Herceptin thus represents a major milestone in the design of rational therapies based on our understanding of the molecular biology of cancer.

Reference

Lebwohl, D. E. and R. Canetta. 1999. New developments in chemotherapy of advanced breast cancer. *Ann. Oncol.* 10: 139–146.

it may contribute to further alterations in oncogenes and tumor suppressor genes during tumor progression.

In addition to mediating cell cycle arrest, p53 is required for apoptosis induced by DNA damage. Unrepaired DNA damage normally induces apoptosis of mammalian cells, a response that is presumably advantageous to the organism because it eliminates cells carrying potentially deleterious mutations (e.g., cells that might develop into cancer cells). Cells lacking p53 fail to undergo apoptosis in response to agents that damage DNA, including radiation and many of the drugs used in cancer chemotherapy. This failure to undergo apoptosis in response to DNA damage contributes to the resistance of many tumors to chemotherapy. In addition, loss of p53 appears to interfere with apoptosis induced by other stimuli, such as growth factor deprivation and oxygen deprivation. These effects of p53 inactivation on cell survival are thought to account for the high frequency of *p53* mutations in human tumors.

Roles of Oncogenes and Tumor Suppressor Genes in Tumor Development

As discussed earlier, the development of cancer is a multistep process in which normal cells gradually progress to malignancy. The complete sequence of events required for the development of any human cancer is not yet known, but it is clear that both the activation of oncogenes and the inactivation of tumor suppressor genes are critical steps in tumor initiation

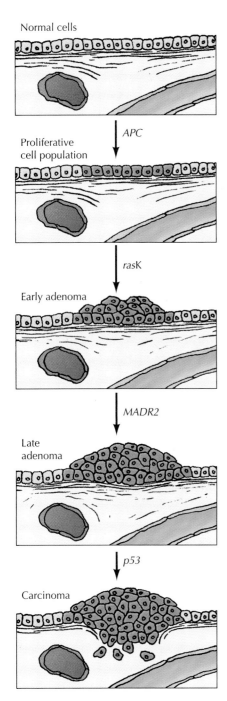

Normal cells

APC

Proliferative cell population

rasK

Early adenoma

MADR2

Late adenoma

p53

Carcinoma

Figure 15.40

Genetic alterations in colon carcinomas Inactivation of *APC* is an early event in tumor development, giving rise to a proliferative cell population. Mutations of *ras*K then frequently occur and are found in early-stage adenomas. Subsequent mutations of *MADR2* and *p53* are associated with later stages of tumor progression.

and progression. Accumulated damage to multiple genes eventually results in the increased proliferation, invasiveness, and metastatic potential that are characteristic of cancer cells.

The role of multiple genetic defects is best understood in the case of colon carcinomas, which have been studied extensively by Bert Vogelstein and his colleagues. These tumors frequently involve mutation of *ras*K oncogenes and inactivation or deletion of three distinct tumor suppressor genes—*APC*, *MADR2*, and *p53*. Lesions representing multiple stages of colon cancer development are regularly obtained as surgical specimens, so it has been possible to correlate these genetic alterations with discrete stages of tumor progression (Figure 15.40).

These studies indicate that inactivation of *APC* is an early event in tumor development. Genetic transmission of mutant *APC* genes in patients with familial adenomatous polyposis results in abnormal colon cell proliferation, leading to the outgrowth of multiple adenomas in the colons of affected patients. Mutations of *APC* also occur frequently in patients with noninherited colon carcinomas and are generally detected at early stages of the disease process. Mutations of *ras*K genes then appear to occur, and *ras*K oncogenes are also frequently present in small and intermediate-size adenomas. In contrast, the *MADR2* and *p53* tumor suppressor genes are inactivated usually at later stages of tumor progression. Mutations in these genes are only rarely found in early-stage adenomas, but they are frequently present in advanced adenomas and malignant carcinomas. The loss of these tumor suppressor genes in colon cancers thus appears to be involved in later stages of progression to malignancy, rather than in the initial stages of tumor formation.

Although these genetic alterations most often occur in the order described here, this is not an obligatory sequence of events. For example, mutations in *p53* are sometimes detected in early adenomas. More important than the order in which mutations occur is the fact that colon cancer ultimately results from accumulated damage to multiple genes. Accumulated damage to both oncogenes and tumor suppressor genes similarly appears to be responsible for the development of other types of cancer, including breast and lung carcinomas. The progressive loss of growth control that is characteristic of cancer cells is thus thought to be the end result of abnormalities in the products of multiple genes that normally regulate cell proliferation, differentiation, and survival.

Applications of Molecular Biology to Cancer Prevention and Treatment

A great deal has been learned about the molecular defects responsible for the development of many human cancers. However, cancer is more than a topic of scientific interest. It is a dread disease that claims the lives of nearly one out of every four Americans. Translating our growing understanding of cancer into practical improvements in cancer prevention and treatment therefore represents a major challenge for current and future research. Fortunately, advances in elucidating the molecular biology of cancer are beginning to contribute to the development of new approaches to its prevention and treatment, which may ultimately yield major advances in dealing with this disease.

Prevention and Early Detection

The most effective way to deal with cancer would be to prevent development of the disease. A second-best, but still effective, alternative would be to reliably detect early premalignant stages of tumor development that can be easily treated. Many cancers can be cured by localized treatments, such as surgery or radiation, if they are detected before they spread through the body by metastasis. For example, early premalignant stages of colon cancer (adenomas) are usually completely curable by relatively minor surgical procedures (Figure 15.41). The cure rate for early carcinomas that remain localized to their site of origin is also high, about 90%. However, survival rates drop to about 50% for patients whose cancers have spread to adjacent tissues and lymph nodes, and to less than 10% for patients with metastatic colon cancer. Early detection of cancer can thus be a critical determinant of the outcome of the disease.

The major application of molecular biology to prevention and early detection may lie in the identification of individuals with inherited susceptibilities to cancer development. Such inherited cancer susceptibilities can result from mutations in tumor suppressor genes, in at least one oncogene (*ret*), and in DNA repair genes, such as the mismatch repair genes responsible for development of hereditary nonpolyposis colon cancer (see Molecular Medicine in Chapter 5). Mutations in these genes can be detected by genetic testing, allowing the identification of high-risk individuals before disease develops.

In addition to contributing to family planning decisions, careful monitoring of high-risk individuals may allow early detection and more effective treatment of some types of cancer. For example, colon adenomas can be detected by colonoscopy and removed prior to the development of malignancy. Patients with familial adenomatous polyposis (resulting from inherited mutations of the *APC* tumor suppressor gene) typically develop hundreds of adenomas within the first 20 years of life, so the colons of these patients are usually removed before the inevitable progression of some of these polyps to malignancy. However, patients with hereditary nonpolyposis colon cancer develop a smaller number of polyps later in life and may therefore benefit from routine colonoscopy and drugs, such as nonsteroidal antiinflammatory drugs (NSAIDs), that inhibit colon cancer development.

The direct inheritance of cancers resulting from defects in known genes is a rare event, constituting about 5% of total cancer incidence. The most common inherited cancer susceptibility is hereditary nonpolyposis colon cancer, which accounts for about 15% of colon cancers and 1 to 2% of all cancers in the United States. Mutations in the *BRCA1* and *BRCA2* tumor suppressor genes are also relatively common, accounting for 5 to 10% of all breast cancers. However, still-unidentified genes conferring cancer susceptibility may contribute to the development of a larger fraction of common adult malignancies, such as breast, colon, and lung cancers. The continuing isolation of cancer susceptibility genes is thus an important undertaking with clear practical implications. Individuals with such inherited susceptibility genes could be appropriately advised to avoid exposure to relevant carcinogens (e.g., tobacco smoke in the case of lung cancer) and carefully monitored to detect tumors at early stages that are more readily treated. The reliable identification of susceptible individuals, if it were followed by appropriate preventive and early detection measures, might ultimately make a significant impact on cancer mortality.

Figure 15.41
Survival rates of patients with colon carcinoma   Five-year survival rates are shown for patients diagnosed with adenomas (polyps), with carcinoma still localized to its site of origin, with carcinoma that has spread regionally to adjacent tissues and lymph nodes, and with metastatic carcinoma.

Molecular Diagnosis

Molecular analysis of the oncogenes and tumor suppressor genes involved in particular types of tumors has the potential of providing information that is useful in the diagnosis of cancer and in monitoring the effects of treatment. Indeed, several applications of such molecular diagnosis are already being put into clinical practice. In some cases, mutations in oncogenes have provided useful molecular markers for monitoring the course of disease during treatment. The translocation of *abl* in chronic myelogenous leukemia is a good example. As previously discussed, this translocation results in the fusion of *abl* with the *bcr* gene, leading to expression of the Bcr/Abl oncogene protein (see Figure 15.25). The polymerase chain reaction (see Figure 3.27) provides a sensitive method of detecting the recombinant *bcr/abl* oncogene in leukemic cells and is therefore used to monitor the response of patients to treatment.

In other cases, the detection of mutations in specific oncogenes or tumor suppressor genes may provide information pertinent to choosing between different therapeutic options. For example, amplification of N-*myc* in neuroblastomas and *erb*B-2 in breast and ovarian carcinomas predicts rapid disease progression. Therefore, it might be appropriate to treat patients with such amplified oncogenes more aggressively. Because of the role of p53 in apoptosis induced by DNA damage, analysis of *p53* mutations may help predict the response of tumors to radiation and chemotherapy. As further information is accumulated on the roles of specific genetic abnormalities in determining tumor behavior, it is likely that molecular diagnosis will become increasingly important in dealing with cancer.

Treatment

The most critical question, however, is whether the discovery of oncogenes and tumor suppressor genes will allow the development of new drugs that act selectively against cancer cells. Most of the drugs currently used in cancer treatment either damage DNA or inhibit DNA replication. Consequently, these drugs are toxic not only to cancer cells but also to normal cells, particularly those normal cells that are undergoing rapid cell division (e.g., hematopoietic cells, epithelial cells of the gastrointestinal tract, and hair follicle cells). The action of anticancer drugs against these normal cell populations accounts for most of the toxicity associated with these drugs and limits their effective use in cancer treatment.

One promising new approach to cancer therapy is the use of drugs that inhibit tumor growth by interfering with angiogenesis (blood vessel formation), rather than acting directly against cancer cells. As noted earlier in this chapter, the formation of new blood vessels is needed to supply the oxygen and nutrients required for tumor growth. Promoting angiogenesis is thus critical to tumor development, and tumor cells secrete a number of growth factors that stimulate the proliferation of capillary endothelial cells, resulting in the outgrowth of new capillaries into the tumor. The importance of angiogenesis was first recognized by Judah Folkman in 1971, and continuing research by Folkman and his colleagues has led to the development of new drugs (endostatin and angiostatin) that inhibit angiogenesis by blocking the proliferation of endothelial cells. Because these drugs act specifically to inhibit the formation of new blood vessels, they are much less toxic to normal cells than standard anticancer agents. Angiogenesis inhibitors have shown very promising results in animal tests, and are currently being tested in clinical trials to evaluate their effectiveness against human cancers.

An alternative strategy for achieving more selective cancer treatment is the development of drugs targeted specifically against the oncogenes that drive tumor growth. Unfortunately, from the standpoint of cancer treatment, oncogenes are not unique to tumor cells. Since proto-oncogenes play important roles in normal cells, general inhibitors of oncogene expression or function are likely to act against normal cells as well as tumor cells. The exploitation of oncogenes as targets for anticancer drugs is therefore not a straightforward proposition, but several promising advances indicate that it may ultimately be possible to develop selective oncogene-based therapies.

The first therapeutic regimen targeted against a specific oncogene is used for the treatment of acute promyelocytic leukemia. This leukemia is characterized by a chromosome translocation in which the gene that encodes the retinoic acid receptor (*RAR*α) is fused to another gene (*PML*) to form the *PML/RAR*α oncogene. The PML/RARα protein is thought to function, at least in part, by blocking cell differentiation. These leukemic cells, however, differentiate in response to treatment with high doses of retinoic acid, which appear to overcome the inhibitory effect of the PML/RARα oncogene protein. Such treatment with retinoic acid results in remission of the leukemia in most patients, although this favorable response is temporary and patients eventually relapse. However, combined treatment with retinoic acid and standard chemotherapeutic agents significantly reduces the incidence of relapse, so the use of retinoic acid is of substantial benefit in the treatment of acute promyelocytic leukemia. The therapeutic activity of retinoic acid was observed prior to identification of the *PML/RAR*α oncogene, so its effectiveness against leukemic cells expressing this oncogene protein was discovered by chance rather than by rational drug design. Nonetheless, the use of retinoic acid for treatment of acute promyelocytic leukemia provides the first example of a clinically useful drug targeted against an oncogene protein.

The development of drugs targeted against Ras proteins is an area of active exploration, consistent with the prevalence of *ras* oncogenes in human tumors. Ras proteins are targeted to the plasma membrane by addition of a farnesyl isoprenoid to their carboxy terminus (see Figure 7.30). Because this is a relatively uncommon type of protein modification, inhibitors of farnesylation have been investigated as potential therapeutic drugs. Such compounds are surprisingly selective inhibitors of cells expressing oncogenic Ras proteins, and their potential clinical utility in the treatment of human cancer is being evaluated.

A major milestone in the rational development of drugs targeted against specific oncogenes was reached in 1998, when the FDA approved Herceptin for treatment of metastatic breast cancer. Herceptin is a monoclonal antibody against the ErbB-2 oncogene protein, which is overexpressed in about 30% of breast cancers as a result of amplification of the *erb*B-2 gene. It was first found that an antibody against the extracellular domain of ErbB-2 (a receptor protein-tyrosine kinase) inhibited the proliferation of tumor cells in which ErbB-2 was overexpressed. These results led to the development and clinical testing of Herceptin, which was found to significantly reduce tumor growth and prolong patient survival in clinical trials involving over 600 women with metastatic breast cancers that overexpressed the ErbB-2 protein. Based on these results, Herceptin became the first drug developed against a specific oncogene to achieve FDA approval for clinical use in cancer treatment.

Another promising area of drug development is the identification of small-molecule inhibitors of oncogene proteins, including the protein kinases that play key roles in signaling the proliferation and survival of cancer cells. An exciting recent advance in this area has been the development of a selective inhibitor of the Bcr/Abl protein-tyrosine kinase, which is generated by the Philadelphia chromosome translocation in chronic myelogenous leukemias (see Figure 15.25). Brian Druker and his colleagues have successfully developed a potent and specific inhibitor of the Bcr/Abl protein kinase and shown that this compound (called STI-571) effectively blocks proliferation of chronic myelogenous leukemia cells. Based on these results, a clinical trial of STI-571 was initiated in June 1998. Remarkably, all of the patients in this initial trial responded to treatment with STI-571, which had only minimal side effects. Continuing trials of STI-571 with larger numbers of patients are in progress, but, if its dramatic activity against chronic myelogenous leukemia continues to be observed, STI-571 will be the first designer drug developed as a specific oncogene inhibitor. Although the eventual impact of molecular biology on the treatment of cancer remains to be seen, the examples of Herceptin and STI-571 clearly suggest that the continuing exploitation of oncogenes and tumor suppressor genes as targets for drug development has the potential of leading to a new generation of drugs selectively targeted against cancer cells.

KEY TERMS

cancer, tumor, benign tumor, malignant tumor, metastasis, carcinoma, sarcoma, leukemia, lymphoma

tumor initiation, tumor progression, adenoma, polyp

carcinogen, tumor promoter

density-dependent inhibition, autocrine growth stimulation, contact inhibition, angiogenesis, programmed cell death, apoptosis

cell transformation

SUMMARY

THE DEVELOPMENT AND CAUSES OF CANCER

Types of Cancer: Cancer can result from the abnormal proliferation of any type of cell. The most important distinction for the patient is between benign tumors, which remain confined to their site of origin, and malignant tumors, which can invade normal tissues and spread throughout the body.

The Development of Cancer: Tumors develop from single altered cells that begin to proliferate abnormally. Additional mutations lead to the selection of cells with progressively increasing capacities for proliferation, survival, invasion, and metastasis.

Causes of Cancer: Radiation and many chemical carcinogens act by damaging DNA and inducing mutations. Other chemical carcinogens contribute to the development of cancer by stimulating cell proliferation. Viruses also cause cancer in both humans and other species.

Properties of Cancer Cells: The uncontrolled proliferation of cancer cells is reflected in reduced requirements for extracellular growth factors and lack of inhibition by cell-cell contact. Many cancer cells are also defective in differentiation, consistent with their continued proliferation *in vivo*. The characteristic failure of cancer cells to undergo apoptosis also contributes substantially to tumor development.

Transformation of Cells in Culture: The development of *in vitro* assays for cell transformation has allowed the conversion of normal cells into tumor cells to be studied in cell culture.

TUMOR VIRUSES

Hepatitis B Viruses: The hepatitis B viruses cause liver cancer in several species, including humans.

SV40 and Polyomavirus: Although neither SV40 nor polyomavirus causes human cancer, they are important models for studying the molecular biology of cell transformation. SV40 T antigen induces transformation by interacting with the cellular Rb and p53 tumor suppressor proteins.

Papillomaviruses: Papillomaviruses induce tumors in a variety of animals, including cervical carcinoma in humans. Like SV40 T antigen, the transforming proteins of papillomaviruses interact with Rb and p53.

Adenoviruses: The adenoviruses do not cause naturally occurring cancers in either humans or other species but are important models in cancer research. Their transforming proteins also interact with Rb and p53.

Herpesviruses: The herpesviruses, which are among the most complex animal viruses, cause cancer in several species, including humans. The molecular biology of transformation by these viruses is not well understood.

Retroviruses: Retroviruses cause cancer in humans and a variety of other animals. Some retroviruses contain specific genes responsible for inducing cell transformation, and studies of these highly oncogenic retroviruses have led to the characterization of both viral and cellular oncogenes.

ONCOGENES

Retroviral Oncogenes: The first oncogene to be identified was the *src* gene of RSV. Subsequent studies have identified more than two dozen distinct oncogenes in different retroviruses.

Proto-Oncogenes: Retroviral oncogenes originated from closely related genes of normal cells, called proto-oncogenes. The oncogenes are abnormally expressed or mutated forms of the corresponding proto-oncogenes.

Oncogenes in Human Cancer: A variety of oncogenes are activated by point mutations, DNA rearrangements, and gene amplification in human cancers. Some of these human tumor oncogenes, such as the *ras* genes, are cellular homologs of oncogenes that were first described in retroviruses.

Functions of Oncogene Products: Most oncogene proteins function as elements of signaling pathways that stimulate cell proliferation. The gene that encodes cyclin D1 is also a potential oncogene, which stimulates cell cycle progression. Other oncogene proteins interfere with cell differentiation, and oncogenes encoding PI 3-kinase, Akt, and Bcl-2 inhibit apoptosis.

TUMOR SUPPRESSOR GENES

Identification of Tumor Suppressor Genes: In contrast to oncogenes, tumor suppressor genes inhibit tumor development. The prototype tumor suppressor gene, *Rb*, was identified by studies of inheritance of retinoblastoma. Loss or mutational inactivation of *Rb* and other tumor suppressor genes, including *p53*, contributes to the development of a wide variety of human cancers.

tumor virus, hepatitis B viruses

simian virus 40 (SV40), polyomavirus

papillomavirus

adenovirus

herpesvirus, Kaposi's sarcoma-associated herpesvirus, Epstein-Barr virus

retrovirus, Rous sarcoma virus (RSV)

oncogene, *src*, *ras*, *raf*

proto-oncogene

***ras*, *c-myc*, *abl*, N-*myc*, *erb*B-2**

Fos, Jun, D1, ErbA, PML/RARα, PI 3-kinase, Akt, Bcl-2

tumor suppressor gene, *Rb*, *p53*

PTEN

Functions of Tumor Suppressor Gene Products: The proteins encoded by most tumor suppressor genes act as inhibitors of cell proliferation or survival. The Rb, INK4, and p53 proteins are negative regulators of cell cycle progression. In addition, p53 is required for apoptosis induced by DNA damage and other stimuli, so its inactivation contributes to enhanced tumor cell survival.

Roles of Oncogenes and Tumor Suppressor Genes in Tumor Development: Mutations in both oncogenes and tumor suppressor genes contribute to the progressive development of human cancers. Accumulated damage to multiple such genes is thought to result in the abnormalities of cell proliferation, differentiation, and survival that characterize the cancer cell.

APPLICATIONS OF MOLECULAR BIOLOGY TO CANCER PREVENTION AND TREATMENT

Prevention and Early Detection: Many cancers can be cured if they are detected at early stages of tumor development. Genetic testing to identify individuals with inherited cancer susceptibilities may allow early detection and more effective treatment of high-risk patients.

Molecular Diagnosis: Detection of mutations in oncogenes and tumor suppressor genes may be useful in diagnosis and in monitoring response to treatment.

Treatment: The development of drugs targeted against specific oncogenes or tumor suppressor genes may lead to the discovery of new therapeutic agents that act selectively against cancer cells.

Questions

1. What is the role of clonal selection in the development of cancer?

2. You have constructed a mutant of SV40 T antigen that fails to induce cell transformation because it no longer binds Rb. Would this mutant T antigen induce transformation if introduced into cells together with a papilloma-virus cDNA that encodes E7? How about E6?

3. The proliferation of fibroblasts is generally inhibited by cAMP. Would *gsp* act as an oncogene in these cells?

4. What effect would overexpression of the *INK4* tumor suppressor gene prod-uct be expected to have on tumor cells in which *Rb* has been inactivated by mutation?

5. Which would you expect to be more sensitive to treatment with irradiation—tumors with wild-type *p53* genes or tumors with mutant *p53* genes?

References and Further Reading

General References

Cooper, G. M. 1992. *Elements of Human Cancer.* Boston: Jones and Bartlett.

Cooper, G. M. 1995. *Oncogenes.* 2nd ed. Boston: Jones and Bartlett.

Varmus, H. and R. A. Weinberg. 1993. *Genes and the Biology of Cancer.* New York: Scientific American Library.

The Development and Causes of Cancer

Christofori, G. and H. Semb. 1999. The role of the cell-adhesion molecule E-cadherin as a tumour-suppressor gene. *Trends Biochem. Sci.* 24: 73–76. [R]

Doll, R. and R. Peto. 1981. *The Causes of Cancer: Quantitative Estimates of Avoidable Risks of Cancer in the United States Today.* New York: Oxford University Press.

Fialkow, P. J. 1979. Clonal origin of human tumors. *Ann. Rev. Med.* 30: 135–143. [R]

Fidler, I. J. 1990. Critical factors in the biology of human cancer metastasis. *Cancer Res.* 50: 6130–6138. [R]

Hanahan, D. and J. Folkman. 1996. Patterns and emerging mechanisms of the angiogenic switch during tumorigenesis. *Cell* 86: 353–364. [R]

Henderson, B. E., R. Ross and L. Bernstein. 1988. Estrogens as a cause of human cancer. *Cancer Res.* 48: 246–253. [R]

Nowell, P. C. 1986. Mechanisms of tumor progression. *Cancer Res.* 46: 2203–2207. [R]

Raff, M. C. 1992. Social controls on cell survival and cell death. *Nature* 356: 397–400. [R]

Sawyers, C. L., C. T. Denny and O. N. Witte. 1991. Leukemia and the disruption of normal hematopoiesis. *Cell* 64: 337–350. [R]

Sporn, M. B. and A. B. Roberts. 1985. Autocrine growth factors and cancer. *Science* 313: 745–747. [R]

Stetler-Stevenson, W. G., S. Aznavoorian and L. A. Liotta. 1993. Tumor cell interactions with the extracellular matrix during invasion and metastasis. *Ann. Rev. Cell Biol.* 9: 541–573. [R]

Temin, H. M. and H. Rubin. 1958. Characteristics of an assay for Rous sarcoma virus and Rous sarcoma cells in culture. *Virology* 6: 669–688. [P]

Thompson, C. B. 1995. Apoptosis in the pathogenesis and treatment of disease. *Science* 267: 1456–1462. [R]

Tumor Viruses

Blumberg, B. S. 1997. Hepatitis B virus, the vaccine, and the control of primary cancer of the liver. *Proc. Natl. Acad. Sci. USA* 94: 7121–7125. [R]

Coffin, J. M., S. H. Hughes and H. E. Varmus, eds. 1997. *Retroviruses.* New York: Cold Spring Harbor Laboratory Press.

Fields, B. N., D. M. Knipe, P. M. Howley, R. M. Chanock, J. L. Melnick, T. P. Monath, B. Roizman and S. E. Straus, eds. 1996. *Fundamental Virology.* 3rd ed. New York: Lippincott-Raven.

Flint, S. J., L. W. Enquist, R. M. Krug, V. R. Racaniello and A. M. Skalka. 1999. *Principles of Virology: Molecular Biology, Pathogenesis, and Control.* Washington, DC: ASM Press.

Ganem, D. 1997. KSHV and Kaposi's sarcoma: The end of the beginning? *Cell* 91: 157–160. [R]

Zur Hausen, H. 1991. Viruses in human cancer. *Science* 254: 1167– 1173. [R]

Oncogenes

Aaronson, S. A. 1991. Growth factors and cancer. *Science* 254: 1146–1153. [R]

Adams, J. M. and S. Cory. 1998. The Bcl-2 protein family: Arbiters of cell survival. *Science* 281: 1322–1326. [R]

Bos, J. L. 1989. *Ras* oncogenes in human cancer: A review. *Cancer Res.* 49: 4682–4689. [R]

Dang, C. V. 1999. c-Myc target genes involved in cell growth, apoptosis, and metabolism. *Mol. Cell. Biol.* 19: 1–11. [R]

Datta, S. R., Brunet, A. and M. E. Greenberg. 1999. Cellular survival: A play in three *Akts. Genes Devel.* 13: 2905–2927. [R]

Der, C. J., T. G. Krontiris and G. M. Cooper. 1982. Transforming genes of human bladder and lung carcinoma cell lines are homologous to the *ras* genes of Harvey and Kirsten sarcoma viruses. *Proc. Natl. Acad. Sci. USA* 79: 3637–3640. [P]

Hunter, T. 1997. Oncoprotein networks. *Cell* 88: 333–346. [R]

Krontiris, T. G. and G. M. Cooper. 1981. Transforming activity of human tumor DNAs. *Proc. Natl. Acad. Sci. USA* 78: 1181–1184. [P]

Leder, P., J. Battey, G. Lenoir, C. Moulding, W. Murphy, H. Potter, T. Stewart and R. Taub. 1983. Translocations among antibody genes in human cancer. *Science* 222: 765–771. [R]

Legauer, C., K. W. Kinzler and B. Vogelstein. 1998. Genetic instabilities in human cancers. *Nature* 396: 643–649. [R]

Look, A. T. 1997. Oncogenic transcription factors in the human acute leukemias. *Science* 278: 1059–1064. [R]

Martin, G. S. 1970. Rous sarcoma virus: A function required for the maintenance of the transformed state. *Nature* 227: 1021–1023. [P]

Sherr, C. J. 1996. Cancer cell cycles. *Science* 274: 1672–1677. [R]

Shih, C., L. C. Padhy, M. Murray and R. A. Weinberg. 1981. Transforming genes of carcinomas and neuroblastomas introduced into mouse fibroblasts. *Nature* 300: 539–542. [P]

Solomon, E., J. Borrow and A. D. Goddard. 1991. Chromosome aberrations and cancer. *Science* 254: 1153–1160. [R]

Stehelin, D., H. E. Varmus, J. M. Bishop and P. K. Vogt. 1976. DNA related to the transforming gene(s) of avian sarcoma viruses is present in normal avian DNA. *Nature* 260: 170–173. [P]

Sugden, B. 1993. How some retroviruses got their oncogenes. *Trends Biochem. Sci.* 18: 233–235. [R]

Tabin, C. J., S. M. Bradley, C. I. Bargmann, R. A. Weinberg, A. G. Papageorge, E. M. Scolnick, R. Dhar, D. R. Lowy and E. H. Chang. 1982. Mechanism of activation of a human oncogene. *Nature* 300: 143–149. [P]

Thompson, C. B. 1995. Apoptosis in the pathogenesis and treatment of disease. *Science* 267: 1456–1462. [R]

Vogt, P. K. 1971. Spontaneous segregation of nontransforming viruses from cloned sarcoma viruses. *Virology* 46: 939–946. [P]

Tumor Suppressor Genes

Cantley, L. C. and B. G. Neel. 1999. New insights into tumor suppression: PTEN suppresses tumor formation by restraining the phosphoinositide 3-kinase/AKT pathway. *Proc. Natl. Acad. Sci. USA* 96: 4240–4245. [R]

Fearon, E. R. 1997. Human cancer syndromes: Clues to the origin and nature of cancer. *Science* 278: 1043–1050. [R]

Friend, S. H., R. Bernards, S. Rogelj, R. A. Weinberg, J. M. Rapaport, D. M. Albert and T. P. Dryja. 1986. A human DNA segment with properties of the gene that predisposes to retinoblastoma and osteosarcoma. *Nature* 323: 643–646. [P]

Harris, H., O. J. Miller, G. Klein, P. Worst and T. Tachibana. 1969. Suppression of malignancy by cell fusion. *Nature* 223: 363–368. [P]

Hartwell, L. H. and M. B. Kastan. 1994. Cell cycle control and cancer. *Science* 266: 1821–1828. [R]

Heldin, C.-H., K. Miyazono and P. ten Dijke. 1997. TGF-β signalling from cell membrane to nucleus through SMAD proteins. *Nature* 390: 465–471. [R]

Hollstein, M., D. Sidransky, B. Vogelstein and C. C. Harris. 1991. *p53* mutations in human cancers. *Science* 253: 49–53. [R]

Kinzler, K. W. and B. Vogelstein. 1996. Lessons from hereditary colorectal cancer. *Cell* 87: 159–170. [R]

Knudson, A. G. 1976. Mutation and cancer: A statistical study of retinoblastoma. *Proc. Natl. Acad. Sci. USA* 68: 820–823. [P]

Lee, W.-H., R. Bookstein, F. Hong, L.-J. Young, J.-Y. Shew and E. Y.-H. P. Lee. 1987. Human retinoblastoma susceptibility gene: Cloning, identification, and sequence. *Science* 235: 1394–1399. [P]

Levine, A. J. 1997. p53, the cellular gatekeeper for growth and division. *Cell* 88: 323–331. [R]

Sherr, C. J. 1996. Cancer cell cycles. *Science* 274: 1672–1677. [R]

Vogelstein, B. and K. W. Kinzler. 1993. The multistep nature of cancer. *Trends Genet.* 9: 138–141. [R]

Weinberg, R. A. 1991. Tumor suppressor genes. *Science* 254: 1138–1146. [R]

Weinberg, R. A. 1995. The retinoblastoma protein and cell cycle control. *Cell* 81: 323–330. [R]

Wicking, C., I. Smyth and A. Bale. 1999. The hedgehog signalling pathway in tumorigenesis and development. *Oncogene* 18: 7844–7851. [R]

Applications of Molecular Biology to Cancer Prevention and Treatment

Degos, L., H. Dombret, C. Chomienne, M.-T. Daniel, J.-M. Miclea, C. Chastang, S. Castaigne and P. Fenaux. 1995. All-*trans*-retinoic acid as a differentiating agent in the treatment of acute promyelocytic leukemia. *Blood* 85: 2643–2653. [R]

Druker, B. J. and N. B. Lydon. 2000. Lessons learned from the development of an Abl tyrosine kinase inhibitor for chronic myelogenous leukemia. *J. Clin. Invest.* 105: 3–7. [R]

Folkman, J. 1995. Clinical applications of research on angiogenesis. *N. Eng. J. Med.* 333: 1757–1763. [R]

Lebwohl, D. E. and R. Canetta. 1999. New developments in chemotherapy of advanced breast cancer. *Ann. Oncol.* 10: 139–146. [R]

Levitzki, A. and A. Gazit. 1995. Tyrosine kinase inhibition: An approach to drug development. *Science* 267: 1782–1788. [R]

Oliff, A. 1999. Farnesyltransferase inhibitors: Targeting the molecular basis of cancer. *Biochim. Biophys. Acta* 1423: C19–C30. [R]

O'Reilly, M. S., T. Boehm, Y. Shing, N. Fukai, G. Vasios, W. S. Lane, E. Flynn, J. R. Birkhead, B. R. Olsen and J. Folkman. 1997. Endostatin: An endogenous inhibitor of angiogenesis and tumor growth. *Cell* 88: 277–285. [P]

Perera, F. P. 1997. Environment and cancer: Who are susceptible? *Science* 278: 1068–1073. [R]

Ponder, B. 1997. Genetic testing for cancer risk. *Science* 278: 1050–1054. [R]

Sidransky, D. 1997. Nucleic acid-based methods for the detection of cancer. *Science* 278: 1054–1058. [R]

Smalley, W. and R. N. DuBois. 1997. Colorectal cancer and nonsteroidal anti-inflammatory drugs. *Adv. Pharmacol.* 39: 1–20. [R]

Answers to Questions

Chapter 1

1. O_2 became abundant in Earth's atmosphere as a result of photosynthesis.

2. Yeasts are surrounded by cell walls; animal cells are not.

3. The refractive index of air is 1.0; the refractive index of oil is approximately 1.4. Since resolution = $0.61\lambda/\eta \sin \alpha$ (η is the refractive index), viewing a specimen through air rather than through oil changes the limit of resolution from about 0.2 μm to about 0.3 μm.

4. You would use velocity centrifugation. In an equilibrium gradient the two organelles would band at the same position.

5. The genome of Rous sarcoma virus (10,000 base pairs) is 100,000 times smaller than that of chicken cells (1.2 billion base pairs).

Chapter 2

1. The formation of stable membranes depends on the amphipathic character of phospholipids.

2. Since phenylalanine residues are hydrophobic, they would probably be located in α helices or β sheets within the interior of the protein. The loop regions connecting these elements of secondary structure would be expected to contain hydrophilic amino acids.

3. Aspartate is an acidic amino acid that interacts with basic amino acids in the substrates of trypsin. Substitution of aspartate with lysine (a basic amino acid) would therefore interfere with substrate binding and catalysis.

4. Substituting the given values into the equation

$$\Delta G = \Delta G + RT \ln \frac{[B][C]}{[A]}$$

gives the following result: $\Delta G = -1.93$ kcal/mol. The reaction will therefore proceed from left to right, with A being converted to B plus C within the cell.

5. Since saturated fatty acids are more highly reduced, more free energy is obtained from their oxidation than from that of unsaturated fatty acids.

6. In order to drive the pathway in the direction of glucose synthesis, the energy-yielding reactions of glycolysis must be bypassed by alternative reactions that are coupled to the utilization of ATP and NADH.

7. The α-helical structure allows the CO and NH groups of peptide bonds to form hydrogen bonds with each other, thereby neutralizing their polar character.

Chapter 3

1. The genetic map is $z \underset{}{\overset{0.5}{\rule{3cm}{0.4pt}}} x \underset{}{\overset{0.2}{\rule{2cm}{0.4pt}}} y$

2. UGC also encodes cysteine, so this mutation would have no effect on enzyme function. UGA, however, is a stop codon, so this mutation would abolish enzyme activity.

3. The restriction map is

4. The haploid sperm contains a single starting copy of the DNA sequence. Each cycle of PCR amplifies the starting material two-fold, so 10 cycles yields 2^{10}, or approximately a thousand, copies. Amplification for 30 cycles yields 2^{30}, or more than a billion, copies.

5. Antibodies against the protein could be raised either by using the predicted amino acid sequence to generate synthetic peptides or by expressing the protein in bacteria. Fluorescence microscopy could then be used to determine subcellular localization.

6. Mutations of the amino acids of interest could be produced by *in vitro* mutagenesis. The effects of these mutations on catalytic activity could then be determined following expression of the mutated protein in bacteria or eukaryotic cells.

Chapter 4

1. DNA reassociation is a bimolecular reaction, so the reaction rate is proportional to the square of the DNA concentration. A sequence repeated 1000 times in the genome would therefore reassociate with a rate 10^6-fold greater than that of single-copy DNA.

2. Given one nucleosome every 200 base pairs, there are 3×10^7 nucleosomes in a diploid human cell. Each nucleosome contains two molecules of the four core histones and one molecule of histone H1, so there are nearly 3×10^8 histone molecules in the cell. Taking the average molecular weight of the histones as approximately 15,000 daltons, the total mass of histones is approximately 0.008 ng—nearly 1% of total cell protein.

3. Because they can accommodate inserts containing thousands of kilobases, YACs can be used for mapping large genomes, which may contain thousands of megabases of DNA. Telomeres maintain the ends of YACs, allowing them to replicate as linear, chromosome-like molecules in yeasts.

4. *C. elegans* and yeast contain about 19,000 and 6,000 genes, respectively, so about two-thirds of *C. elegans* genes are expected to specifically function in multicellular organisms.

5. Approximately 100 kilobases.

Chapter 5

1. This property of DNA polymerases is necessary for proofreading because it enables the polymerase to recognize and excise mismatched bases that are not hydrogen-bonded to the template strand.

2. Primase, like other RNA polymerases, is not capable of proofreading, so errors occur at a comparatively high frequency in RNA primers. However, since the primers are later removed, the overall fidelity of replication is not compromised.

3. The high frequency of skin cancer results from DNA damage induced by solar UV irradiation, which is subject to repair by the nucleotide-excision repair system. The lack of elevated incidence of other cancers may suggest that similar types of damage are not frequent in internal organs and that most cancers of these organs result from other types of mutations (e.g., the incorporation of mismatched bases during DNA replication).

4. RecA mutants are sensitive to UV irradiation because they are deficient in recombinational repair of DNA damage.

5. The mouse would be immunodeficient, lacking both B and T lymphocytes, as a result of being unable to rearrange its immunoglobulin and T cell receptor genes.

6. The cellular reverse transcriptases that might be affected by these drugs include telomerase and the reverse transcriptases encoded by LINE sequences. Inhibition of the LINE reverse transcriptases would not be toxic to the cell, but inhibition of telomerase might interfere with the replication of chromosome ends.

Chapter 6

1. The promoter containing the sequence TATGAT, which more closely resembles the consensus sequence, will be transcribed more efficiently.

2. The wild-type gene will be regulated normally, but the temperature-sensitive gene will be expressed constitutively. β-Galactosidase will therefore be produced at the permissive but not the nonpermissive temperature.

3. The promoter containing the TATA box can be transcribed *in vitro* in the presence of either TBP or TFIID. However, the Inr promoter requires TFIID, since the Inr sequence is recognized by TAFs rather than by TBP.

4. The sequence element would be a candidate binding site for a tissue-specific repressor.

5. Two different DNA-binding domains could be joined to the factor's activation domain by tissue-specific alternative splicing.

Chapter 7

1. A Shine-Delgarno sequence is needed.

2. Polyadenylation is an important translational regulatory mechanism in early development. Its inhibition would block the translation of many oocyte mRNAs following fertilization.

3. An Hsp70 chaperone is expected to be required for the selective uptake of proteins into lysosomes from the cytosol, since this process involves the transfer of unfolded polypeptide chains across the lysosomal membrane. Such a chaperone would not be required for the fusion of membrane vesicles with lysosomes during autophagy.

4. Phospholipase treatment would release a GPI-anchored protein, but not a transmembrane protein, from the cell surface.

5. The mutation from serine to threonine would be expected to have no effect, since both serine and threonine residues are phosphorylated by the same protein kinases. However, the mutation from serine to alanine would prevent phosphorylation and block activation of the enzyme.

Chapter 8

1. The transcription factor could no longer be phosphorylated at these sites, so it would be constitutively imported into the nucleus and would activate target gene expression.

2. Inactivating the nuclear export signal would result in retention of the protein in the nucleus.

3. Such a mutation would prevent proteins destined for nuclear import from being released from the importins within the nucleus.

4. An insulator prevents enhancers from acting on promoters in separate chromosomal domains. Insertion of an insulator between E and P1 would therefore inhibit transcription from both P1 and P2, whereas insertion of an insulator between P1 and P2 would only inhibit transcription from P2.

5. Because formation of a nucleolus requires transcription of the rRNA genes, it is blocked by treatment with actinomycin D. However, because the rRNA genes are transcribed by RNA polymerase I, α-amanitin has no effect.

Chapter 9

1. The signal sequence would direct the growing polypeptide chain to the ER and the protein would then be secreted via the bulk flow pathway.

2. These proteins are unable to enter the ER and therefore remain in the cytosol.

3. Carbohydrate groups are added within the lumen of the ER and Golgi apparatus, which are topologically equivalent to the exterior of the cell.

4. The normally cytosolic protein lacks a signal sequence and does not enter the ER. Therefore, addition of a lysosome-targeting signal would have no effect. In contrast, such an addition would direct a normally secreted protein to lysosomes from the Golgi apparatus.

5. In the absence of mannose-6-phosphate formation in the Golgi apparatus, normally lysosomal proteins would be secreted.

Chapter 10

1. The voltage component of the electrochemical gradient drives membrane insertion and translocation of the positively charged presequences of mitochondrial proteins.

2. Under these conditions, ΔG is approximately -1.4 kcal/mol.

3. They are topologically equivalent compartments.

4. In contrast to mitochondria, there is no electric potential across the chloroplast membrane. Therefore, the charge of transit peptides does not contribute to protein translocation.

5. Two high-energy electrons are required to split each molecule of H_2O, so 24 high-energy electrons are required for the synthesis of each molecule of glucose. The passage of these electrons through the two photosystems generates 12 molecules of NADPH and between 12 and 18 molecules of ATP, depending on the stoichiometry of proton pumping at the cytochrome *bf* complex. Since 18 molecules of ATP are required for the Calvin cycle, the synthesis of glucose may require additional ATPs produced by cyclic electron flow.

6. Three out of four carbons converted to glycolate are returned to choroplasts and reenter the Calvin cycle.

Chapter 11

1. Cytochalasin inhibits actin polymerization. It would therefore inhibit the movement of cells by pseudopodia but would not affect the microtubule-based movement of flagella.

2. The polarity of actin filaments defines the direction of myosin movement. If actin filaments were not polar, the unidirectional movement of myosin that results in the sliding of actin and myosin filaments could not take place.

3. No. Keratins are expressed only in epithelial cells.

4. Colchicine inhibits microtubule polymerization, so it would affect chromosome segregation. It would not affect cytokinesis, which is driven by an actin-myosin contractile ring.

5. By linking the microtubule doublets of cilia together, nexin converts the sliding of individual microtubules to a bending motion that leads to the beating of cilia. If it were eliminated, the microtubules would simply slide past one another.

Chapter 12

1. The presence of porins in the plasma membrane would allow the free diffusion of ions and small molecules between the cytosol and extracellular fluids—a disaster for the cell.

2. Given $C_o/C_i = 10$, the K^+ equilibrium potential calculated from the Nernst equation is –58 mV. The actual resting membrane potential differs from the K^+ equilibrium potential because the plasma membrane is not completely impermeable to other ions.

3. The plant cell wall prevents cell swelling and allows the buildup of turgor pressure.

4. Because gap junctions allow ions to diffuse freely between adjacent cells, expression of normal CFTR in one epithelial cell would provide a functional Cl^- channel to its neighbors.

5. The correct localization of transporters mediating active transport and facilitated diffusion is necessary for the polarized function of epithelial cells in transferring glucose from the intestinal lumen to the blood supply. Tight junctions prevent the diffusion of these transporters between domains of the plasma membrane, as well as sealing the spaces between cells of the epithelium.

Chapter 13

1. Inhibition of cAMP phosphodiesterase would result in elevated levels of cAMP, which would stimulate cell proliferation.

2. The recombinant molecule would function as an epinephrine receptor coupled to G_i. Epinephrine would therefore inhibit adenylyl cyclase, lowering intracellular cAMP levels. Acetylcholine would have no effect, since it would not bind to the recombinant receptor.

3. PDGF monomers would not induce receptor dimerization. Since this is the first critical step in signaling from receptor protein-tyrosine kinases, they would be unable to stimulate the PDGF receptor.

4. Protein phosphatase 1 dephosphorylates serine residues that are phosphorylated by protein kinase A. Cyclic AMP-inducible genes are activated by CREB, which is phosphorylated by protein kinase A, so overexpression of protein phosphatase 1

would inhibit their induction. However, protein phosphatase 1 would not affect the activity of cAMP-gated ligand channels, since these channels are opened directly by cAMP binding rather than by protein phosphorylation.

5. Hydrolysis of PIP_2 by phospholipase C yields both diacylglycerol and IP_3, which signals the release of Ca^{2+} from the endoplasmic reticulum. PIP_2 hydrolysis can therefore activate both PKC-α and PKC-ε. Hydrolysis of phosphatidylcholine yields diacylglycerol but not IP_3; consequently, phosphatidylcholine hydrolysis is sufficient to activate PKC-ε but not PKC-α.

6. Raf acts downstream of Ras in the MAP kinase pathway. Activated Raf can therefore bypass the effects of dominant negative Ras, but activated Ras cannot overcome the inhibitory effects of dominant negative Raf. MEK acts downstream of Raf, so activated MEK can overcome the effects of dominant negative Ras or Raf.

Chapter 14

1. G_1 = 16 hours; S = 9 hours; G_2 = 4 hours; M = 1 hour.

2. One daughter cell would receive two copies of the misaligned chromosome; the other daughter cell would receive none.

3. In a normal cell, overexpression of p16 would inhibit cell cycle progression at the restriction point in G_1. Because Rb is the principal target of Cdk4/cyclin D complexes, a tumor cell lacking functional Rb would be unaffected by p16 overexpression.

4. The nuclear lamina would fail to break down.

5. Anaphase would initiate normally. However, MPF would remain active, so re-formation of nuclei, chromosome decondensation, and cytokinesis would not occur.

6. Oocytes of these mice would fail to arrest at metaphase II.

Chapter 15

1. Clonal selection drives tumor progression by favoring the outgrowth of cells within the tumor population that are proliferating more rapidly.

2. E7 binds Rb, so it would be able to induce transformation in combination with the mutant T antigen. E6 would be unable to do so, since it interacts with p53 but not with Rb.

3. The *gsp* gene encodes the α subunit of G_s, which stimulates adenylyl cyclase. It would therefore not act as an oncogene in cells whose proliferation is inhibited by cAMP.

4. *INK4* encodes the Cdk inhibitor p16, which inhibits Cdk4/cyclin D complexes. Since phosphorylation

of Rb is the critical target of Cdk4/cyclin D, overexpression of p16 would have no effect on cells in which Rb has already been inactivated.

5. Cells with wild-type p53 would be more sensitive because p53 is required for apoptosis induced by DNA damage.

Glossary

α helix A coiled secondary structure of a polypeptide chain formed by hydrogen bonding between amino acids separated by four residues.

ABC transporters A large family of membrane transport proteins characterized by a highly conserved ATP binding domain.

actin An abundant 43-kd protein that polymerizes to form cytoskeletal filaments.

actin bundle Actin filaments that are crosslinked into closely packed arrays.

actin network Actin filaments that are crosslinked into loose three-dimensional meshworks.

activation energy The energy required to raise a molecule to its transition state to undergo a chemical reaction.

active site The region of an enzyme that binds substrates and catalyzes an enzymatic reaction.

active transport The transport of molecules in an energetically unfavorable direction across a membrane coupled to the hydrolysis of ATP or other source of energy.

adaptin A protein that binds to membrane receptors and mediates the formation of clathrin-coated vesicles.

adenine A purine that base-pairs with either thymine or uracil.

adenoma A benign tumor arising from glandular epithelium.

adenylyl cyclase An enzyme that catalyzes the formation of cyclic AMP from ATP.

adherens junction A region of cell-cell adhesion at which the actin cytoskeleton is anchored to the plasma membrane.

Akt A protein-serine/threonine kinase that is activated by PIP$_3$ and plays a key role in signaling cell survival.

allele One copy of a gene.

allosteric regulation The regulation of enzymes by small molecules that bind to a site distinct from the active site, changing the conformation and catalytic activity of the enzyme.

alternative splicing The generation of different mRNAs by varying the pattern of pre-mRNA splicing.

amino acid Monomeric building blocks of proteins, consisting of a carbon atom bound to a carboxyl group, an amino group, a hydrogen atom, and a distinctive side chain.

aminoacyl tRNA synthetase An enzyme that joins a specific amino acid to a tRNA molecule carrying the correct anticodon sequence.

amphipathic A molecule that has both hydrophobic and hydrophilic regions.

anaphase The phase of mitosis during which sister chromatids separate and move to opposite poles of the spindle.

anaphase A The movement of daughter chromosomes toward the spindle poles during mitosis.

anaphase B The separation of the spindle poles during mitosis.

anaphase-promoting complex A ubiquitin ligase that triggers progression from metaphase to anaphase by signaling the degradation of cyclin B and cohesins.

angiogenesis The formation of new blood vessels.

antibody   A protein produced by B lymphocytes that binds to a foreign molecule.

anticodon   The nucleotide sequence of transfer RNA that forms complementary base pairs with a codon sequence on messenger RNA.

antigen   A molecule against which an antibody is directed.

antiport   The transport of two molecules in opposite directions across a membrane.

AP endonuclease   A DNA repair enzyme that cleaves next to apyrimidinic or apurinic sites in DNA.

apical domain   The exposed free surface of a polarized epithelial cell.

apoptosis   An active process of programmed cell death, characterized by cleavage of chromosomal DNA, chromatin condensation, and fragmentation of both the nucleus and the cell.

Arabidopsis thaliana   A small flowering plant used as a model for plant molecular biology and development.

archaebacteria   One of two major groups of prokaryotes; many species of archaebacteria live in extreme conditions similar to those prevalent on primitive Earth.

ARF   A GTP-binding protein required for vesicle budding from the *trans*-Golgi network.

astral microtubules   Microtubules of the mitotic spindle that extend to the cell periphery.

ATP (adenosine 5′-triphosphate)   An adenine-containing nucleoside triphosphate that serves as a store of free energy in the cell.

ATP synthase   A membrane spanning protein complex that couples the energetically favorable transport of protons across a membrane to the synthesis of ATP.

autocrine signaling   A type of cell signaling in which a cell produces a growth factor to which it also responds.

autophagy   The degradation of cytoplasmic proteins and organelles by their enclosure in vesicles from the endoplasmic reticulum that fuse with lysosomes.

autoradiography   The detection of radioisotopically labeled molecules by exposure to X-ray film.

axonemal dynein   The type of dynein found in cilia and flagella.

axoneme   The fundamental structure of cilia and flagella composed of a central pair of microtubules surrounded by nine microtubule doublets.

β sheet   A sheetlike secondary structure of a polypeptide chain, formed by hydrogen bonding between amino acids located in different regions of the polypeptide.

bacteriophage   A bacterial virus.

basal body   A structure similar to a centriole that initiates the growth of axonemal microtubules and anchors cilia and flagella to the surface of the cell.

basal lamina   A sheetlike extracellular matrix that supports epithelial cells and surrounds muscle cells, adipose cells, and peripheral nerves.

base-excision repair   A mechanism of DNA repair in which single damaged bases are removed from a DNA molecule.

basement membrane   See basal lamina.

basolateral domain   The surface region of a polarized epithelial cell that is in contact with adjacent cells or the extracellular matrix.

benign tumor   A tumor that remains confined to its site of origin.

cadherins   A group of cell adhesion molecules that form stable cell-cell junctions at adherens junctions and desmosomes.

Caenorhabditis elegans   A nematode used as a simple multicellular model for development.

calmodulin   A calcium-binding protein.

Calvin cycle   A series of reactions by which six molecules of CO_2 are converted into glucose.

cAMP phosphodiesterase   An enzyme that degrades cyclic AMP.

cAMP-dependent protein kinase   See protein kinase A.

cancer   A malignant tumor.

carbohydrate   A molecule with the formula $(CH_2O)_n$. Carbohydrates include both simple sugars and polysaccharides.

carcinogen   A cancer-inducing agent.

carcinoma   A cancer of epithelial cells.

cardiolipin   A phospholipid containing four hydrocarbon chains.

carrier proteins   Proteins that selectively bind and transport small molecules across a membrane.

caspases   A family of proteases that bring about programmed cell death.

catalase   An enzyme that decomposes hydrogen peroxide.

caveolae   Small invaginations of the plasma membrane that may be involved in endocytosis.

Cdc2   A protein-serine/threonine kinase that is a key regulator of mitosis in eukaryotic cells.

Cdk inhibitor (CKI)   A family of proteins that bind Cdks and inhibit their activity.

Cdks   Cyclin dependent protein kinases that control the cell cycle of eukaryotes. See also Cdc2.

cDNA library   A collection of recombinant cDNA clones.

cell adhesion molecules   Transmembrane proteins that mediate cell-cell interactions.

cell cortex   The actin network underlying the plasma membrane.

cell wall   A rigid, porous structure forming an external layer that provides structural support to bacteria, fungi, and plant cells.

cellulose   The principal structural component of the plant cell wall, a linear polymer of glucose residues linked by $\beta(1 \rightarrow 4)$ glycosidic bonds.

central dogma   The concept that genetic information flows from DNA to RNA to proteins.

centriole   A cylindrical structure consisting of nine triplets of microtubules in the centrosomes of most animal cells.

centromere   A specialized chromosomal region that connects sister chromatids and attaches them to the mitotic spindle.

centrosome   The microtubule-organizing center in animal cells.

cGMP phosphodiesterase   An enzyme that degrades cGMP.

channel proteins   Proteins that form pores through a membrane.

chaperone   A protein that facilitates the correct folding or assembly of other proteins.

chaperonin   A family of heat-shock proteins within which protein folding takes place.

chemiosmotic coupling   The generation of ATP from energy stored in a proton gradient across a membrane.

chiasmata   Sites of recombination that link homologous chromosomes during meiosis.

chitin   A polymer of *N*-acetylglucosamine residues that is the principal component of fungal cell walls.

chlorophyll   The major photosynthetic pigment of plant cells.

chloroplast   The organelle responsible for photosynthesis in the cells of plants and green algae.

cholesterol   A lipid consisting of four hydrocarbon rings. Cholesterol is a major constituent of animal cell plasma membranes and the precursor of steroid hormones.

chromatin   The fibrous complex of eukaryotic DNA and histone proteins. See histones, nucleosome, and chromatosome.

chromatosome   A chromatin subunit consisting of 166 base pairs of DNA wrapped around a histone core and held in place by a linker histone.

chromosomes   The carriers of genes, consisting of long DNA molecules and associated proteins.

cilium   A microtubule-based projection of the plasma membrane that moves a cell through fluid or fluid over a cell.

***cis*-acting control element**   A regulatory DNA sequence that serves as a protein binding site and controls the transcription of adjacent genes.

citric acid cycle   A series of reactions in which acetyl CoA is oxidized to CO_2. The central pathway of oxidative metabolism.

clathrin   A protein that coats the cytoplasmic surface of cell membranes and assembles into basketlike lattices that drive vesicle budding.

clone   See recombinant molecule.

codon   The basic unit of the genetic code; one of the 64 nucleotide triplets that code for an amino acid or stop sequence.

coenzyme A (CoA)   A coenzyme that functions as a carrier of acyl groups in metabolic reactions.

coenzyme Q   A small lipid-soluble molecule that carries electrons between protein complexes in the mitochondrial electron transport chain.

coenzymes   Low-molecular-weight organic molecules that work together with enzymes to catalyze biological reactions.

cohesins   A complex of proteins that maintain the connection between sister chromatids.

colcemid   A drug that inhibits the polymerization of microtubules.

colchicine   A drug that inhibits the polymerization of microtubules.

collagen   The major structural protein of the extracellular matrix.

collenchyma   Plant cells characterized by thick cell walls; they provide structural support to the plant.

complementary DNA (cDNA)   A DNA molecule that is *complementary* to an mRNA molecule, synthesized *in vitro* by reverse transcriptase.

contact inhibition   The inhibition of movement or proliferation of normal cells that results from cell-cell contact.

contractile ring   A structure of actin and myosin II that forms beneath the plasma membrane during mitosis and mediates cytokinesis.

corticosteroids   Steroid hormones produced by the adrenal gland.

cosmid   A vector that contains bacteriophage λ sequences, antibiotic resistance sequences, and an origin of replication. It can accomodate large DNA inserts of up to 45 kb.

crista   A fold in the inner mitochondrial membrane extending into the matrix.

cyanobacteria   The largest and most complex prokaryotes in which photosynthesis is believed to have evolved.

cyclic AMP (cAMP)   Adenosine monophosphate in which the phosphate group is covalently bound to both the 3' and 5' carbon atoms, forming a cyclic structure; an important second messenger in the response of cells to a variety of hormones.

cyclic electron flow   An electron transport pathway associated with photosystem I that produces ATP without the synthesis of NADPH.

cyclic GMP (cGMP)   Guanosine monophosphate in which the phosphate group is covalently bound to both the 3′ and 5′ carbon atoms, forming a cyclic structure; an important second messenger in the response of cells to a variety of hormones and in vision.

cyclins   A family of proteins that regulate the activity of Cdks and control progression through the cell cycle.

cytochalasin   A drug that blocks the elongation of actin filaments.

cytochrome oxidase   A protein complex in the electron transport chain that accepts electrons from cytochrome *c* and transfers them to O_2.

cytokines   Growth factors that regulate blood cells and lymphocytes.

cytokinesis   Division of a cell following mitosis or meiosis.

cytosine   A pyrimidine that base-pairs with guanine.

cytoskeleton   A network of protein filaments that extends throughout the cytoplasm of eukaryotic cells. It provides the structural framework of the cell and is responsible for cell movements.

density gradient centrifugation   A method of separating particles by centrifugation through a gradient of a dense substance, such as sucrose or cesium chloride.

deoxyribonucleic acid (DNA)   The genetic material of the cell.

desmosome   A region of contact between epithelial cells at which keratin filaments are anchored to the plasma membrane. See also hemidesmosome.

diacylglycerol   A secondary messenger formed from the hydrolysis of PIP_2 that activates protein kinase C.

diakinesis   The final stage of the prophase of meiosis I during which the chromosomes fully condense and the cell progresses to metaphase.

Dictyostelium discoideum   A unicellular eukaryote used for studies of cell movement and cell-cell signaling.

diploid   An organism or cell that carries two copies of each chromosome.

diplotene   The stage of mieosis I during which homologous chromosomes separate along their length but remain associated at chiasmata.

DNA glycosylase   A DNA repair enzyme that cleaves the bond linking a purine or pyrimidine to the deoxyribose of the backbone of a DNA molecule.

DNA ligase   An enzyme that seals breaks in DNA strands.

DNA polymerase   An enzyme catalyzing the synthesis of DNA.

dolichol phosphate   A lipid molecule in the endoplasmic reticulum upon which oligosaccharides are assembled for the glycosylation of proteins.

domains   Compact, globular regions of proteins that are the basic units of tertiary structure.

dominant   The allele that determines the phenotype of an organism when more than one allele is present.

dominant inhibitory mutant   A mutant that interferes with the function of the normal allele of the gene.

Drosophila melanogaster   A species of fruit fly commonly used for studies of animal genetics and development.

dynein   A motor protein that moves along microtubules towards the minus end.

E. coli (*Escherichia coli*)   A species of bacteria used as a model for biochemistry and molecular biology.

E2F   A family of transcription factors that regulate the expression of genes involved in cell cycle progression and DNA replication.

ecdysone   An insect steroid hormone that triggers metamorphosis.

ectoderm   The outer germ layer; gives rise to tissues that include the skin and nervous system.

eicosanoid   A class of lipids, including prostaglandins, prostacyclins, thromboxanes, and leukotrienes, that act in autocrine and paracrine signaling.

electrochemical gradient   A difference in chemical concentration and electric potential across a membrane.

electron microscopy   A type of microscopy that uses an electron beam to form an image. In transmission electron microscopy, a beam of electrons is passed through a specimen stained with heavy metals. In scanning electron microscopy, electrons scattered from the surface of a specimen are analyzed to generate a three-dimensional image.

electron transport chain   A series of carriers through which electrons are transported from a higher to a lower energy state.

embryonal stem (ES) cells   Cells cultured from early embryos.

endocrine signaling   A type of cell-cell signaling in which endocrine cells secrete hormones that are carried by the circulation to distant target cells.

endocytosis   The uptake of extracellular material in vesicles formed from the plasma membrane.

endoderm   The inner germ layer; gives rise to internal organs.

endoplasmic reticulum (ER)   An extensive network of membrane-enclosed tubules and sacs involved in protein sorting and processing as well as in lipid synthesis.

endosome   A vesicular compartment involved in the sorting and transport to lysosomes of material taken up by endocytosis.

endosymbiosis   A symbiotic relationship in which one cell resides within a larger cell.

enhancer   A transcriptional regulatory sequence that can be located at a site distant from the promoter.

enzymes   Proteins or RNAs that catalyze biological reactions.

epidermal cells   Cells forming a protective layer on the surfaces of plants and animals.

epidermal growth factor (EGF)   A growth factor that stimulates cell proliferation.

epithelial cells   Cells forming sheets (epithelial tissue) that cover the surface of the body and line internal organs.

equilibrium centrifugation   The separation of particles on the basis of density by centrifugation to equilibrium in a gradient of a dense substance.

erythrocytes   Red blood cells.

estrogen   A steroid hormone produced by the ovaries.

eubacteria   One of two major groups of prokaryotes, including most common species of bacteria.

euchromatin   Decondensed, transcriptionally active interphase chromatin.

eukaryotic cells   Cells that have a nuclear envelope, cytoplasmic organelles, and a cytoskeleton.

exon   A segment of a gene that contains a coding sequence.

exonuclease   An enzyme that hydrolyzes DNA molecules in either the 5′ to 3′ or 3′ to 5′ direction.

exportin   A receptor protein that recognizes nuclear export signals and directs transport from the nucleus to the cytosol.

expression vector   A vector used to direct expression of a cloned DNA fragment in a host cell.

extracellular matrix   Secreted proteins and polysaccharides that fill spaces between cells and bind cells and tissues together.

facilitated diffusion   The transport of molecules across a membrane by carrier or channel proteins.

FAK (focal adhesion kinase)   A nonreceptor protein-tyrosine kinase that plays a key role in integrin signaling.

fats   See triacylglycerols.

fatty acids   Long hydrocarbon chains usually linked to a carboxyl group (COO^-).

feedback inhibition   A type of allosteric regulation in which the product of a metabolic pathway inhibits the activity of an enzyme involved in its synthesis.

fibroblast   A cell type found in connective tissue.

fibronectin   The principal adhesion protein of the extracellular matrix.

filiopodium   A thin projection of the plasma membrane supported by actin bundles.

flagellum   A microtubule-based projection of the plasma membrane that is responsible for cell movement.

flavin adenine dinucleotide ($FADH_2$)   A coenzyme that functions as an electron carrier in oxidation/reduction reactions.

flow cytometer   An instrument that measures the fluoresence intensity of individual cells.

fluid mosaic model   A model of membrane structure in which proteins are inserted in a fluid phospholipid bilayer.

fluid-phase endocytosis   The nonselective uptake of extracellular fluids during endocytosis.

fluorescence *in situ* hybridization (FISH)   A method used to localize genes on chromosomes using fluorescent probes.

fluorescence microscopy   Type of microscopy in which molecules are detected based on the emission of flourescent light.

focal adhesion   A site of attachment of cells to the extracellular matrix at which integrins are linked to bundles of actin filaments.

freeze fracture   Method of electron microscopy in which specimens are frozen in liquid nitrogen and then fractured to split the lipid bilayer, revealing the interior faces of cell membranes.

G-protein   A family of cell signaling proteins regulated by guanine nucleotide binding.

G protein-coupled receptor   A receptor characterized by seven membrane-spanning α helices. Ligand binding causes a conformational change that activates a G protein.

G_0   A quiescent state in which cells remain metabolically active but do not proliferate.

G_1 phase   The phase of the cell cycle between the end of mitosis and the begining of DNA synthesis.

G_2 phase   The phase of the cell cycle between the end of S phase and the begining of mitosis.

gap junction   A plasma membrane channel forming a direct cytoplasmic connection between adjacent cells.

gene   A segment of DNA that encodes a polypeptide chain or an RNA molecule.

gene amplification   An increase in the number of copies of a gene resulting from the repeated replication of a region of DNA.

gene transfer   The introduction of foreign DNA into a cell.

general homologous recombination   The exchange of segments between DNA molecules that share extensive sequence homology.

general transcription factors Transcription factors that are part of the general transcription machinery.

genetic code The correspondence between nucleotide triplets and amino acids in proteins.

genomic imprinting The regulation of genes whose expression depends on whether they are maternally or paternally inherited, apparently controlled by DNA methylation.

genomic library A collection of recombinant DNA clones that collectively contain the genome of an organism.

genotype The genetic composition of an organism.

Gibbs free energy (*G*) The thermodynamic function that combines the effects of enthalpy and entropy to predict the energetically favorable direction of a chemical reaction.

gluconeogenesis The synthesis of glucose.

glycerol phospholipids Phospholipids consisting of two fatty acids bound to a glycerol molecule.

glycocalyx A carbohydrate coat covering the cell surface.

glycogen A polymer of glucose residues that is the principal storage form of carbohydrates in animals.

glycolipid A lipid consisting of two hydrocarbon chains linked to a polar head group containing carbohydrates.

glycolysis The anaerobic breakdown of glucose.

glycoprotein A protein linked to oligosaccharides.

glycosaminoglycan (GAG) A gel-forming polysaccharide of the extracellular matrix.

glycosidic bond The bond formed between sugar residues in oligosaccharides or polysaccharides.

glycosylation The addition of carbohydrates to proteins.

glycosylphosphatidylinositol (GPI) anchor Glycolipids containing phosphatidylinositol that anchor proteins to the external face of the plasma membrane.

glyoxylate cycle The conversion of fatty acids to carbohydrates in plants.

Golgi apparatus A cytoplasmic organelle involved in the processing and sorting of proteins and lipids. In plant cells, it is also the site of the synthesis of cell wall polysaccharides.

green fluorescent protein (GFP) A protein from jellyfish that is commonly used as a marker for fluorescence microscopy.

growth factors Polypeptides that control animal cell growth and differentiation.

guanine A purine that base-pairs with cytosine.

guanylyl cyclase An enzyme that catalyzes the formation of cyclic GMP from GTP.

haploid An organism or cell that has one copy of each chromosome.

heat-shock proteins A highly conserved group of chaperone proteins expressed in cells exposed to elevated temperatures or other forms of environmental stress.

helicase An enzyme that catalyzes the unwinding of DNA.

hemicellulose A polysaccharide that crosslinks cellulose microfibrils in plant cell walls.

hemidesmosome A region of contact between cells and the extracellular matrix at which keratin filaments are attached to integrin.

heterochromatin Condensed, transcriptionally inactive chromatin.

high-energy bonds Chemical bonds that release a large amount of free energy when they are hydrolyzed.

histones Proteins that package DNA in eukaryotic chromosomes.

HMG-14 and HMG-17 Nonhistone chromosomal proteins associated with decondensed transcriptionally active chromatin.

Holliday junction The central intermediate in recombination, consisting of a crossed-strand structure formed by homologous base pairing between strands of two DNA moleucles.

homeobox Conserved DNA sequences of 180 base pairs that encode homeodomains.

homeodomain A type of DNA binding domain found in transcription factors that regulate gene expression during embryonic development.

homologous recombination Recombination between segments of DNA with homologous nucleotide sequences.

hormones Signaling molecules produced by endocrine glands that act on cells at distant body sites.

hydrophilic Soluble in water.

hydrophobic Not soluble in water.

immunoblotting A method that uses antibodies to detect proteins separated by SDS-polyacrylamide gel electrophoresis.

immunoglobulin See antibody.

immunoprecipitation The use of antibodies to isolate proteins.

importin A receptor protein that recognizes nuclear localization signals and directs nuclear import.

in situ **hybridization** The use of radioactive or flourescent probes to detect RNA or DNA sequences in cell extracts, chromosomes, or intact cells.

in vitro **mutagenesis** The introduction of mutations into cloned DNA *in vitro*.

in vitro **translation** Protein synthesis in a cell-free extract.

inositol 1,4,5-trisphosphate (IP$_3$)   A second messenger, formed from the hydrolysis of PIP$_2$, that signals the release of calcium ions from the endoplasmic reticulum.

integral membrane proteins   Proteins embedded within the lipid bilayer of cell membranes.

integrin   A transmembrane protein that mediates the adhesion of cells to the extracellular matrix.

intermediate filament   A cytoskeletal filament about 10 nm in diameter that provides mechanical strength to cells in tissues. See also keratins and neurofilaments.

interphase   The period of the cell cycle between mitoses that includes G$_1$, S, and G$_2$ phases.

intracellular signal transduction   A chain of reactions that transmits chemical signals from the cell surface to their intracellular targets.

intron   A noncoding sequence that interrupts exons in a gene.

ion channel   A protein that mediates the rapid passage of ions across a membrane by forming open pores through the phospholipid bilayer.

ion pump   A protein that couples ATP hydrolysis to the transport of ions across a membrane.

Janus kinase (JAK)   A family of nonreceptor protein-tyrosine kinases associated with cytokine receptors.

keratin   A type of intermediate filament protein of epithelial cells.

kilobase (kb)   One thousand nucleotides or nucleotide base pairs.

kinesin   A motor protein that moves along microtubules towards the plus end.

kinetochore   A specialized structure consisting of proteins attached to a centromere that mediates the attachment and movement of chromosomes along the mitotic spindle.

kinetochore microtubules   Microtubules of the mitotic spindle that attach to condensed chromosomes at their centromeres.

Krebs cycle   See citric acid cycle.

lagging strand   The strand of DNA synthesized opposite to the direction of movement of the replication fork by ligation of Okazaki fragments.

lamellipodium   A broad, actin-based extension of the plasma membrane involved in the movement of fibroblasts.

laminin   The principal adhesion protein of basal laminae.

lamins   Intermediate filament proteins that form the nuclear lamina.

leading strand   The strand of DNA synthesized continuously in the direction of movement of the replication fork.

leptotene   The initial stage of the extended prophase of meiosis I during which homologous chromosomes pair before condensation.

leucine zipper   A protein dimerization domain containing repeated leucine residues; found in many transcription factors.

leukemia   Cancer arising from the precursors of circulating blood cells.

ligand   A molecule that binds to a receptor.

light reactions   The reactions of photosynthesis in which solar energy drives the synthesis of ATP and NADPH.

lignin   A polymer of phenolic residues that strengthens secondary cell walls.

lipids   Hydrophobic molecules that function as energy storage molecules, signaling molecules, and the major components of cell membranes.

long terminal repeat (LTR)   DNA sequences found at the ends of retroviral and retrotransposon DNA that are direct repeats of several hundred nucleotides resulting from reverse transcriptase activity.

lymphocyte   A blood cell that functions in the immune response. B lymphocytes produce antibodies and T lymphocytes are responsible for cell mediated immunity.

lymphoma   A cancer of lymphoid cells.

lysogeny   A viral infection leading to integration of an inactive copy of viral DNA into the cell genome.

lysosome   A cytoplasmic organelle containing enzymes that break down biological polymers.

M phase   The mitotic phase of the cell cycle.

macrophage   A type of white blood cell specialized for phagocytosis.

malignant tumor   A tumor that invades normal tissue and spreads throughout the body.

MAP kinases   A family of mitogen-activated protein-serine/threonine kinases that are ubiquitous regulators of cell growth and differentiation.

matrix   The inner mitochondrial space.

maturation promoting factor (MPF)   A complex of Cdc2 and cyclin B that promotes entry into the M phase of either mitosis or meiosis.

megabase (Mb)   One million nucleotides or nucleotide base pairs.

meiosis   The division of diploid cells to haploid progeny, consisting of two sequential rounds of nuclear and cellular division.

membrane-anchored growth factors   Growth factors associated with the plasma membrane that function as signaling molecules during cell-cell contact.

mesoderm   The middle germ layer; gives rise to connective tissues and the hematopoietic system.

messenger RNA (mRNA)   An RNA molecule that serves as a template for protein synthesis.

metaphase   The phase of mitosis during which the chromosomes are aligned on a metaphase plate in the center of the cell.

metastasis   Spread of cancer cells through the blood or lymphatic system to other organ sites.

5' methylguanosine cap   A structure consisting of GTP and methylated sugars that is added to the 5' ends of eukaryotic mRNAs.

microfilament   A cytoskeleton filament composed of actin.

microspike   See filopodium.

microtubule   A cytoskeletal component formed by the polymerization of tubulin into rigid, hollow rods about 25 nm in diameter.

microtubule-organizing center   An anchoring point near the center of the cell from which most microtubules extend outward.

microvillus   An actin-based protrusion of the plasma membrane, abundant on the surfaces of cells involved in absorption.

mismatch repair   A repair system that removes mismatched bases from newly synthesized DNA strands.

mitochondria   Cytoplasmic organelles responsible for synthesis of most of the ATP in eukaryotic cells by oxidative phosphorylation.

mitosis   Nuclear division.

mitotic spindle   An array of microtubules extending from the spindle poles that is responsible for separating daughter chromosomes during mitosis. See also kinetochore microtubules, polar microtubules, and astral microtubules.

molecular motor   A protein that generates force and movement by converting chemical energy to mechanical energy.

monocistronic   Messenger RNAs that encode a single polypeptide chain.

monoclonal antibody   An antibody produced by a clonal line of B lymphocytes.

monosaccharides   Simple sugars with the basic formula of $(CH_2O)_n$.

mutagen   A chemical that induces a high frequency of mutations.

mutation   A genetic alteration.

myosin   A protein that interacts with actin as a molecular motor.

N-myristoylation   The addition of myristic acid (a 14-carbon fatty acid) to the N-terminal glycine residue of a polypeptide chain.

Na^+-K^+ pump   An ion pump that transports Na^+ out of the cell and K^+ into the cell.

Nernst equation   The relationship between ion concentration and membrane potential.

neurofilament   A type of intermediate filament that supports the axons of nerve cells.

neuron   A nerve cell specialized to receive and transmit signals throughout the body.

neurotransmitter   A small, hydrophilic molecule that carries a signal from a stimulated neuron to a target cell at a synapse.

nexin   A protein that links microtubule doublets to each other in the axoneme.

nicotinamide-adenine dinucleotide (NAD^+)   A coenzyme that functions as an electron carrier in oxidation/reduction reactions.

nitrogen fixation   The reduction of atmospheric nitrogen (N_2) to NH_3.

nonreceptor protein-tyrosine kinase   An intracellular protein-tyrosine kinase.

nuclear envelope   The barrier separating the nucleus from the cytoplasm, composed of an inner and outer membrane, a nuclear lamina, and nuclear pore complexes.

nuclear export signal   An amino acid sequence that targets proteins for transport from the nucleus to the cytosol.

nuclear lamina   A meshwork of lamin filaments providing structural support to the nucleus.

nuclear localization signal   An amino acid sequence that targets proteins for transportation from the cytoplasm to the nucleus.

nuclear membranes   Membranes forming the nuclear envelope; the outer nuclear membrane is continuous with the endoplasmic reticulum and the inner nuclear membrane is adjacent to the nuclear lamina.

nuclear pore complex   A large structure forming a transport channel through the nuclear envelope.

nucleic acid hybridization   The formation of double stranded DNA and/or RNA molecules by complementary base pairing.

nucleolus   The nuclear site of rRNA transcription, processing, and ribosome assembly.

nucleoside   A purine or pyrimidine base linked to a sugar (ribose or deoxyribose).

nucleosome   The basic structural unit of chromatin consisting of DNA wrapped around a histone core.

nucleosome remodeling factors   Proteins that disrupt chromatin structure, allowing transcription factors to bind nucleosomal DNA.

nucleotide   A phosphorylated nucleoside.

nucleotide excision repair A mechanism of DNA repair in which oligonucleotides containing damaged bases are removed from a DNA molecule.

nucleus The most prominent organelle of eukaryotic cells; contains the genetic material.

Okazaki fragments Short DNA fragments that are joined to form the lagging strand of DNA.

oligonucleotide A short polymer of only a few nucleotides.

oligosaccharide A short polymer of only a few sugars.

oncogene A gene capable of inducing one or more characteristics of cancer cells.

operator A regulatory sequence of DNA that controls transcription of an operon.

operon A group of adjacent genes transcribed as a single mRNA.

origin of replication A specific DNA sequence that serves as a binding site for proteins that initiate replication.

origin replication complex (ORC) A protein complex that initiates DNA replication at yeast origins.

oxidative phosphorylation The synthesis of ATP from ADP coupled to the energetically favorable transfer of electrons to molecular oxygen as the final acceptor in an electron transport chain.

p53 A transcription factor (encoded by the *p53* tumor suppressor gene) that arrests the cell cycle in G_1 in response to damaged DNA and is required for apoptosis induced by a variety of stimuli.

pachytene The stage of meiosis I during which recombination takes place between homologous chromosomes.

palmitoylation The addition of palmitic acid (a 16-carbon fatty acid) to cysteine residues of a polypeptide chain.

paracrine signaling Local cell-cell signaling in which a molecule released by one cell acts on a neighboring target cell.

parenchyma cell A type of plant cell responsible for most metabolic activities.

passive diffusion The diffusion of small hydrophobic molecules through a phospholipid bilayer.

passive transport The transport of molecules across a membrane in the energetically favorable direction.

pectin A gel-forming polysaccharide in plant cell walls.

peptide bond The bond joining amino acids in polypeptide chains.

peptidoglycan The principal component of bacterial cell walls consisting of linear polysaccharide chains crosslinked by short peptides.

peptidyl prolyl isomerase An enzyme that facilitates protein folding by catalyzing the *cis-trans* isomerization of prolyl peptide bonds.

pericentriolar material The material in the centrosome that initiates microtubule assembly.

peripheral membrane proteins Proteins indirectly associated with cell membranes by protein-protein interactions.

peroxisome A cytoplasmic organelle specialized for carrying out oxidative reactions.

phagocytosis The uptake of large particles, such as bacteria, by a cell.

phalloidin A drug that binds to actin filaments and prevents their disassembly.

phenotype The physical appearance of an organism.

phorbol esters A class of tumor promoters that stimulate protein kinase C by acting as analogs of diacylglycerol.

phosphatidylinositide 3-kinase (PI 3-kinase) An enzyme that phosphorylates PIP_2 yielding the second messenger phosphatidylinositol 3,4,5-trisphosphate (PIP_3).

phosphatidylinositol 4,5-bisphosphate (PIP_2) A minor phospholipid component of the inner leaflet of the plasma membrane. Hormones and growth factors stimulate its hydrolysis by phospholipase C, yielding the second messengers diacylglycerol and inositol trisphosphate.

phosphodiester bond A bond between the 5′-phosphate of one nucleotide and the 3′-hydroxyl of another.

phospholipase C An enzyme that hydrolyzes PIP_2 to form the second messengers diacylglycerol and inositol trisphosphate.

phospholipid bilayer The basic structure of biological membranes, in which the hydrophobic tails of phospholipids are buried in the interior of the membrane and their polar head groups are exposed to the aqueous solution on either side.

phospholipid transfer protein A protein that transports phospholipid molecules between cell membranes.

phospholipids The principal components of cell membranes, consisting of two hydrocarbon chains (usually fatty acids) joined to a polar head group containing phosphate.

phosphorylation The addition of a phosphate group to a molecule.

photoreactivation A mechanism of DNA repair in which solar energy is used to split pyrimidine dimers.

photosynthesis The process by which cells harness energy from sunlight and synthesize glucose from CO_2 and water.

pinocytosis The uptake of fluids or molecules into a cell by small vesicles.

plant hormones   A group of small molecules that coordinate the responses of plant tissues to environmental signals.

plasma membrane   A phospholipid bilayer with associated proteins that surrounds the cell.

plasmalogens   A family of phospholipids that have an ether bond and an ester bond.

plasmid   A small, circular DNA molecule capable of independent replication in a host cell.

plasmodesma   A cytoplasmic connection between adjacent plant cells formed by a continuous region of the plasma membrane.

plastids   A family of plant organelles including chloroplasts, chromoplasts, leucoplasts, amyloplasts, and elaioplasts.

platelet-derived growth factor (PDGF)   A growth factor released by platelets during blood clotting to stimulate the proliferation of fibroblasts.

polar microtubules   Microtubules of the mitotic spindle that overlap in the center of the cell and push the spindle poles apart.

poly-A tail   A tract of about 200 adenine nucleotides added to the 3′ ends of eukaryotic mRNAs.

polyadenylation   The process of adding a poly-A tail to a pre-mRNA.

polycistronic   Messenger RNAs that encode multiple polypeptide chains.

polymerase chain reaction (PCR)   A method for amplifying a region of DNA by repeated cycles of DNA synthesis *in vitro*.

polynucleotide   A polymer containing up to millions of nucleotides.

polyp   A benign tumor projecting from an epithelial surface.

polypeptide   A polymer of amino acids.

polysaccharide   A polymer containing hundreds or thousands of sugars.

polysome   A series of ribosomes translating a messenger RNA.

pre-mRNA   The primary transcripts that are processed to form messenger RNAs in eukaryotic cells.

pre-rRNA   The primary transcript, which is cleaved to form individual ribosomal RNAs (the 28S, 18S, and 5.8S rRNAs of eukaryotic cells).

pre-tRNA   The primary transcript, which is cleaved to form transfer RNAs.

prenylation   The addition of specific types of lipids (prenyl groups) to C terminal cysteine residues of a polypeptide chain.

primary structure   The sequence of amino acids in a polypeptide chain.

primase   An RNA polymerase used to intiate DNA synthesis.

processed psuedogene   A psuedogene that has arisen by reverse transcription of mRNA.

progesterone   A steroid hormone produced by the ovaries.

programmed cell death   A normal physiological form of cell death characterized by apoptosis.

prokaryotic cells   Cells lacking a nuclear envelope, cytoplasmic organelles, and a cytoskeleton (primarily bacteria).

prometaphase   A transition period between prophase and metaphase during which the microtubules of the mitotic spindle attach to the kinetochores and the chromosomes shuffle until they align in the center of the cell.

promoter   A DNA sequence to which RNA polymerase binds to initiate transcription.

pronuclei   Two haploid nuclei in a newly fertilized egg.

proofreading   The selective removal of mismatched bases by DNA polymerase.

prophase   The beginning phase of mitosis, marked by the appearance of condensed chromosomes and the development of the mitotic spindle.

prosthetic groups   Small molecules bound to proteins.

proteasome   A large protease complex that degrades proteins tagged by ubiquitin.

protein disulfide isomerase   An enzyme that catalyzes the formation and breakage of disulfide (S–S) linkages.

protein kinase   An enzyme that phosphorylates proteins by transferring a phosphate group from ATP.

protein kinase A   A protein kinase regulated by cyclic AMP.

protein kinase C   A family of protein-serine/threonine kinases that are activated by diacylglycerol and Ca^{2+} and function in intracellular signal transduction.

protein phosphatase   An enzyme that reverses the action of protein kinases by removing phosphate groups from phosphorylated amino acid residues.

protein-serine/threonine kinase   A protein kinase that phosphorylates serine and threonine residues.

protein-tyrosine kinase   A protein kinase that phosphorylates tyrosine residues.

protein-tyrosine phosphatase   An enzyme that removes the phosphate groups from phosphotyrosine residues.

proteins   Polypeptides with a unique amino acid sequence.

proteoglycan   A protein linked to glycosaminoglycans.

proteolysis   Degradation of polypeptide chains.

proto-oncogene   A normal cell gene that can be converted into an oncogene.

pseudogene   A nonfunctional gene copy.

pseudopodium   An actin-based extension of the plasma membrane responsible for phagocytosis and amoeboid movement.

PTEN A lipid phosphatase that dephosphorylates PIP_3 and acts as a tumor suppressor.

pyrimidine dimer A common form of DNA damage caused by UV light in which adjacent pyrimidines are joined to form a dimer.

quaternary structure The interactions between polypeptide chains in proteins consisting of more than one polypeptide.

Raf A protein-serine/threonine kinase (encoded by the *raf* oncogene) that is activated by Ras and leads to activation of MAP kinase.

Ran A small GTP-binding protein involved in nuclear import and export.

Ras A family of small GTP binding proteins (encoded by the *ras* oncogenes) that couple growth factor receptors to intracellular targets, including the Raf protein-serine/threonine kinase and the MAP kinase pathway.

Rb A transcriptional regulatory protein encoded by a tumor suppressor gene that was identified by the genetic analysis of retinoblastoma.

receptor protein-tyrosine kinase Membrane-spanning protein-tyrosine kinases that are receptors for extracellular ligands.

receptor-mediated endocytosis The selective uptake of macromolecules that bind to cell surface receptors that concentrate in clathrin-coated pits.

recessive An allele that is masked by a dominant allele.

recombinant DNA library A collection of genomic or cDNA clones.

recombinant molecule A DNA insert joined to a vector.

recombination The exchange of genetic material.

replication fork The region of DNA synthesis where the parental strands separate and two new daughter strands elongate.

repressor A regulatory molecule that blocks transcription.

restriction endonuclease An enzyme that cleaves DNA at a specific sequence.

restriction map The locations of restriction endonuclease cleavage sites on a DNA molecule.

restriction point A regulatory point in animal cell cycles that occurs late in G_1. After this point, a cell is committed to entering S and undergoing one cell division cycle.

retroposon A transposable element that moves via reverse transcription of an RNA intermediate.

retrovirus A virus that replicates by making a DNA copy of its RNA genome by reverse transcription.

reverse transcriptase A DNA polymerase that uses an RNA template.

rhodopsin A G protein-coupled photoreceptor in retinal rod cells that activates transducin in response to light absorption.

ribonucleic acid (RNA) A polymer of ribonucleotides.

ribosomal RNA (rRNA) The RNA component of ribosomes.

ribosomes Particles composed of RNA and proteins that are the sites of protein synthesis.

ribozyme An RNA enzyme.

RNA editing RNA processing events other than splicing that alter the protein coding sequences of mRNAs.

RNA polymerase An enzyme that catalyzes the synthesis of RNA.

RNA splicing The joining of exons in a precursor RNA molecule.

RNase H An enzyme that degrades the RNA strand of RNA-DNA hybrid molecules.

RNase P A ribozyme that cleaves the 5′ end of pre-tRNAs.

rough endoplasmic reticulum The region of the endoplasmic reticulum covered with ribosomes and involved in protein metabolism.

Rous sarcoma virus (RSV) An acutely transforming retrovirus, in which the first oncogene was identified.

ryanodine receptors Calcium channels in muscle and nerve cells that open in response to changes in membrane potential.

S phase The phase of the cell cycle during which DNA replication occurs.

sarcoma A cancer of cells of connective tissue.

sarcomere The contractile unit of muscle cells composed of interacting myosin and actin filaments.

sarcoplasmic reticulum A specialized network of membranes in muscle cells that stores a high concentration of Ca^{2+}.

scanning electron microscopy See electron microscopy.

sclerenchyma cells Plant cells characterized by thick cell walls that provide structural support to the plant.

SDS-polyacrylamide gel electrophoresis (SDS-PAGE) A commonly used method to separate proteins by gel electrophoresis on the basis of size.

second messenger A compound whose metabolism is modified as a result of a ligand-receptor interaction; it functions as a signal transducer by regulating other intracellular processes.

secondary structure The regular arrangement of amino acids within localized regions of a polypeptide chain. See α helix and β sheet.

secretory vesicles Membrane-enclosed sacs that transport proteins from the Golgi apparatus to the cell surface.

selectins Cell adhesion molecules that recognize oligosaccharides exposed on the cell surface.

self-splicing The ability of some RNAs to catalyze the removal of their own introns.

SH2 domain A protein domain of approximately 100 amino acids that binds phosphotyrosine-containing peptides.

Shine-Delgarno sequence The sequence prior to the initiation site that correctly aligns bacterial mRNAs on ribosomes.

signal patch A recognition determinant formed by the three-dimensional folding of a polypeptide chain.

signal peptidase An enzyme that removes the signal sequence of a polypeptide chain by proteolysis.

signal recognition particle (SRP) A particle composed of proteins and 7SL RNA that binds to signal sequences and targets polypeptide chains to the endoplasmic reticulum.

signal sequence A hydrophobic sequence at the amino terminus of a polypeptide chain that targets it for secretion in bacteria or incorporation into the endoplasmic reticulum in eukaryotic cells.

site-specific recombination Recombination mediated by proteins that recognize specific DNA sequences.

smooth endoplasmic reticulum The major site of lipid synthesis in eukaryotic cells.

Southern blotting A method in which radioactive probes are used to detect specific DNA fragments that have been separated by gel electrophoresis.

spectrin A major actin-binding protein of the cell cortex.

sphingomyelin A phospholipid consisting of two hydrocarbon chains bound to a polar head group containing serine.

spliceosomes Large complexes of snRNAs and proteins that catalyze the splicing of pre-mRNAs.

Src A nonreceptor protein-tyrosine kinase encoded by the oncogene (*src*) of Rous sarcoma virus.

starch A polymer of glucose residues that is the principal storage form of carbohydrates in plants.

START A regulatory point in the yeast cell cycle that occurs late in G_1. After this point a cell is committed to entering S and undergoing one cell division cycle.

STAT proteins Trancription factors that have an SH2 domain and are activated by tyrosine phosphorylation, which promotes their translocation from the cytoplasm to the nucleus.

stem cell A cell that divides to produce daughter cells that can either differentiate or remain as stem cells.

stereocilium A specialized microvillus of auditory hair cells.

steroid hormones A group of hydrophobic hormones that are derivatives of cholesterol.

steroid receptor superfamily A family of transcription factors that regulate gene expression in response to steroids and related hormones.

substrate A molecule acted upon by an enzyme.

symport The transport of two molecules in the same direction across a membrane.

synapse The junction between a neuron and another cell, across which information is carried by neurotransmitters.

synapsis The association of homologous chromosomes during meiosis.

synaptic vesicle A secretory vesicle that releases neurotransmitters at a synapse.

synaptomenal complex A zipperlike protein structure that forms along the length of paired homologous chromosomes during meiosis.

T cell receptor A T lymphocyte surface protein that recognizes antigens expressed on the surface of other cells.

TATA box A regulatory DNA sequence found in the promoters of many eukaryotic genes transcribed by RNA polymerase II.

TATA-binding protein (TBP) A basal transcription factor that binds directly to the TATA box.

taxol A drug that binds to and stabilizes microtubules.

telomerase A reverse transcriptase that synthesizes telomeric repeat sequences at the ends of chromosomes from its own RNA template.

telomeres Repeats of simple-sequence DNA that maintain the ends of linear chromosomes.

telophase The final phase of mitosis, during which the nuclei re-form and chromosomes decondense.

temperature-sensitive mutant A cell expressing a protein that is functional at one temperature but not at another, whereas the normal protein is functional at both temperatures.

tertiary structure The three-dimensional folding of a polypeptide chain that gives the protein its functional form.

testosterone A steroid hormone produced by the testis.

thylakoid membrane The innermost membrane of chloroplasts that is the site of electron transport and ATP synthesis.

thymine A pyrimidine found in DNA that base-pairs with adenine.

thyroid hormone A hormone synthesized from tyrosine in the thyroid gland.

tight junction A continuous network of protein strands around the circumference of epithelial cells, sealing the space between cells and forming a barrier between the apical and basolateral domains.

topoisomerase An enzyme that catalyzes the reversible breakage and rejoining of DNA strands.

transcription factor   A protein that regulates the activity of RNA polymerase.

transcription   The synthesis of an RNA molecule from a DNA template.

transcriptional attenuation   A regulatory mechanism that controls elongation of an mRNA molecule past a certain site on the DNA template.

transcytosis   The sorting and transport of proteins to different domains of the plasma membrane following endocytosis.

transducin   A G protein that stimulates cGMP phosphodiesterase when it is activated by rhodopsin.

transfection   The introduction of a foreign gene into eukaryotic cells.

transfer RNA (tRNA)   RNA molecules that function as adaptors between amino acids and mRNA during protein synthesis.

transforming growth factor β (TGF-β)   A polypeptide growth factor that generally inhibits animal cell proliferation.

transgenic mouse   A mouse that carries foreign genes incorporated into the germ line.

translation   The synthesis of a polypeptide chain from an mRNA template.

transmembrane proteins   Integral membrane proteins that span the lipid bilayer and have portions exposed on both sides of the membrane.

transmission electron microscopy   See electron microscopy.

transposition   The movement of DNA sequences throughout the genome.

transposon   A DNA sequence that can move to different positions in the genome.

triacylglycerol   Three fatty acids linked to a glycerol molecule.

tropomyosin   A fibrous protein that binds actin filaments and regulates contraction by blocking the interaction of actin and myosin.

tubulin   A cytoskeletal protein that polymerizes to form microtubules.

tumor   Any abnormal proliferation of cells.

tumor necrosis factor (TNF)   A polypeptide growth factor that induces programmed cell death.

tumor promoter   A compound that leads to tumor development by stimulating cell proliferation.

tumor suppressor gene   A gene whose inactivation leads to tumor development.

tumor virus   A virus capable of causing cancer in animals or humans.

turgor pressure   The internal hydrostatic pressure within plant cells.

ubiquinone   See coenzyme Q.

ubiquitin   A highly conserved protein that acts as a marker to target other cellular proteins for rapid degradation.

uniport   The transport of a single molecule across a membrane.

uracil   A pyrimidine found in RNA that base-pairs with adenine.

vacuole   A large membrane-enclosed sac in the cytoplasm of eukaryotic cells. In plant cells, vacuoles function to store nutrients and waste products, to degrade macromolecules, and to maintain turgor pressure.

vector   A DNA molecule used to direct the replication of a cloned DNA fragment in a host cell.

vinblastine   A drug that inhibits microtubule polymerization.

vincristine   A drug that inhibits microtubule polymerization.

Western blotting   See immunoblotting.

X-ray crystallography   A method in which the diffraction pattern of X rays is used to determine the arrangement of individual atoms within a molecule.

Xenopus laevis   An African clawed frog used as a model system for developmental biology.

yeast artificial chromosome (YAC)   A vector that can replicate as a chromosome in yeast cells and can accomodate very large DNA inserts (hundreds of kb).

yeasts   The simplest unicellular eukaryotes. Yeasts are important models for studies of eukaryotic cells.

zebrafish   A species of small fish used for genetic studies of vertebrate development.

zinc finger domain   A type of DNA binding domain consisting of loops containing cysteine and histidine residues that bind zinc ions.

zygote   A fertilized egg.

zygotene   The stage of meiosis I during which homologous chromosomes become closely associated.

Index

About the Author

Geoffrey M. Cooper is Professor and Chairman of the Department of Biology at Boston University. He received his B.S. from the Massachusetts Institute of Technology in 1969 and his Ph.D. in Biochemistry from the University of Miami in 1973. He then pursued postdoctoral work with Howard Temin at the University of Wisconsin, where he developed gene transfer assays to characterize the proviral DNAs of Rous sarcoma virus and related retroviruses.

In 1975, Dr. Cooper joined the faculty of the Dana-Farber Cancer Institute and Harvard Medical School, where he extended his research studies to the identification of oncogenes in human tumors. Since moving to Boston University in 1998, he has both continued his research and participated in a major expansion of the university's life sciences programs.

Dr. Cooper's current research is focused on understanding the roles of oncogene proteins in the signaling pathways that regulate cell proliferation, differentiation, and programmed cell death. He has published more than 100 research papers in journals such as *Science*, *Nature*, and *Cell*, and in 1984 he received the U.S. Steel Award from the National Academy of Sciences for "the identification and characterization of cellular oncogenes." Dr. Cooper is the author of two textbooks on cancer, *Elements of Human Cancer* and *Oncogenes*, as well as *The Cancer Book*, written for the general public. He also edited, together with Rayla Greenberg Temin and Bill Sugden, the commemorative volume *The DNA Provirus: Howard Temin's Scientific Legacy*.

About the Book

Editor: Andrew D. Sinauer

Project Editor: Joyce Zymeck

Production Manager: Christopher Small

Book Production: Jefferson Johnson, Joan Gemme and Janice Holabird

Illustration Program: Patrick Lane of J/B Woolsey Associates

Photo Research: David McIntyre

Book Design: Susan Brown Schmidler/Jefferson Johnson

Cover Design: Jefferson Johnson

Color Separations: Vision Graphics, Inc.

Cover Manufacture: Henry N. Sawyer Company, Inc.

Book Manufacture: Courier Companies, Inc.